Tutorials in Complex Photonic Media

Library of Congress Cataloging-in-Publication Data

Tutorials in complex photonic media / editors, Mikhail A. Noginov ... [et al.].
p. cm. -- (Press monograph ; 194)
ISBN 978-0-8194-7773-6
1. Photonics. 2. Photonic crystals. 3. Metamaterials. I. Noginov, Mikhail A.
TA1520.T88 2009
621.36--dc22

2009035663

Published by

SPIE
P.O. Box 10
Bellingham, Washington 98227-0010 USA
Phone: +1 360.676.3290
Fax: +1 360.647.1445
Email: Books@spie.org
Web: http://spie.org

Cover art from figures in Chapter 7, courtesy of authors K. O'Holleran, M. R. Dennis, and M. J. Padgett.

Printed in the United States of America.

Tutorials in Complex Photonic Media

Editors:
Mikhail A. Noginov
Graeme Dewar
Martin W. McCall
Nikolay I. Zheludev

SPIE PRESS
Bellingham, Washington USA

Contents

Foreword

Classical optics has been with us for some considerable time, yet the past decade has produced a cornucopia of new research, often revealing unsuspected phenomena hidden like nuggets of gold in the rich lode of optical materials. The key has often been complexity. The range of optical properties available in natural materials is limited, but by adding manmade structure to nature's offerings we can extend our reach, sometimes to achieve properties not seen before. I pick one example from the many included in this volume: negative refraction. Years ago it had been realised that a material with negative magnetic and electric responses would also have a negative refractive index. There, the idea languished for nearly half a century, lacking the naturally occurring materials to realise the effect. However by internally structuring a medium on a scale less than the relevant wavelength, it was proved possible to make a new form of material, a 'metamaterial,' which had the required negatively refracting properties. This concept alone has given rise to thousands of papers. There are other examples I could cite from the chapters in this book: exploitation of near-field properties of nanoparticle arrays, photonic band gap waveguides, metallic nanostructures for sensing proteins, and so on. All of these examples have in common that man adds complexity to the offerings of nature.

In the face of these myriad advances, how are students or other new entrants to the field to educate themselves in the new technology? This book provides the answer, collecting together a definitive series of tutorials, each provided by an expert in the field. It is published at a time when there are many such new entrants and will be of great value.

J. B. Pendry
Imperial College London

Preface

An increasingly large number of high- and low-tech technologies and devices benefit from employing optics and photonics phenomena, the latter originally being termed photon-based electronics. Progress in the research fields of optics and photonics, which have both experienced continuously strong growth over the last few decades, critically depends on the understanding and utilization of the physical, chemical and structural properties of optical materials. The optical materials used in traditional optics technology were macroscopically homogeneous in that their scale of inhomogeneity was much less than the wavelength. In more recent years, multiple breakthroughs have involved inhomogeneous, composite, and multiphase materials, whose structures are either photoinduced or determined by synthesis or fabrication. Examples include holography, optics of scattering media, and metamaterials. These breakthroughs make photonic materials inherently complex. The broad range of physical phenomena underlying complex photonic media makes it difficult for scientists, engineers, and students entering the field to navigate through the range of topics and to understand clearly how they relate to each other.

The purpose of this book is to provide the necessary coverage and inspire the reader's curiosity about the fascinating subject of complex photonic media. All of the tutorial chapters are designed to start with the basics and gradually move toward discussion of more advanced topics. We thus envisage that students and scholars with diverse backgrounds and levels of expertise will find this text interesting and useful. The book can be used as a supplemental text in courses on nanotechnology or optical materials, or a variety of other courses. It can also be used as the main text in a more focused course aimed at fundamental properties of scattering media and metamaterials. The anticipated level of preparation is equivalent to advanced senior undergraduate level, beginning graduate level, or higher. The book covers the topics in the following (rather loose) categorization:

Negative index materials (NIMs). One of the most exciting developments in complex photonic media in recent years is the realization that the basic parameters describing the electromagnetics of simple, isotropic media can take simultaneously negative values. This leads to all kinds of interesting phenomena, from a revised understanding of Snell's law, to lenses that defeat the conventional diffraction resolution limit. In "Negative Refraction" (Chapter 1), Martin W. McCall and Graeme Dewar describe the basic theory and impetus for negative refraction research. In "Optical Hyperspace: Negative Refractive Index and Subwavelength Imaging" (Chapter 2), Leonid V. Alekseyev, Zubin Jacob, and Evgenii Narimanov explore nonmagnetic routes that exploit materials with hyperbolic dispersion relations.

Magneto-optics. The term magneto-optics is used when the direction and polarization state of light are controlled by the application of external magnetic fields, offering opportunities for optical storage and isolation in optical systems. In "Magneto-optics and the Kerr Effect with Ferromagnetic Materials" (Chapter 3), Allan D. Boardman and Neil King introduce the magneto-optics derived from air-ferroelectric interfaces and glass/ferromagnetic film/air multilayer systems. "Nonlinear Magneto-Optics" (Chapter 4) by Yutaka Kawabe gives emphasis to the relationship between the tensors describing the nonlinearity and the underlying crystal point group symmetry. In "Optical Magnetism in Plasmonic Metamaterials" (Chapter 5), Gennady Shvets and Yaroslav A. Urzhumov describe some of the difficult challenges that lie ahead for achieving magnetic activity at optical frequencies.

Chiral media and vortices. Light, being composed of unit spin photons, is inherently chiral. However, chirality in optical systems can also be engaged at structural and macroscopic electromagnetic levels. Structural chirality is covered by Ian Hodgkinson and Levi Bourke in "Chiral Photonic Media" (Chapter 6), which describes the multilayer matrix formalism for novel elliptically polarized filters. When optical beams interfere, phase singularities occur; in "Optical Vortices" (Chapter 7) Kevin O'Holleran, Mark R. Dennis, and Miles J. Padgett describe some of the remarkable topological knots and 3D twists that result.

Scattering in periodic and random media. Scattering of light is fundamental to complex photonic media. Structures that are periodic are generally referred to as photonic crystals. In "Photonic Crystals: From Fundamentals to Functional Photonic Materials" (Chapter 8), Durga P. Aryal, Kosmas L. Tsakmakidis, and Ortwin Hess describe how photonic bandstructure emerges in both 1- and 2D structures, and how optical switching is achievable in inverse-opal structures. When the material inhomogeneity is random, different methods must be employed. In "Wave Interference and Modes in Random Media" (Chapter 9), Azriel Z. Genack and Sheng Zhang describe photon transport in a medium in terms of the interference of multiply scattered partial waves as well as by considering the different spatial, spectral, and temporal characters of the electromagnetic modes.

Photonic media with gain and lasing phenomena. Photonic media with gain and lasing phenomena represents the generic class of active photonic media. "Chaotic Behavior of Random Lasers" (Chapter 10) by Diederik Wiersma, Sushil Mujumdar, Stefano Cavalieri, Renato Torre, Gian-Luca Oppo, and Stefano Lepri examines the irreproducibility of experimental emission spectra and the change of statistics at near threshold. "Lasing in Random Media" (Chapter 11) by Hui Cao provides a detailed review of the concepts and advances in the field of random lasers. "Feedback in Random Lasers" (Chapter 12) by Mikhail A.

Noginov emphasizes the significance of the strength of scattering and/or feedback in determining the properties of random lasers. In "Optical Metamaterials with Zero Loss and Plasmonic Nanolasers" (Chapter 13), Andrey Sarychev discusses how nano-horseshoe inclusion in an active host medium results in a plasmonic nanolaser.

Fundamentals. In "Resonance Energy Transfer: Theoretical Foundations and Developing Applications" (Chapter 14), David L. Andrews explores how the intricate interplay between quantum mechanical and electromagnetic medium properties leads to schemes for energy transfer and all-optical switching. In "Optics of Nanostructured Materials from First Principle Theories" (Chapter 15) Vladimir I. Gavrilenko provides a tutorial on the microscopic modelling of optical response functions using density functional theory and related approaches.

Organic photonic materials. Materials whose nonlinear coefficients often exceed their inorganic counterparts both in magnitude and response rate are examined in "Organic Photonic Materials" (Chapter 16) by Larry R. Dalton, Philip A. Sullivan, Denise H. Bale, Scott R. Hammond, Benjamin C. Olbricht, Harrison Rommel, Bruce Eichinger, and Bruce H. Robinson. These authors explore organic optical material design in terms of critical structure/function relationships. "Charge Transport and Optical Effects in Disordered Organic Semiconductors" (Chapter 17) by Harry H. L. Kwok, You-Lin Wu, and Tai-Ping Sun highlights how, as with regular semiconductors, charge transport can be modified by doping in organic materials, which possess enhanced carrier mobilities.

Holographic media. "Holography and Its Applications" (Chapter 18) by H. John Caulfield and Chandra S. Vikram discusses holograms used as parts of complex light-controlled or light-defined systems that manipulate the direction, spectrum, polarization, or speed of pulse propagation of light in a medium.

Slow and fast light. Slow and fast light is an intriguing topic demystified by Joseph E. Vornehm, Jr. and Robert W. Boyd in the final chapter "Slow and Fast Light" (Chapter 19). The authors show how manipulation of the material dispersion can lead to very slow, halted, or even backward propagating optical pulses.

The conception of *Tutorials in Complex Photonic Media* lies in an effort to consolidate the conference series, Complex Mediums: Light and Complexity, a subconference of the annual SPIE Optics and Photonics meeting held over the years 2003–2006[1]. Incentive for this book was also largely compelled by

---

[1] In 2003 the conference was titled Complex Mediums IV: Beyond Linear Isotropic Dielectrics; in 2006 it was titled Complex Photonic Media.

Introduction to Complex Mediums for Optics and Electromagnetics, edited by Werner S. Weiglhofer and Akhlesh Lakhtakia, SPIE Press (2003), which is a consolidation of the Complex Mediums conferences from 1999 to 2002. We have taken special emphasis in this book to avoid the somewhat disjointed presentation that often accompanies books based on conferences. To this end, all of the chapters underwent round-robin reviews by several editors and coauthors who were frequently not directly involved in the research area. Much "back and forth" has hopefully ironed out the specialist's tendency to dive headlong into details that can only be appreciated once sufficient underpinning motivational material has been presented. Another issue is notation. Eventually, we decided that keeping a consistent notation throughout the book would be self-defeating, as anyone entering a new area must, to a certain extent, be flexible to individual authors' preferences. Nevertheless, we went to some lengths to ensure that the notation within each chapter is consistent.

The four editors who undertook this project have had a unique opportunity to work with some of the leading specialists in the field. Of course, there have been frustrations, but in the end, we hope that that this book presents a broad and balanced summary that will lead many others to take up the exciting challenges of working in complex photonic media. In the introduction to the predecessor volume noted above, Akhlesh Lakhtakia wrote 'I shall be delighted if a companion volume were published after another two or three editions of this conference.' Well, here it is.

Mikhail A. Noginov
Graeme Dewar
Martin W. McCall
Nikolay I. Zheludev
September 2009

List of Contributors

Leonid V. Alekseyev
Princeton University, USA and
Purdue University, USA

David L. Andrews
University of East Anglia
Norwich, UK

Durga P. Aryal
University of Surrey, UK

Denise H. Bale
University of Washington, USA

Allan D. Boardman
University of Salford, UK

Levi Bourke
University of Otago, New Zealand

Robert W. Boyd
The Institute of Optics,
University of Rochester, USA

Hui Cao
Yale University, USA

H. John Caulfield
Fisk University, USA

Stefano Cavalieri
University of Florence, Italy

Larry R. Dalton
University of Washington, USA

Mark R. Dennis
University of Bristol, UK

Graeme Dewar
University of North Dakota, USA

Bruce Eichinger
University of Washington, USA

Vladimir I. Gavrilenko
Norfolk State University, USA

Azriel Z. Genack
Queens College, City University
of New York, USA

Scott R. Hammond
University of Washington, USA

Ortwin Hess
University of Surrey, UK

Ian Hodgkinson
University of Otago, New Zealand

Zubin Jacob
Purdue University, USA

Yutaka Kawabe
Chitose Institute of Science and
Technology, Japan

Neil King
University of Salford, UK

Harry H. L. Kwok
University of Victoria, Canada

Stefano Lepri
Institute of Complex Systems, CNR, Italy

Martin W. McCall
Imperial College London, UK

Sushil Mujumdar
Tata Institute of Fundamental Research, India

Evgenii Narimanov
Purdue University, USA

Mikhail A. Noginov
Norfolk State University, USA

Kevin O'Holleran
University of Glasgow, UK

Benjamin C. Olbricht
University of Washington, USA

Gian-Luca Oppo
University of Strathclyde, UK

Miles J. Padgett
University of Glasgow, UK

Bruce H. Robinson
University of Washington, USA

Harrison Rommel
University of Washington, USA

Andrey K. Sarychev
Institute of Theoretical and Applied Electrodynamics, Russia

Gennady Shvets
University of Texas at Austin, USA

Philip A. Sullivan
University of Washington, USA

Tai-Ping Sun
National Chi-Nan University, Taiwan

Renato Torre
University of Florence, Italy

Kosmas L. Tsakmakidis
University of Surrey, UK

Yaroslav A. Urzhumov
COMSOL, Inc. USA

Chandra S. Vikram (deceased)
Fisk University, USA

Joseph E. Vornehm, Jr.
The Institute of Optics, University of Rochester, USA

Diederik S. Wiersma
LENS—European Laboratory for Non-Linear Spectroscopy, BEC-INFM, Italy

You-Lin Wu
National Chi-Nan University, Taiwan

Sheng Zhang
Queens College, City University of New York, USA

List of Abbreviations

AFM	atomic force microscopy
APC	amorphous polycarbonate
APTE	*addition de photons par transfers d'énergie*
ASE	amplified spontaneous emission
ATR	attenuated total reflection
BCOG	binary chromophore organic glass
BEC	Bose-Einstein condensate
BER	bit-error rate
BZ	Brillouin zone
CCD	charge-coupled device
CCW	coupled-cavity waveguide
CDM	correlated disorder model
CGH	computer-generated hologram
CGS	centimeter-gram-second
CP	circularly polarized
CPO	coherent population oscillation
CQED	cavity QED
CROW	coupled-resonator optical waveguide
CT	charge transfer
CVD	chemical vapor deposition
DBP	delay–bandwidth product
DFB	distributed feedback
DFT	density functional theory
DIOPC	double-inverse-opal PC
DOS	density of states
DPCM	double phase-conjugate mirror
DSC	differential scanning calorimetry
ECP	effective core potential
EE	electrostatic eigenvalue
EET	electronic energy transfer
EFISH	electric-field-induced second harmonic
EIT	electromagnetically induced transparency
EM	electromagnetic
EO	electro-optic
fcc	face-center cubic
FEFD	finite element frequency domain

FRET	fluorescence RET
FWHM	full width at half maximum
FWM	four-wave mixing
GDM	Gaussian disorder model
GEDE	generalized eigenvalue differential equation
GLC	geometric LC
GMR	gap-to-midgap ratio
GVD	group velocity dispersion
hcp	hexagonal close-packed
HOE	holographic optical element
HOMO	highest occupied molecular orbit
HRS	hyper-Raleigh scattering
IR	infrared
IVR	intramolecular vibrational redistribution
JCM	Jaynes-Cummings model
LAP	laser-assisted poling
LCP	left-circular polarization
LDA	local density approximation
LED	light-emitting diode
LF	local field
LHM	left-handed material
LO	longitudinal optical
LUMO	lowest unoccupied molecular orbit
ME	magneto-electric
MO	magneto-optic
MOCVD	metalorganic CVD
MPR	magnetic plasmon resonance
MSHG	magnetization-induced SHG
MTHG	magnetization-induced THG
NA	numerical aperture
NIM	negative index material
NLO	nonlinear optical
NPV	negative phase velocity
OCRET	optically controlled RET
OCT	optical coherence tomography
OEO	optical-electrical-optical
OLED	organic light-emitting diode
OPD	optical path length distance
OPO	optical parameter oscillator
PC (PhC)	photonic crystal
PEC	perfect electric conductor
PFT	power Fourier transform
PGB	photonic band gap
PMT	photomultiplier tubes
PWE	plane wave expansion

QED	quantum electrodynamics
QD	quantum dot
QP	quasi-particle
QW	quantum well
RCP	right-circular polarization
RET	resonance energy transfer
RF	radiofrequency
RIE	reaction ion etching
rms	root mean square
RPA	random-phase approximation
SBS	stimulated Brillouin scattering
SE	stimulated emission
SEIRA	surface-enhanced IR absorption
SEM	scanning electron microscope
SFG	sum frequency generation
SERS	surface-enhanced Raman scattering
SHG	second-harmonic generation
SLM	spatial light modulator
SOA	semiconductor optical amplifier
SP	surface plasmon
SPD	square of the polarizability derivative
SPOF	strip pair-one film
SPP	spiral phase plate
SPR	surface plasmon resonance
SR	slit ring
SRS	stimulated Raman scattering
SRR	split-ring resonator
TD-DFT	time-dependent DFT
TE	transverse electric
TF	Thomas-Fermi
THG	third-harmonic generation
TLS	two-level amplifying system
TM	transverse magnetic
UV	ultraviolet
VCSEL	vertical-cavity surface-emitting laser
WDM	wavelength division multipling
XC	exchange and correlation

Chapter 1
Negative Refraction

Martin W. McCall
Imperial College London, UK

Graeme Dewar
University of North Dakota, Grand Forks, ND, USA

1.1 Introduction

The aim of this tutorial chapter is to cover the elements of the new optical science of negative refraction. We make no claim to exhaustive coverage. Our aim is rather to cover the material in such a way as to make this exciting field accessible to enthusiastic and well prepared undergraduates, as well as to new researchers in the field. Modern-day interest in negative refraction was sparked by Smith et al.,[1] who showed in 2000 that it was possible to make a structure that exhibited a negative index of refraction for microwaves. The novel property of their metamaterial was that the crests of the microwaves travelled in the direction opposite to which the energy was flowing. This led to the term "negative phase velocity" being applied to such media. Previously Schuster,[2] Mandelshtam,[3,4] and Veselago[5] had explored the implications of light's phase velocity being negative. The key characteristics of a medium that allows light to propagate freely with a negative phase velocity are that both the permittivity ϵ and the permeability μ must be negative. The property of having the crests of an electromagnetic wave move from the receiver toward the source is the result of the intervening medium having a negative magnetic permeability. Such waves associated with microwave fields in magnetic materials are called backward waves and have been known for more than half a century. The improvement offered by also having the permittivity negative is that these backward waves propagate without appreciable attenuation.

The metamaterial that Smith et al. fabricated consisted of sets of posts or wires responsible for $\epsilon < 0$ and an array of rings responsible for $\mu < 0$. Their initial experiment demonstrated a pass-band in a frequency range for which ϵ was negative, a condition that could only arise if μ were also negative. Subsequent experiments[6,7] demonstrated that this metamaterial refracted microwaves according to Snell's law, provided the metamaterial was characterized by a negative index of refraction.

Shortly after Smith et al.'s paper appeared, Pendry[8] described how a planar lens made of material with a refractive index of -1 could produce a "perfect" image. The lens was doubly perfect. First, the geometrical optics aberrations were zero. But the second and most intriguing aspect of this proposal was that an image could be produced that had detail much smaller than the wavelength of light used to create the image. Loss in the material used for the lens ultimately limited the subwavelength detail in the image.

1.2 Background

To begin, let us wind back the clock on our understanding of light. Even a child knows that a pencil looks bent when dipped in a glass of water. Later on those who begin studying physics learn how to quantify the bending via Snell's law connecting the incident angle θ_i, and the refracted angle θ_r, according to

$$\sin\theta_i = n\sin\theta_r\ , \tag{1.1}$$

where n is a property of the water called its *refractive index*. If light were particles that were somehow 'pulled' into the medium by the interface, then Eq. (1.1) is explained by postulating that n is given by $c_\perp/v_\perp$, the ratio of the velocity components in air and in the medium perpendicular to the interface. Since we know that $\theta_i > \theta_r$, then this would require $v_\perp > c_\perp$, implying that the speed of light in the medium is *greater* than in air. This was Newton's corpuscular theory of light propagation in media, made at a time when it was not technically possible to measure the speed of light. Of course even beginning physics students learn that the correct explanation, derived from Fig. 1.1, is found from a wave theory of light propagation. By matching wavefronts across the boundary, the refractive index is actually given by

$$n = \frac{c}{v}, \tag{1.2}$$

where c is the vacuum speed of light, and v is the speed of light in the medium. This predicts precisely the opposite of Newton's theory: namely, $v < c$, and light slows down as it crosses into a medium.

Undergraduate physics gives a completely new perspective. Light is the propagation of electromagnetic waves described by Maxwell's equations, which, in linear, homogeneous, isotropic media (e.g., water), is characterized by its permittivity ϵ, and its permeability μ. The theory shows that the speed of wave propagation in such a medium is given by $v = (\epsilon\mu)^{-1/2}$. Since vacuum is an example of such a medium we can write $c = (\epsilon_0\mu_0)^{-1/2}$ and

$$n = \pm\left(\frac{\epsilon\mu}{\epsilon_0\mu_0}\right)^{1/2}, \tag{1.3}$$

where for now we defer discussion of which sign in the square root should be taken. The amplitude of the electric field of a plane wave of frequency ω propagating along the z direction in the medium is given by the real part of

$$E = E_0 e^{i(kz-\omega t)}, \tag{1.4}$$

where $k = n\omega/c$ is the wavenumber. The quantities ϵ and μ relate respectively to the medium's response to excitation by electric and magnetic fields. At the high frequencies associated with optical radiation, media are characterized by $\mu = \mu_0$, so that $n = (\epsilon/\epsilon_0)^{1/2}$.

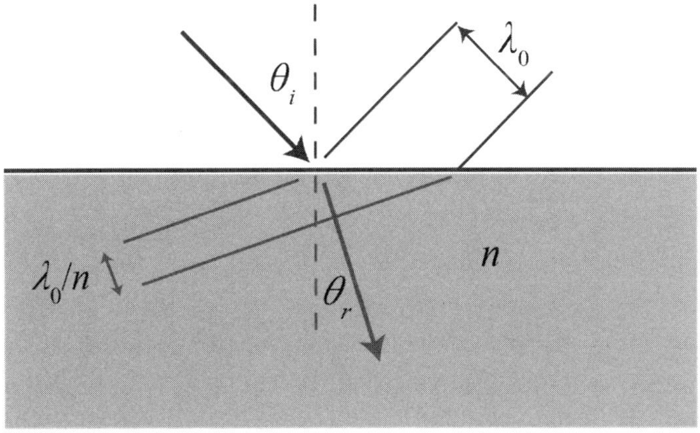

Figure 1.1 Deriving Snell's law of refraction on the basis of wave propagation. Assuming that the wavelength changes by a factor of n in passing from vacuum to the medium, Eq. (1.1) results from matching the projection of the wavelength along the interface.

The behavior of ϵ as a function of frequency is illustrated through the classical Lorentz–Drude model. The equation of a bound electron of mass m driven by the field E is given (at $z = 0$) by

$$\ddot{x} = -\omega_0^2 x - \Gamma \dot{x} - \frac{eE_0}{m} e^{-i\omega t} , \tag{1.5}$$

where ω_0 is the single resonance frequency and Γ is the damping rate per unit mass. For N electrons per unit volume, the polarization P is then given by

$$P = -eNx = \frac{Ne^2 E/m}{\omega_0^2 - \omega^2 - i\Gamma\omega} . \tag{1.6}$$

From $\epsilon E = \epsilon_0 E + P$ we find that

$$\epsilon(\omega) = \epsilon_0 \left(1 + \frac{\omega_p^2}{\omega_0^2 - \omega^2 - i\Gamma\omega} \right) , \tag{1.7}$$

where $\omega_p^2 = Ne^2/m\epsilon_0$. The resulting resonant behavior is illustrated in Fig. 1.2. The imaginary part is always positive as dictated by causality and the choice of the sign of the time dependence in Eq. (1.4). The real part has a negative region on the high-frequency side of the resonance, corresponding to the dielectric polarization being in antiphase with the driving field. Complex ϵ leads, via Eq. (1.3), to a complex refractive index which in turn, via Eq. (1.4), leads to an exponentially decaying field.

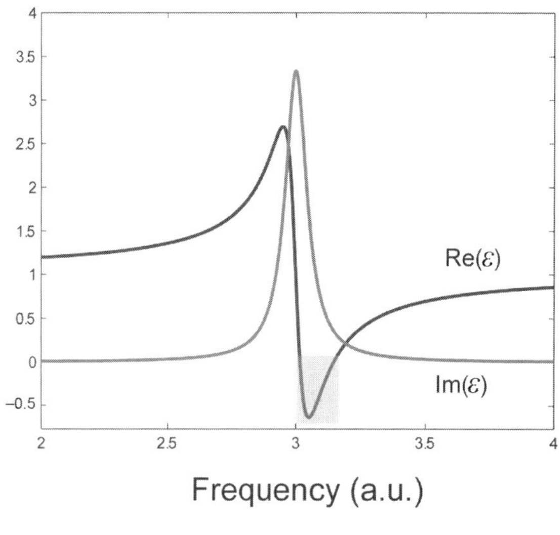

Figure 1.2 Illustrating the resonant behavior of $\epsilon(\omega)$. The shaded region indicates where $\mathrm{Re}(\epsilon) < 0$.

At frequencies just above the resonance apparent in Fig. 1.2, the medium's response, as characterized by the polarization, is just a bit more than 90 deg out of phase with the driving field and, as the frequency goes to infinity, the response approaches 180 deg out of phase. However, for ϵ to be negative, the response of the medium has to be bigger than the driving field, $P > \epsilon_0 E$, so that as the response weakens far from resonance, a point is reached where the response and drive just cancel ($E = 0$). This is called the plasma frequency; in other systems it is called an antiresonance. For a system of unbound charges, the resonance frequency is zero.

A similar sort of behaviour for magnetic moments, in place of electric charge, characterizes the permeability μ of media. Indeed, Fig. 1.2 is qualitatively similar to a description of $\mu(\omega)$. There is a resonance frequency for magnetic moments in an external magnetic field: nuclear magnetic resonance for nuclei, electron paramangetic resonance (or electron spin resonance) for conduction electrons, and ferromagnetic resonance for systems exhibiting a collective magnetization. Only the response of a ferromagnet or ferrimagnet is large enough that the drive can be overwhelmed, with the permeability above the resonance going negative. The point at which the response weakens so that $\mu = 0$ is called ferromagnetic antiresonance.

The field intensity decays according to $I = I_0 \exp(-\alpha z)$, where

$$\alpha = \frac{2\omega}{c}\mathrm{Im}(n) \ . \tag{1.8}$$

Real media can contain several resonance frequencies, and quantum mechanical calculations provide expressions for the individual oscillator strengths.

The presence of Γ in the Lorentz–Drude model indicates the presence of dissipation in the medium, which then absorbs a fraction of the incident radiation. Provided the dissipation is not too great, there exist bulk oscillations (plasmons) of the free electron plasma wherein Eq. (1.7), in the absence of dissipation, becomes[9]

$$\epsilon(\omega) = \epsilon_0 \left(1 - \frac{\omega_p^2}{\omega^2}\right) \ . \tag{1.9}$$

Thus $\epsilon(\omega)$ is negative when ω is less than the plasma frequency ω_p. In this lossless case the refractive index is purely imaginary, corresponding to stationary evanescent fields in the medium. The intensity in the medium still decays according to Eq. (1.8), though in the steady state no energy is transported into the medium, and all of the incident radiation is reflected.

The elucidation of $\epsilon(\omega)$ would appear to complete the description of electromagnetic radiation in linear, homogeneous, isotropic media. Radiation incident to an interface between air ($n \approx 1$) and the medium will be refracted according to Snell's law, Eq. (1.1), whilst fractions of the radiation are reflected and absorbed, according to the frequency relative to any resonance, and the presence of damping. There would appear to be little that could change this picture. The only possibility would be to somehow 'switch on' the permeability μ. However, the magnetic dipole reorientation times are such that natural materials are not magnetically active at the high frequencies at which the same materials are electrically interesting. As Landau and Lifschitz[10] put it:

> "Unlike ϵ, μ ceases to have any physical meaning above a few GHz. To take account of $\mu(\omega)$ would be an unwarrantable refinement ... there is certainly no meaning in using the magnetic susceptibility from optical frequencies onwards and in discussion of such phenomena we must put $\mu = 1$."

It would appear that this section has therefore summarized the electromagnetic behavior of media above a few tens of GHz through its description via $\epsilon(\omega)$ alone. However, physicists like to speculate, which is precisely what Viktor Vesalago did, as will be described in the next section.

1.3 Beyond Natural Media: Waves That Run Backward

In 1968 Viktor Veselago[5] cast aside the restrictions implied by Landau and Lifschitz and speculated on the behavior of a medium that was simultaneously electrically and magnetically responsive at higher frequency (e.g., $\sim$ GHz). Although the most general description would account for complex values of ϵ and μ, the key issues are exposed for media for which ϵ and μ are both real, but their signs are *unrestricted*. We are thus led to consider the quadrant diagram in Fig. 1.3. The familiar behavior of dielectric media is represented in the first quadrant wherein ϵ and μ are both positive. Electromagnetic waves propagate with phase speed $(\epsilon\mu)^{-1/2} = n^{-1}(\epsilon_0\mu_0)^{-1/2}$. In the second quadrant ($\epsilon < 0$, $\mu > 0$) waves are evanescent as described in the previous section. We conclude that in the fourth quadrant ($\epsilon > 0$, $\mu < 0$) waves are also evanescent from the symmetry between ϵ and μ in the expression $n = (\epsilon\mu/\epsilon_0\mu_0)^{1/2}$. The interesting question is what happens in the third quadrant where ϵ and μ are simultaneously negative.

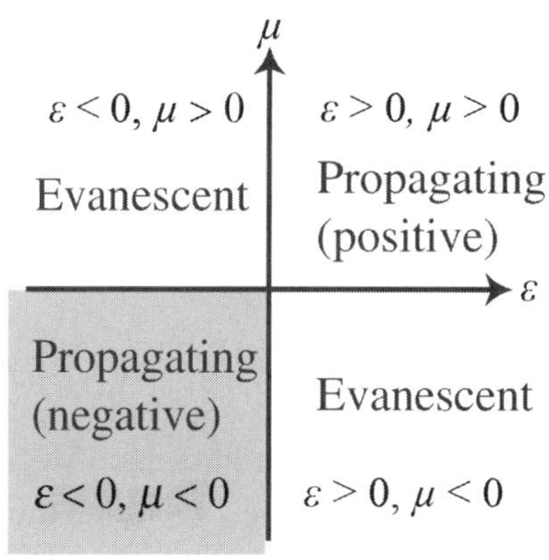

Figure 1.3 The "ϵ–μ" quadrant diagram.

In order to answer this we must consider Maxwell's equations in the frequency domain. The field $\mathbf{E}(\mathbf{r},t)$ can in general be written as a Fourier expansion of plane waves

$$\mathbf{E}(\mathbf{r},t) = \int \mathbf{E}(\mathbf{k},\omega)e^{i(\mathbf{k}\cdot\mathbf{r}-\omega t)}d^3\mathbf{k}\,d\omega\,, \tag{1.10}$$

with similar expansions for the other fields. Faraday's law $\nabla\times\mathbf{E}(\mathbf{r},t) = -\partial\mathbf{B}(\mathbf{r},t)/\partial t$ can then be written as

$$\mathbf{k}\times\mathbf{E}(\mathbf{k},\omega) = \omega\mathbf{B}(\mathbf{k},\omega)\,. \tag{1.11}$$

Similarly, in the absence of free charges, Ampere's law $\nabla\times\mathbf{H}(\mathbf{r},t) = \partial\mathbf{D}(\mathbf{r},t)/\partial t$, becomes

$$\mathbf{k}\times\mathbf{H}(\mathbf{k},\omega) = -\omega\mathbf{D}(\mathbf{k},\omega)\,. \tag{1.12}$$

The quantities $\epsilon(\omega)$ and $\mu(\omega)$ discussed in the previous section are frequency domain quantities. They relate the field variables according to the constitutive relations

$$\mathbf{D} = \epsilon(\omega)\mathbf{E}\,, \tag{1.13}$$

$$\mathbf{B} = \mu(\omega)\mathbf{H}\,, \tag{1.14}$$

where the $(\mathbf{k},\omega)$ dependence of the field variables has been omitted for brevity. Substituting these relations into Eqs. (1.11) and (1.12) yields

$$\mathbf{k}\times\mathbf{E} = \omega\mu\mathbf{H}\,, \tag{1.15}$$

and

$$\mathbf{k}\times\mathbf{H} = -\omega\epsilon\mathbf{E}\,. \tag{1.16}$$

If ϵ and μ are both negative, the disposition of the vectors $\mathbf{k}$, $\mathbf{E}$, $\mathbf{B}$, and $\mathbf{H}$ must be as shown in Fig. 1.4. In particular, we note that the vectors $\mathbf{k}$, $\mathbf{E}$, and $\mathbf{H}$ must form a *left-handed* triad. The power flow is still given by the conventional expression

$$\mathbf{S} = \frac{1}{2}\mathrm{Re}(\mathbf{E}\times\mathbf{H}^*)\,. \tag{1.17}$$

We see from Fig. 1.4 that $\mathbf{S}$ *is anti-parallel to* $\mathbf{k}$, or

$$\mathbf{S}\cdot\mathbf{k} < 0\,. \tag{1.18}$$

Just as is the case for natural media, the dispersion relation adduced from combining Eqs. (1.15) and (1.16) is

$$k^2 = \epsilon\mu\omega^2\,. \tag{1.19}$$

It is now clear that in order to make the direction of $\mathbf{k}$ anti-parallel to $\mathbf{S}$, as required by Eqs. (1.15) and (1.16), *it is necessary to take the negative root of Eq. (1.19).* Hence for this situation

$$n = -\left(\epsilon\mu/\epsilon_0\mu_0\right)^{1/2}\,. \tag{1.20}$$

For a medium described by simultaneously negative (real) values of ϵ and μ, the refractive index is negative.* The necessity to take a negative value of the refractive index may seem trivial and inconsequential, but is not. The immediate implication takes us right back to the pencil in the glass of water experiment, where the pencil appears bent because of Snell's law. Consider refraction at the interface between vacuum $n = 1$ and a medium with refractive index $n = -1$. (See Fig. 1.5.) According to Snell's law, Eq. (1.1), we have $\theta_r = -\theta_i$, *i.e., the refracted wave is refracted on the same side of the normal as the incident wave.* It is the possibility of achieving this so-called *negative refraction* that has caused such excitement in recent years. In fact through negative refraction and related phenomena, it turns out to be possible to achieve effects such as a 'perfect' lens, and an electromagnetic cloak.

Various names have been attributed to the phenomenon discussed in this section without a clear winner emerging. The disposition of $\mathbf{k}$, $\mathbf{E}$, and $\mathbf{H}$ in Fig. 1.4 has prompted some to call media that support this situation *left handed*. The trouble with this is that the media are *not* chiral. The left-handed triad of Fig. 1.4 emerges ultimately from the arbitrary choice (out of two possibilities) of what direction should be associated with the cross product between two vectors. Others have used *double negative media* in recognition of the idea that historically, these issues emerged from consideration of idealized media in which both ϵ and μ are real and negative. The trouble with this name is that actual media are described by complex ϵ and μ, and, as we shall see, the crucial feature of $\mathbf{S} \cdot \mathbf{k} < 0$ can be met by media for which (the real parts of) ϵ and μ are *not* both negative. One of the authors

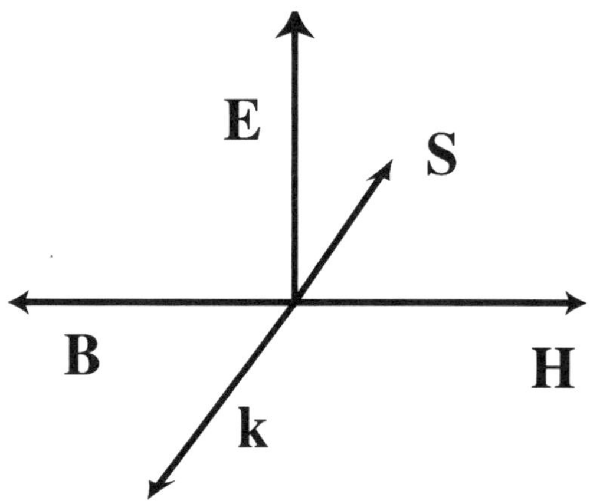

Figure 1.4 The disposition of the vectors $\mathbf{k}$, $\mathbf{E}$, $\mathbf{B}$, $\mathbf{H}$, and $\mathbf{S} = \frac{1}{2}\mathrm{Re}\,(\mathbf{E} \times \mathbf{H}^*)$ when ϵ and μ are both real and negative.

*An informal mnemonic for remembering this result is to say that the square root of the product of two negative numbers is negative. Of course, it is Maxwell's equations and the assumed direction of power flow that dictate which branch of the square root to choose.

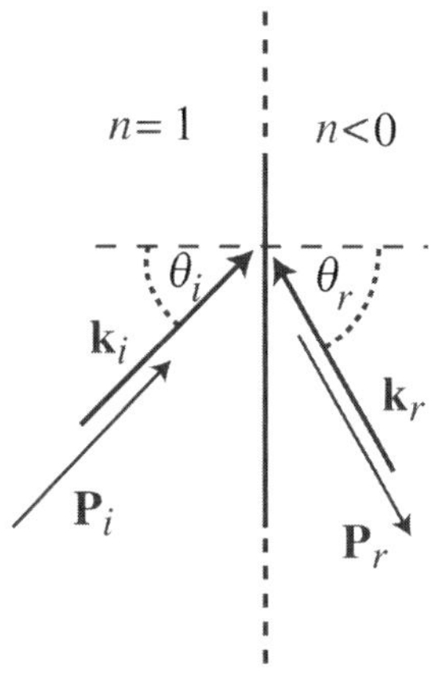

Figure 1.5 Refraction at the interface between vacuum and a medium described by a negative refractive index.

(McCall) and his co-workers have suggested the name Negative Phase Velocity media,[11] which at least captures the essential defining property. The term negative refraction appears to be used most widely. This is somewhat unfortunate since birefringent materials can exhibit negative refraction without the phase velocity being negative.

But how much of this pertains to reality? The intriguing 'double-negative' quadrant in Fig. 1.3 would appear to be inaccessible on account of the lack of a medium that possesses a negative permeability at frequencies above a few GHz. Does including absorption [neglected in Fig. (1.3)] change anything? This is an interesting question to be answered formally in Sec. 1.7. However, we can reach a partial answer by looking at the dispersion of the refractive index according to the Lorentz–Drude model developed earlier. From Eq. (1.7) we calculate the wave number $k = (\epsilon\mu_0)^{1/2}\omega$, the real and imaginary parts being displayed as a function of ω in Fig. 1.6. The wave group velocity $v_g = \partial\omega/\partial k$ is identified as the gradient of the curve $\omega\,(\mathrm{Re}(k))$. We note that v_g is negative for a small frequency region on either side of the resonance frequency.[†] Since the group velocity is associated with the direction of power flow, we conclude that over this frequency range the wave is backward; i.e., whilst the power flows away from the source, the wave's phase advances toward it. However, it is also important to note that this region is necessarily associated with significant absorption indicated by $\mathrm{Im}(k)$. Thus, although the wave runs backward, it is attenuated quite strongly. All of this was noted by Schuster[2] in 1904, who said of this region:

[†]In addition, the group velocity may exceed the vacuum speed of light for frequencies near an absorption resonance. However, the more physically relevant *signal* velocity, a space and time domain concept, is neither negative nor faster than light speed. See Léon Brillouin's book,[12] especially pages 74–79.

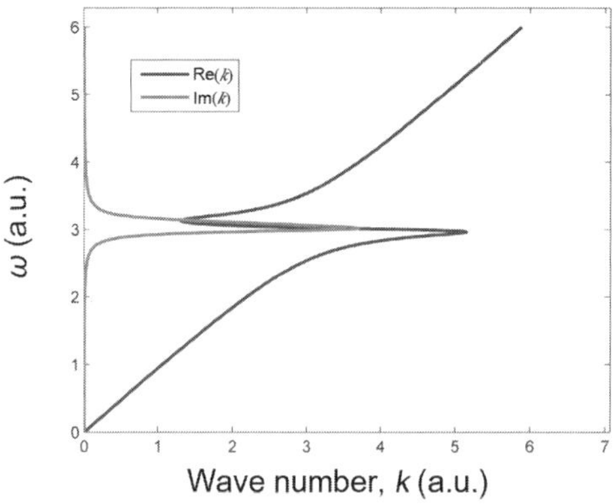

Figure 1.6 Showing the region of anomalous dispersion near a dielectric resonance.

> "If there is a convection of energy forward, the waves must therefore move backwards. In all optical media where the direction of the dispersion is reversed, there is a very powerful absorption ... under these circumstances it is doubtful how far the above results have any application ... One curious result follows: the deviation of the wave on entering a medium is greater than the angle of incidence, so that the wave normal is bent over to the other side of the normal."

So the title of this section is actually a misnomer; naturally occurring media *do* support backward waves. However, as we showed earlier in this section, only if the permittivity and permeability are simultaneously real and negative, are backward waves propagated *without* attenuation.

1.4 Wires and Rings

The ideas in Veselago's prescient paper could not be realized at the time. However, in 2000 a number of factors combined to make Veselago's ideas a reality.

The interesting characteristics of negative refraction are principally associated with spectral regions in which the real parts of both $\epsilon(\omega)$ and $\mu(\omega)$ are simultaneously negative. How can this be achieved? Ostensibly achieving negative $\epsilon(\omega)$ is easy. We have already seen that negative ϵ occurs for plasmas excited below the plasma frequency ω_p. However, there is a problem that we have not yet considered. Real materials always have some loss, so that a small damping coefficient Γ should be included. Copper, for example, has $\Gamma \sim 4 \times 10^{13}\ \mathrm{rad\,s^{-1}}$, $\omega_p \sim 10^{16}\ \mathrm{rad\,s^{-1}}$. For frequencies such that $\Gamma << \omega << \omega_p$, $\mathrm{Re}(\epsilon)$ is negative. However, for low frequencies $\omega << \Gamma$, Eq. (1.7) (with $\omega_0 = 0$) shows that imaginary contribution to the dielectric constant is given by

$$\epsilon(\omega << \Gamma) \approx i\epsilon_0 \frac{\omega_p^2}{\omega\Gamma}\,, \tag{1.21}$$

and the imaginary part of ϵ rises rapidly as the frequency is lowered. The real and imaginary parts are plotted in Fig. 1.7. The attenuation coefficient for this case is determined from Eq. (1.8) as

$$\alpha = \left(\frac{2\omega}{\Gamma}\right)^{1/2} \frac{\omega_p}{c}, \tag{1.22}$$

so that in order for the field to penetrate the material significantly, a low plasma frequency is desirable. However, it cannot be too low, since ω_p precisely determines the frequency at which $\mathrm{Re}(\epsilon) < 0$. In order to lower the plasma frequency, a lower free electron number density N is required. This leads to the idea of embedding thin conducting wires within a host dielectric, as illustrated in Fig. 1.8.

Another benefit of using the thin wire array is that the inductance of the wires enhances the effective electron mass,[13] so that ω_p is further reduced. In fact,

$$m_{eff} = \frac{\mu_0 e^2 \pi r^2 N}{2\pi} \ln\left(\frac{a}{r}\right), \tag{1.23}$$

where r is the wire radius and a is the lattice spacing of the array.

Equation (1.23) was somewhat controversial when it first appeared.[14–16] Some researchers had difficulty with the notion that an electron had an effective mass greater than that of an atom. Of course the physics is clear. Besides accelerating and gaining kinetic energy from the electric field, the electrons must also build up magnetic field energy. The energy associated with the magnetic field of the currents (moving electrons) in the wires is orders of magnitude greater than the electrons' kinetic energy. Hiding this near-zone magnetic field energy in the kinetic energy term by invoking an effective mass for the electron is what accounts for Eq. (1.23). This immediately raises the question of why this does not affect electrons in bulk

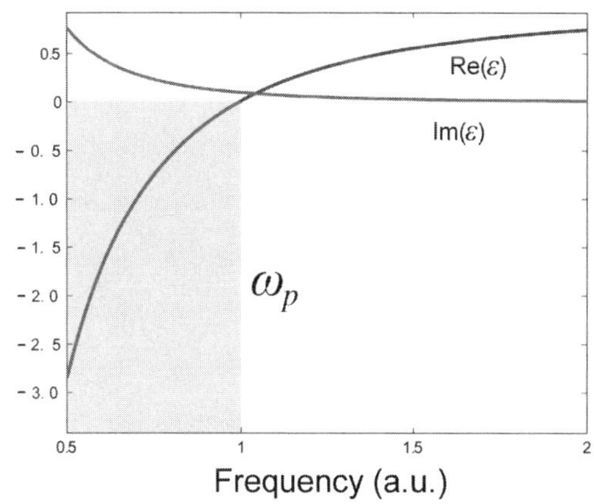

Figure 1.7 Plot of damped plasma permittivity. See Eq. (1.7) (with $\omega_0 = 0$). The shaded region indicates where $\mathrm{Re}(\epsilon) < 0$.

media. The currents associated with moving charges are completely ignored in the standard treatment of the plasma frequency. This is a consequence of the assumed uniformity of the medium under consideration. The magnetic field from currents near any electron to which Newton's second law is applied [as embodied in Eq. (1.5)] cancel by symmetry. The size of the medium is assumed to be large but of such a shape that the magnetic fields arising from distant currents are negligible. Thus magnetic effects in bulk media do not play a role in determining the plasma frequency. However, with an array of wires, this continuous translation symmetry is destroyed, and one must consider the magnetic forces arising from the collective motion of nearby charges.

The three-dimensional (3D) simple cubic grid shown in Fig. 1.8 has the advantage that the dielectric response is isotropic. However, recently, issues of spatial dispersion have been brought to light in both 2D[17] and 3D[18] wire arrays that may mitigate against producing this desirable isotropy in practice.[19]

The general principle of producing an artificial plasma in which $\mathrm{Re}\,[\epsilon(\omega)] < 0$ with small loss using wire grid structures embedded in a dielectric host has been confirmed experimentally in the GHz frequency range.[20] A more detailed description of how the wire array depresses the plasma frequency as well as the attenuation is in Sec. 1.11.1. It is indeed feasible to lower the plasma frequency of a wire array to the GHz range where natural materials with $\mu < 0$ exist. It is then natural to ask what would happen if a wire array were placed in a magnetic medium with $\mu < 0$. The key effect is that the $\epsilon < 0$ property of the wire array would be destroyed.[21,22] The problem is that the negative μ causes $\mathbf{B}$ and $\mathbf{H}$ to be antiparallel and the inductive energy, essentially the volume integral of the $\mathbf{B} \cdot \mathbf{H}$, is negative, so the wire

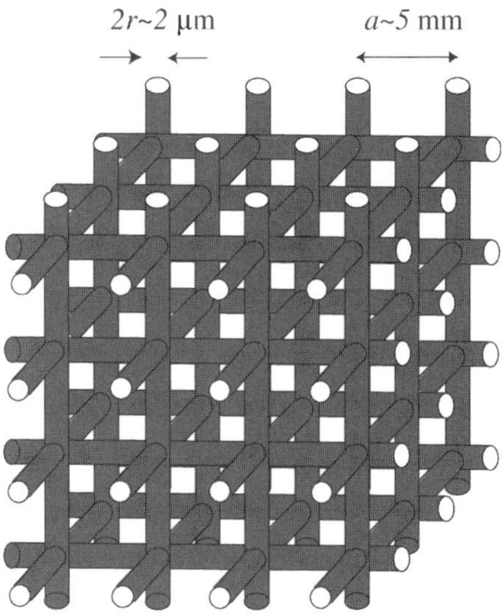

Figure 1.8 Thin wire structure for producing low-frequency plasmons.

array behaves capacitively. This causes m_{eff} of Eq. (1.23) to be negative, hence $\omega_p^2 < 0$ and $\epsilon(\omega) > 0$. However, this can be overcome by decreasing the coupling between the wires and the surrounding magnetic medium by cladding the wires with a nonconducting, nonmagnetic material.[21,23]

Let us turn now to the problem of producing a material that has an effectively negative permeability (real part) at frequencies of tens of GHz and above, where there are no naturally occurring magnetic media. Magnetic monopoles are not found in nature, so an equivalent "magnetic plasma" cannot be used to create a negative permeability. To overcome this, the concept of *artificial magnetism* has been introduced. Whereas naturally magnetic activity originates in the reorientation of magnetic moments within the material's atoms, magnetic activity can be synthesized artificially via conducting elements that produce a magnetic moment in response to excitation. These resonators, first described by Pendry et al.,[24] are rings that interact with microwaves, much as would a magnetic-dipole-type loop antenna. However, the rings have a gap, or split, cut into them. (See Fig. 1.9.)

The splits introduce some capacitance at the ends of the gap, and two rings are placed in close proximity to further increase the capacitance. The net result is an L-C resonator composed of two tightly coupled C-shaped rings, which has a resonance frequency so low that the dimensions of the rings are much smaller than the wavelength for resonant microwaves. The fact that the structure is resonant means that the effective permeability has a similar form to that for the permittivity given in Eq. (1.7). Thus, with appropriate design, a metamaterial in which an array of such split ring resonators is embedded in a dielectric can display magnetic activity at GHz frequencies, with negative values of the effective permeability occurring just above resonance. For the design of Fig. 1.9, the resonance is about 4.7 GHz.[1]

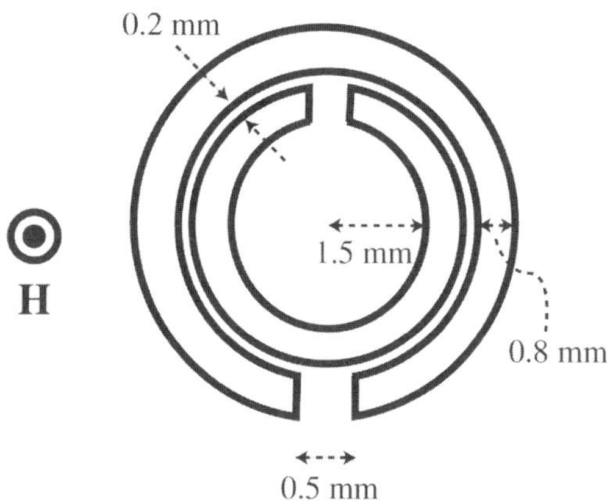

Figure 1.9 The split-ring resonator.

1.5 Experimental Confirmation

The experiment reported by Smith et al.[1] was the first experimental verification that an artificial medium, a metamaterial, could have simultaneously negative ϵ and μ while transmitting electromagnetic waves without overwhelming attenuation. This paper sparked a flurry of activity involving negative refraction in metamaterials. The key innovation they introduced was to combine a wire array constituting an effective electrical plasma ($\epsilon < 0$) with an array of split ring resonators ($\mu < 0$). Initially the microwave transmission through a metamaterial consisting of just an array of split ring structures was measured, and a stop band of several hundred megahertz bandwidth near 5 GHz was noted. This stop band was attributed to the effective negative μ between the array's resonance frequency of about 4.7 GHz and the antiresonance frequency of 5.2 GHz. Subsequently, an array of wires having a plasma frequency of 12 GHz was inserted into the array of split ring resonators and the microwave transmission was again measured. The two interpenetrating arrays of wires and rings had a pass band in the frequency range that was a stop band for the split ring resonators alone. This result was interpreted as evidence for simultaneously negative ϵ and μ.

The question of whether or not Smith et al.'s metamaterial exhibited negative refraction immediately arose. The experimental verification of negative refraction for microwaves falling on a prism made of Smith et al.'s metamaterial came in 2001.[6] In this experiment 10.5-GHz microwaves, normally incident on one side of a prism, refracted upon exiting the prism such that the ray they followed did not cross the normal to the interface. This is the behavior expected of a negative index of refraction metamaterial, as indicated in Fig. 1.5. A similar experiment, with a teflon prism replacing the negative phase velocity metamaterial, showed the ray refracting in the usual manner and crossing the interface normal. An issue arose concerning the propagation distances involved since these distances were only on the order of several wavelengths. Any concern about detectors not being in the prism's radiative far-zone were eliminated by Houck et al.[7]

1.6 The "Perfect" Lens

Conceptually, it is very simple to make a lens out of slab of a medium supporting negative refraction. Consider the imaging geometry in Fig. 1.10. Simple ray tracing shows that if the object is placed a distance d_1 from the front surface of a slab of thickness d, then the image is formed a distance $d_2 = d - d_1$ behind the rear face. It's a far cry from the elementary lens formula, $1/u + 1/v = 1/f$. For one thing, in order to form a real image, the object must be placed a distance $d_1 < d$, i.e., it must be placed in front of the slab a distance that is smaller than the thickness of the slab in order to form a real image. However, a deeper aspect arises as a result of analyzing how images are formed from the perspective of wave theory.

The time-independent part of the electric field emerging from a 2D object placed at some position on the z-axis may be described by its 2D Fourier transform

$$\mathbf{E}(x, y, z) = \int_{k_x, k_y} \mathbf{E}(k_x, k_y) e^{i(k_x x + k_y y + k_z z)} dk_x dk_y \ . \tag{1.24}$$

The integration is over the wave vector components in the $x - y$ plane. Since $n = 1$ in the object space, we also have the free space dispersion relation

$$k_x^2 + k_y^2 + k_z^2 = \frac{\omega^2}{c^2} = \left(\frac{2\pi}{\lambda_0}\right)^2 \ . \tag{1.25}$$

Hence the partial waves with real k_z are restricted to those for which $k_x^2 + k_y^2 < \omega^2/c^2 = k_{\rm max}^2$. Since the transverse wave vectors are restricted in this way, the smallest feature Δ that can be reconstructed from the partial waves up to $k_{\rm max}$ is given by

$$\Delta = \frac{2\pi}{k_{\rm max}} = \lambda_0 \ . \tag{1.26}$$

However, the remarkable thing about the slab lens (or 'Vesalago lens' as it is sometimes called), is that imaging *is not restricted to transverse wave numbers up to* $k_{\rm max}$. This was the major insight of Pendry.[8] Wave numbers beyond $k_{\rm max}$ correspond to waves for which k_z is imaginary. These are the nonpropagating, near-field components of the dipole fields associated with the object. These fields, which are evanescent and do not transport energy, decay very rapidly with distance from the object. Nevertheless they carry the high spatial frequency information that is lost in conventional imaging.

One way to see what happens is to solve the problem of propagating a unit amplitude vacuum electromagnetic field incident to a slab characterized by ϵ and

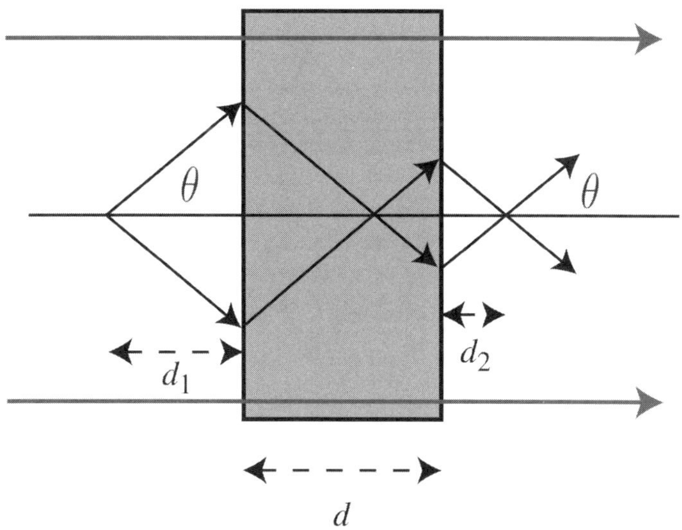

Figure 1.10 A near-field lens made out of a slab of negatively refracting medium.

μ. (See Fig. 1.11.) For s-polarization the electric field component E_y in each of the three indicated regions may be written as

$$E_{y1} = e^{ik_x x}\left(e^{ik_{0z}z} + re^{-ik_{0z}z}\right), \tag{1.27}$$

$$E_{y2} = e^{ik_x x}\left(A^+ e^{ik_z z} + A^- e^{-ik_z z}\right), \tag{1.28}$$

$$E_{y3} = te^{i(k_x x + k_{0z}z)}, \tag{1.29}$$

where r, t, A^+, and A^- are unknown field amplitudes and k_0 is the free space wave number. Note that the transverse wave vector k_x is the same in all three regions, whilst

$$k_{0z} = +\left(k_0^2 - k_x^2\right)^{1/2}, \tag{1.30}$$

$$k_z = -\left[\left(\frac{\epsilon\mu}{\epsilon_0\mu_0}\right)k_0^2 - k_x^2\right]^{1/2}. \tag{1.31}$$

Taking the positive root in Eq. (1.30) ensures that the field decays away from the source for $k_x > k_0$. The sign of the root in Eq. (1.31) actually does not matter in this case, as there are waves in both directions in the slab, although taking the negative root is consistent with negative phase velocity propagation for fields propagating away from the interface. The magnetic field components H_x are then obtained from Eq. (1.15) as

$$\omega H_{x1} = \frac{k_{0z}}{\mu_0}e^{ik_x x}\left(-e^{ik_{0z}z} + re^{-ik_{0z}z}\right), \tag{1.32}$$

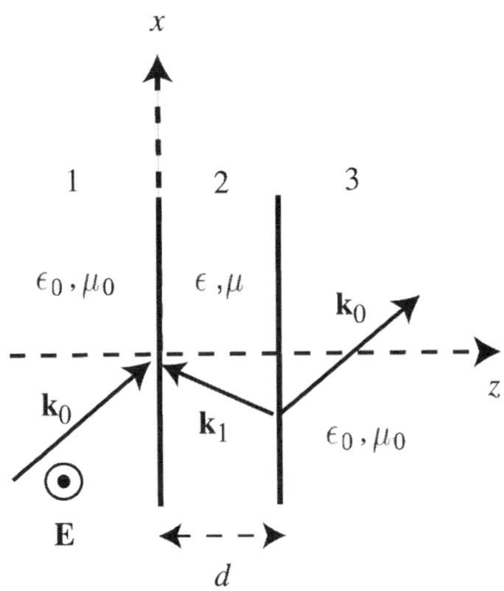

Figure 1.11 Geometry for examining fields in a dielectric slab. Negative refraction is illustrated when, for example, $\mathrm{Re}(\epsilon, \mu) < 0$.

$$\omega H_{x2} = \frac{k_z}{\mu} e^{ik_x x}\left(-A^+ e^{ik_z z} + A^- e^{-ik_z z}\right), \tag{1.33}$$

$$\omega H_{x3} = -\frac{k_z}{\mu_0} t e^{i(k_x x + k_{0z} z)}. \tag{1.34}$$

Matching the tangential fields E_y and H_x at each boundary yields

$$1 + r = A^+ + A^- \tag{1.35}$$

$$\frac{k_{0z}}{\mu_0}(-1 + r) = \frac{k_z}{\mu}(-A^+ + A^-) \tag{1.36}$$

$$A^+ e^{ik_z d} + A^- e^{-ik_z d} = t e^{ik_{0z} d} \tag{1.37}$$

$$\frac{k_z}{\mu}(-A^+ e^{ik_z d} + A^- e^{-ik_z d}) = -\frac{k_{0z}}{\mu_0} t e^{ik_{0z} d}. \tag{1.38}$$

Eliminating $A^{\pm}$ yields the slab reflection and transmission coefficients as

$$r = \rho\left(\frac{1 - e^{2ik_z d}}{1 - \rho^2 e^{2ik_z d}}\right), \tag{1.39}$$

and

$$t = (1 - \rho^2)\left(\frac{e^{i(k_z - k_{0z})d}}{1 - \rho^2 e^{2ik_z d}}\right), \tag{1.40}$$

where

$$\rho = \frac{\mu k_{0z} - \mu_0 k_z}{\mu k_{0z} + \mu_0 k_z} \tag{1.41}$$

is the amplitude reflection coefficient for light passing from vacuum into the medium [note from Eqs. (1.40) and (1.41) that $t(k_z) = t(-k_z)$, so that the sign of the square root in Eq. (1.31) does not matter]. Thus far the theory applies for general ϵ and μ. For current considerations the most interesting case occurs when $\epsilon = -\epsilon_0$ and $\mu = -\mu_0$. We then have from Eq. (1.31) that $k_z = -k_{0z}$ and consequently $\rho = 0$ in Eq. (1.41). This is quite reasonable since a medium with $\epsilon = -\epsilon_0$ and $\mu = -\mu_0$ is impedance matched to vacuum (i.e., $(\mu/\epsilon)^{1/2} = (\mu_0/\epsilon_0)^{1/2}$). However, look what happens to the transmission for a field for which $k_x > k_0$. Setting $k_z = -i\kappa$ (or $k_z = i\kappa$), we find from Eqs. (1.29), (1.40), and (1.41) that

$$|E_{y3}(z = d)|^2 = e^{2\kappa d}. \tag{1.42}$$

The field transmitted through the slab increases exponentially with slab thickness. There is no violation of energy conservation as evanescent waves do not transport energy; the resultant field profile (for $k_x = 1.1k_0$) is illustrated in Fig. 1.12. Although we see that the transmission coefficient is large, the fields decay rapidly on either side of the interface at $z = d$. For an object placed at $z = -\lambda/2$, the field decays to precisely the value it had at the object at a location $z = d + \lambda/2$. So not only does the slab image real waves that are associated with object spatial

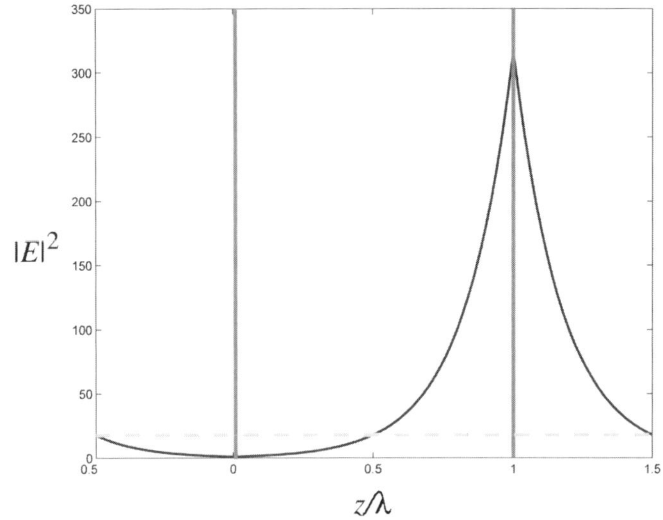

Figure 1.12 Illustrating the amplification of evanescent fields nearby a negative index slab. The slab is one wavelength thick, and the field shown is one for which $k_x = 1.1k_0$. Note that the field decays away from the object location at $z = -\lambda/2$, but is restored to precisely its object value at the image location.

frequencies k_x up to k_0, it also 'images' evanescent waves for which $k_x > k_0$! In principle there is no limit to the resolution obtainable, although a lens made with exactly $\epsilon = -\epsilon_0$, $\mu = -\mu_0$ suffers from a divergence. Surface modes carry off so much energy[25] that, for certain positions of the source, the amount of light reaching the image is vanishingly small.[26] Losses that are present in all media destroy the divergence but, along with dispersion, also limit performance.[27,28] Indeed, as we have seen, the construction of metamaterials that support backward waves is necessarily dispersive in order to access unusual values of ϵ and μ. Nevertheless, the pursuit of a 'perfect' lens has been reduced to one of technology rather than one precluded by diffraction physics.

1.7 The Formal Criterion for Achieving Negative Phase Velocity Propagation

We showed in Sec. 1.3 how a putative material with real negative values of ϵ and μ led to the curious situation in which the direction of power flow of a plane electromagnetic wave opposed the direction of phase advance: $\mathbf{S} \cdot \mathbf{k} < 0$. We now use this criterion as the basis of establishing a rigorous criterion for the occurrence of, let us say, Negative Phase Velocity (NPV) propagation. In recognition that all media will inevitably contain some loss, we allow complex values for ϵ and μ and derive the general conditions for achieving NPV. Since $n = \pm\left(\epsilon\mu/\epsilon_0\mu_0\right)^{1/2}$ is also complex, we have that the wave vector

$$\mathbf{k} = n\frac{\omega}{c}\hat{\mathbf{k}} \tag{1.43}$$

is now a 'complex vector' in the sense that its component in the direction of the unit vector $\hat{\mathbf{k}}$ can be complex valued. Since we know that it is the real part of n that relates to the phase of the wave, we will write the NPV condition as

$$\mathbf{S} \cdot \mathrm{Re}(\mathbf{k}) < 0 \,, \tag{1.44}$$

where $\mathrm{Re}(\mathbf{k})$ means $\mathrm{Re}(n)\frac{\omega}{c}\hat{\mathbf{k}}$. Substituting Eqs. (1.4) and (1.15) into (1.17) we obtain

$$\mathbf{S} = \frac{1}{2}\mathrm{Re}\left(\frac{n}{\mu}\right) E_0^2 \exp(-2\mathrm{Im}(n)\omega z/c)\hat{\mathbf{k}} \,. \tag{1.45}$$

Now let us examine the complex numbers ϵ, μ, and n in detail by writing them as

$$\epsilon = \epsilon' + i\epsilon'' = |\epsilon| \exp i\phi_\epsilon \,, \tag{1.46}$$

$$\mu = \mu' + i\mu'' = |\mu| \exp i\phi_\mu \,, \tag{1.47}$$

$$n = n' + in'' = |n| \exp i\phi_n \,. \tag{1.48}$$

Choosing the positive root in $n = n_+ = +\left(\epsilon\mu/\epsilon_0\mu_0\right)^{1/2} = c\left(\epsilon\mu\right)^{1/2}$ we have

$$n_+ = c\,|\epsilon|^{1/2}|\mu|^{1/2} \exp\left[\frac{i}{2}\left(\phi_\epsilon + \phi_\mu\right)\right] \,, \tag{1.49}$$

and

$$\frac{n_+}{c\mu} = \frac{|\epsilon|^{1/2}}{c|\mu|^{1/2}} \exp\left[\frac{i}{2}\left(\phi_\epsilon - \phi_\mu\right)\right] \,. \tag{1.50}$$

The causality condition that ϵ'' and μ'' are both positive implies that $0 \leq \phi_{\epsilon,\mu} \leq \pi$, so that the argument of n_+ must also lie between these limits. This also means that

$$\mathrm{Re}\left(\frac{n_+}{\mu}\right) > 0 \,, \tag{1.51}$$

and thus according to Eq. (1.45), the power flow for this root is in the direction of $\hat{\mathbf{k}}$. Similar considerations show that the argument of the negative root lies between 0 and $-\pi$, and the power flow is in the direction of $-\hat{\mathbf{k}}$. We can now formalize the conditions for NPV propagation. From Eqs. (1.44) and (1.45) we have that either

$$\mathrm{Re}(n_+) < 0 \,, \text{ or } \mathrm{Re}(n_-) > 0 \,. \tag{1.52}$$

In fact these two conditions go together since imposition of one automatically implies the other. It is clear from Eq. (1.51) that it is the permeability μ that dictates the sign of $n_\pm$; the negative phase velocity is a magnetic effect. From consideration of the definition $n = \pm c\left(\epsilon\mu\right)^{1/2}$ it is readily shown that NVP propagation occurs whenever

$$\left(|\epsilon| - \epsilon'\right)\left(|\mu| - \mu'\right) > \epsilon''\mu'' \,. \tag{1.53}$$

This is the general condition for NPV propagation shown first by McCall et al.[11]

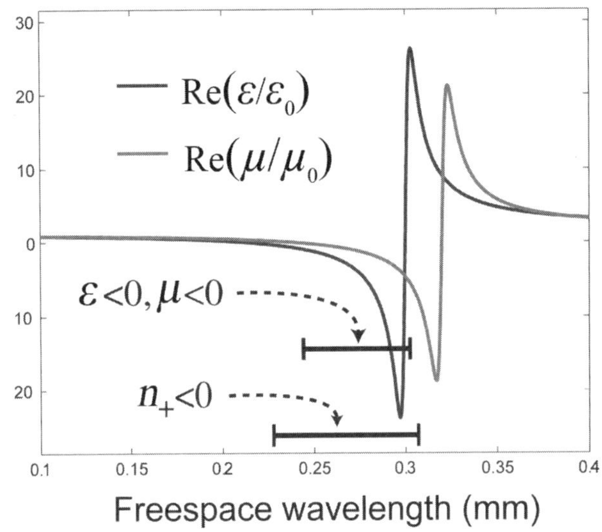

Figure 1.13 Illustrating the occurrence of NPV propagation according to Eq. (1.53). Note that the defined NPV propagation region ($\mathrm{Re}(n_+) < 0$) extends beyond the central "double negative" region $\epsilon < 0$, $\mu < 0$.

As an illustration, single-resonance Lorentz–Drude models can be applied for both $\epsilon(\omega)$ and $\mu(\omega)$ where the resonance frequencies for ϵ and μ are close but distinct, as illustrated in Fig. 1.13. The figure shows ϵ' and μ' together with the region for which $\mathrm{Re}(\epsilon, \mu)$ are simultaneously negative as well as the region for which $n_+ < 0$, according to the criterion of Eq. (1.53). The defined NPV propagation region ($\mathrm{Re}(n_+) < 0$) extends beyond the central "double negative" region. Thus it is possible to achieve negative refraction in regimes distinct from negative real parts of ϵ and μ.

1.8 Fermat's Principle and Negative Space

Classical optics is often couched in terms of Fermat's principle of least time, which asserts that the variation of the optical path length is zero

$$\delta \int_1^2 n(\mathbf{r})dl = 0\,. \tag{1.54}$$

The variation is carried out over all possible optical paths with dl being the physical infinitesimal physical path length. Will this principle be modified if n is allowed to be negative in some parts of the possible paths? We can take the example of imaging through a negative index slab (Fig. 1.10) as canonical. It is not hard to see that the total optical path traversed by any ray from object to image is actually zero. Compare this situation with conventional imaging where the total optical path of different rays is a positive constant. Both cases result in zero variation as we move from one path to the next; however, the zero path length in the negative imaging case yields additional insight. One can regard the intervening region where the index is negative as constituting a region of *negative space*. The imaging geometry determines that light rays traverse through as much positive space as negative, making the total path length zero. But notice that one can regard the interposition of a region of negative space as exactly compensating for (or 'annihilating') the positive distance travelled by *any* ray emerging from the object. Thus, when the image is reconstituted, in a sense it *is* the object. This gives an additional insight as to how the imaging can be perfect when a negative index slab is used.

1.9 Cloaking

A recent and rather exciting suggestion on the application of negative index media is the concept of the 'invisibility cloak,' of Harry Potter fame.[29,30] Although some of the hype is perhaps due to this connection with a popular fictional idea, current research is making a serious and concerted attempt at creating at least a limited form of electromagnetic cloak. We will describe in this section roughly how it works.

The idea is based on coordinate transformations. Consider transforming cylindrical polar coordinates (r, θ, z) according to

$$r' = \left(1 - \frac{a}{b}\right) r + a\,, \tag{1.55}$$

$$\theta' = \theta\,, \tag{1.56}$$

$$z' = z\,, \tag{1.57}$$

where a and b are constants with $b > a$. Since $r > 0$, $r' > a$, and in the transformed space a 'hole' of radius a has effectively appeared. The straight lines for which $y = r\sin\theta$ is constant in the original coordinate system become the lines shown in Fig. 1.14. If the lines of constant $y = r\sin\theta$ represent straight ray paths going from left to right in the original system, the curved lines represent light rays bending around an obstacle in the new system. After curving around the obstacle, the rays eventually become parallel again so that an observer looking toward the object from the right would not detect its presence, and the object is hidden. This is quite different from technologies that seek to render an object undetectable by minimizing the backscatter when illuminated by radar, such as for the stealth bomber.

The transformation could be effected in physical (r, θ, z) space by replacing vacuum with a dielectric medium with appropriate properties. The coordinate transformation of Eqs. (1.55)–(1.57) provides the required recipe. The vacuum electromagnetic quantities ϵ_0, μ_0 become, after the stretching and hole-punching of Eq. (1.55)–(1.57), anisotropic. The resultant medium then responds in the directions of increasing r, θ, z according to

$$(\epsilon,\mu)_r = \left(1-\frac{a}{r}\right)(\epsilon,\mu)_0 , \tag{1.58}$$

$$(\epsilon,\mu)_\theta = \left(1-\frac{a}{r}\right)^{-1}(\epsilon,\mu)_0 , \tag{1.59}$$

$$(\epsilon,\mu)_z = \left(1-\frac{a}{b}\right)^{-2}\left(1-\frac{a}{r}\right)(\epsilon,\mu)_0 . \tag{1.60}$$

The medium surrounds a cylindrical object of radius $r = a$. Details of how the electromagnetic constitutive parameters are transformed from one coordinate system to another may be found in the book by Post.[31] The recipe of Eqs. (1.58)–(1.60) is quite demanding since it requires that the medium's response in each of the three orthogonal directions must depend on the radial distance r. The first cloaking experiments[32] relaxed this constraint by fixing the electric field to be along the z-direction. The only relevant components for a plane wave propagating normal to the z-axis are then ϵ_z, μ_r, and μ_θ. The next simplification was to eliminate the r-dependence of μ_θ and ϵ_z by re-scaling the constitutive parameters to

$$\mu_r = \left(1-\frac{a}{r}\right)^{2}\mu_0 , \tag{1.61}$$

$$\mu_\theta = \mu_0 , \tag{1.62}$$

$$\epsilon_z = \left(1-\frac{a}{b}\right)^{-2}\epsilon_0 . \tag{1.63}$$

In the chosen configuration, the constitutive parameters of Eqs. (1.61)–(1.63) actually yield the same dispersion characteristics as those given by Eqs. (1.58)–(1.60), so that the ray trajectories[33] determined by the latter parameters are, under the geo-

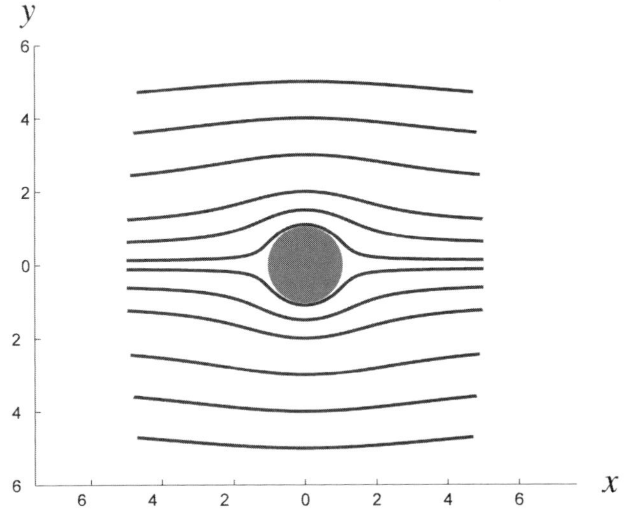

Figure 1.14 The coordinate transformation of Eq. (1.55) for the lines of constant $y = r\sin\theta$.

metrical optics approximation, the same as those of the former. The penalty is that some light is reflected by the cloak, so that the observer to the right notices a drop in the overall intensity.

The tailoring of a metamaterial to yield the effective constitutive relations approaching those of Eqs. (1.61)–(1.63) was achieved using metallic inclusions on concentric dielectric rings, as shown in Fig. 1.15. Simulation of the performance and experimental realization of the cloak are shown in Fig. 1.16.

It is seen that the cloak in these experiments is highly restricted in its operation. It works only at one frequency; the cloak and the object it shrouds (which, in order to preserve the effective medium approximation, cannot be too large) must be stationary (since the constitutive parameters of a medium depend on its velocity relative to the observer's reference frame[31]); only guided waves polarized parallel and propagating normal to the axis of a given cylinder are allowed, and the observer must ignore any changes in overall intensity. Nevertheless, the first experimental demonstrations are impressive, and it is likely that technological improvements and theoretical insights will further advance the concept.

1.10 Conclusion

The aim of this chapter has been to introduce the elements of negative refraction. We have shown how the concept of a metamaterial, using subwavelength metallic inclusions, has enabled us to now access (effective) values of the material parame-

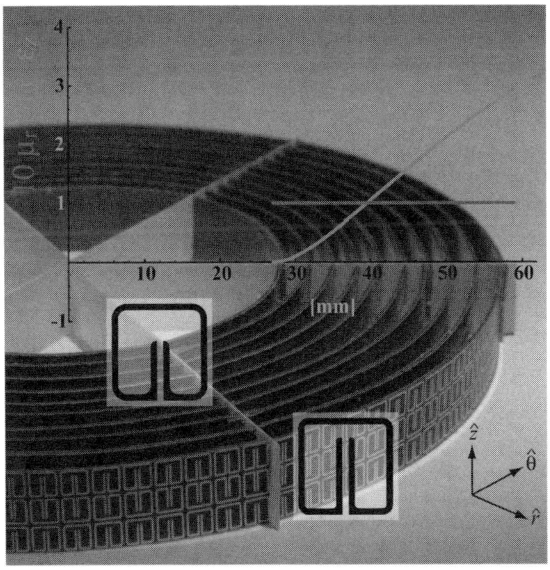

Figure 1.15 The experimental realization of the metamaterial with 'cloak parameters' defined by Eqs. (1.61)–(1.63). (Reprinted from Ref. 32 with permission from the American Association for the Advancement of Science.)

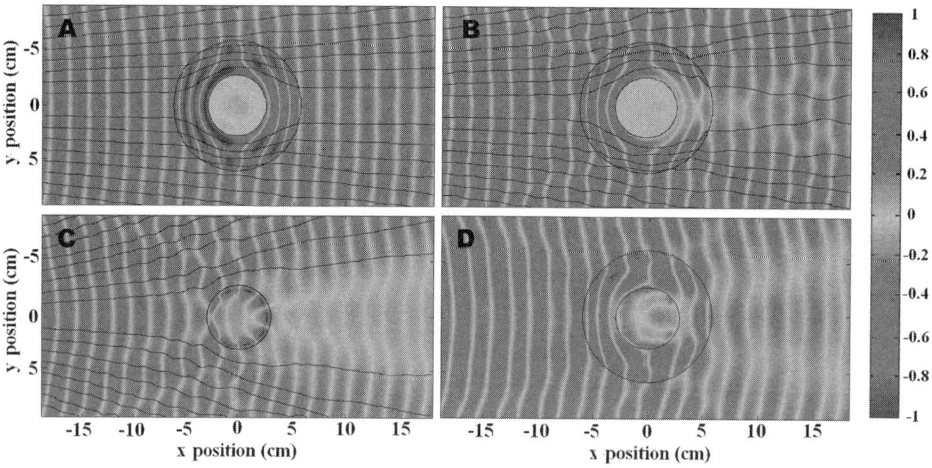

Figure 1.16 Theoretical simulation and experimental realization of the electromagnetic cloak. The instantaneous value of the electric field (color bar) is shown for (A) calculations using the exact material properties prescribed by Eqs. (1.58)–(1.60), (B) calculations using the reduced material properties prescribed by Eqs. (1.61)–(1.63), (C) measurements with a copper disc without the surrounding cloak, and (D) measurements with a copper disc surrounded by the metamaterial cloak of Fig. 1.15. Note that the wavefront kinking in the geometric shadow of the object in (C) is largely absent in (D). [Reprinted with permission from Ref. 32. © (2006) by the American Association for the Advancement of Science.] (See color plate section.)

ters that Vesalago speculated about all those years ago. After some controversies, the idea that certain values of these parameters can lead to the negatively refracting form of Snell's law has been conclusively confirmed experimentally. Although some issues are still being debated in the literature, the science of negative refraction has now firmly taken root. Much of the interest has been due to the possible applications in producing a lens that is beyond the diffraction limit, and is an electromagnetic cloak. It is important to remember that these and other applications fundamentally rely on the dispersive nature of the metamaterial. Perfect lensing and electromagnetic cloaking strictly only work at one frequency. In the foundation experiments reviewed here, that frequency has been in the microwave regime at a few tens of GHz. However, much progress has been made over the last two or three years in pushing the working range of metamaterials to much higher frequencies: $\sim$ 200 THz. (See Ref. 34 for a recent review.) This already corresponds to a wavelength of 1.5 μm, so it is clear that the community is fast closing in on the optical range. Metamaterials offer tremendously exciting possibilities, and no doubt more applications will emerge over the next few years.

Although we do not claim to have been exhaustive in this review, it is hoped that by focusing on the underlying principles of negative refraction, we have prepared the reader to appreciate these developments.

1.11 Appendices

1.11.1 Appendix I: The $\epsilon(\omega)$ of a square wire array

The notion that the electromagnetic response of the wire array of Fig. 1.8 can even be described by an $\epsilon(\omega)$ requires some explanation. The main objection is that the wavelength for waves of interest within the structure are not much greater than the lattice constant of the wire array. The concept of a permittivity requires a well defined and slowly varying average of electric field over a volume of linear dimensions much smaller than any wavelength considered. This criterion is violated for waves with frequencies near the plasma frequency of the wire array. It turns out that the $\epsilon(\omega)$ calculated in the long-wavelength limit ($\lambda >> a$ or $ka << 1$) is quite accurate down to wavelengths approaching the first Bragg reflection of the wire lattice.

The $\epsilon(\omega)$ for a wire array can be calculated analytically in the long-wavelength limit $ka << 1$. Since the three interpenetrating wire arrays shown in Fig. 1.8 are mutually perpendicular, one can simplify the analysis to a single square array of wires along the x-axis and consider only electromagnetic waves propagating in the plane perpendicular to the wires and with the electric field polarized along the x-axis. The permittivity for this case is found to be:[21]

$$\epsilon(\omega) = \epsilon_0 \left\{ 1 - \frac{\sigma^{eff}}{\omega\epsilon_0 \left[i + \left(\frac{\mu_0 \omega a^2 \sigma^{eff}}{2\pi} \right) \left(\ln \frac{a}{r} - \frac{3 + \ln 2 - \pi/2}{2} \right) \right]} \right\}, \tag{1.64}$$

where $\sigma^{eff} = (\pi r^2/a^2)\sigma$ is the wire conductivity scaled by the ratio of the wire volume to the unit cell volume. If the skin depth δ of the wire is taken into account, then πr^2 should be replaced by $2\pi r\delta$. Only the area of the wire that actually carries current needs to be accounted for in determining the effective conductivity. Note that $\epsilon(\omega)$ is more properly expressed as a function of ω and k as $\epsilon(\omega,\, k << 1/a)$.

How accurate is this $\epsilon(\omega)$? One approach to answering this is to dispense with the concept of a permittivity and solve Maxwell's equations directly for electromagnetic waves within the wire structure. It turns out[34] that one can determine k in the limit in which the wavelength is much smaller than the wire radius, i.e., $kr << 1$. For typical structures, this is a 10^2 to 10^3 times more stringent condition than the $ka << 1$ approximation used to calculate $\epsilon(\omega)$. One can get a sense of how accurate the $\epsilon(\omega)$ of Eq. (1.64) is by using it to calculate k from Eq. (1.19) and comparing it to the more exact k. Both approaches to calculating ka were used to produce Fig. 1.17; the difference between the calculations is less than the thickness of the lines used to plot ka. Indeed, $\epsilon(\omega)$ gives a fairly precise representation of the electromagnetic response of the wire array up to frequencies approaching the first Bragg reflection. Although beyond the scope of this chapter, it is possible to calculate the reflection coefficient of the wire array with $\epsilon(\omega)$ and in a more exact formalism with $kr << 1$. These different calculations of the reflection coefficient were found[35] to agree well beyond the plasma frequency and up to a frequency 1/3 of that of the first Bragg reflection.

If one were to extend the plot of ka in Fig. 1.17 to lower frequencies than shown, one would see that the real and imaginary parts of k approach equality and are proportional to $\omega^{1/2}$. This is the behavior one expects for a simple, homogeneous metal. For wavelengths $> 50a$, the wire array behaves as a poor conductor with a large skin depth, and the $\epsilon(\omega)$ representing it shows no effect of the (spatially inhomogeneous) lattice. It is fortuitous that the calculation of $\epsilon(\omega)$, nominally accurate for low frequencies at which the wire array behaves as an uninteresting poor metal, is in fact quite accurate up to and beyond the plasma frequency.

1.11.2 Appendix II: Physics of the wire array's plasma frequency and damping rate

The reduction of the plasma frequency of a square array of wires relative to the plasma frequency of the metal forming the wires is attributable to two sources: the dilution of the number density of mobile charges and the inductance of the wires. The damping rate for the wire array is also reduced; however, it is only the inductance of the wires that is responsible for this reduction.

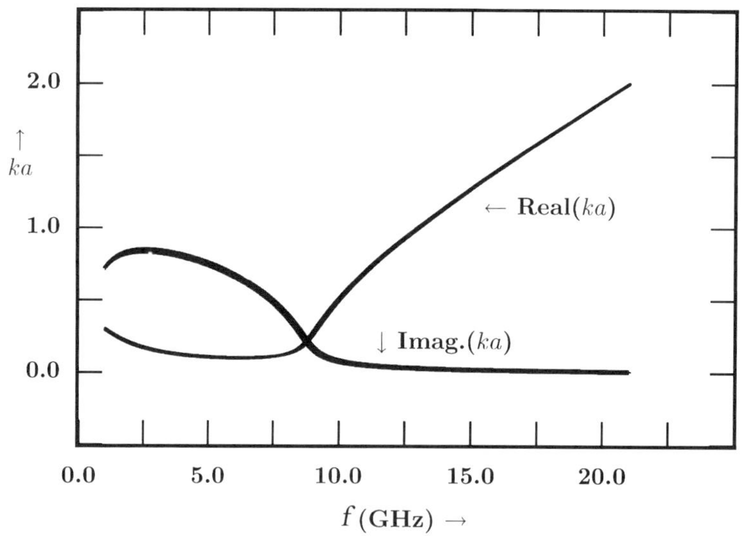

Figure 1.17 The scaled propagation constant ka is plotted versus frequency. Two versions of ka are shown: one is a near-exact calculation and the other an approximation based on the $\epsilon(\omega)$ of Eq. (1.64). The difference between these two calculations of ka is less than the width of the lines drawn in the figure. The crossing of the real and imaginary part of the dispersion relation occurs at the plasma frequency of 8.86 GHz in the near-exact calculation and 8.71 GHz for the calculation based on $\epsilon(\omega)$. Parameters used in the calculation were: $a = 5.0 \times 10^{-3}$ m, $r = 1.0 \times 10^{-6}$ m, and $\sigma = 3.65 \times 10^{7}$ siemen/m.

The plasma frequency for a bulk conductor is defined in the derivation of Eq. (1.7), namely

$$\omega_p^2 = \frac{Ne^2}{m\epsilon_0} , \tag{1.65}$$

where N is the number density, and m the mass of electrons in the conductor. Filling space with a wire array reduces this number density drastically since there are no mobile charges between the wires. If the wires have a radius r and are arranged in a square array of lattice constant a, then the effective number density is

$$N_{eff} = N\frac{\pi r^2}{a^2} . \tag{1.66}$$

The inductance of the wires also contributes to the depression of the plasma frequency. In effect, one augments the momentum associated with a free electron by a term proportional to the vector potential arising from currents in the wires. An electric field applied to a wire not only works to speed up electrons, thereby increasing their kinetic energy, but also increases the energy stored in the magnetic field resulting from the moving electrons. One can incorporate this near-zone magnetic field energy into the electrons' kinetic energy by invoking an effective mass m_{eff} for the electrons, i.e.,

$$W_{elec} = \Delta\left(\frac{1}{2}mv^2 + \frac{1}{2}\int \mathbf{B}\cdot\mathbf{H}\,dV\right) = \Delta\left(\frac{1}{2}m_{eff}v^2\right) , \tag{1.67}$$

where W_{elec} is the work done by the imposed electric field. Some care should be exercised in determining the volume over which the magnetic field energy is integrated. A unit cell of the 2D array, centered on a wire, and of length such that the total volume equals the volume per electron, is the appropriate choice.[13,21] The result for a square array is

$$m_{eff} = m + \frac{\mu_0 e^2 \pi r^2 N}{2\pi}\left[\ln\left(\frac{a}{r}\right) - \frac{3 + \ln 2 - \pi/2}{2}\right] . \tag{1.68}$$

For reasonable choices of a and r, i.e., a on the order of millimeters and r on the order of microns, the part of the effective mass associated with the inductive energy is about three orders of magnitude greater than the electron mass. The constant term following the logarithmic term is dependent on the choice of a square lattice; a hexagonal lattice's constant term would be different. Equation (1.23) captures the essence of the enhancement of the electron's effective mass resulting from near-zone magnetic fields within the array.

Thus, the plasma frequency is suppressed by two effects: the dilution of the number density of free electrons and the effective mass enhancement which accounts for inductance. The damping rate, however, pertains to the rate at which each free electron loses momentum; number density is irrelevant. Equation (1.5), an expression of Newton's second law for electrons in the conductor, must include the effect of inductive electric fields acting on the electrons. This diminishes the effectiveness of the damping rate in removing kinetic energy from the electrons. Thus, on physical grounds, one expects the damping rate in the wire array to be depressed relative to the bulk metal by a factor proportional the enhanced effective mass of the electrons.

We now turn to calculations based on the long-wavelength limit to verify that 1) the plasma frequency for a wire array is indeed much lower that the plasma frequency of the conductor of which the wire is made and 2) that the damping rate in the wire array is also lower than that for the bulk conductor. The $\epsilon(\omega)$ described in Appendix I can be set to zero and the plasma frequency extracted from the resulting quadratic equation. Neglecting the small imaginary part of this result, one obtains

$$\omega_p^2 = \frac{2\pi}{(\mu_0\epsilon_0 a^2)\left[\ln(\frac{a}{r}) - \frac{3+\ln 2-\pi/2}{2}\right]} = \frac{2\omega_B^2}{\pi\left[\ln(\frac{a}{r}) - \frac{3+\ln 2-\pi/2}{2}\right]}, \tag{1.69}$$

where $\omega_B = \pi c/a$ is the frequency of the first Bragg reflection. This is in agreement with Eqs. (1.65), (1.66), and (1.68). Note that the plasma frequency is always comparable to that of the first Bragg reflection since $\ln(a/r)$ is typically less than 10.

The loss in a wire array is also described by $\epsilon(\omega)$. It is instructive to note how this loss compares with that for a bulk conductor as described by the Drude result for $\epsilon(\omega)$, Eq. (1.7). Incorporating the plasma frequency of Eq. (1.69) into the $\epsilon(\omega)$ for the wire array, Eq. (1.64), one obtains

$$\epsilon(\omega) = 1 - \frac{\omega_p^2}{\omega^2 + i\omega\epsilon_0\omega_p^2/\sigma_0^{eff}}. \tag{1.70}$$

Comparison of this with Eq. (1.7) implies that the damping rate for the wire array is

$$\Gamma_{array} = \frac{\epsilon_0\omega_p^2}{\sigma_0^{eff}} = \frac{m}{m_{eff}\tau}, \tag{1.71}$$

where τ in the intrinsic scattering rate for electrons in the wire characterized by the conductivity $\sigma = ne^2\tau/m$. It is clear from this expression for the dissipation rate for the wire array that this rate is suppressed by the ratio by which the electron's effective mass is enhanced by inductive effects. In contrast to the case with the plasma frequency, the dilution of the number density of electrons in the wire array versus bulk conductor, does not affect the damping rate. Furthermore, since Eq. (1.68) shows that the electron effective mass increases with the area of wire carrying current, one can conclude that an array made of wires of larger diameter results in less loss than an array made of wires of smaller diameter. This result has been verified by explicit calculations, including the skin effect in the wires.[35]

References

1. R. D. Smith, W. J. Padilla, D. C. Vier, S. C. Nemat-Nasser, and S. Schultz, "Composite medium with simultaneously negative permeability and permittivity," *Phys. Rev. Lett.* **84**, 4184–4187 (2000).

2. A. Schuster, *Theory of Optics*, Edward Arnold, London (1904).

3. L. I. Mandelshtam, "Group velocity in crystalline arrays," *Zh. Eksp. Teor. Fiz.* **15**, 475–478 (1945).

4. L. I. Mandelshtam, *Complete Collected Works*, Akad. Nauk SSSR, Moscow (1947).

5. V. G. Veselago, "The electrodynamics of substances with simultaneously negative values of ϵ and μ," *Sov. Phys. USPEKHI* **10**, 517–526 (1968).

6. R. A. Shelby, D. R. Smith, and S. Schultz, "Experimental verification of a negative index of refraction," *Science* **292**, 77–79 (2001).

7. A. A. Houck, J. B. Brock, and I. L. Chuang, "Experimental observations of a left-handed material that obeys Snell's law," *Phys. Rev. Lett.* **90**, 137401 (2003).

8. J. B. Pendry, "Negative refraction makes a perfect lens," *Phys. Rev. Lett.* **85**, 3966–3969 (2000).

9. N. W. Ashcroft and N. D. Mermin, *Solid State Physics*, Holt, Rinehart and Winston, Philadelphia, PA (1976).

10. L. D. Landau and E. M. Lifschitz, *Electrodynamics of Continuous Media*, Pergamon Press, Oxford (1963).

11. M. W. McCall, A. Lakhtakia, and W. S. Weiglhofer, "The negative index of refraction demystified," *Eur. J. Phys.* **23**, 353–359 (2002).

12. L. Brillouin, *Wave Propagation and Group Velocity*, 1st ed., Academic Press, New York (1960).

13. J. B. Pendry, A. J. Holden, W. J. Stewart, and I. Youngs, "Extremely low frequency plasmons in metallic mesostructures," *Phys. Rev. Lett.* **76**, 4773–4776 (1996).

14. S. A. Mikhailov, "Comment on 'Extremely low frequency plasmons in metallic mesostructures'," *Phys. Rev. Lett.* **78**, 4135 (1997).

15. J. B. Pendry, A. J. Holden, W. J. Stewart, and I. Youngs, "Pendry et al. reply:," *Phys. Rev. Lett.* **78**, 4136–4136 (1997).

16. R. M. Walser, A. P. Valanju, and P. M. Valanu, "Comment on 'Extremely low frequency plasmons in metallic mesostructures'," *Phys. Rev. Lett.* **87**, 119701 (2001).

17. P. A. Belov, R. Marqués, S. I. Maslovski, I. S. Nefedov, M. Silveirinha, C. R. Simovski, and S. A. Tretyakov, "Strong spatial dispersion in wire media in the very large wavelength limit," *Phys. Rev. B* **67**, 113103 (2003).

18. G. Shvets, A. K. Sarychev, and V. M. Shalaev, "Electromagnetic properties of three-dimensional wire arrays: photons, plasmons, and equivalent circuits," in *Complex Mediums IV: Beyond Linear Isotropic Dielectrics*, M. W. McCall and G. Dewar, Eds., *Proc. SPIE* **5218**, 156–165 (2003).

19. M. Shapiro, G. Shvets, J. R. Sirigiri, and R. J. Temkin, "Spatial dispersion in metamaterials with negative dielectric permittivity and its effect on surface waves," *Opt. Lett.* **31**, 2051–2053 (2006).

20. J. B. Pendry, A. J. Holden, D. J. Robbins, and W. J. Stewart, "Low frequency plasmons in thin-wire structures," *J. Phys.: Condens. Matter* **10**, 4785–4809 (1998).

21. G. Dewar, "The applicability of ferrimagnetic hosts to nanostructured negative index of refraction (left-handed) materials," in *Complex Mediums III: Beyond Linear Isotropic Dielectrics*, A. Lakhtakia, G. Dewar, and M. W. McCall, Eds., *Proc. SPIE* **4806**, 156–166 (2002).

22. A. L. Pokrovsky and A. L. Efros, "Electrodynamics of metallic photonic crystals and the problem of left-handed materials," *Phys. Rev. Lett.* **89**, 093901 (2002).

23. G. Dewar, "Transverse electromagnetic waves in periodic wire arrays," in *Complex Mediums IV: Beyond Linear Isotropic Dielectrics*, M. W. McCall and G. Dewar, Eds., *Proc. SPIE* **5218**, 140–144 (2003).

24. J. B. Pendry, A. J. Holden, D. J. Robbins, and W. J. Stewart, "Magnetism from conductors and enhanced nonlinear phenomena," *IEEE Trans. Microwave Theory Tech.* **47**, 2075–2084 (1999).

25. F. D. M. Haldane, "Electromagnetic surface modes at interfaces with negative refractive index makes a "not-quite-perfect" lens," *arXiv:cond-mat/0206420v3* (2002).

26. G. W. Milton, N. P. Nicorovici, and R. C. McPhedran, "Opaque perfect lenses," *arXive:physics/0608225v1* (2006).

27. N. Garcia and M. Nieto-Vesperinas, "Left-handed materials do not make a perfect lens," *Phys. Rev. Lett.* **88**, 207403 (2002).

28. I. A. Larkin and M. I. Stockman, "Imperfect perfect lens," *Nano Letters* **5**, 339–343 (2005).

29. J. B. Pendry, D. Schurig, and D. R. Smith, "Controlling electromagnetic fields," *Science* **312**, 1780–1782 (2006).

30. U. Leonhardt and T. G. Philbin, "General relativity in electrical engineering," *New Journal of Physics* **8**, 247 (2006).

31. E. J. Post, *Formal Structure of Electromagnetics*, Dover, Mineola, NY (1997).

32. D. Schurig, J. J. Mock, B. J. Justice, S. A. Cummer, J. B. Pendry, A. F. Starr, and D. R. Smith, "Metamaterial electromagnetic cloak at microwave frequencies," *Science* **314**, 977–980 (2006).

33. M. W. McCall and D. Censor, "Relativity and mathematical tools: Waves in moving media," *American Journal of Physics* **75**, 1134–1140 (2007).

34. G. Dewar, "Transverse electromagnetic waves in periodic wire arrays," in *Complex Mediums V: Light and Complexity*, M. W. McCall and G. Dewar, Eds., *Proc. SPIE* **5508**, 158–166 (2004).

35. G. Dewar, "Strategies for minimizing losses in negative phase velocity metamaterials," in *Complex Photonic Media*, G. Dewar, M. W. McCall, M. A. Noginov, and N. I. Zheludev, Eds., *Proc. SPIE* **6320**, 63200D (2006).

Biographies

Martin W. McCall graduated in Physics from Imperial College London in 1983. His doctoral thesis, completed while he worked for an electronics company, concerned development of vector coupled-wave theory describing anisotropic diffraction in photorefractives for use in real-time image processing. After a spell at the University of Bath, UK, where he worked on nonlinear dynamics in semiconductor lasers, McCall returned to Imperial College in 1988 to work on a range of optoelectronic themes, including nonlinear coupling and mixing in semiconductor amplifiers and laser arrays, optical interconnects, and Bragg grating physics. Sometimes referring to himself as a "reformed experimentalist," McCall's research is now purely theoretical. Broadly within the remit of describing the electromagnetics of complex media, he has specifically worked on chiral photonic films and negative index metamaterials. Recently he has specialized in the use of covariant methods in electromagnetism.

Graeme Dewar earned his Ph.D. in Physics from Simon Fraser University in 1980. After stints on the faculties of Princeton University and the University of Miami, he joined the University of North Dakota in 1989 where he is currently Professor and Chair of the Department of Physics and Astrophysics. Most of his research projects have involved the interaction of electromagnetic radiation with complex media. These have included experimental investigations of the radio frequency properties of ferromagnetic metals, with an emphasis on magnetoelastic effects and photonic crystals. His current interests are primarily in metamaterials having a tailored permittivity and permeability.

Chapter 2
Optical Hyperspace: Negative Refractive Index and Subwavelength Imaging

Leonid V. Alekseyev[a,b], Zubin Jacob[b], and Evgenii Narimanov[b]
[a]Princeton University, Princeton, NJ, USA
[b]Purdue University, West Lafayette, IN, USA

2.1 Introduction

The art and science of optics is centered upon our ability to control the refractive index of materials, thereby directing the flow of light. From the stained-glass windows of Gothic cathedrals to modern LCD projectors, from Galileo's telescope to terabit optical communication systems, devices made possible by skillful manipulation of the refractive index have resulted in countless technological and cultural breakthroughs. For centuries, the refractive index has been regarded as a strictly positive quantity — such was the universal experience. Recent advances in fabrication and processing techniques, however, have enabled the creation of materials with a *negative* refractive index. This development opens many new chapters in the fields of optical physics and device engineering. Negative index greatly expands the parameter space accessible for manipulating light, opening the way for devices with unprecedented capabilities — for example, imaging systems with subwavelength resolution and ultrasmall waveguides. The novel systems made possible by negative index materials (NIMs) may bring about revolutionary technological changes.[1]

In the present chapter we describe a method to achieve negative refraction via materials with a *hyperbolic* dispersion relation. Both natural materials and metamaterials can exhibit this property. We show that in addition to providing a simple path to nonmagnetic negative refraction, the hyperbolic dispersion relation enables novel devices for waveguiding and subwavelength imaging.

The present-day interest in NIMs started in the early 2000s.[2–4] The origins of the subject, however, date back many decades. Indeed, as a general wave propagation phenomenon, negative refraction has been known since the early 20th century.[5,6] It was noted, in particular, that negative refraction naturally occurs at the interface with a medium characterized by negative phase velocity. No such materials were known in the electromagnetic domain, and so the early discussions involved only mechanical oscillations. The first detailed treatment of negative refraction in electromagnetism was provided by Veselago in 1968.[7] He showed that to attain negative phase velocity for electromagnetic (EM) waves, the material response must be of the form $\epsilon < 0$, $\mu < 0$. When this condition is satisfied, the $\mathbf{E}$, $\mathbf{H}$, and $\mathbf{k}$ vectors form a left-handed triplet. As a result, the wave vector $\mathbf{k}$ and the Poynting vector $\mathbf{S}$ are oriented in opposite directions; the system has negative phase velocity, which is the condition for negative refraction. Indeed, negative phase velocity serves as a definition of negative index materials.[8] While mechanical and radio frequency devices exhibiting such effective negative indices were known at the time of Veselago's writing, bulk materials with negative phase velocity were not found in nature and were not readily attainable.[8]

The once-fledgling field of negative refraction has experienced a major surge in the past decade, owing to major theoretical and experimental advances. On the theoretical side, Pendry has proposed negative refractive media as a platform for subwavelength resolution and aberration-free imaging.[2] In particular, Pendry showed that a slab of Veselago's "left-handed" material with $\epsilon = \mu = -1$ acts as a perfect lens: it does not suffer from aberrations and is not subject to the diffraction limit. The proposed "superlens" stimulated enormous interest in NIMs, but generated some initial controversy regarding their experimental realizability. This controversy was soon resolved by Smith and colleagues, who fabricated a material with $\epsilon < 0$, $\mu < 0$ in the microwave band and explicitly demonstrated negative refraction.[3] The required response was attained by artificially structuring the material on a scale smaller than the operational wavelength, thereby creating a *metamaterial*. Utilizing the latest nanofabrication techniques, material patterning can be done on a submicron scale. This opens the way for NIMs operating at infrared and even visible wavelengths. Indeed, negative refraction was demonstrated experimentally with wavelengths as short as 772 nm.[9]

Fabricating structures that exhibit negative refraction at such high frequencies presents many difficulties. The most challenging aspect of the engineered electromagnetic response is the required negative magnetic permeability. Negative permeability is a result of a resonant response by a miniature conductive structure. For an effective negative permeability response, these microresonators must reside in subwavelength unit cells. Thus, to attain negative permeability for THz and higher frequencies, one must resort to lithographic methods in structuring the materials. For the optical frequencies, fully three-dimensional (3D) subwavelength patterning is currently unfeasible.

Aside from the manufacturing difficulties, negative magnetic response presents another significant challenge. The resonance in the real component of magnetic permeability which leads to negative values of μ is necessarily accompanied by a spike in its imaginary component. This leads to high absorption at the operating frequencies of magnetic NIMs, which can significantly impair NIM devices.[10]

In the quest to minimize losses, it becomes prudent to examine ways of obtaining negative refraction without resorting to optical magnetism. It was shown by several groups that negative refraction can arise for light in suitably designed photonic crystals.[11–14] From the standpoint of losses, photonic crystal materials are generally superior to magnetic NIMs.[14] However, photonic crystals present many of the same fabrication challenges as magnetic metamaterials, especially for 3D structures. While the characteristic features of photonic crystals are simpler and larger (and hence easier to produce), the photonic band behavior is strongly sensitive to disorder, necessitating high manufacturing precision.

For an alternative approach to nonmagnetic negative refraction, we start with the observation that for appropriately cut surfaces of anisotropic crystals, negative refraction occurs (for a limited range of angles) due to Poynting vector walk-off.[15] This effect generalizes to *all-angle* negative refraction for a particular class of strongly anisotropic materials — those in which the components of the dielectric tensor have opposite signs.[16,17] Metamaterials designed to satisfy this condition are vastly simpler than typical magnetic metamaterials, and are therefore potentially more amenable to bulk fabrication. In addition, these metamaterials are not sensitive to disorder and operate far from resonances, thus helping minimize absorption losses. Finally, for certain frequencies, materials with the prescribed anisotropy can be found in nature.[18]

It should be noted that like all nonmagnetic negative refraction systems, this approach cannot be utilized to create the originally proposed superlens; the superlens relies on the excitation of high-wavenumber surface plasmon-polariton modes, which is only possible for simultaneously negative ϵ and μ. However, negatively refracting materials based on strong anisotropy lead to an entirely new class of exciting devices.

In subsequent sections we describe in detail the role of anisotropy in creating all-angle negative refraction and discuss natural and artificial materials that can be used to demonstrate this phenomenon. We then describe potential applications enabled by such materials. These applications include negative phase velocity waveguides, slow light waveguides, and *the hyperlens* — a novel device that enables far-field subwavelength-resolved imaging.

2.2 Nonmagnetic Negative Refraction

For a plane wave with wave vector $\mathbf{k}$, incident on some surface, translational invariance demands that $k_{\|}$, the component of $\mathbf{k}$ along the surface, be preserved for the refracted wave. So long as the direction of the energy flow (given by the Poynting vector $\mathbf{S}$) and the direction of the wave vector $\mathbf{k}$ are the same, negative refraction cannot occur. Thus, negative refraction is only possible in media where the unit vectors $\hat{\mathbf{k}}$ and $\hat{\mathbf{S}}$ do not coincide. More specifically, for the transmitted wave we must have $S_{\|} < 0$ when $k_{\|} > 0$ and vice versa. For a medium with negative phase velocity, $\hat{\mathbf{S}} = -\hat{\mathbf{k}}$ holds, and the condition $\{S_{\|} < 0 \text{ and } k_{\|} > 0\}$ is then satisfied automatically. Material parameters $\epsilon < 0$, $\mu < 0$ lead to exactly this scenario. More generally, however, we may inquire as to what material parameters lead to negative refraction without requiring negative phase velocity.

The simplest answer to this question comes from considering wave propagation in anisotropic crystals and noting that the directions of $\mathbf{S}$ and $\mathbf{k}$ are, generally, different. To see how this comes about, we consider plane wave propagation in a uniaxial medium. Depending on polarization, the waves can be characterized as *ordinary* or *extraordinary*. For extraordinary waves, the electric field vector has a nonvanishing component along the optical axis; therefore, the different components of the electric field $\mathbf{E}$ experience different dielectric constants. Furthermore, the relationship between $\mathbf{E}$ and $\mathbf{D}$ (the electric displacement vector) depends on the propagation direction. Ordinary waves, on the other hand, are not affected by the anisotropy and are of no special interest. For this reason, in the subsequent discussion we treat only the extraordinary polarization.

Taking $\hat{\mathbf{x}}$ as the direction of the optical axis, we may characterize the extraordinary wave in a uniaxial crystal by the dispersion relation[19]

$$\frac{k_x^2}{\epsilon_z} + \frac{k_{y,z}^2}{\epsilon_x} = \frac{\omega^2}{c^2}. \tag{2.1}$$

For sufficiently weak absorption, the direction of the Poynting vector is identical to the direction of the group velocity vector $\mathbf{v}_g = \nabla_{\mathbf{k}}\omega(\mathbf{k})$.[19] This means that $\mathbf{S}$ is normal to the isofrequency curves given by Eq. (2.1).

What does this imply for the relative angle between $\mathbf{S}$ and $\mathbf{k}$? In the isotropic case, the wave vector surfaces are circles, and therefore $\mathbf{S} \propto \nabla_{\mathbf{k}}\omega(\mathbf{k}) \propto \mathbf{k}$, i.e., $\mathbf{S}$ and $\mathbf{k}$ are collinear, as can be seen in Fig. 2.1(a). Consider now the situation

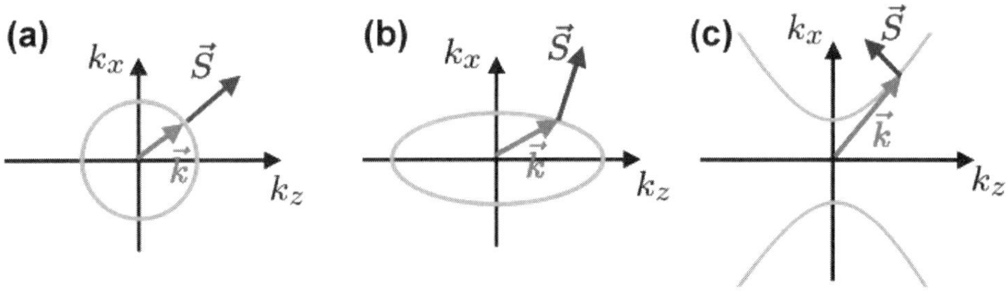

Figure 2.1 Isofrequency curve and relative direction of the wave vector $\vec{k}$ and the Poynting vector $\vec{S}$ for (a) isotropic material, (b) material with $\epsilon_x, \epsilon_z > 0$, and (c) material with $\epsilon_x < 0$, $\epsilon_z > 0$. (Reprinted from Ref. 20.)

in Fig. 2.1(b), where $\epsilon_x \neq \epsilon_z$ and $\epsilon_{x,z} > 0$. The wave vector surfaces become ellipsoidal; as a consequence, the angle between $\mathbf{S}$ and $\mathbf{k}$ is nonzero. (Its exact value depends on the direction of propagation and the degree of anisotropy.) This implies that it is possible to pick a coordinate system $x'z'$ in which $S_{z'} < 0$, $k_{z'} > 0$. If the material is cut such that $\hat{\mathrm{x}}'$ defines the surface normal, negative refraction occurs. Note, however, that this situation is only realizable for a finite range of $\mathbf{k}$ values, and hence a finite set of incidence angles.

Finally, we consider the case shown in Fig. 2.1(c), where ϵ_x and ϵ_z are not only non-equal but also possess different *signs*. This drastically changes the nature of the dispersion relation in Eq. (2.1). More specifically, for a material with negative transverse dielectric permittivity ($\epsilon_x < 0$) and positive in-plane permittivity ($\epsilon_z > 0$), Eq. (2.1) describes a hyperbola. Constructing the vectors $\mathbf{S}$ and $\mathbf{k}$, as before, we see that the signs of S_z and k_z are opposite for all admissible values of $\mathbf{k}$.

This result can also be obtained by examining the $\hat{\mathbf{z}}$ component of the Poynting vector for the extraordinary wave:

$$S_z = \frac{k_z}{\epsilon_x \, \omega/c} E_0^2, \tag{2.2}$$

where E_0 is the electric field amplitude (in CGS units). Evidently, if $\epsilon_x < 0$, S_z is negative, i.e., opposite to the direction of the wave vector component k_z.

If we now consider a beam incident on an interface of a material exhibiting a hyperbolic dispersion relation, we find that the sign of the tangential component of the wave vector (k_z) is preserved as usual upon transmission through the boundary, while the Poynting vector (and thus the energy flux) undergoes negative refraction, as in Figs. 2.2 (a) and (b). Furthermore, a slab of such material can function as a planar lens, as shown in Fig. 2.2(c). In this sense, the $\epsilon_x < 0$, $\epsilon_z > 0$ material mimics the behavior of negative refraction systems with $\epsilon < 0$, $\mu < 0$. We should keep in mind, however, that the hyperbolic dispersion relation in Eq. (2.1) has a profound impact not only on refraction behavior at the interface, but also on the

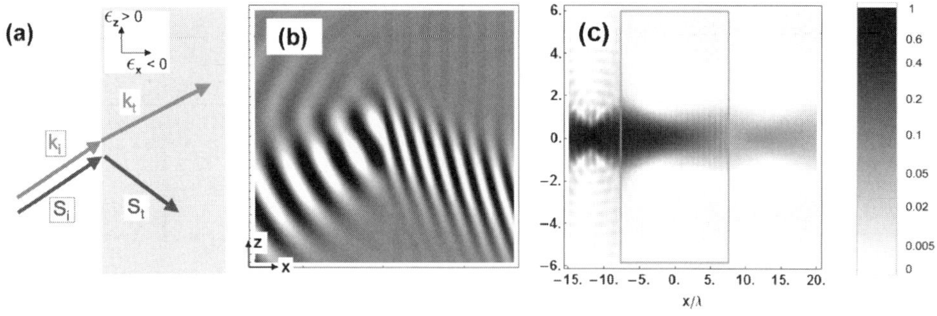

Figure 2.2 (a) The ray diagram and (b) the electric field for the refraction of a light beam at the boundary of air with an $\epsilon_x < 0$, $\epsilon_z > 0$ material. Note the negative refraction of the beam and the direction of the wavefronts ($\epsilon_z = 3$, $\epsilon_x = -1.5$). (c) The intensity distribution of a beam propagating through a slab made of such material. This slab functions as a planar lens. (Adapted from Ref. 20.)

general properties of wave propagation. (Indeed, we shall see in Section 2.4.1 that this dispersion relation enables devices with negative phase velocity and near-zero group velocity.)

One hyperbolic dispersion effect that is of particular interest in imaging applications involves directionality constraints on propagating radiation. Fig. 2.1(c) shows that the allowed directions of the wave vector and the Poynting vector are restricted by the asymptotes of the hyperbola. The locus of the allowed **S** vectors is a cone, with the half-angle θ_c given by

$$\tan\theta_c = \sqrt{\frac{\epsilon_z}{|\epsilon_x|}}. \tag{2.3}$$

Such beam-like directional radiation patterns have indeed been observed for sources embedded in strongly anisotropic plasmas.[21,22] It is interesting to note that in the case of an ideal point source and with zero losses, all of the energy is concentrated at the boundary of the propagation cone, since there are infinitely many wave vectors — solutions of Eq. (2.1) — that accumulate close to the asymptotes of the hyperbola, and therefore share the same direction. Furthermore, for $|\epsilon_x| \gg \epsilon_z$, the beam divergence angle approaches zero. Thus, in this so-called *channeling regime*,[23] subdiffraction-limited imaging can be performed.

It should be emphasized that the absence of the conventional diffraction limit is a general feature for wave propagation in $\epsilon_x < 0$, $\epsilon_z > 0$ materials. For positive dielectric constants, Eq. (2.1) implies a finite spatial frequency bandwidth limit, which causes diffraction. For instance, in the isotropic case, we have

$$k_{x,y}^2 > \epsilon\frac{\omega^2}{c^2} \implies k_z = i\kappa. \tag{2.4}$$

In other words, for large values of $k_{x,y}$ the wave equation solutions $\exp(i\,\mathbf{k}\cdot\mathbf{r})$ are the evanescent waves that exponentially decay away from the source. At the same time, it is precisely these waves that carry information about structure on a subwavelength scale, as scattering from subwavelength features results in large $k_{x,y}$. For a hyperbolic dispersion relation, however, Eq. (2.1) can be satisfied for *arbitrarily large* values of $k_{x,y}$ and k_z. These high spatial frequency waves propagate through the $\epsilon_x < 0,\ \epsilon_z > 0$ structure and enable subdiffraction-limited imaging.

2.3 Hyperbolic Dispersion: Materials

Clearly, the special nature of the $\epsilon_x < 0,\ \epsilon_z > 0$ systems leads to a multitude of exotic effects. Here arises a natural question: how can we elicit such a response from physical materials?

Perhaps surprisingly, the $\epsilon_x < 0,\ \epsilon_z > 0$ behavior is observed in a number of natural materials where structural anisotropy strongly affects the dielectric response. Examples of such materials can easily be found in the infrared and THz spectral bands. For instance, in the far-infrared/low-THz domain, this behavior can be found in triglycine sulfate (TGS), a compound widely used in fabricating infrared photodetectors. In TGS, a strong phonon anisotropy leads to a large anisotropy in the dielectric tensor. In particular, dielectric response for the field polarized along the crystal's monoclinic C_2 axis features a resonance at 268 μm, which is absent for light polarized transverse to C_2.[24] Measured dielectric functions[17,24,25] reveal that $\epsilon_x < 0$, while $\epsilon_z > 0$ in the region $250 \leq \lambda \leq$ 268 μm (here and subsequently we let $\hat{\mathrm{x}}$ lie along the appropriate crystallographic axis). Furthermore, the imaginary part of ϵ becomes small away from the resonance, minimizing absorption.

Whereas the phonon anisotropy of TGS exists in the low-THz domain, for other materials, it may occur in a different spectral band. The strong anisotropy of the dielectric response of sapphire (Al_2O_3) is also due to excitation of different phonon modes (polarized either parallel or perpendicular to the c axis of the rhombohedral structure), but occurs around 20 μm. A region of $\epsilon_x < 0$, $\epsilon_z > 0$ for wavelengths of 19.5 to 21 μm has been experimentally observed.[26]

Anisotropic phonon excitation is not the only mechanism that can lead to strong dielectric anisotropy. Bismuth, a group V semimetal, exhibits such anisotropy due to a substantial difference in its effective electron masses along different directions in the crystal. Measurements of bismuth plasma frequencies[27,28] can be used to reconstruct its dielectric tensor. The $\epsilon_x < 0,\ \epsilon_z > 0$ anisotropy is revealed between 54 and 63 μm. It should be noted that pure bismuth samples exhibit much lower absorption than most metals, due to long electron relaxation times (a conservative estimate is $\tau = 0.1$ ns at 4 K[28]). The typical ratio of imaginary and real parts of the dielectric function in bismuth is thus expected to be on the order of 0.1% in the frequency interval of interest.

For spectral domains where natural effects do not result in differing signs of the dielectric tensor components, such anisotropy may be attained in metamaterials. To satisfy the requirement $\epsilon_x < 0$ and $\epsilon_z > 0$, the metamaterials must combine plasmonic or polar materials (with $\epsilon < 0$) with conventional dielectrics in an appropriate geometry.

The $\epsilon < 0$ components of such nanocomposites may come from a variety of sources. For instance, these negative permittivity materials can be created artificially. One approach involves strongly doping a semiconductor, thereby creating a plasmon resonance. Another possible technique to induce negative permittivity is engineering quantum wells with appropriate intrasubband transitions. Negative permittivity is also quite common in naturally occurring materials. In the visible spectrum, plasmon resonances result in $\epsilon < 0$ for a number of metals. Silver is one example of a relatively low-loss plasmonic material. At longer wavelengths, phonon resonances can yield $\epsilon < 0$, with losses typically lower than those in silver. One such low-loss material, well-suited for studying negative-index phenomena in the mid-IR, is silicon carbide,[29,30] with $\epsilon < 0$ between 10.3 and 11 μm.

Metamaterials can be structured in many different ways. For instance, the plasmonic inclusions can take the form of aligned nanowires. Alternatively, these inclusions can be anisotropically distributed in a dielectric host. The simplest arrangement that yields the desired dielectric properties is a layered medium with alternating permittivities in the x direction.[18,29,31] This medium consists of a sequence of "dielectric" layers ($\epsilon_1 > 0$) and "conductive" layers ($\epsilon_2 < 0$).[32] The effective dielectric tensor of such a structure (with the volume fraction of the conducting layers N_c) is given by[33]

$$\begin{aligned} \epsilon_x &= \frac{\epsilon_1 \epsilon_2}{N_c\, \epsilon_1 + (1 - N_c)\epsilon_2} \\ \epsilon_z &= (1 - N_c)\epsilon_1 + N_c\, \epsilon_2. \end{aligned} \tag{2.5}$$

Provided that $\epsilon_1 > 0$ and $\epsilon_2 < 0$ in a certain frequency range, these equations lead to a well-defined frequency interval with $\epsilon_x < 0,\ \epsilon_z > 0$ (the exact values of the interval are determined from the dispersive characteristics of ϵ_1 and ϵ_2). Such a layered system can be fabricated using epitaxial semiconductor growth, with selective doping used to attain $\epsilon_2 < 0$ in the "metallic" regions.

2.4 Applications

2.4.1 Waveguides

As discussed above, the $\epsilon_x < 0,\ \epsilon_z > 0$ materials enable all-angle negative refraction for incident plane waves. However, for *guided modes*, this form of the dielectric tensor results in negative phase velocities and even negative group delays — phenomena primarily associated with magnetic ($\epsilon_x < 0,\ \mu < 0$) NIMs. To see how this comes about, let us consider guided-mode solutions for a planar waveguide of thickness d with perfectly conducting walls. Suppose that the boundaries of the waveguide lie at $x = 0$ and $x = d$, and that guided modes propagate in the z direction. We assume that the waveguide is filled with a uniaxial anisotropic material characterized by dielectric constants $\epsilon_x \equiv \epsilon_\perp$ (for field components transverse to the waveguide) and $\epsilon_{y,z} \equiv \epsilon_{\parallel}$. The solution for transverse magnetic (TM) modes propagating in such a waveguide is[19]

$$\mathbf{E}(\mathbf{r}, t) = E_0 \left[-i\frac{\beta}{\epsilon_{\perp}} \cos(\kappa x)\hat{\mathbf{x}} + \frac{\kappa}{\epsilon_{\parallel}} \sin(\kappa x)\hat{\mathbf{z}} \right] \exp[-i(\beta z - \omega t)], \tag{2.6}$$

where $\kappa = m\pi/d$, and κ and β satisfy the dispersion relation similar to Eq. (2.1):

$$\frac{\beta^2}{\epsilon_{\perp}} + \frac{\kappa^2}{\epsilon_{\parallel}} = \frac{\omega^2}{c^2}. \tag{2.7}$$

We note that in the isotropic case ($\epsilon_{\parallel} = \epsilon_{\perp}$), the above expressions reduce to the familiar solutions for a metallic waveguide with the maximum supported mode $m_{\max}$ derived from the condition $\kappa \leq \sqrt{\epsilon}\omega/c$:

$$m_{\max} = \left\lfloor \frac{d\sqrt{\epsilon}\omega/c}{\pi} \right\rfloor \tag{2.8}$$

($\lfloor\cdot\rfloor$ and $\lceil\cdot\rceil$ denote floor and ceiling functions).

When both $\epsilon_{\parallel}$ and $\epsilon_{\perp} > 0$, this expression generalizes readily to the anisotropic case [in fact, we only need to replace ϵ with $\epsilon_{\parallel}$ in Eq. (2.8)]. However, if the *signs* of $\epsilon_{\parallel}$ and $\epsilon_{\perp}$ differ, the situation changes dramatically. Consider, for instance, the case $\{\epsilon_{\perp} < 0,\ \epsilon_{\parallel} > 0\}$. The condition for Eq. (2.7) to be satisfied now reads $\kappa \geq \sqrt{\epsilon}\omega/c$, leading to

$$m_{\min} = \left\lceil \frac{d\sqrt{\epsilon_{\parallel}}\omega/c}{\pi} \right\rceil. \tag{2.9}$$

Rather than having a maximum mode cutoff, the guided modes are now bounded *from below*. By adjusting the values of d and $\epsilon_{\parallel}$, it is possible to allow *all* modes to propagate in a waveguide, or to elevate the minimum cut-off threshold $m_{\min}$ to admit only high-order modes.

This result has interesting potential applications. First, the optical power in a given mode is proportional to β, which, asymptotically, is linear in the mode number m. Thus, it might be possible to concentrate unusually high fields in a subwavelength waveguide, an impossible feat with conventional materials. Such a capability would be of great interest in developing nonlinear devices.

Secondly, it should be noted that mode profiles for high-m solutions exhibit rapid oscillations, i.e., correspond to high spatial frequencies. Such high-order modes would be able to couple to evanescent fields of finely structured objects, which are also characterized by high transverse spatial frequencies. These high spatial frequencies carry the information about the object's subwavelength features — the information typically lost as a consequence of the diffraction limit. This ability to guide waves that would exponentially decay in an ordinary medium is of great interest in constructing subwavelength imaging devices, and will be discussed in more detail in Section 2.4.2.

Let us now consider the group velocity of the guided modes, $v_g = \partial\omega/\partial\beta$. Differentiating Eq. (2.7), we obtain

$$\frac{\partial\omega}{\partial\beta} = \frac{c^2}{\epsilon_\perp}\frac{1}{\omega/\beta} = \frac{c^2}{\epsilon_\perp}\frac{1}{v_\phi}, \tag{2.10}$$

where v_ϕ is the phase velocity. For $\epsilon_\perp < 0$, we see immediately that the phase velocity and the group velocity are of different signs. This implies that the Poynting vector $\mathbf{S}$ is directed opposite the wave vector $\mathbf{k}$.

It is worth noting that this conclusion can be made from the simple geometrical argument if we represent the mode of a metallic waveguide by a plane wave with wave vector $\mathbf{k}$ bouncing between the two waveguide boundaries. Due to the $\{\epsilon_\perp < 0,\ \epsilon_\| > 0\}$ anisotropy, the components of $\mathbf{S}$ and $\mathbf{k}$ along the waveguide, S_z and k_z, differ in sign (as was shown in an earlier section). However, in the process of constructing a waveguide mode out of the multiply reflecting plane wave, it can be seen that S_z represents the net energy flow in the mode, while k_z coincides with the mode propagation constant β. We therefore arrive at the same conclusion — that the direction of the phase fronts is opposite to the direction of the energy flow.

The guided modes therefore mimic the refractive behavior of magnetic ($\epsilon < 0$, $\mu < 0$) NIMs. Indeed, if we consider the waveguide shown in Fig. 2.3(a), filled with a regular dielectric on the left and with an $\epsilon_\perp < 0$ anisotropic material on the right, and a mode with propagation vector $\beta = \beta_y\hat{\mathbf{y}} + \beta_z\hat{\mathbf{z}}$ incident on this boundary, the phase fronts of the mode reveal negative refraction.

Yet another counterintuitive phenomenon is associated with propagation in anisotropic waveguides. Recall that for a waveguide with perfectly conducting walls, as above, the energy flux in the core is antiparallel to the wave vector. The same is true if the core is bounded by a cladding made from a regular, isotropic dielectric. However, for a dielectric waveguide, a portion of the energy flux exists in the cladding. In this region, the energy flux is, as usual, collinear with the wave vector [Fig. 2.3(b)]. For a particular value of the light frequency ω and the waveguide thickness d, the negative energy flux inside the waveguide can be nearly balanced by the positive energy flux outside. This leads to a dramatic reduction in the group velocity.

The frequency-dependent group velocity of a single slow mode is plotted in Fig. 2.3(c). It is evident that $v_g \lesssim 0.004\,c$ is attainable over a 1.1-THz frequency range. Such a wide bandwidth suggests the possibility of using the proposed system as an optical buffer.

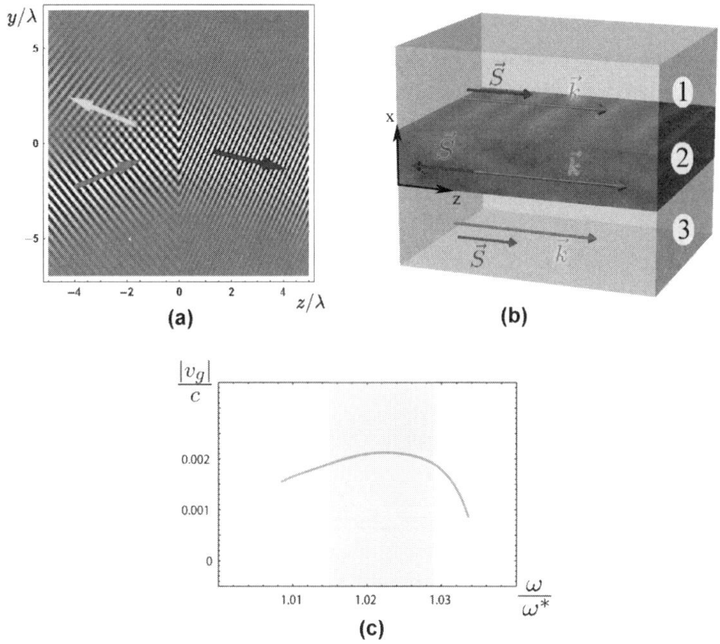

Figure 2.3 (a) Negative refraction exhibited by wavefronts in a 2D slab waveguide with metallic walls, filled with an isotropic dielectric on the left, and $\{\epsilon_\perp < 0,\ \epsilon_\| > 0\}$ material on the right. Arrows indicate the direction of the power flow. (b) Schematics of a waveguide supporting slow group velocity modes: dielectric cladding in regions 1 and 3; $\{\epsilon_\perp < 0,\ \epsilon_\| > 0\}$ material in region 2. (c) Group velocity as a function of frequency for the waveguide in (b). Note that $v_g \lesssim 0.004\,c$ throughout the shaded region. (Adapted from Refs. 20 and 31 with permission from Optical Society of America and Taylor & Francis Ltd.)

2.4.2 The hyperlens

2.4.2.1 Theoretical description

We saw in Section 2.2 that a medium with a hyperbolic dispersion relation allows propagation of high spatial frequency waves that would decay in a conventional dielectric. This phenomenon, however, is of limited utility in stand-off subwavelength imaging, as the high-k modes start exponentially decaying outside the material. It turns out, however, that hyperbolic dispersion implemented in curvilinear coordinates can yield devices that convert the high-k modes to propagating waves by essentially magnifying subwavelength structures.

A *hyperlens* is a hollow core cylinder (or half cylinder), made of a strongly anisotropic material, that can function as a far-field subdiffraction lens.[34–37] To understand the origin of subwavelength resolution in the hyperlens, it is useful to consider the imaging problem in the context of detecting a wave, scattered by a subwavelength object.

Waves scattered by the illuminated object can be examined in a monochromatic plane wave basis with a wide spectrum of spatial frequencies. The choice of basis, however, is dictated by the symmetry of the object under consideration and/or by convenience. Mathematically, the problem can be equivalently treated in a basis of cylindrical waves. In particular, any plane wave illuminating an object can be expanded in a basis of cylindrical waves as

$$\exp(ikx) = \sum_{m=-\infty}^{\infty} i^m J_m(kr) \exp(im\phi), \tag{2.11}$$

where $J_m(kr)$ denotes the Bessel function of the first kind and m is the angular momentum mode number of the cylindrical wave. This decomposition is illustrated schematically in Fig. 2.4(a). In this representation, reconstructing an image is equivalent to retrieving the scattering amplitudes and phase shifts of the various constituent angular momentum modes. The resolution limit in the cylindrical wave basis can be restated as the limit to the number of retrieved angular momentum modes with appreciable amplitude or phase change after scattering from the object.

We may think of the scattered angular momentum modes as distinct information channels through which the information about the object at the origin is conveyed to the far-field. However, even though the number of these channels is infinite [m is unbounded in Eq. (2.11)], very little information is carried over the high-m channels. Figure 2.4(b) shows the exact radial profile of the electric field for $m = 1$ and $m = 14$. For high values of m, the field exponentially decays toward the origin. This suggests that the interaction between a high-m mode and an object

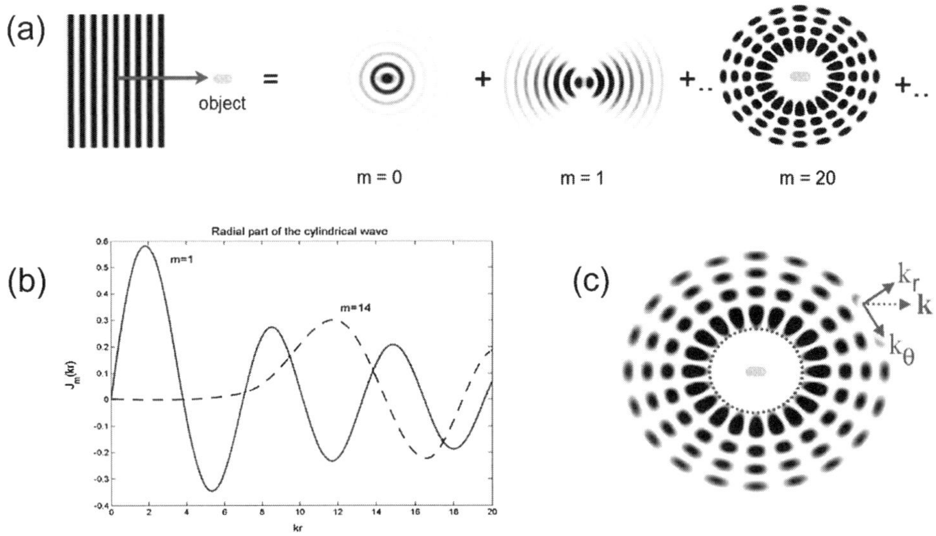

Figure 2.4 (a) The scattering of an incident plane wave by a target can be represented as scattering of various angular momentum modes. The regions of high intensity are shown in black, and low intensity in white. (b) Higher-order modes are exponentially small at the center. (c) The attenuation of high-order modes results from an upper bound on values of k_θ and the formation of the caustic shown as a dashed circle in the panel. (Reprinted from Ref. 34.)

placed at the origin is exponentially small; i.e., the scattering of such modes from the object is negligible. Classically, this corresponds to the parts of an illuminating beam that have a high *impact parameter* and therefore miss the scatterer and carry no information about the object into the far field.

The high-m modes are evanescent within a circle of critical radius called *the caustic*. This is because conservation of angular momentum implies that the tangential wave vector of a high-angular-momentum mode increases toward the center ($k_\theta r = m = \text{const}$). In a medium such as vacuum characterized by a circular isofrequency curve [see Fig. 2.1(a)], this increase in the tangential component is not supported, as both the tangential and radial wave vectors are bounded (see Eq. (2.4) and related discussion). These incident high-angular-momentum modes simply reflect without ever reaching the scatterer. As such, they do not contribute to the retrieval of information about the object's structure. However, if there existed a way to drive these states to the center, whereupon they could interact with the object, then these high-angular-momentum states would act as extra information channels for subwavelength structure retrieval.

It turns out that this scenario is possible for cylindrical systems with a hyperbolic dispersion relation. Consider wave propagation in a bulk medium with strong cylindrical anisotropy where dielectric permittivities have different signs in the tangential and radial directions ($\epsilon_\theta > 0$, $\epsilon_r < 0$). Since there exist no natural materials with such an anisotropy, we assume that the required dielectric response could be implemented using metamaterials. In particular, the desired anisotropy may be attained in a cylinder composed of "slices" of metal and dielectric or alternating concentric layers of metal and dielectric (see Fig. 2.5). The layer thickness h in each of these structures is much less than the wavelength λ, and when $h \ll \lambda \leq r$, the effective medium expressions in Eq. (2.5) (with $\epsilon_x, \epsilon_z \rightarrow \epsilon_r, \epsilon_\theta$) can be used for dielectric permittivities. A low-loss cylindrically anisotropic material can also be achieved by metallic inclusions in a hollow core dielectric cylinder.

It should be noted that the polar dielectric permittivities are ill defined at the center, and any practical realization of cylindrical anisotropy using metamaterials can only closely approximate the desired dielectric permittivities away from the center (when $r \geq \lambda$). However, numerical simulations show that the effective medium description is adequate and that the hyperlens functions even in the case where the inner radius is no greater than a wavelength.[34] The hyperlens functions in the channeling regime where a smaller inner radius aids in higher resolution.

As before, we focus on extraordinary waves (TM modes, with the magnetic field along the axis of the cylinder). These waves obey a hyperbolic dispersion relation similar to Eq. (2.1), namely,

$$\frac{k_r^2}{\epsilon_\theta} - \frac{k_\theta^2}{|\epsilon_r|} = \frac{\omega^2}{c^2}, \tag{2.12}$$

which allows for very high values of k, limited only by the patterning scale of the metamaterial medium. As the tangential component of the wave vector increases

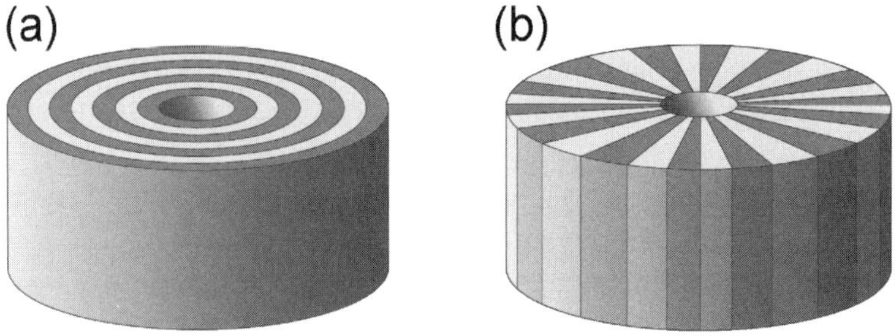

Figure 2.5 Possible realizations of metacylinders. (a) Concentric alternate metallic layers and dielectric layers or (b) radially symmetric "slices" of metal and dielectric produce ($\epsilon_\theta > 0$, $\epsilon_r < 0$) anisotropy. This results in a hyperbolic dispersion relation necessary for penetration of the field close to the center. (Reprinted from Ref. 34.)

toward the center, the radial component also increases; Eq. (2.12) can be satisfied for any radius and any value of m. Thus, as long as the effective medium description is valid, the field of high-angular-momentum states has appreciable magnitude close to the center.

This can be verified by solving Maxwell's equations for the TM mode in the cylindrical geometry for the ($\epsilon_\theta > 0$, $\epsilon_r < 0$) anisotropy

$$B_z \propto J_{m\sqrt{\epsilon_r/\epsilon_\theta}}\left(\frac{\omega}{c}\sqrt{\epsilon_\theta}\, r\right)\exp(im\phi). \tag{2.13}$$

This mode is plotted in Fig. 2.6(b). Note that the cylindrical anisotropy causes a high-angular-momentum state to penetrate toward the center — in contrast to the behavior of high-m modes in regular dielectrics [see Fig. 2.6(a)].

We now consider a hollow core cylinder of inner radius $R_{\text{inner}} \approx \lambda$ and outer radius R_{outer}, made of a cylindrically anisotropic homogeneous medium. The high-angular-momentum states with caustic radius $R_c \leq R_{\text{outer}}$ are captured by the device and guided toward the core. In this case, cylindrical symmetry implies that the distance between the field nodes at the core is less than the vacuum wavelength (see Fig. 2.6). Therefore, such high-angular-momentum states can act as a subwavelength probe for an object placed inside the core. Furthermore, since in the medium under consideration these states are propagating waves, they can carry information about the detailed structure of the object to the far field. The hyperlens thus enables extra information channels for retrieving the object's subwavelength structure. In the absence of the device, the high-angular-momentum modes representing these channels do not reach the core and, as such, carry no information about the object.

2.4.2.2 Imaging simulations

To confirm the subwavelength imaging capabilities of the hyperlens, we consider placing two point sources in the vicinity of the hollow cylinder's inner boundary.

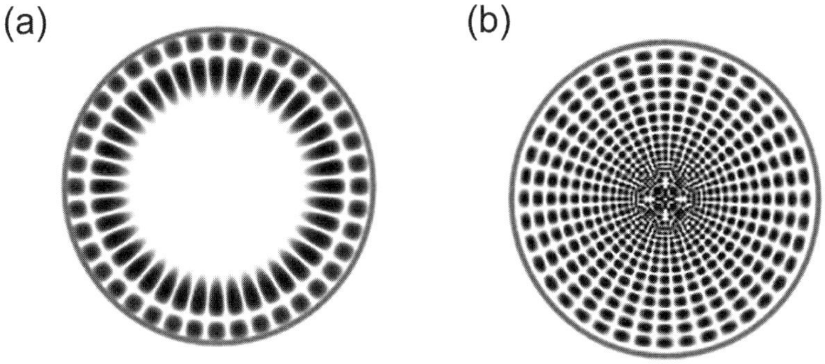

Figure 2.6 (a) High-angular-momentum states in an isotropic dielectric cylinder. (b) High-angular-momentum states in a cylinder made of $\epsilon_\theta > 0$, $\epsilon_r < 0$ metamaterial (in the effective medium approximation); note that the field penetrates to the center. (Reprinted from Ref. 34.)

To improve the coupling of high spatial frequency Fourier components to the high-angular-momentum modes, we assume that the inner layer of the hyperlens has $\text{Re}[\epsilon] \approx -1$. The two sources are placed at a distance of $\lambda/4.5$ apart (with $\lambda = 365$ nm), and we assume that the hyperlens is made of 160 alternating layers of silver ($\epsilon = -2.4012 + 0.2488i$) and dielectric ($\epsilon \approx 2.7$), each 10 nm thick. Exact numerical simulations can be used to study the imaging characteristics of this device. The resulting intensity pattern is shown in Fig. 2.7(a). The highly directional nature of the beams from the two sources allows for the resolution at the outer surface of the hyperlens. The separation between the two output beams at the boundary of the device is 5 times the distance between the sources and is larger than the diffraction limit, thereby allowing for subsequent processing by conventional optics. This magnification corresponds to the ratio of the outer and inner radii, and is a consequence of cylindrical symmetry, together with the directional nature of the beams.

The intensity distribution at the source is shown in Fig. 2.7(b), whereas the intensity distribution just outside the hyperlens is shown in Fig. 2.7(c). The two sources are clearly resolved, even though the distance between them is below the diffraction limit. It should be noted that realistic losses do not significantly affect the subdiffraction resolution capabilities of the hyperlens. Furthermore, due to the optical magnification in the hyperlens (by a factor of 5 in the simulation of Fig. 2.7), even for the subwavelength object, the scale of the image can be substantially larger than the wavelength, thus allowing for further optical processing of the image (e.g., further magnification) by conventional optics.

2.4.2.3 Semiclassical treatment

The above results were obtained by numerically propagating fields through the cylindrical layered structure. There exists, however, an analytic approach to analyzing light propagation in the hyperlens. Owing to the hyperbolic form of the

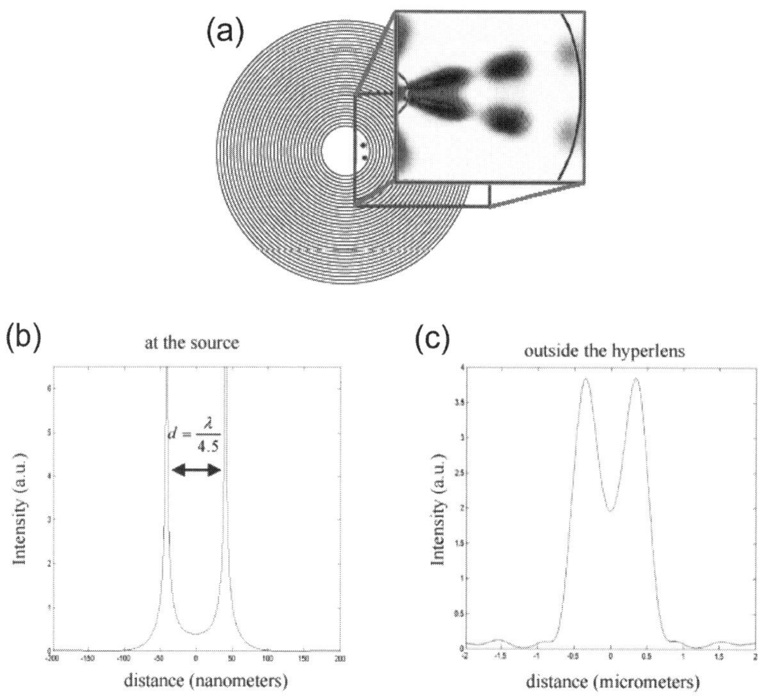

Figure 2.7 (a) Schematics of imaging by the hyperlens. Two point sources separated by $\lambda/4.5$ are placed within the hollow core of the hyperlens. The hyperlens consists of 160 alternating layers of metal and dielectric, each of 10 nm thickness. Intensity plot in the region bounded by the rectangle shows the highly directional nature of the beams from the two point sources. The separation between the beams at the outer boundary of the device is greater than λ, due to magnification. (b) and (c) Demonstration of subwavelength resolution in the composite hyperlens containing two sources placed a distance $\lambda/4.5$ apart inside the core: (b) field at the source; (c) field outside the hyperlens. (Adapted from Ref. 34.)

dispersion relation in Eq. (2.12), the radial and tangential momenta of the fields increase as light approaches the core of the device. This leads to a substantial decrease of wavelength, which suggests a semiclassical description of field propagation using Hamiltonian ray optics.

With the key assumption that the dielectric permittivity does not vary significantly over the scale of the wavelength, we can obtain the ray-optical Hamiltonian for a cylindrically anisotropic medium such as the hyperlens:

$$H = c\sqrt{\frac{p_r^2}{\epsilon_\theta} + \frac{p_\theta^2}{r^2\epsilon_r}}, \tag{2.14}$$

where c is the velocty of light in vacuum, p_r and p_θ are the radial and angular momenta, and ϵ_θ, ϵ_r are the tangential and radial dielectric permittivities.

Solving for the ray dynamics, the equation of the ray trajectory inside the hyperlens is seen to be[38]

$$r(\theta) = \frac{p_\theta}{\xi\sqrt{|\epsilon_r|}\sinh[\eta(\theta - \theta_0)]}. \tag{2.15}$$

This is the equation of a spiral where θ_0 is a parameter related to the initial conditions and

$$\eta = \sqrt{\frac{|\epsilon_r|}{\epsilon_\theta}} \tag{2.16}$$

critically determines the ray dynamics inside the hyperlens. The implications of this analytical solution beyond the ray approach will be presented in the subsequent section.

For a ray of light impinging on the hyperlens (outer radius $r_{\max}$) from vacuum with an impact parameter ρ, we can use the conservation of angular momentum ($p_\theta = \rho\frac{\omega}{c}$) upon refraction to evaluate the constant θ_0. The above equation then becomes

$$r(\theta) = \frac{\rho}{\sqrt{|\epsilon_r|}\sinh[\eta(\theta - \theta_0)]}, \tag{2.17}$$

with

$$\theta_0 = \sin^{-1}\left(\frac{\rho}{r_{\max}}\right) - \frac{1}{\eta}\sinh^{-1}\left(\frac{\rho}{r_{\max}\sqrt{|\epsilon_r|}}\right). \tag{2.18}$$

We plot the analytical result of Eq. (2.17) in Figs. 2.8 (a) and (b) for small values of the parameter η which explicitly shows the spiraling behavior. The negative refraction of the ray is consistent with the known negative refraction of the Poynting vector in strongly anisotropic materials. For large values of the parameter η, we are in the channeling regime, where the ray moves in a straight line inside the hyperlens.

If we visualize a Gaussian beam impinging on the layered hyperlens with impact parameter ρ [Fig. 2.9(a)] as a pencil of parallel rays, then Eq. (2.17) predicts that the distance between the rays will decrease as it approaches the core, where the rays bounce off the inner hollow region. This is seen by plotting the analytical expression inside the hyperlens for $\eta = 1$, $\epsilon_\theta = 1$, $\epsilon_r = -1$, and considering specular reflection at the inner radius, as shown in Fig. 2.9(b). By choosing an appropriate metal ($\epsilon_m \approx -0.4$) and dielectric ($\epsilon_d \approx 2.4$) we can achieve the layered hyperlens yielding the desired dielectric response, which is $\epsilon_\theta = 1, \epsilon_r = -1$ according to Eq. (2.5). We choose an inner radius of λ, outer radius 7λ, thickness of layers $\lambda/100$, $N = 600$ layers, and impact parameter $\rho = 2.4\lambda$ at an operating wavelength of 700 nm.

The magnitude of the field is plotted in Fig. 2.9(c), and the ray trajectory calculated from Eq. (2.17) is in white, superimposed on the field plot. Black denotes regions of high intensity. The two circles denote the inner and outer boundaries of the device. The ray is clearly seen to move along the center of the Gaussian beam. The narrowing effect obtained from the ray equations is evident in the width of

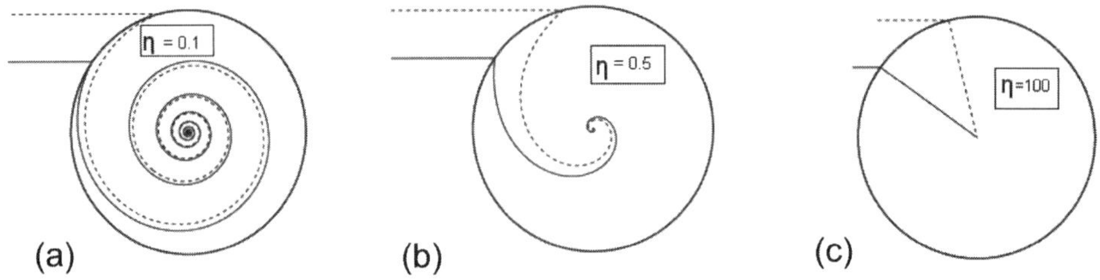

Figure 2.8 Trajectories of two rays incident on the hyperlens with different impact parameters, calculated using the analytical expression in Eqs. (2.17) and (2.18). (a) η = 0.1. (b) η = 0.5. Note the strong spiraling behavior. (c) "Channeling regime" for large η (η = 100), where rays travel in straight lines radially. Note that all rays travel toward the center. (Reprinted from Ref. 38.)

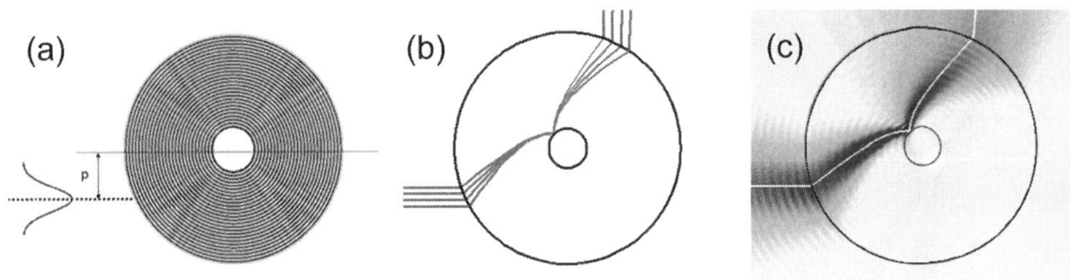

Figure 2.9 (a) Schematic of a Gaussian beam with impact parameter ρ impinging on the layered hyperlens (top view) consisting of alternating layers of metal and dielectric. The inner hollow region and the region outside the hyperlens is vacuum. (b) Ray trajectories representing the Gaussian beam calculated for the effective medium parameters of the hyperlens using Eq. (2.17). Note the narrowing of the Gaussian beam toward the core of the hyperlens, as predicted by the semiclassical theory. We consider specular reflection at the inner core. (c) Absolute value of the field for a Gaussian beam scattering from the layered hyperlens with parameters $\rho \approx 4\lambda$, $r_{\min} \approx \lambda$, $r_{\max} \approx 7\lambda$, $h \approx \lambda/100$, $\epsilon_m \approx -0.4$, and $\epsilon_d \approx 2.4$. The ray trajectory shown in white is calculated using Eq. (2.17) and specular reflection at the inner boundary. Note the narrowing of the Gaussian beam and also the motion of the center of the beam along the calculated ray trajectory. (Adapted from Ref. 38.)

the Gaussian beam near the core. This validates the semiclassical description presented, as well as the adequacy of the effective medium approximation in describing the hyperlens. Note that the narrowing effect opens up the possibility of using the hyperlens for subdiffraction lithography in the channeling regime where the Gaussian beam is expected to travel radially to the core with reduced beam width.

The semiclassical description can bring further insight into the hyperlens imaging setup of Fig. 2.7. Recall from earlier discussion that energy carried by waves in media with negative transverse permittivity is constrained to a cone. In the case of cylindrical anisotropy, the half-angle of the cone [see also Eq. (2.3)] is given by

$$\tan(\theta_c) = \sqrt{\frac{\epsilon_\theta}{|\epsilon_r|}} = \frac{1}{\eta}, \tag{2.19}$$

where η is the parameter entering the Eqs. (2.17) and (2.18) that determines the pitch of the ray spirals. For large values of η (the channeling regime), the energy cone divergence angle tends to zero, i.e., radiation from a point source propagates as a narrow beam. This is the condition that enables subdiffraction-limited imaging. We note from Fig. 2.8(c) that in the channeling regime, rays of light move in the hyperlens in straight lines, which is essential for a narrow beam divergence angle.

We verify this fact using the analytical expression for rays inside the hyperlens in the case of two point sources kept inside the hyperlens. The point source is represented as a source of rays in all directions as shown in the inset of Fig. 2.10(a). Note that even though we have assumed isotropic emission in the core, the density of rays is high in two cone-like regions within the hyperlens. The rays of light are negatively refracted at the inner curved surface of the hyperlens, which helps in the formation of a beam. Inside the hyperlens, the rays then move in straight lines, almost radially, traveling to the outer interface. These rays arrive at normal incidence, and the beam-like nature in the hyperlens is preserved as they emerge into vacuum. Thus, the two point sources give rise to two distinct beams in the far-field, even though they are separated by less than the diffraction limit inside the hyperlens. Furthermore, due to the cylindrical geometry and almost radial nature of propagation, the distance between the point sources is magnified and above the diffraction limit. We verify this behavior by considering a practical realization of the hyperlens made of alternating layers of metal ($\epsilon_m \approx -1$) and dielectric ($\epsilon_d \approx 1.1$) to achieve a dielectric response in the effective medium approximation ($\epsilon_\theta = 0.05$, $\epsilon_r = -22$). This gives a large value of $\eta \approx 20$, and hence we are in the channeling regime. The magnification due to the radial nature of light propagation is the ratio of the radii, which is approximately 5 in this case. The two beams emanating from the point sources that carry information to the far-field can clearly be seen in Fig. 2.10(b), consistent with the plot obtained from the analytical expression for the rays in the hyperlens.

2.5 Conclusion

Anisotropic metamaterials with hyperbolic dispersion relations were originally proposed as a simple alternative to negatively refractive media operating via magnetic resonances. We have seen, however, that this class of metamaterials goes far beyond geometric negative refraction. Its properties enable a multitude of novel systems with applications in imaging and wave guiding.

In the coming years, we expect to see the emergence of many metamaterial-enabled devices. At radio frequencies, negative index metamaterials have already found applications in reflectors and radio antennas,[39] as well as in magnetic resonance imaging.[40] Optical domain metamaterials remain the subject of intense research. The goal of creating a medium with customized spatial and spectral variation of its dielectric tensor is ambitious but not far-fetched. After all, the ability to tailor electromagnetic response of materials by nanoscale patterning has become

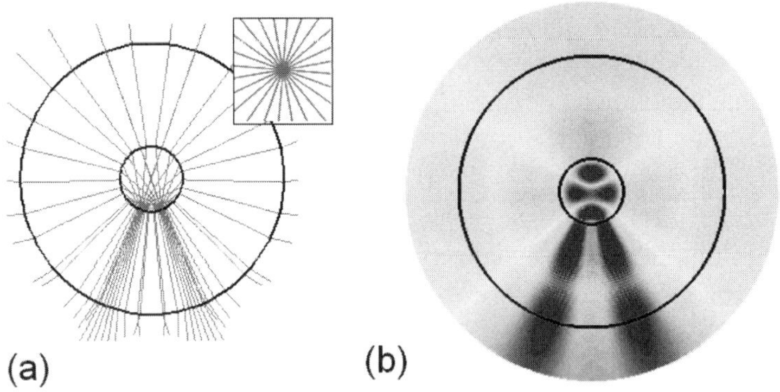

Figure 2.10 Subdiffraction imaging in the hyperlens. (a) Beam-like radiation obtained from Eq. (2.17) for two point sources kept near the inner boundary of the hyperlens for large η (channeling regime). The rays are negatively refracted at the inner surface and proceed radially outward, leading to magnification at the outer surface. The point source is represented as a source of rays in all directions (inset). (b) Numerical confirmation of the beam-like radiation using a layered metamaterial hyperlens made of alternating layers of metal ($\epsilon_m \approx -1$) and dielectric ($\epsilon_d \approx 1.1$) and two point sources near the inner boundary. The regions of high intensity are dark. (Reprinted from Ref. 38.)

common in active optoelectronic devices (such as quantum cascade lasers), as well as in photonic crystal and plasmonic systems. Great opportunities exist for constructing and interfacing useful devices based on hyperbolic dispersion.

References

1. V. M. Shalaev, "Optical negative-index metamaterials," *Nature Photonics* **1**, 41–48 (2007).
2. J. B. Pendry, "Negative refraction makes a perfect lens," *Phys. Rev. Lett.* **85**, 3966–3969 (2000).
3. D. R. Smith, D. Schurig, and J. B. Pendry, "Negative refraction of modulated electromagnetic waves," *Appl. Phys. Lett.* **81**, 2713–2715 (2002).
4. J. B. Pendry and D. R. Smith, "Reversing light with negative refraction," *Physics Today* **57**, 37–43 (2004).
5. H. Lamb, "On group-velocity," *Proc. Lond. Math. Soc.* **1**, 473–479 (1904).
6. A. Schuster, *An Introduction to the Theory of Optics*, Arnold, London (1904).
7. V. G. Veselago, "The electrodynamics of substances with simultaneously negative value of ϵ and μ," *Sov. Phys. Usp.* **10**, 509–514 (1968).
8. V. G. Veselago and E. E. Narimanov, "The left hand of brightness: past, present and future of negative index materials," *Nature Materials* **5**, 759–762 (2006).

9. U. K. Chettiar, A. V. Kildishev, H.-K. Yuan, W. Cai, S. Xiao, V. P. Drachev, and V. M. Shalaev, "Dual-band negative index metamaterial: double negative at 813 nm and single negative at 772 nm," *Opt. Lett.* **32**, 1671–1673 (2007).
10. V. A. Podolskiy and E. E. Narimanov, "Near-sighted superlens," *Opt. Lett.* **30**, 75–77 (2005).
11. M. Notomi, "Theory of light propagation in strongly modulated photonic crystals: Refraction-like behavior in the vicinity of the photonic band gap," *Phys. Rev. B* **62**, 10696–10705 (2000).
12. C. Luo, S. G. Johnson, J. D. Joannopoulos, and J. B. Pendry, "All-angle negative refraction without negative effective index," *Phys. Rev. B* **65**, 201104 (2002).
13. P. V. Parimi, W. T. Lu, P. Vodo, and S. Sridhar, "Imaging by flat lens using negative refraction," *Nature* **426**, 404 (2003).
14. E. Schonbrun, T. Yamashita, W. Park, and C. J. Summers, "Negative-index imaging by an index-matched photonic crystal slab," *Phys. Rev. B* **73**, 195117 (2006).
15. Y. Zhang, B. Fluegel, and A. Mascarenhas, "Total negative refraction in real crystals for ballistic electrons and light," *Phys. Rev. Lett.* **91**, 157404 (2003).
16. P. A. Belov, "Backward waves and negative refraction in uniaxial dielectrics with negative dielectric permittivity along the anisotropy axis," *Microwave Opt. Technol. Lett.* **37**, 259–263 (2003).
17. T. Dumelow, J. A. P. da Costa, and V. N. Freire, "Slab lenses from simple anisotropic media," *Phys. Rev. B* **72**, 235115 (2005).
18. V. A. Podolskiy and E. E. Narimanov, "Strongly anisotropic waveguide as a nonmagnetic left-handed system," *Phys. Rev. B* **71**, 201101(R) (2005).
19. L. D. Landau, E. M. Lifshitz, and L. P. Pitaevskii, *Electrodynamics of Continuous Media*, 2nd ed., Reed Ltd., Oxford (1984).
20. L. V. Alekseyev and E. Narimanov, "Slow light and 3D imaging with non-magnetic negative index systems," *Opt. Express* **14**, 11184–11193 (2006).
21. R. K. Fisher and R. W. Gould, "Resonance cones in the field pattern of a short antenna in an anisotropic plasma," *Phys. Rev. Lett.* **22**, 1093–1095 (1969).
22. E. Arbel and L. B. Felsen, *Electromagnetic Theory and Antennas*, Pergamon Press, New York (1963).
23. P. A. Belov and Y. Hao, "Subwavelength imaging at optical frequencies using a transmission device formed by a periodic layered metal-dielectric structure operating in the canalization regime," *Phys. Rev. B* **73**, 113110 (2006).

24. A. Hadni and X. Gerbaux, "Far IR excitation of longitudinal optical phonons in triglycine sulfate," *Ferroelectrics* **248**, 15–26 (2000).

25. X. Gerbaux, M. Tazawa, and A. Hadni, "Far IR transmission measurements on triglycine sulfate (TGS), at 5 K," *Ferroelectrics* **215**, 47–63 (1998).

26. M. Schubert, T. E. Tiwald, and C. M. Herzinger, "Infrared dielectric anisotropy and phonon modes of sapphire," *Phys. Rev. B* **61**, 8187–8201 (2000).

27. W. S. Boyle, A. D. Brailsford, and J. K. Galt, "Dielectric anomalies and cyclotron absorption in the infrared: observations on bismuth," *Phys. Rev.* **109**, 1396 (1958).

28. V. D. Kulakovskii and V. D. Egorov, "Plasma reflection in bismuth and bismuth-antimony alloys," *Sov. Phys. Solid State* **15**(7), 1368 (1974).

29. G. Shvets, "Photonic approach to making a material with a negative index of refraction," *Phys. Rev. B* **67**, 035109 (2003).

30. G. Shvets, Y. Urzhumov, and D. Korobkin, "Enhanced near-field resolution in mid-infrared using metamaterials," *J. Opt. Soc. Am. B* **23**, 468–478 (2006).

31. V. A. Podolskiy, L. V. Alekseyev, and E. E. Narimanov, "Strongly anisotropic media: the THz perspectives of left-handed materials," *J. Mod. Opt.* **52**, 2343–2349 (2005).

32. S. A. Ramakrishna, J. B. Pendry, M. C. K. Wiltshire, and W. J. Stewart, "Imaging the near field," *J. Mod. Opt.* **50**, 1419–1430 (2003).

33. R. Wangberg, J. Elser, E. E. Narimanov, and V. A. Podolskiy, "Non-magnetic nano-composites for optical and infrared negative-refractive-index media," *J. Opt. Soc. Am. B.* **23**, 498–505 (2006).

34. Z. Jacob, L. V. Alekseyev, and E. Narimanov, "Optical hyperlens: Far-field imaging beyond the diffraction limit," *Opt. Express* **14**, 8247–8256 (2006).

35. A. Salandrino and N. Engheta, "Far-field subdiffraction optical microscopy using metamaterial crystals: Theory and simulations," *Phys. Rev. B* **74**, 075103 (2006).

36. I. I. Smolyaninov, Y.-J. Hung, and C. C. Davis, "Magnifying superlens in the visible frequency range," *Science* **315**, 5819 (2007).

37. Z. Liu, H. Lee, Y. Xiong, C. Sun, and X. Zhang, "Far-field optical hyperlens magnifying sub-diffraction-limited objects," *Science* **315**, 1686 (2007).

38. Z. Jacob, L. V. Alekseyev, and E. Narimanov, "Semiclassical theory of the hyperlens," *J. Opt. Soc. Am. A* **24**, A52–59 (2007).

39. C. G. Parazzoli, R. B. Greegor, J. A. Nielsen, M. A. Thompson, K. Li, A. M. Vetter, M. H. Tanielian, and D. C. Vier, "Performance of a negative index of refraction lens," *Appl. Phys. Lett.* **84**, 3232–3234 (2004).

40. M. C. K. Wiltshire, J. B. Pendry, I. R. Young, D. J. Larkman, D. J. Gilderdale, and J. V. Hajnal, "Microstructured magnetic materials for RF flux guides in magnetic resonance imaging," *Science* **291**, 849–851 (2001).

Biographies

Leonid Alekseyev received a B.S. in Physics and an M.S. in Electrical Engineering from Stanford University in 2003. He is currently working toward a Ph.D. in electrical engineering at Princeton University. His research focuses on theory and applications of novel metamaterial devices.

Zubin Jacob completed his B.Tech. in electrical engineering at the Indian Institute of Technology-Bombay. He received M.A.E.E. and M.S.E.E degrees from Princeton University and is currently a Ph.D. candidate in the department of Electrical and Computer Engineering at Purdue University. He was the recipient of the best student paper award at ETOPIM 7 and the SPIE student award for potential long-range contributions to optical engineering, and was the finalist for the Theodor Maiman best student paper award at IQEC/CLEO in 2009. His research interests are in the area of metamaterials and nano-optics.

A biography for Evgenii Narimanov was not available.

Chapter 3
Magneto-optics and the Kerr Effect with Ferromagnetic Materials

Allan D. Boardman and Neil King
Institute for Materials Research, Joule Physics Laboratory, University of Salford, UK

3.1 Introduction to Magneto-optical Materials and Concepts

As implied by its name, magneto-optics embraces all activities that concern the interaction of light with a magnetic material.[1–4] A vast range of substances can be accessed for such investigations, and insulating materials such as YIG (yttrium iron garnet) are popular for isolator applications. It is metallic materials, however, that are discussed in this tutorial, so that the basics of the all-important Kerr effect can be exposed in a manner that exploits popular ferromagnetic materials that yield effects that are of practical use.[2] In this spirit, the discussion will always refer to metallic conductors, even though, of course, all magneto-optic materials can display the full range of magneto-optic properties to a greater or lesser degree.[2–4] Of the five ferromagnetic elements, cobalt, iron, and nickel are by far the most important. In bulk form, each material contains what are called domains. Domains are magnetised subregions that, although quite often chaotically related to each other, can be brought into alignment by the application of a modest applied magnetic field. In this way, a magnetic material and its magnetic state can be organised to have an electromagnetic impact simply by reflecting p-polarised light from it. In fact, the magnetisation of the material produces a small rotation of the plane of polarisation of this type of incident light. Upon reflection from a magnetised material, an incident plane p-polarised light beam can become elliptically polarised, with the principal axis moving a few degrees away from the axis of the incoming light. Such a rotation is known as the Kerr effect, named after the Rev. John Kerr, who announced his discovery in 1888. Kerr could not have imagined just how significant this discovery was; its impact on data storage technologies is still holding out fantastic prospects.[1–4] As discussed below, magnetic materials can be magnetised in a number of ways. A popular orientation for the magnetisation is to be perpendicular to the optical reflection surface.[1–4] It should also be possible to observe such a magneto-optic effect using only a nanomagnet that is a member of an array of nanomagnets. The aim is to deposit a massive number of isolated magnetised units in the multiterabit range over a surface so that each unit can be interrogated by bouncing a laser beam off of it. The 'bits' are digital data that can be read optically through the Kerr effect. This is not the only use for Kerr effect; it is generically useful as a tool for monitoring all types of magnetic states. The data storage aspect is amazing, however, when it is realised that the print collection in the U.S. Library of Congress is about 10 terabytes. The fact that 1 terabyte represents about 50,000 trees turned into printed paper makes the development of magneto-optic storage even more imperative and exciting.

Ferromagnetic elements such as iron and cobalt exhibit the maximum magnetisation that they can attain. This is technically known as the saturation magnetisation and is very high. Such elements also have a high Curie point, the temperature at which the magnetisation is destroyed; e.g., for cobalt this temperature is 1120° C. For nickel, this value drops to 360° C. It is against the background of these attractive properties of ferromagnetic metals that the rest of the chapter develops a working discussion of the Kerr effect.

3.2 Reflection of Light from a Plane Ferromagnetic Surface

In this section, the Kerr effect will first be investigated for a single plane surface. In this way, the basic properties can be revealed in a manner that will inform any search for enhancement and address the question of whether any connection to surface waves exists. The latter, discussed in the concluding part of the chapter, is informative in its own right and could form the basis for modern implementations of Kerr effect technologies.

When p-polarised light is reflected from the surface of a magnetised ferromagnetic material, the reflected beam acquires an induced s component, known as the Kerr component, which is out of phase with the reflected p-polarised component. Hence, the reflected wave is generally elliptically polarised. Experimentally, there are three globally favoured orientations for applying a magnetic field to a magneto-optical material. These orientations are polar, transverse, and longitudinal, each of which refers to the orientation of the propagation direction of an electromagnetic wave to an applied magnetic field.

For a reflecting/transmitting surface of a magneto-optical material, if the magnetisation **M** created by an applied magnetic field is parallel to both the surface and the plane containing the incident light beam, the arrangement is called a longitudinal configuration. If **M** is perpendicular to the surface, it is called polar; if, however, **M** lies in the surface plane but is also perpendicular to the incident plane containing the light beam, it has a transverse orientation.

For the longitudinal and polar cases, a Kerr rotation of the polarisation of the incoming *p*-polarised light beam takes place, thereby creating a small *s*-polarised component in the reflected wave. For the transverse case, no rotation occurs; instead, there is magnetic change to the reflected beam intensity. Given the high suitability of the polar orientation for data storage systems, this is the one that is now favoured for devices, over the longitudinal orientation.[1,2]

3.2.1 Single-surface polar orientation

To begin, let us consider Fig. 3.1, which illustrates the existence of a plane wave $\exp\left(i\left[\gamma \sin\theta x + \gamma \cos\theta z - \omega t\right]\right)$ propagating with a wave vector $\boldsymbol{\gamma}$ lying in the (x,z) plane of an infinite magneto-optic material. The wave vector is orientated at angle θ to the z axis. The magnetisation vector **M** is set to lie along the z axis. Neglecting spatial dispersion, the dielectric tensor $\boldsymbol{\varepsilon}$ of the magneto-optic material is

$$\boldsymbol{\varepsilon} = \begin{pmatrix} \varepsilon_{xx} & \varepsilon_{xy} & 0 \\ \varepsilon_{yx} & \varepsilon_{yy} & 0 \\ 0 & 0 & \varepsilon_{zz} \end{pmatrix}. \tag{3.1}$$

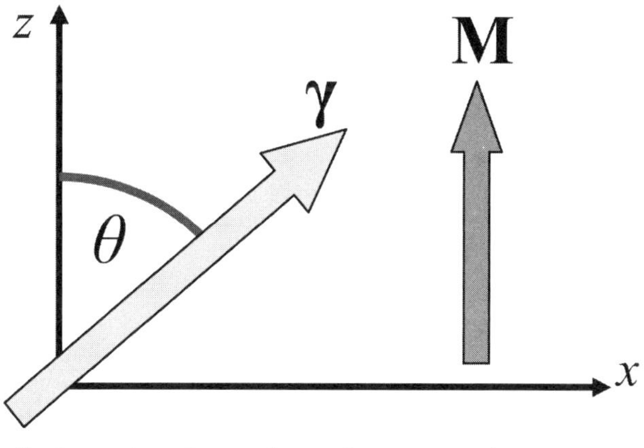

Figure 3.1 Coordinate system for a plane electromagnetic wave propagating with a complex wave number *γ* in the (*x*,*z*) plane inside a bulk ferromagnetic material that is magnetised along the *z* axis.

This dielectric tensor accounts for the fact that a magneto-optical material, such as a ferromagnetic metal, exhibits a form of forced gyrotropy. Specifically, it is this character that produces off-diagonal elements. This can be quickly appreciated by considering, as an example, the behaviour of a free charge q with velocity $\mathbf{v}$ in the presence of an applied magnetic field $\mathbf{H}$. It will experience a force $q\mathbf{v} \times \mathbf{H}$, which, when added into the current density in Maxwell's equations, immediately creates the off-diagonal elements in question. The permittivity tensor of the magneto-optic materials then has cylindrical symmetry about the z axis, which in this case accounts for the zeros in the tensor. Switching off the applied magnetic field causes the off-diagonal elements to disappear and the diagonal elements to become equal. The generic conclusion is that all types of magneto-optic materials that are exposed to a magnetic field have a relative permittivity with off-diagonal elements and can have appropriately positioned zero elements that depend on the direction of the applied field.

In the Fourier domain, the transforms of the electric field are $\mathbf{E}(r,\omega)$, which satisfy

$$\nabla \times \left(\nabla \times \mathbf{E}\right) - \frac{\omega^2}{c^2}\boldsymbol{\varepsilon} \cdot \mathbf{E} = 0 . \tag{3.2}$$

This equation can be rewritten as

$$\nabla\left(\nabla \cdot \mathbf{E}\right) - \nabla^2\mathbf{E} - \frac{\omega^2}{c^2}\boldsymbol{\varepsilon} \cdot \mathbf{E} = 0 . \tag{3.3}$$

As stated earlier, $\mathbf{E}(r,\omega)$ varies as $\exp\left(i\left[\gamma \sin\theta x + \gamma \cos\theta z - \omega t\right]\right)$, so that substitution into Eq. (3.3) leads to the component equations

$$\left(\gamma^2 \cos^2\theta - \frac{\omega^2}{c^2}\varepsilon_{xx}\right)E_x - \frac{\omega^2}{c^2}\varepsilon_{xy}E_y + \gamma^2 \sin\theta\cos\theta E_z = 0 \tag{3.4}$$

$$\left(\gamma^2 - \frac{\omega^2}{c^2}\varepsilon_{yy}\right)E_y - \frac{\omega^2}{c^2}\varepsilon_{yx}E_x = 0 \tag{3.5}$$

$$\gamma^2 \sin\theta\cos\theta E_x + \left(\gamma^2 \sin^2\theta - \frac{\omega^2}{c^2}\varepsilon_{zz}\right)E_z = 0 . \tag{3.6}$$

The field components can now be eliminated to yield

$$\left(\gamma^2 - \frac{\omega^2}{c^2}\varepsilon_{yy}\right)\left[\gamma^2\left(\varepsilon_{xx}\sin^2\theta + \varepsilon_{zz}\cos^2\theta\right) - \frac{\omega^2}{c^2}\varepsilon_{xx}\varepsilon_{zz}\right] + \gamma^2\frac{\omega^2}{c^2}\varepsilon_{xy}\varepsilon_{yx}\sin^2\theta - \frac{\omega^4}{c^4}\varepsilon_{xx}\varepsilon_{yy}\varepsilon_{zz} = 0. \tag{3.7}$$

Up until now in this discussion, the relative dielectric permittivity tensor has been kept deliberately general; however, an inspection of the literature[1,2] shows that it is practical to set its diagonal elements equal to each other.[2] This can be expressed with the following notation: $\varepsilon_{xx} = \varepsilon_{yy} = \varepsilon_{zz} = \varepsilon_1$, $\varepsilon_{xy} = -\varepsilon_{yx} = \varepsilon_2$. Note that, in addition, $|\varepsilon_2| \ll |\varepsilon_1|$, keeping in mind that the *tensor elements are complex*. Technically speaking, Eq. (3.7), being a quartic, has four roots identified with a suitable distribution of plus and minus signs. The roots required here are

$$\gamma_2 = \frac{\omega}{c}\sqrt{\varepsilon_1}\left(1 + i\frac{\varepsilon_2}{2\varepsilon_1}\cos\theta_2\right) \tag{3.8}$$

$$\gamma_3 = \frac{\omega}{c}\sqrt{\varepsilon_1}\left(1 - i\frac{\varepsilon_2}{2\varepsilon_1}\cos\theta_3\right), \tag{3.9}$$

where the labelling (2,3) is used for convenience. At this stage, the polar configuration is adopted; this is sketched in Fig. 3.2 to show what occurs when a p-polarised beam encounters a magnetic material interface. The results that flow from Fig. 3.1 are placed in context, and the orientations of the complex wave vectors are schematically displayed.

A further refinement of the notation is achieved at this stage by writing[3,4]

$$\varepsilon_1 \equiv n^2; \qquad \varepsilon_2 = -in^2Q; \qquad \frac{\varepsilon_2}{\varepsilon_1} = -iQ, \tag{3.10}$$

where n is the complex refractive index of the magneto-optic material, and Q, which is also complex, is a measure of the strength of the magnetisation. Hence, the final convenient form for the roots of Eq. (3.7) to be used here is

$$\gamma_2 = \frac{\omega}{c}n\left(1 + i\frac{Q}{2}\cos\theta_2\right) \tag{3.11}$$

$$\gamma_3 = \frac{\omega}{c}n\left(1 - i\frac{Q}{2}\cos\theta_3\right). \tag{3.12}$$

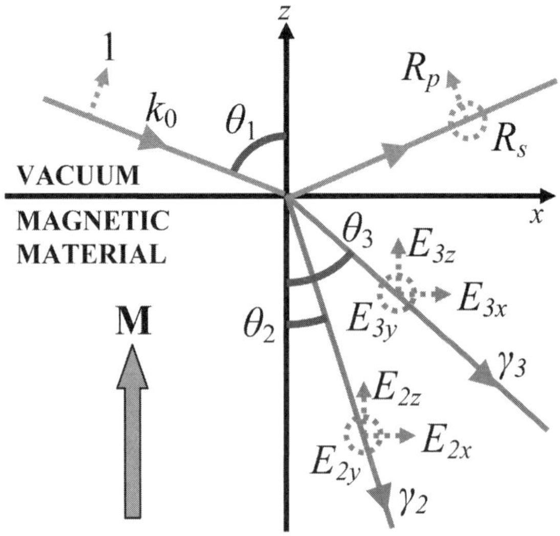

Figure 3.2 Sketch of the polar configuration showing that **M** is perpendicular to the interface. The refraction and reflection at a single interface are sketched to show the existence of the complex amplitudes R_s and R_p, and the vectors **E** and $\boldsymbol{\gamma}$.

Furthermore, the relationships between the electric field components (E_x, E_y, E_z) are

$$E_y = \frac{\frac{\omega^2}{c^2}\varepsilon_{yx}}{\gamma^2 - \frac{\omega^2}{c^2}\varepsilon_{yy}} E_x \tag{3.13}$$

$$E_z = -\frac{\gamma^2 \sin\theta\ \cos\theta}{\gamma^2 \sin^2\theta\ - \frac{\omega^2}{c^2}\varepsilon_{zz}} E_x . \tag{3.14}$$

These will be developed below with the subscripts introduced in Fig. 3.2. In the latter, a plane wave is incident on a plane boundary between a vacuum and a magneto-optic material, and a polar orientation of the magnetisation is adopted. However, even though here the vacuum is selected for convenience, the results readily generalise to any unmagnetised, insulating, dielectric upper bounding medium. The mathematical spatial dependences of the rays involved in Fig. 3.2, after factoring out the common time-dependence $\exp(-i\omega t)$ are

- ***Incident*** $\exp\left(i\left[k_0 \sin\theta_1 x - k_0 \cos\theta_1 z\right]\right)$
- ***Reflected*** $\exp\left(i\left[k_0 \sin\theta_1 x + k_0 \cos\theta_1 z\right]\right)$

- ***Transmitted*** (type 1) $\exp\left(i\left[\gamma_2 \sin\theta_2 x - \gamma_2 \cos\theta_2 z\right]\right)$
  (type 2) $\exp\left(i\left[\gamma_3 \sin\theta_3 x - \gamma_3 \cos\theta_3 z\right]\right)$.

It is interesting that the polarly configured magneto-optic surface creates two types of transmitted waves. Using the (2,3) notation, these waves are associated with electric field components (E_{2x}, E_{2y}, E_{2z}) and (E_{3x}, E_{3y}, E_{3z}), and the magnetic field components are H_{2x}, H_{3x}.

Using the Maxwell equation $\nabla \times \mathbf{E} = i\omega\mu_0\mathbf{H}$, the relationships of the y components of the electric field to the x components of the magnetic field are

$$-k_0 \cos\theta_1 E_{1y} = \omega\mu_0 H_{1x} \tag{3.15}$$

$$\gamma_2 \cos\theta_2 E_{2y} = \omega\mu_0 H_{2x} \tag{3.16}$$

$$\gamma_3 \cos\theta_3 E_{3y} = \omega\mu_0 H_{3x}, \tag{3.17}$$

where the first equation refers to the vacuum upper half-space, and the next two equations refer to the magneto-optic material.

At the interface, the boundary conditions require the continuity of the tangential components of the electric and magnetic fields as well as the continuity of the normal component of the displacement vector. Upon applying any boundary condition, the outcome is that a given quantity X on one side of the boundary is the same as its value on the other side of the boundary (in this case, located at $z = 0$). Hence, the difference between the X values on each side of the boundary is zero. This result can be compactly written as $[\![X]\!]_{z=0} = 0$. For the type of surface interrogation shown in Fig. 3.2, it is envisaged that an optical beam is incident upon the surface and that the reflected beam from that surface experiences a change of polarisation. In other words, if a p-polarised optical beam is fired onto the surface, then the reflected wave is not simply p-polarised, but has acquired an extra polarisation that is perpendicular to the incident plane. The net reflectivity is therefore a combination of a p-polarised beam and an induced s-polarised component. In Fig. 3.2, R_p and R_s represent the complex amplitudes of the reflected p-polarised and s-polarised waves.

Applying the boundary conditions gives the following four equations:

$$[\![E_x]\!]_{z=0} = 0 \Rightarrow \quad \cos\theta_1 - R_p \cos\theta_1 = E_{2x} + E_{3x} \tag{3.18}$$

$$[\![D_z]\!]_{z=0} = 0 \Rightarrow \quad \sin\theta_1 + R_p \sin\theta_1 = \varepsilon_{zz} E_{2z} + \varepsilon_{zz} E_{3z} \tag{3.19}$$

$$[\![H_x]\!]_{z=0} = 0 \Rightarrow \quad -R_s k_0 \cos\theta_1 = E_{2y}\gamma_2 \cos\theta_2 + E_{3y}\gamma_3 \cos\theta_3 \tag{3.20}$$

$$\left[\!\left[E_y \right]\!\right]_{z=0} = 0 \Rightarrow \qquad R_s = E_{2y} + E_{3y}\,. \tag{3.21}$$

Using the notation established in Fig. 3.2, the relationships for Eqs. (3.13) and (3.14) become

$$E_{2y} = \frac{\dfrac{\omega^2}{c^2}\varepsilon_{yx}}{\gamma_2^2 - \dfrac{\omega^2}{c^2}\varepsilon_{yy}} E_{2x} = A_2 E_{2x} \tag{3.22}$$

$$E_{2z} = -\frac{\gamma_2^2 \sin\theta_2 \cos\theta_2}{\gamma_2^2 \sin^2\theta_2 - \dfrac{\omega^2}{c^2}\varepsilon_{zz}} E_{2x} = -B_2 E_{2x} \tag{3.23}$$

$$E_{3y} = \frac{\dfrac{\omega^2}{c^2}\varepsilon_{yx}}{\gamma_3^2 - \dfrac{\omega^2}{c^2}\varepsilon_{yy}} E_{3x} = A_3 E_{3x}\,;\; E_{3z} = -\frac{\gamma_3^2 \sin\theta_3 \cos\theta_3}{\gamma_3^2 \sin^2\theta_3 - \dfrac{\omega^2}{c^2}\varepsilon_{zz}} E_{3x} = -B_3 E_{3x}\,. \tag{3.24}$$

3.2.2 Kerr rotation

One of the main drivers of magneto-optics investigations is to determine what is known as the Kerr rotation. The foregoing analysis establishes a set of equations from which the rotation of the electric field vector of a p-polarised wave (that is out of the plane of incidence due to the acquisition of an s-polarised contribution) can be calculated. If the reflected p-polarised electric field amplitude is designated as R_p and the induced s-polarised electric field amplitude is designated as R_s, the complex Kerr rotation angle can be defined through the complex ratio[1,2,4]

$$\frac{R_s}{R_p} = \tan\left(\Theta_K + i\Theta_E\right). \tag{3.25}$$

Θ_K is called the Kerr rotation, with Θ_E being the measure of the ellipticity introduced. In practical cases, these angles are small enough for the tangent to be replaced by the complex angle itself.

The calculation of R_s and R_p is considerably expedited by expressing the basic boundary condition equations in the following form:

$$\begin{bmatrix} \cos\theta_1 & 1 & 1 & 0 \\ \sin\theta_1 & \varepsilon_{zz}B_2 & \varepsilon_{zz}B_3 & 0 \\ 0 & \gamma_2 A_2\cos\theta_2 & \gamma_3 A_3\cos\theta_3 & k_0\cos\theta_1 \\ 0 & -A_2 & -A_3 & 1 \end{bmatrix}\begin{bmatrix} R_p \\ E_{2x} \\ E_{3x} \\ R_s \end{bmatrix} = \begin{bmatrix} \cos\theta_1 \\ -\sin\theta_1 \\ 0 \\ 0 \end{bmatrix}, \quad (3.26)$$

in which the incident p-polarised wave has a unit amplitude, thus enabling Cramer's rule to be elegantly applied to the quantification of the elements in the left-hand column vector. For example,

$$R_s = \frac{\left\|\begin{matrix} \cos\theta_1 & 1 & 1 & \cos\theta_1 \\ \sin\theta_1 & \varepsilon_{zz}B_2 & \varepsilon_{zz}B_3 & -\sin\theta_1 \\ 0 & \gamma_2 A_2\cos\theta_2 & \gamma_3 A_3\cos\theta_3 & 0 \\ 0 & -A_2 & -A_3 & 0 \end{matrix}\right\|}{\left\|\begin{matrix} \cos\theta_1 & 1 & 1 & 0 \\ \sin\theta_1 & \varepsilon_{zz}B_2 & \varepsilon_{zz}B_3 & 0 \\ 0 & \gamma_2 A_2\cos\theta_2 & \gamma_3 A_3\cos\theta_3 & k_0\cos\theta_1 \\ 0 & -A_2 & -A_3 & 1 \end{matrix}\right\|}. \quad (3.27)$$

The quantities A_2, A_3, B_2, and B_3 can be easily manipulated into the forms

$$A_2 = \frac{i}{\cos\theta_2};\; A_3 = -\frac{i}{\cos\theta_3} \quad (3.28)$$

$$B_2 = \frac{(1+Q\cos\theta_2)\sin\theta_2}{Q\sin^2\theta_2 - \cos\theta_2};\; B_3 = -\frac{(1-Q\cos\theta_3)\sin\theta_3}{Q\sin^2\theta_3 - \cos\theta_3}. \quad (3.29)$$

These forms are more suitable for numerical processing, which yields the results displayed in Fig. 3.3. This figure illustrates the behaviour of a single interface and shows very dramatically the expected Brewster effect induced into the p-polarised reflectivity. Reflection intensity drops smoothly to a minimum at what can now be called the pseudo-Brewster angle because this is a metal surface, and $|R_p|$ does not drop to zero. This contrasts nicely with a glass interface for which the p-polarised reflectivity would drop to zero at the Brewster angle. The metallic surface may be absorbing but there is still an identifiable Brewster angle that can be exploited to yield, in combination with the induced s-polarized reflectivity, a measurable Kerr effect. In the next section, this pseudo-Brewster effect will be put into competition with the simultaneous creation of surface plasmon-polaritons that are created with a system that uses attenuated total reflection (ATR).

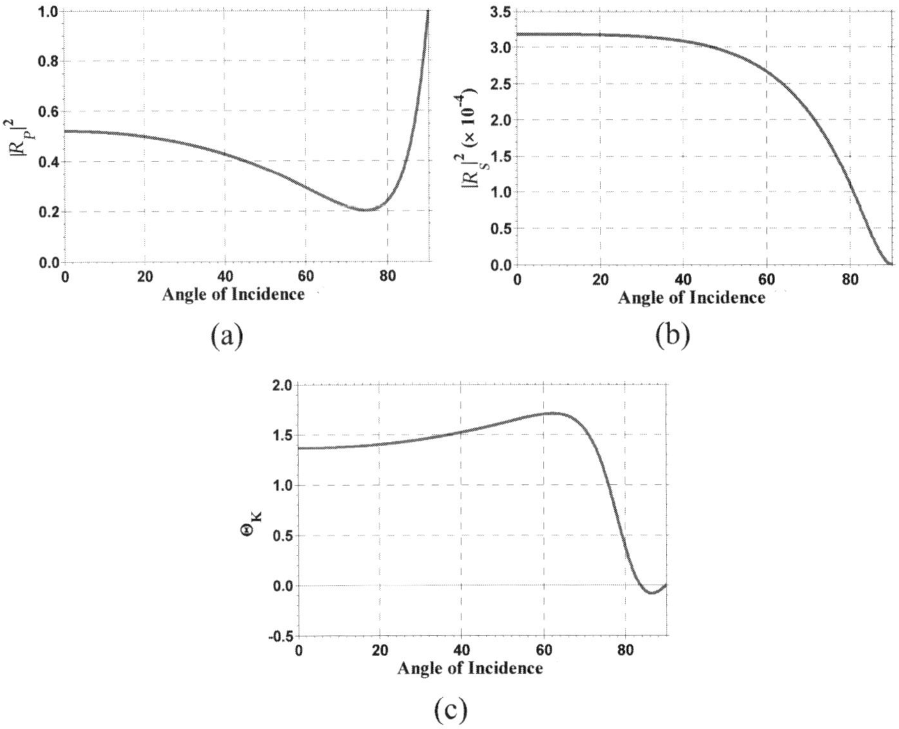

Figure 3.3 (a) *p*-polarised and (b) induced *s*-polarised reflection intensity at a vacuum/MnBi interface. (c) The Kerr rotation: λ = 578 nm, n = 2.44 + 2.92i, Q = –0.089 + 0.034i, where λ is the wavelength of the incident *p*-polarised light beam.

3.3 Enhancing the Kerr Effect with Attenuated Total Reflection

During recent decades, considerable attention has been paid to the launching of surface waves on metallic interfaces. For example, a surface wave can exist at the interface between a metal (such as sodium) and air.[5] Using the high electromagnetic frequencies associated with the optical range, a metal can be viewed as essentially a free-electron gas, electrically balanced by the positively charged lattice background that is too inert to move. Hence, it can be said that a metal is an electron plasma, behaving like a wobbling jelly. A bulk plasma of this kind can sustain jelly-like oscillations that possess an angular frequency ω_p. Since this is the case, $\hbar\omega_p$ is the quantum of energy associated with it, and it is called a plasmon. $\hbar = h/2\pi$, where h is Planck's constant. A metal can also sustain surface plasmons with an angular frequency $\omega_p / \sqrt{2}$. This description is not enough, however, because if a surface wave is created by an incoming electromagnetic wave, the plasma oscillations are stimulated by photons, and associated polarising effects result.[5] Hence, such a surface excitation is actually a combination called a surface plasmon-polariton. When an electromagnetic wave passes through a dielectric, it excites internal polarisation of the dielectric, thus causing the wave to become a hybrid, called a polariton. For an electron plasma, such a mode has both plasmon and photon content, and the total energy is shared over the whole system. This physical feature means that it is possible to examine plasmon-polaritons in their extreme states when they have either a strong photon or strong plasmon content. The question of how to create such polaritons (such as surface modes on an unmagnetised metal surface) has attracted a lot of attention in the literature, and the fundamentals have been addressed in detail.[5] The popular conclusion is that the way to generate polaritons is by deploying a prism hovering above or attached to a metal surface and to use incoming *p*-polarised rays beyond the onset of total internal reflection inside the prism.[5] *s*-polarised light rays are unable to satisfy the boundary conditions required to maintain a surface mode. The well-known Kretschmann-Raether[5] configuration, designed to achieve this outcome, is shown in Fig. 3.4; however, the Otto method,[5] in which there is a gap between the base of the prism and the metal surface, is another acceptable, well-known choice. Both methods rely on the appearance of a minimum in the reflected intensity $|R_p|^2$ of the *p*-polarised light incident through the prism at an angle greater than θ_c. This is, at first sight, surprising since the latter is associated with the total internal reflection. In this total internal reflection regime, the reflectivity would be unity for a prism standing alone. However, if a metal surface is held parallel and near the base of the prism, an angle of incidence is encountered at which the reflectivity drops considerably from unity, sometimes to a very low value. A large amount of energy is then channelled into a surface plasmon-polariton wave. Hence the name *attenuated total reflection* (ATR) is appropriate for this situation.

A configuration involving a thin metal film attached to the base of a prism is sketched in Fig. 3.4. The figure illustrates the appearance of the type of surface mode discussed above; however, it should be re-emphasised that the discussion at this point does not yet involve a magneto-optic material.

In order to understand why an ATR configuration is needed, it is useful to consider the dispersion curve associated with the expected surface modes. First, these surfaces are localised at the metal surface in Fig. 3.4. This means that a surface plasmon-polariton mode carries an electromagnetic field that exponentially decays away from the metal surface in either z direction. Assuming for this part of the discussion that the metal is lossless, its relative dielectric permittivity is

$$\varepsilon(\omega) = 1 - \frac{\omega_p^2}{\omega^2} = 1 - \frac{1}{\Omega^2}, \tag{3.30}$$

where ω is the angular frequency of the surface wave (mode), ω_p is the plasma frequency introduced earlier, and $\Omega = \omega / \omega_p$ is a convenient normalisation to

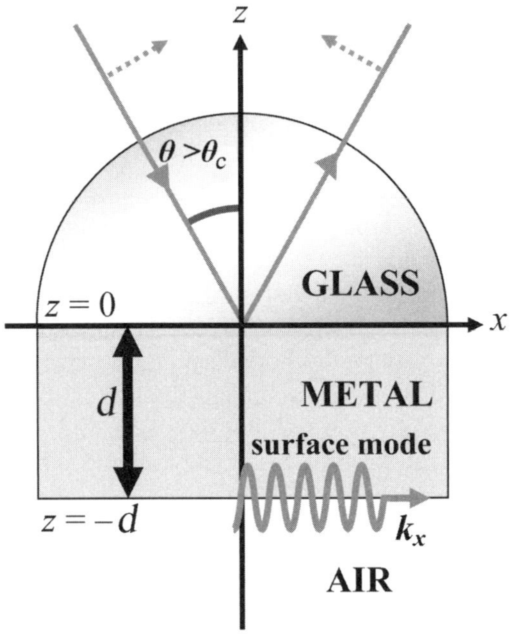

Figure 3.4 Sketch illustrating that a reflected *p*-polarised light beam incident in the (*x*,*z*) plane beyond the critical angle θ_C onto a metal film of width *d* attached to the base of a hemispherical glass prism can generate a surface mode on the lower film surface with wave number k_x. The hemispherical shape of the prism is selected only to avoid the unnecessary complication of dealing with the refraction at the prism surface.

adopt, together with a normalised wave number $K_x = ck_x / \omega_p$. Applying the usual boundary conditions that the tangential electric field and magnetic field components are continuous at the metal/air boundary, and leaving aside the presence of the prism, the dispersion equation of the surface plasmon-polariton for an air/metal boundary is[5]

$$K_x^2 = \Omega^2 \left[\frac{1 - 1/\Omega^2}{2 - 1/\Omega^2} \right]. \tag{3.31}$$

The numerical properties of Eq. (3.31) are displayed in Fig. 3.5 and can be interpreted in the following manner: $\Omega = 1/\sqrt{2}$ is the asymptotic value reached as $K_x \to \infty$; this is the regime of a pure surface plasmon oscillation. For small K_x, the photons become strongly involved, and two branches for the dispersion are revealed, namely, $\Omega_+ = \sqrt{1 + K_x^2}$ and $\Omega_- = K_x$. The two branches are also shown in Fig 3.5 with the lower branch corresponding to a nonradiative surface excitation, known as the Fano mode. The upper radiative branch is called the Brewster mode.[5] The exciting question to ask about these dispersion curves is how to connect them back to the idea that a prism can be used to generate the localised modes at the metal/air interface. First of all, do the dispersion curves indicate whether simply shining a beam of light onto the surface of the metal in the absence of a prism can generate surface modes? To answer this question it should first be noticed that for light in air or vacuum, the dispersion curve is simply $\Omega = K_x$. In other words, there is no dispersion, so this curve is often called the *light line*. However, this line lies above the Fano curve, so the wave numbers for the surface plasmon-polariton are inaccessible by only directing a beam of light onto a free metal surface. This is universally true, no matter what the bounding medium of the metal happens to be. Even in the case of Fig. 3.4, the surface modes are on the boundary of the metal and access the bounding medium, which happens to be air in this case. The conclusion at this stage is that the momenta of the surface waves are unreachable unless a new light line can be introduced. This barrier can be overcome, however, as shown in Figs. 3.4 and 3.5. What is required is a reduction in the slope of the light line. This is achieved by decreasing the slope of the light line to that of the light line in glass, through the placement of a prism, in the manner shown in Fig. 3.4. The operation must be beyond the prism critical angle to take advantage of exponentially decaying fields, hence attenuating, or frustrating, the total reflection of the prism. A portion of the surface plasmon-polariton between the air and glass light lines then becomes accessible to measurement by transferring the horizontal momentum through the prism, acting under total internal reflection, across to the metal surface. Both the Otto and the Kretschmann-Raether systems work in exactly this way.

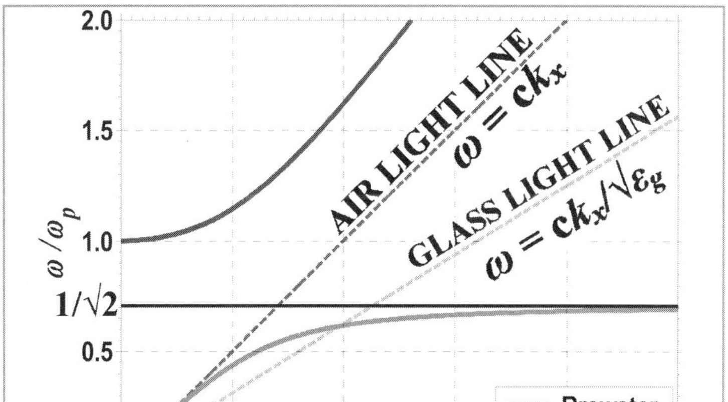

Figure 3.5 Dispersion curves relating the frequency of the incoming light to k_x, the wave number of a surface excitation. The Brewster and Fano modes are clearly displayed. Two light lines are also included to illustrate how a part of the surface plasmon-polariton dispersion is made available to external excitation through a prism.

The reason for providing the set-up shown in Fig. 3.4 and for analysing the dispersion curves in Fig. 3.5 is to anticipate the possibility that this set-up could be used to create an enhancement of the Kerr effect, if the metal film is replaced by a suitably magnetised ferromagnetic material. Given the fact that the ATR system relies on a sudden drop in the p-polarised reflectivity, i.e., on a resonance, the question arises as to what is the nature of the Kerr rotation if a ferromagnetic material, instead of an ordinary metal film, is attached to the base of a prism. The fundamental desire is to enhance the Kerr rotation. However, will this be achieved by arranging for the p-polarised reflectivity to drop suddenly because of a surface-plasmon-generated resonance? Or, will it do so because of a drop in the p-polarised reflectivity due to a kind of pseudo-Brewster effect? The net outcomes will depend on the magnetic material selected, and in all cases it has to be determined whether sufficient energy is transferred into the induced Kerr-driven s-polarised reflection to give a reasonable expectation of observation. In other words, is the figure of merit acceptable? There is a global interest in improving the Kerr effect in this way, so this tutorial will investigate a detailed analysis of the arrangement shown in Fig. 3.6, which is inspired by the early surface plasmon investigations.[5,6]

For the ATR system shown in Fig. 3.6, p-polarised light is incident upon a glass prism that has a magnetic film attached to its base. Given that two types of wave will be initiated within the magnetic film, subscript 1 will apply to fields within the prism and, as before, subscripts 2 or 3 will be used to designate field

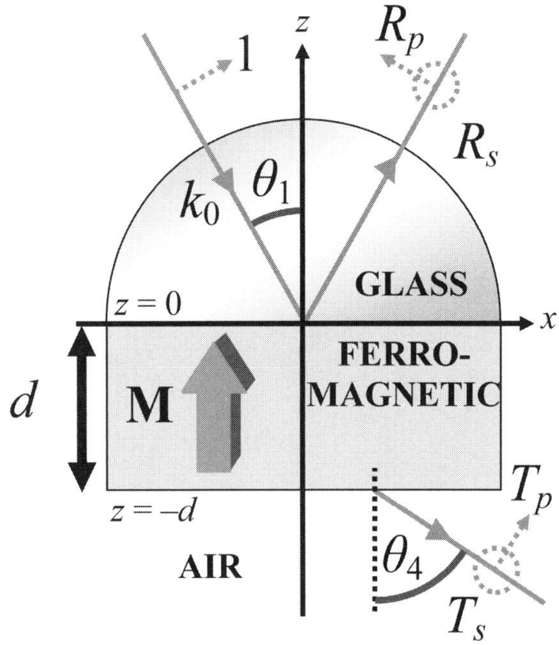

Figure 3.6 ATR arrangement deploying a magnetic film attached to the base of a prism, using a polar configuration. R_p, R_s and T_p, T_s are the complex amplitudes used in the analysis.

components within the film. Subscript 4 designates field components in the emergent medium, which in this case is air. Superscripts i and r are used to distinguish between waves in the metal film that head toward, or away, from the lower boundary. Again, for simplicity, the prism is assumed to be hemispherical so that it will not produce the added complication of refraction at the point of entry of any incident wave. The plane of incidence is the (x,z) plane, and the magnetic material is magnetised in the z direction, thus making this a polar orientation system. If the subscript 1 is used for the prism, (2,3) for the modes in the metal film, and 4 for the air into which the wave is transmitted, then the form of the electromagnetic waves is now as follows:

Glass prism

- Incident wave: $\exp\left(i\left[k_0 \sin\theta_1 x - k_0 \cos\theta_1 z - \omega t\right]\right)$
- Reflected wave: $\exp\left(i\left[k_0 \sin\theta_1 x + k_0 \cos\theta_1 z - \omega t\right]\right)$.

Magnetic film

- Incident wave: $\exp\left(i\left[\gamma_2 \sin\theta_2 x - \gamma_2 \cos\theta_2 z - \omega t\right]\right)$
- Reflected wave: $\exp\left(i\left[\gamma_2 \sin\theta_2 x + \gamma_2 \cos\theta_2 z - \omega t\right]\right)$
- Incident wave: $\exp\left(i\left[\gamma_3 \sin\theta_3 x - \gamma_3 \cos\theta_3 z - \omega t\right]\right)$
- Reflected wave: $\exp\left(i\left[\gamma_3 \sin\theta_3 x + \gamma_3 \cos\theta_3 z - \omega t\right]\right)$.

Air substrate

- Transmitted wave: $\exp\left(i\left[\frac{\omega}{c}\sin\theta_4 x - \frac{\omega}{c}\cos\theta_4 z - \omega t\right]\right)$.

The angles can be easily understood from Fig. 3.2, except that in this case there is an angle θ_4 that accounts for the transmitted ray into the air substrate.

The boundary conditions are the continuity of E_x, D_z, H_x, and E_y at every interface of the ATR system. Upon application they yield the following equations:

Glass/metal boundary

$$[\![E_x]\!]_{z=0} = 0 \Rightarrow \quad \cos\theta_1 - R_p \cos\theta_1 = E_{2x}^i + E_{3x}^i + E_{2x}^r + E_{3x}^r \tag{3.32}$$

$$[\![D_z]\!]_{z=0} = 0 \Rightarrow \quad \begin{aligned} & n_g^2 \sin\theta_1 + n_g^2 R_p \sin\theta_1 \\ & = \varepsilon_{zz}\left(E_{2z}^i + E_{3z}^i + E_{2z}^r + E_{3z}^r\right), \end{aligned} \tag{3.33}$$

where n_g is the refractive index of the glass prism.

$$[\![H_x]\!]_{z=0} = 0 \Rightarrow \begin{aligned} -R_s k_0 \cos\theta_1 &= E_{2y}^i \gamma_2 \cos\theta_2 - E_{2y}^r \gamma_2 \cos\theta_2 \\ &\quad + E_{3y}^i \gamma_3 \cos\theta_3 - E_{3y}^r \gamma_3 \cos\theta_3 \end{aligned} \tag{3.34}$$

$$[\![E_y]\!]_{z=0} = 0 \Rightarrow \qquad R_s = E_{2y}^i + E_{3y}^i + E_{2y}^r + E_{3y}^r . \tag{3.35}$$

Metal/air boundary

$$[\![E_x]\!]_{z=-d} = 0 \Rightarrow \begin{aligned} & E_{2x}^i \exp\left(i\gamma_2 \cos\theta_2 d\right) + E_{3x}^i \exp\left(i\gamma_3 \cos\theta_3 d\right) \\ & \quad + E_{2x}^r \exp\left(-i\gamma_2 \cos\theta_2 d\right) + E_{3x}^r \exp\left(-i\gamma_3 \cos\theta_3 d\right) \\ & \quad = T_p \cos\theta_4 \exp\left(i\frac{\omega}{c}\cos\theta_4 d\right) \end{aligned} \tag{3.36}$$

$$[\![D_z]\!]_{z=-d} = 0 \Rightarrow \begin{aligned} & \varepsilon_{zz} E_{2z}^i \exp\left(i\gamma_2 \cos\theta_2 d\right) + \varepsilon_{zz} E_{3z}^i \exp\left(i\gamma_3 \cos\theta_3 d\right) \\ & \quad + \varepsilon_{zz} E_{2z}^r \exp\left(-i\gamma_2 \cos\theta_2 d\right) \\ & \quad + \varepsilon_{zz} E_{3z}^r \exp\left(-i\gamma_3 \cos\theta_3 d\right) \\ & \quad = T_p \sin\theta_4 \exp\left(i\frac{\omega}{c}\cos\theta_4 d\right) \end{aligned} \tag{3.37}$$

$$[\![H_x]\!]_{z=-d} = 0 \Rightarrow \begin{aligned} & E_{2y}^i \gamma_2 \cos\theta_2 \exp\left(i\gamma_2 \cos\theta_2 d\right) \\ & \quad + E_{3y}^i \gamma_3 \cos\theta_3 \exp\left(i\gamma_3 \cos\theta_3 d\right) \\ & \quad - E_{2y}^r \gamma_2 \cos\theta_2 \exp\left(-i\gamma_2 \cos\theta_2 d\right) \\ & \quad - E_{3y}^r \gamma_3 \cos\theta_3 \exp\left(-i\gamma_3 \cos\theta_3 d\right) \\ & \quad = T_s \frac{\omega}{c} \cos\theta_4 \exp\left(i\frac{\omega}{c}\cos\theta_4 d\right) \end{aligned} \tag{3.38}$$

$$[\![E_y]\!]_{z=-d} = 0 \Rightarrow \begin{aligned} & E_{2y}^i \exp\left(i\gamma_2 \cos\theta_2 d\right) + E_{3y}^i \exp\left(i\gamma_3 \cos\theta_3 d\right) \\ & \quad + E_{2y}^r \exp\left(-i\gamma_2 \cos\theta_2 d\right) + E_{3y}^r \exp\left(-i\gamma_3 \cos\theta_3 d\right). \\ & \quad = T_s \exp\left(i\frac{\omega}{c}\cos\theta_4 d\right). \end{aligned} \tag{3.39}$$

Note that T_p and T_s are the transmission analogues of the R_p and R_s reflectivities, and that Snell's law implies that $n_g \sin\theta_1 = (c/\omega)\gamma_2 \sin\theta_2 = (c/\omega)\gamma_3 \sin\theta_3 = \sin\theta_4$.

In order to progress with the calculation, it is useful to recall Eqs. (3.22)–(3.24) and use the following relationships that the various field components have to one another:

$$E^i_{2y} = A_2 E^i_{2x}\,; \quad E^r_{2y} = A_2 E^r_{2x}\,; \quad E^i_{3y} = A_3 E^i_{3x}\,; \quad E^r_{3y} = A_3 E^r_{3x} \tag{3.40}$$

$$E^i_{2z} = -B_2 E^i_{2x}\,; \quad E^r_{2z} = B_2 E^r_{2x}\,; \quad E^i_{3z} = -B_3 E^i_{3x}\,; \quad E^r_{3z} = B_3 E^r_{3x}\,. \tag{3.41}$$

Note that A_2, A_3, B_2 and B_3 have the same definitions as those given previously and can be substituted directly into the boundary conditions for the E_z and E_y components. After this is done, a more simplified appearance for the equations can be achieved by defining the following quantities before the whole system is expressed in matrix form

$$C_1 = \gamma_2 \cos\theta_2\,; \quad C_2 = \gamma_3 \cos\theta_3 \tag{3.42}$$

$$C_3 = \exp\left(iC_1 d\right); \quad C_4 = \exp\left(iC_2 d\right);\, C_5 = \exp\left(i\frac{\omega}{c}\cos\theta_4 d\right). \tag{3.43}$$

Gathering all of the equations that have emerged from the application of the boundary conditions gives the following matrix equation:

$$\mathbf{D}\cdot\mathbf{V} = \mathbf{G}\,, \tag{3.44}$$

in which the matrix **D** is

$$\begin{bmatrix} \cos\theta_1 & 0 & 1 & 1 & 1 & 1 & 0 & 0 \\ -n_g^2 \sin\theta_1 & 0 & -\varepsilon_{zz}B_2 & \varepsilon_{zz}B_2 & -\varepsilon_{zz}B_3 & \varepsilon_{zz}B_3 & 0 & 0 \\ 0 & k_0\cos\theta_1 & C_1A_2 & -C_1A_2 & C_2A_3 & -C_2A_3 & 0 & 0 \\ 0 & -1 & A_2 & A_2 & A_3 & A_3 & 0 & 0 \\ 0 & 0 & C_3 & \dfrac{1}{C_3} & C_4 & \dfrac{1}{C_4} & -C_5\cos\theta_4 & 0 \\ 0 & 0 & -\varepsilon_{zz}B_2C_3 & \varepsilon_{zz}\dfrac{B_2}{C_3} & -\varepsilon_{zz}B_3C_4 & \varepsilon_{zz}\dfrac{B_3}{C_4} & -C_5\sin\theta_4 & 0 \\ 0 & 0 & C_1C_3A_2 & -\dfrac{C_1A_2}{C_3} & C_2C_4A_3 & -\dfrac{C_2A_3}{C_4} & 0 & -\dfrac{\omega}{c}\cos\theta_4 C_5 \\ 0 & 0 & C_3A_2 & \dfrac{A_2}{C_3} & C_4A_3 & \dfrac{A_3}{C_4} & 0 & C_5 \end{bmatrix},$$

and this operates on the vectors

$$\mathbf{V} = \begin{bmatrix} R_p & R_s & E_{2x}^i & E_{2x}^r & E_{3x}^i & E_{3x}^r & T_p & T_s \end{bmatrix}^T \tag{3.45}$$

$$\mathbf{G} = \begin{bmatrix} \cos\theta_1 & n_g^2 \sin\theta_1 & 0 & 0 & 0 & 0 & 0 & 0 \end{bmatrix}^T . \tag{3.46}$$

If the determinant of the matrix **D** is D, then this set of linear equations can be readily solved using Cramer's rule,[7] which states that the ith element in the vector **V** is the ratio D_i/D, in which D_i is created by replacing the elements of the ith column of D with the elements in the vector **G**. It is interesting that this rule is attributed to Gabriel Cramer, who lived during the period 1704–1752.

3.4 Numerical Investigations of Attenuated Total Reflection

In this section, an example is designed to illustrate what is possible for a particular type of material in terms of what can be visibly expected when the *p*-polarised and *s*-polarised reflectivities are calculated. In order to be very specific and to use commonly discussed ferromagnetic materials, the figures shown are the outcomes of using a glass prism and manganese bismuth or nickel.

Figures 3.7(a) and (b) compare the *p*-polarised reflectivity that arises when using MnBi, or Ni, respectively. It can be seen in Fig. 3.7(a) that the reflectivity declines slightly until the total internal reflection angle of the prism is reached. It then rises sharply only to suffer a further drop to a minimum between 60 and 80 deg. The calculations use data for Ni and MnBi magneto-optical materials, given in terms of refractive indices.[2,3] In fact, what is needed for calculations is the off-diagonal element of the permittivity tensor expressed in the form n^2Q, where n is the refractive index. This step introduces a convenient magneto-optic parameter Q and leads to the diagonal elements being simply expressed as n^2. The data[2,3] for MnBi shows that the real part of the relative permittivity has a modest negative value of –2.5. On the other hand, the real part of the relative permittivity of Ni has the dominant value of –10.7. Hence, in comparing Fig. 3.7(a) to Fig. 3.7(b), it should be expected that the sharp drop in R_p for Ni is clearly associated with surface plasmon-polariton generation. This is confirmed by the fact that the angle of incidence θ_p associated with the minimum in the case of Ni approximately satisfies the formula[5]

$$\sin\theta_p = \frac{1}{n_p} \operatorname{Re}\left[\frac{\varepsilon(\omega)}{\varepsilon(\omega)+1}\right]^{\frac{1}{2}}, \tag{3.47}$$

where n_p is the refractive index of the prism and $\varepsilon(\omega)$ is the relative permittivity defined by the diagonal element of the permittivity tensor of the magnetic film. There is competition between the surface plasmon generation and the pseudo-

Brewster effect but, for Ni, the surface plasmon mode plays a very strong role. This can be seen more easily by considering Fig. 3.8, which shows the variation in incident angle with film thickness. In Fig. 3.8(b) the scale shows that a deep minimum range occurs at quite thin film thicknesses and that its onset lies significantly below the onset of the pseudo-Brewster type of angle. Although in this case it can be asserted that surface plasmon generation is a major effect, the formula given by Eq. (3.47) is not precisely satisfied. By contrast, Fig. 3.8(a) shows that the pseudo-Brewster effect sets in with equal clarity over a wide range of thicknesses. It must be deduced that, even as the film thickness declines, the pseudo-Brewster reflection effect dominates over the surface plasmon-polariton resonance.

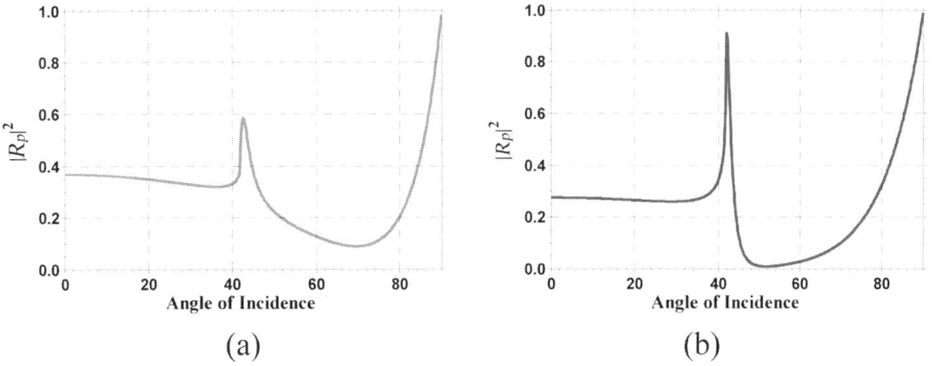

Figure 3.7 Variation of the *p*-polarised reflectivity of the ATR system. (a) MnBi: λ = 578 nm, n = 2.44 + 2.92i, Q = –0.089 + 0.034i. (b) Ni: λ = 600 nm, n = 2.09 + 3.89i, Q = 0.01 – 0.0052i, where λ is the wavelength of an incident light beam. The refractive index of the glass prism is 1.55.

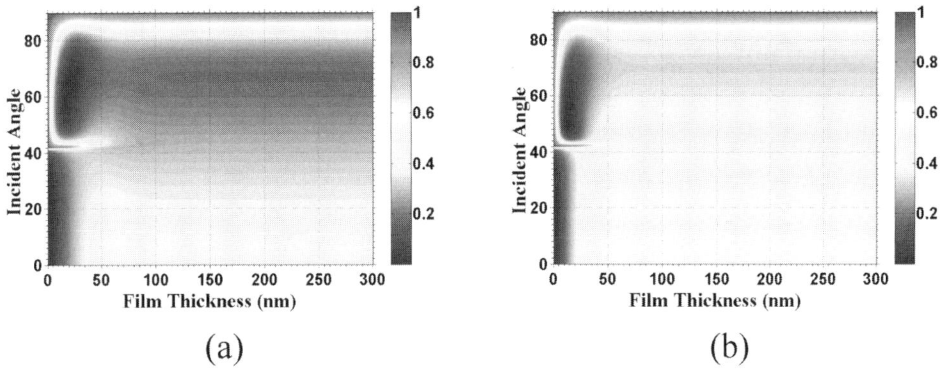

Figure 3.8 Variation of the *p*-polarised reflectivity of the ATR system over the plane defined by the incident angle and the film thickness. (a) MnBi: λ = 578 nm, n = 2.44 + 2.92i, Q = –0.089 + 0.034i. (b) Ni: λ = 600 nm, n = 2.09 + 3.89i, Q = 0.01 – 0.0052i, where λ is the wavelength of an incident light beam. The refractive index of the glass prism is 1.55. (See color plate section.)

The conclusion is that for Ni a surface plasmon effect can be expected, but for MnBi it may not be possible to exploit anything other than the pseudo-Brewster effect. Nevertheless, for both materials, it is still worth pursuing the ATR configuration to see if there is some form of Brewster mode, or surface plasmon resonant, or another enhancement of the Kerr effect over that obtained by creating a simple reflection from an unadorned plane surface. The main point to focus on is that the Kerr rotation is measured as the ratio of the induced *s*-polarised reflectivity to the *p*-polarised reflectivity. Hence, it might be reasonably expected that as the *p*-polarised reflectivity drops at a resonant frequency (for example resulting from the creation of a surface plasmon), the Kerr rotation will increase significantly. This expected outcome has to be balanced, however, with the presence of absorption that will significantly alter the width of the resonance, making it much less useful. In the figures used here, only magnetic metal materials are used for enhancement systems, and for the *s*-polarised reflectivity, the Kerr rotations do show an enhancement over the outcome achievable by reflection of *p*-polarised light off an uncovered plane surface of a magnetic material. Figure 3.9 shows the *s*-polarised reflectivity for MnBi and Ni, respectively, together with the Kerr rotations that can be expected. Further steps toward improving the magnitude of the Kerr rotation could be taken[6,8] by adding layers of a metal such as gold to make specific use of purely metallic-driven plasmonic enhancement. This is straightforward to do, but leads to a more complicated matrix equation, so will not be pursued in this tutorial.

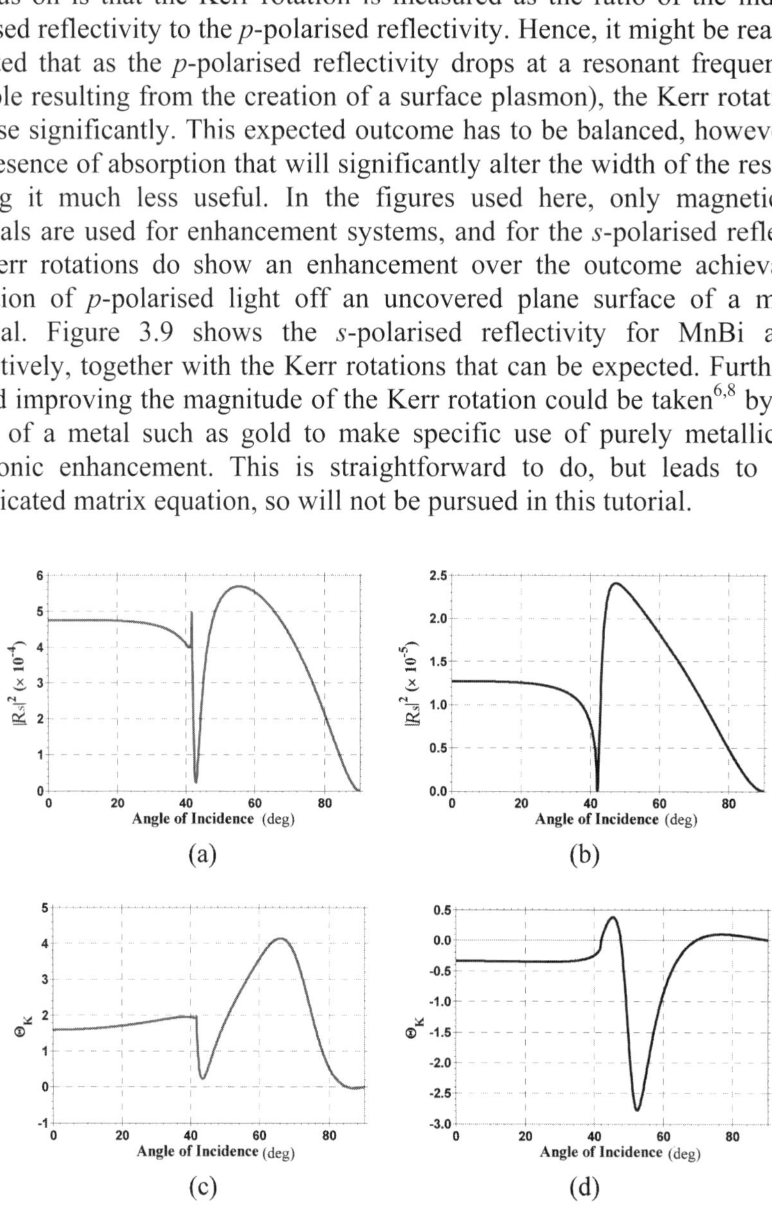

Figure 3.9 *s*-polarised reflectivity of (a) MnBi: film thickness 35 nm and (b) Ni: film thickness 15 nm. (c) and (d) show Kerr rotations of MnBi and Ni, respectively, with the same data as is presented in Fig. 3.8.

Although there can be an enhancement of the Kerr rotation, care needs to be taken if the results are to be translated into downstream, practical device domains. The golden rule is that generating a large Kerr rotation is not a sufficient condition for an application to be undertaken if the magnitude of the *s*-polarised signal is too low, or the absorption is too high. To go into more detail, the light reflected off the magnetic film is actually elliptically polarised. In fact, the *p* and *s* components of the reflected light have a phase difference $\Delta\varphi$. This aspect will not be pursued any further here except to note that striving for an enhanced Kerr rotation based on an ATR system is ultimately quantified by what is known as a figure of merit.[8] Essentially, the figure of merit is proportional to $|R_p|^2 |R_s|^2 \cos(\Delta\varphi)$. This can be broadly understood in the following manner: If $\Delta\varphi$ tends toward π, as opposed to zero, then the figure of merit goes to zero and the effect cannot be exploited. A high value of $|R_p|^2$ means that a large signal is being observed, while a high value of $|R_s|^2$ implies a strong rotation.

Now that the essentials of enhanced *p*-to-*s* conversion have been covered, it is interesting to display the two-dimensional outcomes shown in Fig. 3.10 of the variation of $|R_p|^2$, $|R_s|^2$, and Θ_K as they vary over the incident angle–film thickness plane.

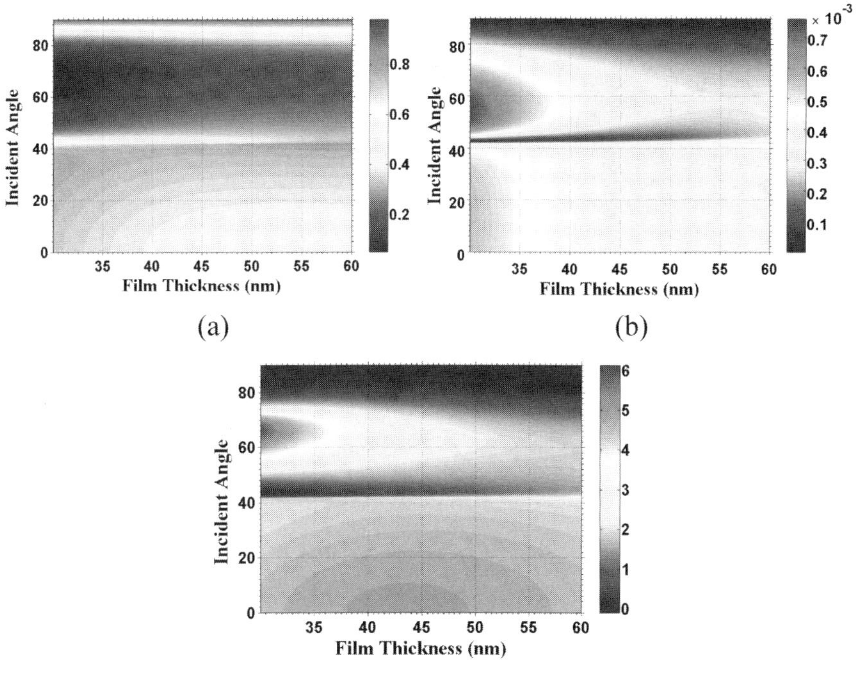

Figure 3.10 Incident angle–film thickness variation of (a) $|R_p|^2$, (b) $|R_s|^2$, and (c) Θ_K, for MnBi. Data is the same as is presented in Figs. 3.8 and 3.9. Note the difference in the color scale for (b). (See color plate section.)

3.5 Conclusions

The purpose of this chapter is to introduce the reader to magneto-optics through the agency of the eponymous Kerr effect introduced by the Rev. John Kerr. It should be appreciated that the topic of magneto-optics is very broad, but the essentials of this kind of electromagnetic constitutive relationship implied by the application of an applied magnetic field to a material are well demonstrated. A further specialisation is to consider only ferromagnetic materials. There is a lot of activity based around insulators, but the concern here is to investigate how a rotation of incoming *p*-polarised light can be readily created. Given this motivation, the chapter takes a step-by-step approach to two distinctive reflection problems. First, a plane surface of a ferromagnetic material is considered in what is known as the polar configuration. The polar configuration involves establishing a net magnetisation that is both perpendicular to the reflection surfaces and lying in the plane of incidence of the exciting electromagnetic wave. After this has been achieved, the so-called attenuated total reflection (ATR) method of exciting a ferromagnetic film is explained. This is followed by a detailed mathematical formulation, and finally, a number of elegant numerical results are given. It is hoped that the work presented here will stimulate interest in the wider field currently embraced by the magneto-optics community.

References

1. S. Sugano and N. Kojima, Eds., *Magneto-optics*, Springer Series in Solid-State Sciences, Berlin (2000).

2. A. K. Zvezdin, and V. A. Kotov, *Modern Magnetooptics and Magnetooptical Materials*, Institute of Physics, Bristol (1997).

   G. Q. Di, and S. Uchiyama, "Optical and magneto-optical properties of MnBi Film," *Phys. Rev.B,* **53**, 3327 (1996).

3. A. D. Boardman, and M. Xie, "Magneto-optics: A Critical Review," in *Introduction to Complex Mediums*, W. S. Weighlhofer and A. Lakhtakia, Eds., SPIE Press, Bellingham, WA (2003).

4. A. D. Boardman, Ed., *Electromagnetic Surface Modes*, John Wiley, Chichester, West Sussex (1982).

5. C. Hermann, V. A. Kosobukin, G. Lampel, J. Peretti, V. I. Safarov and P. Bertrand, "Surface-enhanced magneto-optics in metallic multilayer films," *Phys. Rev. B,* **64**, 235422 (2001).

6. T. S. Blyth, and E. F. Robertson, *Basic Linear Algebra*, Springer, New York (2002).

7. V. I. Safarov, V. A. Kosobukin, C. Hermann, G. Lampel, J. Peretti, "Magneto-optical effects enhanced by surface plasmons in metallic multilayer films," *Phys. Rev. Lett.* **73**, 3584, (1994).

Biographies

Allan D. Boardman is Professor of Applied Physics at the University of Salford, Greater Manchester, UK. This was the home of James Joule, whose house is still on the campus. In addition to possessing a Ph.D., Boardman was awarded the Doctor of Science degree by the University of Durham, UK, his *alma mater*. He has organized conferences and has been a director of a number of NATO Advanced Study Institutes and gives many invited conference presentations. He is Vice-President of the UK Consortium for Optics and Photonics (UKCPO) that services the industrial and business photonics sectors. He is also the founder and Director of the UK North West Photonics Association. He serves as Chair of the Optics and Photonics Division of the UK Institute of Physics and is Secretary of the Quantum Electronics and Optics Board of the European Physical Society that includes representatives from the whole of the EU and Russia. Boardman works on negative index materials from microwaves to the optical regime. He is especially interested in magneto-optical problems. He has served, for several years, as a Topical Editor for *Journal of the Optical Society of America B* on metamaterials, magneto-optics, and nonlinear electromagnetic wave theory and has recently accepted an invitation to serve as the Topical Editor for metamaterials and photonic structures for the UK Institute of Physics' *Journal of Optics A*: *Pure and Applied.*

Neil King completed his Ph.D. in Physics at the University of Salford in 2007 under the direction of Professor A. D. Boardman. He is the co-author of five journal papers and one book chapter. His research interests are in metamaterials as well as the theory and simulation of magneto-optic phenomena and solitons. His current research involves laser physics and nonlinear materials.

Chapter 4
Symmetry Properties of Nonlinear Magneto-optical Effects

Yutaka Kawabe
Chitose Institute of Science and Technology, Chitose, Japan

4.1 Introduction

Nonlinear optical phenomena have been extensively studied theoretically, experimentally, and practically after second-harmonic generation (SHG) was first observed by Franken et al.[1] and the theoretical framework had been established, mainly by Bloembergen and his colleagues.[2] Most nonlinear optical phenomena are successfully described by nonlinear optical tensors with adequate rank, whether the involved processes are resonant or nonresonant, parametric or dissipative, slow or fast, temporally dependent or not. The magnitudes of nonlinearities are determined by complex physical processes, including real or virtual electronic transitions, electron-phonon interactions, and even the thermal properties of materials.[2–5] Therefore, the quantitative studies of optical nonlinearities require precise experimental techniques to extract intrinsic data to be compared by theoretical values obtained from large amounts of numerical computation. However, there are simple symmetry relationships among the tensor components of the nonlinear optical coefficients determined only by the point group to which materials belong. For example, it is well known that the second-order optical nonlinearity vanishes in any material with spatial inversion but does not vanish in noncentrosymmetric systems.[3–5] Understanding the symmetry properties of the relevant tensors is critical to distinguishing and assigning the observed nonlinear effects of various physical origins, especially when small signals from magnetic effects are investigated.

Regarding atomic positions, all crystals can be classified into 32 point groups and 230 space groups. According to von Neumann's principle, the symmetry properties of nonlinear optical phenomena, as well as any other physical property, are determined by the types of symmetries each point group respects.[6,7] These classifications are, however, not complete when magnetism affects the properties of materials. Since angular momentum vectors, such as spin vectors, are defined as the vector product of two conventional vectors, the behavior under a coordinate transformation is different from that of the usual vectors. For example, spatial inversion reverses the direction of conventional vectors, while it does not change the direction of spins. Instead, the time-reversal operation flips the direction of the spin vector. If we consider a material with spins aligned in an antiferromagnetic order on a lattice having centrosymmetry, the conventional spatial inversion cannot be a symmetry element; however, the combination of spatial inversion and temporal reversal reproduces the equivalent alignment. In order to describe the correct symmetry of magnetic materials, we need to employ the time-reversal operation as a transformation element.

When time-reversal symmetries are taken into account, crystals are classified into 122 magnetic point groups, that is, 32 conventional groups, 32 groups composed of the elements of each group and their products with the time-reversal operation, and a new class of 58 groups with a combination of spatial and temporal operations.[6–11]

For the case of second-order optical nonlinear effects, such as the SHG process—the main subject of this chapter—if a crystal belongs to one of the new class of groups with spatial centrosymmetry, it does not show an SHG signal in the conventional manner, but may give a weak nonlinear optical response originating from the magnetic ordering. The details of the symmetry of nonlinear optical tensors for magnetic point groups are given in the next section of this chapter.

In the third section, magnetization-induced SHG (MSHG) will be described; this effect is the most popular topic in the field of nonlinear magneto-optics. When an external magnetic field is applied to materials, it induces magnetization. Alternatively, some materials may have a spontaneous magnetization without external fields (ferromagnetism). For such cases, we can re-assign the material from the original structural point group to one of the 122 magnetic point groups by taking into account the spin ordering. The symmetry properties of SHG from magnetized bulk materials and from their surfaces can be found from the properties of the newly assigned magnetic groups given in Sec. 4.2. In Sec. 4.3, however, we will introduce a more convenient calculation procedure to describe MSHG effects. We express the effects as a modulation of tensors caused by external fields. In the linear case, Faraday and Kerr effects of conventional magneto-optics are usually explained as a change of the dielectric constant tensor due to magnetization and are expressed by additional terms of a power series in the fields. In Sec. 4.3, we will discuss MSHG (which is sometimes called the nonlinear Kerr effect) as a nonlinear version of magneto-optics; that is, SHG tensors are modulated by magnetization. In this formulation, we can describe the MSHG effects in a unified and simplified manner, since we need not consider complex magnetic symmetry, which depends on the applied magnetic field direction.

The next topic, presented in Sec. 4.4, is related to the effects caused by the magnetic dipole transition process. When an optical field is incident upon an atomic or ionic system embedded in a space with a highly symmetric environment, transitions of magnetic dipole moment origin can sometimes be observed when the electric dipole transition is forbidden.[12] When optical absorption or an emission process occurs due to a magnetic dipole moment induced by the incident electric field, these effects can be considered as an electro-magnetic effect in the optical frequency range.[13] This concept can be extended into the nonlinear case. When the magnetic polarization oscillating at the second-harmonic frequency is induced by virtual excitation to a two-photon resonant level, it will be a source of SHG. In such a case, we must consider the axial properties of the magnetic field and the induced magnetization, which result in axial nonlinear optical tensors, as discussed in Sec. 4.4.

Finally, some experimental remarks are presented in Sec. 4.5. The author gratefully acknowledges Profs. E. Hanamura and Y. Tanabe for fruitful discussions on the symmetry of physical tensors.

4.2 Nonlinear Optics in Magnetic Materials

When high-power monochromatic light is incident on materials, polarization with a magnitude proportional to a higher order of the electric field beyond linear is induced. For simplicity, in this tutorial, we concentrate on the case in which the incident wave has a single frequency. But the extension to multiwavelength cases can be made without too much effort by referring to the standard textbooks on nonlinear optics.[2–5] In general, the polarization vector **P** can be expanded as a power series of electric field **E** as follows:

$$\mathbf{P} = \chi^{(1)}\mathbf{E} + \chi^{(2)}\mathbf{EE} + \chi^{(3)}\mathbf{EEE} + \cdots. \tag{4.1}$$

Because the coefficient of the nth term $\chi^{(n)}$, the nth order nonlinear optical coefficient (or susceptibility), is an $(n + 1)$-th rank tensor, the equation can be expressed in another way as

$$P_i = \chi^{(1)}_{ij}E^j + \chi^{(2)}_{ijk}E^jE^k + \chi^{(3)}_{ijkl}E^jE^kE^l + \cdots. \tag{4.2}$$

With sub- and superscripts indicating the Cartesian coordinates, we employ Einstein's rule, so the summation over coordinates (x, y, z) is made for the indices appearing twice in each term.

In principle, the $(n + 1)$-th rank tensor in three-dimensional space has 3^{n+1} components. However, due to the crystalline symmetry of materials, several components of the tensors vanish, and many of the nonzero components are not independent. The relations among them are determined by the point group to which the material belongs.

As long as we consider the symmetry of atomic sites, any crystal belongs to one of the 32 point groups. Due to von Neumann's principle, the symmetry properties of tensors for physical phenomena are determined by the point groups. Each point group is composed of elements corresponding to symmetry operations that move atomic positions into equivalent sites in the crystal. The operations can be formulated as corresponding coordinate transformations. Therefore, the elements of a group can be represented by 3 × 3 orthogonal matrices. All symmetry relations of the tensors can be deduced by applying so-called generator matrices, which generate all members of the group by multiplication among them, to the equations shown below.[6,7]

The coordinate transformation representing an element of the group can be expressed by an orthogonal matrix (a_{ij}). The change of the second-, third- and n^{th} rank tensor components under the coordinate transformation are given as

$$\chi_{ij}^{(1)} = a_{ip} a_{iq} \chi_{pq}^{(1)}, \tag{4.3.a}$$

$$\chi_{ijk}^{(2)} = a_{ip} a_{jq} a_{kr} \chi_{pqr}^{(2)}, \text{ and} \tag{4.3.b}$$

$$\chi_{ijk\cdots m}^{(n-1)} = a_{ip} a_{jq} a_{kr} \cdots a_{mu} \chi_{pqr\cdots u}^{(n-1)}. \tag{4.3.c}$$

The term on the left side, giving a component in the 'new' coordinate system, should be the same as the corresponding component in the old coordinate, due to their equivalence after transformation. Therefore, a set of 3^n equations (n being the rank of the tensor) is derived that will give all of the relationships among components. In order to obtain the results for one of the point groups, one need not make the calculation for all of the transformation matrices of the group; only the calculations for generators give the complete results.[9]

When we consider magnetic symmetries in crystals, the symmetry classification of materials is more complex. In addition to the conventional point groups (so-called colorless or mono-colored groups), there must be two classes, one of which is a gray group that is made of all elements of a colorless group and their products with the time-reversal operation. Therefore, the number of these groups is also 32. The other class of groups includes 58 members; these are composed of the usual spatial operations and the products of some of these operations and temporal inversion. These groups are usually called black and white groups (sometimes referred to as bicolored groups).

In order to explain the details of these groups and their relationship to nonlinear optical tensors, we give simple examples by using the most primitive five groups representing the triclinic system. The groups and their elements are listed in Table 4.1.

Two groups [1] and [$\bar{1}$] are members of the colorless groups, of which structures are the same as the corresponding conventional point groups. Rigorously speaking, however, they do not represent paramagnetic or diamagnetic materials exactly, because they do not include the time-reversal operation as an element. Instead, the materials with permanent magnetization (ferromagnets) can be expressed by these groups. The paramagnetic counterparts correspond to the two gray groups denoted by [$\underline{1}$] and [$\bar{1}$] × [$\underline{1}$], respectively. In conventional nonlinear optics, the time-reversal operations are not taken into account explicitly, although these gray groups correctly describe the symmetry of such nonmagnetic materials. For the last example [$\underline{\bar{1}}$], the spatial inversion operation followed by the temporal inversion is the unique nontrivial element of the group. Such combinations are characteristic of the black and white groups, which describe antiferromagnetic spin ordering in crystals.

Symmetry properties of nonlinear optical tensors for these point groups depend on the types of tensor. We must introduce classifications regarding the nonlinear optical tensors, that is, *i* type and *c* type, and polar and axial. The *i* type and *c* type can be characterized by their behavior under the time-reversal operation; the *i*-type tensor does not change its sign, while the *c*-type tensor changes its sign under temporal inversion. For conventional nonlinear optics, only *i*-type tensors are considered because of their dominance over others and their practical importance. In this section, we discuss both types of tensors in relationship to SHG.

Here, we choose five triclinic systems that have the lowest symmetry among the crystal groups. In these systems, all 27 (actually 18 for the case of SHG due to wavelength degeneracy) components of the tensor would be independent, or all of them would be zero.[4,5] Although for simplicity we sometimes use the symbol $\chi^{(2)}$ instead of $\chi^{(2)}{}_{ijk}$, keep in mind that $\chi^{(2)}$ represents every component.

First, we discuss the *i*-type tensor of these systems. This tensor is transformed according to Eq. (4.3.b) under coordinate transformations. When substituting the identity matrix into Eq. (4.3.b), the resulting equation $\chi^{(2)}{}_{ijk} = \chi^{(2)}{}_{ijk}$ does not give

Table 4.1 Groups, generators, and elements included in five magnetic point groups in the triclinic system and their classification. In this table and text, 1 represents the identity transformation, $\bar{1}$ represents the spatial inversion, and $\underline{1}$ represents the temporal inversion operation. The symbol × indicates a direct product of groups.

Group	Generators	Elements	Classification
[1]	1	1	colorless
[$\bar{1}$]	$\bar{1}$	1, $\bar{1}$	colorless
[$\underline{1}$]	$\underline{1}$	1, $\underline{1}$	gray
[$\bar{1}$] × [$\underline{1}$]	$\bar{1}$, $\underline{1}$	1, $\bar{1}$, $\underline{1}$, $\underline{\bar{1}}$	gray
[$\underline{\bar{1}}$]	$\underline{\bar{1}}$	1, $\underline{\bar{1}}$	black and white

any information on intercomponent relations. Therefore, all of $\chi^{(2)}$'s components of group [1] are found to be independent and nonzero. On the other hand, group $[\bar{1}]$, has spatial inversion as a generator denoted by the matrix

$$\bar{1} = \begin{pmatrix} -1 & 0 & 0 \\ 0 & -1 & 0 \\ 0 & 0 & -1 \end{pmatrix}. \tag{4.4}$$

In this case, we obtain $\chi^{(2)}{}_{ijk} = -\chi^{(2)}{}_{ijk}$, which means that all components vanish.

The symmetry relations of the tensor for $[\underline{1}]$ are the same as those for [1], because the temporal inversion operation does not affect the *i*-type tensor. Likewise, the relations for $[\bar{1}] \times [\underline{1}]$ and $[\underline{\bar{1}}]$ are the same as those for $[\bar{1}]$. For the *c*-type tensor, which changes its sign under the time-reversal operation, we must multiply by −1 on the right side of Eq. (4.3.b), as in Eq. (4.5) below, when the considered generator includes the time-reversal operation[9]

$$\chi^{(2)}_{ijk} = -a_{ip}a_{jq}a_{kr}\chi^{(2)}_{pqr}. \tag{4.5}$$

Therefore, for the groups having the time-reversal operation (gray groups) such as $[\underline{1}]$ and $[\bar{1}] \times [\underline{1}]$, *c*-type $\chi^{(2)}$ values vanish, because the element of the groups gives the relation $\chi^{(2)} = -\chi^{(2)}$. For the two colorless groups, since the transformation of *c*-type $\chi^{(2)}$ obeys the same rule as that for the *i* tensor for all generators, [i.e., Eq. (4.3.b)], the *c*-type $\chi^{(2)}$ for [1] is nonzero and that for $[\bar{1}]$ is zero.[9] Finally, the group $[\underline{\bar{1}}]$ has the nonzero *c*-type $\chi^{(2)}$ because its unique generator $\underline{\bar{1}}$ gives the relation $\chi^{(2)} = \chi^{(2)}$, due to the cancellation of minus signs in matrix Eq. (4.4) and Eq. (4.5). All of these results are summarized in Table 4.2 (See the rows of polar tensors; polar and axial characteristics are discussed in the next section.)

Table 4.2 The symmetry relations of the polar and axial $\chi^{(2)}$ values of *i*-type and *c*-type tensors in the five triclinic magnetic point groups.

		[1]	$[\bar{1}]$	$[\underline{1}]$	$[\bar{1}] \times [\underline{1}]$	$[\underline{\bar{1}}]$
polar	*i* tensor	nonzero	0	nonzero	0	0
	c tensor	nonzero	0	0	0	nonzero
axial	*i* tensor	nonzero	nonzero	nonzero	nonzero	nonzero
	c tensor	nonzero	nonzero	0	0	0

For group [1], there are both *i*-type and *c*-type nonzero components. Because the magnitude of a *c*-type tensor is usually several orders of magnitude smaller than that of the *i*-type tensor, it is difficult to separate them experimentally. But the signature of a *c*-type tensor will change by reversing the magnetization direction (equivalent to time inversion). Therefore, careful experimental study can distinguish the two components by changing the direction of the external magnetic field. This phenomenon can be also described in terms of MSHG, which will be discussed in the next section.

The tensor symmetry for the other 30 point groups (or 117 magnetic point groups) can be obtained by similar but tedious calculations. First, one makes a list of generator matrices of the selected group that are given in several books; one can also find them from the international notation of the group.[9] Then, one makes 27 equations for each matrix by substituting them into Eq. (4.3.b) or Eq. (4.5). For the derivation for *i*-type tensors, Eq. (4.3.b) is always employed; however, for *c*-type tensors, Eq. (4.5) must be used when a generator is accompanied by the time-reversal operation. All of the relationships among tensor components can be obtained by solving the simultaneous linear equations. Generator matrices and the derived tensor symmetries as well as the details of magnetic point groups are provided in the book by Birss.[9]

If materials experience an antiferromagnetic phase transition, the symmetry will change from a gray group to a black and white group. Below the transition temperature, a *c*-type tensor may be observed because additional nonzero tensor components are allowed by the results of lowered symmetry due to spin ordering. Several experimental results are obtained from antiferromagnetic materials. Here, we briefly introduce the results of Fiebig et al.,[14–16] who investigated SHG using single crystalline Cr_2O_3, which has a centrosymmetric corundum structure of the point group $\bar{3}m$. The spin directions of chromium ions align in an antiferromagnetic manner below the Néel temperature T_N. Optical characteristics of the crystal are determined by the chromium ions with three *d* electrons embedded at quasi-octahedral symmetry sites. For example, several absorption peaks due to *d-d* transitions have been observed, and their origin has been explained by the representation of the group. The material has not shown conventional SHG due to a (polar) *i*-type tensor above T_N. However, its magnetic point group changes to $\underline{\bar{3}m}$ below T_N, which gives several nonzero *c*-type $\chi^{(2)}$ components.[15]

Actually, in this material there are also *i*-type axial $\chi^{(2)}$ components (discussion of the axial tensor is given in the following sections) both below and above T_N. Fiebig et al. observed the interference between these two components by employing circularly polarized beams for incident light. By varying the excitation wavelength over a wide range, they observed several SHG peaks resonant to *d-d* transitions in chromium ions. Utilizing the contrast between constructive and destructive interference obtained with the right- and left-circularly polarized incident beams, they succeeded in visualizing antiferromagneic domains of Cr_2O_3 defined by the direction of sublattice magnetization.[16]

In another example, the same team also studied the SHG from the hexagonal crystals of rare-earth manganite $RMnO_3$ (R: Y, Ho, Er, etc.) which have a cystallographic noncentrosymmetry due to ferro- or pyro-electricity as well as an antiferromagnetic spin ordering below T_N. In these cases, there is an *i*-type tensor due to ferroelectricity and a *c*-type tensor due to antiferromagnetism. The team distinguished the two components by selecting an appropriate polarization direction for the exciting light beam. Ordering configurations of the manganese spin (i.e., magnetic point groups of the material) depend on the type of rare-earth ion and temperature. Fiebig et al. determined the magnetic structures for materials that are otherwise difficult to clarify by other methods except neutron diffractometry.[14,17,18] They visualized the contrast caused by the interference between SHG from the samples and references and defined the ferroelectric and antiferromagnetic domains as well as the coupling between these two effects.[19] For microscopic theory regarding the resonant nonlinear optical coefficients of these materials, refer to several papers and the book written by Hanamura et al. and references therein.[20–22]

4.3. Magnetic-Field-Induced Second-Harmonic Generation

Magnetization, either spontaneous or induced by an external field, sometimes causes weak changes to linear and nonlinear polarizabilities. The case of the linear effect is known as the magneto-optical effect and is widely used for magneto-optical recording devices. On the other hand, nonlinear optical susceptibility also experiences changes under the magnetic field or magnetization, which causes new nonlinear optical effects such as MSHG. MSHG, a second harmonic observed only when a macroscopic magnetization exists in the medium, has been the most popular topic in nonlinear magneto-optics. For a discussion of the symmetry properties of this effect, we start by examining the analogy among the electro-optic (EO) effect, the magneto-optic (MO) effect, and the electric-field-induced SHG (EFISH).

The EO effect is sometimes described as the modulation of refractive index (or polarizability) by an external electric field $\mathbf{E}_0$ and is sometimes referred to as the second- or third-order nonlinear optical effect because it can be formulated in terms of nonlinear optics.[23] When a strong static electric field $\mathbf{E}_0$ is applied on a material (typically the order is ~ MV/m), its polarization and polarizability are functions of the static field and can be expanded by a power series of the field as

$$\begin{aligned} \mathbf{P} &= \chi^{(1)}(\mathbf{E}_0)\mathbf{E} = \left(\chi^{(1)} + \chi^{(1)e}\mathbf{E}_0 + \chi^{(1)ee}\mathbf{E}_0\mathbf{E}_0 + \cdots\right)\mathbf{E} \\ &= \chi^{(1)}\mathbf{E} + \chi^{(1)e}\mathbf{E}_0\mathbf{E} + \chi^{(1)ee}\mathbf{E}_0\mathbf{E}_0\mathbf{E} + \cdots. \end{aligned} \tag{4.6}$$

The first term is the optical polarization without the external field, and the second term indicates the Pockels effect (the first-order EO effect). The coefficient $\chi^{(1)e}$ is a third-rank tensor with three subscripts *i*, *j*, and *k*, so its symmetry relation is the same as that of the SHG tensor. However, one may not commute the subscripts i, j, and k, corresponding to the polarization, static field, and alternating optical field, respectively. Since the contraction of the tensor due to Einstein's rule reduces the rank of the products $\chi_{ijk}^{(1)e} E_0^j$ from three to two, this term works as an additional modulated part of the polarizablity $\chi^{(1)}$, which is proportional to the static field. Likewise, the third term that causes the Kerr effect (the second-order EO effect that changes polarization as the square of the static field) consists of the contracted products of the third-order optical nonlinearity (fourth-rank tensor) and three vectors (each a first-rank tensor). Therefore, the part $\chi_{ijkl}^{(2)ee} E_0^j E_0^k$ is also the change of polarization due to the static field.

Because the MO effect can be also described as the change of polarizability due to magnetization, the same treatment as applied to the EO case is possible.[24] As given in Eq. (4.6), the polarization can be expanded to

$$\begin{aligned}\mathbf{P} = \chi^{(1)}(\mathbf{M}_0)\mathbf{E} &= \left(\chi^{(1)} + \chi^{(1)m}\mathbf{M}_0 + \chi^{(1)mm}\mathbf{M}_0\mathbf{M}_0 + \cdots\right)\mathbf{E} \\ &= \chi^{(1)}\mathbf{E} + \chi^{(1)m}\mathbf{M}_0\mathbf{E} + \chi^{(1)mm}\mathbf{M}_0\mathbf{M}_0\mathbf{E} + \cdots. \end{aligned} \tag{4.7}$$

The second term gives the conventional MO effect called the Faraday, or Kerr effect, depending on the alignment of the material and the optical beam. The very weak third term—the Cotton-Mouton effect—is sometimes observed. In these cases, the polarizability is expanded in terms of the magnetization instead of in terms of the electric field, so we must take into account the difference between electric field and magnetization.

Since vectors regarding magnetic effects, such as magnetization **M**, magnetic field **H**, or magnetic flux density **B** are defined as the vector products of conventional vectors or as the rotation of vectors (i.e., **B** = Curl**A**, where **A** is a vector potential), these vectors do not experience a change of direction under spatial inversion. This property can be recognized by applying spatial inversion, i.e., $\mathbf{a} \to -\mathbf{a}$ and $\mathbf{b} \to -\mathbf{b}$, to the definition of the vector product $\mathbf{c} = \mathbf{a} \times \mathbf{b}$. These types of vectors are called axial, while conventional vectors are called polar.

For tensors as well as for vectors, there is also a distinction between axial and polar types of characteristics. The algebraic properties among {polar, axial} is isomorphic to {1, −1}; that is, the product between polar and polar is polar, between axial and polar is axial, and between axial and axial is polar. Therefore, the coefficient $\chi^{(1)m}$ must be axial, and $\chi^{(1)mm}$ must be polar, considering that polarization and electric field are polar vectors.

Changes of axial tensors under the coordinate transformations are given by Eq. (4.3.b) when the determinant of transformation matrices is 1 (proper transformation), while these changes are given by Eq. (4.5) when the determinant is −1 (improper transformation).[9] Applying this rule to the five magnetic point groups for triclinic systems, the (i type) axial third-rank tensor [for example, $\chi^{(1)m}$ in Eq. (4.7)] is known to be nonzero for both [1] and [$\bar{1}$], as well as for the other three magnetic groups.

There is also a classification of i type and c type for axial tensors; the transformation rule for axial c-type tensors is not very complex.[9] When a generator includes both the temporal inversion and a mirror symmetry or point inversion (the determinant is −1), we must multiply by −1 twice. Therefore, the relation is given by Eq. (4.3.b). All transformation relations can be summarized in the following equation:

$$\chi_{ijk}^{(2)} = (-1)^c (-1)^a a_{ip} a_{jq} a_{kr} \chi_{pqr}^{(2)}. \tag{4.8}$$

Here, the symbol c means that only in the calculation for c tensors, the minus sign must be present as a multiplier when a generator includes the temporal inversion. The symbol a means that only in the calculation for axial tensors, the minus sign must be multiplied when the determinant of the generator matrix is −1. For example, considering the axial c tensor for the group [$\underline{\bar{1}}$], the transformation for the element $\underline{\bar{1}}$ gives the relation $\chi^{(2)} = -\chi^{(2)}$, resulting in zero, due to five −1 factors on the right side. All of the results for the five groups were given in Table 4.2 in the previous section.

MSHG can be described by $\chi^{(3)}$, the fourth-rank tensor with axial properties. Before discussing MSHG, let us introduce the electric analogue which is called electric-field-induced SHG (EFISH).[25] Usually, an isotropic medium such as a liquid or an amorphous polymer does not show SHG due to its macroscopic centrosymmetry. However, an external electric field always induces noncentrosymmetry, due to molecular orientation or electron redistribution. We can describe $\chi^{(2)}$ of such a system by selecting the adequate group for a newly emerged material having the symmetry reduced from the original material. In the case of the liquid, its symmetry is reduced from isotropic to ∞mm by the external field. However, it is possible to treat this phenomenon as a modulation of $\chi^{(2)}$ using the higher-rank tensor contracted with the external field. Here, we describe it by the latter method, because we can obtain the equivalent results as long as its nonlinear coefficient can be expanded in a power series of the field. Under the external field, the second-order nonlinear polarization can be expressed as

$$\mathbf{P}^{(2)} = \chi^{(2)}(\mathbf{E}_0)\mathbf{E}\mathbf{E} = \left(\chi^{(2)} + \chi^{(2)e}\mathbf{E}_0 + \chi^{(2)ee}\mathbf{E}_0\mathbf{E}_0 + \cdots\right)\mathbf{E}\mathbf{E}. \tag{4.9}$$

If the medium is isotropic, or SHG inactive ($\chi^{(2)} = 0$), the second term in Eq. (4.9) is dominant for SHG, and the term $\chi^{(2)e}\mathbf{E}_0$ functions as the second-order nonlinear optical tensor. Considering that this term has a contracted third-rank form made by the fourth-rank tensor $\chi^{(2)e}$ and the vector $\mathbf{E}_0$, EFISH can be described as a third-order nonlinear optical effect. Because the third-order nonlinear optical tensor (in this case, polar i type) has nonzero components for materials with any symmetry, we can observe EFISH signals for any materials.

MSHG is the magnetic version of EFISH and has been observed in materials with large magnetization such as ferromagnets, i.e., iron or nickel, and so on. In the literature, discussion of tensor symmetry for MSHG has been based on the polar *c*-type second-order nonlinear optical tensors for the magnetic point groups of the materials, including time-reversal symmetry, as given in the previous section. However, the equivalent results can be obtained by applying the same procedure as was applied with EFISH. For this description, we express the modulation of $\chi^{(2)}$ by a power series in the magnetization of the material as

$$\mathbf{P}^{(2)} = \chi^{(2)}\left(\mathbf{M}_0\right)\mathbf{EE} = \left(\chi^{(2)} + \chi^{(2)m}\mathbf{M}_0 + \chi^{(2)mm}\mathbf{M}_0\mathbf{M}_0 + \cdots\right)\mathbf{EE}\,. \quad (4.10)$$

Here, the term proportional to the magnetization has a contracted form. However, $\chi^{(2)m}$ should be an axial tensor because of the axial property of the magnetization and the polar properties of the polarization. Therefore, the symmetry relationship for MSHG can be described by fourth-rank axial tensors. Generally, invariance of the tensors under the space-time transformation gives the following relationship for the fourth-rank tensor $\chi^{(3)}$

$$\chi^{(3)}_{ijkl} = \left(-1\right)^c\left(-1\right)^a a_{ip}a_{jq}a_{kr}a_{ls}\chi^{(3)}_{pqrs}\,. \quad (4.11)$$

Although Eq. (4.11) is parallel to Eq. (4.8), the relationships among components are different from the odd-rank case. Instead of rehashing the discussion above, we summarize the results for the five triclinic magnetic point groups in Table 4.3. Because the even-rank axial *i* tensor is zero for centrosymmetric materials, as is the odd-rank polar *i* tensor, we must choose noncentrosymmetric samples in order to observe the MSHG effect.

Until now most of the experiments have been conducted on the surface of magnetized metals with crystalline centrosymmetry where the spatial symmetry is locally broken. These surfaces, as well as having the MSHG optical effect, also generate second harmonics with electronic origin given by the conventional $\chi^{(2)}$ tensor. However, it is sometimes possible to separate the tensors by changing the polarization of the excitation light and/or selecting a specific polarization component of the generated light.

Table 4.3 The symmetry relations of the even-rank $\chi^{(3)}$ tensors for the five triclinic magnetic point groups.

		[1]	$[\bar{1}]$	$[\underline{1}]$	$[\bar{1}]\times[\underline{1}]$	$[\underline{\bar{1}}]$
polar	*i* tensor	nonzero	nonzero	nonzero	nonzero	nonzero
	c tensor	nonzero	nonzero	0	0	0
axial	*i* tensor	nonzero	0	nonzero	0	0
	c tensor	nonzero	0	0	0	nonzero

Here, we discuss the symmetry of the MSHG tensor for the system similar to that employed in the first proposal of Shen's group in which they discussed the possibility of MSHG at several surfaces of face-centered cubic crystals.[26] Let us choose the (001) surface of the crystal. Because spatial inversion symmetry is broken at the surface, we must use the 4-*mm* group to describe the surface nonlinear optical effects. First, we consider $\chi^{(2)}$ when there is no magnetization. Generators of this group are the 90-deg rotation around the z axis and the reflection plane perpendicular to the surface. These are expressed by the following two matrices:

$$4_z = \begin{pmatrix} 0 & 1 & 0 \\ -1 & 0 & 0 \\ 0 & 0 & 1 \end{pmatrix} \tag{4.12.a}$$

and

$$\overline{2}_y = \begin{pmatrix} 1 & 0 & 0 \\ 0 & -1 & 0 \\ 0 & 0 & 1 \end{pmatrix}. \tag{4.12.b}$$

The SHG tensor can be expressed by a 3 × 6 matrix form instead of one with 27 components because of the degeneracy of the incident light wavelengths for the **EE** term in Eq. (4.1). In order to obtain relationships among components, we substitute these two matrices into Eq. (4.3.b). If the medium has the lowest symmetry (that is the group [1]), the tensor form is given by

$$\begin{pmatrix} xxx & xyy & xzz & xyz & xzx & xxy \\ yxx & yyy & yzz & yyz & yzx & yxy \\ zxx & zyy & zzz & zyz & zzx & zxy \end{pmatrix}. \tag{4.13}$$

Here, we show only the subscript parts of the $\chi^{(2)}$ tensor, because that is the useful notation adopted by some references.[4,5] For surface SHG (that is, polar i type), the substitution of Eq. (4.12.a) gives the following relationship among the components:

$$\begin{pmatrix} 0 & 0 & 0 & xyz & xzx & 0 \\ 0 & 0 & 0 & xzx & -xyz & 0 \\ zxx & zxx & zzz & 0 & 0 & 0 \end{pmatrix}. \tag{4.14}$$

The results from the application of Eq. (4.12.b) are easily given to the residual components in Eq. (4.14) because the components vanish if the subscripts include an odd number of y terms. Hence, the tensor form is finally reduced to

$$\begin{pmatrix} 0 & 0 & 0 & 0 & xzx & 0 \\ 0 & 0 & 0 & xzx & 0 & 0 \\ zxx & zxx & zzz & 0 & 0 & 0 \end{pmatrix}. \quad (4.15)$$

One method for obtaining the MSHG tensor is to calculate the polar c-type tensor by substituting the generators appropriate for the system with magnetization. We calculate using two different cases: (a) when the magnetization is perpendicular to the surface (**M** // z), the generators are 4_z and $\underline{\bar{2}_y}$; (b) when the magnetization is parallel to one axis on the surface (**M** // x), generators are $\bar{2}_x$ and $\underline{\bar{2}_y}$, as shown in Fig. 4.1. The adequacy of these generators can be understood from the behavior of axial vectors under reflection by mirror planes; that is, the component parallel to the mirror plane changes its direction but the perpendicular part does not. For instance, $\underline{\bar{2}_y}$ in the first case changes spin direction twice, so is known to be a generator of the corresponding group.

Here, instead of applying the calculation rule for c-type tensors, we will show another method based on the fourth-rank axial tensors, which also gives the correct results for MSHG. In this method, we need not take into account the time-reversal symmetry of materials themselves; these effects are introduced as a modulation of nonlinear tensors caused by an external field or spontaneous magnetization. Hence, we need to consider the axial i-type tensors.

The fourth-rank tensor describing MSHG appeared as the second term in Eq. (4.10). Among the four subscripts of $\chi^{(2)m}{}_{ijkl}$, the letter i indicates that the coordinate component of the generated harmonics, j and k (these two are

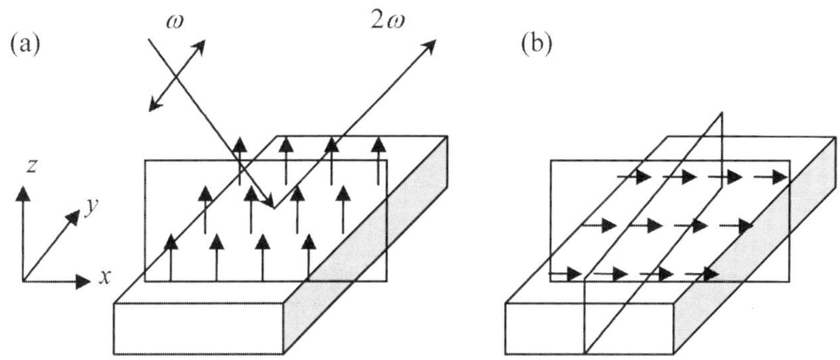

Fig. 4.1 Schematic diagram for a face-centered cubic metal surface with magnetization (a) perpendicular or (b) parallel to the surface. From these figures, we can understand that 4_z and $\underline{\bar{2}_y}$ are the generators for (a) and $\bar{2}_x$ and $\underline{\bar{2}_y}$ are the generators for (b).

commutable) apply to the incident light field, and l to the direction of the magnetization. All components of the tensor can be given by a three-layered stack of 3 × 6 matrices, so the total MSHG tensor can be expressed as

$$\begin{pmatrix} xxxx & xyyx & xzzx & xyzx & xzxx & xxyx \\ yxxx & yyyx & yzzx & yyzx & yzxx & yxyx \\ zxxx & zyyx & zzzx & zyzx & zzxx & zxyx \\ xxxy & xyyy & xzzy & xyzy & xzxy & xxyy \\ yxxy & yyyy & yzzy & yyzy & yzxy & yxyy \\ zxxy & zyyy & zzzy & zyzy & zzxy & zxyy \\ xxxz & xyyz & xzzz & xyzz & xzxz & xxyz \\ yxxz & yyyz & yzzz & yyzz & yzxz & yxyz \\ zxxz & zyyz & zzzz & zyzz & zzxz & zxyz \end{pmatrix}. \qquad (4.16)$$

The three rows at the top, middle, and bottom of the matrix give the MSHG tensor when the magnetization is parallel to the x, y, and z directions, respectively (see the last index of all of the elements). Applying the generators of the 4-*mm* group to Eq. (4.11) for the axial *i*-type case, these components can be reduced to the following form:

$$\begin{pmatrix} 0 & 0 & 0 & 0 & 0 & xxyx \\ yxxx & yyyx & yzzx & 0 & 0 & 0 \\ 0 & 0 & 0 & zyzx & 0 & 0 \\ -yyyx & -yxxx & -yzzx & 0 & 0 & 0 \\ 0 & 0 & 0 & 0 & 0 & -xxyx \\ 0 & 0 & 0 & 0 & -zyzx & 0 \\ 0 & 0 & 0 & xyzz & 0 & 0 \\ 0 & 0 & 0 & 0 & -xyzz & 0 \\ 0 & 0 & 0 & 0 & 0 & 0 \end{pmatrix}. \qquad (4.17)$$

This matrix gives all of the symmetry information for the MSHG tensor, which is equivalent to the results obtained from the third-rank *c*-tensor calculation done for the magnetic point groups. If the magnetization is parallel to the x axis, for example, we can use the 3 × 6 SHG tensor given in the upper three rows of the matrix as

$$\begin{pmatrix} 0 & 0 & 0 & 0 & 0 & xxy \\ yxx & yyy & yzz & 0 & 0 & 0 \\ 0 & 0 & 0 & zyz & 0 & 0 \end{pmatrix}. \tag{4.18}$$

If the magnetization is parallel to the z axis, the MSHG tensor must be

$$\begin{pmatrix} 0 & 0 & 0 & xyz & 0 & 0 \\ 0 & 0 & 0 & 0 & -xyz & 0 \\ 0 & 0 & 0 & 0 & 0 & 0 \end{pmatrix}. \tag{4.19}$$

These results are known to correspond to the components calculated by Pan et al.[26] This method appears to require considerably more calculation than the previous method described, but can be performed by automatically applying the calculation rules. Therefore, this method is more convenient than finding the corresponding generators, especially in complex cases such as those in which the direction of the field deviates from the crystal axes, or both electric and magnetic fields have an influence. It is also advantageous that the calculation can be done by symbolic manipulation software systems such as Maple™ or Mathematica™.

Among the several active components in Eqs. (4.15) and (4.18), let us consider the SHG generated by the incident light with the polarization parallel to the y axis (s polarization) when the surface is magnetized parallel the to x axis, as shown in Fig. 4.1(b). In this case, the $\chi^{(2)}_{zyy}$ term ($= \chi^{(2)}_{zxx}$) of Eq. (4.15) contributes to SHG with the p polarization, while the $\chi^{(2)}_{yyy}$ term of Eq. (4.18) gives SHG with the s polarization. Fortunately, even small magnitudes of the susceptibility components are dominant due to the lack of conventional bulk nonlinear effects. It is possible to observe the superposition of two components as the rotation of polarization of the SHG. The direction of the polarization axis is controllable by changing the direction of magnetization. This phenomenon is called the nonlinear Kerr effect; its application to new magneto-optical memories is being studied because the rotation angle of the nonlinear Kerr effect is much larger than that of the linear Kerr effect.

After the first paper by Pan et al., several pioneering studies on theories and experiments on MSHG were undertaken.[27–30] The main purpose of the early MSHG studies was to probe the spin alignment at the surfaces of magnetic materials. The first MSHG was observed by Reif et al. from the surface of an iron crystal.[28] Later, the same author obtained clearer MSHG signals from a PtMnSb alloy.[29] Stimulated by the growing interest in recording applications, the MSHG from thin layered films was studied, and very large nonlinear Kerr rotation was reported for films composed of Fe and Cr.[30]

Recently, magnetic semiconductors have received attention for use in functional integrated devices for the next generation. Most of the materials in such devices are composed of a semiconductor matrix doped with magnetic ions.

For spectroscopic studies, MSHG has been employed as a method to probe the magnetic structure and the physical origin of magnetism. For example, Sänger et al. made detailed MSHG spectroscopic studies of the system composed of a II–VI semiconductor doped with manganese ions. They measured a broad range of MSHG spectra and observed their dependences on angle, magnetic field, and temperature as well as discussed the physical origins.[31] These researchers not only observed MSHG due to the fourth-rank tensor, but also assigned the contribution from nonlinear magneto-optical spatial dispersion that is expressed by the axial fifth-rank tensor. Magnetization in highly correlated electronic systems, such as manganese perovskite, exhibiting the colossal magnetoresistance effect, was also studied by the MSHG method.[32]

In addition to MSHG, magnetization-induced third-harmonic generation (MTHG) has also been observed.[33,34] Symmetry properties of this phenomenon can be described by fifth-rank axial *i*-type nonlinear optical tensors. The parities of this tensor for the triclinic five groups are the same as those of other odd-rank tensors (see Table 4.2), according to the relationship corresponding to Eqs. (4.8) and (4.11). Therefore, all of the materials show MTHG, so a bulk region of the crystals will contribute to the generation. Discrimination from the conventional third-harmonic generation is also possible by observing the polarization direction and change of ellipticity and their dependence on the magnitude and direction of the external magnetic field, as discussed in Ref. 34. Details of recent studies are available in the articles contained within a special issue of the *Journal of the Optical Society of America*.[35–37]

4.4 Effects Due to an Optical Magnetic Field or Magnetic Dipole Moment Transition

If an electric field is applied to some media, magnetization is sometimes induced, as in a dielectric polarization. Conversely, dielectric polarization can be induced by an external magnetic field. The former effect is electromagnetic (EM) and the latter is magneto-electric (ME). These effects have been studied for a long time, usually with the application of static or low-frequency fields.[38]

When the induced electric or magnetic polarizations are proportional to the external fields of the counterpart, the relationships can be described by the second-rank tensors as

$$M_i = \chi_{ij}^{me} E^j \tag{4.20.a}$$

and

$$P_i = \chi_{ij}^{em} H^j . \tag{4.20.b}$$

The tensors in Eq. (4.20.a) giving the EM effect and those in Eq. (4.20.b) giving the ME effect have axial characteristics due to the axial properties of the **M** and **H** vectors. These tensors are frequency dependent, as is well known for dielectric functions. In an optically resonant frequency region, the electric field in light induces electronic transitions by which an electron situated in the ground state is excited into a higher energy state. Its transition probability can be given by a matrix element of the perturbation Hamiltonian calculated from the wave functions of the initial and final states. If the matrix element for the electric dipole moment is nonzero, then the transition induced by the electric field of light, called an electric-dipole-allowed transition, is dominant and has a relatively large oscillator strength. If atoms or ions are located in highly symmetric sites, the electric dipole transition probability sometimes vanishes. In such a case, transition due to a higher-order perturbation, that is, a magnetic-dipole-allowed transition, or an electric quadrupole transition, can be observed.[12]

In the quantum mechanical description, the SHG process can be given by continual three-step transitions; a typical example is depicted in Fig. 4.2.[3–5] A quantum description of the total nonlinear optical process can be given as a superposition of the relatives of Fig. 4.2 corresponding to each perturbation term (or Feynman diagram). However, this description would be outside the scope of this tutorial. For centrosymmeric crystals, all wave functions can be classified into odd or even functions, and the electric transition dipole moment between the functions with the same parity must be zero. Then, it is impossible to return to the initial state by the three-step allowed transitions; this fact is consistent with the conclusion given by tensor symmetry that the conventional SHG is forbidden in centrosymmetric crystals. However, if we consider the magnetic- or quadrupole-allowed transitions, the three-step process can be observed even in centrosymmetric systems.

In some SHG processes, a virtual transition to a dipole forbidden state (one with the same parity of the ground state) is caused by the two-photon absorption process by strong incident light. The subsequent second-harmonic emission is consequentially generated by a magnetic dipole transition to the initial state. In

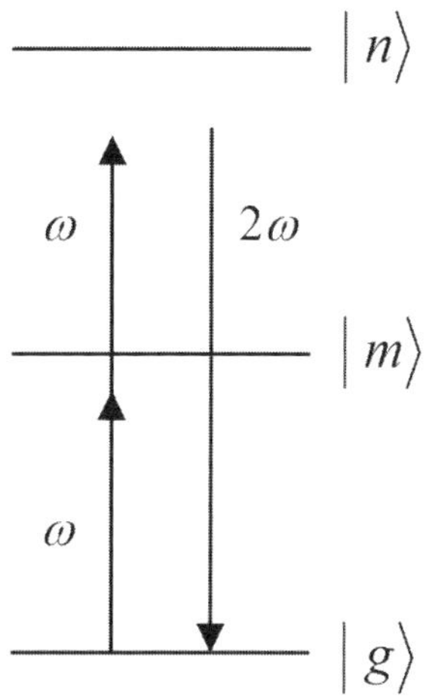

Figure 4.2. Diagram showing the SHG process. The intermediate states denoted by *n* and *m* need not be resonant to the incident light. The transition probability of each step is determined by the matrix element and detuning of levels.

this case, the second-harmonic magnetization can be described by the relationship

$$M_i = \chi_{ijk}^{mee} E^j E^k . \tag{4.21}$$

Comparing this equation with Eq. (4.20), the process can be considered to be the 'nonlinear' and 'optical' version of the EM effect where the coefficient tensor has axial properties. Other than this example, many types of SHG may exist in which the external magnetic field or induced magnetizations play a role in different transition steps. These details are discussed in Ref. 13.

Here, we introduce several examples of SHG caused by such processes. As we explained in Sec. 4.2, Cr_2O_3 shows SHG due to the axial *i*-type tensor both below and above T_N. A theory based on microscopic analysis was provided by Hanamura's group, who effectively reproduced the spectroscopic results obtained by Fiebig et al.[13,14,20,21]

Another example is the orthorhombic $YCrO_3$ crystal, which has a distorted perovskite structure belonging to the *mmm* point group, and shows antiferromagnetic and weak ferromagnetic spin ordering below the Néel temperature $T_N = 141$ K. In the ordered phase, its magnetic point group is $m\underline{mm}$. Because the nonmagnetic *mmm* crystal has a centrosymmetic structure, its polar *i*-type tensor vanishes, while there exist nonzero axial *i*-type tensor components as given by

$$\begin{pmatrix} 0 & 0 & 0 & xyz & 0 & 0 \\ 0 & 0 & 0 & 0 & yzx & 0 \\ 0 & 0 & 0 & 0 & 0 & zxy \end{pmatrix} . \tag{4.22}$$

In its antiferromagnetic $m\underline{mm}$ phase, in addition to Eq. (4.22), the axial *c*-type tensor is also available, for which components are shown in Eq. (4.23), while the polar *c*-type tensor is zero even in this phase:

$$\begin{pmatrix} 0 & 0 & 0 & 0 & xzx & 0 \\ 0 & 0 & 0 & xzx & 0 & 0 \\ zxx & zxx & zzz & 0 & 0 & 0 \end{pmatrix} . \tag{4.23}$$

We studied the SHG spectra and temperature dependence of SHG in this material.[39,40] When we irradiated the sample with an incident beam for which the electric field was directed between two crystalline axes (for example, *x* and *y*), we observed SHG with the polarization parallel to the incident light in the whole temperature range. As known from Eq. (4.22), induced nonlinear magnetic oscillation should be perpendicular to the electric field of the incident beam, so the polarization (electric field) of the harmonic light must be parallel to that of the incident beam. The SHG spectrum obtained showed a resonant peak located at the absorption peak corresponding to the transition from the ground $^4A_{2g}$ to magnetic-dipole-allowed $^4T_{2g}$ levels. Therefore, it has been concluded that the observed SHG was emitted by a downward magnetic dipole transition.

Below the Néel temperature, an axial *c*-tensor is supposed to give a contribution via the nonzero $\chi^{(2)}_{zxx} = \chi^{(2)}_{zyy}$ components in our experimental setup. This contribution has the same polarization as that of the axial *i* tensor and is, hence, indistinguishable from this source of SHG. In our experimental results, we found that the SHG intensity changed around the Néel temperature. The intensity change of the SHG at the Néel temperature indicates the observation of the SHG signal due to the axial *c*-tensor component, although more precise measurements are required to confirm it.

In our other example, we studied the SHG from the pyroelectric and weakly ferromagnetic $GaFeO_3$ crystal. This material shows relatively strong SHG due to the polar *i*-type tensor (with the conventional electronic origin) and also has a nonzero axial *i* tensor. Below the Curie temperature, around 300 K, it has also polar and axial *c*-type tensors. Basically it is possible to partially separate these components experimentally. The details of tensor symmetry, experimental results, and a theoretical microscopic model are given in Ref. 41.

4.5 Experiments

In order to observe nonlinear optical effects, usually a strong optical field is required because the generated signal is proportional to the square or higher order of the incident light intensity. This is especially significant because the magnitude of nonlinear optical effects due to a magnetic origin is much smaller than in the case where the same effects are due to conventional nonlinearities; for this reason, an ingenious experimental set-up is required. For example, detectors with high sensitivity, such as photomultiplier tubes (PMTs) or cooled CCD systems, and monochromators for signal separation are recommended for experimental use. Usually magnetic materials have absorption due to free electrons for metals, or due to *d-d* transitions for oxides, so the influence from the absorption can be crucial, and signal enhancement by phase matching is difficult to achieve. The effects of absorption and coherence length must be taken into account for data analysis.

The nonlinear magneto-optical effect is now used as a method of spectroscopy that is complementary to the conventional linear, or nonlinear, spectroscopies. For such applications widely tunable laser sources are preferred, and their line width is required to be narrow. Several types of laser systems are available for this purpose.

In order to obtain SHG signals, a mode-locked Ti:sapphire femtosecond laser system with high peak power and high repetition frequency is the most useful because the optical nonlinearity of magnetic origins is usually small. Lasers with high peak power and low total energy limit thermal damage caused by optical absorption at the fundamental, or SHG, wavelength. If wavelength dependence is not significant, or not of interest, these sources are very appropriate. Indeed, the studies of metallic materials have been conducted mainly with such lasers.[33,34]

For spectroscopic studies, the wavelength regions must be expanded by adding regenerative amplifier and optical parametric amplifier systems. Such systems are appropriate for nonlinear optical studies because they make it possible to do experiments in any wavelength region from the near UV to mid-IR, although they are sometimes expensive and unstable. However, it is often preferable to employ nanosecond laser systems for this purpose, since the spectral resolution is limited in femtosecond lasers. The wavelength range of Nd^{3+}:YAG, the most popular nanosecond laser source, can be expanded by using an accompanying dye laser or an optical parametric oscillator (OPO). The OPO pumped by the second or third harmonic of the YAG laser yields nanosecond pulses in the region 400 nm to 2 μm, typically with mJ energy.[14,37,40,41]

The peak power of nanosecond systems is weaker than that of femtosecond systems, and the total energy per pulse is relatively high. Therefore, caution is needed to eliminate damage to samples due to residual absorption. It is also necessary to use high-sensitivity detectors such as PMTs. Compactness and ease of use of systems composed of a Nd^{3+}:YAG laser, harmonic generators, and an OPO make it so convenient that we have used such a laser system for the studies of $YCrO_3$ and other materials. A block diagram of the experimental set-up is sketched in Fig. 4.3.[40]

An OPO system pumped by the third harmonic of a Q-switched YAG laser was used at a repetition rate of 10 Hz. High-energy (~ mJ) light pulses from 400 to 700 nm were obtained as a signal beam, and from 800 to 1700 nm as an idler. The idler beam was deployed by blocking visible light with color filters. In order to keep the incident energy constant, the pulse energy was adjusted to 0.2 mJ by a variable neutral-density filter positioned in front of the samples. Emitted SHG was separated by heat absorption filters or interference filters in front of the monochromator. Harmonic signals were detected by a PMT, accumulated in a boxcar integrator, and recorded by a digital multimeter and PC. The adequacy of the set-up has been confirmed over the whole wavelength region by measuring the Maker fringes obtained from an α-quartz crystal plate. We eliminated artifacts caused by changing the wavelength. Because temperature dependence was needed for the study of the effects due to the phase transition, a closed cycle

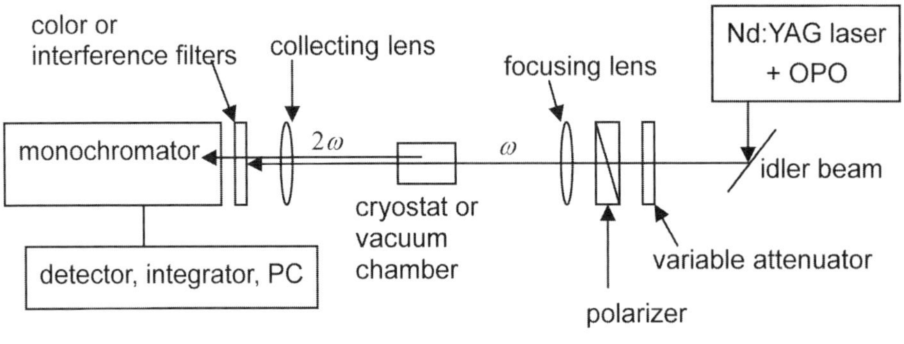

Figure 4.3 Block diagram of an experimental set-up for SHG spectroscopy.

cryostat was used to control temperature from room temperature to 10 K, as shown in Fig. 4.3.

In contrast to conventional nonlinear optics, which is generally studied over a wavelength range in which the samples are transparent, nonlinear magneto-optical studies are often affected by absorption because most magnetic materials have absorption in the visible wavelength range. In the resonant wavelength range, not only does the imaginary part of linear susceptibility appear as absorption, but the nonlinear optical coefficients have complex values. The imaginary part of nonlinear optical coefficients manifests itself in different ways, depending on the types of phenomena. For harmonic generation, the imaginary part also gives harmonic oscillation with the quadrature phase, so it is impossible to separate these two components with a measurement of the harmonic intensity. However, it is possible to separate them by utilizing interference with another harmonic signal having a known phase. In our study, the SHG spectrum of $YCrO_3$ had a broad peak at 600 nm. Therefore, there is strong dispersion of $\chi^{(2)}$ for both the real and the imaginary parts in the region. We have superposed the SHG from the α-quartz and the SHG from the sample in a vacuum and have adjusted both intensities to nearly equal the other but with their polarization perpendicular to each other, by changing the position and direction of the quartz plate. The phase of $\chi^{(2)}$ of the sample can be calculated from the degree of circular polarization of the resulting SHG.[40]

Another area of interest is the magnitude of $\chi^{(2)}$ values that have a magnetic origin. Because this value is very small compared to that of the conventional $\chi^{(2)}$, which is about 10^{-8} to 10^{-9} (esu), no precise investigations have been performed. Some discussion of the $YCrO_3$ case is given below. Generally, the SHG intensity can be given by the absolute square of $\chi^{(2)}$ and some characteristic length l as

$$I_{2\omega} \propto l^2 \left|\chi^{(2)}\right|^2 . \tag{4.24}$$

The characteristic length is the shortest length among the coherence length, the absorption length, and the crystal dimension. For simplicity, we neglect effects such as walk-off and others. For a transparent medium, the intensity is determined by the coherence length if the phases of the fundamental and harmonic do not match and if the highest intensity appears as peaks of Maker fringes. If the refractive indices for both wavelengths are tuned to be the same (phase matching) by choosing the appropriate incident angle on the crystals, the harmonic intensity is proportional to the square of the crystal length.

When the absorption at the SHG wavelength is non-negligible, we must take absorption into account. If the absorption is very strong, as in the cases of some organic materials or metals (typically absorption length is less than 1 μm), the characteristic length l in Eq. (4.24) might be the absorption length. However, if the sample is very thin (less than 100 nm), and thinner than the absorption length, the SHG intensity is proportional to the square of the sample thickness. Because the absorption length is not so short in the case of a typical magnetic oxide, the intensity is determined by the coherence length, which can be measured by the Maker fringe pattern. When direct measurement of the coherence length is difficult due to absorption, we must refer to dispersion of the dielectric constant or refractive indices. Unfortunately, the availability of experimental data on these quantities is limited. Sometimes we must make an estimate for the refractive indices of the magnetic oxides or related materials being studied. We estimated the dispersion of the refractive index of $YCrO_3$ from that of different crystals with similar composition. The $\chi^{(2)}$ value is estimated at 10^{-10} (esu) at the peak wavelength,[39] which is one order of magnitude smaller than the proper $\chi^{(2)}$, even though it is a resonantly enhanced quantity. This is very reasonable, considering that the magneto-optical effects are given by a higher-order perturbation.

Although there are few quantitative estimates of the magneto-optical nonlinear susceptibilities, most of these effects are very weak. It is probably difficult to apply these phenomena to a practical use other than magnetic structural analysis. However, new concepts such as photonic crystals, metamaterials, slow light, and radially polarized light are now being widely investigated. Some of these topics are included in this book. We hope that these emerging subfields of optical physics may incorporate nonlinear magneto-optical effects.

References

1. P. A. Franken, A. E. Hill, C. W. Peters, and G. Weinreich, "Generation of optical harmonics," *Phys. Rev. Lett.* **7**(4), 118–119 (1961).
2. N. Bloembergen, *Nonlinear Optics*, W. A. Benjamin, Inc. Reading, MA (1965).
3. Y. R. Shen, *Principles of Nonlinear Optics*, John Wiley and Sons, Inc., New York (1984).
4. P. N. Butcher and D. Cotter, *The Elements of Nonlinear Optics*, Cambridge University Press, Cambridge (1990).
5. R. W. Boyd, *Nonlinear Optics*, Academic Press, San Diego (1992).
6. M. Tinkham, *Group Theory and Quantum Mechanics*, McGraw-Hill, New York (1964).
7. G. Burns, *Introduction to Group Theory with Applications*, Academic Press, New York (1977).
8. A. V. Shubnikov and N. V. Belov, *Colored Symmetry*, Pergamon, Oxford (1964).
9. R. R. Birss, *Symmetry and Magnetism*, North-Holland, Amsterdam (1966).
10. D. B. Litvin, "Point group symmetries" in *Introduction to Complex Mediums for Optics and Electromagnetics*, W. S. Weiglhofer and A. Lakhtakia, Eds., SPIE Press, Bellingham, WA, 79–98 (2003).

11. H. Schmid, "Magnetoelectric effects in insulating magnetic materials" in *Introduction to Complex Mediums for Optics and Electromagnetics*, W. S. Weiglhofer and A. Lakhtakia, Eds., SPIE Press, Bellingham, WA, 167–188 (2003).

12. A. Messiah, *Quantum Mechanics*, North-Holland, Amsterdam (1964).

13. R. Zawodny, "Nonlinear magneto-optics of magnetically ordered crystals" in *Modern Nonlinear Optics* Part 1, M. Evans and S. Kielich, Eds., Wiley, New York (1993).

14. M. Fiebig, V. V. Pavlov, and R. V. Pisarev, "Second-harmonic generation as a tool for studying electronic and magnetic structures of crystals: review," *J. Opt. Soc. Am. B* **22**(1), 96–118 (2005).

15. M. Fiebig, D. Frohlich, B. B. Krichevtsov, and R. V. Pisarev, "Second harmonic generation and magnetic-dipole-electric-dipole interference in antiferromagnetic Cr_2O_3," *Phys. Rev. Lett.* **73**, 2127–2130 (1994).

16. M. Fiebig, D. Frohlich, L. G. v. Sluyterman, and R. V. Pisarev, "Domain topography of antiferromagnetic Cr_2O_3 by second-harmonic generation," *Appl. Phys. Lett.* **66**, 2906–2908 (1995).

17. M. Fiebig, D. Frohlich, K. Kohn, St. Leute, Th. Lottermoser, V. V. Pavlov, and R. V. Pisarev, "Determination of the magnetic symmetry of hexagonal manganites by second harmonic generation," *Phys. Rev. Lett.* **84**, 5620–5623 (2000).

18. D. Frohlich, St. Leute, V. V. Pavlov, R. V. Pisarev, and K. Kohn, "Determination of the magnetic structure of hexagonal manganites $RMnO_3$ (R = Sc, Y, Ho, Er, Tm, Yb) by second-harmonic spectroscopy," *J. Appl. Phys.* **85**, 4762–4764 (1999).

19. T. Lottermoser, M. Fiebig, D. Frohlich, S. Kallembach, and M. Maat, "Coupling of ferroelectric and antiferromagnetic order parameters in hexagonal $RMnO_3$," *Appl. Phys. B* **74**, 759–764 (2002).

20. M. Muto, Y. Tanabe, T. Iizuka-Sakano, and E. Hanamura, "Magnetoelectric and second-harmonic spectra in antiferromagnetic Cr_2O_3," *Phys. Rev. B* **57**, 9586–9607 (1998).

21. Y. Tanabe, M. Muto, M. Fiebig, and E. Hanamura, "Interference of second harmonics due to electric and magnetic dipoles in antiferromagnetic Cr_2O_3," *Phys. Rev. B* **58**, 8654–8666 (1998).

22. E. Hanamura, Y. Kawabe, and A. Yamanaka, *Quantum Nonlinear Optics*, Springer, Berlin (2007).

23. A. Yariv, *Optical Electronics*, Holt, Rinehart and Winston, New York (1985).

24. S. Sugano and N. Kojima, Eds., *Magneto-optics,* Springer-Verlag, Berlin (2000).

25. D. S. Chemla and J. Zyss, Eds., *Nonlinear Optical Properties of Organic Molecules and Crystals*, Academic Press, Orlando (1987).

26. R.-P. Pan, H. D. Wei, and Y. R. Shen, "Optical second-harmonic generation from magnetized surfaces," *Phys. Rev. B* **39**, 1229–1234 (1989). (Actually, there are mistakes in the tensor components, due to an incorrect choice of symmetry element; this is another motivation to provide a detailed description in Sec. 4.3 of this chapter.)

27. W. Hübner and K.-H. Benneman, "Nonlinear magneto-optical Kerr effects on a nickel surface," *Phys. Rev. B* **40**, 5973–5979 (1989).

28. J. Reif, J. C. Zink, C.-M. Schneider, and J. Kirschner, "Effects of surface magnetism on optical second harmonic generation," *Phys. Rev. Lett.* **67**, 2878–2881 (1991).

29. J. Reif, C. Rau, and E. Matthias, "Influence of magnetism on second harmonic generation," *Phys. Rev. Lett.* **71**, 1931–1934 (1993).

30 B. Koopmans, M. C. Koerkamp, T. Rasing, and H. van den Berg, "Observation of large Kerr angles in the nonlinear optical response from magnetic multilayers," *Phys. Rev. Lett.* **74**, 3692–3695 (1995).

31. I. Sänger, D. R. Yakovlev, B. Kaminski, R. V. Pisarev, V. V. Pavlov, and M. Bayer, "Orbital quantization of electronic states in a magnetic field as the origin of second-harmonic generation in diamagnetic semiconductor," *Phys. Rev. B* **74**, 165208 (2006).

32. Y. Ogawa, H. Yamada, T. Ogasawara, T. Arima, H. Okamoto, M. Kawasaki, and Y. Tokura, "Nonlinear magneto-optical Kerr rotation of an oxide superlattice with artificially broken symmetry," *Phys. Rev. Lett.* **90**, 217403 (2003).

33. O. A. Akitsipetrov, T. V. Murzina, E. M. Kim, R. V. Kapra, A. A. Fedyanin, M. Inoue, A. F. Kravets, S. V. Kuznetsova, M. V. Ivanchenko, and V. G. Lifshits, "Magnetization-induced second- and third-harmonic generation in magnetic thin films and nanoparticles," *J. Opt. Soc. Am. B* **22**, 138–147 (2005).

34. S. Ohkoshi, J. Shimura, K. Ikeda, and K. Hashimoto, "Magnetization-induced second- and third-harmonic generation in transparent magnetic films," *J. Opt. Soc. Am. B* **22**, 196–203 (2005).

35. L. C. Sampaio, J. Hamrle, V. V. Pavlov, J. Ferre, P. Georges, A. Brun, H. Le Gall, and J. Ben Youssef, "Magnetization-induced second-harmonic generation of light by exchange-coupled magnetic layers," *J. Opt. Soc. Am. B* **22**, 119–127 (2005).

36. A. Kirilyuk and T. Rasing, "Magnetization-induced-second-harmonic generation from surfaces and interfaces," *J. Opt. Soc. Am. B* **22**, 148–167 (2005).

37. V. V. Pavlov, A. M. Kalashinikova, R. V. Pisarev, I. Sanger, D. R. Yakovlev, and M. Bayer, "Second-harmonic generation in the magnetic semiconductor (Cd, Mn)Te," *J. Opt. Soc. Am. B* **22**, 168–175 (2005).

38. T. H. O'Dell, *The Electrodynamics of Magneto-electric Media*, North-Holland, Amsterdam (1970).

39. K. Eguchi, Y. Kawabe, and E. Hanamura "Second harmonic generation of yttrium orthochromite with magnetic origin," *J. Phys. Soc. Jpn.* **74**, 1075–1076 (2005).

40. K. Eguchi, Y. Katayama, Y. Kawabe, and E. Hanamura "Nonlinear magneto-optical spectroscopy of $YCrO_3$ and $GaFeO_3$," *Proc. SPIE* **5924**, 59240X (2005).

41. K. Eguchi, Y. Tanabe, T. Ogawa, M. Tanaka, Y. Kawabe and E. Hanamura "Second-haromonic generation from pyroelectric and ferrimagnetic $GaFeO_3$," *J. Opt. Soc. Am. B* **22**, 128–137 (2005).

Biography

Yutaka Kawabe is a professor at Chitose Institute of Science and Technology (CIST) in Japan. He received his B.S. and M.S. degrees in physics from Kyoto University in 1982 and 1984, respectively, and his Ph.D. degree in electrical engineering from Osaka University in 1993. From 1984 to 1995, he was a researcher for Central Research Laboratories of Idemitsu Kosan Co. Ltd. After working at the Optical Sciences Center, University of Arizona for two years as an assistant research scientist, he moved to CIST as an associate professor in 1998. Prof. Kawabe's current research interests include nonlinear magneto-optics and bio- or organic optical materials for laser applications. Dr. Kawabe is a member of SPIE.

Chapter 5
Optical Magnetism in Plasmonic Metamaterials

Gennady Shvets
Department of Physics, University of Texas at Austin, TX, USA

Yaroslav A. Urzhumov
COMSOL, Inc., Los Angeles, CA, USA

5.1 Introduction

In this review we describe the challenges and opportunities for creating magnetically active metamaterials in the optical part of the spectrum. Several techniques for extracting the effective parameters of plasmonic metamaterials are introduced and compared. The emphasis is on the periodic metamaterials whose unit cell is much smaller than the optical wavelength. The conceptual differences between microwave and optical metamaterials are demonstrated. We show that concepts for microwave metamaterials can be extended to the optical domain in a rather limited way. Whenever a metamaterial that exhibits magnetic response in the microwave part of the spectrum is scaled down in size to exhibit similar electromagnetic behavior in the optical part of the spectrum, plasmonic effects play a major role.

Specifically, we demonstrate that for every unit cell's geometry of a magnetically active metamaterial, there exists the shortest wavelength λ_{res} beyond which the magnetic response disappears. In general, the more elaborate is the unit cell (e.g., split rings with narrow inter-ring gaps, etc.), the longer is the λ_{res}.

A new area of electromagnetics has recently emerged: electromagnetic metamaterials. The emergence of this new field occurred in response to the demand for materials with the electromagnetic properties that are not available in naturally occurring media. One of the best known properties unattainable without significant metamaterials engineering is a negative refractive index. The main challenge in making a negative index metamaterial (NIM) is that both the effective dielectric permittivity ϵ_{eff} and magnetic permeability μ_{eff} must be negative.[1] Numerous applications exist for NIMs in every spectral range, from microwave to optical. These include "perfect" lenses, transmission lines, antennas, electromagnetic cloaking, and many others.[2–5] Recent theoretical[6,7] and experimental[8] work demonstrates that for some applications such as electromagnetic cloaking it may not even be necessary to have a negative index; just controlling the effective magnetic permeability suffices.

The first realizations of NIMs were made in the microwave part of the spectrum.[9] The unit cell consisted of a metallic split-ring resonator (SRR)[10] (responsible for the negative permeability $\mu_{\text{eff}} < 0$) and a continuous thin metal wire[11] (responsible for the negative permittivity $\epsilon_{\text{eff}} < 0$). Remarkably, even in the first microwave realizations of the NIM its unit cell was strongly subwavelength: $a/\lambda \approx 1/7$, where a is the lattice constant and λ is the vacuum wavelength. In fact, the condition that $a \ll \lambda$ must be satisfied in order for the effective description using ϵ_{eff} and μ_{eff} to be sensible. If the electromagnetic structure consists of larger unit cells with $a \geq \lambda/2n_d$, where n_d is the refractive index of a substrate onto which metallic elements are deposited (e.g., Duroid in some of the recent microwave experiments[8]), they cannot be described by the averaged quantities such as permittivity and permeability. It is the high λ/a ratio that distinguishes a true metamaterial from its more common cousin, the photonic crystal.[12,13]

Developing NIMs for optical frequencies, however, has proven to be much more challenging than for microwaves. Microwave structures can be made extremely subwavelength using several standard microwave approaches to making a subwavelength resonator: enhancement of the resonator's capacitance by making its aspect ratio (e.g., ratio of the SRR's radius to gap size) high, inserting high-permittivity materials into the SRR's gap, etc. These microwave techniques are briefly illustrated in Sec. 5.2 using a simple SRR shown in Fig. 5.1. Simply scaling down existing microwave NIM designs from microwave to optical wavelengths does not work for two reasons. First, to develop a $\lambda/10$ unit cell requires much smaller (typically, another factor 10) subcellular features such as metallic linewidths and gaps. For $\lambda = 1$ μm, that corresponds to 10-nm features, which are presently too difficult to reliably fabricate. For example, the classic SRR has been scaled down to $\lambda = 3$ μm, but further wavelength reduction using the same design

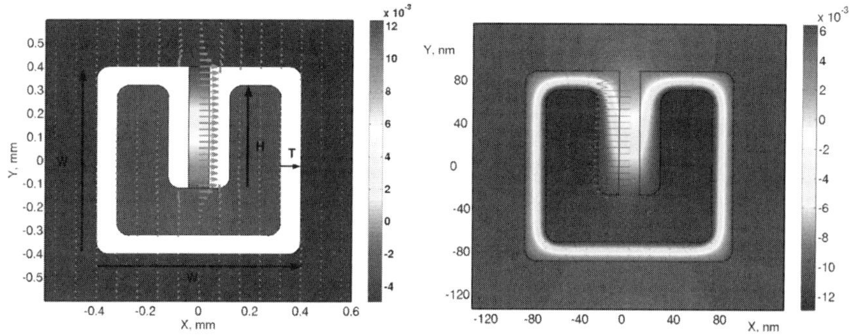

Figure 5.1 Magnetic field distribution inside a lattice of split-ring resonators. Color: magnetic field H_z. Arrows: electric field vectors. Left: SRR made of a perfect electric conductor (PEC) material. The gap is filled with a high-permittivity ($\epsilon_d = 4$) dielectric. Resonator parameters: periods $a_x = a_y = 1.2$ mm, ring size $W = 0.8$ mm, ring thickness $T = 80$ µm, gap height $H = 0.44$ mm, gap width $G = 80$ µm. The gap is filled with a high-permittivity dielectric $\epsilon_d = 4$. At the magnetic cutoff ($\mu_{\text{eff}} = 0$) shown here the vacuum wavelength $\lambda = 1.57$ cm. Right: the scaled-down by a factor ≈ 4400 plasmonic version of the silver SRR operating at $\lambda = 3.44$ µm. (Reprinted from Ref. 29.) (See color plate section.)

paradigm seems impractical. Second, as the metal linewidth approaches the typical skin depth $l_{\text{sk}} \approx 25$ nm, the metal no longer behaves as an impenetrable perfect electric conductor (PEC). Optical fields penetrate into the metal, and the response of the structure becomes *plasmonic*.

These difficulties have not deterred the research community from trying to fabricate and experimentally test magnetically active and even negative index structures in the infrared[14–17] and visible[18,19] spectral regions. Because fabricating intricate metallic resonators on a nanoscale is not feasible, much simpler magnetic resonances such as pairs of metallic strips or wires[14,16,17] or metallic nanoposts[18] have been used in experiments. Unfortunately, so far there has been no success in producing *subwavelength* magnetically active metamaterials in the optical range satisfying $a < \lambda/2n_d$, where n_d is the refractive index of a dielectric substrate or filler onto which magnetic materials are deposited. The reason for this is very simple:[20] in the absence of *plasmonic effects*, simple geometric resonators (such as pairs of metal strip or wires) resonate at the wavelength $\lambda \sim 2n_dL$, where L is the characteristic size of the resonator. In other words, "simple" metallic resonators are naturally not sub-λ resonators.

Fortunately, metallic resonators can be miniaturized using *plasmonic effects*.[20–24] In the optical regime metals can no longer be described as PECs. Instead, they are best described by a frequency-dependent plasmonic dielectric permittivity $\epsilon(\omega) \equiv \epsilon' + i\epsilon''$. For low-loss metals such as silver (for most frequencies) or gold (for infrared frequencies), $\epsilon'' \ll |\epsilon'|$ and $\epsilon' \ll -1$. Therefore, there is a significant field penetration into metallic structures that are thinner than or comparable to the skin depth $l_{\text{sk}} = \lambda/2\pi\sqrt{-\epsilon'} \approx 25$ nm. In metals $\epsilon(\omega)$ is determined by the

Drude response of the free electron to the optical fields. When the kinetic energy of the oscillating free electrons becomes comparable to the energy of the electric field, one can refer to the structure as being operated in a plasmonic regime. One of the most serious issues in the plasmonic regime is resistive loss; even a small amount of loss can drastically reduce the magnetic response and prevent access to the $\mu_{\text{eff}} < 0$ regime. The main differences between the microwave and optical regimes are analyzed in Sec. 5.2.

Presently, the only theoretical method of characterizing plasmonic metamaterials is by carrying out fully electromagnetic scattering simulations, obtaining complex transmission t and reflection r coefficients, and then calculating the effective parameters ϵ_{eff} and μ_{eff} of the metamaterial from r and t.[25,26] Such a direct approach lacks the intuitive appeal and rigor of the earlier microwave work that provided semi-analytic expressions for both ϵ_{eff}[11] and μ_{eff}.[10] Moreover, the extracted parameters of a periodic structure exhibit various artifacts such as antiresonances[27] that make their interpretation even less intuitive. In Sec. 5.3 we describe recent progress in rigorously calculating the quasistatic dielectric permittivity $\epsilon_{\text{qs}} \approx \epsilon_{\text{eff}}$ of plasmonic nanostructures.[22,23,28] Three approaches are described: the capacitor model, the electrostatic homogenization model, and the generalized eigenvalue differential equation (GEDE) model.

5.2 Why is Optical Magnetism Difficult to Achieve?

Before trying to understand how one can make a subwavelength magnetic material in the optical part of the spectrum, it is useful to review how magnetism is accomplished in microwave metamaterials. Consider one of the most basic design elements of NIMs: the split ring resonator (SRR). The particular design shown in Fig. 5.1 has been recently used[8] for making an invisibility cloak. For simplicity, we've simulated an infinite array of two-dimensional (infinite in the z direction) SRRs, with all sizes given in the caption to Fig. 5.1. Metal resistivity has been neglected, and PEC boundary conditions have been used at the metal surface. The nonvanishing components of the electromagnetic field are H_z, E_x, and E_y. The following eigenvalue equation has been solved:

$$-\vec{\nabla} \cdot \left(\frac{1}{\epsilon} \vec{\nabla} H_z \right) = \frac{\omega^2}{c^2} H_z, \tag{5.1}$$

where $\epsilon(x, y)$ is the spatially nonuniform dielectric permittivity. Finite element frequency domain (FEFD) code COMSOL Multiphysics[30] was used for solving Eq. (5.1).

Periodic boundary conditions have been applied to the cell boundaries. Therefore, the calculated eigenfrequency $f = \omega/2\pi = c/\lambda = 19$ GHz ($\lambda = 1.56$ cm) corresponds to the magnetic cutoff $\mu_{\text{eff}}(f_0) = 0$. Assuming that the wave is propagating in the x direction, the dispersion relation for the entire propagation band can be computed by setting the phase shift $\phi_x \equiv k_x a$ in the phase-shifted boundary conditions between the $x = -a_x/2$ and $x = +a_x/2$ sides of the unit cell. In this particular case, the propagation band extends from $f = f_0$ upward in frequency and corresponds to $\mu_{\text{eff}} \geq 0$ and $\epsilon_{\text{eff}} > 0$. The $\mu_{\text{eff}} < 0$ region is just below f_0. Because the ratio of the SRR size to wavelength is approximately $1/20$, and because its magnetic response is so strong, it can be characterized as a subwavelength magnetic resonator.

Why is this structure subwavelength? The natural resonance frequency of a resonator scales as $1/\sqrt{L_R C_R}$. Therefore, its size can be reduced from its "natural" $\lambda/2$ value by increasing either its inductance L_R or capacitance C_R. For this particular design the largest increase comes from the capacitance, which is increased by a factor $\epsilon_d H/G \approx 36$. Note that both the narrow gap and high permittivity of the dielectric placed in the gap enhance the capacitance and reduce the resonant frequency. The role of the dielectric filling is verified by an additional numerical simulation (not shown) in which the $\epsilon_d = 4$ dielectric filler is removed from the gap. The resonant frequency increases by a factor 1.8, confirming that most of the capacitance comes from the gap region. Note that at the resonant frequency the average electric and magnetic energies are equal to each other because $L_R \langle I^2 \rangle = \langle Q^2 \rangle / C_R$. This intuitive result is confirmed by the numerical simulations. The fact that the average magnetic energy constitutes a significant fraction of the total energy (one-half) explains the strong magnetic response of the structure. The results of this dissipation-free simulation are not significantly affected by the finite metallic losses because microwaves penetrate by only a fraction of a micron into a typical metal (e.g., 0.45 μm into copper at $\lambda = 1.56$ cm).

We now investigate whether this structure can be naturally scaled down to optical wavelengths. Because high-ϵ_d dielectrics are not available at optical frequencies, it is assumed that the SRR's gap is air-filled. Instead of using PEC boundary conditions at the metal surface, the metal is modelled as a lossless negative-ϵ dielectric with $\epsilon(\omega)$ taken from the standard tables.[32] By scanning the value of the plasmonic ϵ (physically equivalent to scanning the frequency), we have calculated the corresponding size of the SRR that has the same geometric proportions as the one shown in Fig. 5.1. Losses have been neglected for this calculation: $\epsilon \equiv \epsilon'$. Nontrivial solutions of Eq. (5.1) corresponding to the magnetic cutoff were found only for $\epsilon < \epsilon_{res} \equiv -330$, corresponding to $\lambda > \lambda_{res} \equiv 3$ μm for silver. Here ϵ_{res} corresponds to the *electrostatic resonance* of an SRR. The electrostatic potential ϕ corresponding to the electrostatic resonance and the corresponding electric field $\vec{E} = -\vec{\nabla}\phi$ are shown in Fig. 5.2(a). Electrostatic resonances depend only on the geometry of the nanostructure and are independent of their physical size. For nanostructures that are not much smaller than the wavelength of light, a nonzero magnetic field is associated with some of the electrostatic resonances.[22,23,33,34]

Magnetic field distribution for $\lambda = 3.44$ μm is presented in Fig. 5.1. Clearly, the magnetic field penetrates deep into the metal. This specific wavelength has been chosen because, for $\lambda = 3.44$ μm, the ratio of the wavelength to the SRR's

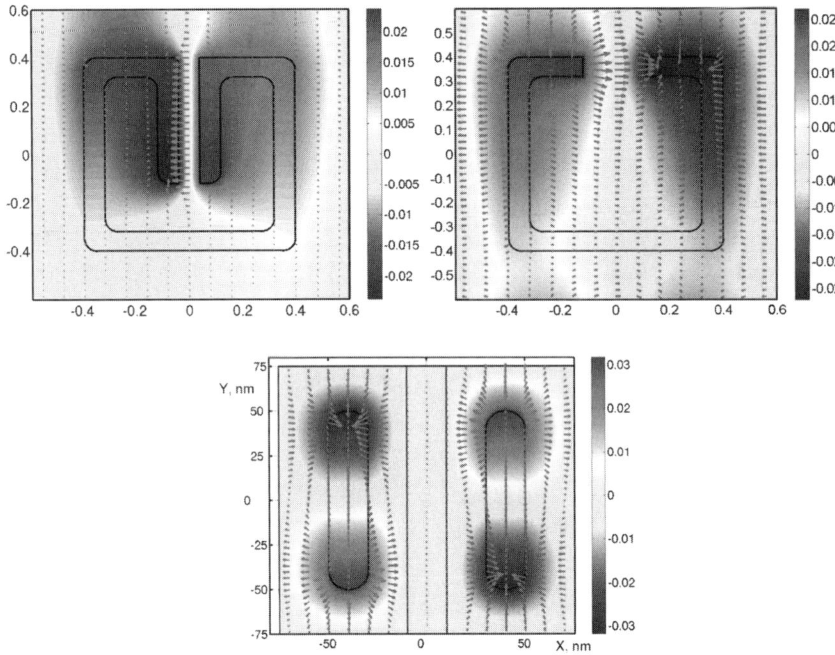

Figure 5.2 Potential distribution ϕ (color-coded) inside a lattice of (a) split-ring resonators (SRRs), (b) split rings (SRs), and (c) metal strips separated by a metal film corresponding to electrostatic resonances responsible for the magnetic response. Arrows: $\vec{E} = -\nabla\phi$. Electrostatic resonances occur at (a) $\epsilon_{res} \equiv -330$ ($\lambda = 3$ μm) for SRR, (b) $\epsilon_{res} \equiv -82$ ($\lambda = 1.5$ μm) for SR, and (c) $\epsilon_{res} \equiv -8.8$ ($\lambda = 0.5$ μm), assuming that the plasmonic material is silver. (Reprinted from Refs. 29 and 31.) (See color plate section.)

width is the same as for the microwave design: $1/20$. One may be tempted to assume that we've demonstrated a successful down-scaling (by a factor $4,400$) of a NIM from microwave to optical frequencies, and that there are no conceptual differences between the $\lambda = 1.56$ cm and $\lambda = 3.44$ μm. To see why this is not the case, consider the implications of the plasmonic ϵ of metals at optical frequencies.

For simplicity assume a collisionless Drude model for the metal: $\epsilon(\omega) = \epsilon_b - \omega_p^2/\omega(\omega + i\gamma)$ with $\gamma = 0$. Then the total energy density is

$$U_{tot} = \int_{V_m+V_v} d^2x \left(\frac{H_z^2}{8\pi} + \frac{\vec{E}^2}{8\pi}\frac{\partial(\omega\epsilon)}{\partial\omega} \right) = \int_{V_v} d^2x \frac{\vec{E}^2}{8\pi}$$

$$+\epsilon_b \int_{V_m} d^2x \frac{\vec{E}^2}{8\pi} + \frac{\omega_p^2}{\omega^2}\int_{V_m} d^2x \frac{\vec{E}^2}{8\pi} + \int_{V_m+V_v} d^2x \frac{H_z^2}{8\pi}, \tag{5.2}$$

where V_v and V_m are the vacuum and metallic volumes, respectively. The physical meaning of the four terms in Eq. (5.2) is as follows: the first two terms represent the energy of the electric field U_E, the third one represents the kinetic energy U_k of plasma electrons, and the fourth one represents the magnetic energy U_m. Because both the kinetic and magnetic energies (with $H_z^2 = c^2/\omega^2|\vec{\nabla}_\perp \times \vec{E}|^2$) scale as ω^{-2}, these two terms can be grouped into $U_L = U_m + U_k$. Again, at the resonance the "capacitive" energy U_E matches the "inductive" energy U_L. For the most relevant cases of $|\epsilon| \gg 1$ the following rules can be established for plasmonic structures: (a) most of the capacitive energy U_E resides in the vacuum region (outside of the plasmonic material); (b) most of the magnetic energy U_m also resides in the vacuum region. Of course, the same is true for the microwave structures. The main difference between the two types of structures comes from the kinetic energy U_k present in plasmonic but absent from microwave structures.

It is instructive to compare the values of U_E, U_m, and U_k for the plasmonic structure. From the results of the numerical simulation shown in Fig. 5.1 (right) it is found that $U_E = 0.5U_{tot}$, $U_k = 0.32U_t$, and $U_m = 0.18U_t$. As noted earlier, for the microwave SRR $U_E = U_m = 0.5U_t$. Therefore, the distinction between the scaled-down plasmonic structure and its microwave counterpart is the contribution of the kinetic energy U_k of the Drude electrons versus that of the magnetic energy U_m. In the present example, U_k exceeds U_m by almost a factor of 2. Thus it is fair to say that the present structure operates in a *strongly plasmonic* regime. This is quite remarkable, given that the operating wavelength $\lambda = 3.44$ μm is fairly long. To quantify the plasmonic effects in a magnetic structure, we introduce the plasmonic parameter $T_p = U_k/U_m$. The structure can be said to operate in a strongly plasmonic regime when $T_p > 1$. The plasmonic SRR can be made even more subwavelength by reducing the operating wavelength to $\lambda = \lambda_{res}$. Of course, this happens at the expense of the magnetic energy; the plasmonic parameter T_p diverges as the structure shrinks and the operating wavelength approaches that of the electrostatic resonance. At the electrostatic resonance, $T_p = \infty$ and $U_E = U_k$. The analogy between the plasmonic energy and inductance energy has been suggested earlier,[24,35] although the relation between the electrostatic resonance at $\lambda = \lambda_{res}$ and the vanishing of the negative magnetic permeability for $\lambda < \lambda_{res}$ has not been previously discussed.

The importance of having a reasonably small plasmonic parameter for making a magnetic resonator becomes apparent when losses are taken into account. Achieving $\mu_{\text{eff}} = 0$ (qualitative threshold for a strong magnetic response) sets a threshold for the magnetic energy U_m in the resonator. The total energy $U_{tot} = 2(1+T_p)U_m$ increases with T_p. This reduces the group velocity of the propagating mode, with the effect of narrowing the propagation band. Even small losses tend to destroy such bands. To get a qualitative estimate of the role of resistive losses we assume a finite γ in the Drude model. The group velocity can be roughly estimated as $v_g/c \sim U_m/U_t$. The propagation band is assumed to be destroyed by losses if the transit time across a single cell is longer than the decay time γ^{-1}, or $a_x/v_g > \gamma^{-1}$. This results in the following condition for achieving optical magnetism in a lossy system:

$$\frac{\gamma}{\omega} \approx \frac{\epsilon''}{\epsilon'} < \left[\frac{\omega a_x}{c}(1+T_p)\right]^{-1}. \tag{5.3}$$

This condition is easily satisfied for any subwavelength structure as long as T_p is of order unity. It becomes increasingly difficult to satisfy for the structures operated in a strongly plasmonic regime. Note that in the absence of losses a strong magnetic response of a subwavelength SRR can be achieved for any $\lambda > \lambda_{res}$.

Several instructive conclusions can be drawn from the above examples of the large (microwave) and small (mid-infrared) rings. First, for a given geometry of the unit cell there exists the shortest wavelength λ_{res} for which a strong magnetic response is expected. This wavelength corresponds to the electrostatic resonance of the structure, and for the SRR geometry shown in Fig. 5.1, $\lambda_{res} = 3$ μm. A strong optical response can be obtained for any $\lambda > \lambda_{res}$, with the actual dimensions of the ring adjusted accordingly. The positive result here is that even for a very subwavelength ring ($\lambda/20$) the plasmonic parameter $T_p \approx 2$ is modest and, therefore, nonzero losses do not destroy the magnetic response. The negative result is that the standard SRR design cannot be used for obtaining a magnetic response for visible/near-infrared frequencies because the electrostatic resonance occurs in the mid-infrared range.

The rule of thumb is that the more elaborate the design of the plasmonic structure, the longer the wavelength of the electrostatic resonance. For example, for a simplified version of the SRR, the SR shown in Fig. 5.2(b) (no vertical capacitor-forming strips, similar to the one used earlier[36]), it is found that the electrostatic resonance occurs at $\lambda_{res} = 1.5$ μm (corresponding to $\epsilon_{res} = -82$ and assuming that the SR material is silver). The electrostatic potential ϕ and the corresponding electric field $\vec{E} = -\nabla\phi$ are shown in Fig. 5.2(b). Using a typical dielectric substrate or a filler with $\epsilon_d = 2.25$ would reduce the resonant permittivity to $\epsilon_{res} = -186$ (corresponding to $\lambda_{res} = 2.25$ μm for Ag). At the same time, this design simplification comes at a cost; for very long wavelengths (microwave) such an SR is only moderately subwavelength, with $\lambda/W = 5.2$. Again, using a filler with $\epsilon_d = 2.25$ would result in $\lambda/W = 7.8$. If a more subwavelength resonator is required, one needs to operate in the vicinity of λ_{res}. The drawback of operating too close to λ_{res} is a high plasmonic parameter T_p and, therefore, high resistive losses.

Of course, the structure can be simplified even further: from a split ring to a pair of metal strips[15,20] or a pair of metal strips separated by a thin metal film.[28] The advantage of this structural simplification is that the magnetically active plasmonic resonance is pushed even further into the visible/near-infrared part of the spectrum: $\lambda_{res} = 0.5$ μm without a dielectric filler and $\lambda_{res} = 0.7$ μm with the $\epsilon_d = 2.25$ filler. The potential distribution for the strip pair-one film (SPOF) structure is shown in Fig. 5.2(c). Because of the promise of the SPOF for NIM development, we have scaled the structure in nanometers. Of course, the results of electrostatic simulations can be plotted with an arbitrary spatial scale [as was done in Figs. 5.2(a) and (b)] because there is no spatial scale in electrostatics. In the PEC limit (long wavelength) the ratio of the wavelength to period was found to be $\lambda/a_x = 1.85$ at the magnetic cutoff for the SPOF structure without a filler, and

$\lambda/a_x = 2.8$ with the $\epsilon_d = 2.25$ filler. Therefore, the SPOF structure presents a unique opportunity for developing a strongly subwavelength ($\lambda/10 - \lambda/6$) optical (visible to near-IR) NIM. As was recently shown,[28] the addition of a thin film modifies the electric response of the double-strip structure and turns this magnetically active metamaterial into a true subwavelength ($a_x = \lambda/6$) NIM. The simple qualitative remarks presented above explain why. Note that no optical NIM with a cell size smaller that $\lambda/2.5$ has been experimentally demonstrated[14–17] to date.

We summarize this section by noting that optical magnetic resonators are conceptually very different from their microwave counterparts. Intricate designs of microwave resonators (such as SRRs) are simply inappropriate for optical frequencies because they lose their strong magnetic response for $\lambda < \lambda_{res}$. This is a *rigorous quantitative result*; for a given resonator geometry, there is no strong magnetic response for wavelengths shorter than the one corresponding to the electrostatic resonance. λ_{res} enters the electrostatic theory parametrically, through the dependence of the metal's dielectric permittivity; $\epsilon(\lambda) < \epsilon_{res} \equiv \epsilon(\lambda_{res})$ is necessary for the strong response. Because intricate SRR designs correspond to extremely large negative values of ϵ_{res}, they do not exhibit a strong magnetic response in the visible/near-infrared parts of the spectrum. Therefore, the transition from SRRs [see Fig. 5.2(a)] to simple pairs of strips [see Fig. 5.2(c)] is not just a matter of fabricational convenience; it is physically necessary for making optical NIMs. The price one pays for using these simplified structures is that they are no longer subwavelength for $\lambda \gg \lambda_{res}$. Therefore, one is forced to operate close to λ_{res}. The price for that is significant plasmonic effects that enhance the plasmonic parameter T_p and make the bandwidth of the magnetic response very narrow. That makes these near-resonant structures highly susceptible to losses according to Eq. (5.3). Thus, plasmonic effects play two roles in optical magnetism. Their positive role is that they turn simple metallic structures into subwavelength resonators. Their negative role is that they enhance resistive losses.

Thus, the role of an optical NIM designer is to choose a structure that (a) has a λ_{res} in the visible/mid-IR, and (b) is sufficiently subwavelength in the PEC limit (long λ) that the structure does not need to be operated too close to λ_{res}. The electrostatic resonance at λ_{res} is the natural starting point for computing the optical response of the subwavelength metamaterial. In Sec. 5.3 we describe several techniques for calculating the effective dielectric permittivity $\epsilon_{\text{eff}}(\omega) \approx \epsilon_{\text{qs}}(\omega)$ in the quasistatic regime corresponding to the unit cell size much smaller than the wavelength, or, more precisely, $\omega^2 a_{x,y}^2/c^2 \ll 1$.

5.3 Effective Quasistatic Dielectric Permittivity of a Plasmonic Metamaterial

Several theoretical techniques are available for calculating $\epsilon_{\text{qs}}(\omega)$ for plasmonic nanostructures. The first one is the capacitor approach described in Sec. 5.3.1: AC voltage is imposed across the unit cell of a structure, and the effective capacitance is evaluated. The second approach, described in Sec. 5.3.2, is to homogenize the electrostatic equations using a multiscale expansion with two independent spatial variables: the intracell (small scale) variable $\vec{\xi}$ and the intercell (large scale) variable $\vec{X}$. The third approach, described in Sec. 5.3.3, uses the method of electrostatic eigenvalues and its extension to plasmonic structures with continuous plasmonic phase.

5.3.1 The capacitor model

The most simple and intuitive method of introducing the effective dielectric permittivity of a complex periodic plasmonic metamaterial is to imagine what happens when a single cell of such a structure is immersed in a uniform electric field. For simplicity, all calculations in Sec. 5.3 will be limited to a two-dimensional case, i.e., the structure is assumed uniform along the z axis; generalizations to three dimensions are straightforward. To further simplify our calculations, it is assumed that the unit cell is a rectangle of the size $a_x \times a_y$ in the xy plane: $-a_x/2 < x < a_x/2$ and $-a_y/2 < y < a_y/2$. The unit cell is assumed to consist of a plasmonic inclusion with a complex frequency-dependent dielectric permittivity $\epsilon(\omega)$ embedded into a dielectric host with the dielectric permittivity ϵ_d. The plasmonic inclusion may intersect the unit cell's boundary. We assume that the structure has an inversion symmetry and two reflection axes (x and y). Thus, the effective permittivity tensor is diagonal: $\epsilon_{\text{qs}} = \text{diag}(\epsilon_{\text{qs}}^{xx}, \epsilon_{\text{qs}}^{yy})$.

Applying a constant electric field $\vec{E}_0 = \vec{e}_x E_{0x} + \vec{e}_y E_{0y}$ is equivalent to solving the Poisson equation for the potential $\phi \equiv \phi_{\vec{E}_0}$:

$$\vec{\nabla} \cdot \left(\epsilon \vec{\nabla} \phi\right) = 0 \tag{5.4}$$

on a rectangular domain $ABCD$ (where AB and CD are parallel to y, and BC and AD are parallel to x). The external electric field $\vec{E}_0$ determines the boundary conditions satisfied by ϕ: (a) $\phi(x + a_x, y) = \phi(x, y) - E_{0x} a_x$, where (x, y) belongs to AB, and (b) $\phi(x, y + a_y) = \phi(x, y) - E_{0y} a_y$, where (x, y) belongs to AD. We now view the unit cell of a metamaterial as a tiny capacitor immersed in a uniform electric field that is created by the voltage applied between its plates. For calculating $\epsilon_{\text{qs}}^{xx}$ assume that the voltage $V_0 = E_{0x} a_x$ is applied between its sides AB and CD, and that $E_{0y} = 0$. From the potential distribution $\phi(x, y)$ the required surface charge density on the "capacitor plate" AB is $\sigma(y) = (\vec{n} \cdot \vec{D}) = -\epsilon_d \partial_x \phi(x = -a_x/2, y)$. The total charge (per unit length in z) on the capacitor is $Q = \int_{-a_y/2}^{+a_y/2} dy\ \sigma(y)$. The opposite capacitor plate CD is oppositely charged. The unit length capacitance of this capacitor, $C \equiv \epsilon_{\text{qs}}^{xx} a_y / a_x$ is thus given by $C = Q/V_0$, or

$$\epsilon_{\text{qs}}^{xx} = -\frac{\epsilon_d}{E_0 a_y} \int_{-a_y/2}^{+a_y/2} dy\ \partial_x \phi(x = -a_x/2, y)$$

$$\equiv \frac{a_y^{-1} \int_{-a_y/2}^{a_y/2} dy D_x(x=-a_x/2, y)}{a_x^{-1} \int_{-a_x/2}^{a_x/2} dx E_x(x, y=-a_y/2)}. \tag{5.5}$$

The $\epsilon_{\rm qs}^{yy}$ component is determined similarly by applying the voltage $V_0 = E_{0y}a_y$ between the capacitor plates AD and BC. Both $\epsilon_{\rm qs}^{xx}$ and $\epsilon_{\rm qs}^{yy}$ depend parametrically on the frequency ω because of the $\epsilon(\omega)$ dependence of the plasmonic permittivity. Therefore, extracting the frequency-dependent components of the $\epsilon_{\rm qs}$ tensor involves scanning the frequency ω, repeatedly solving Eq. (5.4) with the described boundary conditions, and applying the capacitance-based definition of $\epsilon_{\rm qs}$ given by Eq. (5.5). In the electrostatic limit of $\omega^2 a_{x,y}^2/c^2 \ll 1$ this quasistatic dielectric tensor accurately approximates the effective dielectric permittivity tensor: $\epsilon_{\rm eff}^{ij}(\omega) \approx \epsilon_{\rm qs}^{i,j}(\omega)$. We refer to this technique of extracting the electrostatic $\epsilon_{\rm eff}$-tensor as the "frequency scan" technique. As it turns out, there is a faster and more physically appealing approach to calculating $\epsilon_{\rm eff}^{ij}(\omega)$, described in Sec. 5.3.3. Note that the definition of the effective permittivity given by Eq. (5.5) is equivalent to the one introduced earlier by Pendry et al.,[10] which was derived[37] from the integral form of Maxwell's equations.

The capacitor model can be shown to be equivalent to another intuitive definition of $\epsilon_{\rm qs}$ based on the dipole moment density. The total dipole moment of a unit cell $\vec{p} = \int dxdy\vec{P}$ (where $\vec{P} = \frac{\epsilon-1}{4\pi}\vec{E}$ is the polarization density) is linearly proportional to the external electric field $\vec{E}_0$. In a homogeneous medium with anisotropic permittivity, vector components of the total dipole moment are $p_i = \frac{1}{4\pi}\left(\epsilon_{\rm qs}^{ij} - \delta^{ij}\right) E_{0j}$, where $\epsilon_{\rm qs}^{ij}$ is the effective permittivity tensor. Therefore, $\epsilon_{\rm qs}$ can be defined by requiring that

$$a_x a_y(\epsilon_{\rm qs}^{ij} - \delta^{ij})E_{0j} \equiv \int dxdy(\epsilon - 1)E_i. \tag{5.6}$$

Because $\int dxdyE_i = a_x a_y E_{0i}$, this *dipole density* definition of $\epsilon_{\rm qs}$ simplifies to

$$a_x a_y \epsilon_{\rm qs}^{ij} E_{0j}^{(k)} \equiv \int dxdy D_i^{(k)} = \int dxdy \epsilon E_i^{(k)}, \tag{5.7}$$

where the external field $E_{0j}^{(k)} \equiv E_0\delta_{jk}$ applied to the unit cell produces the total internal field $\vec{E}^{(k)}$, and $k = 1, 2$. The internal electric fields are computed by solving Eq. (5.4) subject to its $\vec{E}_0^{(k)}$-dependent boundary conditions. Equation (5.7) is equivalent to the capacitance-based definitions of $\epsilon_{\rm qs}$ given by Eq. (5.5). Owing to an identity $\int dxdy\ D_i = \oint ds\ x_i(\vec{D}\cdot\vec{n})$, where $\vec{n}$ is the unit normal to the closed contour of integration, Eq. (5.7) is expressed through a contour integral of the first kind:

$$\epsilon_{\rm qs}^{ij} E_{0j}^{(k)} = \frac{1}{a_x a_y} \oint ds\ x_i(\vec{D}^{(k)} \cdot \vec{n}). \tag{5.8}$$

This contour integral indeed reduces to the surface charge on the capacitor plates. For example, for the x-polarized external field ($k = 1$), we obtain $\oint ds\, x(\vec{D}\cdot \vec{n}) = (-a_x/2)(-Q) + (a_x/2)Q = a_x Q$, thereby completing the proof of equivalence between the definitions given by Eqs. (5.5) and (5.6). All formulas of this section can be generalized to three dimensions. Also, note that Eq. (5.8) can be used for determining *all* (diagonal and off-diagonal) elements of tensor ϵ_{qs} of an arbitrary nanostructure (with or without inversion symmetry of a unit cell) because the number of unknown components of ϵ_{qs} is equal to the number of equations.

5.3.2 Effective medium description through electrostatic homogenization

While the capacitance model developed in Sec. 5.3.1 is simple and intuitive, it needs to be justified in the context of the rigorous homogenization theory. In this section we review a multiscale approach[38,39] to calculating the effective permittivity of a metamaterial and show its equivalence with the capacitance model. Also known as the homogenization theory of differential operators with periodic coefficients,[38,39] it is the most vigorous approach to homogenizing a periodic metamaterial with a unit cell size a being much smaller than that of the typical variation scale Λ of the dominant electrostatic potential Φ. As in the previous sections, the key assumption leading to the electrostatic approximation is that $\omega a/c \ll \omega\Lambda/c \ll 1$. Under this set of assumptions, the frequency ω enters only as a parameter determining the plasmonic dielectric permittivity $\epsilon(\omega)$. The basis of the method is the two-scale expansion. Let $\vec{X}$ be the macroscopic coordinate enumerating the cells (corresponding to the large spatial scale Λ), and $\vec{\xi}$ be the local coordinate inside the unit cell (corresponding to the small spatial scale a). The potential $\phi(\vec{X}, \vec{\xi})$ and the local permittivity $\epsilon(\vec{\xi})$ are periodic in $\vec{\xi}$.

Using $\tau = a/\Lambda$ as the small parameter, we expand

$$\phi(x) = \phi_0(X, \xi) + \tau\phi_1(X, \xi) + \tau^2\phi_2(X, \xi) + O(\tau^3) \tag{5.9}$$

and use $\nabla_x = \nabla_\xi + \tau\nabla_X$ to solve the Laplace equation $\nabla_x\epsilon(x)\nabla_x\phi(x) = 0$ order-by-order in τ. The goal of the homogenization theory is to show that there exists a macroscopic potential $\phi_{macro}(X)$ that obeys a Laplace-Poisson equation in a certain homogeneous, possibly anisotropic, medium. In doing so, both the rigorous definitions of $\phi_{macro}(X)$ and $\epsilon_{\text{eff}}^{ij}$ are recovered. This goal is achieved by expanding the Poisson equation in powers of τ. Terms with different powers of τ must vanish independently, resulting in three equations:

$$\nabla_\xi\epsilon(\xi)\nabla_\xi\phi_0(X, \xi) = 0, \tag{5.10}$$

$$\nabla_\xi\epsilon(\xi)\nabla_\xi\phi_1(X, \xi) = -\left(\nabla_\xi\epsilon(\xi) + \epsilon(\xi)\nabla_\xi\right)\nabla_X\phi_0(X, \xi), \tag{5.11}$$

and

$$\begin{aligned}&\nabla_X\epsilon(\xi)\nabla_X\phi_0(X, \xi) + \\ &\left(\epsilon(\xi)\nabla_\xi + \nabla_\xi\epsilon(\xi)\right)\nabla_X\phi_1(X, \xi) + \nabla_\xi\epsilon(\xi)\nabla_\xi\phi_2(X, \xi) = 0.\end{aligned} \tag{5.12}$$

Equation (5.10) is identically satisfied by $\phi_0(X, \xi) \equiv \phi_0(X)$, and $\phi_0(X)$ assumes the role of the macroscopic potential. Next, $\phi_1(X, \xi)$ is expressed through the macroscopic gradients of ϕ_0:

$$\phi_1(X, \xi) = -\frac{\partial \phi_0(X)}{\partial X_i} \phi_{sc}^{(i)}(\xi), \tag{5.13}$$

where $\phi_{sc}^{(i)}(\xi)$ ($i = 1, 2$ in two dimensions) are periodic basis functions satisfying the local Poisson equation

$$\vec{\nabla}_\xi \cdot \epsilon(\xi) \vec{\nabla}_\xi \phi_{sc}^{(i)}(\xi) = \left(\vec{\nabla}_\xi \epsilon(\xi) \right) \cdot \vec{e}_i, \tag{5.14}$$

where $\vec{e}_i$ is the ith basis vector of a Cartesian coordinate system. Note that the periodic potentials $\phi_{sc}^{(i)}$ are linearly related to the $\phi_{\vec{E}_0}$ defined by Eq. (5.4) through $\phi_{\vec{E}_0} \equiv \left(E_{0i} \phi_{sc}^{(i)}(\xi) - \vec{E}_0 \cdot \vec{\xi} \right) / |\vec{E}_0|$. The difference between Eq. (5.14) and Eq. (5.4) is that the macroscopic electric field $\vec{E}_0 \equiv -\vec{\nabla}_X \phi_0$ is explicitly included in the rhs in (5.14), but embedded in the boundary conditions in (5.4).

Finally, the macroscopic equation for $\phi_0(X)$ is obtained by substituting ϕ_1 from Eq. (5.13) into Eq. (5.12) and averaging over the local variable ξ: $\nabla_{X_i} \epsilon_{\text{qs}}^{ij} \nabla_{X_j} \phi_{macro}(X) = 0$, where $\phi_{macro}(X) \equiv \phi_0(X) \equiv \langle \phi(X, \xi) \rangle$ and

$$\epsilon_{\text{qs}}^{ij} = \left\langle \epsilon(\xi) \delta_{ij} - \epsilon(\xi) \nabla_{\xi_i} \phi_{sc}^{(j)}(\xi) \right\rangle \equiv \left\langle \epsilon(\xi) \nabla_{\xi_i} \left(\xi_j - \phi_{sc}^{(j)}(\xi) \right) \right\rangle; \tag{5.15}$$

angle brackets denote averaging over the unit cell and $\langle F \rangle \equiv \int F d^2\xi / \int d^2\xi$. It can be shown that this definition of ϵ_{qs} is equivalent to the one obtained from the capacitor model. Indeed, note that $\vec{E}^{(i)}(\xi) = \vec{\nabla}_\xi(-\phi_{sc}^{(i)}(\xi) + \xi_i)$ is the total electric field excited by external electric field $\vec{E}_0 = \vec{e}_i$ with the unit amplitude. Therefore, Eq. (5.15) is equivalent to $\epsilon_{\text{qs}}^{ij} = \langle \epsilon(\xi) E_i^{(j)}(\xi) \rangle$, in exact agreement with Eq. (5.7).

5.3.3 The eigenvalue expansion approach

The frequency scan technique described in Sec. 5.3.1 is a simple yet time-consuming approach to calculating the quasistatic response of subwavelength metamaterials. The electrostatic eigenvalue (EE) approach[22,40,41] enables calculating this response for a wide range of frequencies by evaluating the position and strength of the electric dipole active plasmonic resonances in that range. As we shall see from the examples below, there are only a few eigenmodes that strongly contribute to ϵ_{qs}, making the EE approach extremely efficient. Additional theoretical insights [such as the Hermitian nature of ϵ_{qs} that is not evident from Eq. (5.7)] can be gained from the EE approach. In addition to reviewing some of the known facts about the EE approach[40–42] to calculating ϵ_{qs}, we extend the original theory of plasmon resonances to include plasmonic metamaterials with continuous plasmonic phase. Such structures have become increasingly important in the field of NIMs since the introduction of the so-called *fishnet* structure,[14,17] as well as the SPOF structure.[28]

One way of obtaining eigenvalue expansions is the *generalized eigenvalue differential equation* (GEDE).[22,40] Another essentially equivalent method is based on a surface integral eigenvalue equation.[43,44] The steps of the GEDE approach are briefly described here, with the details appearing elsewhere.[22,23] We assume that a periodic nanostructure consists of two dielectric, nonmagnetic materials: one with a frequency-dependent permittivity $\epsilon(\omega) < 0$ and another with permittivity ϵ_d. The local permittivity of such a structure is $\epsilon(\vec{r}, \omega) = \epsilon_d\left[1 - \frac{1}{s(\omega)}\theta(\vec{r})\right]$, where $\theta(\vec{r}) = 1$ inside the plasmonic material and $\theta(\vec{r}) = 0$ elsewhere. Note that $s(\omega) = (1 - \epsilon(\omega)/\epsilon_d)^{-1}$ can be thought of as the "frequency label" that selects the correct ϵ for a given ω. Conversely, once s is obtained, the appropriate ω can be estimated for any low-loss metal.

First, the GEDE

$$\vec{\nabla} \cdot \left[\theta(\vec{r})\vec{\nabla}\phi_n\right] = s_n \nabla^2 \phi_n \tag{5.16}$$

is solved for the real eigenvalues s_n. Spectral properties of the GEDE are discussed in detail in Refs. 40 and 42 and references therein. Second, the solution of Eq. (5.4) is expressed as an eigenmode expansion[40]

$$\phi(\vec{r}) = \phi_0(\vec{r}) + \sum_n \frac{s_n}{s(\omega) - s_n} \frac{(\phi_n, \phi_0)}{(\phi_n, \phi_n)} \phi_n(\vec{r}), \tag{5.17}$$

where the scalar product is defined as $(\phi, \psi) = \int dxdy\theta\vec{\nabla}\phi^* \cdot \vec{\nabla}\psi$, and $\phi_0 = -E_0 x$ represents the x-polarized external field. Third, the quasistatic permittivity is calculated by substituting $\phi(\vec{r})$ from Eq. (5.17) into any of the equivalent definitions of ϵ_{qs}. For example, the dipole moment definition, Eq. (5.6), leads to the following analytical expression for ϵ_{qs}:

$$\epsilon_{\text{qs}}^{ij}(\omega) = 1 - \frac{f_0^{ij}}{s(\omega)} - \sum_{n>0} \frac{f_n^{ij}}{s(\omega) - s_n}, \tag{5.18}$$

where $f_n^{ij} = A^{-1}(\phi_n, x_i)(\phi_n, x_j)/(\phi_n, \phi_n)$ are the electric dipole strengths of the nth resonance,

$$f_0^{ij} = (A_p/A)\delta_{ij} - \sum_{n>0} f_n^{ij} \tag{5.19}$$

is the measure of the electric response of the continuous plasma phase, and $A_p = \int dxdy\theta(\vec{r})$ is the area of the plasmonic phase contained within the area A of a unit cell. From the expression for f_n^{ij} we note that only dipole-active resonances having a nonvanishing dipole moment (ϕ_n, x_j) contribute to the dielectric permittivity. Examples of such resonances are shown in Figs. 5.2(a) and (b) for the SRR and

SR. But, for example, the electrostatic resonance of the SPOF structure shown in Fig. 5.2(c) is not electric dipole active. It does not contribute to ϵ_{qs} but contributes to the magnetic permeability.

Application of the equivalent capacitance definition given by Eq. (5.8) leads to the same Eq. (5.18) with the same f_n^{ij}. However f_0^{ij} is obtained in a different (and more instructive) form:

$$f_0^{ij} = q^{ij} - \frac{1}{A}\sum_n \frac{(\phi_n, x_j)}{(\phi_n, \phi_n)} \oint ds\theta x_i \frac{\partial \phi_n}{\partial n}, \tag{5.20}$$

where $\partial/\partial n$ is the normal derivative, $q^{ij} = \frac{1}{A}\oint ds\theta x_i \frac{\partial x_j}{\partial n}$, and integration is carried out over a closed contour encapsulating the plasmonic phase in one unit cell. Note that, if there is no continuous plasmonic phase inside the unit cell of a nanostructure, its boundaries can be chosen such that they are not intersected by the plasmonic inclusion and, therefore, $f_0^{ij} \equiv 0$. Combining Eqs. (5.19) and (5.20) results in a generalized sum rule for plasmonic oscillators in nanostructures that contain a continuous plasmonic phase.

To illustrate this method, we chose the two-dimensional SPOF[28] shown in Fig. 5.2(c). The real and imaginary parts of the yy-component of ϵ_{qs} corresponding to the electric field along the film are plotted as dashed and dash-dotted lines, respectively, in Fig. 5.3. The gray and black dashed lines show Re $\epsilon_{\text{qs}}^{yy}$ with and without retardation correction to the frequency of the plasmon resonances. It is generally known[44] that frequencies of the optical resonances of finite-sized nanoparticles (with the typical spatial dimension w) are red-shifted from their electrostatic values because of the retardation effects proportional to $\eta^2 \equiv \omega^2 w^2/c^2$. As shown in the Appendix, these shifts can be expressed as corrections to the frequency labels s_n: $s_n = s_n^{(0)} + s_n^{(2)}\eta^2$, where $s_n^{(0)}$ are the electrostatic resonances and $s_n^{(2)}$ are the retardation corrections computed in the Appendix. To obtain any meaningful comparison between the electromagnetically extracted values of ϵ_{eff} and the electrostatic ϵ_{qs}, these corrections must be included even for structures as small as $\lambda/10$.

In the chosen range of frequencies, there are only two dipolar resonances that contribute to ϵ^{yy}; quasistatic curves in Fig. 5.3 are computed from Eq. (5.18) with the following numerical coefficients: conduction pole residue $f_0^{yy} = 0.043$, electric resonance strengths $f_1^{yy} = 0.0045$, $f_2^{yy} = 0.0005$, and pole positions $s_n = s_n^{(0)} + s_n^{(2)}\eta^2$ with $s_1^{(0)} = 0.0426$, $s_2^{(0)} = 0.1630$, and $s_1^{(2)}\eta^2 = -0.007$, $s_2^{(2)}\eta^2 = -0.004$. The other component of ϵ_{qs} has no resonances between $\lambda = 500 - 800$ nm and remains approximately constant, $\epsilon_{\text{qs}}^{xx} \approx 1.2$ in the $\lambda = 500 - 800$ nm frequency. Quasistatic calculations of $\epsilon_{\text{qs}}^{yy}$ are compared with the $\epsilon_{em} \equiv \epsilon_{\text{eff}}$ extracted from the first-principles electromagnetic scattering simulations[25–27] of a single layer of SPOF. Electromagnetic simulations are performed using the FEFD method implemented in the software package COMSOL Multiphysics.[30] Overall, agreement be-

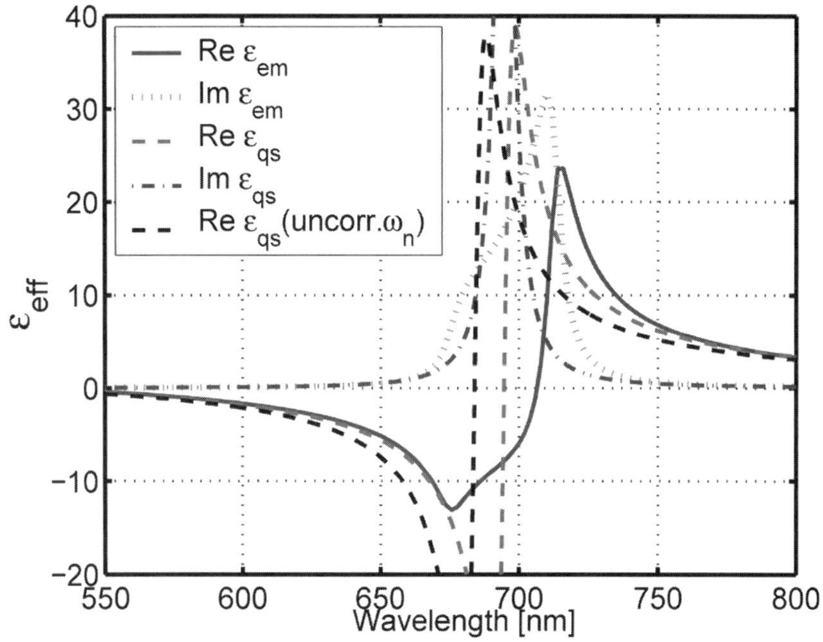

Figure 5.3 Effective dielectric permittivity $\epsilon_{\text{eff}}^{yy}$ of the SPOF structure (with the film in the yz plane) calculated using two methods: electromagnetic scattering through a single layer (solid and dotted curves), and the quasistatic formula (5.18) (dashed and dash-dotted). The red-shifted gray dashed curve differs from the black dashed one by the frequency shift given by Eq. (5.24) discussed in the Appendix. Structure parameters: periods $a_x = 100$ nm, $a_y = 75$ nm, strip width $w = 50$ nm, strip thickness $t_s = 15$ nm, film thickness $d_f = 5$ nm, strip separation in a pair $h = 15$ nm; plasmonic component: silver (Drude model); immersion medium: $\epsilon_d = 1$. (Reprinted from Ref. 31.)

tween ϵ_{em} and ϵ_{qs} is very good everywhere except near the strong absorption line associated with electric resonance at ≈ 700 nm. Inside that band ($680-720$ nm) the shape of ϵ_{em} strongly deviates from Lorentzian. We speculate that this irregularity of ϵ_{em} is related to the large phase shift per cell $\theta \equiv k_x a_x = \text{Re}[\sqrt{\epsilon_{\text{eff}}^{yy}}]\omega a_x/c$ neglected in the quasistatic approximation based on periodic electrostatic potentials. A more accurate description of ϵ_{eff} should include spatial dispersion.[45,46] Development of an adequate theory of this phenomenon in plasmonic crystals is underway.

5.4 Summary

In this tutorial review we have discussed several physics issues important for designing subwavelength plasmonic metamaterials in the optical part of the spectrum. We have demonstrated that the major difference between subwavelength metamaterials in the *microwave* and *optical* frequency ranges is the plasmonic nature of the optical metamaterials. We have shown that plasmonic effects enable miniaturization of simple nanostructures in which optical magnetism can be observed.

The contribution of the plasmonic effects has been quantified by introducing a plasmonic parameter T_p which characterizes the ratio of the kinetic energy of the conduction electrons and the magnetic energy stored by the nanostructure. High values of T_p enhance resistive losses that can suppress magnetically active optical resonances. Because of the importance of electrostatic resonances in determining the electromagnetic properties of subwavelength metamaterials, we have also presented several approaches to calculating the electrostatic effective permittivity $\epsilon_{qs}(\omega)$. The accuracy of the electrostatic calculations can be greatly improved by taking into account retardation effects as described in the Appendix. Our theory is supported by first-principles electromagnetic simulations.

Acknowledgments

This work is supported by the ARO MURI W911NF-04-01-0203, AFOSR MURI FA9550-06-01-0279, and the DARPA contract HR0011-05-C-0068.

5.5 Appendix: Electromagnetic Red Shifts of Plasmonic Resonances

In this Appendix we describe the effect of retardation (relevant when the unit cell size of a plasmonic metamaterial is a significant fraction of the optical wavelength) on the frequencies of plasmonic resonances. The frequency corrections calculated here scale as $\eta^2 = \omega^2 w^2/c^2$, where w is the characteristic size of the plasmonic element.

First, we sketch the perturbation theory of the selfconsistent optical response of a nanostructure to a propagating light beam. To isolate the role of electrostatic resonances, it is convenient to decompose electric and magnetic fields into "incident" and "scattered," i.e., $\vec{E} = \vec{E}_{in} + \vec{E}_{sc}$ and $\vec{H} = \vec{H}_{in} + \vec{H}_{sc}$, such that $\vec{E}_{sc}$ and $\vec{H}_{sc}$ vanish in a homogeneous structure. To achieve this, we use the plane wave ansatz: $\vec{E}_{in} = \vec{E}_0 e^{i\vec{k}\cdot\vec{r}}$ and $\vec{H}_{in} = \vec{H}_0 e^{i\vec{k}\cdot\vec{r}}$. After the effective medium parameters ϵ_{eff} and μ_{eff} are expressed through $\vec{k}$, we use the dispersion relation of transverse waves in a homogenized medium: $k = \sqrt{\epsilon_{\text{eff}}}\sqrt{\mu_{\text{eff}}}\omega/c$ to determine $\epsilon_{\text{eff}}(\omega)$ and $\mu_{\text{eff}}(\omega)$. This calculation is reminiscent of the Maxwell-Garnett homogenization theory, where individual particles to be homogenized are assumed to be immersed inside an effective medium with some unknown (and selfconsistently determined) ϵ_{eff} and μ_{eff}. To simplify the calculation, we assume that $\vec{k}||\hat{x}$, $\vec{E}_0||\hat{y}$, and $\vec{H}_0||\hat{z}$. Therefore, $\mu_{\text{eff}} \equiv \mu_{\text{eff}}^{zz}$. Additionally, because the incident fields satisfy Maxwell's equations in the homogenized medium, we have $\vec{H}_0 = Z_e^{-1}[\vec{n}_k \times \vec{E}_0]$, where $Z_e = \sqrt{\mu_{\text{eff}}}/\sqrt{\epsilon_{\text{eff}}}$ is the effective impedance and $\vec{n}_k = \vec{k}/|\vec{k}|$ is the direction of the phase velocity. We further assume that $\epsilon_{\text{eff}} = \epsilon_{\text{qs}} + O(\eta^2)$, where ϵ_{qs} is given by Eq. (5.18).

The scattered electric field $\vec{E}_{sc}$ is decomposed into the potential and solenoidal parts, $\vec{E}_{sc} = \vec{E}_{pot} + \vec{E}_{sol} = -\vec{\nabla}\Phi_{sc} + ik_0\vec{A}_{sc}$, where $\vec{\nabla}\cdot\vec{A}_{sc} = 0$ and $k_0 \equiv \omega/c$.

Note that $\vec{A}_{sc}$ is related to the scattered magnetic field: $\vec{H}_{sc} = (\mu_{\text{eff}} - 1)\vec{H}_{in} + \vec{\nabla} \times \vec{A}_{sc}$. It can be demonstrated that $\vec{A}_{sc}$ is first order in η, making the contribution of $\vec{A}_{sc}$ to $\vec{E}_{sc}$ second order in η. Therefore, the lowest-order (η^2) expression for $\mu_{\text{eff}} - 1$ can be found without directly computing $\vec{A}_{sc}$ or magnetic fields.

The potential part of $\vec{E}_{sc}$ is determined from $\vec{\nabla} \cdot \vec{D} = 0$, resulting in

$$\vec{\nabla}\epsilon\vec{E}_{pot} = -\vec{\nabla}\epsilon\vec{E}_{in} - \vec{\nabla}\epsilon\vec{E}_{sol} \equiv -\left(\vec{E}_{in} + ik_0\vec{A}_{sc}\right) \cdot \vec{\nabla}\epsilon, \tag{5.21}$$

where $k_0\vec{A}_{sc}$ can be neglected to order η^2. For completeness, we note that $\vec{A}_{sc}$ is computed, to the lowest order in η, as

$$-\nabla^2\vec{A}_{sc} - ik_0\epsilon\vec{\nabla}\Phi_{sc}^{(0)} = ik_0\left(\epsilon_{\text{qs}} - \epsilon\right)\vec{E}_{in}. \tag{5.22}$$

We are now in a position to calculate retardation shifts of the plasmonic resonances. These frequency shifts are always red and can be understood physically as follows. Quasistatic currents associated with electric fields of electrostatic resonances induce magnetic fields via Ampere's law. These magnetic fields generate secondary electric fields according to Faraday's law. The secondary electric fields contribute to the Poisson equation $\vec{\nabla} \cdot \epsilon\vec{E} = 0$, causing shifts in the electrostatic eigenvalues. While such frequency shifts have been calculated earlier for isolated plasmonic nanoparticles, they have never been calculated for periodic plasmonic metamaterials.

To the lowest order in η and in the close vicinity of the nth plasmonic resonance [i.e., at $s \approx s_n^{(0)}$, where $s_n^{(0)}$ is the purely electrostatic eigenvalue of the Eq. (5.16)], the vector potential induced by the nth electrostatic resonance is found from Eq. (5.22): $\vec{A}_{sc}^{(1)}(\vec{r}) = ik_0 \int dV'G(\vec{r}-\vec{r}')\epsilon(\vec{r}')\nabla\phi_n(\vec{r}')$, where $\epsilon \approx \epsilon_d\left(1 - \frac{1}{s_n^{(0)}}\theta\right)$, and $G(\vec{r}-\vec{r}')$ is the modified Green's function of the Poisson equation with periodic boundary conditions originally calculated[47] in the context of solid state physics. Thus, computed vector potential $\vec{A}_{sc}^{(1)}$ contributes to the Poisson equation:

$$\vec{\nabla}\theta\vec{\nabla}\Phi + k_0^2\epsilon_d\vec{\nabla}\theta \cdot \int dV'G(r,r')\left(1 - \frac{1}{s_n^{(0)}}\theta\right)\vec{\nabla}\Phi = s_n\nabla^2\Phi. \tag{5.23}$$

This is a generalized linear eigenvalue problem with an integro-differential operator. Treating the integral term as a perturbation, corrections to electrostatic eigenvalues $s_n^{(0)}$ can be shown to be

$$s_n^{(2)} = k_0^2\epsilon_d s_n^{(0)} \oint dS\phi_n(r)\vec{n} \cdot \int G(\vec{r}-\vec{r}')\left(1 - \frac{1}{s_n^{(0)}}\theta(\vec{r}')\right)\vec{\nabla}\phi_n(\vec{r}')dV' \times \left(\int |\nabla\phi_n|^2\theta dV\right)^{-1}, \tag{5.24}$$

where $\oint dS$ is a surface integral over a closed surface S of the plasmonic inclusion (which reduces to a contour integral for two-dimensional crystals). The renormalized s_n is calculated as $s_n = s_n^{(0)} + s_n^{(2)}$. The volume integral in Eq. (5.24) can be reduced to a surface integration over the surface S by introducing an auxiliary vector $\vec{a}(\vec{r}, \vec{r}') = -\vec{\nabla} \int G(\vec{r} - \vec{r}'')G(\vec{r}'' - \vec{r}')dV''$:

$$s_n^{(2)} = k_0^2 \epsilon_d \oint dS \phi_n(\vec{r})\vec{n} \cdot \oint dS' \vec{a}(\vec{r}, \vec{r}') \times [\vec{n}' \times \nabla \phi_n(\vec{r}')] \times \left(\oint \phi_n \frac{\partial \phi_n}{\partial n} dS \right)^{-1}, \tag{5.25}$$

where the normal derivative $\partial \phi_n / \partial n$ is evaluated on the plasmonic side of the surface. A particular case of this formula has been previously reported for isolated three-dimensional particles.[44]

Despite substantial progress in calculations of periodic Green's functions,[47] closed-form expressions for double- or triple-periodic G in two or three dimensions are not known. However, there exists one simple yet exact result for a two-dimensional Green's function in the limit $a_y \gg a_x$:[47]

$$G_2(\vec{r} - \vec{r}') = \frac{1}{2a_x} \left(\frac{(y - y')^2}{a_y} - |y - y'| + \frac{a_y}{6} \right) - \frac{1}{4\pi} \ln \left[1 - 2e^{-2\pi|y-y'|/a_x} \cos(2\pi|x - x'|/a_x) + e^{-4\pi|y-y'|/a_x} \right]. \tag{5.26}$$

The function (5.26) is periodic only in the x direction. It is therefore applicable only for $|y - y'| \ll a_y$, i.e., when plasmonic inclusions are much thinner in the y direction than the period ($w_y \ll a_y$). For periodic metamaterials based on "current loops" (strip pairs, horse shoes, etc.), interaction between consecutive layers of resonators is usually insignificant, and the function (5.26) provides a reasonable approximation. When the condition $w_x \ll a_x$ is satisfied, one can use expression (5.26) with interchanged variables $x \leftrightarrow y$, $a_x \leftrightarrow a_y$. When both dimensions are small, $w_{x,y} \ll a_{x,y}$, a symmetrized (in x, y) version of Eq. (5.26) is used.

References

1. V. G. Veselago, "The electrodynamics of substances with simultaneously negative values of ϵ and μ," *Soviet Physics – Uspekhi* **10**, 509 (1968).

2. J. B. Pendry, "Negative refraction makes a perfect lens," *Phys. Rev. Lett.* **85**, 3966 (2000).

3. Y. Horii, C. Caloz, and T. Itoh, "Super-compact multilayered left-handed transmission line and diplexer application," *IEEE Trans. Microwave Th. Tech.* **53**, 1527 (2005).

4. A. Alu and N. Engheta, "Guided modes in a waveguide filled with a pair of single-negative (sng), double-negative (dng), and/or double-positive (dps) layers," *IEEE Trans. Microw. Th. Tech.* **52**, 199 (2004).

5. R. W. Ziolkowski and A. Erentok, "Metamaterial-based efficient electrically small antennas," *IEEE Trans. Antennas Propag.* **54**, 2113 (2006).

6. J. B. Pendry, D. Schurig, and D. R. Smith, "Controlling electromagnetic fields," *Science* **312**, 1780 (2006).

7. U. Leonhardt, "Optical conformal mapping," *Science* **312**, 1777 (2006).

8. D. Schurig, J. J. Mock, B. J. Justice, S. A. Cummer, J. B. Pendry, A. F. Starr, and D. R. Smith, "Metamaterial electromagnetic cloak at microwave frequencies," *Science* **314**, 977 (2006).

9. D. R. Smith, W. J. Padilla, D. C. Vier, S. C. Nemat-Nasser, and S. Schultz, "Composite medium with simultaneously negative permeability and permittivity," *Phys. Rev. Lett.* **84**, 4184 (2000).

10. J. B. Pendry, A. J. Holden, D. J. Robbins, and W. J. Stewart, "Magnetism from conductors and enhanced nonlinear phenomena," *IEEE Trans. Microwave Theory Tech.* **47**, 2075 (1999).

11. J. B. Pendry, A. J. Holden, W. J. Stewart, and I. Youngs, "Extremely low frequency plasmons in metallic mesostructures," *Phys. Rev. Lett.* **76**, 4773 (1996).

12. E. Yablonovitch, "Inhibited spontaneous emission in solid-state physics and electronics," *Phys. Rev. Lett.* **58**, 2059 (1987).

13. S. John, "Strong localization of photons in certain disordered dielectric superlattices," *Phys. Rev. Lett.* **58**, 2486 (1987).

14. S. Zhang, W. Fan, N. C. Panoiu, K. J. Malloy, R. M. Osgood, and S. R. J. Brueck, "Experimental demonstration of near-infrared negative-index metamaterials," *Phys. Rev. Lett.* **95**, 137404 (2005).

15. G. Dolling, C. Enkrich, M. Wegener, J. F. Zhou, C. M. Soukoulis, and S. Linden, "Cut-wire pairs and plate pairs as magnetic atoms for optical metamaterials," *Opt. Lett.* **30**, 3198 (2005).

16. V. M. Shalaev, W. Cai, U. K. Chettiar, H.-K. Yuan, A. K. Sarychev, V. P. Drachev, and A. V. Kildishev, "Negative index of refraction in optical metamaterials," *Opt. Lett.* **30**, 3356 (2005).

17. G. Dolling, C. Enkrich, M. Wegener, C. M. Soukoulis, and S. Linden, "Simultaneous negative phase and group velocity of light in a metamaterial," *Science* **312**, 892 (2006).

18. A. N. Grigorenko, A. K. Geim, H. F. Gleeson, Y. Zhang, A. A. Firsov, I. Y. Khrushchev, and J. Petrovic, "Nanofabricated media with negative permeability at visible frequencies," *Nature* **438**, 335 (2005).

19. W. Cai, U. K. Chettiar, H.-K. Yuan, V. C. de Silva, A. V. Kildishev, V. P. Drachev, and V. M. Shalaev, "Metamagnetics with rainbow colors," *Opt. Exp.* **15**, 3333 (2007).

20. G. Shvets and Y. A. Urzhumov, "Negative index meta-materials based on two-dimensional metallic structures," *J. Opt. A: Pure Appl. Opt.* **8**, S122 (2006).

21. G. Shvets, "Photonic approach to making a material with a negative index of refraction," *Phys. Rev. B.* **338**, 035109 (2003).

22. G. Shvets and Y. Urzhumov, "Engineering the electromagnetic properties of periodic nanostructures using electrostatic resonances," *Phys. Rev. Lett.* **93**, 243902 (2004).

23. G. Shvets and Y. Urzhumov, "Electric and magnetic properties of sub-wavelength plasmonic crystals," *J. Opt. A: Pure Appl. Opt.* **7**, S23–S31 (2005).

24. J. Zhou, T. Koschny, M. Kafesaki, E. N. Economou, J. B. Pendry, and C. M. Soukoulis, "Saturation of the magnetic response of split-ring resonators at optical frequencies," *Phys. Rev. Lett.* **95**, 223902 (2005).

25. D. R. Smith, S. Schultz, P. Markos, and C. M. Soukoulis, "Determination of effective permittivity and permeability of metamaterials from reflection and transmission coefficients," *Phys. Rev. B* **65**, 195104 (2002).

26. P. Markos and C. M. Soukoulis, "Transmission properties and effective electromagnetic parameters of double negative metamaterials," *Opt. Exp.* **11**, 649 (2003).

27. T. Koschny, P. Markos, D. R. Smith, and C. M. Soukoulis, "Resonant and antiresonant frequency dependence of the effective parameters of metamaterials," *Phys. Rev. E* **68**, 065602 (2003).

28. V. Lomakin, Y. Fainman, Y. Urzhumov, and G. Shvets, "Doubly negative metamaterials in the near infrared and visible regimes based on thin nanocomposites," *Opt. Exp.* **14**, 11164 (2006).

29. Y. Urzhumov and G. Shvets, "Optical magnetism and negative refraction using plasmonic metamaterials," *Solid State Communications* **146**, 208–220 (2008).

30. Comsol AB, Burlington, MA, *COMSOL Multiphysics User's Guide, Version 3.3* (2006).

31. Y. Urzhumov and G. Shvets, "Quasistatic effective medium theory of plasmonic nanostructures," *Proc. SPIE* **6642**, 66420X (2007).

32. E. D. Palik, Ed., *Handbook of Optical Constants of Solids*, vol. 1, Academic Press, Orlando, FL (1985).

33. A. K. Sarychev, G. Shvets, and V. M. Shalaev, "Magnetic plasmon resonance," *Phys. Rev. E* **73**, 036609 (2006).

34. A. K. Sarychev and V. M. Shalaev, "Magnetic resonance in metal nanoantennas," in *Complex Mediums V: Light and Complexity*, *Proc. SPIE* **5508**, 128 (2004).

35. N. Engheta, A. Salandrino, and A. Alu, "Circuit elements at optical frequencies: Nanoinductors, nanocapacitors, and nanoresistors," *Phys. Rev. Lett.* **95**, 095504 (2005).

36. S. Linden, C. Enkrich, M. Wegener, J. Zhou, T. Koschny, and C. M. Soukoulis, "Magnetic response of metamaterials at 100 terahertz," *Science* **306**, 1351 (2004).

37. D. R. Smith and J. B. Pendry, "Homogenization of metamaterials by field averaging," *J. Opt. Soc. Am. B* **23**, 391 (2006).

38. G. W. Milton, *The Theory of Composites*, Cambridge University Press (2002).

39. V. V. Zhikov, *Homogenization of Differential Operators and Integral Functionals*, Springer-Verlag, New York/Berlin (1994).

40. D. Bergman and D. Stroud, "The physical properties of macroscopically inhomogeneous media," *Solid State Phys.* **46**, 147 (1992).

41. M. I. Stockman, S. V. Faleev, and D. J. Bergman, "Localization versus delocalization of surface plasmons in nanosystems: Can one state have both characteristics?," *Phys. Rev. Lett.* **87**, 167401 (2001).

42. M. I. Stockman, D. J. Bergman, and T. Kobayashi, "Coherent control of nanoscale localization of ultrafast optical excitation in nanosystems," *Phys. Rev. B* **69**, 054202 (2004).

43. D. R. Fredkin and I. D. Mayergoyz, "Resonant behavior of dielectric objects (electrostatic resonance)," *Phys. Rev. Lett.* **91**, 253902 (2003).

44. I. D. Mayergoyz, D. R. Fredkin, and Z. Zhang, "Electrostatic (plasmon) resonances in nanoparticles," *Phys. Rev. B* **72**, 155412 (2005).

45. V. M. Agranovich and Y. N. Gartstein, "Spatial dispersion and negative refraction of light," *Physics – Uspekhi* **49**, 1029 (2006).

46. M. A. Shapiro, G. Shvets, J. R. Sirigiri, and R. J. Temkin, "Spatial dispersion in metamaterials with negative dielectric permittivity and its effect on surface waves," *Opt. Lett.* **31**, 2051 (2006).

47. S. L. Marshall, "A periodic Green function for calculation of Coulomb lattice potentials," *J. Phys.: Condens. Mat.* **12**, 4575 (2000).

Biographies

Gennady Shvets received his Ph.D. in Physics from MIT in 1995 and has been on the Physics faculty at the University of Texas at Austin since 2004. Previously, he held research positions at the Princeton Plasma Physics Laboratory and the Fermi National Accelerator Laboratory, and was on the faculty of the Illinois Institute of Technology. He is one of the pioneers in the emerging field of negative index metamaterials (NIMs). He also pioneered the concept of subwavelength plasmonic crystals and developed their applications to subwavelength NIMs in the optical part of the spectrum. With his collaborators, he developed novel techniques for analyzing optical properties of plasmonic nanostructures, including band-structure calculations of periodic nanostructures and quasi-static calculations of plasmonic resonances.

Dr. Shvets' research interests include nanophotonics, metamaterials with exotic optical properties (especially negative index), near-field optics, laser processing of materials on a nanoscale, and advanced particle accelerators. He is the author or coauthor of more than 90 papers in refereed journals, including *Science, Nature Physics, Nature Materials, Physical Review Letters*, and *Applied Physics Letters*. Dr. Shvets was a Department of Energy Postdoctoral Fellow from 1995-96 and was a recipient of the Presidential Early Career Award for Scientists and Engineers in 2000. His research is supported by DOE, NSF, DARPA, AFOSR, and ARO.

Yaroslav Urzhumov received his Ph.D. in Physics from the University of Texas at Austin in 2007. His research has focused on theoretical design and engineering of optical metamaterials, plasmonic and photonic nanotechnology, subwavelength resolution in optical imaging devices, and exotic electromagnetic phenomena in metamaterials. He is particularly interested in the development of novel and efficient numerical techniques for modeling of nanophotonic components. Dr. Urzhumov served as a Research Assistant for Illinois Institute of Technology and the University of Texas at Austin. Currently, he is an Applications Engineer with COMSOL, an international multiphysics software company.

Chapter 6
Chiral Photonic Media

Ian Hodgkinson and Levi Bourke
Department of Physics, University of Otago, New Zealand

6.1 Introduction

Since the discovery of optical activity[1] in crystalline quartz in 1811, and then in liquids, gases, and liquid crystals, materials that effectively rotate the plane of polarization of a beam of light have continued to be of great interest in many disciplines including biology, chemistry, and physics. Quartz crystals were found to exist in two structural forms that are mirror images of each other. In right-handed (d-rotary) quartz, the silicon and oxygen atoms form a right-handed helix, and an observer looking along the axis toward a source of light sees a clockwise rotation of the vibration direction, whereas the left-handed (l-rotary) version produces anticlockwise rotation. Handedness of optically active substances is a critical issue in biology and chemistry; thus, sugar dextrose (d-glucose) is the most important carbohydrate in human metabolism, and in general the amino acids that form proteins are l-rotary.

In many optical devices such as computer and TV displays, it is necessary to control or modify the reflectance spectrum and the polarization spectrum of a beam of light. An emerging industrial trend is to combine passive dielectric anisotropic films and active liquid crystal layers.[2] Both can act as birefringent slabs with aligned microstructural columns or aligned cigar-shaped molecules, and both can behave as chiral materials with a helical structure and large rotary power.

In this chapter we are concerned in particular with the optical properties of dielectric structurally chiral media. In 1959 the pioneering work of Young and Kowal[3] prepared the way for fabrication of chiral media by vacuum deposition of dielectric material onto a rotating substrate.[4] Such a medium is locally birefringent in form,[5,6] with axes that twist steadily with distance into the medium. The effect on light is small except near the so-called circular Bragg resonance, where light that matches the handedness and pitch of the chiral structure is reflected strongly. Given that the material is handed and periodic, we refer to it as a chiral photonic medium. Other types of chiral media have been fabricated, including planar arrays of metallic and nonmetallic chiral structures,[7,8] but they are not considered here.

In practice, interfacial reflections change the handedness of a fraction of the light reflected from a chiral photonic medium, and then maximum reflectance occurs for slightly elliptical, rather than circular, light. Following such observations we speculate that elliptical Bragg resonators may be designed for any polarization, possibly as a composite material in which a chiral medium is threaded through a birefringent medium with fixed axes.

With such optical applications in mind we survey the properties of several chiral architectures, including standard-chiral material, thickness-modulated chiral material, chiral material threaded through an isotropic medium and through a birefringent medium, and chiral material threaded through a second chiral material with the same or different pitch and handedness. Characterization is initially by polarization response maps and then by reflectance spectra calculated for the polarization of maximum response. Also, in some examples an abrupt central twist is added to establish a photonic defect mode.[9–11] Low threshold lasing has been observed from such dye-doped chiral material.[12,13]

In tutorial fashion we begin with an outline of appropriate mathematical methods for describing propagation in chiral photonic materials. Matrix methods[14] and the MATLAB[15] environment are convenient for computation of reflectances and transmittances, and here we introduce elliptically polarized basis fields for the cover (usually the medium of incidence) and for the substrate, so that copolarized and cross-polarized remittances appropriate to any elliptical polarization may be generated directly. Overall, the current work brings together necessary equations and properties of chiral architectures that were discussed at recent SPIE conferences on complex mediums.[16–19]

6.2 Stratified Anisotropic Media

In this section we give a brief outline of the Berreman matrix method[14] that we use for computing the optical properties of inorganic chiral films. We begin generally by discussing the matrix method for a stack of thin birefringent layers that are assumed to be lossless and list all equations needed for the computation of reflection coefficients expressed in terms of elliptical basis vectors.

In the discussion, we characterize each biaxial layer by three principal refractive indices (n_1, n_2, n_3) and three rotation angles (η, ψ, ξ) that tilt the material axes (1, 2, 3) with respect to the fixed propagation axes (x, y, z). Figure 6.1(a) shows the material axes aligned with the propagation axes ($\eta = \psi = \xi = 0$), and Fig. 6.1(b) illustrates an azimuthal rotation of the material by angle ξ. A general angular position of the material is achieved by rotating in turn by η about the x axis, by ψ about the z axis, and by ξ about the x axis.

We assume that the azimuthal angle of a chiral film is ξ_1 at the cover-film interface and twists uniformly by a total angle (not modulo 2π) of $\xi_2 - \xi_1$ through the film thickness d. The structure is said to be right handed if $h \equiv \mathrm{sign}(\xi_2 - \xi_1) = +1$, isotropic if $h = 0$, and left handed if $h = -1$. Equivalently, the handedness may be defined by the sign of the differential twist $d\xi/dx$. As well, we make use of the basic notation of Ref. 14, for which propagation is in the x-y plane and a 4×4 characteristic matrix $\hat{M}$ is used to transfer total-field vectors between pairs of y-z planes:

$$\begin{bmatrix} E_y \\ H_z \\ E_z \\ H_y \end{bmatrix}_{x=0} = \hat{M} \cdot \begin{bmatrix} E_y \\ H_z \\ E_z \\ H_y \end{bmatrix}_{x=d} . \tag{6.1}$$

6.2.1 Biaxial material

We start with the diagonal matrix that represents the relative permittivity of the dielectric substance in material frame coordinates,

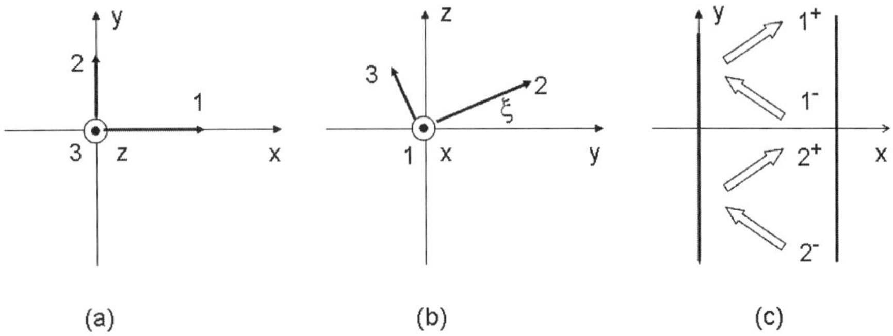

Figure 6.1 Material axes 1, 2, and 3 for a biaxial material shown aligned with the fixed propagation axes x, y, and z in (a) and following an azimuthal rotation ξ of the material in (b). The four basis fields that propagate in such a layer are represented in (c).

$$\hat{\varepsilon}_{123} = \begin{bmatrix} n_1^2 & 0 & 0 \\ 0 & n_2^2 & 0 \\ 0 & 0 & n_3^2 \end{bmatrix}. \tag{6.2}$$

Then rotation matrices

$$\hat{R}_x(\phi) = \begin{bmatrix} 1 & 0 & 0 \\ 0 & \cos\phi & -\sin\phi \\ 0 & \sin\phi & \cos\phi \end{bmatrix} \tag{6.3}$$

for rotation by an angle ϕ about the x axis, and

$$\hat{R}_z(\phi) = \begin{bmatrix} \cos\phi & -\sin\phi & 0 \\ \sin\phi & \cos\phi & 0 \\ 0 & 0 & 1 \end{bmatrix} \tag{6.4}$$

for rotation about the z axis are used to position the dielectric material with respect to the propagation. In this way, the symmetric relative permittivity matrix $\hat{\varepsilon}$ for the propagation plane is determined:

$$\hat{\varepsilon} = \begin{bmatrix} \varepsilon_{xx} & \varepsilon_{xy} & \varepsilon_{xz} \\ \varepsilon_{xy} & \varepsilon_{yy} & \varepsilon_{yz} \\ \varepsilon_{xz} & \varepsilon_{yz} & \varepsilon_{zz} \end{bmatrix} = \hat{R}_x(\xi)\hat{R}_z(\psi)\hat{R}_x(\eta)\hat{\varepsilon}_{123}\hat{R}_x(-\eta)\hat{R}_z(-\psi)\hat{R}_x(-\xi). \tag{6.5}$$

6.2.2 Propagation and basis fields

Propagation in a layer of dielectric material, including all multiple reflections, can be described in terms of a set of four traveling-wave basis fields,[14] as illustrated in Fig. 6.1(c). A + sign is used here to indicate left-to-right propagation and a minus sign indicates right-to-left. Also we have used the labels 1 and 2 for association with orthogonal polarizations.

The four basis fields propagate at different angles θ to the x axis and see a different effective refractive index n. We make use of two propagation parameters that relate to the x and y components of a dimensionless unit vector that is parallel to the propagation vector,

$$\begin{aligned} \alpha &= n\cos\theta &= ns_x \\ \beta &= n\sin\theta &= ns_y. \end{aligned} \tag{6.6}$$

The parameter β is the Snell's law invariant. It has the same value for each basis field and is constant throughout the structure. In most applications the value of β is set by the angle of incidence and refractive index of the cover medium. However, in general the four basis waves have different values of α.

The field components E_x, E_y, E_z, H_x, H_y, and H_z of a basis wave are required to satisfy Maxwell's equations for plane harmonic waves in a biaxial medium. This requirement can be written in column-vector form[14] as

$$\begin{aligned} n\hat{s}\,\mathbf{E} &= z_0\mathbf{H} \\ n\hat{s}\,\mathbf{H} &= -(\hat{\varepsilon}/z_0)\mathbf{E} \end{aligned} \tag{6.7}$$

where

$$n\hat{s} = \begin{bmatrix} 0 & 0 & \beta \\ 0 & 0 & -\alpha \\ -\beta & \alpha & 0 \end{bmatrix}. \tag{6.8}$$

Substitution of Eq. (6.8) into Eq. (6.7) and elimination of the field components E_x and H_x that are normal to interfaces and not required for boundary condition matching, yields the eigenequation

$$\hat{L}\begin{bmatrix} E_y \\ H_z \\ E_z \\ H_y \end{bmatrix} = \alpha \begin{bmatrix} E_y \\ H_z \\ E_z \\ H_y \end{bmatrix}, \tag{6.9}$$

where

$$\hat{L} = \begin{bmatrix} -\dfrac{\beta\varepsilon_{xy}}{\varepsilon_{xx}} & z_0 - \dfrac{z_0\beta^2}{\varepsilon_{xx}} & -\dfrac{\beta\varepsilon_{xz}}{\varepsilon_{xx}} & 0 \\ \dfrac{\varepsilon_{yy}}{z_0} - \dfrac{\varepsilon_{xy}^2}{z_0\varepsilon_{xx}} & -\dfrac{\beta\varepsilon_{xy}}{\varepsilon_{xx}} & \dfrac{\varepsilon_{yz}}{z_0} - \dfrac{\varepsilon_{xy}\varepsilon_{xz}}{z_0\varepsilon_{xx}} & 0 \\ 0 & 0 & 0 & -z_0 \\ -\dfrac{\varepsilon_{yz}}{z_0} + \dfrac{\varepsilon_{xy}\varepsilon_{xz}}{z_0\varepsilon_{xx}} & \dfrac{\beta\varepsilon_{xz}}{\varepsilon_{xx}} & \dfrac{\beta^2}{z_0} + \dfrac{\varepsilon_{xz}^2}{z_0\varepsilon_{xx}} - \dfrac{\varepsilon_{zz}}{z_0} & 0 \end{bmatrix}, \tag{6.10}$$

and $z_0 \approx 377\,\text{ohm}$ is the impedance of free space.

Next we define a field matrix of basis vectors,

$$\hat{F} = \begin{bmatrix} E^+_{y1} & E^-_{y1} & E^+_{y2} & E^-_{y2} \\ H^+_{z1} & H^-_{z1} & H^+_{z2} & H^-_{z2} \\ E^+_{z1} & E^-_{z1} & E^+_{z2} & E^-_{z2} \\ H^+_{y1} & H^-_{y1} & H^+_{y2} & H^-_{y2} \end{bmatrix}, \tag{6.11}$$

and a diagonal matrix that holds the corresponding α's,

$$\hat{\alpha} = \begin{bmatrix} \alpha_1^+ & 0 & 0 & 0 \\ 0 & \alpha_1^- & 0 & 0 \\ 0 & 0 & \alpha_2^+ & 0 \\ 0 & 0 & 0 & \alpha_2^- \end{bmatrix}, \tag{6.12}$$

and generalize Eq. (6.9) so that it applies to the four basis waves,

$$\hat{L}\hat{F} = \hat{F}\hat{\alpha}. \tag{6.13}$$

Then solutions for the four basis fields and the four α's can be obtained by a single MATLAB[15] call,

$$[\hat{F}, \hat{\alpha}] = \text{eig}\, \hat{L}. \tag{6.14}$$

The basis fields are linearly polarized, and in the case of an isotropic medium, commonly the cover or the substrate, the field matrix can be put in the form

$$\hat{F} = \begin{bmatrix} 1 & 1 & 0 & 0 \\ \gamma_p & -\gamma_p & 0 & 0 \\ 0 & 0 & 1 & 1 \\ 0 & 0 & \gamma_s & -\gamma_s \end{bmatrix}, \tag{6.15}$$

where $\gamma_p = n/z_0 \cos\theta$ and $\gamma_s = -n\cos\theta/z_0$. Here p and s are the normal polarization descriptors for transverse magnetic (TM) waves and transverse electric (TE) waves, respectively. Note that we have organized the columns of $\hat{F}$ so that the basis vectors 1^+, 1^- on the left side have the TM polarization, and 2^+, 2^- on the right side have the TE polarization.

In an isotropic medium the p and s waves travel at the same speed and superpose to form elliptically polarized waves. We form a set of elliptically polarized basis vectors as linear sums of the p and s waves (columns of $\hat{F}$) and construct a new matrix of field vectors

$$\hat{F} = \begin{bmatrix} \cos\theta\, \cos\chi & \cos\theta\, \cos\chi & -i\cos\theta\, \sin\chi & i\cos\theta\, \sin\chi \\ \gamma_p\cos\theta\, \cos\chi & -\gamma_p\cos\theta\, \cos\chi & -i\gamma_p\cos\theta\, \sin\chi & -i\gamma_p\cos\theta\, \sin\chi \\ -i\sin\chi & i\sin\chi & \cos\chi & \cos\chi \\ -i\gamma_s\sin\chi & -i\gamma_s\sin\chi & \gamma_s\cos\chi & -\gamma_s\cos\chi \end{bmatrix} \tag{6.16}$$

that is appropriate to an isotropic cover or substrate. The labeling scheme shown in Fig. 6.2 is used to distinguish cover and substrate fields.

The parameter χ in Eq. (6.16) is the auxiliary angle that is sometimes used to specify the shape and handedness of the polarization ellipse of a beam of light.[5] Specifically, $\chi = \pm\tan^{-1}(b/a)$, where b/a is the ratio of the minor and major axes of the ellipse. The light is left handed if $-\pi/4 \le \chi < 0$ and right handed if $0 < \chi \le \pi/4$. Left-circular polarization (LCP) is represented by $\chi = -45$ deg, linear polarization by $\chi = 0$ deg, and right-circular polarization (RCP) by $\chi = 45$ deg. All independent states of elliptical polarization are represented by this restricted range of the auxiliary angle χ, but for our purposes here it is convenient to allow a full 360-deg range for χ in Eq. (6.16). The polarizations of the cover and substrate basis vectors are listed in Table 6.1.

In a typical experimental situation, elliptical light may be generated and analyzed using a linear polarizer oriented at angle χ, and a quarter-wave plate oriented with its fast axis horizontal. In Eq. (6.16) the principal axes of the polarization ellipses are aligned horizontally and vertically, parallel to the y and z axes for normal incidence. Relative azimuthal orientation of the ellipses and the film may be achieved by setting the azimuthal angle ξ of the film. Given that dielectric properties repeat when the film is rotated about its normal axis by half a full turn, we see that all independent values for an outcome such as a reflectance are covered by the ranges $-\pi/2 < \xi \le \pi/2$ and $-\pi/4 \le \chi \le \pi/4$.

6.2.3 Field transfer matrices

The *phase matrix*[14]

$$\hat{A}_d = \begin{bmatrix} \exp[-i\phi_1^+] & 0 & 0 & 0 \\ 0 & \exp[-i\phi_1^-] & 0 & 0 \\ 0 & 0 & \exp[-i\phi_2^+] & 0 \\ 0 & 0 & 0 & \exp[-i\phi_2^-] \end{bmatrix}, \qquad (6.17)$$

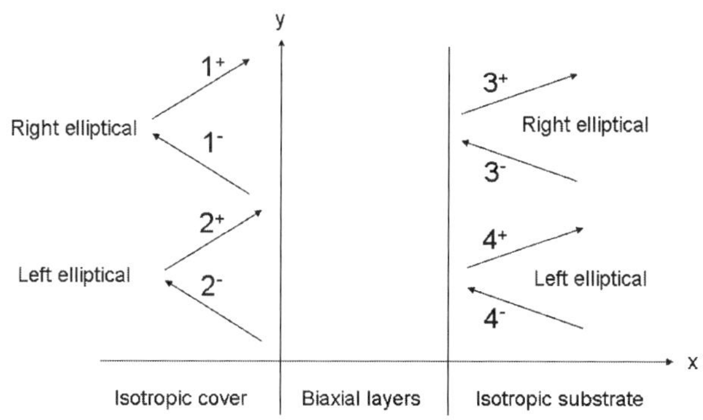

Figure 6.2 Labeling of elliptically polarized basis vectors. The polarization labels apply to a case such as $0 < \chi < \pi/4$.

Table 6.1 Polarization of the cover and substrate basis vectors.

Auxiliary angle χ (deg)	Basis vectors 1^+, 1^- and 3^+, 3^-	Basis vectors 2^+, 2^- and 4^+, 4^-
180	Horizontal linear	Vertical linear
	Left elliptical	Right elliptical
135	Left circular	Right circular
	Left elliptical	Right elliptical
90	Vertical linear	Horizontal Linear
	Right elliptical	Left elliptical
45	Right circular	Left circular
	Right elliptical	Left elliptical
0	Horizontal linear	Vertical linear
	Left elliptical	Right elliptical
−45	Left circular	Right circular
	Left elliptical	Right elliptical
−90	Vertical linear	Horizontal linear
	Right elliptical	Left elliptical
−135	Right circular	Left circular
	Right elliptical	Left elliptical
−180	Horizontal linear	Vertical linear

where

$$\phi_{1,2}^{\pm} = k\alpha_{1,2}^{\pm} d \tag{6.18}$$

transforms traveling wave fields across a layer. The *characteristic matrix*[14]

$$\hat{M} = \hat{F}\hat{A}_d\hat{F}^{-1} \tag{6.19}$$

transforms total fields across the layer; in turn $\hat{F}^{-1}$ transforms total fields to traveling waves on the right-hand side of the layer, $\hat{A}_d$ transforms the traveling waves across the layer, and $\hat{F}_c^{-1}$ transforms back to total fields on the left-hand side. Similarly, the *system matrix*[14]

$$\hat{A} = \hat{F}_c^{-1}\hat{M}\hat{F}_s, \tag{6.20}$$

where $\hat{F}_c$ is the field matrix for the cover medium and $\hat{F}_s$ is the field matrix for the substrate, transforms elliptically polarized traveling waves from a plane just inside the substrate to a plane just inside the cover.

6.2.4 Reflectance and transmittance

Expressions for the reflection and transmission coefficients can be determined from the system matrix.[14] For convenience, the eight amplitude reflection and transmission coefficients and the eight reflectance and transmittance coefficients are organized into a pair of 4×4 arrays,

$$\hat{r} \equiv \begin{bmatrix} r_{11} & r_{12} & t_{13} & t_{14} \\ r_{21} & r_{22} & t_{23} & t_{24} \\ t_{31} & t_{32} & r_{33} & r_{34} \\ t_{41} & t_{42} & r_{43} & r_{44} \end{bmatrix}$$

$$= \begin{bmatrix} 0 & 0 & -A_{11} & -A_{13} \\ 1 & 0 & -A_{21} & -A_{23} \\ 0 & 0 & -A_{31} & -A_{33} \\ 0 & 1 & -A_{41} & -A_{43} \end{bmatrix}^{-1} \begin{bmatrix} -1 & 0 & A_{12} & A_{14} \\ 0 & 0 & A_{22} & A_{24} \\ 0 & -1 & A_{32} & A_{34} \\ 0 & 0 & A_{42} & A_{44} \end{bmatrix}, \tag{6.21}$$

and

$$\hat{R} \equiv \begin{bmatrix} R_{11} & R_{12} & T_{13} & T_{14} \\ R_{21} & R_{22} & T_{23} & T_{24} \\ T_{31} & T_{32} & R_{33} & R_{34} \\ T_{41} & T_{42} & R_{43} & R_{44} \end{bmatrix}$$

$$= \begin{bmatrix} |r_{11}|^2 \frac{|p_1^-|}{|p_1^+|} & |r_{12}|^2 \frac{|p_1^-|}{|p_2^+|} & |t_{13}|^2 \frac{|p_1^-|}{|p_3^-|} & |t_{14}|^2 \frac{|p_1^-|}{|p_4^-|} \\ |r_{21}|^2 \frac{|p_2^-|}{|p_1^+|} & |r_{22}|^2 \frac{|p_2^-|}{|p_2^+|} & |t_{23}|^2 \frac{|p_2^-|}{|p_3^-|} & |t_{24}|^2 \frac{|p_2^-|}{|p_4^-|} \\ |t_{31}|^2 \frac{|p_3^+|}{|p_1^+|} & |t_{32}|^2 \frac{|p_3^+|}{|p_2^+|} & |r_{33}|^2 \frac{|p_3^+|}{|p_3^-|} & |r_{34}|^2 \frac{|p_3^+|}{|p_4^-|} \\ |t_{41}|^2 \frac{|p_4^+|}{|p_1^+|} & |t_{42}|^2 \frac{|p_4^+|}{|p_2^+|} & |r_{43}|^2 \frac{|p_4^+|}{|p_3^-|} & |r_{44}|^2 \frac{|p_4^+|}{|p_4^-|} \end{bmatrix}. \tag{6.22}$$

In Eq. (6.22) the p's are Poynting fluxes associated with the basis vectors in Fig. 6.1(c), and a reflectance coefficient such as R_{21} refers to 1^+ input waves and 2^- output waves.

6.3 Chiral Architectures and Characteristic Matrices

6.3.1 Five chiral architectures

Five chiral architectures formed from thin layers of the same normal-columnar birefringent material with in-plane refractive indices n_2 and n_3 are illustrated in Fig. 6.3. Ω is the dielectric period (half a structural period), and the bars and dots mark the fast axes of the sublayers. The standard-chiral material in 6.3(a) is shown as a discontinuous structure, but in practice with forty or so sublayers per dielectric period it behaves as continuous chiral material.

The remaining four architectures can be described as perturbed versions of the standard-chiral material. In Fig. 6.3(b) (modulated-chiral) the perturbation is a

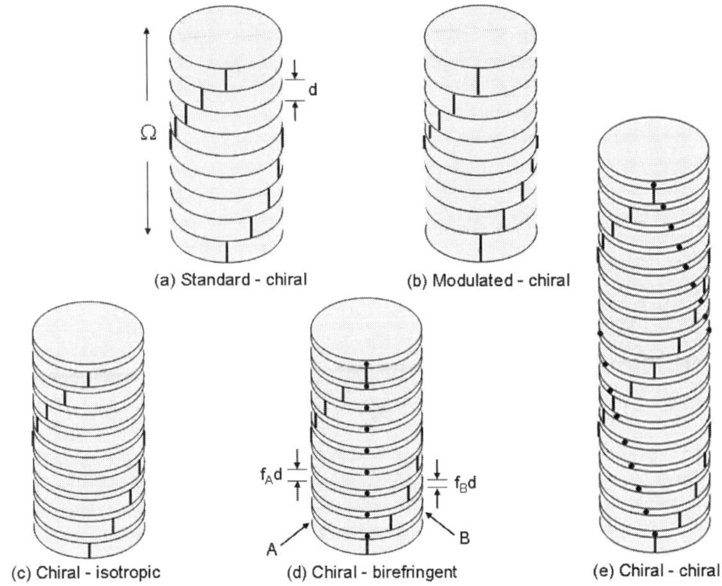

Figure 6.3 Chiral architectures formed from thin layers of the same birefringent material. The bars and dots mark the fast axes of the sublayers.

sinusoidal thickness modulation of the sublayers, such that the sublayer with axial twist ξ has thickness $d\,[1 + a\sin(2\xi + \xi_0)]$, where d is the unperturbed thickness, a is the amplitude of the modulation, and ξ_0 is an angular offset used to position the peak modulation with respect to the start of the dielectric period. In Fig. 6.3(b) the modulated-chiral structure is illustrated with $\xi_0 = \pi/2$.

For architectures (c) chiral-isotropic, (d) chiral-birefringent, and (e) chiral-chiral, the perturbation consists of replacement or realignment of a part of each layer from (a), so that the new architectures appear as a chiral medium A (marked with bars) threaded through a background medium B that may be isotropic (no markers), birefringent (marked with dots), or chiral (marked with dots).

Structures (a), (b), (c), and (d) have the same dielectric pitch Ω, and multiple-period stacks of these materials resonate to the same Bragg wavelength,

$$\lambda_{Br} = 2\,n_{av}\,\Omega, \tag{6.23}$$

where

$$n_{av} = \frac{n_2 + n_3}{2}, \tag{6.24}$$

independent of the material fractions f_A and f_B.

The chiral-chiral architecture (e) is drawn as a right-handed material of pitch Ω_A threaded though a left-handed material of pitch $\Omega_B = 2\Omega_A$. Such a material has two Bragg resonances, with $\lambda_{BrB} = 2\lambda_{BrA}$. In general for such a medium, the Bragg wavelengths may be set via the incremental layer-to-layer twist angles. Thus,

$$\Delta\xi_A = \frac{2\, n_{av}\pi\, d}{\lambda_{BrA}} \tag{6.25}$$

and

$$\Delta\xi_B = \frac{2\, n_{av}\pi\, d}{\lambda_{BrB}}. \tag{6.26}$$

The number of half turns is larger for the smaller resonance, but the strength of the resonances can be balanced via choice of the material fractions.

6.3.2 Matrix for a continuous chiral film

A continuous chiral film formed by the serial bideposition technique[20] has a normal-chiral nanostructure. The characteristic matrix $\hat{M}$ for such a film illuminated at normal incidence can be expressed in terms of an auxiliary matrix

$$M_a = \exp\left\{ |\xi_2 - \xi_1| \begin{bmatrix} 0 & -i\frac{z_0}{n_{av}}G & -h & 0 \\ -i\frac{n_2^2}{z_0\, n_{av}}G & 0 & 0 & h \\ h & 0 & 0 & i\frac{z_0}{n_{av}}G \\ 0 & -h & i\frac{n_3^2}{z_0\, n_{av}}G & 0 \end{bmatrix} \right\} \tag{6.27}$$

and a rotation matrix

$$\hat{S}(\xi) = \begin{bmatrix} \cos\xi & 0 & -\sin\xi & 0 \\ 0 & \cos\xi & 0 & \sin\xi \\ \sin\xi & 0 & \cos\xi & 0 \\ 0 & -\sin\xi & 0 & \cos\xi \end{bmatrix}. \tag{6.28}$$

Thus,

$$M = \hat{S}(\xi_1) \cdot M_a \cdot \hat{S}(-\xi_2)\,. \tag{6.29}$$

In Eq. (6.27), ξ_1, ξ_2 are the cover-side and substrate-side azimuthal angles of the fast axis of the material, and $G = \lambda_{Br}/\lambda$ is a dimensionless parameter.[16]

6.3.3 Matrix for a biaxial film

We note that Eq. (6.29) yields the 4×4 matrix for a normal-columnar biaxial film that is illuminated at normal incidence when $\xi_2 = \xi_1$ (no twisting). However, for a film that is known to be normal-columnar biaxial and illuminated normally, it is more convenient to use the explicit form

$$\hat{M} = \begin{bmatrix} \cos\phi_2 & -i\frac{z_0}{n_2}\sin\phi_2 & 0 & 0 \\ -i\frac{n_2}{z_0}\sin\phi_2 & \cos\phi_2 & 0 & 0 \\ 0 & 0 & \cos\phi_3 & i\frac{z_0}{n_3}\sin\phi_3 \\ 0 & 0 & i\frac{n_3}{z_0}\sin\phi_3 & \cos\phi_3 \end{bmatrix}, \tag{6.30}$$

where $\phi_2 = 2\pi n_2 d/\lambda$ and $\phi_3 = 2\pi n_3 d/\lambda$.

6.3.4 Matrix for an isotropic film

The 4×4 matrix for an isotropic film can be obtained from Eq. (6.29) by substituting $n = n_2 = n_3$ (no birefringence) and $\xi_2 = \xi_1$ (no twisting). However, for a film that is known to be isotropic and illuminated normally, it is more convenient to use

$$\hat{M} = \begin{bmatrix} \cos\phi & -i\frac{z_0}{n}\sin\phi & 0 & 0 \\ -i\frac{n}{z_0}\sin\phi & \cos\phi & 0 & 0 \\ 0 & 0 & \cos\phi & i\frac{z_0}{n}\sin\phi \\ 0 & 0 & i\frac{n}{z_0}\sin\phi & \cos\phi \end{bmatrix}, \qquad (6.31)$$

where $\phi = 2\pi n\, d/\lambda$.

6.3.5 Matrix for a stack of films

The characteristic matrix $\hat{M}$ of a stack of films is the product of the matrices of the individual films.[14] If a stack has N identical half turns then usually it is convenient to compute $\hat{m}$ for one period and raise it to the power of N, i.e., $\hat{M} = \hat{m}^N$. In a second example, for material that is locally normal columnar and light that is incident normally, a spacerless Fabry-Perot filter fabricated as N half turns, but with a central, abrupt change in phase of 90 deg may be represented by $\hat{M} = \hat{m}^{N/2}\,\hat{S}(\pi/2)\,\hat{m}^{N/2}\,\hat{S}(-\pi/2)$.[17]

6.3.6 Matrices for discontinuous and structurally perturbed films

As illustrated in Fig. 6.3, basic chiral films can be nanoengineered as a stack of birefringent sublayers of constant thickness and steadily increasing azimuthal angle. Similarly, complex chiral media based on deposition of mixed architectures (isotropic, birefringent, and chiral) and modulation of a material property can be realized. For such cases the characteristic matrix of the stack is formed as the product of the matrices that represent the sublayers. When light is incident normally, computation time can be reduced for normal-columnar sublayers that are identical apart from an azimuthal rotation by first using Eq. (6.30) to determine the matrix for the sublayer in the standard position, with principal axis-2 parallel to the y-axis, and then applying rotation matrices to allow for subsequent changes in azimuthal angle. When light is incident at an oblique angle, the elements of matrices depend on β and should be calculated using Eq. (6.19); rotation matrices cannot be used in such cases.

6.3.7 Herpin effective birefringent media

One focus of this article is the design of handed Bragg media for elliptically polarized light. Such a medium exhibits maximum resonant reflectance for a given axially propagating elliptical state with the same handedness and pitch as the structure and nearly zero reflectance for the orthogonal state. An in-plane anisotropy is introduced to standard-chiral media to facilitate this goal. In practice this has been achieved by depositing thin birefringent layers A that twist to form a standard-chiral medium so that they thread through a secondary birefringent medium B with fixed axes.

To obtain physical insight into such perturbations, we express the local properties of the resulting medium as the effective principal refractive indices and effective principal axes of a Herpin period $(A/2)B(A/2)$.[14] Some cosmetic rearrangements of the 'as deposited' structures that are depicted in Fig. 6.3 may be required for mathematical equivalence, but in the end the effect on reflectances is negligible. However, with ease of fabrication in mind and to allow the derivation of approximate expressions, we make simplifications at this point. Specifically we consider only normal optical incidence, so that the light doesn't "see" n_{1A} or n_{1B}. Also we assume that the average of the in-plane refractive indices is the same for A and B, to avoid spurious reflections at boundaries. Thus,

$$n_{av} = \frac{n_{2A} + n_{3A}}{2} = \frac{n_{2B} + n_{3B}}{2}. \tag{6.32}$$

Figure 6.4 illustrates the combination of A and B, represented by n_{2A}, n_{3A}, ξ_A and n_{2B}, n_{3B}, and ξ_B, respectively, to form an effective medium represented by n_{2C}, n_{3C}, and ξ_C. The fraction of A in the complex medium is f_A, and the fraction of B is $f_B = 1 - f_A$. In the program that we use, the effective indices are labeled so that $n_{2C} < n_{3C}$. Then for the three media we have in-plane birefringences of

$$\Delta n_A = n_{3A} - n_{2A}, \tag{6.33}$$

$$\Delta n_B = n_{3B} - n_{2B}, \tag{6.34}$$

and

$$\Delta n_C = n_{3C} - n_{2C}. \tag{6.35}$$

Simulations show that the refractive indices n_{2C} and n_{3C} of the effective medium oscillate about n_{av} as a function of the difference in the angles ξ_A and ξ_B. Thus,

$$n_{2C} \approx n_{av} - \frac{1}{2} f_A \,\Delta n_A - \frac{1}{2} f_B \,\Delta n_B \,\cos 2(\xi_A - \xi_B), \tag{6.36}$$

$$n_{3C} \approx n_{av} + \frac{1}{2} f_A \,\Delta n_A + \frac{1}{2} f_B \,\Delta n_B \,\cos 2(\xi_A - \xi_B), \tag{6.37}$$

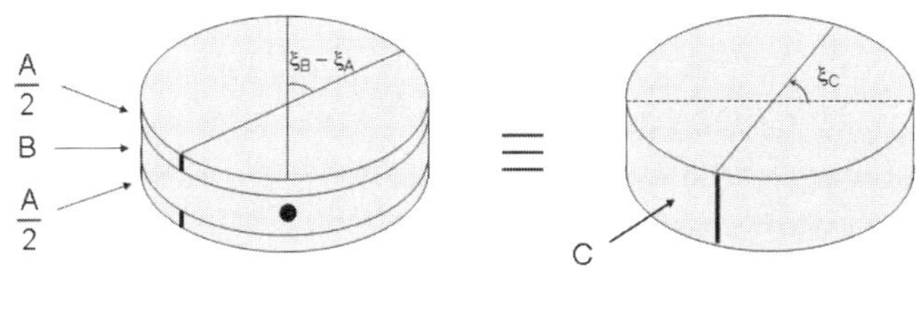

Figure 6.4 Herpin period.

and hence

$$\Delta n_C \approx f_A \Delta n_A + f_B \Delta n_B \cos 2(\xi_A - \xi_B). \tag{6.38}$$

These modulations, together with associated modulations in the effective angle ξ_C, provide in-plane asymmetries that we can exploit to develop handed Bragg materials for elliptical states. However, in the examples that follow we use a more general technique that reverses the computational path for the characteristic matrix of the Herpin period (defined by the parameter sets n_{1A}, n_{2A}, n_{3A}, η_A, ψ_A, and ξ_A and n_{1B}, n_{2B}, n_{3B}, η_B, ψ_B, and ξ_B) and yields the set of effective principal refractive indices and effective rotation angles n_{1C}, n_{2C}, n_{3C}, η_C, ψ_C, and ξ_C for the local effective medium.[14] In each case independent calculations made using the discontinuous film model with 40 or so sublayers per chiral period gave essentially the same remittances.

6.4 Reflectance Spectra and Polarization Response Maps

6.4.1 Film parameters

The same parameter values are used throughout for the simulations. Apart from noted exceptions, we use $n_c = 1$ for the refractive index of the cover, $n_s = 1.5$ for the substrate, and $n_1 = 2.00$, $n_2 = 1.75$, and $n_3 = 1.85$ for the principal refractive indices of the birefringent material. Axis 1 is perpendicular to the substrate, and hence axis 2 (the fast axis) and axis 3 (the slow axis) are in the plane of the substrate. At the start and finish of a chiral film with an integral number N half turns, axis 2 is horizontal, aligned with the y axis. The light is assumed to be incident normally, so that $\theta = 0$ and $\beta = 0$. It follows that the local in-plane birefringence seen by the light is $\Delta n = 0.10$, and the average refractive index is $n_{av} = 1.80$. Also with the conditions as listed, rotation matrices can be employed to account for azimuthal rotations of a thin birefringent layer, part of a chiral coating, or an entire chiral coating.

6.4.2 Standard-chiral media

In this section we consider the question "Is the standard-chiral Bragg mirror formed as several periods of the structure in Fig. 6.3(a) an ideal structure for circularly polarized (CP) light?" But first we need to ask, what do we mean by ideal? Several properties, not all independent, come to mind. We might reasonably expect that an ideal right-handed chiral mirror would exhibit at the Bragg resonant wavelength (i) strong copolarized reflectance R_{11}, (ii) zero copolarized reflectance R_{22}, (iii) zero cross-polarized reflectances R_{12} and R_{21}, and the reflectance R_{11} should (iv) increase toward unity as the number of half turns is increased, and (v) be invariant to rotation of the mirror about the x axis.

The polarizing properties of a chiral mirror at the Bragg wavelength λ_{Br} can be characterized by a response map of copolarized reflectance versus ξ and χ.[18] Such a map is shown for a standard-chiral mirror with $N = 30$ in Fig. 6.5. Notice

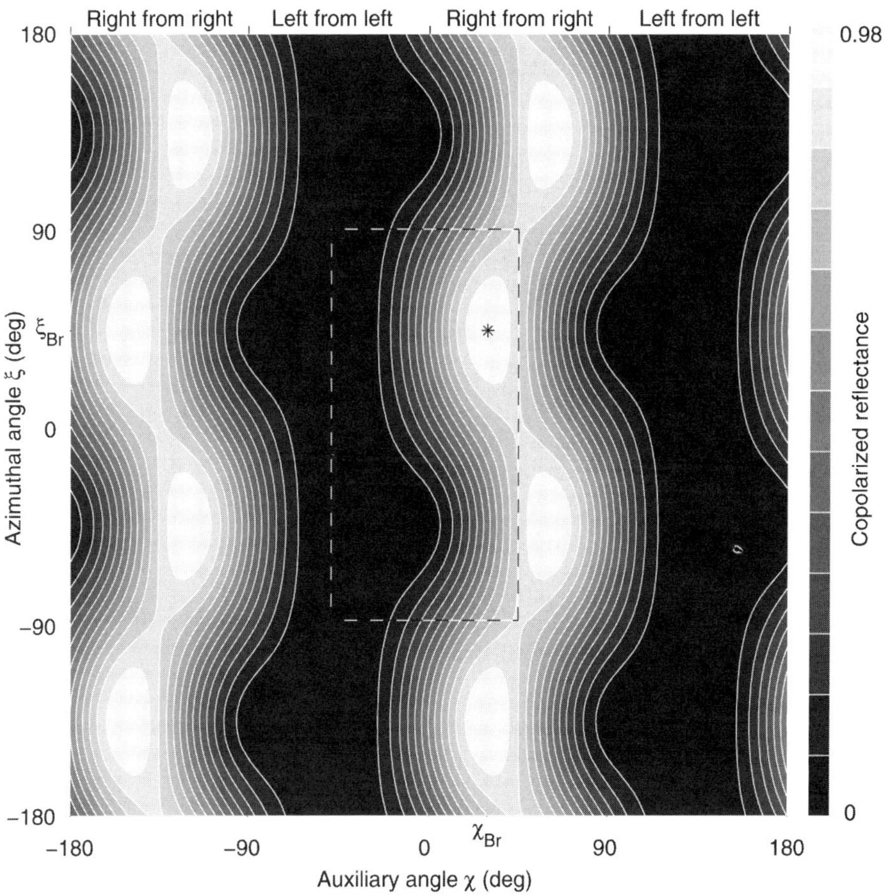

Figure 6.5 Response map of a right-handed standard-chiral reflector on a glass substrate.

that the map is plotted for the full range of ξ and χ values, and that the rectangular area defined by the broken lines contains all independent values of the copolarized reflectance.

First, we note that R_{11} is constant along the line $\chi = 45$ deg on Fig. 6.5. Thus, $R_{11}^{\xi,\frac{\pi}{4}}$ is independent of ξ, and hence condition (v) is satisfied. In turn, this invariance means that a characteristic spectral response to CP light can be calculated using any value of ξ. The spectra shown in Fig. 6.6(a) for the unmatched standard-chiral coating on glass were computed using $\chi = \pi/4$ (circularly polarized basis vectors) and $\xi = 0$; any other value of ξ would give the same result. The spectra show that requirements (i) and (ii) for an ideal chiral reflector are satisfied, but the cross-polarized reflectances are significant (and equal). Additional calculations show that R_{11} tends to a value that is considerably smaller than 1 as N is increased from small values. Thus, the requirements (iii) and (iv) are not satisfied by the standard-chiral reflector on glass.

The response map in Fig. 6.5 contains eight peaks of the same height, onc of which is located within the highlighted rectangle and labeled by a * marker. We use the coordinates of this peak to define the polarization of the Bragg resonance. Thus, the standard-chiral film on glass behaves as an elliptical Bragg resonator characterized by three parameters: the Bragg resonant wavelength λ_{Br}, the azimuthal angle ξ_{Br} that positions the sample with respect to the ellipse, and the auxiliary angle χ_{Br} that defines the ellipticity.

We have found that for both standard and perturbed chiral resonators, ξ_{Br} and χ_{Br} can be computed from elements of the amplitude reflection matrix $\hat{r}^{0,\frac{\pi}{4}}$.[19] Summarizing, we have

$$\lambda_{Br} = 2\, n_{av} \Omega, \tag{6.39}$$

$$\xi_{Br} = \frac{1}{2} \arctan\left[\frac{\Re\{(r_{11}^{0,\frac{\pi}{4}} + r_{22}^{0,\frac{\pi}{4}})/(r_{12}^{0,\frac{\pi}{4}} + r_{21}^{0,\frac{\pi}{4}})\}}{\Re\{i\,(r_{11}^{0,\frac{\pi}{4}} - r_{22}^{0,\frac{\pi}{4}})/(r_{12}^{0,\frac{\pi}{4}} + r_{21}^{0,\frac{\pi}{4}})\}}\right], \tag{6.40}$$

and

$$\chi_{Br} = \frac{1}{2} \arctan\left[\frac{r_{11}^{0,\frac{\pi}{4}} e^{-2i\xi_{Br}} + r_{22}^{0,\frac{\pi}{4}} e^{2i\xi_{Br}}}{i\,(r_{12}^{0,\frac{\pi}{4}} + r_{21}^{0,\frac{\pi}{4}})}\right]. \tag{6.41}$$

Figure 6.6(b) shows spectra calculated for the unmatched standard-chiral coating on glass, similar to 6.6(a), but computed for the elliptically polarized basis vectors of the Bragg resonance.

6.4.3 Remittance at the Bragg wavelength

An audit of circularly polarized light remitted at the resonant wavelength from a highly reflecting standard-chiral mirror yields simple physical explanations (Fig. 6.7)

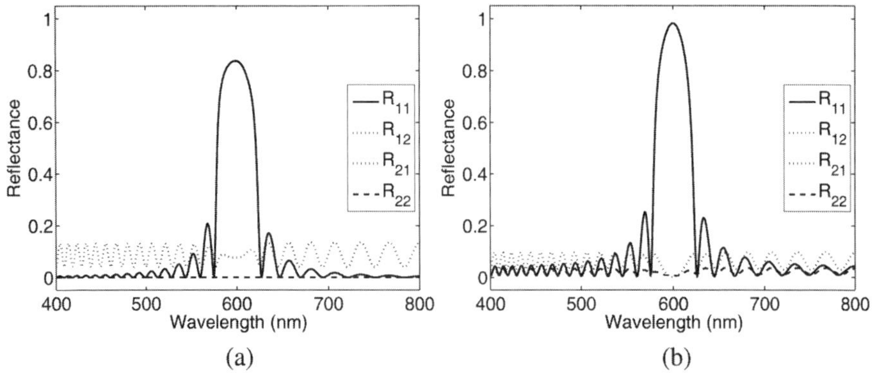

(a) (b)

Figure 6.6 (a) Reflectance of an unmatched standard-chiral coating on a glass substrate computed with circularly polarized basis vectors. (b) Reflectance of the same coating but computed with the elliptically polarized basis vectors of the Bragg resonance.

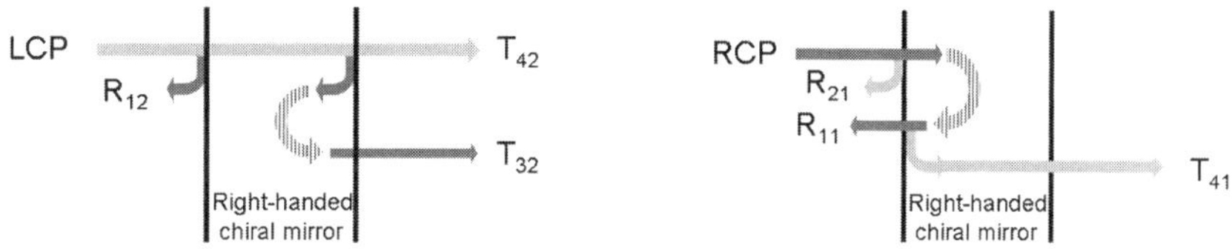

Figure 6.7 Remittance from a right-handed chiral mirror illuminated with circularly polarized light at normal incidence, showing the origin of the copolarized and cross-polarized reflectances and transmittances.

and approximate expressions [Eqs. (6.42)–(6.49)] for the polarized reflectances and transmittances in terms of Fresnel reflection and transmission coefficients for the cover-film and film-substrate interfaces.

$$R_{11} \approx \left[\frac{4n_c n_{av}}{(n_c+n_{av})^2}\right]^2 \tag{6.42}$$

$$R_{12} \approx \left(\frac{n_c-n_{av}}{n_c+n_{av}}\right)^2 \tag{6.43}$$

$$R_{21} \approx \left(\frac{n_c-n_{av}}{n_c+n_{av}}\right)^2 \tag{6.44}$$

$$R_{22} \approx 0 \tag{6.45}$$

$$T_{31} \approx 0 \tag{6.46}$$

$$T_{32} \approx \frac{4n_c n_{av}}{(n_c+n_{av})^2}\,\frac{(n_{av}-n_s)^2}{(n_{av}+n_s)^2}\,\frac{4n_{av}n_s}{(n_{av}+n_s)^2} \tag{6.47}$$

$$T_{41} \approx \frac{4n_c n_{av}}{(n_c+n_{av})^2}\,\frac{(n_c-n_{av})^2}{(n_c+n_{av})^2}\,\frac{4n_{av}n_s}{(n_{av}+n_s)^2} \tag{6.48}$$

$$T_{42} \approx \frac{4n_c n_{av}}{(n_c+n_{av})^2}\,\frac{4n_{av}n_s}{(n_{av}+n_s)^2} \tag{6.49}$$

The above approximations satisfy the principle of conservation of energy, *viz.*,

$$R_{11}+R_{21}+T_{31}+T_{41}=1 \tag{6.50}$$

for right-handed incident light, and

$$R_{12}+R_{22}+T_{32}+T_{42}=1 \tag{6.51}$$

for left-handed incident light.

Thus, the origins of the cross-polarized reflectances R_{12} and R_{21} from a highly reflecting chiral mirror are the handedness-reversing Fresnel reflections at the cover-film interface. In general, index matching a chiral coating to its bounding media is an effective strategy for eliminating cross-polarized reflections. Another strategy is to apply antireflection coatings at the interfaces. In both cases the energy

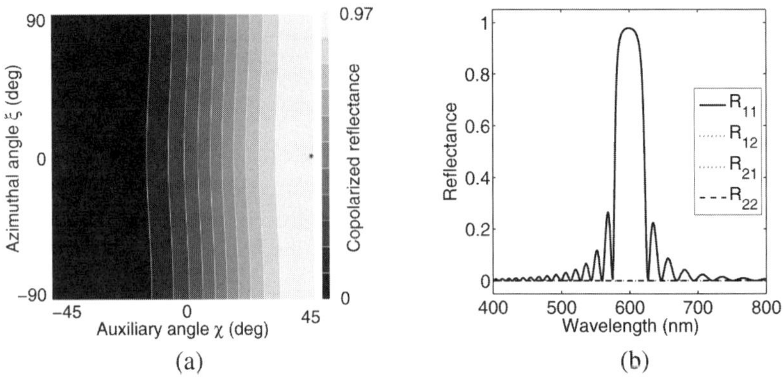

Figure 6.8 (a) Response map for an index-matched, standard-chiral reflector. (b) Reflectance spectra for circularly polarized light.

saved adds in part to the strength of R_{11} in the case of a right-handed reflector. Figure 6.8(a) shows the response map computed for the index-matched standard-chiral reflector, and Fig. 6.8(b) shows reflectance spectra computed for circularly polarized light.

6.4.4 Modulated-chiral media

Figures 6.9 and 6.10 use examples with $\xi_0 = \pi/2$ and $N = 20, 30, 40$ to illustrate the optical performance of thickness-modulated chiral reflectors. The left side of Fig. 6.9 shows that the effect of the modulation on ξ_{Br} is small. In the right side the 30-deg value of χ_{Br} for zero modulation amplitude represents the contribution of the cover-coating index mismatch to the ellipticity. As the amplitude of the modulation is increased from 0 to 0.5, the medium effect that is generated reduces χ_{Br} by a further 10 deg or so, and hence provides a modest increase in the ellipticity. Additional calculations show that a low level of the copolarized reflectance R_{22} is maintained as the amplitude of the modulation is increased and that the cross-polarized reflectance remains negligible, confirming that the modulated-chiral coatings on glass operate as elliptically polarized Bragg reflectors.

Reflectances computed for a right-handed, index-matched, modulated-chiral filter with parameter values $a = 0.5$ and $N = 30$ are shown in Fig. 6.10(a). This spectrum shows a primary resonance with parameter values $\lambda_{Br} = 800\,\text{nm}$, $\xi_{Br} = 43\,\text{deg}$, and $\chi_{Br} = 32\,\text{deg}$, and a secondary resonance with $\lambda_{Br} = 400\,\text{nm}$, $\xi_{Br} = 40\,\text{deg}$, and $\chi_{Br} = 11\,\text{deg}$. Secondary resonances do not occur with a standard-chiral reflector, but have been predicted and observed for modulated-chiral films.[19,21] Basically, they relate to harmonics of the ξ versus x profile. When an abrupt twist of 90 deg is applied at the center of the filter, defect modes appear in both the principal peak and the secondary peak, as shown in Fig. 6.10(b).[19]

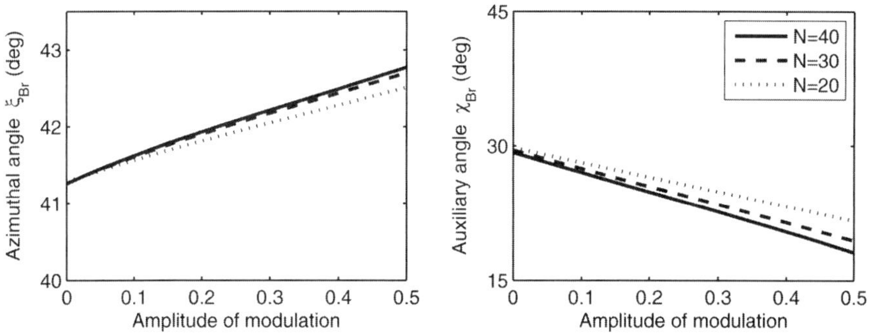

Figure 6.9 Dependence of the parameters ξ_{Br} and χ_{Br} of the elliptical Bragg resonance on amplitude for a modulated-chiral reflector.

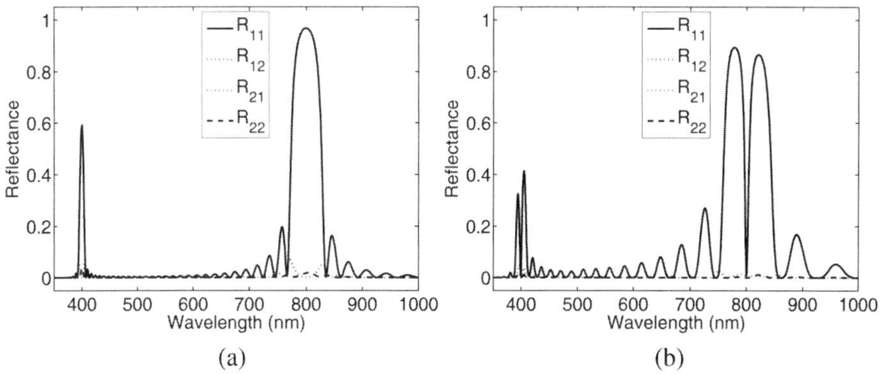

Figure 6.10 (a) Reflectance from an index-matched, modulated-chiral reflector. (b) The same architecture but with defect modes caused by an abrupt 90-deg central twist.

6.4.5 Chiral-isotropic media

With $\Delta n_B = 0$, the approximation Eq. (6.38) reduces to $n_c = f_A \Delta n_A$. Thus, a chiral medium A threaded through an isotropic medium B is equivalent to a chiral medium C with reduced local linear birefringence.[18] C has the same circular Bragg wavelength as A, but the bandwidth of the reflectance versus wavelength peak is smaller, and more half turns are needed for the same peak reflectance.

6.4.6 Chiral-birefringent media

Next we consider chiral media threaded through birefringent media.[18, 19] In the examples that we present here, the background medium B has the same birefringence as the chiral medium A ($\Delta n_A = \Delta n_B = 0.1$), but whereas A twists regularly with thickness, the axes of B remain fixed with $\xi_B = 0$.

The left and right sides of Fig. 6.11 show the effective birefringence and the effective azimuthal angle of the Herpin equivalent medium for the case in which A is right handed and $f_B = 0.25$.

The response map for 30 half-turns of medium C illuminated in turn by all elliptical polarizations is shown in Fig. 6.12(a). The elliptical polarization that gives maximum response is marked with the * symbol, and the reflectance spectra computed for the polarization of maximum response are shown in Fig. 6.12(b). Clearly, medium C behaves as a right-handed Bragg medium for elliptically polarized light.

Figure 6.13 illustrates the optical performance of a range of chiral-birefringent reflectors with $N = 20, 30, 40$ and as a function of the fraction f_B of aligned birefringent material. The left side of Fig. 6.13 shows that ξ_{Br} is nearly constant and close to 45 deg. As for the modulated-chiral reflector, the 30-deg value of χ_{Br} shown in the right side of Fig. 6.13 is due to mismatch of the cover and coating

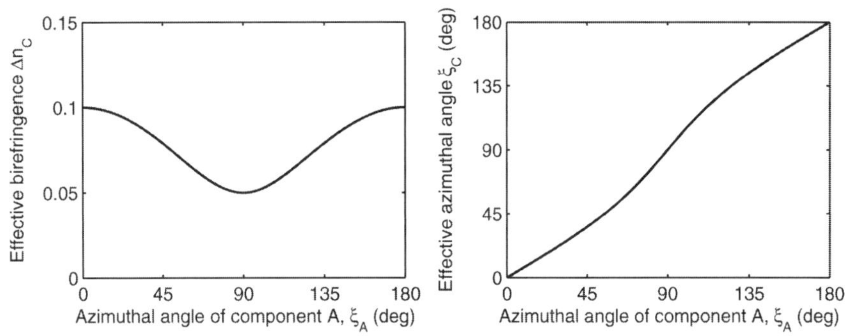

Figure 6.11 Effective birefringence and effective azimuthal angle of a Herpin medium equivalent to a chiral-birefringent medium.

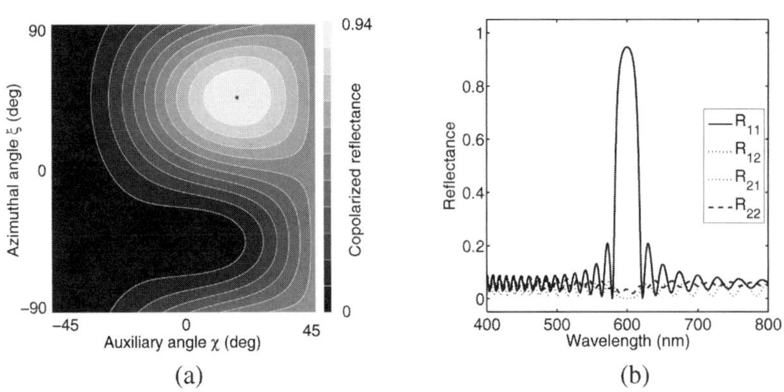

Figure 6.12 (a) Response map for a chiral-birefringent reflector. (b) Spectra computed for the elliptical polarization that gives maximum reflectance.

refractive indices. The same part of the figure shows that χ_{Br} decreases from 30 deg to zero as f_B is increased from 0 to 0.5. Essentially, this demonstrates the possibility of designing Bragg reflectors for any polarization, from circular through elliptical to linear. Further calculations, however, show that the wide range of χ_{Br} comes at the expense of reduced coreflectance for a given number of half-turns N, due to the 'dilution' of the twisting birefringence by the aligned birefringence. As with the standard-chiral coating on glass and the modulated-chiral coating, the cross-polarized reflectance remains at a negligible level, particularly when index matching is used.

Typical sensitivities of the parameters of the Bragg resonance to change in the angle of incidence are shown in Fig. 6.14 for the reflector with $N = 30$. The blue shift of λ_{Br} with increasing θ is a well-known phenomenon for interference filters. Only minor changes occurred in the polarization parameters ξ_{Br} and χ_{Br}.

6.4.7 Chiral-chiral media

Finally, we consider two index-matched architectures in which pairs of chiral structures each with $N \approx 50$ thread through each other.[18,22,23] In the first case, which

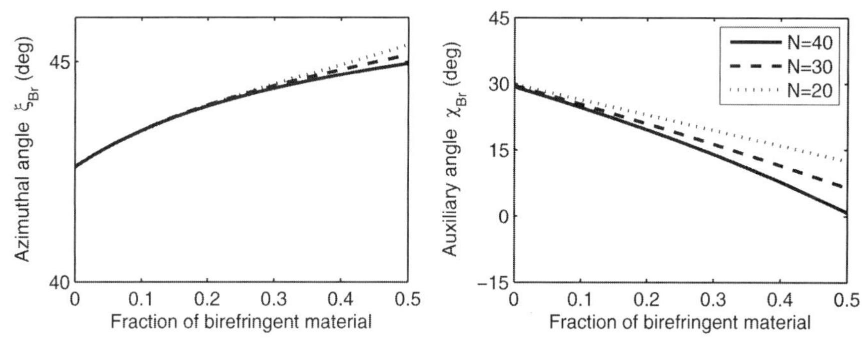

Figure 6.13 Dependence of the parameters ξ_{Br} and χ_{Br} of the elliptical Bragg resonance of a chiral-birefringent reflector on material fraction.

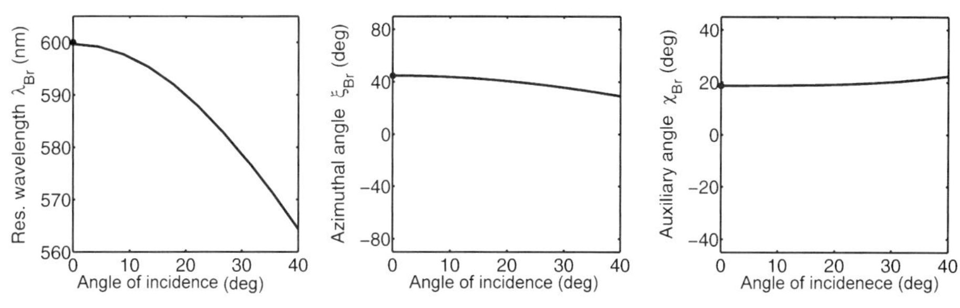

Figure 6.14 Dependence of the parameters λ_{Br}, ξ_{Br} and χ_{Br} of the elliptical Bragg resonance of a chiral-birefringent reflector on angle of incidence.

is illustrated in Figs. 6.15 and 6.16, the chiral structures are both right handed. Figure 6.15(a) shows a material designed to give balanced right-circular Bragg resonances at $\lambda_{BrA} = 500$ nm and $\lambda_{BrB} = 700$ nm. Equations (6.25) and (6.26) were used here, and balancing was achieved by setting $f_B = 0.58$. A central 90-deg twist in one chiral structure, A or B, establishes a defect mode in the corresponding resonance, and twists in both A and B cause defect modes in both resonances, as shown in Figure 6.15(b).

Figures 6.16(a) and 6.16(b) are similar to Figs. 6.15(a) and 6.15(b), except that the Bragg wavelengths are closer together, now at 595 nm and 605 nm. At this separation the light reflected from the two structures interferes and diffuses the shapes of the resonances and defect modes.

Similarly, Figs. 6.17(a) and 6.17(b) and Figs. 6.18(a) and 6.18(b) show principal resonances and defect modes for a material in which right-handed and left-

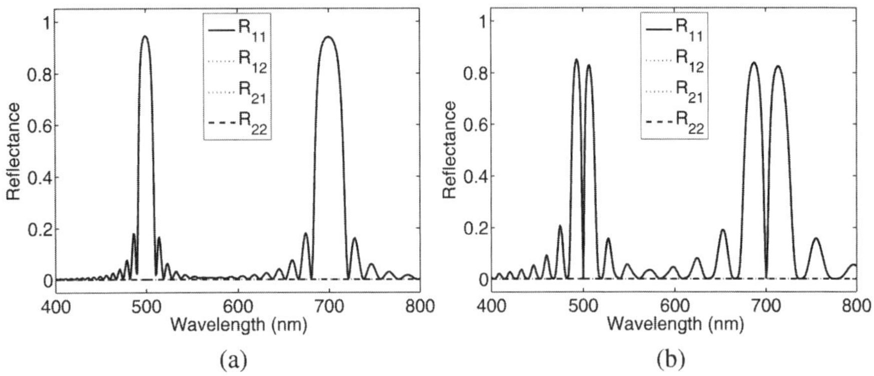

Figure 6.15 (a) Reflectance from a chiral-chiral material in which two right-handed structures with Bragg resonances at 500 nm and 700 nm thread through each other. (b) The same architecture but with defect modes caused by central 90-deg twists.

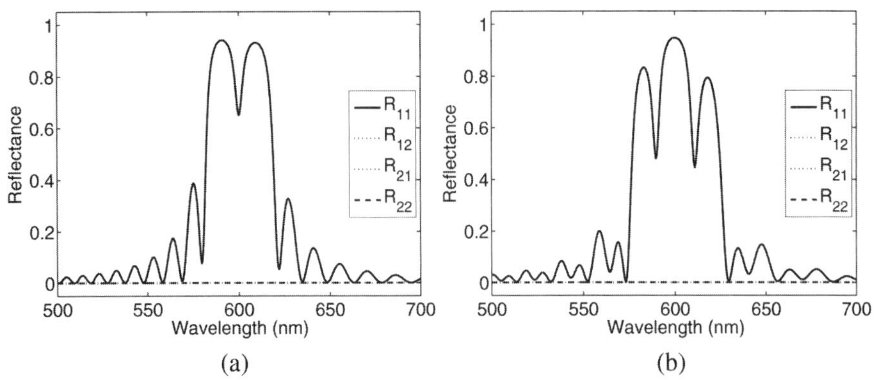

Figure 6.16 (a) Reflectance from an architecture similar to that described in Fig. 6.15(a) but with Bragg resonances at 595 nm and 605 nm. (b) The same architecture but with central 90-deg twist defects.

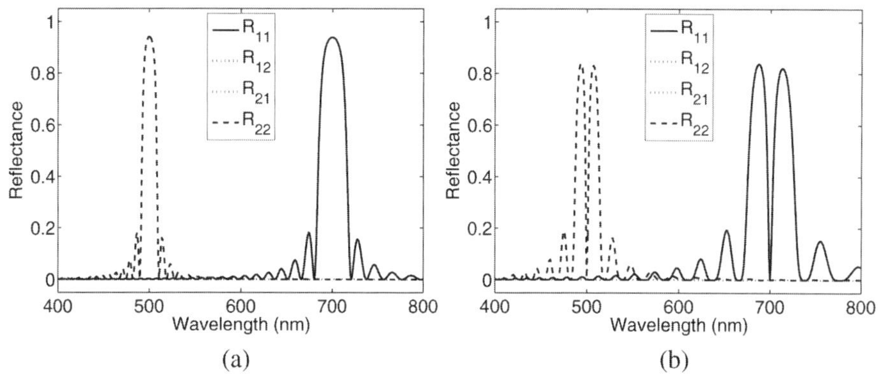

Figure 6.17 (a) Reflectance from a chiral-chiral material in which right-handed and left-handed structures with Bragg resonances at 500 nm and 700 nm thread through each other. (b) The same architecture but with defect modes caused by central 90-deg twists.

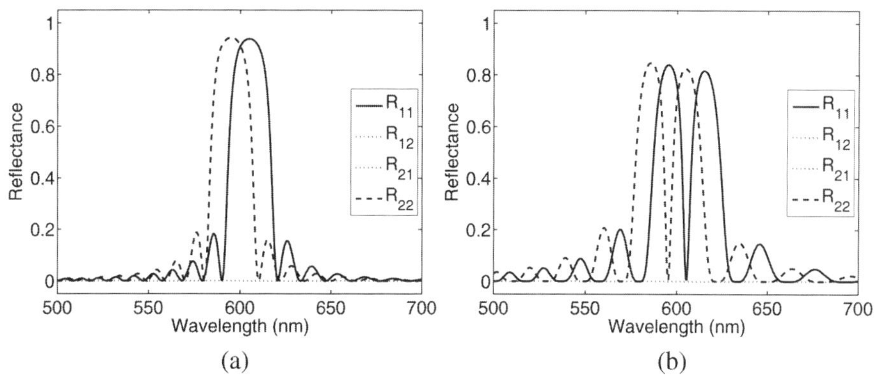

Figure 6.18 (a) Reflectance from an architecture similar to that described in Fig.6.17(a) but with Bragg resonances at 595 nm and 605 nm. (b) The same architecture but with central 90-deg twist defects.

handed structures thread through each other. The main difference is that the right-circular and left-circular waves do not interact, as a consequence of their mutual orthogonality. Thus, in Fig. 6.18(a) and 6.18(b) the principal resonances and the defect modes overlap without interfering. A potential application is a narrowband optical filter with overlapping passbands.

6.5 Summary

We have listed basic equations for computing the optical properties of chiral photonic media. Five chiral architectures formed by layers of the same normal-columnar birefringent media are considered in detail. The factors that cause the standard chiral medium to reflect slightly elliptical rather than circular light at the Bragg resonance are explained, and we show that perturbations to the standard architecture can enhance the ellipticity. Finally, we have demonstrated chiral architectures that support two or more Bragg resonances with the same or opposite handedness and at the same or different wavelengths; experimental realization and characterization of these architectures is described elsewhere.[22,23]

Acknowledgments

The authors acknowledge financial support from the New Zealand Foundation for Research, Science and Technology (FoRST) and from the MacDiarmid Institute for Advanced Materials and Nanotechnology.

References

1. E. Hecht, *Optics*, Addison Wesley, San Francisco (2002).
2. M. K. Tilsch, K. Hendrix, K. Tan, D. Shemo, R. Bradley, R. Erz, and J. Buth, "Production scale deposition of multilayer film structures for birefringent optical components," presented at Metallurgical Coatings and Thin Films, 23–27 April 2007, San Diego, USA, Paper C-13.
3. N. O. Young and J. Kowal, "Optically active fluorite films," *Nature* **183**, 104–105 (1959).
4. K. Robbie, M. J. Brett, and A. Lakhtakia, "Chiral sculptured thin films," *Nature* **384**, 616 (1996).
5. M. Born and E. Wolf, *Principles of Optics*, Pergamon Press, New York (1980).
6. I. J. Hodgkinson and Q. H. Wu, "Inorganic chiral optical materials," *Adv. Mat.* **13**, 889–897 (2001).
7. A. Potts, A. Papakostas, D. M. Bagnall, and N. I. Zheludev, "Planar chiral meta-materials for optical applications," *Microelectron. Eng.* **73–74**, 367–371 (2004).
8. M. W Horn, M. D. Pickett, R. Messier, and A. Lakhtakia, "Blending of nanoscale and microscale in uniform large-area sculptured thin-film architectures," *Nanotechnology* **15**, 303–310 (2004).
9. Y-C. Yang, C-S. Kee, J-E. Kim, and H. Y. Park, "Photonic defect modes of cholesteric liquid crystals," *Phys. Rev. E* **60**, 6852–6854 (1999).
10. I. J. Hodgkinson, Q. H. Wu, K. E. Thorn, A. Lakhtakia, and M. W. McCall, "Spacerless circular-polarization spectral-hole filters using chiral thin films: theory and experiment," *Opt. Commun.* **184**, 57–66 (2000).
11. V. I. Kopp and A. Z. Genack, "Twist defect in chiral photonic structures," *Phys. Rev. Lett.* **89**, 033901 (2002).

12. V. I. Kopp, B. Fan, H. K. M. Vithana, and A. Z. Genack, "Low-threshold lasing at the edge of a photonic stop band in cholesteric liquid crystals," *Opt. Lett.* **23**, 1707–1709 (1998).

13. J. Schmidtke, W. Stille, and H. Finkelmann, "Defect mode emission of a dye doped cholesteric polymer network," *Phys. Rev. Lett.* **90**, 83902 (2003).

14. I. J. Hodgkinson and Q. H. Wu, *Birefringent Thin Films and Polarizing Elements*, World Scientific, Singapore (1998).

15. MATLAB is a trade mark of The MathWorks, Inc., 24 Pine Park Way, Natick, MA 01760, USA.

16. I. J. Hodgkinson, Q. H. Wu, and L. De Silva, "Layered and continuous handed materials for chiral optics," in *Complex Mediums III: Beyond Linear Isotropic Dielectrics*, A. Lakhtakia, G. Dewar, and M. W. McCall, Eds., *Proc. SPIE* **4806**, 118–128 (2002).

17. I. J. Hodgkinson, Q. H. Wu, L. De Silva, and M. D. Arnold, "Chiral supercavities," in *Complex Mediums IV: Beyond Linear Isotropic Dielectrics*, M. W. McCall and G. Dewar, Eds., *Proc. SPIE* **5218**, 40–50 (2003).

18. I. J. Hodgkinson, Q. H. Wu, L. De Silva, and M. D. Arnold, "Threaded-chiral media: reflectors for elliptically polarized light," in *Complex Mediums V: Light and Complexity*, M. W. McCall and G. Dewar, Eds., *Proc. SPIE* **5508**, 47–56 (2004).

19. I. J. Hodgkinson and L. De Silva, "Sculptured thin film handed mirrors," in *Complex Mediums VI: Light and Complexity*, M. W. McCall, G. Dewar, and M. A. Noginov, Eds., *Proc. SPIE* **5924**, 147–158 (2005).

20. I. J. Hodgkinson and Q. H. Wu, "Serial bideposition of anisotropic thin films with enhanced linear birefringence," *Appl. Opt.* **38**, 3621–3625 (1999).

21. M. W. McCall, "Combination morphologies in sculptured thin films," in *Complex Mediums V: Light and Complexity*, M. W. McCall and G. Dewar, Eds., *Proc. SPIE* **5508**, 77–84 (2004).

22. L. Bourke, "Threaded Chiral Media and Flake," MSc Thesis, University of Otago, New Zealand (2008).

23. L. Bourke, I. J. Hodgkinson, L. De Silva, and J. Leader, "Chiral photonic film and flake," *Opt. Exp.* **16**, 16889–16894 (2008).

Biographies

Ian Hodgkinson is an emeritus professor at the University of Otago and a researcher in the MacDiarmid Institute for Advanced Materials and Nanotechnology. He is the author of a book on birefringent films and more than 100 papers in optics journals. His current research interests include natural and nanoengineered chiral films.

Levi Bourke is a graduate student at the University of Otago. His MSc research dissertation introduces chiral media with multiple Bragg resonances and describes the realization and characterization of such media in the forms of film and flake.

Chapter 7
Optical Vortices

Kevin O'Holleran
Department of Physics & Astronomy, University of Glasgow, UK

Mark R. Dennis
H. H. Wills Physics Laboratory, University of Bristol, UK

Miles J. Padgett
Department of Physics & Astronomy, University of Glasgow, UK

7.1 Introduction

To fully describe even a monochromatic light field, one needs to specify its magnitude, phase, and polarization state, all as a function of spatial position. The requirement to specify the polarization is commensurate with the light being a vector field. However, in many optical systems and associated phenomena, the polarization is uniform throughout the field, and it is sufficient to consider the light as a scalar field with only its magnitude and phase being spatially dependent. In such systems, if one examines any cross-section through the complicated light field, one aspect of the light's phase distribution is immediately apparent. Throughout the cross-section there are positions of phase singularity where all phases meet, and around which the phase changes by 2π, in either a clockwise or anticlockwise direction.

In the area immediately surrounding the phase singularity, the 2π phase variation gives rise to a helicity in the phase front with a pitch equal to the optical wavelength λ. At all positions in an isotropic medium, the Poynting vector is perpendicular to the phase fronts and hence, a helical phase front results in azimuthal components of the Poynting vector around the singularity.[1] Since the Poynting vector indicates the direction of both the energy and momentum flow, the helical phase fronts have associated with them an azimuthal energy and momentum flow, leading to this type of phase singularity also being called an "optical vortex." Similarly, the azimuthal momentum flow gives an angular momentum around the beam axis. Indeed since 1992, it has been clearly understood that beams possessing perfect helical phase fronts with a scalar field amplitude $u(r, \theta)$, described by

$$u(r, \theta) = A(r)e^{i\ell\theta}, \tag{7.1}$$

carry an orbital angular momentum of $\ell\hbar$ per photon,[2] where θ is the azimuthal angle around the beam axis. Most typically, helically phased beams with finite aperture are described by Laguerre-Gaussian polynomials. Figure 7.1 shows examples of the phase fronts, intensity, and phase cross-section of different Laguerre-

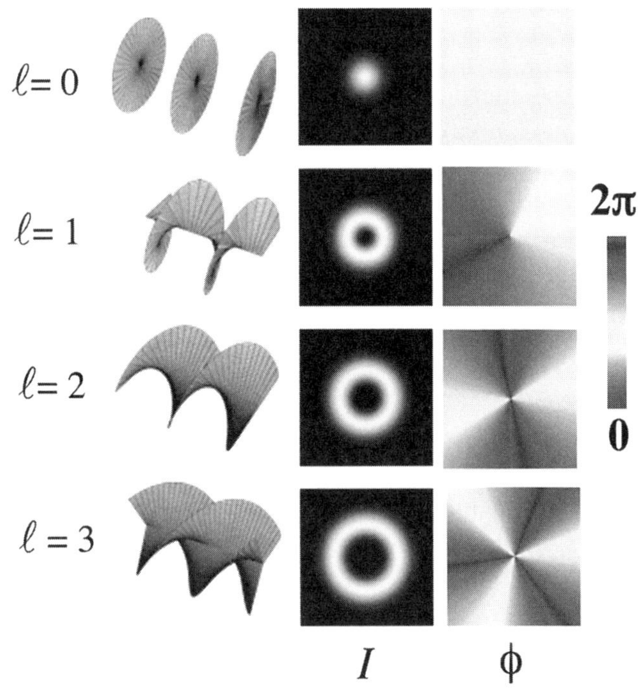

Figure 7.1 Examples of Laguerre-Gaussian beams illustrating the form of the helical phase fronts with various values of ℓ. The left-hand column shows the surfaces of constant phase, the middle column the intensity cross-sections, and the right-hand column the phase cross-sections.

Gaussian modes with various values of ℓ. Since the mid-1990s, their orbital angular momentum has been a topic of both significant theoretical and experimental study.[3]

Independently of their momentum properties, the study of the optical vortices themselves has been of great interest since a paper by Nye and Berry introduced the concept into wave theory.[4] Although the positions of optical vortices within an arbitrary beam cross-section are points, it is essential to appreciate that the light field is fully three-dimensional (3D) and that on moving from one cross-section to the next, the positions of the vortices map out lines. These lines of phase singularity are lines of complete darkness embedded within the 3D light field.

Experimentally, the structure and hence topology of the vortex lines can in principle be determined by imaging either the intensity or phase distribution in a series of neighboring beam cross-sections. Linking these vortex points in successive cross-sections allows the complete 3D structure to be deduced. As the viewing cross-section is moved through the light field, one frequently observes that on moving from one plane to the next, two oppositely signed vortices may be spontaneously created or may annihilate each other. However, the vortex lines are fixed within the beam and do not move. Hence, the use of temporally sequenced terms is misleading. In two dimensions, the creation or annihilation of a pair of oppositely signed vortices is manifestation of a 3D hairpin in a single vortex line.[5] If two vortices are seen to be created in one cross-section, then in a subsequent plane annihilate each other, the true 3D structure is that of a closed-loop vortex line, as seen in Fig. 7.2. The apparent change in handedness occurring at the extremes of such a loop arises from defining handedness with respect to the viewing direction. The sense of phase circulation around the vortex line, and hence continuity of energy flow, is maintained at all points around the loop.

As mentioned above, in a monochromatic field, these vortex lines are stationary. Perhaps the most obvious example is that of laser speckle where the stationary individual black specks are the intersections of the vortex lines with the viewing cross-section, hence the tiny bright regions of light are in fact surrounded by a complex network of dark vortex lines.[6] For polychromatic fields, the situation is more complicated; any interference pattern created from beams of different fre-

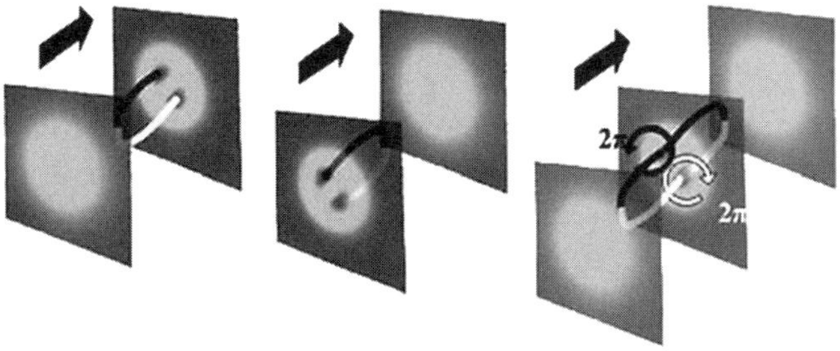

Figure 7.2 When examining successive cross-sections in the field, a pair of oppositely signed vortices may appear and then annihilate. This is a signature of a vortex loop.

quencies is not stationary, the speed of its movement being related to the range of frequencies it contains.

The underlying questions that this chapter and the work it reports attempt to elucidate are: "What is the topology of these monochromatic vortex lines within random light fields? Do they all form vortex loops and are any of these loops linked, or even knotted?"

The structure and topology of vortices in monochromatic fields have been analyzed in a number of specific cases ranging from diffraction catastrophes[7] to the specific superpostions of Bessel and Laguerre-Gaussian beams that result in linked and knotted vortex loops[8–10] (see Fig. 7.3). In addition to this work on specific structures, there is also interest in the general properties of vortex lines leading to results for the density of vortex points within a cross-section, vortex line curvature, and even the vortex line velocity (for polychromatic fields) for various types of random wavefields.[11,12]

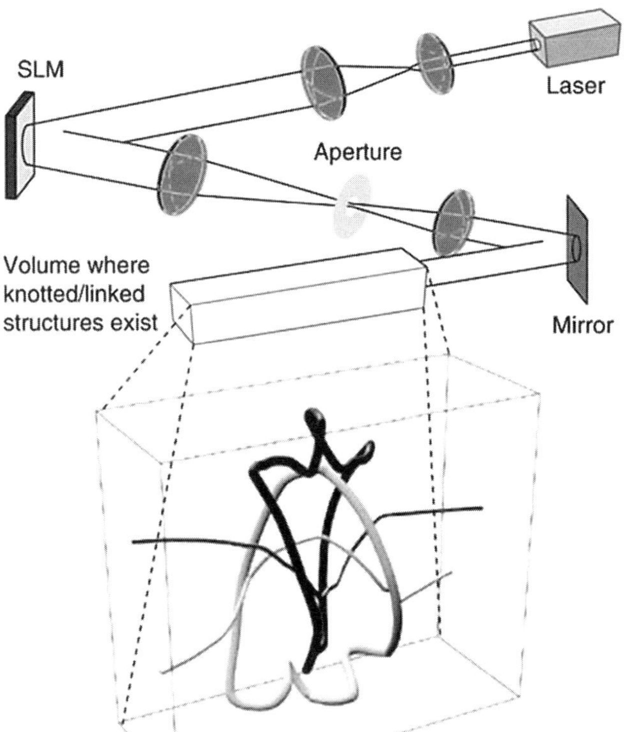

Figure 7.3 Experimental configuration used to produce the linked vortex loops. The expanded beam from a He-Ne laser is incident on a hologram designed to give the required beam superposition. A second lens system is used to recollimate the beam, and a moveable CCD used to record successive cross-sections from which the vortex line structure is deduced.

7.2 Locating Vortex Lines

Figure 7.4 shows a cross-section through the intensity and corresponding phase distribution created by the interference between a random set of plane waves. As discussed in the previous section, the optical vortices are points of zero intensity in the cross-section of the resulting optical field at which the phase is singular.[4] We note here that all of the vortices are $l = \pm 1$; indeed, random light fields do not contain any higher vortices, i.e., $|l| > 1$, since any $l = N$ vortex when subject to a perturbation will break up into $N \times (l = 1)$ vortices.

When locating the positions of the vortex in any particular cross-section, one can look for the phase singularity, but mathematically (and as we will see later, experimentally) it is convenient to express the scalar field amplitude as a complex number separated into its real and imaginary parts. One recognizes that at the positions of the vortices, both the real and imaginary parts are zero. At a vortex position within a complex scalar field, one can therefore write

$$u(r, \phi) = Ae^{i\phi} = \xi + i\eta = 0, \tag{7.2}$$

where A, ϕ, ξ, and η are the magnitude, phase, real part, and imaginary part of the field. The handedness of the 2π phase change around the vortex position determines whether $l = +1$ or -1, the sign often being referred to as the topological charge of the vortex or phase singularity. Throughout the entire field, both ξ and η can take any value ranging from $+A$ to $-A$. Where ξ is zero, $\phi = \pi/2$ or $3\pi/2$, and where $\eta = 0$, $\phi = 0$ or π.

Within the beam cross-section, one can plot lines of zero ξ and η, where these lines cross mark the vortex positions. Within the 3D light field, these lines of zero ξ and η are surfaces; where the two intersect marks the lines of the optical vortices.

Plotting the surfaces of zero ξ and η gives a clear visualization of how the dimension and topology of a vortex line arise. For instance, if the peak of one surface just passes through the other, the perimeter of this 'island' would be a vortex loop.

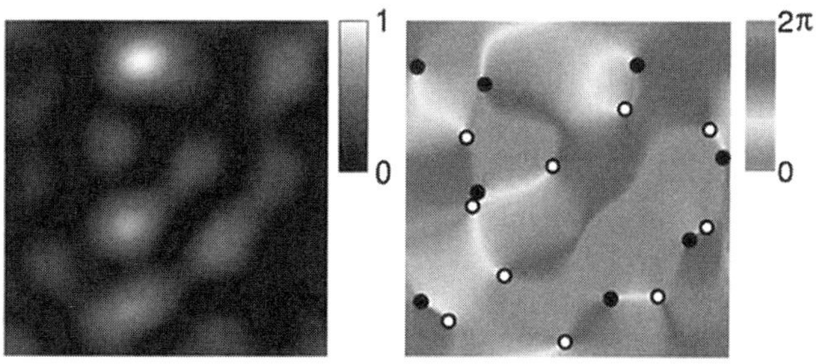

Figure 7.4 A cross-section of a superposition of beams consisting of 25 randomly weighted and directed plane waves. Left: normalized intensity of field. Right: phase of field with grayscale representing the phase $(0 - 2\pi)$. The vortices are marked here with their topological charge $l = \pm 1$ indicated by the black and white spots, respectively.

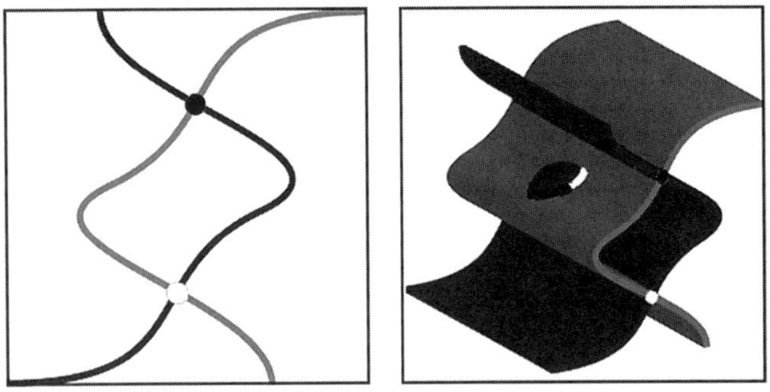

Figure 7.5 Illustration of how the dimension of the vortex arises from the field contours. In two dimensions the $\xi = 0$ (gray) and $\eta = 0$ (black) contours are lines and therefore intersect at points ($l = +1$ and $l = -1$ indicated in black and white, respectively). In three dimensions the contours are surfaces, intersecting along lines. The intersections shown here illustrate how loops may arise.

The formation of such a loop is shown in Fig. 7.5. If the two surfaces intersect at grazing angles, then the precise lines of intersections depend critically on the field parameters. This means that the vortex lines are subject to dramatic movements in response to even small perturbations of the generating light fields.[13]

7.3 Making Beams Containing Optical Vortices

The generation of vortices within what are initially nearly plane-wave laser beams relies on the introduction of a phase term of $\exp(il\theta)$ into the beam. The most obvious way to introduce this term is to pass a beam with a flat phase front through a dielectric with an azimuthal optical thickness equal to $\theta\lambda/2\pi$. Such an optical component is known as a spiral phase plate (SPP) and is illustrated in Fig. 7.6 (a). This method was first successfully implemented by immersing the SPP into a liquid with very similar refractive index that could be varied through heating.[14] This allowed the SPP to be fine tuned such that the step height exactly matched the optical wavelength. More recently, SPPs have been precisely micromachined for the optical regime without need for fine tuning.[15] If the step height of the SPP delays the optical path by $l\lambda$, then an optical vortex is generated with an azimuthal phase dependence of $\exp(il\theta)$. However, if the step height is mismatched with the wavelength, a more complex vortex structure will emerge.[16,17] This will be discussed later in the chapter.

An alternative to using optics based on refractive materials is to use diffractive optics. For example, for operation at a single wavelength, a glass lens can be replaced by a Fresnel zone plate. More generally, diffractive optics can replace any glass optic, and spiral phase plates are no exception. An equivalent description of a diffractive optic considers it as a hologram of the component it is replacing. This description is sometimes referred to as computer-generated holograms (CGHs). The first CGH used to create a vortex beam was a spiral Fresnel zone plate.[18]

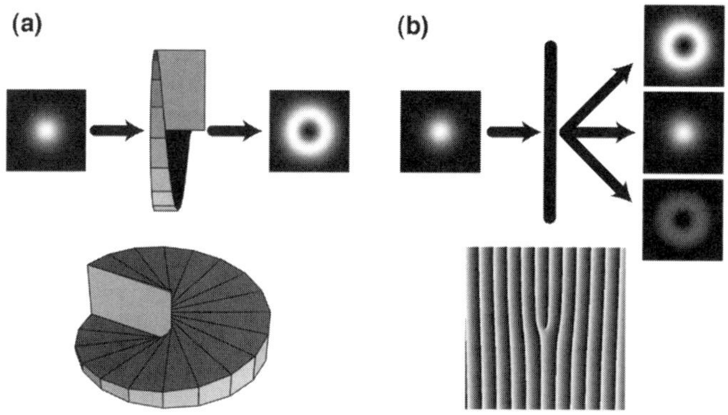

Figure 7.6 Illustration of a Gaussian beam incident on (a) an SPP of height $\lambda\theta/2\pi$ and (b) a hologram with phase modulation $\mathrm{mod}2\pi|\lambda\theta/2\pi + \alpha x|$, where the second term adds a blazed diffraction grating to preferentially diffract light into the positive first order.

More commonly, to generate a pure helically phased beam, the CGH is a simple diffraction grating with an l-pronged fork in the middle; see Fig. 7.6 (b). Like all holograms, obtaining a diffraction efficiency in excess of even 10% requires the hologram to produce a phase modulation, not amplitude modulation, of the illumination light. Although having a lower optical efficiency (i.e., the diffraction efficiency) than the glass equivalent, CGHs have the advantage of being completely general in the phase structure that they impose on the diffracted beam, meaning that a single CGH can create a complex superposition of beams. Originally, the production of CGHs required either complicated etching within an expensive etching facility or a time-consuming and highly complicated photographic process. Since 2000, however, the commercial availability of spatial light modulators (SLMs) has made holographic generation of vortices the most popular method used in current experiments. SLMs are computer controled, pixelated liquid-crystal devices that allow the spatially dependent phase modulation of the illumination light, effectively producing a CGH that can be updated at video frame rates. Many different algorithms for designing holograms have been developed and due to the power of modern desktop computers, a hologram can be calculated and displayed on the SLM in tens of milliseconds.

The methods discussed so far have the aim of generating a vortex by introducing a global $\exp(il\theta)$ phase term. However, optical vortices are not introduced into an optical field solely as a result of specialized optical components. The simplest way to generate a vortex is to superpose three (or more) plane waves.[19] Hence, rather than originating from specific scientific experiments, optical vortices are ubiquitous throughout nature, arising whenever polarized light is scattered or reflected from a rough surface. In most cases of plane wave interference, the result is a lattice of vortices such that there are equal numbers of $l = +1$ and $l = -1$ singularities, i.e., no global $\exp(il\theta)$ phase and hence, no total orbital angular momentum.

7.4 Topology of Vortex Lines

Both the SSP and its CGH equivalent produce a perfect helically phased beam in which a single vortex line runs along the beam axis. However, the topologies of random fields are much more intricate, the vortex lines being curved and frequently curved sufficiently to form loops. In this section we will consider the interference between a finite number of plane waves, with random amplitudes, phases, and directions.

First let us look at interference between two- and three-wave superpositions. When there are only two waves, the destructive interference condition can be satisfied only if the amplitudes of the waves are equal to each other and the wavevectors are pointing in different directions. This results in the well-known sinusoidal interference pattern with a series of maxima and minima. In three dimensions, however, these minima are planes of zero intensity. Although the intensity is zero, these planes do not contain vortices, rather they are a perfect phase step of π. This is a very special and unstable case. The smallest perturbation (i.e., the introduction of a weak third wave) results in these planes breaking to form a skewed honeycomb lattice of vortex lines (see Fig. 7.7 for a cross-section through such a superposition). Provided that no one wave is greater than the sum of the others, all combinations of three waves result in this regular array of straight vortex lines. When more than three waves are present, the topology is dependent on the complex amplitudes of the waves.[19]

When considering the superposition of multiple plane waves, the condition for a vortex, i.e., Eq. 7.2 becomes

$$u(r,\theta) = \sum_{\mathbf{k}} a_{\mathbf{k}} e_{\mathbf{k}}^{i\phi(\mathbf{k},\mathbf{r})} = 0, \tag{7.3}$$

where $a_{\mathbf{k}}$ and $\phi(\mathbf{k},\mathbf{r})$ are the magnitude and phase of the plane wave of wavevector $\mathbf{k}$.

With the addition of a fourth wave, various topologies of the resulting vortex lines are possible. Numbering the wave amplitudes in Eq. 7.3, such that $a_n \geq a_{n+1}$,

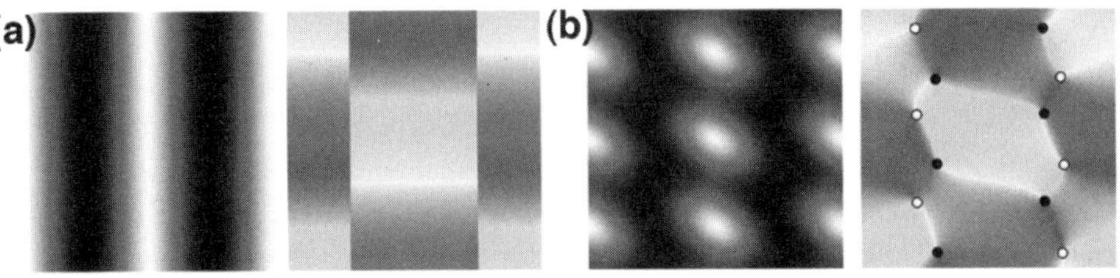

Figure 7.7 The shadings used above are the same as in Fig. 7.4. The intensity and phase cross-section shown in (a) is that of two plane waves with equal amplitude, while (b) shows the effect of adding a third wave of small amplitude.

the vortex line topology is determined by the following inequalities:

$$a_1 + a_4 \leq a_2 + a_3 \qquad \text{(lines)} \tag{7.4}$$
$$a_1 + a_4 = a_2 + a_3 \qquad \text{(reconnections)} \tag{7.5}$$
$$a_1 + a_4 \geq a_2 + a_3 \qquad \text{(loops)}. \tag{7.6}$$

An illustration of these topologies and the transition between them (reconnections) is shown in Fig. 7.8. Note that it is possible to form both open (but periodic) vortex lines and closed vortex loops. The reason that these inequalities determine the topology follows from a consideration of the phasors at a vortex point in the field. At a vortex position, the addition of the four phasors on an Argand diagram must form a quadrilateral (i.e., the resulting field is zero). As the position on the vortex line is followed in x, y, z, the phasors rotate at a rate given by the projections of their wavevector onto the direction of the vortex line, but the quadrilateral must remain *closed.* The quadrilaterals can be split into two types: those that can have full rotations between phasors (Eq. 7.4) and those that have a maximum angle between phasors (Eq. 7.6). In the latter case, none of the wavevectors are free to rotate through a full 2π, meaning that the vortex line cannot indefinitely continue in any one direction, hence, the line must curve back on itself to form a loop. The transition between these topologies corresponds to the quadrilateral having the ability to become flat (which occurs exactly at the reconnection point). Once the magnitudes of the four waves are set, adjusting any of the relative phases simply results in a translation of the vortex structure, leaving its topology unchanged. Extension of these phasor ideas to five or more plane waves is more complicated (as irregular pentagons have many more degrees of freedom in terms of rotating vertices). Even with five waves, the relative amplitudes alone are not sufficient to define the resulting topology; rather, an adjustment of the phase of any one wave can dramatically change the topology of the vortex structure (see Fig. 7.9).

Before examining the topologies of vortex lines obtained within random light fields, it is useful to consider some vortex topologies that have already been created using specific combinations of laser beams. As mentioned previously, in 2001 Berry and Dennis realized that combining modes that already had helical phase

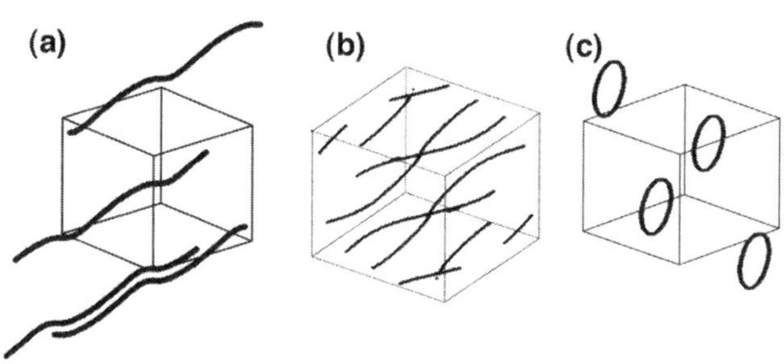

Figure 7.8 The possible vortex topologies resulting from four-wave interference: (a) twisted lines $a_1 + a_4 \leq a_2 + a_3$, (b) reconnections $a_1 + a_4 = a_2 + a_3$, and (c) loops $a_1 + a_4 \geq a_2 + a_3$.

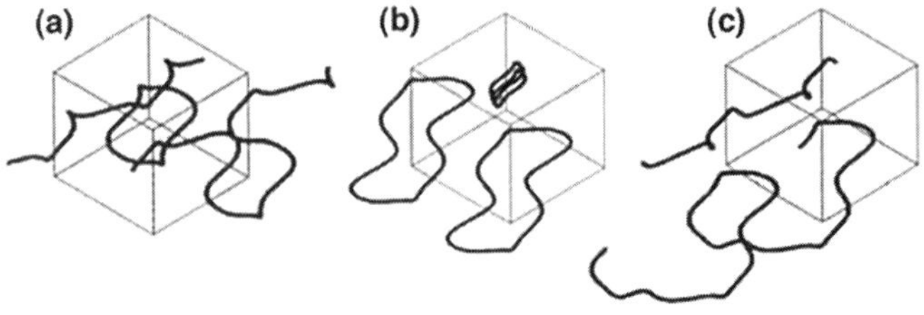

Figure 7.9 Illustration of how vortex topology is sensitive to phase values for five or more waves. In (a) the vortex lines are formed by a wave superposition, and there are two loops and two lines. In (b) the phase of one wave is changed by $\pi/2$, resulting in four loops. In (c) the phase of the same wave is now adjusted to $\pi/3$, resulting in two lines.

fronts allowed complex topologies to be created from a small number of beams. Using four Bessel or Laguerre-Gaussian modes of the correct weighting and waist size results in vortex loops becoming linked or even knotted.[8,20] Both of these constructions rely on perturbation of a high-charge vortex ($l > 1$). As the perturbation is increased from zero, the high-charge vortex splits into multiple single-charge vortices of the same sign (totaling the original l). The loops begin to twist and other loops emerge in the field. Eventually, at the right perturbation amplitude, the loops connect with each other and the topology changes through these reconnections. For a finite window of perturbation amplitude, the vortex loops exist in a linked or knotted state before breaking, as increased perturbation causes more reconnections. The influence of the increasing perturbation amplitude on the formation then causes destruction of linked vortex loops, as shown in Fig. 7.10. As mentioned earlier, linked and knotted vortex lines were experimentally realized using Laguerre-Gaussian modes created with a single SLM, displaying a precalculated hologram to produce the correct superposition of modes in the diffracted beam.[10]

Another interesting vortex structure is formed by an imperfect SPP. If the step height is exactly matched to the illumination wavelength being used, then, as discussed previously, the vortex line is straight along the beam axis. However, if the step is not correctly matched to the optical wavelength, then l ceases to be an integer and the symmetric annular ring of the Laguerre-Gaussian beam is broken by a radial discontinuity appearing as a dark radial line in the intensity cross-section. Examination of the phase cross-section shows that additional vortex lines have been

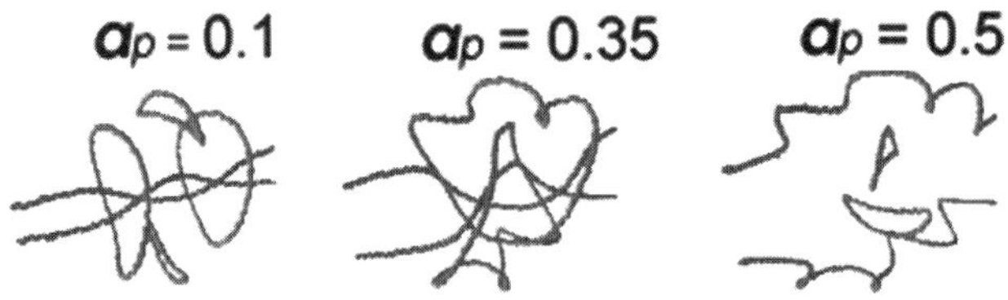

Figure 7.10 Perturbation of high-order vortex and loops leading to formation of a link between two loops. The perturbation is a Gaussian beam with amplitude a_p.

created along the dark line that cuts through the original symmetric ring of intensity. The most dramatic departure from the single straight vortex is when the step height is an odd number of half wavelengths.[16,17] Within the 3D light field, a series of vortex hairpins is formed. Within the beam cross-section, these hairpins are apparent as a chain of alternating charge vortices appearing along the radial line, illustrated in Fig. 7.11.

It has also been shown that vortices may form braids.[21] This last prediction has not yet been achieved in real beams, as it requires counterpropagating beams, leading to a vortex structure on a wavelength scale and hence, has not yet lent itself to experiment verification. Whether similar structures are present within random light fields, such as laser speckle patterns, is the subject of the remaining sections of this chapter.

7.5 Computer Simulation of Vortex Structures

The interference between optical beams of known form and the corresponding formation of optical vortices is a deterministic problem and therefore can be modeled precisely, if somewhat laboriously, by computer. For modeling random waves, an obvious choice of basis set in which to work is that of plane waves traveling in various directions. Any finite number of plane waves with random magnitude, phase, and direction can be combined, and the resulting interference pattern calculated over a finite volume, the scale of the calculation being limited only by the computer's processing power and memory capacity, in particular. The next step is to search element by element (voxcel by voxcel) for vortices. As singularities of the phase, the vortices can be located with very small loop integral paths, consisting of eight elements (the perimeter of a 3×3 grid centered on the voxcel in question). This search routine is illustrated in Fig. 7.12. If one restricts this vortex search to a particular orientation of beam cross-sections, then the search routine frequently fails to locate vortex lines when their direction is tangential to the section. However, as the generated array of interference data is 3D, the integral can be performed within cross-sections of various orientations. Using the Cartesian directions, i.e., the xy, xz, and yz planes, three separate integrals for one voxcel are sufficient to determine if a vortex line passes through it. For example, if a vortex is perpendic-

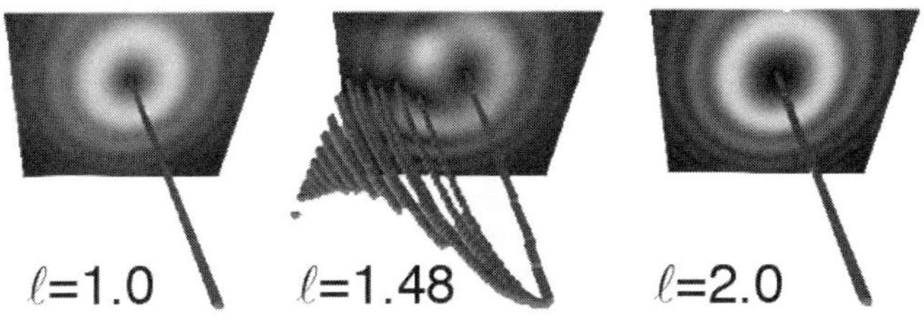

Figure 7.11 Vortex structure with increasing height of an SPP, where step height is $h = \ell\lambda$. The back plane of the box shows the intensity at that plane.

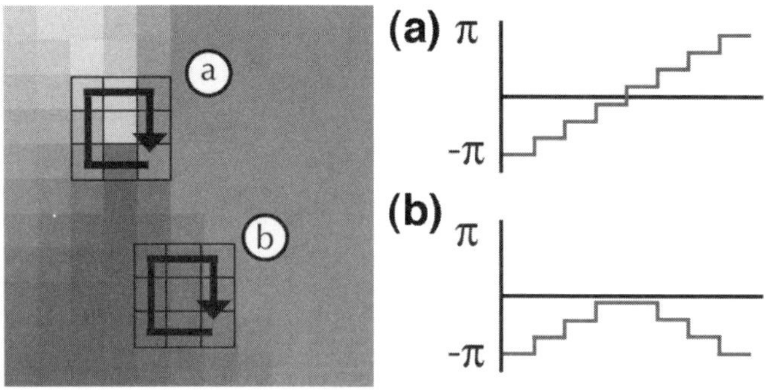

Figure 7.12 Two examples of integration paths on computed phase values. Path (a) encloses a vortex that can be identified by the overall change in phase by 2π as the path returns to its starting point. Path (b) does not enclose a vortex, and the phase can be seen to return to its original value with no discontinuity.

ular to the z axis, the numerical limitations of performing the integral in xy may result in it being missed, but it will be identified by the integration in the other two planes; a vortex line cannot be simultaneously tangential to three orthogonal planes!

The end result is a 3D array with certain voxcels being 'flagged' as containing a vortex. These vortex positions form connected lines through the 3D array, either connecting back onto themselves or terminating at the edge of the modeled volume. The vortex positions can then be sorted according to their individual structures. However, topology of the vortex lines that connect to the edge of the volume is ambiguous and cannot be resolved for any finite calculation.

One approach to solving this ambiguity can be applied if the interference is both laterally and axially periodic and the calculated volume is the repeating cell. In this case, the lines can be 'wrapped' back through the array and ultimately connect back to themselves. Any vortex line feature can be traced out of the initial volume through the neighboring (identical) cell and ultimately be traced back to its starting point, with a path either a) wholly within the initial cell, b) entering neighboring cells but then returning to the initial cell (illustrated in Fig. 7.13), or c) at the corresponding position in a neighboring cell. Paths a) and b) correspond to closed vortex loops. Path c) is an infinite vortex line, albeit one whose structure is periodically repeated.

To obtain an interference pattern that is periodic, there must be a rational relation between the projection of the wavevectors that make up the field components. This is achieved by restricting the directions of waves within k-space to a square grid of spacing Δk. Combined with the paraxial approximation, this results in a field that is periodic in x, y, and z. The transverse and axial repeat distances are

$$d_t = \frac{2\pi}{\Delta k} \tag{7.7}$$

$$d_a = \frac{4\pi k_0}{(\Delta k)^2}. \tag{7.8}$$

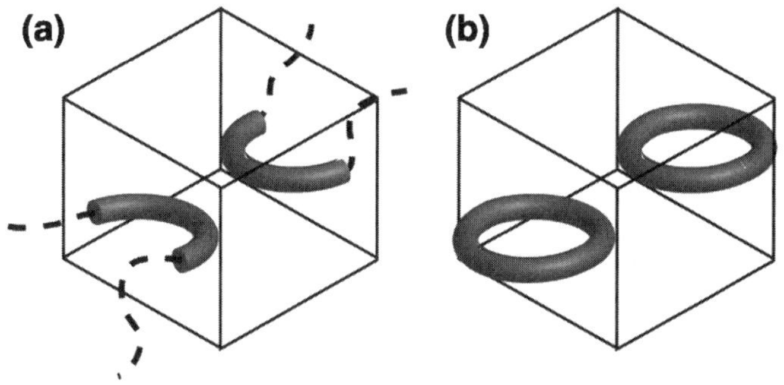

Figure 7.13 Illustration of how a periodic cell can be used to determine topology. In (a) the topology of the two vortex lines is ambiguous as the edge of the calculated volume intersects them. In (b) the cell is periodic, and the topology of the vortex lines can be determined by wrapping the faces onto each other.

This repeating interference pattern is an example of the Talbot effect[22]; hence, we refer to these calculated volumes as "Talbot Cells."

Finding a loop that is itself threaded is slightly more complicated, since large loops may themselves have a very intricate structure. The difficultly is compounded by the fact that the phase of the interference pattern is only actually known at discrete points corresponding to the calculated voxcels; higher-order derivatives cannot be reliably calculated. One approach is based on the recognition that if threaded by a single vortex line, the loop must have a 2π phase variation around its inner perimeter. This winding of the phase around the loop is the signature of a single threading and can be used as a route to identification. Double threadings by vortex lines of opposite charge are missed by this routine, but in any event seem even less frequent than single threadings.[23]

7.6 Vortex Structures in Random Fields

As discussed, anyone who has used a laser will be familiar with the speckle pattern that is observed when an expanded beam is scattered from a surface. In such speckle fields, each of the black specks is a point where a vortex line intersects a plane. The analytically derived point density in the plane is known[12] and is related to the numerical aperture (NA) of the system. The NA is a measure of the maximum angle that light within the system makes with respect to the optical axis. A high NA, i.e., a rapidly diverging or converging light field, will result in a high vortex point density and is closely related to the spatial resolving power of the optical system. However, this tells us only about the intersections of the vortex lines with the plane and nothing of the 3D structure or topology of the vortex lines themselves.

The methods described in the previous section can be used to obtain insights about the topology of vortex lines in random fields of the type typified by the example of laser speckle. A speckle field can be approximated by a serics of plane waves having a Gaussian distribution for their real and imaginary amplitudes (giving a randomized magnitude and a randomly distributed $0 - 2\pi$ phase value) and further multiplied with a Gaussian envelope in the angular power spectrum. Furthermore, by restricting the directions of the plane waves to lie on a regular grid in k-space, as in the previous section, the resulting interference pattern is both laterally and axially periodic in real space.

Of concern is that by restricting the interference pattern to being periodic, one may be imposing some subtle constraints on the vortex topology. To minimize this possibility, one can deliberately run the simulations for various values of Δk and for each of these compare the average of the modeled vortex density within many cross-sections to that analytically predicted by Berry and Dennis.[12]

Typically we use k-space grids as large as 27×27 and calculate the Talbot cell over a Cartesian volume of $500 \times 500 \times 4000$ voxcels, giving sufficient spatial resolution to resolve even the smallest structures. A typical Talbot cell for a superposition of 25×25 plane waves is shown in Fig. 7.14. By generating and analyzing hundreds of randomly generated Talbot cells, various statements can be made concerning the number and properties of the vortex line loops and their topology.

From examination of several hundred randomly generated Talbot cells, we find that approximately 27% of the total vortex line length is made up of closed loops (of various sizes), with the remaining 73% made up of periodic lines. Therefore, it seems likely that within real, nonperiodic optical fields of finite extent, a similar

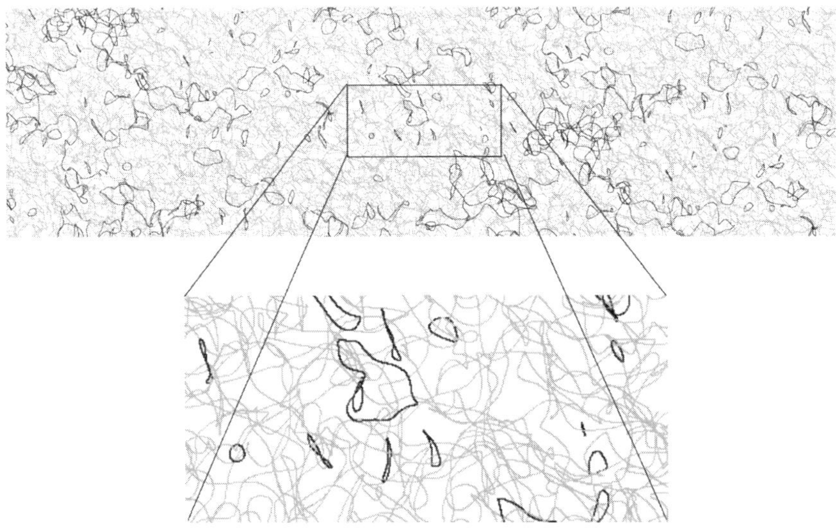

Figure 7.14 A Talbot cell vortex structure, created using 625 randomly selected plane waves, projected into the xz plane. Closed vortex loops are plotted in black and periodic vortex lines plotted in gray.

proportion of lines traverses the entire bright region. Previously, Berry and Dennis have pointed out that in regions of low intensity, the residual field is dominated by the background vacuum fluctuations.[24] In such regions, because of the time-varying nature of the field components, the field is no longer monochromatic, hence, the vortex lines are not stationary and the topology is also time varying. Consequently, although the vortex structures and associated topology are stationary within the bright regions of the beam, as the vortex lines leave the edge of the beam, their path becomes time varying, perturbed by random fluctuations.

When plotting or analyzing the vortex lines, it is natural to rescale the axial and lateral dimensions such that the rescaled coherence length Λ is the same in all directions. The results of our numerical simulations strongly suggest that vortex lines, on the large scale, have fractal structure (given by power-law scalings).[23] Fig. 7.15 shows a log-log plot of the Pythagorean distance between a fixed point and a varying point on an infinite periodic line as a function of the vortex line length between them; the straight-line trend (over 3 orders of magnitude), with gradient approximately 0.5, suggests that the large-scale structure of the vortex line is a Brownian random fractal. The distribution of closed loops also shows a fractal structure. In Fig. 7.16 we plot a log-log histogram of closed loops against loop length. The fitted gradient of -2.46 is consistent with Brownian fractality combined with scale invariance over the fitted range. (The distribution of closed loops is unaffected by a rescaling of length.) The shoulder of this distribution suggests that small vortex loops follow a different scaling. Our results for both infinite periodic lines and closed loops are surprisingly similar to those found for

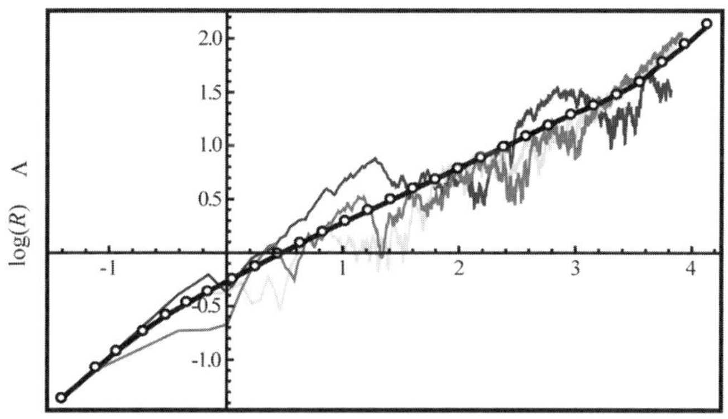

Figure 7.15 The gray lines show the Pythagorean distance from one point on an individual line as the length of the line is followed. These lines have different periods (on the log scale) of 2.7, 3.3, and 3.5 for the gray lines. After these lengths, the plots converge on a linear displacement (gradient 1). The thick black line shows the average based on many pairs of points from 100 different lines from different superpositions and has a gradient of 0.52 ± 0.01 over three decades.

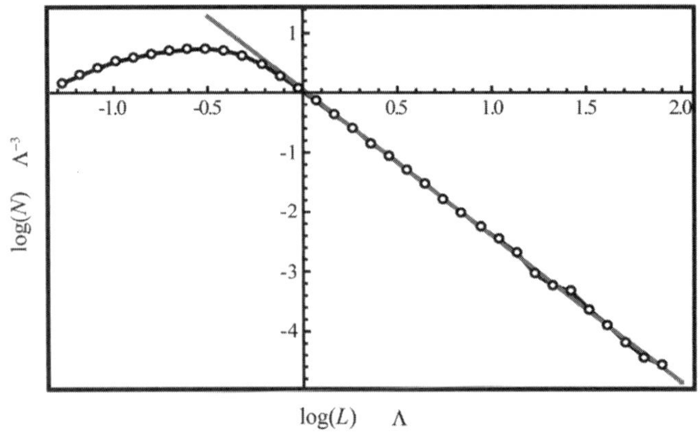

Figure 7.16 The loop size distribution for vortex loops in simulated speckle fields using the rescaled length Λ. The gray line is the linear fit for the straight section of the distribution and has a gradient of -2.46 ± 0.02, consistent with Brownian fractality.

random lattice models representing the distribution of cosmic strings in the early universe.[25]

Topologically, we find that fewer than 1% of the vortex line loops are threaded by another vortex line and we have not found linked loops or knotted loops in any simulation. The reasons for this are unclear, but it is possible that topological features such as loops and knots require the beam superposition to have a nonzero orbital angular momentum (as is the case with known links and knots), which is highly unlikely within a random speckle field.

7.7 Experiments for Visualizing Vortex Structures

There are two obvious ways to look for vortices in experimentally produced light fields. One can either overexpose a CCD array and look for localized dark spots or one can perform an interferometric experiment and look for singularities in the phase. The first method is easier but has disadvantages. Turning points and loops form in very dark sections of the field and the available dynamic range of the CCD array may be insufficient to unambiguously resolve the vortex structure. By using an interferometer to measure the phase of a beam, the vortices can be found to the nearest pixel. A schematic of a general experiment to create and record interference patterns formed between a vortex field and a plane wave reference beam is shown in Fig. 7.17.

By interfering the object beam with a reference beam, fringes are formed across the acquired image, with distinctive 'forks' at the vortex points. This can be seen in Fig. 7.18. In order to pin down the location of a vortex, one beam is phase stepped and the intensity at each pixel of a CCD array is recorded. Any pixel that is not on a vortex will show some modulation in intensity and a relative phase value for the underlying phase of the vortex carrying field can be extracted by Fourier analysis.

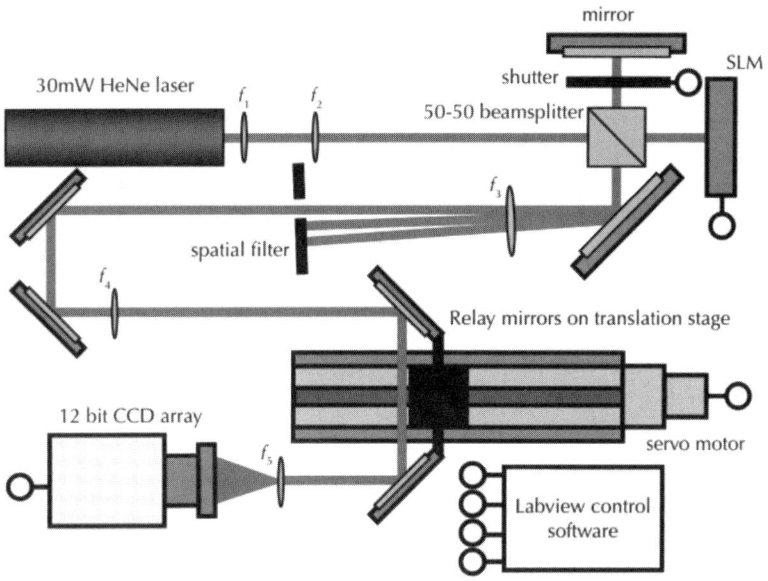

Figure 7.17 Setup of a general vortex experiment using an SLM to create an arbitrary beam and a camera mounted on a motorized stage to image cross-sections of the resulting intensity and phase distribution.

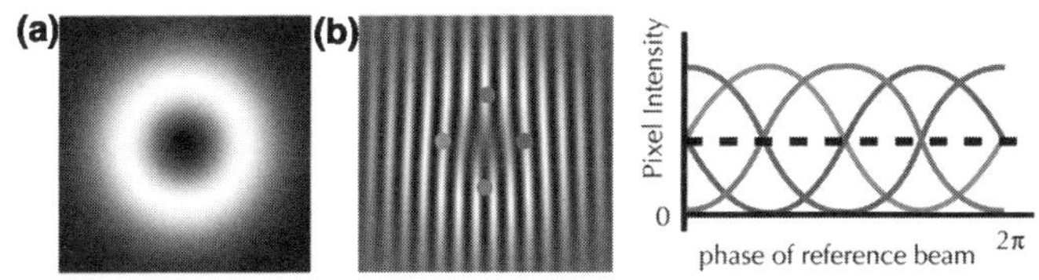

Figure 7.18 (a) Intensity of a beam containing an on-axis vortex. (b) Same beam with an added reference beam; note the distinctive 'fork.' The grayscale lines indicate how the intensity of various pixels changes as the reference beam phase is modulated. The dashed line shows the constant intensity measured at the vortex position.

In contrast, a pixel located at a vortex will remain at constant intensity (that of the reference beam) and will be identified as a phase singularity by the integration process described earlier. Once an image set has been analyzed and vortex locations logged, the camera can be translated along the beam axis and another of the neighboring planes analyzed. After the entire volume has been scanned, the accumulated vortex positions can be displayed, giving a map of the vortex lines embedded in the sample volume (see Fig. 7.19).

7.8 Conclusions

We have explained how phase singularities and their associated optical vortices are present through most light fields, occurring whenever three or more plane wave

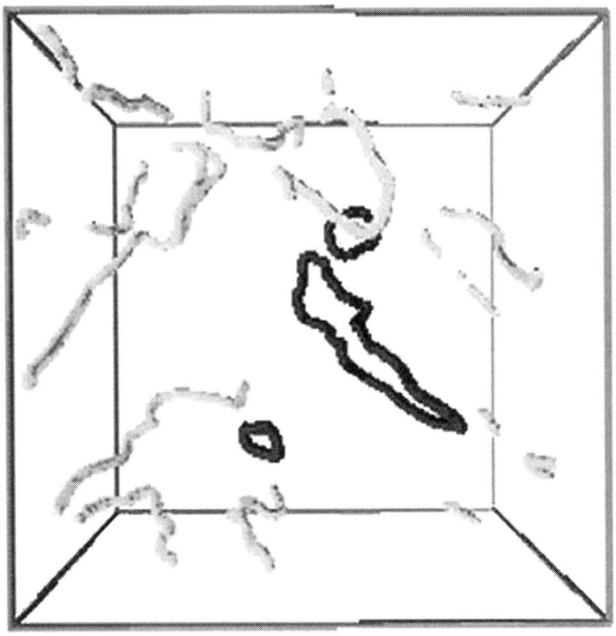

Figure 7.19 A projected view of a volume within an experimentally observed speckle pattern containing a number of vortex line loops (black) and some vortex lines (gray) of unknown topology.

components interfere. For monochromatic fields, these lines of complete darkness are fixed in space and time; one manifestation is the observation of laser speckle where each of the black specks corresponds to a crossing of a vortex line with the viewing plane. For small numbers of interfering plane waves, the resulting vortex patterns are well understood and the resulting topologies can be inferred. For larger numbers of plane waves, the problem remains deterministic but is complicated, the vortex structure being highly and nonlinearly dependent on the precise values of the interfering fields. Numerical modeling of random light fields is capable of generating representative interference patterns as well as the corresponding vortex structures. We see that vortex lines naturally form loops; a small number of these loops (<1%) are themselves threaded by other vortex lines, and rarer still, vortex loops can be linked to other loops. The linking and knotting of vortex loops predicted and subsequently observed to be possible using specific combinations of high-charge helically phased beams seems to be a special case, rarely encountered within random light fields.

References

1. L. Allen and M. J. Padgett, "The Poynting vector in Laguerre-Gaussian beams and the interpretation of their angular momentum density," *Opt. Commun.* **184**, 67–71 (2000).

2. L. Allen, M. W. Beijersbergen, R. J. C. Spreeuw, and J. P. Woerdman, "Orbital angular momentum and the transformation of Laguerre-Gaussian laser modes," *Phys. Rev. A.* **45**, 8185–8189 (1992).

3. L. Allen, S. M. Barnett, and M. J. Padgett, *Optical Angular Momentum*, Institute of Physics Publishing and CRC Press (2003).

4. J. F. Nye and M. V. Berry, "Dislocations in wave trains," *Proc. R. Soc. Lond. A* **336**, 165–190 (1974).

5. M. V. Berry, "Much ado about nothing: optical dislocation lines (phase singularities, zeros, vortices...)," in *Int. Conf. on Singular Optics*, M. S. Soskin, Ed., *SPIE Proc.* **3487**, 1–5 (1998).

6. M. R. Dennis, "Local phase structure of wave dislocation lines: twist and twirl," *J. Opt. A: Pure Appl. Opt.* **6**, S202–S208 (2004).

7. M. Berry, J. F. Nye, and F. J. Wright, "The elliptic umbilic diffraction catastrophe," *Phil. Tran. R. Soc.* **291**, 454–483 (1979).

8. M. V. Berry and M. R. Dennis, "Knotted and linked phase singularities in monochromatic waves," *Proc. R. Soc. Lond. A* **457**, 2251–2263 (2001).

9. J. Leach, M. R. Dennis, J. Courtial, and M. J. Padgett, "Knotted threads of darkness," *Nature* **432**, 165 (2004).

10. J. Leach, M. R. Dennis, J. Courtial, and M. J. Padgett, "Vortex knots in light," *New J. Phys.* **7**, 55 (2005).

11. M. V. Berry, "Disruption of wavefronts: statistics of dislocations in incoherent Gaussian random waves," *J. Phys. A: Math. Gen.* **11**(1), 27–37 (1978).

12. M. V. Berry and M. R. Dennis, "Phase singularities in isotropic random waves," *Proc. R. Soc. Lond. A* **456**, 2059–2079 (2000).

13. M. V. Berry and M. R. Dennis, "Topological events on wave dislocation lines: birth and death of loops, and reconnection," *J. Phys. A: Math. Theor.* **40**, 65–74 (2007).

14. M. W. Beijersbergen, R. P. C. Coerwinkel, M. Kristensen, and J. P. Woerdman, "Helical-wave-front laser beams produced with a spiral phaseplate," *Opt. Commun.* **112**, 321–327 (1994).

15. S. S. R. Oemrawsingh, J. A. W. van Houwelingen, E. R. Eliel, J. P. Woerdman, E. J. K. Verstegen, J. G. Kloosterboer, and G. W. 't Hooft, "Production and characterization of spiral phase plates for optical wavelengths," *Appl. Opt.* **43**, 688–694 (2004).

16. M. V. Berry, "Optical vortices evolving from helicoidal integer and fractional phase steps," *J. Opt. A: Pure Appl. Opt.* **6**, 259–268 (2004).

17. J. Leach, E. Yao, and M. J. Padgett, "Observation of the vortex structure of a non-integer vortex beam," *New J. Phys.* **6**, 71 (2004).

18. N. R. Heckenberg, R. McDuff, C. P. Smith, and A. G. White, "Generation of optical phase singularities by computer-generated holograms," *Optics Letters* **17**, 221–223 (1992).

19. K. O'Holleran, M. J. Padgett, and M. R. Dennis, "Topology of optical vortex lines formed by the interference of three, four, and five plane waves," *Opt. Express* **14**, 3039–3044 (2006).

20. M. V. Berry and M. R. Dennis, "Knotting and unknotting of phase singularities: Helmholtz waves, paraxial waves and waves in 2+1 dimensions," *J. Phys. A: Math. Gen.* **34**, 8877–8888 (2001).

21. M. R. Dennis, "Braided nodal lines in wave superpositions," *New J. Phys.* **5**, 134.1–134.8 (2003).

22. W. H. F. Talbot, "Facts relating to optical science," *Philos. Mag.* **9**, 401–407 (1836).

23. K. O'Holleran, M. R. Dennis, F. Flossmann, and M. J. Padgett, "Fractality of light's darkness," *Phys. Rev. Lett* **100**, 053902 (2008).

24. M. V. Berry and M. R. Dennis, "Quantum cores of optical phase singularities," *J. Opt. A: Pure Appl. Opt.* **6**, S178–S180 (2004).

25. T. Vachaspati and A. Vilenkin, "Formation and evolution of cosmic strings," *Phys. Rev. D.* **30**, 2036–2044 (1984).

Biographies

Kevin O'Holleran is a Ph.D. student in the Optics Group in the University of Glasgow's Department of Physics & Astronomy. His research interests center around optical vortices and their associated angular momentum. He is funded by the Engineering and Physical Science Research Council.

Mark Dennis is a Research Fellow and Lecturer in the Theory Group in the Physics Department of the University of Bristol. He has wide interests in mathematical optical physics, particularly Singular and Topological Optics. He is supported by a University Research Fellowship from the Royal Society of London.

Miles Padgett is Professor of Optics in the Department of Physics & Astronomy at the University of Glasgow. He heads a 15-strong research team covering the full spectrum of blue-sky research to applied commercial development, funded by a combination of government, charity, and industry. In 2001 he was elected to Fellowship of the Royal Society of Edinburgh. From 2007 to 2008 he was supported by a Royal Society Leverhulme Trust Senior Research Fellowship.

Chapter 8
Photonic Crystals: From Fundamentals to Functional Photonic Opals

Durga P. Aryal, Kosmas L. Tsakmakidis, and Ortwin Hess
Advanced Technology Institute and Department of Physics, Faculty of Engineering and Physical Sciences, University of Surrey, UK

8.1 Introduction

Controlling the flow of light is one of the major challenges in modern optics. With optical telecommunication and computing technologies becoming increasingly important, there is an ever-growing need for devices that will be able to control and manipulate lightwave signals. Guiding of light over large distances with ultralow losses has revolutionized the communications industry, allowing for fiber optic transmission of information. Therefore, it is certainly conceivable that the control of light flow on a microscopic scale may equally well open a new era in the realms of computation, quantum electronics, photonics, optical chips, and functional devices. Classic means for controlling light signals are Bragg mirrors, waveguides, resonators, and beam splitters. However, considering that the diversity of modern optical devices has dramatically increased, there is now a plethora of new challenges in our quest for new ways of controlling light. An example of desired lightwave-based functionality is optically switchable windows, whose appearance can be switched on demand (e.g., from opaque to totally transparent and vice-versa). It is thus clear that such sorts of applications fundamentally entail an exploration and pursuit of new ideas, designs, and photonic devices that will enable us to mould the flow of light beyond current constraints.

Photonic crystals (PCs) are engineered structures that have a photonic functionality on the materials level, enabling the complete prohibition or allowance of the propagation of light in certain directions and at certain frequencies. They accomplish this feat by means of a periodic modulation of the refractive index of a suitable host medium. Within these three-dimensionally periodic structures, the distribution of electromagnetic modes and their accompanying dispersion relations differ dramatically from those of bulk media. PCs are, in this regard, highly attractive because they allow the design and manipulation of their photonic properties based on a so-called "band-structure engineering." In particular, it swiftly turns out from a pertinent modal analysis that PCs possess photonic band gap (PBG) regions, i.e., regions in which the propagation of photons is forbidden and the density of allowed electromagnetic states vanishes. These regions can be designed to exist in one-, two- or three-dimensional structures, depending on whether the dielectric constant is periodic along one direction and homogeneous in the others (1D PCs), periodic in a plane and homogeneous in the third direction (2D PCs), or periodic in all three directions (3D PCs).

Although 1D PCs have been known and well-studied for decades in the form of highly reflecting dielectric (Bragg) mirrors, the idea of constructing a 2- or 3D PC is no more than about two decades old. From the start, 3D PCs have attracted enormous attention by scientists, owing to the prediction that they posses highly unusual features, such as full 3D PBGs, and also because of the conceivable applications of these structures.[1,2] Some of the best known 3D PCs are the Yablonovite structure,[3] the "layer-by-layer" structure,[4] the silicon woodpile structure,[5] and the opal[6] and inverse-opal[7] PC structures. Among the several 3D

PC structures, opal-based PCs are the most extensively studied, owing to the fact that they can be synthesized relatively easily by colloidal self-assembly.[8–12]

Current and foreseeable applications of PCs can be divided according to their principle of operation. Some rely solely on the existence (or nonexistence) of a complete PBG, while others rely on the peculiar properties of the individual photonic bands and their dispersion. Yet another allure of PCs lies in the possibility of dynamically (actively) tuning their optical properties, which may allow for the realization of controllable and functional nanophotonic devices. Since the photonic band structure mainly depends on the spatial (geometric) arrangement of the crystal and the refractive indices of the materials used, there are currently two main approaches toward tunable PBGs. The first approach is based on changing the lattice constants or the spatial symmetry by means of external forces, such as mechanical forces,[13–15] electrical/magnetic fields,[16] or light.[17] Although large shifts in the existing PBGs have been demonstrated with these techniques, the required structural changes, which are of the order of micrometre dimensions, may limit the practical deployment of such schemes in real-life devices. The second of the aforesaid methodologies is based on controlling the refractive indices of the materials. Liquid crystals have been widely used in this connection due to their inherent anisotropy and their different phases that allow for considerable variation in their optical properties.[18–22]

However, a number of more challenging, emerging applications, such as the deployment of PCs to functional surfaces or windows, require tunable structures that can allow for a complete switching of the PBG. To this end, promising new types of inverse-opal-based PCs have been experimentally realized,[23] wherein each air void of a conventional inverse opal is replaced by a hybrid sphere made of a core and a shell region of different materials. As will be shown in detail in the following pages, this arrangement opens up new perspectives for achieving a complete switching of the PBG.

It is the purpose of this tutorial chapter to provide a thorough introduction to the principles of PC operation, as well as to explore more advanced concepts, such as the different tuning possibilities inherent in contemporary opal-based PC structures. The work presented here is organized as follows. We begin our study with a review of the basic properties of PCs. Upon introducing the characteristic or 'master' equation giving the optical modes in such structures, we proceed with a review of Bloch's theorem, Bragg's law, and the concept of the 'band structure' in, successively, one to three dimensions. This is followed by a detailed description of the physical origins of the band gaps appearing, in particular, in 3D inverse-opal PCs. As a next step in our study, we introduce and study in detail a novel PC structure, which we call a double-inverse-opal photonic crystal (DIOPC); we show that this structure allows for complete PBG switching and may have applications in structural color and functional windows. We conclude with a summary of the presented results and an appendix that concisely explains the plane wave expansion (PWE) method frequently used in the analyses of PCs.

8.2 Principles of Photonic Crystals

Concepts and fundamental principles from solid state physics and electrodynamics concerning the structure of crystals and involving solution methodologies to Maxwell's equations are essential in understanding the physics describing the interaction of light with periodic PCs. In this section we will, thus, analyze the basic principles and underlying theory of 1D PCs. After having laid the foundations for a more involved analysis, we will proceed with succinctly extending the previous principles to the case of 2D and 3D PCs.

8.2.1 Electromagnetism of periodic dielectrics

The intriguing optical properties of PCs can be formally described by combining Maxwell's equations from electromagnetism with Bloch's theorem from solid state physics. Consequently, it is of crucial importance for the understanding of PC theory to have an unambiguous view of how these two theories are combined in periodic dielectric structures. This is discussed in the following two sections.

8.2.2 Maxwell's equations

All of the macroscopic electromagnetic phenomena, including the propagation of light in PCs, are governed by Maxwell's equations. In centimetre-gram-second (CGS) units, these equations take the form:

$$\nabla \cdot \mathbf{B}(\mathbf{r},t) = 0\,, \tag{8.1}$$

$$\nabla \cdot \mathbf{D}(\mathbf{r},t) = 4\pi\rho(\mathbf{r},t)\,, \tag{8.2}$$

$$\nabla \times \mathbf{E}(\mathbf{r},t) + \frac{1}{c}\frac{\partial \mathbf{B}(\mathbf{r},t)}{\partial t} = 0\,, \tag{8.3}$$

$$\nabla \times \mathbf{H}(\mathbf{r},t) - \frac{1}{c}\frac{\partial \mathbf{D}(\mathbf{r},t)}{\partial t} = \frac{4\pi}{c}\mathbf{J}(\mathbf{r},t)\,. \tag{8.4}$$

Here, $\mathbf{E}(\mathbf{r}, t)$ is the time- and space-dependent electric field strength, $\mathbf{H}(\mathbf{r}, t)$ is the magnetic field strength, $\mathbf{D}(\mathbf{r}, t)$ is the electric displacement, $\mathbf{B}(\mathbf{r}, t)$ is the magnetic flux density, $\rho(\mathbf{r}, t)$ and $\mathbf{J}(\mathbf{r}, t)$ are the charge and current densities, respectively, and c is the speed of light in vacuum.

When an engineered medium, i.e., a suitably designed dielectric structure composed of homogeneous regions, does not contain free electric charges, as well as currents or sources of light inside it, one can set $\rho(\mathbf{r}, t) = \mathbf{J}(\mathbf{r}, t) = 0$. Additionally here, we will make the following four assumptions:

(a) The field strengths are small, so that nonlinear effects can be ignored;
(b) The material is fully isotropic at the macroscopic scale, so that $\mathbf{E}(\mathbf{r}, t)$ and $\mathbf{D}(\mathbf{r}, t)$ can be related by a scalar dielectric constant $\varepsilon(\mathbf{r})$;

(c) There is no explicit dependence of the dielectric constant on frequency;
(d) The magnetic permeability is equal to unity.

Hence, we can write **D**(**r**, t) = ε(**r**) **E**(**r**, t), where ε(**r**) is the spatially varying dielectric constant, and **B**(**r**, t) = **H**(**r**, t). With these assumptions in place, Maxwell's equations become:

$$\nabla \cdot \mathbf{H}(\mathbf{r},t) = 0\,, \tag{8.5}$$

$$\nabla \cdot \varepsilon(\mathbf{r})\mathbf{E}(\mathbf{r},t) = 0\,, \tag{8.6}$$

$$\nabla \times \mathbf{E}(\mathbf{r},t) + \frac{1}{c}\frac{\partial \mathbf{H}(\mathbf{r},t)}{\partial t} = 0\,, \tag{8.7}$$

$$\nabla \times \mathbf{H}(\mathbf{r},t) - \frac{\varepsilon(\mathbf{r})}{c}\frac{\partial \mathbf{E}(\mathbf{r},t)}{\partial t} = 0\,. \tag{8.8}$$

We further assume, without loss of generality, that the electromagnetic fields have a time-harmonic nature. Harmonic modes can be expressed in the following form:

$$\mathbf{E}(\mathbf{r},t) = \mathbf{E}(\mathbf{r})e^{-i\omega t}\,, \tag{8.9}$$

$$\mathbf{H}(\mathbf{r},t) = \mathbf{H}(\mathbf{r})e^{-i\omega t}\,. \tag{8.10}$$

Substituting these expressions for **E**(**r**, t) and **H**(**r**, t) into Eqs. (8.5) and (8.6), we arrive at the following, time-independent, divergence equations:

$$\nabla \cdot \mathbf{H}(\mathbf{r}) = 0\,, \tag{8.11}$$

$$\nabla \cdot \mathbf{D}(\mathbf{r}) = 0\,. \tag{8.12}$$

Equations (8.11) and (8.12) indicate that the supported electromagnetic waves are transverse in nature. Further, after some algebraic manipulations, one can straightforwardly derive the following characteristic (or master) equation for the (continuous throughout) magnetic field components[24]

$$\nabla \times \left\{ \frac{1}{\varepsilon(\mathbf{r})} \nabla \times \mathbf{H}(\mathbf{r}) \right\} = \left(\frac{\omega}{c} \right)^2 \mathbf{H}(\mathbf{r})\,. \tag{8.13}$$

Equation (8.13) is solved in the frequency domain. For a given frequency and refractive index distribution, we identify its possible algebraic solutions, i.e., determine the spatial distribution of the magnetic field components; the physically acceptable solutions are those that also satisfy the transversality condition Eq. (8.11). Therefore, we start by considering Eq. (8.13) in its eigenvalue form

$$\Theta\mathbf{H}(\mathbf{r}) = \left(\frac{\omega}{c}\right)^2 \mathbf{H}(\mathbf{r}), \tag{8.14}$$

where

$$\Theta\mathbf{H}(\mathbf{r}) = \nabla \times \left\{ \frac{1}{\varepsilon(\mathbf{r})} \nabla \times \mathbf{H}(\mathbf{r}) \right\}. \tag{8.15}$$

Expression (8.14) indicates that the sought-after field patterns of the supported harmonic eigenmodes are just the eigensolutions $\mathbf{H}(\mathbf{r})$ of the eigenproblem at hand; the eigenvalues $(\omega/c)^2$ are proportional to the squared frequencies of those modes. Further, one may note that the operator Θ is linear and contains all of the information concerning the spatial distribution of the dielectric constant ε as can be readily inferred by inspection of Eq. (8.15). To a very good approximation, particularly for wavelengths at or below the infrared regime, most dielectric materials used in PCs are lossless, i.e., ε is real and positive throughout. Thus, Θ is real valued. Importantly, it also turns out that Θ is symmetric.[24] Accordingly, Θ is Hermitian, i.e., $\Theta = \Theta^\dagger$, where $\Theta^\dagger$ denotes the conjugate transpose of Θ. Of interest to us here is that Θ, being Hermitian, has real eigenvalues, which are also orthogonal and form a complete set of basis functions. This plays a crucial role for the efficient numerical analysis of PCs using, e.g., the PWE method[25] or other popular mode-solving methodologies.

One may note here that another approach for obtaining the required mode patterns could be to resort to the fully vectorial **E**-field equation, namely,

$$\frac{1}{\varepsilon(\mathbf{r})} \nabla \times \{\nabla \times \mathbf{E}(\mathbf{r})\} = \left(\frac{\omega}{c}\right)^2 \mathbf{E}(\mathbf{r}). \tag{8.16}$$

A difficulty with this approach is that the electric field components not tangential to a dielectric interface are discontinuous. As a result, Eq. (8.16) cannot be cast in a simple eigenvalue problem. Although it can still be solved, its solution procedure is complicated by the fact that the operator $\Xi = [1/\varepsilon(\mathbf{r})]\nabla\times\nabla\times$ in Eq. (8.16) is now non-Hermitian. Indeed, the different position of $\varepsilon(\mathbf{r})$ in Ξ compared with the position of $\varepsilon(\mathbf{r})$ in Θ destroys the hermiticity. For this reason, this approach is generally avoided.

It is appropriate at this stage to comment on a few properties of PCs that can be deduced from Eq. (8.13). First, we see that the field solution in Eq. (8.13) has a vector nature and should, therefore, also satisfy the constraint $\nabla \cdot \mathbf{H}(\mathbf{r}) = 0$ that is embraced by Maxwell's equations. This forces the solutions to Eq. (8.13) to be *transverse*; longitudinal modes are not allowed (at finite frequencies). Moreover, it is clear from the form of Eq. (8.13) that Maxwell's equations are not characterized by a particular length scale, i.e., they are fully scalable, at least until, in the very small nanometre scales, quantum mechanical description of the physical world prevails. This lack of an absolute length scale makes the physics of PCs scalable. Accordingly, a band structure for a system with a lattice constant a will be the same as the band structure of another system having lattice constant a/x, as long as we also scale the frequencies $\omega \rightarrow x\omega$.

8.2.3 Bloch's theorem

The dielectric function of a PC is made of a unit cell, which is repeated in space according to a well-defined periodic pattern. Exploiting the analogy to solid state physics, this information can be reduced to only two concepts: basis and lattice. The lattice defines the spatial arrangement of the unit cell; the basis specifies the content of the unit cell. For instance, the basis for a 3D PC can be a dielectric sphere in air or a dielectric cube in air. The lattice is then generated by a linear combination of primitive vectors $\mathbf{a}_i$, determined by the minimum translations that leave the dielectric function unchanged. Choosing a reference frame and placing a lattice point at its origin indicates that any other lattice point has a one-to-one correspondence with a vector $\mathbf{R}$, which is a linear combination of the primitive lattice vectors $\mathbf{a}_i$

$$\mathbf{R} = \sum_{i=1}^{N} n_i \mathbf{a}_i , \tag{8.17}$$

where the numbers n_i are integers. Recalling the analogy with solid state physics, we see that the dielectric constant $\varepsilon(\mathbf{r})$ in Eq. (8.13) acts as a "potential" for the eigenfunction $\mathbf{H}(\mathbf{r})$. Since the "potential" $\varepsilon(\mathbf{r})$ is periodic in one or more dimensions in PCs, we are able to write

$$\varepsilon(\mathbf{r}) = \varepsilon(\mathbf{r} + \mathbf{R}) . \tag{8.18}$$

The real space lattice (or Bravais lattice) defines an infinite collection of points generated by a set of discrete translation operations. The set of all wave vectors $\mathbf{k}$ that yield plane waves corresponding to the aforementioned periodicity is known as the reciprocal lattice. More formally, $\mathbf{k}$ belongs to the reciprocal lattice of a Bravais lattice of points $\mathbf{R}$, provided that the relation $e^{i\mathbf{k}\cdot(\mathbf{r}+\mathbf{R})} = e^{i\mathbf{k}\cdot\mathbf{r}}$ holds for any $\mathbf{r}$ and for all $\mathbf{R}$ in the Bravais lattice. Equivalently, we can identify the reciprocal lattice as the set of wave vectors $\mathbf{k}$ satisfying $e^{i\mathbf{k}\cdot\mathbf{R}} = 1$ for all $\mathbf{R}$ in the Bravais lattice. The Wigner-Seitz primitive cell of the reciprocal lattice, which designates the irreducible full symmetry of the lattice, is defined as the Brillouin zone (BZ). The BZ also corresponds to the set of points in the reciprocal space that can be reached from the origin without crossing any Bragg plane.

The real space primitive lattice vectors $\mathbf{a}_i$ and reciprocal space primitive lattice vectors $\mathbf{b}_j$ are related by

$$\mathbf{a}_i \cdot \mathbf{b}_j = 2\delta_{ij}, \tag{8.19}$$

where δ_{ij} is the Kronecker delta function, i.e., $\delta_{ij} = 1$ for $i = j$ and 0 otherwise. Moreover, it is well known that the reciprocal lattice is itself a Bravais lattice; its primitive vectors can be generated from the vectors of the direct lattice. Let $\mathbf{a}_1$, $\mathbf{a}_2$, $\mathbf{a}_3$ be a set of primitive vectors. Then the reciprocal lattice can be generated by the three primitive vectors

$$\begin{aligned}\mathbf{b}_1 &= 2\pi \frac{\mathbf{a}_2 \times \mathbf{a}_3}{\mathbf{a}_1 \cdot (\mathbf{a}_2 \times \mathbf{a}_3)} \\ \mathbf{b}_2 &= 2\pi \frac{\mathbf{a}_3 \times \mathbf{a}_1}{\mathbf{a}_1 \cdot (\mathbf{a}_2 \times \mathbf{a}_3)}, \\ \mathbf{b}_3 &= 2\pi \frac{\mathbf{a}_1 \times \mathbf{a}_2}{\mathbf{a}_1 \cdot (\mathbf{a}_2 \times \mathbf{a}_3)}\end{aligned} \tag{8.20}$$

where $\mathbf{b}_1$, $\mathbf{b}_2$ and $\mathbf{b}_3$ are primitive reciprocal lattice vectors.

We now consider a 1D periodic system that has discrete translational symmetry along the x direction, $\mathbf{R} = n\mathbf{a}$ (n integer), $\mathbf{a} = a\mathbf{x}_0$, and we assume that there is continuous translational symmetry in the other two directions. The discrete translational symmetry can be expressed in terms of an operator O, which creates a spatial shift a along the x direction. Owing to the structure of the dielectric potential $\varepsilon(\mathbf{r})$, it can be shown that O commutes with Θ:

$$\left[O, \Theta\right] = 0. \tag{8.21}$$

This means that we can construct simultaneous eigenfunctions of O and Θ, and therefore classify the eigenfunctions of Θ by the eigenvalues of O. The eigenfunctions of O are easily determined as plane waves with wave vector $\mathbf{k}$, because the translational symmetry allows the eigenfunctions to differ only by a phase shift that corresponds to the eigenvalue φ

$$Oe^{ik_x x} = e^{ik_x(x+a)} = e^{ik_x x} e^{ik_x a} = \varphi e^{ik_x x}. \tag{8.22}$$

In fact, all of the eigenfunctions with wave vectors of the form $k_x + m(2\pi/a)$, m being an integer number, form a degenerate set; they all have the same eigenvalue φ. This suggests that φ is not unique for all wave vectors. Mathematically, all plane waves corresponding to $k_x + mG$, with $G = 2\pi/a$, form a degenerate set of eigenfunctions, and every superposition of plane waves with wave vectors $k_x + mG$ is also an eigenfunction with eigenvalue φ. Since φ depends on k only as a free parameter for a given structure, we can now take a step back (to the master equation) and state the following:

The eigenfunctions of Θ can be classified by a wave vector **k** and have the form

$$\mathbf{H}_k(x) = e^{ik_x x} \sum_m \mathbf{H}_m e^{imGx} = e^{ik_x x} \mathbf{u}_k(x), \tag{8.23}$$

with the lattice periodic function $u_k(x) = u_k(x + ma)$ and the plane wave amplitudes $\mathbf{H}_m$. In solid state physics, the form of Eq. (8.23) is known as Bloch's theorem.[26] A key fact concerning this theorem is that a Bloch state with wave vector k_x and a Bloch state with wave vector $k_x + mG$ are identical. The k_x's that differ by integral multiples of $G = 2\pi/a$ are not different from a physical point of view. In fact, we need only consider k_x existing in the range $-\pi/a \leq k_x \leq +\pi/a$. This region of important, nonredundant values of k_x is referred to as the first BZ.

Let us go back to the Bloch theorem and generalize it for a periodic system with N dimensions. A given N-dimensional periodic structure with dielectric constant $\varepsilon(\mathbf{r}) = \varepsilon(\mathbf{r} + \mathbf{R})$, where $\boldsymbol{R}$ is given by Eq. (8.17), has eigenfunctions that can be labeled by a wave vector **k** and expressed in the form

$$\mathbf{H}_{\mathbf{k}}(\mathbf{r}) = e^{i\mathbf{k}\cdot\mathbf{r}} \sum_G \mathbf{H}_{\mathbf{k},\mathbf{G}} e^{i\mathbf{G}\cdot\mathbf{r}} = e^{i\mathbf{k}\cdot\mathbf{r}} \mathbf{u}_{\mathbf{k}}(\mathbf{r}), \tag{8.24}$$

where **G** denotes reciprocal lattice vectors and is given by

$$\mathbf{G} = \sum_{i=1}^{N} n_i \mathbf{b}_i, \tag{8.25}$$

with primitive reciprocal lattice vectors $\mathbf{b}_i$ and integer numbers n_i. In Eq. (8.24), $\mathbf{u}_{\mathbf{k}}(\mathbf{r})$ is a periodic function with the periodicity of the lattice: $\mathbf{u}_{\mathbf{k}}(\mathbf{r}) = \mathbf{u}_{\mathbf{k}}(\mathbf{r} + \mathbf{R})$ for all vectors **R**.

8.2.4 Photonic band structure

The concept of the band structure is crucial, since most of the optical properties of PCs are based on this feature. Among the parameters influencing the band structure, the PC lattice and its corresponding BZ are of primary importance.

Since a PC corresponds to a periodic dielectric function [Eq. (8.18)], the solutions to Eq. (8.13), using the Bloch theorem, can be chosen in the form

$$\mathbf{H}^{\upsilon}_{\mathbf{k}}(\mathbf{r}) = e^{i\mathbf{k}\cdot\mathbf{r}} \sum_G \mathbf{H}^{\upsilon}_{\mathbf{k},\mathbf{G}} \mathrm{e}^{i\mathbf{G}\cdot\mathbf{r}} = \mathrm{e}^{i\mathbf{k}\cdot\mathbf{r}} \mathbf{u}^{\upsilon}_{\mathbf{k}}(\mathbf{r}), \tag{8.26}$$

with eigenvalues $\omega_\upsilon(\mathbf{k})$ yielding a different Hermitian eigenproblem over the primitive cell of the lattice for each Bloch wave vector **k**. This primitive cell is a finite domain in the directions where the structure is periodic, leading to discrete eigenvalues, which are labeled by $\upsilon = 1, 2\ldots$. The eigenvalues $\omega_\upsilon(\mathbf{k})$ are continuous functions of **k**, forming discrete bands when plotted versus the latter. The plot $\omega_\upsilon(\mathbf{k})$ is called a band structure, wherein each $(\omega_\upsilon, \mathbf{k})$ mode is a Bloch mode.

In general, the band structure is plotted only along the characteristic path of the irreducible part of the BZ, i.e., a line following the edges of the irreducible BZ. In practice,[26] all of the maximum and minima of the band structure lie on this characteristic path. Hence, the existence and frequency range of a PBG can be deduced from a plot of the band structure along the characteristic path. The frequencies are usually displayed in units of $2\pi c/a$. Indeed, since Maxwell's equations are linear, the geometry and eigenvalues/eigenfunctions also scale linearly, and it is thus reasonable to normalize the frequencies by the fundamental length of the crystal, the lattice constant. Normalization of $2\pi c/a$ is assumed throughout this work when no explicit frequency units are given. The wave vector **k** is also expressed in the aforementioned normalized units.

In a uniform material, the dispersion equation characterizing the propagation of an electromagnetic plane wave is $\omega = ck/n$. For a homogeneous material, this relation is a line whose slope is proportional to the inverse of the refractive index. However, as we highlighted above, in a periodic material, the frequency as a function of the wave vector forms bands that can be separated by band gaps. A photonic band gap (PBG) is a frequency range in which no state exists for any **k**. A 1D band gap is shown in Fig. 8.1. The photons are altogether forbidden from propagating along the direction in which the PBG appears. In the case of a partial PBG, this forbidden range is limited to one (or a few) directions within the PC. In the case of a complete (3D) PBG, the forbidden frequency range extends to all directions of propagation within the PC. In the direction where a PBG occurs, the PC acts as a perfectly reflecting mirror for all waves having frequencies within the PBG.

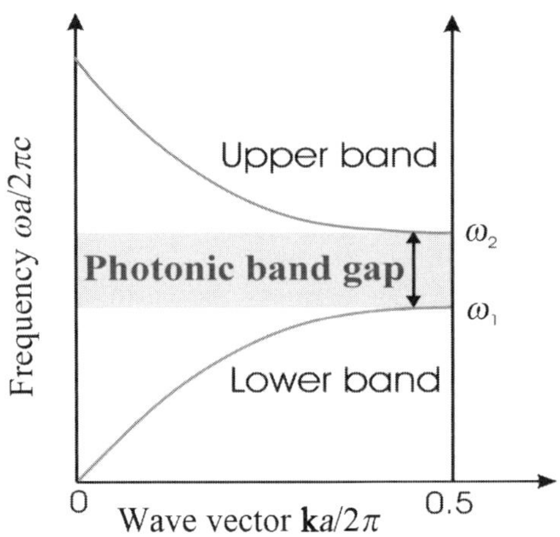

Figure 8.1 Schematic representation of a photonic band structure.

In most practical applications, one aims to make the PBG as large as possible. Therefore, the size of the PBG is an important feature. The PBG size is usually described in terms of the gap-to-midgap ratio (GMR), which is defined as the ratio of the absolute band gap width ($\Delta\omega = |\omega_1 - \omega_2|$) to the midgap frequency [$\omega_c = (\omega_1 + \omega_2)/2$], with ω_1 and ω_2 being the upper edge frequency of lower band and lower edge frequency of upper band limiting the PBG, respectively. This can be expressed as

$$\mathrm{GMR} = \frac{\Delta\omega}{\omega_c} = \frac{|\omega_1 - \omega_2|}{\omega_c}. \tag{8.27}$$

The reason for choosing the GMR instead of the absolute PBG is to make use of a metric (GMR) that is independent of frequency. The size of the band gap depends on the refractive index contrast of the materials constituting the PC. Moreover, geometrical parameters are also important for the size and position of the band gap. The origin of the PBG and its dependence on the PC geometrical parameters are discussed in the sections below.

8.3 One-Dimensional Photonic Crystals

In this section, we present Bragg's law for an ordinary crystal, which will then be extended to photonic crystals. We present in one dimension the concept of the band structure for a multilayer system, and finally, we discuss its novel features.

Despite being the simplest relevant system, 1D PCs convey most of the physical features of the more complex 2D and 3D PCs. Therefore, in order to gain an understanding of photonic band structures and the origin of PBGs, we here consider the case of the 1D PC in somewhat more detail. Two different approaches will be presented below: the first one is based on Bragg's law, and the second on the evolution of the photonic band structure. We will show how both approaches are complementary in explaining the origin of PBGs.

Figure 8.2 illustrates a multilayer structure of a 1D PC with lattice constant a. It is periodic along the z direction and homogeneous along the other two directions. Here, we assume that light is propagating along the z direction. The wave vector $\mathbf{k}$ can assume discrete values only along z, since the structure is periodic along that direction. We stress that the reflection of light from this structure should not be the only means by which one determines whether a PBG is present or not. Because of the symmetry or polarization mismatching between the input mode (pulse/wave) and the supported modes in a dielectric structure, we may still observe high reflection spectra from a structure not possessing a PBG.

8.3.1 Bragg's law

Particles scatter incident rays in all directions. In some of these directions the scattered beams are in phase and reinforce each other, giving rise to enhanced

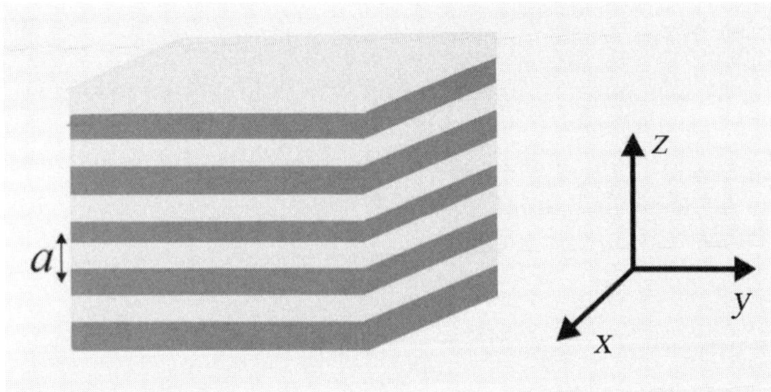

Figure 8.2 One-dimensional PCs consist of alternating layers of materials (light gray and dark gray) with different refractive indices, spaced by a distance a, the lattice constant.

diffracted beams and a constructive interference. The mathematical description of diffraction was first written down by von Laue in 1912[26], and his equations are still useful. However, a simpler way to describe the geometry of diffraction can be obtained by using Bragg's law.

Figure 8.3 schematically illustrates the interference between waves scattered from two adjacent rows of atoms in a crystal. The net effect of scattering from a single row is equivalent to partial reflection from a hypothetical mirror imagined to be aligned with the row. Thus, the angle of "reflection" equals the angle of incidence for each row. Interference then occurs between the beams reflecting off of different rows of atoms in the crystal.

A large intensity will be detected at this angle if the reflected rays from each successive layer add up constructively. Obviously, the successive layers will interfere constructively if the path difference is an integer multiple of the wavelength λ: $AB + BC = m\lambda$. Since $AB = BC = d\sin\theta$, constructive interfere-ence will occur when

$$2d\sin\theta = m\lambda\,, \tag{8.28}$$

where d is the spacing between the subsequent planes, and θ is the angle between the incident rays and the plane surface. Solids have structural features given by the inter-atomic spacing, on the order of 2 Å. To probe the structure of solids, we thus need light with wavelength less than or equal to 2 Å, or x rays. The periodicity of structures can be probed particularly effectively through diffraction; light waves reflected from the individual electrons can interfere constructively or destructively, and an intense reflection is proof of constructive interference. Furthermore, if only two rows are involved, the transition from constructive to destructive interference is gradual as θ changes. However, if interference from many rows occurs, then the constructive interference peaks become very sharp with mostly destructive interference in between.

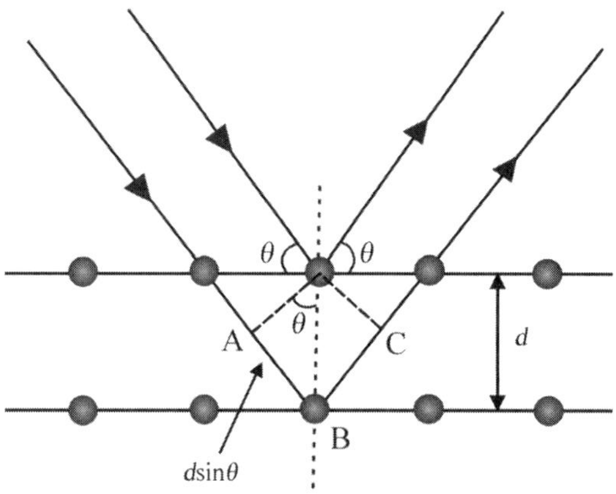

Figure 8.3 Schematic representation of Bragg's diffraction. Maximal diffraction occurs at $2d\sin\theta = m\lambda$, where λ is the wavelength of electromagnetic wave and m is an integer.

Although Bragg's law was used to explain the interference pattern of x rays scattered by crystals, diffraction has been developed to study the structure of all states of matter with any beam, e.g., ions, electrons, neutrons, protons, and photons, with a wavelength similar to the distance between the atomic, molecular, or dielectric layer structures of interest.

A periodic dielectric structure, either in one, two, or three dimensions, diffracts photons in a fashion analogous to the way a crystal diffracts x rays. Alternating regions of high-n and low-n materials create a periodic structure of differing dielectric material densities, like the atomic planes in crystals. If the phases of the waves scattered from each dielectric layer coincide, the structure will achieve maximum reflectivity. Due to the existence of dielectric crystalline planes in PCs, waves within some frequency regions will be diffracted according to Bragg's law[6]

$$\lambda_{hkl} = 2d_{hkl}\sqrt{n_{\text{eff}}^{\ 2} - \sin^2\theta_{hkl}}\ , \tag{8.29}$$

where λ_{hkl} is the wavelength of the electromagnetic wave, d_{hkl} is the interplanar distance for the hkl crystallographic direction, n_{eff} is the effective refractive index of the PC, and θ_{hkl} is the angle between the incident radiation and the normal to the set of crystalline planes determined by the $[hkl]$ indices. An important difference between diffraction in solids and diffraction in PCs is the bandwidth of the Bragg peaks. The aforementioned Eq. (8.29) is derived below.

Equation (8.28) can be expressed in terms of the wave vector $\mathbf{k}_{hkl}$ and the reciprocal lattice vector $\mathbf{G}_{hkl}$ as[26]

$$2\mathbf{k}_{hkl} \cdot \mathbf{G}_{hkl} = G_{hkl}^{\ 2}. \tag{8.30}$$

Furthermore, we can also write

$$\mathbf{k}_{hkl} \cdot \mathbf{G}_{hkl} = k_{hkl} G_{hkl} \cos\theta_{hkl} \, . \tag{8.31}$$

Now, we may introduce an effective index of refraction n_{eff}^2, which can be estimated by an effective medium approach as $n_{\text{eff}}^2 = f_1 n_1^2 + f_2 n_2^2 + f_3 n_3^2 + \ldots + f_n n_n^2$, with f_i and n_i representing different filling fractions for the various constituents and their refractive indices, respectively. For the magnitudes of the vectors $\mathbf{k}_{hkl}$ and $\mathbf{G}_{hkl}$, we can then write $k_{hkl} = n_{eff}(2\pi/\lambda_{hkl})$ and $G_{hkl} = 2\pi/d_{hkl}$. Introducing these definitions in Eq. (8.30), we obtain the wavelength of the electromagnetic radiation

$$\lambda_{hkl} = 2 d_{hkl} n_{\text{eff}} \cos\theta_{hkl}. \tag{8.32}$$

Using Snell's law, we can write

$$n \sin\theta = n_{\text{eff}} \sin\theta_{hkl} \, , \tag{8.33}$$

and by applying this law in Eq. (8.32), we arrive at the previous Bragg condition, Eq. (8.29)

$$\lambda_{hkl} = 2 d_{hkl} \sqrt{n_{\text{eff}}^2 - \sin^2\theta_{hkl}} \, .$$

In one dimension, Eq. (8.30) becomes

$$\mathbf{k} = \pm \left(\frac{m \cdot \pi}{a} \right), \tag{8.34}$$

where $m = 0, 1, 2\ldots$.

If we now divide both terms of Eq. (8.2.30) by 4, we obtain:

$$2\mathbf{k} \cdot \left(\frac{\mathbf{G}}{2} \right) = \left(\frac{G}{2} \right)^2 . \tag{8.35}$$

This relation has a simple geometrical interpretation, shown in Fig. 8.4. The vectors **k** satisfying the maximum diffraction condition are actually those that lie on the edge of the BZ. Therefore, the edge of the BZ plus its center (Γ) $G = 0$ satisfy the maximum diffraction condition. Consequently, the band structures are calculated along the high-symmetry points of the BZ.

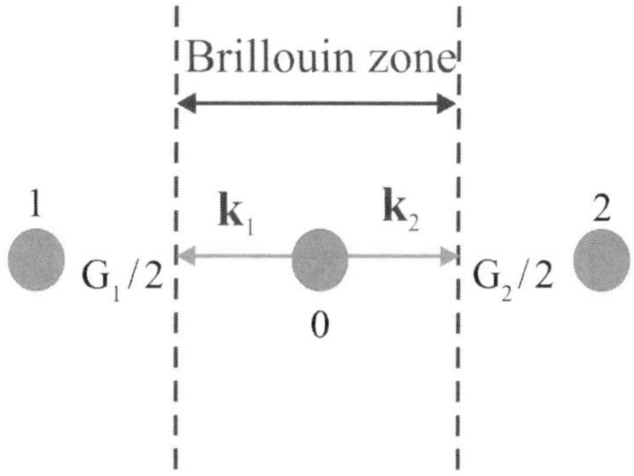

Figure 8.4 Graphical solution of Eq. (8.35): The dashed lines indicate the points where the reciprocal lattice vectors **G***i*, *i* = 1, 2 are halved. Each vector **k***i*, *i* = 1, 2 (originating from the point 0 and being parallel to the line connecting the points 0 and 1, or 0 and 2) with its tip on the dashed line is a solution to Eq. (8.35).

8.3.2 One-dimensional photonic band structure

The photonic band structure gives us information about the propagation properties of electromagnetic radiation within the photonic crystal. It is a representation in which the available energy states are plotted as a function of propagation direction.

In order to understand how the photonic band structure is constructed, a 1D dielectric multilayer system will be studied and compared with the case of a homogeneous dielectric system. The band structures of three different multilayer films are plotted in Fig. 8.5. Figure 8.5(a) reports the plot for the band structure of multilayer films where each layer has the same dielectric constant, 11.56 (silicon). This is a homogeneous structure to which we have artificially assigned a periodicity a. Moreover, the photons do not react to the periodic dielectric layers' presence and behave as they would in a homogeneous dielectric structure; therefore, no PBG will appear. Replacing one of the two materials comprising the multilayer by a material having a slightly lower dielectric constant, 10.24 (tin disulfide), introduces a perturbation of the homogeneous system. The resulting band structure is shown in Fig. 8.5(b). This plot looks like the dispersion curves of a homogeneous system, with one important difference: there is a gap frequency between the upper and lower branches of the bands—i.e., a frequency gap in which no mode, regardless of wave vector **k**, can exist in the crystal. The reason behind the PBG in PCs is a lift of degeneracy of the two states existing at $\pm\pi/a$. Indeed, for π/a the modes are standing waves with a wavelength of $2a$, twice the crystal's lattice constant. In Fig. 8.5(a), these two modes are degenerate, since the system is homogeneous. However, in the system corresponding to Fig. 8.5(b), the two materials are not identical; there are only two ways to center a standing wave of this type.

We can position the nodes of the standing wave either in each low-dielectric layer, as in Fig. 8.6(a), or in each high-dielectric layer, as in Fig. 8.6(b). Any other position would violate the symmetry of the unit cell about its center.

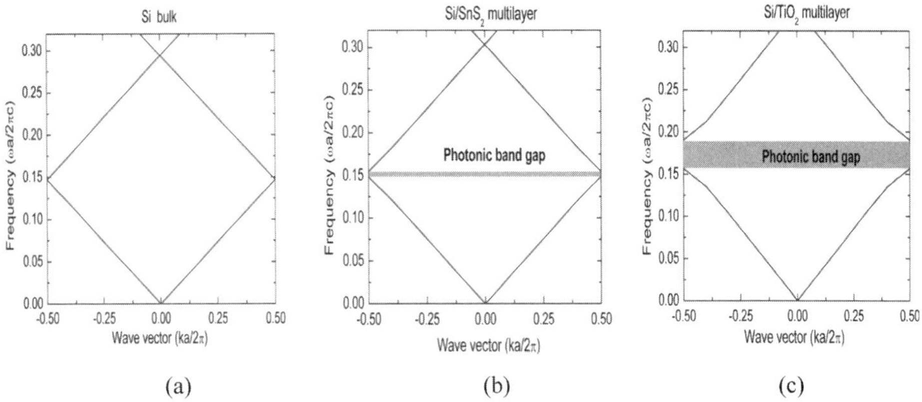

Figure 8.5 Photonic band structures of 1D multilayer PCs. Each layer has a width 0.5*a*, where *a* is the lattice constant. Here, we consider an electromagnetic wave propagating along the *z* direction (see Fig. 8.2).

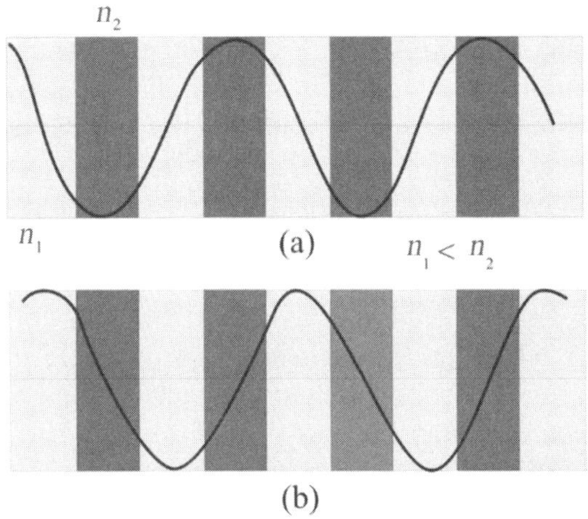

Figure 8.6 Schematic representation of the electric field in 1D PCs at (a) low-dielectric and (b) high-dielectric multilayers.

On the other hand, according to the electromagnetic variational theorem,[24] modes having their energy concentrated in high-refractive-index dielectric regions have a lower frequency than modes having their energy in the low-index dielectric regions. This gives rise to the frequency difference between the two cases, and therefore a PBG appears. In other words, as the electromagnetic energy

$$\text{Energy} = \int \left[\varepsilon(x) \left| E(x) \right|^2 + \left| H(x) \right|^2 \right] dx \tag{8.36}$$

is different in these two cases, the modes have different frequency eigenvalues, resulting in a band gap.

How does the band structure agree with Bragg's law? To answer this question let us consider an example of a periodic array of dielectric layers, as displayed in Figs. 8.7 and 8.8. When an incident wave enters a periodic array of dielectric layers, it is partially reflected and partially refracted at the boundaries of the dielectric layers. This phenomenon is strongly dependent on the geometry and refractive index contrast. According to Bragg's law, if the partially reflected waves are in phase and superimposed, the incident wave is totally back-reflected and is unable to enter the medium, as illustrated in Fig. 8.7. The range of wavelengths in which the incident waves are totally reflected corresponds to a band gap. On the other hand, when the wavelength of an incident wave does not lie within the PBG, destructive interference occurs, and the partially reflected waves cancel one other. Consequently, total reflection from the periodic structure does not occur, and part of the light is transmitted through the PC, as shown in Fig. 8.8.

The interaction between electromagnetic waves and PCs causes the splitting of degenerate bands for wave vectors on the surface of the BZ and the appearance of frequency gaps. Waves with frequencies within these stop gaps are Bragg diffracted and cannot propagate. The widths of PBGs increase with the interaction strength between light and the crystal.

This condition provides an intuitive idea of what kind of structure may become a PC. That is, the structure should be periodic so that scattered waves are superimposed and in phase at any point of the structure. Moreover, the structure should possess symmetry in as many directions as possible, so that waves are scattered in a similar manner from 'equivalent' points in the lattice, i.e., from points that are translationally invariant.

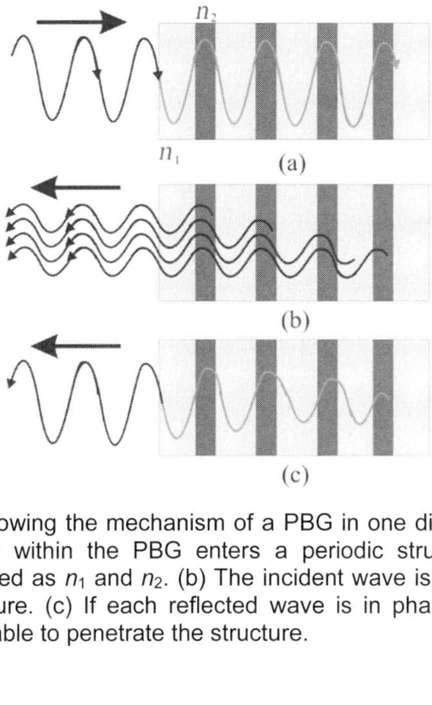

Figure 8.7 Diagram showing the mechanism of a PBG in one dimension. (a) An incident wave at a wavelength within the PBG enters a periodic structure with two different refractive indices denoted as n_1 and n_2. (b) The incident wave is partially reflected by the boundary of the structure. (c) If each reflected wave is in phase, the incident wave is totally reflected and unable to penetrate the structure.

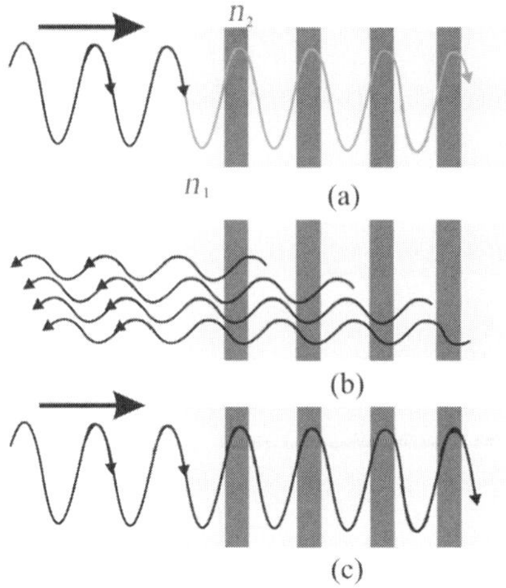

Figure 8.8 Diagram showing destructive interference. (a) An incident wave at a wavelength outside the PBG enters a periodic structure. (b) The incident wave is partially reflected by the boundary of the structure, but the reflected waves are out of phase and interfere destructively with each other. (c) Reflection does not take place, and the incident wave penetrates the structure.

8.4 Generalization to Two- and Three-Dimensional Photonic Crystals

Armed with the knowledge of the basic principles of photonic crystals in one dimension, we will in the following two sections briefly describe 2D and 3D PCs.

8.4.1 Two-dimensional photonic crystals

Two-dimensional PCs can possess different types of lattices. Square and triangular lattices are the most popular. These lattices, in real space, are illustrated in Fig. 8.9(a). There, $\mathbf{a}_1$ and $\mathbf{a}_2$ denote lattice vectors in real space. The square and triangular lattices in reciprocal space are displayed in Figs. 8.9(b) and (c), respectively, where $\mathbf{b}_1$ and $\mathbf{b}_2$ represent reciprocal lattice vectors. The inset shows the BZ constructed from $\mathbf{b}_1$ and $\mathbf{b}_2$ as the Wigner-Seitz cell of the reciprocal lattice.

The BZ is quadratic, and therefore has additional symmetries, which reduce the part of $\mathbf{k}$ space that has to be considered. The triangle (ΓMX) is the smallest area that can be mapped to the whole BZ by mirror or rotation operations, and contains all of the nonredundant information. This smallest possible region is the irreducible BZ. Therefore, it is sufficient to calculate the band structure for a closed path along the lines connecting the high symmetry points of the first (irreducible) BZ.

Now consider a square lattice with spacing α, the lattice vectors $\mathbf{a}_1 = \alpha\mathbf{x}_0$ and $\mathbf{a}_2 = \alpha\mathbf{y}_0$ being along the x and y directions, respectively ($\mathbf{x}_0$ and $\mathbf{y}_0$ are the corresponding unit vectors). Relational Eq. (8.19) can be used to determine which are the reciprocal lattice vectors and which are the real ones. Hence, the reciprocal lattice vectors for this 2D square lattice are $\mathbf{b}_1 = (2\pi/a)\mathbf{y}_0$ and $\mathbf{b}_2 = (2\pi/a)\mathbf{x}_0$. We note that the reciprocal lattice is also a square lattice, but with spacing $(2\pi/a)$ instead of a. Similarly, reciprocal lattice vectors for a given triangular lattice can be calculated using Eq. (8.19).

For light propagating in the plane of periodicity, the modes can be separated into two independent polarizations, namely TE (transverse electric, in which the electric field lies in the plane of periodicity and the magnetic field is perpendicular to it) and TM (transverse magnetic, in which the magnetic field lies in the plane of periodicity and the electric field is perpendicular to it). For a proper choice of lattice, the PC can have a PBG in the plane of periodicity. As an example, consider a square arrangement of rods with dielectric constant 10.24 (tin disulfide) and radius $0.2a$ in air. We already know from our 1D example that a PBG will appear at the border of the BZ. Figure 8.10 shows a calculation of the band structure for the TE and TM polarizations. As expected, a TM gap appears (for normalized frequencies ranging from 0.304 to 0.433), while there is no TE gap. The TM gap has a GMR of 34.99% and is limited by the first band at the point M and by the second band at the point X. There is also a partial PBG in the Γ–X direction for TE modes.

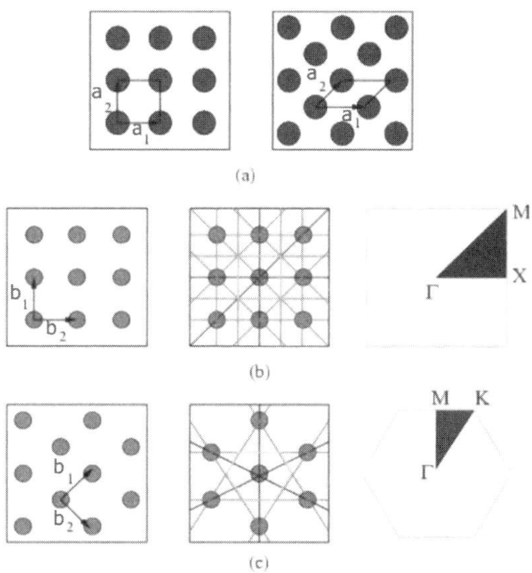

Figure 8.9 (a) 2D rectangular and triangular lattices in real space. The vectors $\mathbf{a}_1$ and $\mathbf{a}_2$ are the primitive lattice vectors. (b) 2D rectangular lattice in reciprocal space and its corresponding Brillouin zone. The darkened triangular region is the irreducible part of the BZ. High-symmetry points are represented by Γ, M and X. (c) 2D triangular lattice in reciprocal space and its corresponding BZ. The vectors $\mathbf{b}_1$ and $\mathbf{b}_2$ are the reciprocal primitive lattice vectors.

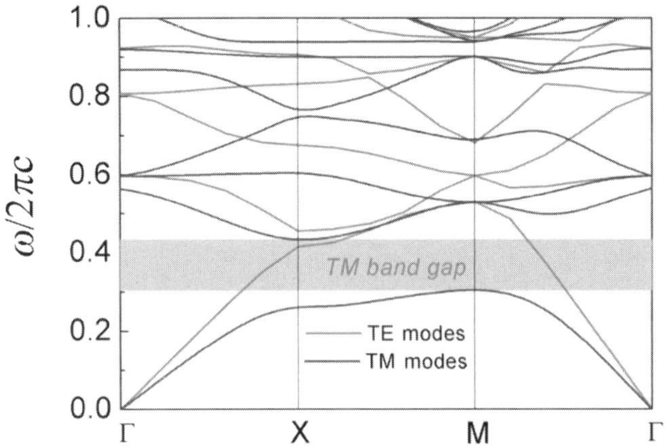

Figure 8.10 Band structure for a square lattice of refractive index 3.2 rods with radius 0.2*a* in air.

To explain the appearance of the PBGs according to Eq. (8.36), we will now analyze the distribution of the electric energy $\varepsilon(\mathbf{r})|\mathbf{E}(\mathbf{r})|^2$ of the first two TE and TM bands at the point X. Figures 8.11(a) and (b) show the energy for the first TM band (here, light grey (rods) indicates high intensity values and dark grey low intensity values). Since the **E** field points along the rod axis and is therefore always parallel to the dielectric interface, the energy can be confined very strongly in the region of high-refractive-index dielectric material. Consequently, this leads to the lowest frequency in the eigenvalue spectrum. A mode of higher order will now require an extra node in the crystal plane because there is no additional degree of freedom, owing to the quasi-scalar character of the z-polarized electric field. This nodal line, as can be seen from Fig. 8.11(a), has to go through the center of the rods because the mode has to be orthogonal to the first one (mode-orthogonality theorem). This pushes a significant part of the energy out of the high dielectric into the air region and causes an energy shift that leads to the large gap.

The situation, however, is different for the TE modes. As the electric field vector lies in the crystal plane, there are additional possibilities for the energy to localize. The electric field vector can be orientated perpendicular or parallel to the interface. For the lowest order [Fig. 8.11(b)] the field lines tend to be parallel. This indicates a perpendicular crossing at the two sides of the rods. The corresponding discontinuous increase in energy is clearly observed in Fig. 8.11(b). This causes a high localization in the low-index dielectric. The second-order band in Fig. 8.11(b) has a more complex structure in the field distribution and even higher low-index localization. However, the difference within the first-order band is obviously much smaller than in the TM case.

We now understand that PBGs arise from the net interferences of scattered incident light waves from the lattice points of a periodic structure. Here, we stress that high refractive index contrasts of the periodic structures play a vital role in order for the PBGs to become more pronounced for a given structure. There are two reasons for the importance of high refractive index contrasts.

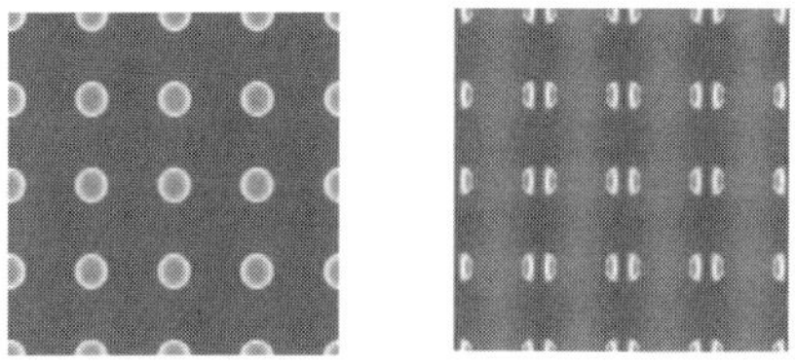

(a)

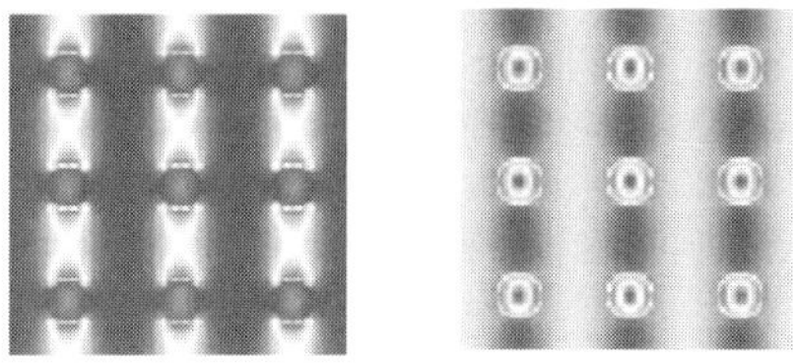

(b)

Figure 8.11 Electric field energy density for a square lattice of refractive index 3.2 rods with radius 0.2*a* in air (a) at the X point for the first and the second TM band and (b) at the X point for the first and the second TE band.

First, each PC structure has a minimum value of refractive index contrast to exhibit a complete PBG. This phenomenon originates from the fact that combinations of two dielectrics with high refractive index contrast tend to more strongly scatter waves (from any direction) compared to low refractive index contrast dielectric mixtures; therefore, both partial and complete PBGs are more likely to occur in structures that contain high refractive index contrast materials. Second, the higher the refractive index contrast, the fewer layers are necessary to have sufficient PBG effects. As previously explained, each layer or lattice in a PC partially reflects the propagating wave. As a consequence, the higher the refractive index contrast, the higher is also the reflection coefficient per layer. Sufficient net reflections can thus be achieved by fewer layers of high refractive index contrast, as compared to a structure with the same configuration but with a lower refractive index contrast.

8.4.2 Three-dimensional photonic crystals

Although 2D PCs display many of the properties of 3D PCs, they lack one very obvious, yet important, capability: they cannot confine light in the third direction. Three-dimensional control of photons can be achieved by a 3D periodic dielectric structure, i.e., a 3D PC.

The idea of a 3D periodic dielectric structure as a means of controlling spontaneous emission was first proposed by Eli Yabolonovitch.[1] The motivation was to create a structure in which the PBG would overlap the electronic gap, thereby making it possible to improve the performance of semiconductor lasers, heterojunction bipolar transistors, and solar cells. This idea was independently proposed by Sajeev John while studying the phenomenon of localization of light in disordered dielectric superlattices.[2] Several structures based on the face-center-cubic (fcc) lattice were experimentally fabricated by Gmitter and Yabolonovitch,[27] and transmission was measured in search for PBGs. We note that the BZ of an fcc lattice has more spherical symmetry and that it is more likely for a full, 3D PBG to open up when the PBGs along individual directions overlap. Following this approach it was found that a suitably designed structure did exhibit a 3D PBG. This was an inverse fcc structure, which is now known as an inverse opal.

A more systematic search for a structure with a full PBG ensued when theorists started looking at this problem.[28–31] The first attempt was to solve the 3D case with the scalar wave approximation by decoupling Maxwell's equations. The results predicted PBGs for the inverse fcc structure made by Yablonovitch, but also for the fcc structure made out of dielectric spheres, which contradicted experimental results. The quantitative agreement between the experimental and theoretical values was not good, suggesting that the vector nature of the electromagnetic field could be crucial and should not be neglected. Calculations were made using the PWE method developed earlier,[28,29] incorporating full-vector waves, and it was discovered that the fcc structure did not, in fact, posses a full PBG, owing to degeneracy of the bands at the W and U points. This degeneracy could not be lifted, even for refractive index contrast as high as 4.0 (germanium) and filling fraction of 96%.[25,32,33] Ho and co-workers proposed a way to lift this degeneracy by choosing the diamond lattice with dielectric atoms.[25,34] For a fixed dielectric constant of 3.6, a full PBG was found to be present for dielectric spheres, as well as for air spheres, and for a good range of filling fractions. The diamond structure was experimentally fabricated by drilling cylindrical holes into a dielectric material.[3] In good agreement with theory, measurements on this structure verified the presence of a full PBG.

Following the success of the diamond structure, experimental fabrication of different structures began. A simple layer-by-layer approach was designed,[35] which also exhibited a full PBG. For micron- and submicron-length-scale materials, the theoretical community turned to the fcc structure. The reason was two-fold: first, further band-structure calculations indicated that the fcc structure did posses a complete PBG between the eighth and ninth bands.[36–37] Second, the colloidal self-organization of monodisperse submicron spheres[14] made it easier for such structures to be fabricated. It has long been known that natural opals have an interesting optical property: the wavelength of the reflected color changes when viewed at different angles under white light illumination. This phenomenon is known as opalescence. In 1964, Sanders, Jones, and Segnit[38] discovered that opals were composed of spherical particles of amorphous silica in the size range of 1.5–3.5 μm and are arranged hexagonally in layers, which are either in a random close-packed fcc, or a hexagonal close-packed (hcp)[39] geometry. Optical properties for probing possible PBGs, such as transmission, reflection, and diffraction spectra, were first measured in artificial opal structures made from monodisperse polystyrene colloids.[40,41]

As the direct fcc structure does not show a full gap, efforts were made to produce an inverted fcc structure using the colloidal crystal as a template. One of the early attempts produced a porous network of titania using emulsion templating.[42] The technique demonstrated that an interconnected network of uniform pores in a titania background could indeed be produced. However, the structure did not show any crystalline order. This approach was followed by a number of techniques in which a template of fcc-ordered silica or microspheres was first synthesized and then infiltrated with different background materials, either as ceramic precursors,[9,43] metals and polymers,[44,45] or semiconducting nanoparticles.[46] The template material was subsequently removed by calcinations in the case of polystyrene, or by etching with HF in the case of silica templates. Samples produced using ceramic precursors and sol gels typically yielded small sample sizes owing to breakage. In yet another approach, colloidal crystallization was shown to occur simultaneously with the introduction of a high-refractive-index background material. This technique was used to produce thin-film inverse fcc-type[47] PCs.

8.5 Physics of Inverse-Opal Photonic Crystals

Basic properties of inverse opals, such as their band structures and variations in geometrical and material engineering, are extremely crucial for understanding their optical characteristics, as well as for conceiving their potential applications. Therefore, in the following sections we will study the influence of the geometrical and material parameters of inverse opals on their photonic band structures and PBGs.

8.5.1 Introduction

Opals are among the few PCs existing in natural states, e.g., in the wings of butterflies and in various minerals. They are also among the most colorful of all gems, despite being composed primarily of silica, a colorless solid with the chemical formula SiO_2. The origin of the colors in opals is an ordered microstructure of closely packed silica spheres, which causes light to diffract from the interface between the SiO_2 balls and the air in the voids between the balls. Since the size of these silica spheres is on the order of hundreds of nanometres, the range of wavelengths of the diffracted light falls within the visible region.

On the other hand, inverse opals are inverse replicas of opals. Instead of consisting of a regular arrangement of uniform spherical particles (as in opals), inverse opals consist of a regular arrangement of spherical void spaces surrounded by solid walls. Moreover, the ordered arrangement of the pore structures leads to diffraction of light in a manner similar to the diffraction observed with opals. These diffraction effects endow inverse opals with optical and photonic crystal properties. This class of crystals has distinguished itself as easier to fabricate by self-organizing processes[48,49] on a relatively large scale.

The geometrical distribution of a dielectric material in an inverse opal can be most easily understood by concisely reviewing the fabrication process: latex spheres of radius R_2 are allowed to sediment slowly in a liquid, e.g., an alcohol.

Owing to self-organizing processes, the spheres settle arranged in an fcc close-packed lattice. Then the liquid is evaporated and the system infiltrated by a high-refractive-index material. Ideally, as will be assumed here, all voids between the spheres are completely filled. In practice, however, the high-refractive-index material is often slightly porous due to inhomogeneous filling. After the infiltration, the system is heated, and the latex spheres are also evaporated, leaving an fcc arrangement of air–spheres in a dielectric matrix. If the refractive index of the remaining material is high enough (> 2.85), a full PBG, located between the eighth and ninth bands, is expected according to theoretical calculations.[14, 33]

Figure 8.12 shows the fcc lattice and its corresponding first BZ, where the circles indicate the high-symmetry points Γ, L, X, U, W, and K. The points L, X, and K correspond to the (111), (100), and (110) directions, respectively. In particular, the (111) direction is of prime interest, since it corresponds to the growth direction of the photonic crystal and is, therefore, the most convenient direction for optical measurements. In Fig. 8.12, the area enclosed by the high-symmetry points can be mapped to the whole BZ by symmetry operations (translations, rotations, inversions, reflections), and contains all nonredundant information. This region of the first BZ enclosed by the set of high-symmetry points is called, as mentioned before, the irreducible BZ. Within it, we already know from 1D examples that a PBG exists for all **k** in a given direction, but the maxima and minima of each band are always lying on the edge of the first BZ. Therefore, it is sufficient to calculate the band structure for a closed path along the lines connecting the high symmetry points, i.e., Γ, L, X, U, W, and K.

Here we will investigate two possibilities in the modeling of inverse opal PCs: model 1, in which overlapping spheres have a radius larger than the closed packed radius (0.35*a*, where *a* is the lattice constant); and model 2, in which spheres are connected by channels or air cylinders (the radius of the spheres is equal to the close-packed radius).

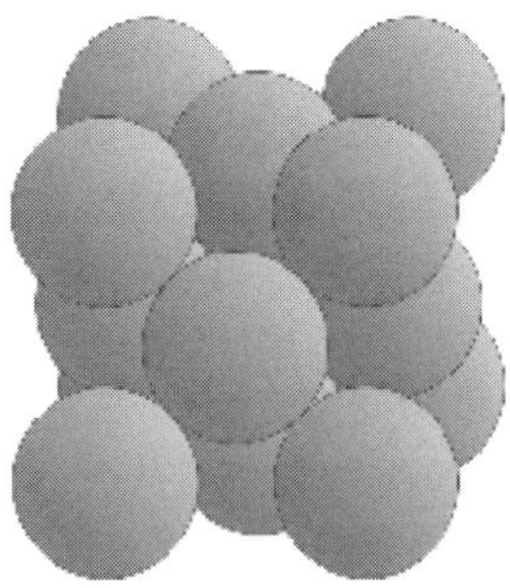

(a)

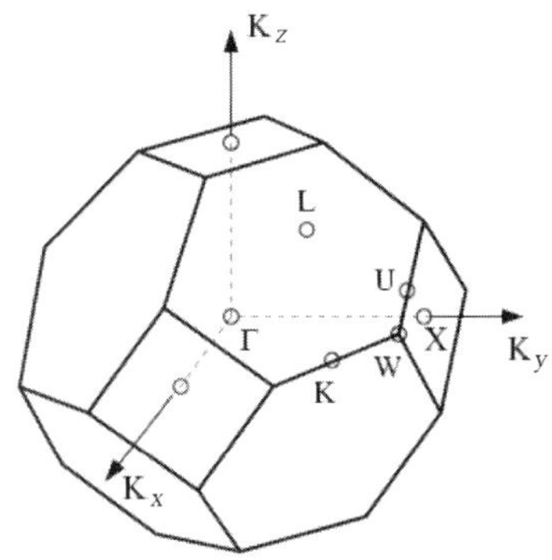

(b)

Figure 8.12 (a) An fcc structure and (b) its corresponding first BZ. Γ indicates the origin of the reciprocal space, while K, L, X, U, and W are high-symmetry points on the edges of the Brillouin zone.

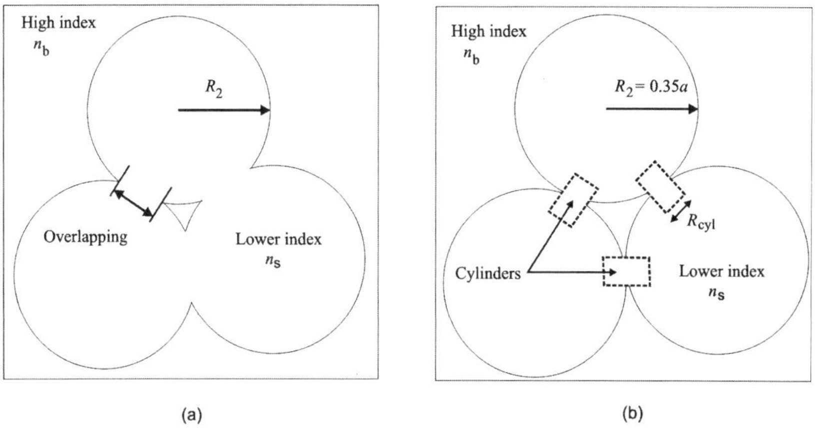

Figure 8.13 Schematic representation of the two different models used in this work. (a) model 1: overlapping spheres and (b) model 2: spheres connected by channels or cylinders. Here, R_2 is the sphere radius, n_b is the background dielectric constant, and n_s is the dielectric constant of the spheres.

A schematic representation of these two models is shown in Fig. 8.13. We should note here that both models produce similar results for similar filling fractions. The reason for choosing this approach is that the experimental structures lie between the two limiting cases of model 1 and model 2. Moreover, model 1 is useful for study of the direct influence of sphere size on the optical properties of inverse opals, whereas model 2 is useful for study of the influence of connecting channels on the optical properties of inverse opals.

For a numerical simulation tool, we used MIT Photonic-Bands (© 2002 Massachusetts Institute of Technology), a frequency domain iterative code, to perform a direct computation of the eigenstates and eigenvalues of Maxwell's equations using a plane wave basis.[50,51] Before starting our studies with that software, we performed a convergence test and found a good convergence of the results; e.g., for a frequency tolerance 10^{-7}, grid size $48 \times 48 \times 48$, and mesh size 15. The mesh size is a subgrid laid on each grid point in order to average the dielectric constant over a "mesh" of points and find an effective dielectric tensor.[50]

8.5.2 Inverse opals with moderate-refractive-index contrast

A schematic illustration of inverse opals, which are investigated in this section, is shown in Fig. 8.14. In this figure, n_s represents the refractive index of the spheres, and R_2 stands for the sphere radius.

We begin our study with the optical properties of inverse opals composed of close-packed air spheres embedded in a titania (TiO_2) background with refractive index 2.5 (Fig. 8.14). It is worth mentioning at this point the reason for choosing titania as the background material. Titania is a good candidate for future applications of PCs (particularly in relation to functional windows in the visible regime) because it is one of the few high-refractive-index materials that exhibits low absorption. Moreover, the infiltration of titania is a more homogeneous process and leads to a less porous material compared to that for higher-refractive-index materials, the latter of which is still at the optimization stage.[7]

The band structures depicted in Figs. 8.15(a) and (b), obtained using model 1, are for an inverse opal of close-packed air (n_s = 1.0) and silica (n_s = 1.5) spheres, respectively, embedded in titania. Both structures possess only partial PBGs between the second and third bands at the points L and X, respectively. They do not have a complete PBG because the dielectric contrast is lower than the threshold refractive index of 2.8 that is necessary for a complete PBG to appear. In Fig. 8.15(a), the magnitude of the GMR between the second and third bands at the L point is found to be 19.39%, with a midgap frequency of 0.60. In Fig. 8.15(b), the corresponding magnitude of the GMR is 11.24%, with a midgap frequency of 0.49, which is smaller than that in the first case. This can be explained by the decrease in dielectric contrast between the two materials constituting the PC.

As mentioned earlier, the presence of PBGs in PCs can be explained through various means. Here, we choose to analyze the electric field energy distributions of the bands limiting the band gap at the relevant high-symmetry points. The electric energy distribution for both the second and third bands at the points L and X, respectively [for the band gap of Fig. 8.15(a)], is shown in Fig. 8.16. We should recall here (from the electromagnetic energy variational theorem) that the lower band tends to concentrate its energy in the high-refractive-index dielectric regions in order to lower its frequency, while the higher band tends to concentrate its energy in the lower dielectric region, leading to a raise in its frequency and to the formation of a frequency gap, i.e., a PBG. In a 3D PC, the field energy distributions are more complicated; in Fig. 8.16(a) both bands seem to concentrate their energies in the higher dielectric regions. However the second-band energy is more concentrated inside the high-dielectric region (70%) than it is in the third band (53%). The difference in energy concentration results in the splitting of the frequency at the L point, giving rise to the band gap between the second and third bands [Fig. 8.15(a)]. On the other hand, at X point, the second- and third-band energies [Fig. 8.16(b)] in the higher dielectric region are found to be 71% and 67%. The small energy difference between these bands results in the correspondingly small band gap width.

Now we consider the case of Fig. 8.15(b). The energy distributions of the second and third bands at the points L and X are shown in Fig. 8.17. In Fig. 8.17(a) the energy of the second and third bands at the L point is continuous in the higher dielectric regions. However, some amount of energy of the third band is slightly pushed into the silica spheres. The quantitative values of energies localized by the second and third bands at the L point are found to be 57% and 33%. This difference in energy of the second and third bands explains the band gap. At the X point, the energy of the second band is concentrated primarily in the higher dielectric region. The energy of the third band is also found to be concentrated in the titania. The energies [Fig. 8.17(b)] of the second and third bands localized in titania are found to be 52% and 50%, which results in a very small band gap width, shown in Fig. 8.15(b). Indeed, the relative energy difference between the second and third bands in the higher dielectric regions at the X point for inverse opal of air spheres in titania is found to be 13%, which is much larger than that found for inverse opal of silica spheres in titania (4%). Therefore, the band gap width between the second and third bands for inverse opal of silica spheres in titania is much smaller than that for inverse opal of air spheres in titania at the X point.

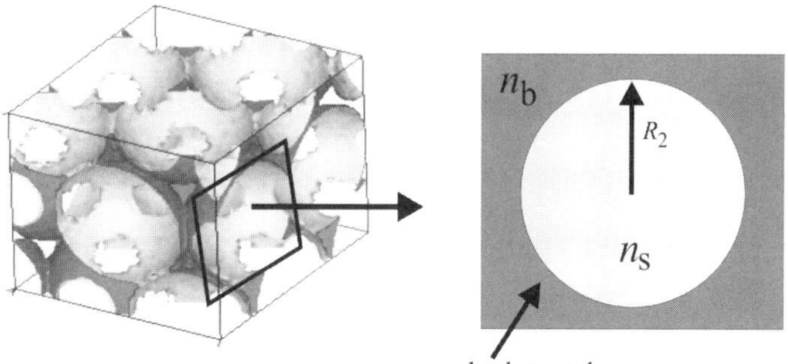

Figure 8.14 Schematic representation of an inverse opal, in which titania (TiO_2) is the background material with refractive index 2.5, and the sphere has a refractive index n_s and a radius R_2.

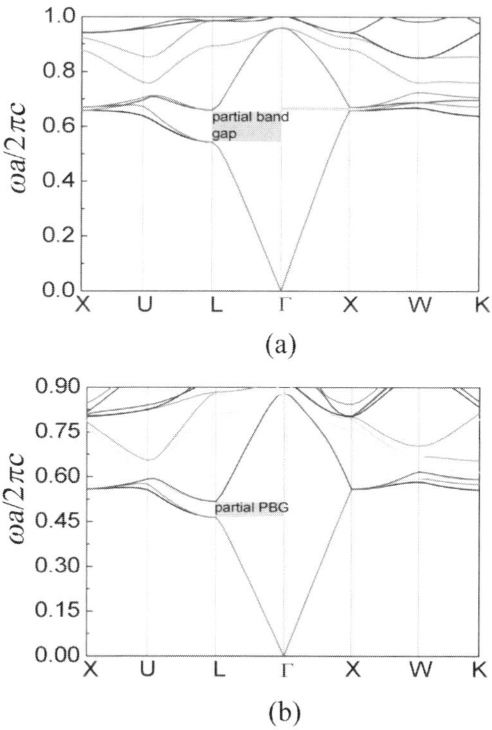

Figure 8.15 Band structures of inverse opals of close packed (a) air and (b) silica spheres, in titania (refractive index 2.5), calculated using model 1.

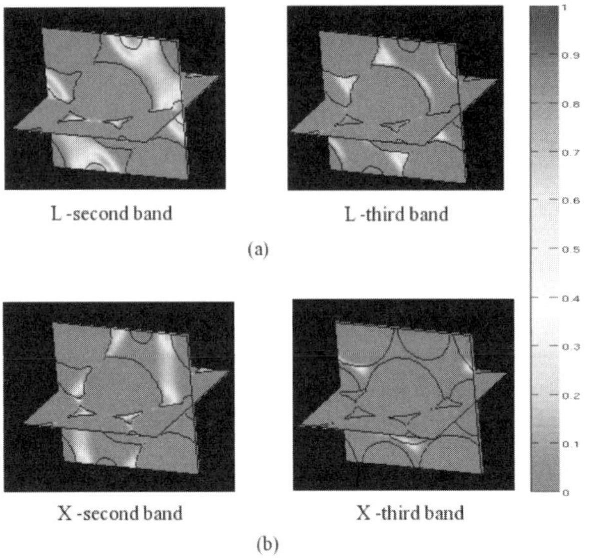

Figure 8.16 Electric field energy density distributions of the second and third bands at the points (a) L and (b) X in an inverse opal of close-packed air spheres (R_2 = 0.35a) in titania (model 1). The black contours indicate horizontal and vertical cross sections of the dielectric structures. The electric field energy density distributions are shown in real space for the modes corresponding to the X and L points of the reciprocal space. The horizontal and vertical cross sections shown are perpendicular to the (111) and (100) crystallographic directions. In each plot, the maximum occurring electric field energy density distribution was normalized to a value of 1 (dark blue). The corresponding reciprocal space band structure is shown in Fig. 8.15(a), and the real space structure is shown in Figs. 8.13(a) and 8.14. (See color plate section.)

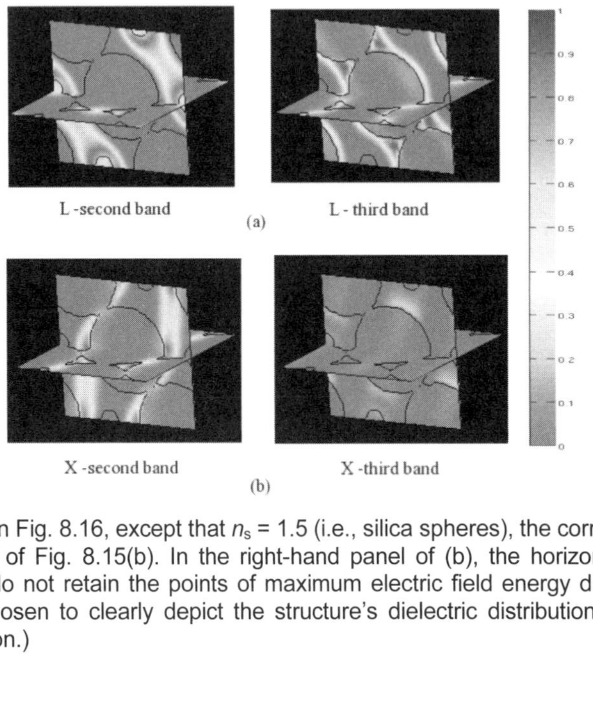

Figure 8.17 As in Fig. 8.16, except that n_s = 1.5 (i.e., silica spheres), the corresponding band structure is that of Fig. 8.15(b). In the right-hand panel of (b), the horizontal and vertical cross sections do not retain the points of maximum electric field energy distributions; they were instead chosen to clearly depict the structure's dielectric distribution contours. (See color plate section.)

The band structure and eigenmode calculations presented so far have assumed an infinitely extended crystal. Obviously, this can never be realized in real applications. An important question is, therefore, how a finite PC responds to an external excitation. This can be addressed by studying transmission spectra of the structures. To verify the band gap width, we calculated the transmission spectra of the inverse opal of close-packed air and silica spheres in a titania background, using model 1, with the finite-difference-time-domain (FDTD) method.

In Figs. 8.18(a) and (b) we compare the band structures of an infinitely extended inverse-opal PC of air and silica spheres in titania (in the Γ–L direction), with the transmission through a finite system of a plane wave propagating along Γ–L with normal incidence. This transmission calculation is performed for a finite crystal size of 20 layers having an ABCABC arrangement (i.e., six lattice constants, as shown in Fig. 8.19) with resolution of 50 grid points per lattice constant. In Fig. 8.18(a) the first band existing in the frequency range

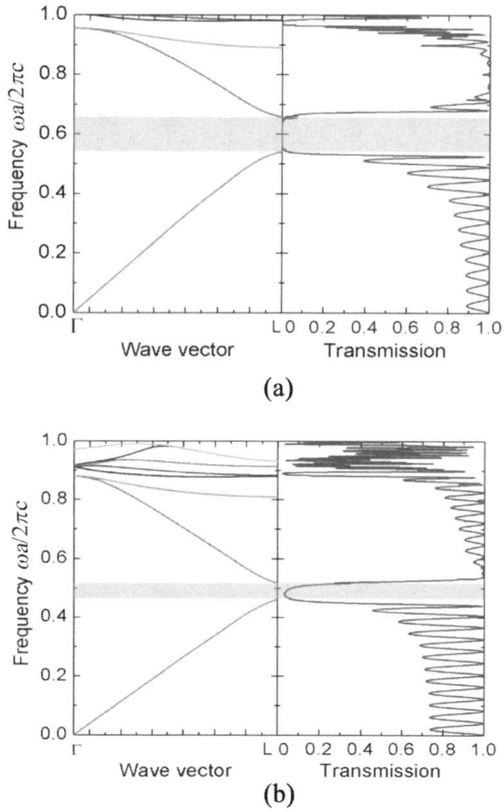

Figure 8.18 Band structure (left) and transmission (right) through inverse opals of close-packed (a) air and (b) silica spheres, respectively, in titania, calculated using model 1. Here the transmission spectra are calculated for electromagnetic waves propagating along the Γ–L direction. 20 crystal layers are considered in these calculations. The resolution per lattice constant (three layers) is 50 grid points.

(0.0–0.542) corresponds to high transmission, modulated only by Fabry-Perot oscillations that are a result of the finite length of the structure. This can be explained by the large spatial overlap of the plane wave's first-order band because the latter makes a large contribution to the lowest-order Fourier components. When reaching the first band gap edge at frequency 0.543, the transmission becomes smaller. The reason for this is the flattening of the band (indicating smaller group velocity) near the band edge, and therefore a poor coupling, owing to impedance mismatch of the plane wave and the photonic crystal mode. Within the band gap frequency range of the band structure calculation, from 0.543 to 0.659, there is no transmission. A perfect fit in the position of the PBG can be observed between the two simulations. As the frequency becomes larger than 0.659, we again observe a high transmission through the structure. In Figs. 8.18(a) and (b) the transmission through the crystal in the case of silica spheres in titania is larger than the transmission through the air spheres in titania. This is due to the difference in index contrasts within the two structures. In the case of the silica/titania PC, the index contrast is smaller, and consequently, the diffraction at the dielectric interfaces is smaller. Therefore, the attenuation per lattice constant within the PBG is smaller. It has thus been found from our previous simulation results that the titania inverse opals are examples of structures exhibiting partial PBGs. Now, it is interesting to investigate how these PBGs vary with the geometrical parameters.

Figure 8.20 shows the dependences of the GMRs of the partial band gap on the sphere radius at points L and X for an inverse opal with $n_s = 1.0$. This figure indicates that the magnitude of the GMR at the L point reaches a maximum value for a close-packed arrangement. On the other hand, the variations in the GMR

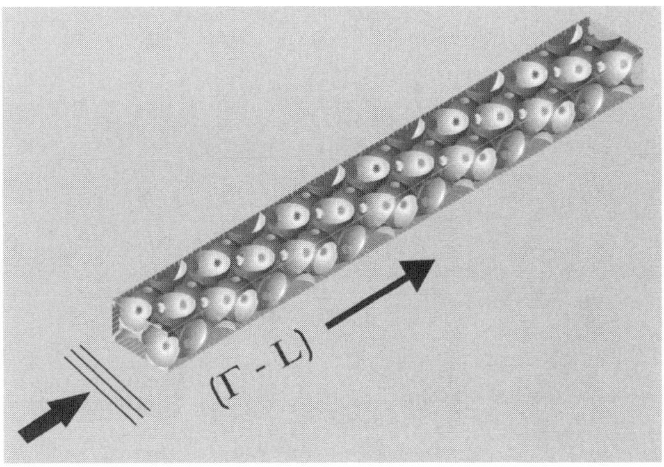

Figure 8.19 Structure used to calculate the transmission spectra in the Γ–L: 20 layers (6 lattice constants) of air spheres in a titania background. The light and darker greys represent the air spheres and the titania background material, respectively. The large arrow shows the incident plane wave used in the calculation, and the three parallel lines schematically represent the corresponding wavefronts.

magnitudes at the X point are more complex because of the several bands that are present at this high-symmetry point. The minima are observed for radii of approximately 0.35*a*. For larger radii, the spheres overlap very strongly, and the volume fraction of titania is very small. Consequently, the PBG for each point of the band structure is strongly reduced until the GMR altogether vanishes (for sphere radii above 0.47*a*).

Let us now investigate the variation of the GMRs with the refractive index of the spheres, exploiting model 1. Figure 8.21 displays the dependences of the GMR on the refractive indices of the spheres (n_s) for the first partial PBG at points L and X of the close-packed inverse opal with a titania background. The refractive index of the spheres is varied from 1.0 to 3.0. To explain the behaviour of the PBG shown in Fig. 8.21, the curves are divided into two regions: inverse opal region for n_s varying from 1.0 to 2.5 (region A) and opal region for n_s varying from 2.5 to 3.0 (region B). Generally, in PCs the size of the PBG depends on the dielectric contrast and, as we have previously pointed out, the PBG arises only when there is a difference in the field energy between the bands. In region A, increasing n_s decreases the dielectric contrast, and consequently, the PBG size decreases. At the particular value n_s = 2.5, the structure has no dielectric periodicity and becomes a bulk titania. Therefore, no PBG exists and the GMR is zero for both points L and X.

In region B, the dielectric contrast increases again with increasing n_s and, as a result, the PBG size increases. The behaviour observed in Fig. 8.21 highlights the strong influence of the index contrast in the inverse opal on the resulting PBGs of the structure. To complete this study, the next section will consider structures with higher-refractive-index contrasts by increasing the background refractive index.

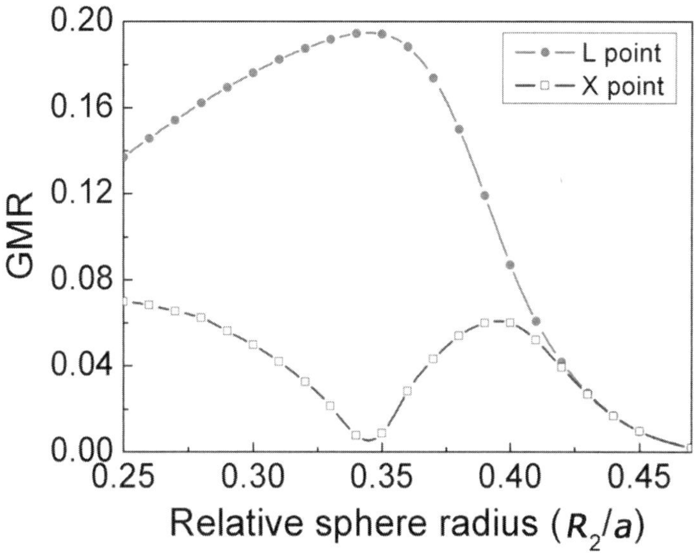

Figure 8.20 Dependence of the GMR on the radius of the air sphere for an inverse opal of air spheres in titania.

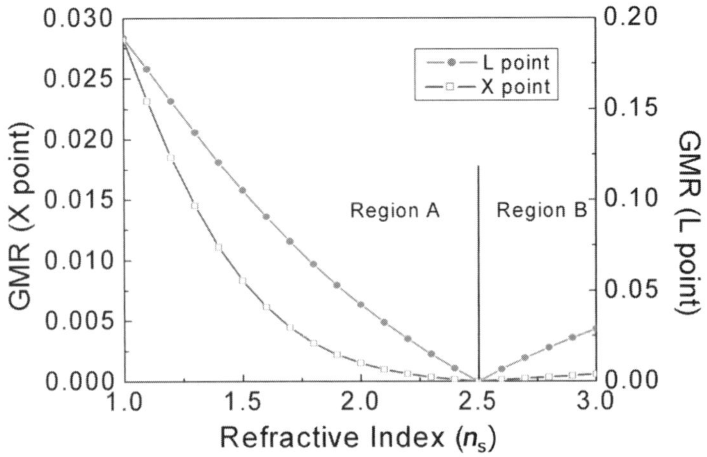

Figure 8.21 Dependence of the GMR at the points L and X of inverse opals composed of close-packed spheres of varying dielectric material in a titania background.

8.5.3 Toward a higher-refractive-index contrast

In this section, we will discuss inverse opals with a higher-refractive-index contrast compared to the inverse opals presented in the previous section. To this end, we consider background materials having higher refractive indices compared to the previously studied titania. We start this study by presenting the band structure of the inverse opal with background refractive index $n_b = 3.2$ and refractive index of the sphere $n_s = 1.0$, illustrated in Fig. 8.22. This structure possesses a partial PBG between the second and third bands at both points L and X. The magnitudes of the GMR at points L and X are found to be 22.52% and 2.69%, respectively—larger than those found for an inverse opal of air spheres in titania. The increase in the width of the partial PBG can be explained by the increase in the dielectric contrast of the materials comprising the structure. Compared to the two previous cases, the dielectric contrast of the system is higher than the threshold value 2.8 needed for a complete PBG to appear. As expected, a complete PBG between the eighth and ninth bands is observed. This complete PBG is limited by the W point at the eighth band and by the X point at the ninth band. The magnitude of the complete GMR is found to be 3.21% with the midgap frequency 0.85.

On the other hand, Fig. 8.23 shows the band structure of an inverse opal of air spheres in a silicon (n_b = 3.4) background. This structure also exhibits a complete PBG of 4.70% between the eighth and ninth bands, limited by the W and X points, with a (normalized) midgap frequency 0.81. Again, the larger complete PBG compared to the tin disulfide case is owing to the increase in the dielectric contrast between the materials comprising the PC. Moreover, a pronounced partial PBG is also observed between the second and third bands at points L and X, respectively. The GMR at the L point is found to be 23.47% with the midgap frequency 0.46, which is much larger than that found in the case of an inverse opal of air spheres in tin disulfide.

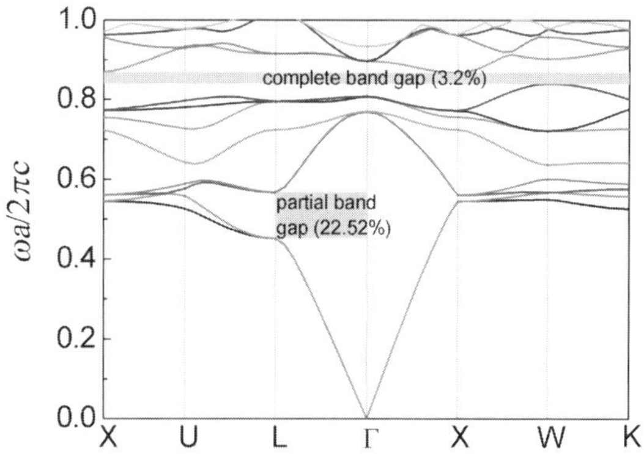

Figure 8.22 Band structure of an inverse opal of close-packed air spheres in tin disulfide (n_b = 3.2), calculated using model 1.

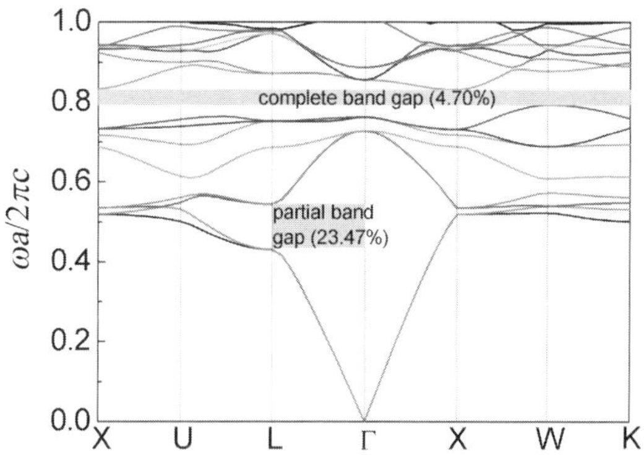

Figure 8.23 Band structure of an inverse opal of close-packed air spheres in silicon (n_b = 3.4), calculated using model 1.

The comparison of the transmission spectrum with the band structure for the inverse opal of air spheres in silicon is shown in Fig. 8.24. This calculation is performed on the basis of model 1, and the electromagnetic wave is considered to be propagating along Γ–L and Γ–X through a crystal of finite size consisting of 20 layers. In the transmission spectrum, the lower zero-transmission region coincides well with the lower band gaps. For higher frequencies in particular, the partial PBG frequencies between the eighth and ninth bands at points W and X overlap, resulting in a complete PBG, as highlighted in Fig. 8.24.

To verify the effect of the variation of the material's dielectric constant on the GMR, we calculated a sequence of band structures for an inverse opal of air spheres in various background refractive indices n_b for close-packed arrangements. As can be seen in Fig. 8.25, the GMR increases with the increasing refractive index of the background material. From this we can verify that the complete PBG appears when the refractive index of the background is greater than 2.8.[14,36] This is the threshold refractive index value required for the complete PBG to appear in an inverse opal with close-packed air spheres.

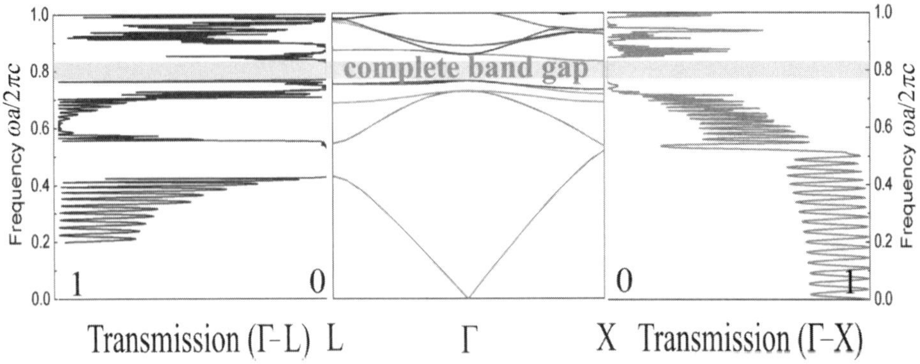

Figure 8.24 Transmission Γ–L (left) and Γ–X (right) and band structure (middle) through an inverse opal of close-packed air spheres in silicon, calculated using model 1. The highlighted region indicates the overlapping of the band gap frequencies between the eighth and ninth bands, at points L and X, resulting in complete PBG. 20 layers of crystals are taken for this calculation. The resolution per lattice constant is 50 grid points.

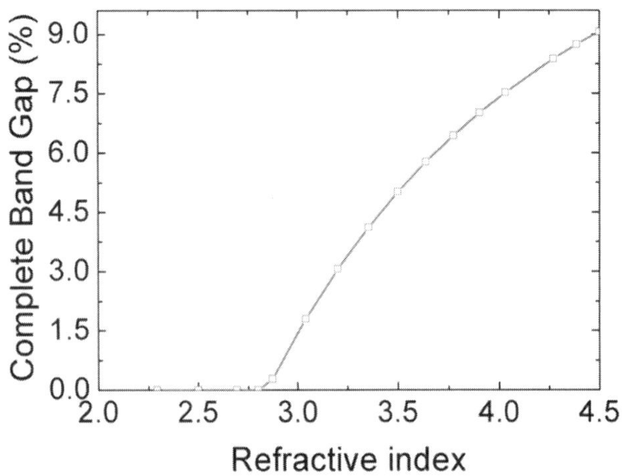

Figure 8.25 GMR as a function of background refractive index for inverse opal of close-packed air spheres in a dielectric background (model 1).

Following detailed computational analysis, we found that the complete PBG of this inverse opal can be optimized by varying the sphere radius (e.g., the size of the voids using model 1). Figure 8.26 summarizes the way the GMR of the complete PBG between the eighth and ninth bands changes with the radius of the spheres. The maximum GMR of 4.92% is found for a sphere radius 0.355*a*, where the spheres slightly overlap, and vanishes for a sphere radius 0.375*a*. For sphere radii below 0.35*a* the spheres are separated by silicon walls, whereas above this value the spheres overlap. Note that while the variation of the radius of the spheres directly affects the existence of the complete PBG between the eighth and ninth bands, we still have the partial PBG at L point between the second and third bands for those radii where the complete PBG vanishes.

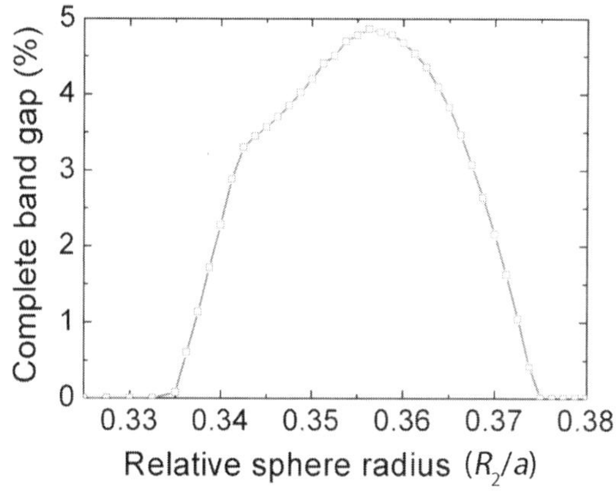

Figure 8.26 GMR as a function of sphere radius for an inverse opal of air spheres in silicon (model 1).

We now consider the influence of the size of the connecting cylinders (model 2) on the optical properties of an inverse opal composed of air spheres embedded in silicon. The results are summarized in Fig. 8.27, where R_{cyl} represents the radius of the cylinders. The maximum PBG is found to be 7.13% for a cylinder radius 0.13*a,* and it vanishes for cylinder radius 0.18*a*.

If we compare the PBG values obtained from the variations of two different parameters (namely, the radius of the spheres and the radius of the cylinders) for the inverse opal of air spheres in silicon background, we may find two different PBG maxima values depending on the type of parameter selected to vary. This result indicates that both sphere radii R_2 and R_{cyl} have to be optimized in order to achieve maximum PBGs. Therefore, the radius of the spheres R_2 and the radius of the cylinders R_{cyl}, as well as the dielectric constant, strongly influence the PBG size. This means that inverse opal structures, in principle, have the potential to yield tunable PBGs via geometrical and material tailoring.

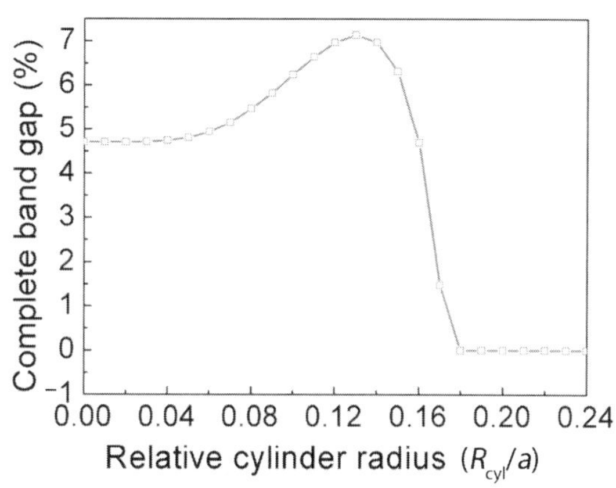

Figure 8.27 GMR as a function of the radius of connecting cylinders for the inverse opal of close-packed air spheres in silicon background.

8.6 Double-Inverse-Opal Photonic Crystals (DIOPCs)

In this section, a novel PC structure is introduced that has additional degrees of freedom in its design when compared to inverse opals. We call this structure a double-inverse-opal photonic crystal (DIOPC). In the next section, we present a comprehensive study of the geometrical and material parameters influencing the PBG in the aforementioned new structure. We conclude by presenting a new, realistic, and promising method for obtaining *complete* PBG *switching* with the DIOPC structure.

8.6.1 Introduction

The DIOPC structure is founded on a 3D periodic arrangement of silica cores with air shells embedded in a high-refractive-index dielectric background (titania). The air shells allow for relative movement of the silica core within them. This distinctive feature of the DIOPC structure provides us with more degrees of freedom. As will be numerically demonstrated below, a change in either the positions of the core spheres within the shells or in their optogeometric characteristics can, indeed, provide a new means of partial PBG tuning, as well as complete PBG switching. Figure 8.28 illustrates a unit cell of a typical DIOPC, composed of a shell sphere with radius R_2 and refractive index $n_s = 1.0$, and a core sphere with radius R_1 and refractive index $n_c = 1.5$, both integrated within a background dielectric material that has refractive index $n_b = 2.5$ or 3.2.

Following a judicious choice of the structure's optogeometric parameters, the work presented here shows that a complete PBG switching of the DIOPC can be achieved by shifting the core spheres in different directions. The principle of the fabrication process and experimental demonstration of such structures have been

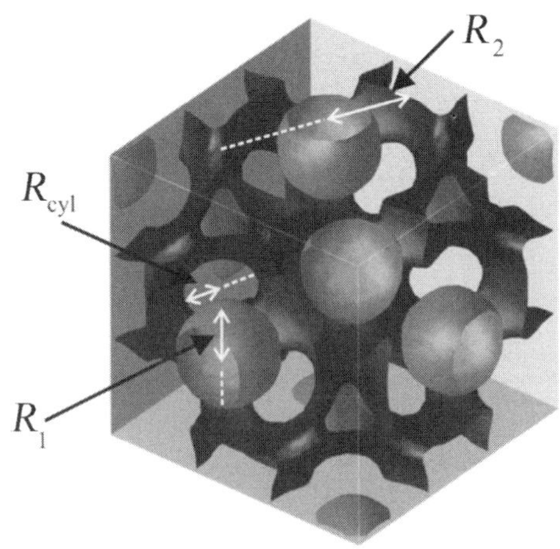

Figure 8.28 Schematic representation of a DIOPC. R_1 and R_2 represent the core and shell spheres' radii, respectively, while R_{cyl} represents the cylinder radius. The refractive indices of the background, core, and shell spheres are considered to be n_b = 2.5 or 3.2, n_c = 1.5, and n_s = 1.0, respectively.

detailed in a series of recent works.[23,50] The silica spheres have inorganic cores (not shown in Fig. 8.28), which in the following calculations are assumed to be considerably smaller than the wavelength of light. These inorganic cores can be addressed by external electric or magnetic fields, thereby allowing for a collective movement of the silica spheres at will.

8.6.2 Photonic band gap switching via symmetry breaking

In DIOPC structures, the air shell allows for movement of the cores within the air pores, opening the door for a completely new method of tuning the optical properties of PC structures. Depending on the position of the cores within the air pores, the fields will be distributed in different ways, and consequently, the optical properties will be affected. As will be shown, this new approach is suitable not only for tuning the partial PBGs, but also for switching the complete PBG, provided the refractive index contrast is high enough in the structure.

Figure 8.29 shows a schematic illustration of the core spheres of the DIOPC shifted in two different directions, namely the (111) and (100) directions. The choice of the (111) direction as a shifting direction is somewhat obvious, since it corresponds to the normal to the DIOPC sample and is therefore the easiest direction to affect by external means. Moreover, the (111) direction corresponds to the "natural" position of the core spheres, provided the interactions with the pore walls are less important compared to the effect of gravity inside the structure. The (100) path is the in-plane direction that is intuitively expected to be primarily affected by the sphere shifting, since the air band of the complete PBG is limited at the X point.

(111) direction

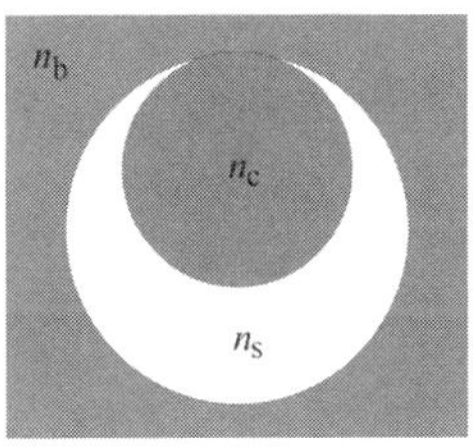

(a)

(100) direction

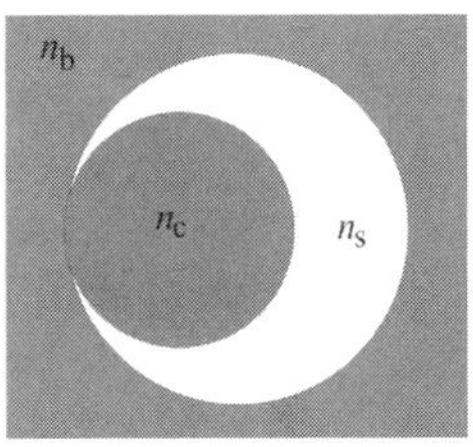

(b)

Figure 8.29 Schematic representation of core spheres of the DIOPC shifted along (a) the (111) direction and (b) the (100) direction.

8.6.3 Tuning of the partial photonic band gap

We now consider the DIOPC structure with off-center core spheres and investigate its potential for PBG tuning. This arrangement may break particular symmetries of the crystal, and therefore, the degeneracy of the high-symmetry points. This has to be taken into account when determining the position of the complete PBG. Regarding the question as to whether it is possible to shift the core spheres in any direction, practical experience suggests that it should be possible to move the spheres at will within the air voids in any direction (e.g., by incorporating small metallic spheres at the centers of the spheres, and then exerting attractive/repulsive magnetic or electric forces). However, focusing on the most recent experimental cases, directions such as the (111) and (100) are normally those of prime interest. Therefore, in our calculations we have chosen the core spheres to be shifted along the aforementioned directions and study the influence of their position and radius on the PBG properties of the DIOPC structure (with titania as a background). The core spheres are shifted in a way that they are always in contact with the titania wall. In the following calculations, the cylinder radius was chosen to be $0.12a$.

Figure 8.30 summarizes the influence of the core spheres' position and radius on the partial PBG between the second and third bands of the DIOPC structure. The core sphere is positioned (within the air void) in three different ways, namely centered cores and shifted cores along the (111) and (100) directions. First, we note from Fig. 8.30(a) that for $R_1 < 0.05a$ the influence of the core sphere radius on the GMR appears to be constant for all three cases. However, as the core radius increases, its influence on the GMR in all three cases is found to be stronger. The centered cores, as well as those shifted along the (100) direction, have a somewhat similar behavior; i.e. the GMR decreases monotonically with increasing values of the core radius. However, in the case of core spheres shifted along the (111) direction, we observe a different behavior compared to the previous two cases. In this case, the GMR first increases with the core radius to a highest value of 19.9% for a core radius $0.15a$, and afterward starts decreasing monotonically. Furthermore, Fig. 8.30(a) indicates that the influence of the core sphere's position and radius on the PBG for the (111) shifting case is stronger compared with the other two cases.

For a quantitative comparison of the GMR values in the three different cases, let us focus on a core radius equal to $R_1 = 0.15a$. We then immediately observe that the GMR value for spheres shifted along the (111) direction is 19.9%, which is larger than the GMR value corresponding to a centered core sphere [17.03%] and a core sphere shifted along the (100) direction (17.86%). Therefore, we may unambiguously conclude that the GMRs are more influenced by core spheres shifted along the (111) direction than in the (100) direction.

The previous conclusions change quite dramatically when we consider the GMR variations at the X point, as shown in Fig. 8.30(b). Unlike the case of the L point, we may now observe that similar behaviour of the GMR dependences on the core radius occurs for the cases where the core spheres are shifted along the

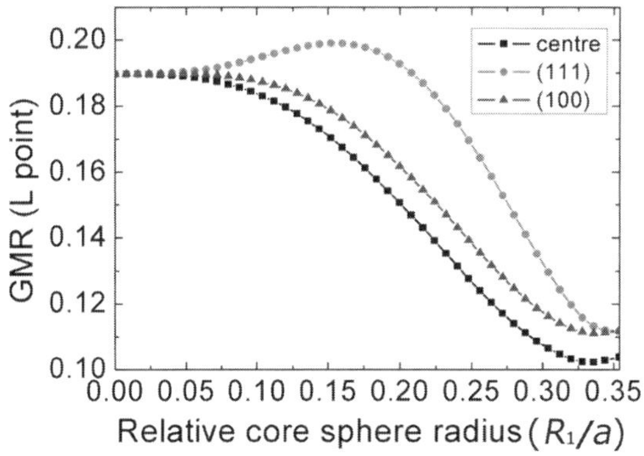

(a)

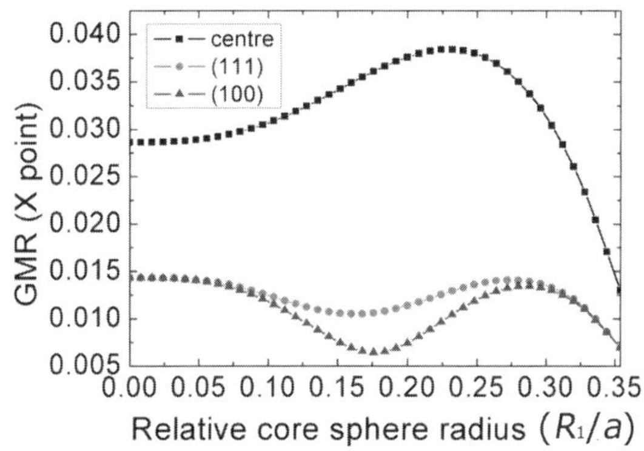

(b)

Figure 8.30 Dependences on R_1 of the GMR between the second and third bands of the DIOPC at (a) the L point and (b) the X point, when the silica core is centered or shifted along the (111) and (100) directions. Here the silica core radius R_1 is varied, while the shell radius is kept constant at $R_2 = 0.35a$ in a titania background, calculated using model 2. The cylinder radius is chosen to be $0.12a$.

(111) and (100) directions. For core spheres shifted along the (111) direction, the GMR at the X point decreases with increasing core radius until it reaches its first smallest value (1.05% for radius $R_1 = 0.168a$). Its global minimum value of 0.69% is reached for a core sphere radius equal to $0.35a$. In the case of core spheres shifted along the (100) direction, we note that the GMR decreases monotonically for core sphere radius $R_1 \leq 0.176a$. It becomes minimum (0.64%) for core radius $0.176a$. Afterward, the GMR begins to increase, reaching its second (global) highest value, then again starts decreasing; it finally assumes the value of 0.69% for core sphere radius equal to exactly $0.35a$.

The case corresponding to the centered core spheres is quite different. Here, the GMR increases with core sphere radius (for $R_1 \leq 0.23a$), taking its highest value of 3.84% for a core radius of $R_1 = 0.23a$. For $R_1 \geq 0.23a$, the GMR values are found to be decreasing. We conclude that the variation with core sphere radius of the GMR at point L is qualitatively the same for the cases where the spheres are shifted along either the (111) or the (100) direction [see Fig. 8.30(a)]. Similarly, the variation of the GMR at point X with core sphere radius [see Fig. 8.30(b)] is, again, qualitatively very similar for the cases where the spheres are shifted along either the (111) or the (100) direction. By contrast, as we see from Figs. 8.30(a) and (b), the variation with the core sphere radius of the GMR at point L (or X) for the case where the core spheres are centered exhibits a distinct and qualitatively different behaviour compared to the aforementioned cases where the spheres are shifted along either the (111) or the (100) direction. It is the interaction between the field and the core sphere that influences the PBG, especially when the core sphere is touching the wall.

8.6.4 Switching of the complete photonic band gap

For applications concerning switchable windows that can be changed dynamically from completely opaque to completely transparent, a partial PBG may not always be sufficient. Therefore, the purpose of this section is to extend the study of the effects of the spheres' position to the case where the PBG is complete. Moreover, we aim to put into evidence the existence of "switching states," which correspond to completely opaque or completely transparent aspects, depending on the positioning of the spheres. In order to obtain a complete PBG of reasonable size, the index contrast in the structure should be increased. Therefore, in the following discussion, we are considering DIOPC structures that have tin disulfide (with $n_b = 3.2$) as a background material.

We recall from Sec. 8.5.3 that an inverse opal of close-packed air spheres in tin disulfide has a complete PBG between the eighth and ninth bands. Similarly, the DIOPC can show a complete PBG between the eighth and ninth band. As discussed in the following paragraphs, the existence or not of this complete PBG strongly depends on the choice of the geometrical parameters, as well as on the actual position of the core sphere.

Figure 8.31 illustrates the band structures of the DIOPC with close-packed air spheres and silica cores of radius $0.186a$ shifted either in the (111) or the (100) direction. In Fig. 8.31(a), the complete PBG between the eighth and ninth band is highlighted by a light grey horizontal zone. We notice that the PBG is open when the silica spheres are shifted along the (111) direction, whereas it is closed when the silica spheres are shifted along the (100) direction. Therefore, we may at this point conclude that the complete PBG is, indeed, strongly influenced by the position of the silica spheres. In order to see how the complete PBG (for both cases) evolves with the geometrical parameters, we next perform a study of the influence of the radii of the connecting cylinders and the silica spheres on the PBG of the DIOPC structure.

Figure 8.32(a) displays the variations of the complete PBG with the relative radius of the cylinder R_{cyl}/a for a core sphere radius $R_1 = 0.186a$. We infer from this figure that there are two regions where the PBG may be completely switched

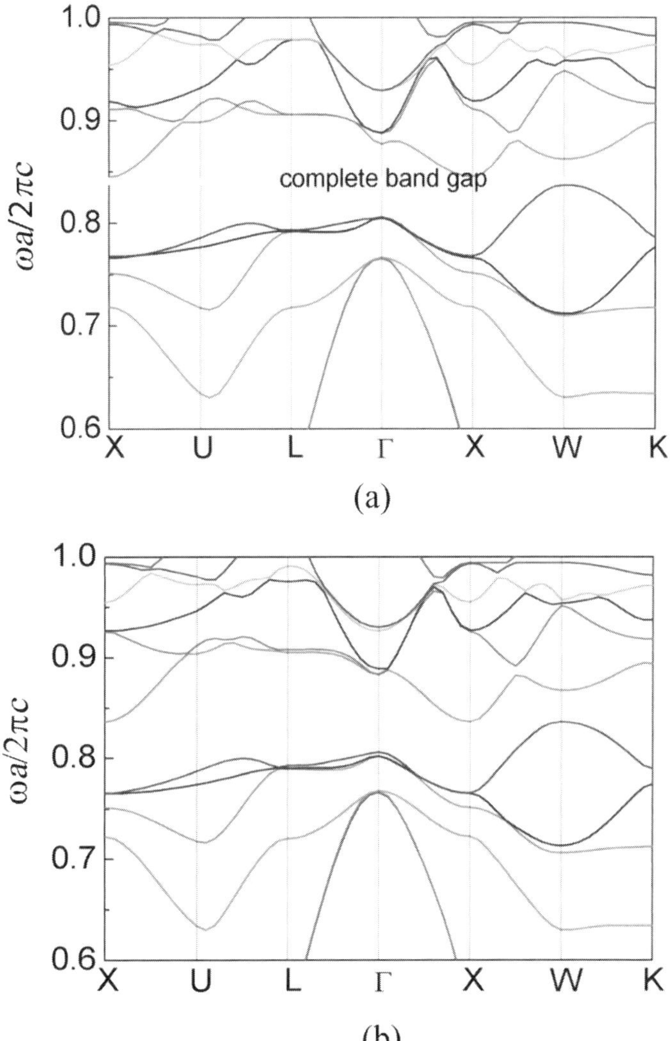

Figure 8.31 Band structures of the DIOPC with silica core spheres shifted along (a) the (111) direction, and (b) the (100) direction, in a tin disulfide, calculated using model 2. These calculations are performed for a core sphere radius of 0.18a, shell sphere radius 0.35a, and cylinder radius 0.035a.

from open to closed, namely: (a) for $R_{cyl} \leq 0.045a$, yielding a complete PBG of about 0.87% and (b) for $0.15a \leq R_{cyl} \leq 0.16a$, yielding a complete PBG of up to 1.4%. The first region is very stable with respect to variations of R_{cyl}. However, it is still not of considerable experimental interest, since the corresponding small values of R_{cyl} are difficult to achieve. In the second region, the range of possible R_{cyl} values is smaller, but these values are experimentally realizable. Furthermore, the PBG size is larger.

Figure 8.32(b) summarizes the influence of the radius of the core spheres on the complete PBG for the two switching states. In these calculations, the cylinder radius was kept constant: $R_{cyl} = 0.0354a$. As expected, the PBG is larger for the (111) shift and decreases with silica core radius. Consequently, we found a small range of R_1 (between $0.185a$ and $0.218a$) where the gap was closed for the (100) shift but remained open for the (111) shift. Therefore, for this range of core radius, the switching of the complete PBG can be realized by means of dynamic control of the core spheres.

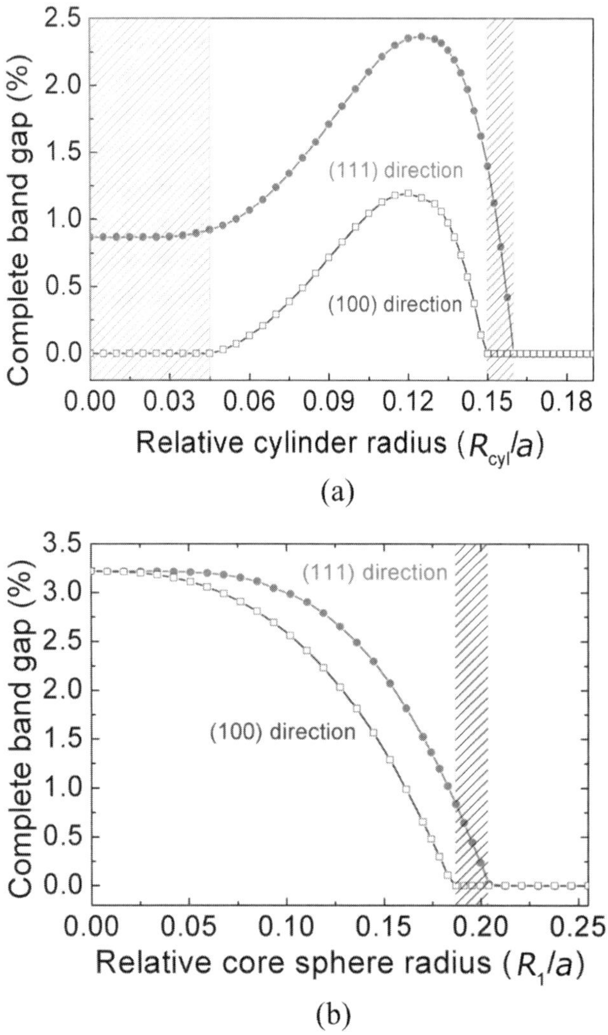

Figure 8.32 GMR for two switching states: sphere shifted in (a) the (111) direction and (b) the (100) direction. The ranges of parameters where a complete switching is possible are marked in shaded boxes. In (a) the core radius is kept constant at $0.186a$, and the cylinder radius is varied. In (b) the cylinder radius is kept constant at $0.0354a$, and the core radius is varied. These calculations were performed using model 2.

8.7 Conclusion

By deploying both the rigorous plane wave expansion (PWE) method and the 3D, full wave, finite-difference time-domain (3D FDTD) algorithm, we have investigated in detail and presented novel results regarding the complex interaction of electromagnetic waves with opal-based photonic crystal structures. Throughout the present work, we have presented a realistic description of opal-based PCs with technological and/or fundamental relevance. The principal PC applications aimed at this work were the realization of functional surfaces and windows. We, therefore, focused on studying the tuning and switching possibilities of opal-based PCs.

As a first step of our study we discussed the so-called "master" characteristic equation, which provides the eigenvalues and eigenmodes of an infinitely extended dielectric structure with discrete translational symmetry, based on a combination of Maxwell's equations and Bloch's theorem. Useful concepts such as the Brillouin zone and band structure were also introduced and explained. By discussing the [illegible] energy localization in 1D and 2D structures, we explained the existence of a photonic band gap (PBG), i.e., a frequency region wherein, under appropriate conditions, no electromagnetic eigenmodes can exist. The influence of the refractive indices and the geometry were discussed by means of several examples. The particular focus on lower-dimensional systems was an excellent first step for the more detailed investigation of higher-dimensionality systems discussed in the succeeding sections.

In the second step of our study, we focused on the optical properties of conventional [illegible] opal PCs. The influence of the geometrical and material parameters was thoroughly studied, paving the way for the discussions in the subsequent sections. After detailed computations, we observed partial band gaps for an [illegible] of [illegible] spheres in titania background (n_b = 2.5) and complete band gaps for [illegible] (n_b = 3.1), as well as for a background in which n_b = 3.4.

Finally, we [illegible] a new type of opal-based photonic crystal, which we called a [illegible] inverse opal photonic crystal (DIOPC). As we explained, this structure [illegible] more degrees of freedom compared to the inverse-opal structure [illegible] a hollow shell, enabling control of the positioning of the core spheres within the hollow shells, and resulting in novel overall optical properties. Depending on the position and size of the core spheres within the hollow shells, as well as on a judicious choice of the core dielectric material, a complete band gap was found for certain positions of the cores. This PBG can close for other positions of the core spheres. As a result, provided that dynamical movement of the cores is maintained, light can efficiently pass through the DIOPC for one position, while being totally reflected for another position. We conclude by proposing the use of this property as a switching mechanism that allows for dynamic changes (i.e., from completely transparent to completely opaque [illegible]) in the appearance of a surface made of a DIOPC.

8.8 Appendix: Plane Wave Expansion (PWE) Method

In this appendix we concisely discuss the numerical method used in the calculations of the PC band structures. Since PCs generally represent complex 2D or 3D vectorial systems, numerical computations form an indispensable part of most theoretical analyses. For our band structure calculations, we used the MIT Photonic-Bands (© 2002 Massachusetts Institute of Technology) online software, which relies on the plane wave expansion (PWE) method. This method operates in the reciprocal space and, in order to analyze a given PC geometry and determine the existence of PBGs, requires the calculation of the allowed mode frequencies for all possible **k** vectors. Fortunately, by utilizing the translational symmetries of the crystals, it is possible to determine the required solutions (eigenfrequencies/eigenvectors of the associated eigenvalue equation) by considering only the **k** vectors that are restricted to the first BZ.

The PWE method for the calculation of photonic band structures has undeniably made substantial contributions to the development of PCs. The similarity between Schrödinger's equation for electrons and the wave equation for light has played an important role in this development. The PWE method is based on Bloch's theorem, in which a periodic function is expanded into appropriate Fourier series. Inserting the so-obtained expansions into the eigenvalue characteristic (or master) Eq. (8.13) results in an infinite matrix eigen-problem. After the matrix is suitably truncated, the solutions to the aforementioned problem provide the eigenfrequencies and expansion coefficients for the eigenfunctions.[4]

The matrix equation can be derived in the following way. The master equation is Eq. (8.13)

$$\nabla \times \left\{ \frac{1}{\varepsilon(\mathbf{r})} \nabla \times \mathbf{H}(\mathbf{r}) \right\} = \left(\frac{\omega}{c} \right)^2 \mathbf{H}(\mathbf{r}) .$$

Since PCs are periodic with regard to the dielectric distribution, the solutions to Eq. (8. 13), using the Bloch's theorem, can be written as:

$$\mathbf{H}(\mathbf{r}) = \mathbf{u}_{\mathrm{k}}(\mathbf{r}) e^{i\mathbf{k}\cdot\mathbf{r}}, \tag{8.37}$$

with $\mathbf{u}_{\mathrm{k}}(\mathbf{r})$ being a periodic function and **k** the wave vector of the solution. The Fourier series expansion of the **H** field in terms of the reciprocal lattice vector **G** reads

$$\mathbf{H}(\mathbf{r}) = \sum_{G} \mathbf{H}_{\mathbf{k}}(\mathbf{G}) e^{i(\mathbf{k}+\mathbf{G})\cdot\mathbf{r}} . \tag{8.38}$$

In reciprocal space, the corresponding wave equation is found by Fourier transformations

$$-(\mathbf{k}+\mathbf{G})\times\left\{\sum_{G'}\varepsilon^{-1}(\mathbf{G}-\mathbf{G}')(\mathbf{k}+\mathbf{G}')\times\mathbf{H}_{\mathbf{k}}(\mathbf{G}')\right\}=\left(\frac{\omega}{c}\right)^2\mathbf{H}_{\mathbf{k}}(\mathbf{G}). \quad (8.39)$$

This equation may be expressed in matrix form and solved using standard numerical routines as in eigenvalue problems. The subscript **k** is used to indicate that the eigenvalue problem is solved for a fixed wave vector **k** to find the angular frequencies ω of all allowed modes. By truncating the summations to N reciprocal lattice vectors, the matrices are of dimensions $3N \times 3N$. However, the matrix equations involving the **H** field may be reduced to $2N \times 2N$ dimensions.[25] This follows from the transverse condition on the **H** fields (i.e., $\nabla\cdot\mathbf{H}(\mathbf{r})=0$), where $\mathbf{H}_k(\mathbf{G})$ may be written as a sum of two vectors orthogonal to the relevant $(\mathbf{k}+\mathbf{G})$

$$\mathbf{H}_{\mathbf{k}}(\mathbf{G})=h_{\mathbf{k},\mathbf{G},1}\,\hat{\mathbf{e}}_1+h_{\mathbf{k},\mathbf{G},2}\,\hat{\mathbf{e}}_2, \quad (8.40)$$

where **k** is a wave vector in the BZ of the lattice, **G** is the reciprocal lattice vector, and $\hat{\mathbf{e}}_1, \hat{\mathbf{e}}_2$ are orthogonal unit vectors that are both perpendicular to wave vectors $(\mathbf{k}+\mathbf{G})$. Hence, Eq. (8.38) may be written as:

$$\mathbf{H}(\mathbf{r})=\sum_{G}\sum_{\lambda=1,2}h_{\mathbf{k},\mathbf{G},\lambda}\,\hat{\mathbf{e}}_{\lambda}\,e^{i(\mathbf{k}+\mathbf{G})\cdot\mathbf{r}}. \quad (8.41)$$

Substituting Eq. (8.41) into Eq. (8.13), we obtain the following linear matrix equations for the **H** field[25]:

$$\sum_{G,\lambda}\mathbf{H}^{\lambda,\lambda'}_{\mathbf{k},\mathbf{G},\mathbf{G}'}\,h_{\mathbf{k},\mathbf{G},\lambda'}=\left(\frac{\omega}{c}\right)^2 h_{\mathbf{k},\mathbf{G},\lambda}, \quad (8.42)$$

where

$$\mathbf{H}_{\mathbf{G},\mathbf{G}'}=|\mathbf{k}+\mathbf{G}||\mathbf{k}+\mathbf{G}'|\varepsilon^{-1}(\mathbf{G}-\mathbf{G}')\times\begin{vmatrix}\hat{\mathbf{e}}_2\cdot\hat{\mathbf{e}}_2' & -\hat{\mathbf{e}}_2\cdot\hat{\mathbf{e}}_1' \\ -\hat{\mathbf{e}}_1\cdot\hat{\mathbf{e}}_2' & \hat{\mathbf{e}}_1\cdot\hat{\mathbf{e}}_1'\end{vmatrix}, \quad (8.43)$$

and $\varepsilon(\mathbf{G}\text{-}\mathbf{G}')=\varepsilon_{\mathbf{G},\mathbf{G}'}$ is the Fourier transform of $\varepsilon(\mathbf{r})$.

Equation (8.42) is the final matrix equation we wanted to derive. The involved matrix is furthermore Hermitian and positive definite, which allows employment of considerably faster numerical routines than when solving for general eigenvalue problems. The reduction from a $3N \times 3N$ to a $2N \times 2N$ dimension matrix equation is naturally of crucial importance in this respect, since it reduces the overall computation time; for this reason, the **H**-field version of the wave equation is the preferred implementation. Indeed, the misplaced position of the dielectric constant $\varepsilon(\mathbf{r})$ within the wave equation for the $\mathbf{E}(\mathbf{r})$ formalism does not destroy the hermiticity, and the problem is left with the normal eigenvalue solution procedures.

In our numerical simulations of 3D PC band structures, we have used the aforedescribed fully vectorial numerical technique. This method uses a frequency-domain iterative approach to perform a direct computation of the eigenstates and eigenvalues of Maxwell's equations, using a plane wave basis.[51] Further details and discretization aspects of the method's actual numerical implementation can be found in Ref. 52.

Acknowledgments

We wish to thank Cécile Jamois, Peter Spahn, Christian Hermann, Klaus Boehringer, and Jeremy Allam for fruitful discussions, and Joahim Hamm for technical assistance. This work was partially supported by the Engineering and Physical Sciences Research Council (EPSRC), UK, and the German Federal Ministry of Research and Education within the project KODO. K. L. Tsakmakidis acknowledges support by the Royal Academy of Engineering through a research fellowship. Finally, we would like to express our sincere gratitude to the anonymous reviewers for their constructive comments that enhanced the quality of the presented work.

References

1. E. Yablonovitch, "Inhibited spontaneous emission in solid-state physics and electronics," *Phys. Rev. Lett.* **58**, 2059–2062 (1987).
2. S. John, "Strong localization of photons in certain disordered superlattices," *Phys. Rev. Lett.* **58**, 2486–2489 (1987).
3. E. Yablonovitch, T. J. Gmitter, and K. M. Leung, "Photonic band structure: the face-centered-cubic case employing nonspherical atoms," *Phys. Rev. Lett.* **67**, 2295–2298 (1991).
4. S. Fan, P. R. Villeneuve, R. D. Meade, and J. D. Joannopoulos, "Design of three-dimensional photonic crystals at submicron lengthscales," *Appl. Phys. Lett.* **65**, 1466–1468 (1994).
5. H. S. Sözüer, and J. P. Dowling, "Photonic band calculations for woodpile structures," *J. Mod. Opt.* **41**, 231–239 (1994).
6. A. Reynolds, F. López-Tejeira, D. Cassagne, F. J. García-Vidal, C. Jouanin, and J. Sánchez-Dehesa, "Spectral properties of opal-based photonic crystals having a SiO_2 matrix," *Phys. Rev. B* **60**, 11422–11426 (1999).
7. E. Palacios-Lidón, A. Blanco, M. Ibisate, F. Meseguer, C. López, and J. Sánchez-Dehesa, "Optical study of the full photonic band gap in silicon inverse opals," *Appl. Phys. Lett.* **81**, 4925–4927 (2002).

8. O. D. Velev, T. A. Jede, R. F. Lobo, and A. M. Lenhoff, "Porous silica via colloidal crystallization," *Nature* **389**, 447–448 (1997).
9. J. E. G. J. Wijnhoven, and W. L. Vos, "Preparation of photonic crystals made of air spheres in titania," *Science* **281**, 802–804 (1998).
10. N. Tétreault, A. Mihi, H. Míguez, I. Rodríguez, G. A. Ozin, F. Meseguer, and V. Kitaev, "Dielectric planar defects in colloidal photonic crystal films," *Adv. Mater.* **16**, 346–349 (2004).
11. P. Jiang, J. F. Bertone, K. S. Hwang, and V. L. Colvin, "Single-crystal colloidal multilayers of controlled thickness," *Chem. Mater.* **11**, 2132–2140 (1999).
12. C. F. Blanford, H. Yan, R. C. Schroden, M. Al-Daous, and A. Stein, "Gems of chemistry and physics: macroporous metal oxides with 3D order," *Adv. Mater.* **13**, 401–407 (2001).
13. K. Sumioka, H. Kayashima, and T. Tsutsui, "Tuning the optical properties of inverse opal photonic crystals by deformation," *Adv. Mater.* **14**, 1284–1286 (2002).
14. K. Busch and S. John, "Photonic band gap formation in certain self-organizing systems," *Phys. Rev. E* **58**, 3896–3908 (1998).
15. S. H. Foulger, P. Jiang, A. Lattam, D. W. Smith, Jr., J. Ballato, D. E. Dausch, S. Grego, and B. R. Stoner, "Photonic crystal composites with reversible high-frequency stop band shifts," *Adv. Mater.* **15**, 685–689 (2003).
16. P. Sheng, W. Wen, N. Wang, H. Ma, Z. Lin, W. Y. Zhang, X. Y. Lei, Z. L. Wang, D. G. Zheng, W. Y. Tam, and C. T. Chan, "Multiply coated microspheres. A platform for realizing fields-induced structural transition and photonic band gap," *Pure Appl. Chem.* **72**, 309–315 (2000).
17. Z. Gu, A. Fujishima, and O. Sato, "Photochemically tunable colloidal crystals." *J. Am. Chem. Soc.* **122**, 12387–12388 (2000).
18. S. W. Leonard, J. P. Mondia, H. M. van Driel, O. Toader, S. John, K. Busch, A. Birner, U. Gösele, and V. Lehmann, "Tunable two-dimensional photonic crystals using liquid-crystal infiltration," *Phys. Rev. B* **61**, R2389–R2392 (2000).
19. H. Takeda and K. Yoshino, "Tunable photonic band schemes of opals and inverse opals infiltrated with liquid crystals," *J. Appl. Phys.* **92**, 5658–5662 (2002).
20. Ch. Schuller, F. Klopf, J. P. Reithmair, M. Kamp, and A. Forchel, "Tunable photonic crystals fabricated in III-V semiconductor slab waveguides using infiltrated liquid crystals," *Appl. Phys. Lett.* **82**, 2767–2769 (2003).
21. G. Mertens, T. Röder, H. Matthias, H. Marsmann, H. R. Kitzerow, S. L. Schweizer, C. Jamois, R. B. Wehrspohn, and M. Neubert, "Two- and three-dimensional photonic crystals made of macroporous silicon and liquid crystals," *Appl. Phys. Lett.* **83**, 3036–3038 (2003).

22. S. Kubo, Z. Gu, K. Takahashi, A. Fujishima, H. Segawa, and O. Sato, "Tunable photonic band gap crystals based on a liquid crystal-infiltrated inverse opal structure," *J. Am. Chm. Soc.* **126**, 8314–8319 (2004).
23. T. Ruhl, P. Spahn, C. Hermann, C. Jamois, and O. Hess, "Double inverse opal photonic crystals: the route to photonic band gap switching," *Adv. Funct. Mater.* **16**, 885–890 (2006).
24. J. D. Joannopoulos, R. D. Meade, and J. N. Winn, *Photonic Crystals: Molding the Flow of Light*, Princeton University Press, Princeton, NJ (1995).
25. K. M. Ho, C. T. Chan, and C. M. Soukoulis, "Existence of a photonic gap in periodic dielectric structures," *Phys. Rev. Lett.* **65**, 3152–3155 (1990).
26. S. Kittel, *Introduction to Solid State Physics*, Ch. 2, John Wiley and Sons (1995).
27. E. Yablonovitch and T. J. Gmitter, "Photon band structure: the face-centered-cubic case," *Phys. Rev. Lett.* **63**, 1950–1953 (1989).
28. S. Satpathy, Ze Zhang, and M. R. Salehpour, "Theory of photon bands in three-dimensional periodic dielectric structures." *Phys. Rev. Lett.* **64**, 1239–1242 (1990).
29. K. M. Leung and Y. F. Liu, "Photonic band structures: the plane-wave method," *Phys. Rev. B* **41**, 10188–10190 (1990).
30. S. John and R. Rangarajan, "Optimal structures for classical wave localization: an alternative to the Ioffe-Regel criterion," *Phys. Rev. B* **38**, 10101–10104 (1988).
31. E. N. Economou and A. Zdetsis, "Classical wave propagation in periodic structures," *Phys. Rev. B* **40**, 1334–1337 (1989).
32. K. M. Leung and Y. F. Liu, "Full vector wave calculation of photonic band structures in face-centered-cubic dielectric media," *Phys. Rev. Lett.* **65**, 2646–2649 (1990).
33. Ze Zhang and S. Satpathy, "Electromagnetic wave propagation in periodic structures: Bloch wave solution of Maxwell's equations," *Phys. Rev. Lett.* **65**, 2650–2653 (1990).
34. C. T. Chan, K. M. Ho, and C. M. Soukoulis, "Photonic band gaps in experimentally realizable periodic dielectric structures," *Europhys. Lett.* **16**, 563–568 (1991).
35. E. Ozbay, A. Abeyta, G. Tuttle, M. C. Tringides, R. Biswas, C. T. Chan, C. M. Soukoulis, and K. M. Ho, "Measurement of a three-dimensional photonic band gap in a crystal structure made of dielectric rods," *Phys. Rev. B* **50**, 1945–1948 (1994).
36. H. S. Sözüer, J. W. Haus and R. Inguva, "Photonic bands: convergence problems with the plane-wave method," *Phys. Rev. B* **45**, 13962–13972 (1992).
37. T. Suzuki and P. K. L. Yu, "Tunneling in photonic band structures," *J. Opt. Soc. Am. B* **12**, 804–820 (1995).

38. J. B. Jones, J. V. Sanders, and E. R. Segnit, "Structure of opal," *Nature* **204**, 990–991 (1964).

39. J. V. Sanders, "Diffraction of light by opals," *Acta Cryst.* A **24**, 427–434 (1968).

40. İ. İ. Tarhan and G. H. Watson, "Photonic band structure of fcc colloidal crystals," *Phys. Rev. Lett.* **76**, 315–318 (1996).

41. W. L. Vos, M. Megens, C. M. van Kats, and P. Bösecke, "Transmission and diffraction by photonic colloidal crystals," *J. Phys.: Condens. Matter* **8**, 9503–9507 (1996).

42. A. Imhof and D. J. Pine, "Ordered macroporous materials by emulsion templating," *Nature* **389**, 948–951 (1997).

43. B. T. Holland, C. F. Blanford, and A. Stein, "Synthesis of macroporous minerals with highly ordered three-dimensional arrays of spheroidal voids," *Science* **281**, 538–540 (1998).

44. P. Jiang, J. Cizeron, J. F. Bertone, and V. L. Colvin, "Preparation of macroporous metal films from colloidal crystals," *J. Am. Chem. Soc.* **121**, 7957–7958 (1999).

45. O. D. Velev, P. M. Tessier, A. M. Lenhoff, and E. W. Kaler, "Materials: a class of porous metallic nanostructures," *Nature* **401**, 548 (1999).

46. Y. A. Vlasov, N. Yao, and D. J. Norris, "Synthesis of photonic crystals for optical wavelengths from semiconductor quantum dots," *Adv. Mater.* **11**, 165–169 (1999).

47. G. Subramania, K. Constant, R. Biswas, M. M. Sigalas, and K. M. Ho, "Optical photonic crystals fabricated from colloidal systems," *Appl. Phys. Lett.* **74**, 3933–3935 (1999).

48. A. Blanco, E. Chomski, S. Grabtchak, M. Ibisate, S. John, S. W. Leonard, C. Lopez, F. Meseguer, H. Miguez, J. P. Mondia, G. A. Ozin, O. Toader, and H. M. van Driel, "Large-scale synthesis of a silicon photonic crystal with a complete three-dimensional bandgap near 1.5 micrometers," *Nature* **405**, 437–439 (2000).

49. H. Míguez, F. Meseguer, C. López, M. Holgado, G. Andreasen, A. Mifsud, and V. Fornés, "Germanium fcc structure from a colloidal crystal template," *Langmuir* **16**, 4405–4408 (2000).

50. T. Ruhl, P. Spahn, H. Winkler, and G. P. Hellmann, "Large area monodomain order in colloidal crystals," *Macromol. Chem. Phys.* **205**, 1385–1393 (2004).

51. S. G. Johnson and J. D. Joannopoulos, "The MIT Photonic-Bands Package," http://ab-initio.mit.edu/mpb.

52. S. G. Johnson and J. D. Joannopoulos, "Block-iterative frequency-domain methods for Maxwell's equations in a planewave basis," *Optics Express* **8**, 173–190 (2001).

Biographies

Durga P. Aryal obtained an M.S. degree in physics from the Tribhuvan University (Nepal) and the University of Stuttgart (Germany) in 1994 and 2003, respectively. He received a Ph.D. degree in 2007 from the University of Surrey (United Kingdom), where he is currently a postdoctoral research fellow. His research interests are in the areas of photonic crystal devices and structures, computational photonics, silicon photonics, and metamaterials.

Kosmas L. Tsakmakidis received a Diploma in electrical and computer engineering in 2002 from the Aristotle University of Thessaloniki (Greece) and Master of Research (M.Res.) and Ph.D. degrees in 2003 and 2009, respectively, from the University of Surrey (United Kingdom). He is currently a fellow of the Royal Academy of Engineering and Physical Sciences Research Council (EPSRC). His research interests are in the areas of metamaterials, slow light, photonic crystals, computational photonics, and solid state physics.

Ortwin Hess obtained M.S. and Ph.D. degrees from the Technical University of Berlin in 1990 and 1993, respectively. From 1990 to 1992 he was a research associate in Edinburgh and served as a postdoctorate researcher at the University of Marburg (1993–1994). From 1995 to 2003 Hess headed the Theoretical Quantum Electronics Group at the German Aerospace Centre (DLR) in Stuttgart. Since 2001 he has served as Docent of Photonics at Tampere University of Technology in Finland. Hess was a visiting professor at Stanford University (1997–1998) and the University of Munich (2000–2001). Since March 2003, he has served at the University of Surrey (United Kingdom) as Chair of Theoretical Condensed Matter and Optical Physics in the Department of Physics and the Advanced Technology Institute, where he heads the Theory and Advanced Computation Group.

His research interests lie in theoretical condensed matter and optical physics, focusing, in particular, on the physics of slow light, optical metamaterials and plasmonics, ultrafast and spatiotemporal dynamics of (quantum dot) semiconductor lasers, and quantum optics of complex nanosystems. Recent interests also lie in nanothermodynamics and the nanorheology linked with biophotonics, as well as in plasmonic nanomaterials and photonic nanodevices. Together with his group he has made pioneering contributions to slow light and light storage in metamaterials (the 'trapped rainbow' effect), the ultrafast and spatiotemporal dynamics of semiconductor and quantum dot lasers, quantum fluctuations of lasers and optical amplifiers, as well as microcavity and optically pumped semiconductor lasers.

Chapter 9
Wave Interference and Modes in Random Media

Azriel Z. Genack and Sheng Zhang
Queens College of CUNY, Queens, NY, USA

9.1 Introduction

Mass and energy are transported via waves. These waves are quantum mechanical for classical particles, such as electrons, and classical for quantum mechanical particles, such as photons. For particles that do not mutually interact, transport through disordered media reduces to the study of wave scattering by inhomogeneities in the phase velocity of a medium. Examples of disturbances within uniform or periodic media are atomic dislocations in resistors, molecules in the atmosphere, or dielectric fluctuations in composite media. When the wave is multiply scattered, but scattering is sufficiently weak that the wave returns only rarely to a coherence volume within the sample through which it has passed, average transport may be described by particle diffusion. However, the wave nature is still strongly exhibited in fluctuations on a wavelength scale and in the statistics of transport. Though details of the scattering process depend on the type of waves and the specific environment, many essential characteristics of wave transport on length scales greater than both the wavelength λ and the transport mean free path ℓ are strikingly similar. Thus, for example, in the limit of particle diffusion, electronic conductance follows Ohm's law, and optical transmission through clouds and white paint correspondingly falls inversely with thickness. In this review, we consider features of multiply scattered waves in samples in which the wave is temporally coherent throughout the sample. This is readily achieved for classical wave scattering from static dielectric structures, but is obtained only for electrons interacting with restless atoms at ultralow temperatures in mesoscopic samples intermediate in size between the microscopic atomic scale and the macroscopic scale.

The superposition of randomly scattered waves in static disordered systems produces a random spatial pattern of field or intensity referred to as the speckle pattern because of its grainy appearance, as shown schematically in Fig. 9.1.[1–4] The speckle pattern at different frequencies provides a complex fingerprint of the interaction of a wave with the sample. However, in general, it is not possible to infer the internal structure of a body, even from a complete set of such patterns for all incident wave vectors. Indeed, even the forward problem of calculating the speckle pattern from a given structure for 3D systems cannot be solved at present, except for samples with dimensions considerably larger than the wavelength scale. Nonetheless, essential elements of a description of wave transport can be inferred from the statistics of the speckle pattern of radiation scattered from ensembles of random samples.[5–23] Here, we examine the statistics of random transmission variables within a random ensemble of sample configurations, as well as the statistics of the evolution of the speckle pattern as a whole.

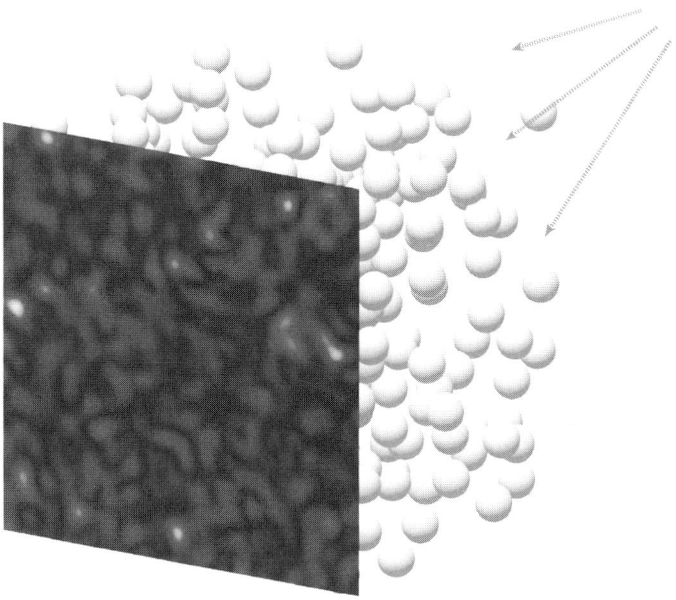

Figure 9.1 Schematic diagram of speckle pattern produced by multiple scattering in a random medium.

The statistics of wave transport depend on the spatial distribution of excitation within the sample. Waves may be either exponentially localized within the sample or may extend throughout the sample.[24–47] The localization transition is a consequence of the change in the nature of modes of classical and quantum mechanical waves in disordered systems and engenders a dramatic change in transport. Wave propagation and localization may be characterized by a host of interrelated parameters expressing different aspects of propagation. Propagation parameters include (1) the degree of spectral overlap of modes of the random medium, which is the ratio of the averages of the spectral width and spacing of the modes $\delta = \delta\nu/\Delta\nu$, (2) the dimensionless conductance g, which is the sum of transmission coefficients over all incident and outgoing transverse propagation channels, a and b, respectively, (3) the variance of total transmission for a single incident channel normalized by the average value of the transmission var($s_a = T_a/\langle T_a \rangle$), (4) the degree of intensity correlation κ between points for which the field correlation function vanishes, and (5) the probability of return P_{return} of a partial wave to a coherence volume within the sample. These parameters are related as follows:

$$\delta = g = 2/3\text{var}(s_a) = 2/3\kappa = 1/P_{\text{return}}. \tag{9.1}$$

The nature of propagation within the sample is seen both in fluctuations of random variables of the scattered wave and in the evolution of the scattered wave with shift in incident frequency. The probability distributions of total transmission and of various measures of transformation of the speckle pattern with frequency shift are described by a single functional form that depends only on the variance of the corresponding distribution. Characteristics of wave propagation in random media are discussed from two perspectives: one emphasizing the interference of partial waves within the medium, the other focusing on the underlying modes excited within the medium.

In this review, we consider this divide from two perspectives with some illustrative examples. We first consider the complex interference within the medium of multiply scattered partial waves. The interference of these waves leads to weak localization, which suppresses transport. We then treat transport by considering the different spatial, spectral, and temporal characters of the electromagnetic modes of the medium. These approaches are closely related, as it is the interference of waves that produces the field distribution in a given mode.

9.2 Wave Interference

9.2.1 Weak localization

Average transport from any region within a random sample is suppressed by constructive interference of waves following paths that return to a coherence volume through which the waves passed and that differ only in the sense in which they are traversed.[48–61] The coherent sum of the complex amplitudes for return for two time-reversed paths may be written as $A = A_{\circlearrowright} + A_{\circlearrowleft}$. Because the amplitudes and phase associated with these partial waves are identical, the probability of return is proportional to

$$|A|^2 = |A_{\circlearrowright} + A_{\circlearrowleft}|^2 = |2A_{\circlearrowright}|^2 = 4|A_{\circlearrowright}|^2, \tag{9.2}$$

which corresponds to twice the return probability obtained for the incoherent sum $|A_{\circlearrowright}|^2 + |A_{\circlearrowleft}|^2 = 2|A_{\circlearrowright}|^2$. In samples in which the probability of return is small, average transport is not appreciably affected by wave interference. The average over an ensemble of random configurations of the intensity at a point is then proportional to the weighted average of the amplitude squared for each of the sinuating partial waves that passes through the point. The probability of return to a point for an ensemble of random configurations is then well approximated by the diffusion equation for intensity or photon density. However, when the probability of return approaches unity, wave transport is strongly suppressed by wave interference, and the wave becomes exponentially localized within the medium. The limits in which the particle diffusion model holds can therefore be established by considering the probability of return of a random walker to a coherence volume in the medium. Since the actual probability of return to a coherence volume P_{return} is twice the probability for the incoherent addition of the associated paths

$$P_{\text{return}} = 2P_{\text{return}}^{\text{inc}}, \tag{9.3}$$

the enhanced return due to wave interference, known as weak localization or coherent backscattering, is equal to $P_{\text{return}}^{\text{inc}}$. This corresponds to a reduction in transport and to a suppression of the bare or Boltzmann diffusion coefficient D_{B} to a renormalized effective diffusion coefficient D. In the limit $P_{\text{return}} << 1$, the fractional reduction in the diffusion coefficient is given by

$$\frac{D_{\text{B}} - D}{D_{\text{B}}} = P_{\text{return}}^{\text{inc}}. \tag{9.4}$$

Here, $D_{\text{B}} = v\ell / d$ is the photon diffusion coefficient in the Boltzmann approximation in which interference of waves returning to a point is neglected and d is the dimensionality of the sample. We take the coherence volume to be $V_{\text{c}} \sim (\lambda/2)^d$, where $\lambda/2$ is the field correlation length that corresponds to the first zero of the field correlation function[5,18,62,63] within the medium.

In an unbounded medium, $P_{\text{return}}^{\text{inc}}$ may be calculated in the diffusion model, and is given by

$$P_{\text{return}}^{\text{inc}} = \frac{V_{\text{c}}}{\tau_{\text{c}}} \int_{2\tau}^{\infty} P(0,t)dt. \tag{9.5}$$

Here,

$$P(r,t) = \frac{1}{(4\pi D_{\text{B}}t)^{d/2}} \exp\left(\frac{-r^2}{4D_{\text{B}}t}\right) \tag{9.6}$$

is the Green's function of the diffusion equation, and τ_{c} is the time needed to travel a coherence length $\tau_{\text{c}} = \lambda/2v$, where v is the transport velocity. The lower limit of the integral is the earliest time of return, which is twice the mean free time between scattering events $\tau = \ell / v$. This corresponds to the time to return for a single scattering event at a distance of a single mean free path from the point $r = 0$, giving

$$P_{\text{return}}^{\text{inc}} = \frac{(\lambda/2)^d}{\tau_{\text{c}}(4\pi D_{\text{B}})^{d/2}} \int_{2\tau}^{\infty} \frac{dt}{t^{d/2}}, \tag{9.7}$$

which is finite only for $d > 2$.[34,60] $P_{\text{return}}^{\text{inc}}$ is unbounded in 1D or 2D media. This suggests that escape from a point is substantially suppressed due to coherent backscattering and that the intensity at the origin may not vanish with increasing delay. Thus, the wave is localized independently of the strength of scattering for $d \leq 2$. Only above the marginal dimension for localization $d = 2$, may P_{return} be less than unity. For $d > 2$,

$$P_{\text{return}} = \frac{(\lambda/2)^d}{\tau_{\text{c}}(4\pi D_{\text{B}})^{d/2}} \frac{1}{\frac{d}{2}-1} (2\tau)^{1-d/2}. \tag{9.8}$$

The interference of waves returning to a point then results in a reduction of D from its Boltzmann value D_{B}. Expressing D_{B} and τ in terms of the mean free path for $d = 3$ gives

$$P_{\text{return}} = \frac{D_{\text{B}} - D}{D_{\text{B}}} = \frac{3\sqrt{6\pi}}{8} \frac{1}{(k\ell)^2} = \frac{1.63}{(k\ell)^2}. \tag{9.9}$$

Thus, $P_{\text{return}} \approx 1$ when $k\ell \approx 1$. The influence of interference is then so strong that D tends to zero and waves are localized. Thus, $k\ell \approx 1$ is the threshold for electron localization transition in 3D, which is the Ioffe-Regel criterion.[26] Once $\ell < \lambda / 2\pi$, the phase of the wave does not change substantially between scattering events, and it is not possible to define a trajectory with a specific direction. The particle diffusion model can therefore no longer describe transport. Thus, limitations on transport are not exclusively the province of quantum mechanics, but arise from the wave nature of the electron. This stands in contrast to the original quantum mechanical perspective of the Anderson model.[24] In that model, an electron interacts with an atomic lattice with disorder in either the diagonal or off-diagonal elements of the Hamiltonian, which represent the site energy and the coupling coefficients, respectively. Anderson showed that a transition in the nature of propagation may occur for $d = 3$ as a function of electron energy or scattering strength. At the mobility edge separating diffusive and localized waves, the correlation length diverges with a critical exponent as the transition is approached from either the diffusive or localized side of the mobility edge.[28,29,32,64–66]

9.2.2 Coherent backscattering

The impact of coherent backscattering can be directly observed as an enhancement of retroreflected light from a random medium.[52–55,60,61] The trajectories of two partial waves that follow time-reversed paths within the sample are then scattered at an angle θ to the normal, as shown in Fig. 9.2. These waves interfere in the far field. When scattering is time-reversal invariant, the accumulated phase shifts for waves traversing the same path within the medium but in opposite senses are the same. However, a phase difference arises due to the different lengths of the trajectories outside the sample

$$\Delta\phi = \frac{2\pi}{\lambda}\Delta r = \mathbf{k}\cdot\boldsymbol{\rho}, \tag{9.10}$$

where Δr is the additional pathlength outside the sample and $\boldsymbol{\rho}$ is the transverse excursion of the wave along the sample surface. The contribution to the field at a point r in the far-field associated with the partial waves that enters the sample normally at r_i and emerges from the sample at $r_i + \rho$, after following a trajectory α within the sample between these points, and which is then scattered with wave vector $\mathbf{k}$, is the real part of the complex field

$$p(r_i, \alpha, \rho, k)e^{i\phi_\alpha}e^{i\mathbf{k}\cdot[\mathbf{r}-(\mathbf{r}_i+\boldsymbol{\rho})]}. \tag{9.11}$$

If we add the fields for the time-reversed pair of partial waves shown in Fig. 9.2, we obtain

$$p(\mathbf{r}_\mathbf{i}, \alpha, \boldsymbol{\rho}, \mathbf{k})e^{\mathrm{i}\phi_\alpha}e^{i\mathbf{k}\cdot[r-(\mathbf{r}_i+\boldsymbol{\rho})]}(1+e^{\mathrm{i}\Delta\phi}). \tag{9.12}$$

Summing all pairs gives the intensity in the direction of k at an angle θ to the normal

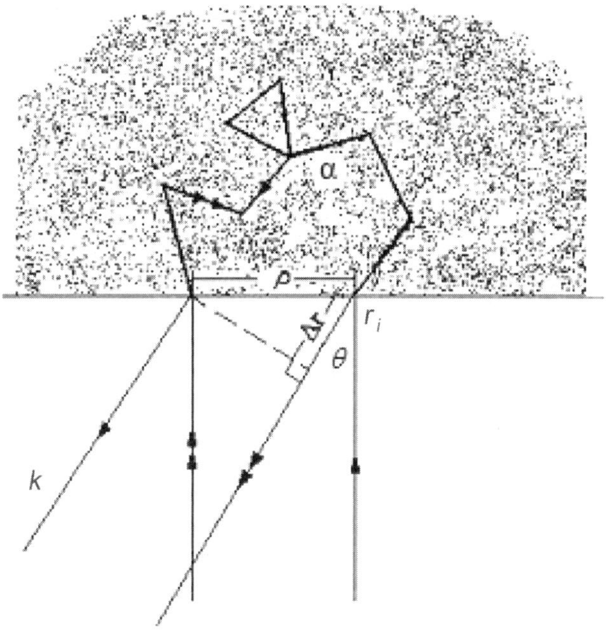

Figure 9.2 Schematic of coherent backscattering for a pair of partial waves traversing along path *α* in opposite directions.

$$I(\theta) = \left| \sum_{\alpha} \sum_{r_i\rho}' p(\mathbf{r}_i, \alpha, \boldsymbol{\rho}, \mathbf{k}) e^{i\phi_\alpha} e^{i\mathbf{k}\cdot[\mathbf{r}-(\mathbf{r}_i+\boldsymbol{\rho})]} (1 + e^{i\mathbf{k}\cdot\boldsymbol{\rho}}) \right|^2, \tag{9.13}$$

where the prime indicates the sum is taken over pairs with reversed points of entry and exit. The average over an ensemble of random configurations, denoted by $\langle ... \rangle$, can be obtained by averaging scattered light over time in a colloidal sample or by spinning or translating a static sample. Cross terms in the square of the sum of the partial waves for different paths do not contribute to the ensemble average, since the phase accumulated within the sample for different configurations is not correlated. This gives

$$\langle I(\theta) \rangle = \langle \sum_{\alpha} \sum_{r_i\rho}' P(\mathbf{r}_i, \alpha, \boldsymbol{\rho}, \mathbf{k})(2 + 2\cos(\mathbf{k}\cdot\boldsymbol{\rho})) \rangle = I_{\text{inc}} + I_{\text{c}}, \tag{9.14}$$

where

$$P(\mathbf{r}_i, \alpha, \boldsymbol{\rho}, \mathbf{k}) = p^2(\mathbf{r}_i, \alpha, \boldsymbol{\rho}, \mathbf{k}), \tag{9.15}$$

and I_{inc} and I_{c} are the intensity values associated, respectively, with the incoherent and coherent sums over paths. The factor of 2 appears because the sum is taken over pairs of time-reversed paths. In the exact backscattered direction, $\mathbf{k}\cdot\boldsymbol{\rho} = 0$, and the intensity is double that of the incoherent background.

This enhancement is seen in the measurement of coherent backscattering shown in Fig. 9.3. The enhanced reflection corresponds to a reduction in transmission associated with a renormalization of D.

The term associated with coherent backscattering may be expressed as the integral

$$\langle I_c(\theta)\rangle = \int\int P(s,\boldsymbol{\rho},\mathbf{k})\cos(\mathbf{k}\cdot\boldsymbol{\rho})\,ds\,d^2\rho\,, \tag{9.16}$$

where $P(s,\boldsymbol{\rho},\mathbf{k})$ is the probability density that a photon incident upon the medium will exit the sample at a position displaced by ρ with wave-vector $\mathbf{k}$ after following a path of length s within the medium. Since the distribution of scattering angles for the last scattering event in the path is broad, we expect that $\mathbf{k}$ will not be correlated with s or ρ, which relate to the entire wave trajectory within the sample. We therefore express $P(s,\boldsymbol{\rho},\mathbf{k})$ as a product

$$P(s,\boldsymbol{\rho},\mathbf{k}) = A(s,\boldsymbol{\rho})B(\mathbf{k})\,, \tag{9.17}$$

where $B(\mathbf{k})$ is the specific intensity for reflected radiation. Integrating $A(s,\boldsymbol{\rho})$ over s gives $I_0(\boldsymbol{\rho})$, the point spread function, which is the transverse

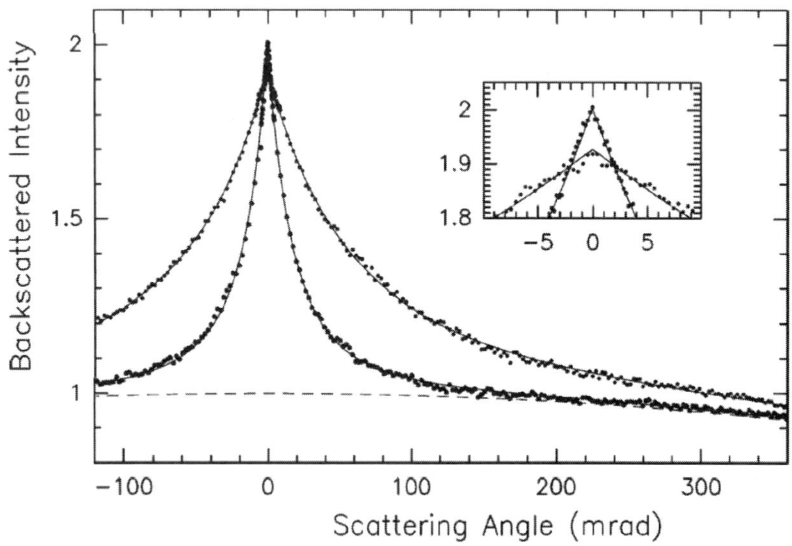

Figure 9.3 Coherent backscattering of light measured with two different values for the transport mean free path ℓ. The typical angular width varies as λ/ℓ. Narrow cone: a sample of $BaSO_4$ powder with $\ell/\lambda = 4$; broad cone: TiO_2 sample with $\ell/\lambda = 1$. The inset confirms the triangular cusp predicted by diffusion theory and also shows that the maximum enhancement factor is lowered for the sample with a small value of ℓ/λ. [Reprinted with permission from Ref. 55. © (1995) by the American Physical Society.]

spatial distribution of intensity on the surface due to excitation of a point on the surface by a normally incident wave. Thus, the coherent backscattering peak is the Fourier transform of the point spread function on the incident surface multiplied by $B(\mathbf{k})$

$$\langle I_c(\theta)\rangle = B(\mathbf{k})\int I_0(\boldsymbol{\rho})\cos(\mathbf{k}\cdot\boldsymbol{\rho})\,d^2\rho.^5 \tag{9.18}$$

In weakly scattering media with $k\ell >> 1$, the coherent backscattering peak is much narrower than $B(\mathbf{k})$. The width of $I_0(\boldsymbol{\rho})$ is on the order of a few mean free path lengths $\Delta\rho \approx \ell$, since most of the incident energy returns to the surface after only a few scattering events. Because components of ρ and the corresponding components of $\mathbf{k}$ along the same direction are conjugate variables in the Fourier transform relation between the coherent backscattering peak and the point spread function, we have

$$\Delta k_{\|} \sim k\Delta\theta \sim \frac{1}{\Delta\rho} \approx \frac{1}{\ell}, \text{ or } \Delta\theta \approx \frac{1}{k\ell}, \tag{9.19}$$

where $\Delta\theta$ is the width of the coherent backscattering peak. Because some of the reflected radiation survives many scattering events within the sample, $I_0(\boldsymbol{\rho})$ has a broad tail that can produce a sharp structure in its Fourier transform. This results in a sharp triangular peak in $\langle I_c(\theta)\rangle$,[54] as seen in Fig. 9.3.[55] For strong scattering, corresponding to small values of $\ell/(\lambda/2\pi) = k\ell$, the coherent backscattering peak broadens and corresponds to a sizable fraction of the total energy. As a result, when $\ell \approx \lambda$, the Fourier transform of $P(\rho)$ goes negative as the reflected radiation falls below the incoherent background level at large angles. This has been observed in recent measurements.[61]

9.3 Modes

Numerous approaches have been taken to treat waves in random media. The methods of wave interference can be used to calculate the average of transmission and reflection, fluctuations, and correlation of scattered waves, all of which can be carried out by calculating the Green's function. This approach is particularly useful when the Green's function can be expressed as a perturbation expansion.[62] However, it is difficult to obtain analytic results for localized waves where no small parameter exists and all of the diagrams must be summed. An approach that has proven useful when scattering is weak is the phenomenological radiative transfer method. The radiative transfer equation is a Boltzmann equation for the specific intensity $I_{\hat{v}}(\mathbf{r},t)$ describing the flow of radiation in a disordered medium at position r and time t in a direction $\hat{v}$.[67,68] This method can be used to derive a diffusion equation for the specific intensity and to sort out the complexities of the wave interaction with the surface. Random matrix theory is another powerful approach that gives important results for the statistics of localized waves in quasi-1D samples.[15–17, 69–74] This method treats the reflection or transmission matrices for propagating waveguide modes whose elements are assumed random except for certain constraints. In the rest of this chapter, we will emphasize the underlying electromagnetic modes of the random medium, which can be directly observed for localized waves. We will see that fundamental characteristics of wave propagation can be described in terms of the statistics of modes.

9.3.1 Quasimodes

The field inside an open random sample may be viewed as a superposition of quasimodes, each of which corresponds to a volume speckle pattern of the field. The spatial and spectral variation of the polarization component j of the field for the nth quasimode is given by

$$A_{n,j}(r,\omega) = a_{n,j}(r)\frac{\Gamma_n/2}{\Gamma_n/2 + i(\omega - \omega_n)}, \tag{9.20}$$

where ω_n and Γ_n are the central frequency and linewidth, respectively. In contrast to modes of a closed system, quasimodes are eigenstates associated with open systems, in which energy may be absorbed within the sample or may leak out through the boundaries. The spatial field distribution of a quasimode $a_n(r)$, satisfies the Helmholtz equation

$$Ha_n(r) = (\omega_n - i\Gamma_n/2)^2 a_n(r), \tag{9.21}$$

where H is the non-Hermitian Helmholtz operator with open appropriate boundary conditions. The imaginary part of the corresponding eigenfrequency characterizes the broadening of the line. If a single one of these quasimodes was excited by a pulse, the local field amplitudes would decay exponentially in a time $2/\Gamma_n$. For the sake of simplicity, we will often refer to quasimodes as modes, without risk of confusion. Although, in general, the eigenstates of a non-Hermitian operator do not form a complete basis, Leung et al. have demonstrated[75] that when the refractive index of materials in a leaky system varies discontinuously and approaches a constant asymptotic value sufficiently rapidly, the quasimodes are complete and orthogonal so that an arbitrary state of the system can be expressed as a superposition of quasimodes.[45,76] In these cases, the Green's function for $\omega^2 - H$ can be expanded with the use of these mutually orthogonal quasimodes to give[77]

$$G(r_1, r_2, \omega) = \sum_n \frac{a_n(r_1)a_n(r_2)}{\omega^2 - (\omega_n - i\Gamma_n/2)^2}. \tag{9.22}$$

This expression encapsulates the equivalence between treatments that emphasize wave interference of multiply scattered trajectories arriving at a point and an analysis that stresses the decomposition of the field into the quasimodes of the wave within the sample.

An alternative way to approach modes of an open system is to consider the waves both inside and outside of the medium. In nondissipative systems, the eigenstates with constant flux for the system as a whole have real eigenvalues. Because these states are complete and orthorgonal,[78] the Green's functions can be expressed in terms of these constant flux modes. This has proved useful in the study of the spatial structure of modes in random lasers and nonlinear systems.[79] In the remainder of this chapter, we will emphasize the quasimode description of the wave inside the random sample, with complex eigenvalues reflecting leakage from the system and absorption within the system.

9.3.2 Localized and extended modes

The nature of wave propagation in disordered samples reflects the spatial extent of the wave within the medium.[24,27,28] When modes are spectrally isolated, the envelope of the field inside the sample is exponentially peaked.[27,29,31] On the other hand, modes that overlap spectrally may extend throughout the sample. In diffusive samples, eigenstates are spread throughout the sample, while in samples in which waves are generally localized, the intensity distribution is singly or multiply peaked. The emergence of a number of intensity peaks within samples that are longer than the localization length was explained by Mott as the hybridization of localized modes due to exponentially small overlap of excitation in neighboring localized states.[80–83] Apart from an overall modulation on the wavelength scale, the spatial intensity distribution of such Mott states exhibits a number of peaks along the length of the sample equal to the number of coupled modes.[45,47,82–84] Overlapping modes are observed even in samples in which the wave is typically localized. When the sample supports a number of roughly equally spaced localization centers—which have been termed necklace states by Pendry—their contribution to transmission is particularly large.[47] Necklace states play a large role in transport since their increased spatial extent facilitates the advance of the wave through the sample, and the resulting increased transmission occurs over a relatively broad frequency range. Because these states are short lived, they are relatively weakly affected by absorption, which strongly attenuates localized states.[45,85,86] Conversely, occasionally the number of overlapping modes in diffusive samples may be small, and the lifetimes of such modes will tend to be larger than is typical. In the presence of gain, lasing will occur first in such long-lived quasi-extended modes.[87–91] An important issue discussed in reviews of lasing in diffusive sample in this volume[92–94] is the extent to which the presence of gain modifies the modes of the sample when modes overlap spectrally and spatially.

The intensity distribution within the interior of a multiply scattering sample is generally inaccessible in 3D samples, but can be examined in 1D and 2D samples.[45,95] The presence of both isolated and overlapping modes within the same frequency range has been observed in measurements of field spectra carried out with a vector network analyzer inside a single-mode waveguide containing randomly positioned dielectric elements. Many of the elements had a binary structure so that a pseudogap is created in which the density of states was particularly low[34] in the frequency range of the band gap of the corresponding periodic structure of binary elements. The waveguide containing the sample was slotted and covered with a movable copper slab so that measurements could be made at any point along the sample length without introducing substantial leakage through the top of the waveguide. Measurements in Fig. 9.4 show that spectrally isolated lines have Lorentzian shapes with the same width at all points within the sample and are strongly peaked in space. On the other hand, when modes overlap spectrally, the line shape varies with position within the sample, and the spatial intensity distribution is multiply peaked. Quasi-extended intensity distributions within a region in which waves are typically localized arise from the superposition modes with Lorentzian lines that are multipeaked in space.[45]

One-dimensional localization has also been observed in optical measurements in single-mode optical fibers[96] and in single-mode channels that guide light within photonic crystals.[46] When the structure bracketing the channel

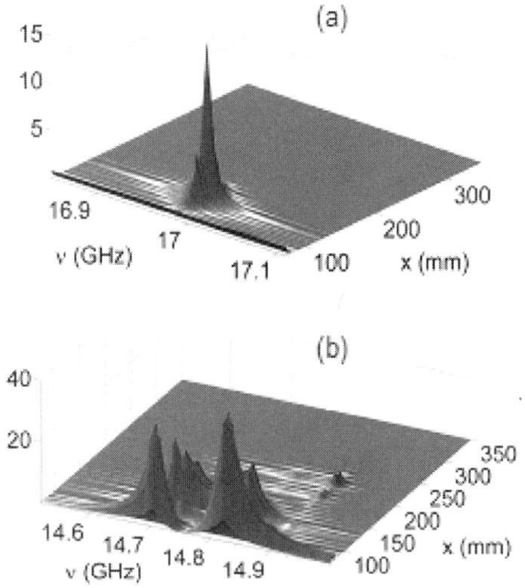

Figure 9.4 Intensity spectra of quasimodes versus positions within a random single-mode waveguide for (a) isolated and (b) overlapping waves. In (a) the intensity of the wave is exponentially localized, whereas in (b) the intensity distribution is multiply peaked and hence quasi-extended. [Reprinted from Ref. 45. © (2006) by the American Physical Society.]

is periodic, the velocity of the wave propagating down the channel experiences a periodic modulation so that a forbidden band is created. When disorder is introduced into the lattice, quasimodes with spatially varying amplitude along the channel are created. Modes near the edge of the band gap are long lived and readily localized by disorder. An example of spectra of vertically scattered light versus frequency for light launched down a channel through a tapered optical fiber is shown in Fig. 9.5.[46] The inset shows the disordered sample of holes with random departure from circularity in silicon-on-insulator substrates at a hole filling fraction of $f \sim 0.30$. The structure is produced using electron-beam lithography and chlorine-based inductively coupled plasma reactive ion etching.

Narrow spectral lines are also observed near the band edge of cholesteric liquid crystals.[97] In such anisotropic liquid crystals, the average molecular orientation, known as the director, rotates with depth into the sample to create a sample with planar anisotropy with constant pitch. A photonic band gap is formed at a wavelength within the medium equal to the pitch for light with the same sense of circular polarization as the handedness of the helical structure. Low-threshold lasing is observed in dye-doped cholesteric liquid crystals in long-lived modes at the band edge. When even small disorder is present, the wavelengths of the lasing modes differ from values of modes for a periodic sample.[97]

We have considered the localization of waves inside samples with reduced dimensionality. Localization can also be achieved in samples that are homogeneous along one direction and inhomogeneous in the transverse plane.[98] For light incident along the direction in which the sample is homogeneous, the

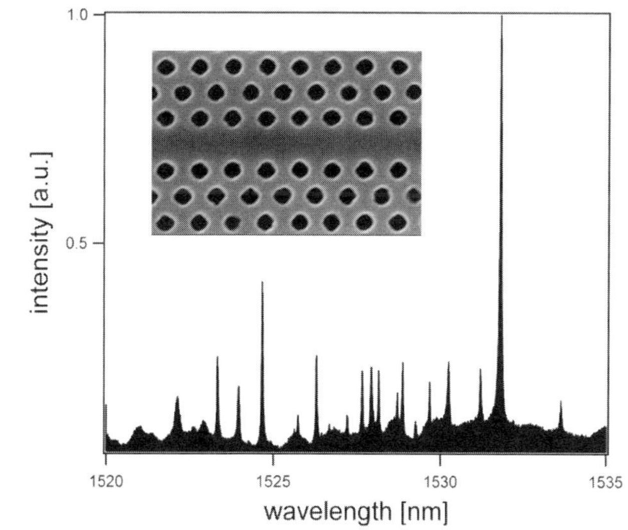

Figure 9.5 Spectrum of wave transmitted to a region within a single-mode photonic crystal waveguide near the short wavelength edge of the first stop band at ~ 1520 nm. The channel surrounded by irregular holes is shown in the inset. [Reprinted with permission from Ref. 46. © (2007) by the American Physical Society.]

velocity of the wave is constant, and so the variation of the waveform along this direction in samples of different lengths provides the time evolution in the transverse directions. The wave vector gains a transverse component as a result of scattering in the transverse direction, which arises from disorder. Since the component of the wave vector in the transverse direction $k_\perp$ is small, the product of the transverse component of the wave vector and the mean free path in the transverse direction $k_\perp \ell$ is also small. This facilitates the observation of localization. Linear and nonlinear localization have been observed in disordered 2D lattices written into photorefractive materials by superimposing a random speckle pattern onto three interfering beams, which alone would produce a hexagonal lattice.[99] Exponential localization in the plane of the disordered lattice is achieved even though the index modulation is small. Transverse localization was also produced in a 1D lattice of coupled optical waveguides patterned on an AlGaAs substrate.[100] Light is injected into one or a few waveguides at the input and tunnels coherently between neighboring waveguides as it propagates along the waveguide axis. Samples with different strengths of disorder are created by changing the width of the distribution of random widths of the waveguides, which are periodic on average. At short lengths, a ballistic component is observed in the transverse direction moving away from the point in the transverse direction at which the light was injected. At greater sample lengths, the ballistic component disappears, while an exponential central peak grows. Nonlinear perturbations enhance localization in states with uniform phase in the transverse direction and tend to delocalize waves in which the phase is modulated by π rad in neighboring waveguides.

A qualitative understanding of the characteristics of modes in localized samples can be obtained by considering the intensity distributions and associated spectra for samples with one or two defects within a periodic background. This produces exponentially peaked states within the band gap. Computations for the cases of a single defect either at the center of the sample [Figs. 9.6(a) and (b)] or displaced from the center [Figs. 9.6(c) and (d)] and a pair of defect states [Figs. 9.6(e) and (f)] placed symmetrically about the center of the sample are plotted in Fig. 9.6. The structure within each sample and the corresponding spatial intensity pattern and transmission spectrum are shown. The underlying periodic structure is a sample of binary elements with indices of refraction of 1 and 2. A quarter-wave defect with index of refraction 1 and a thickness of one half the period is introduced. This places the transmission peak for the single mode at the center of the band gap.

For a single defect, the intensity falls exponentially from the defect, apart from a modulation on the scale of $\lambda/2$. The modulation is characteristic of a standing wave since the components of the wave propagating in opposite directions are nearly of the same intensity except near the sample boundaries. The difference in flux in the forward and backward directions is constant and equal to the transmission. The exponential decay length is the same as that for the evanescent wave excited at the same frequency within the band gap in a defect-free structure. The transmission coefficient depends only on the exponential decay length and the shortest distance from the defect to the boundary.[31,101] When the defect is in the first half of the sample, the intensity rises from the input to the defect site and then falls exponentially to the output surface. On the other hand,

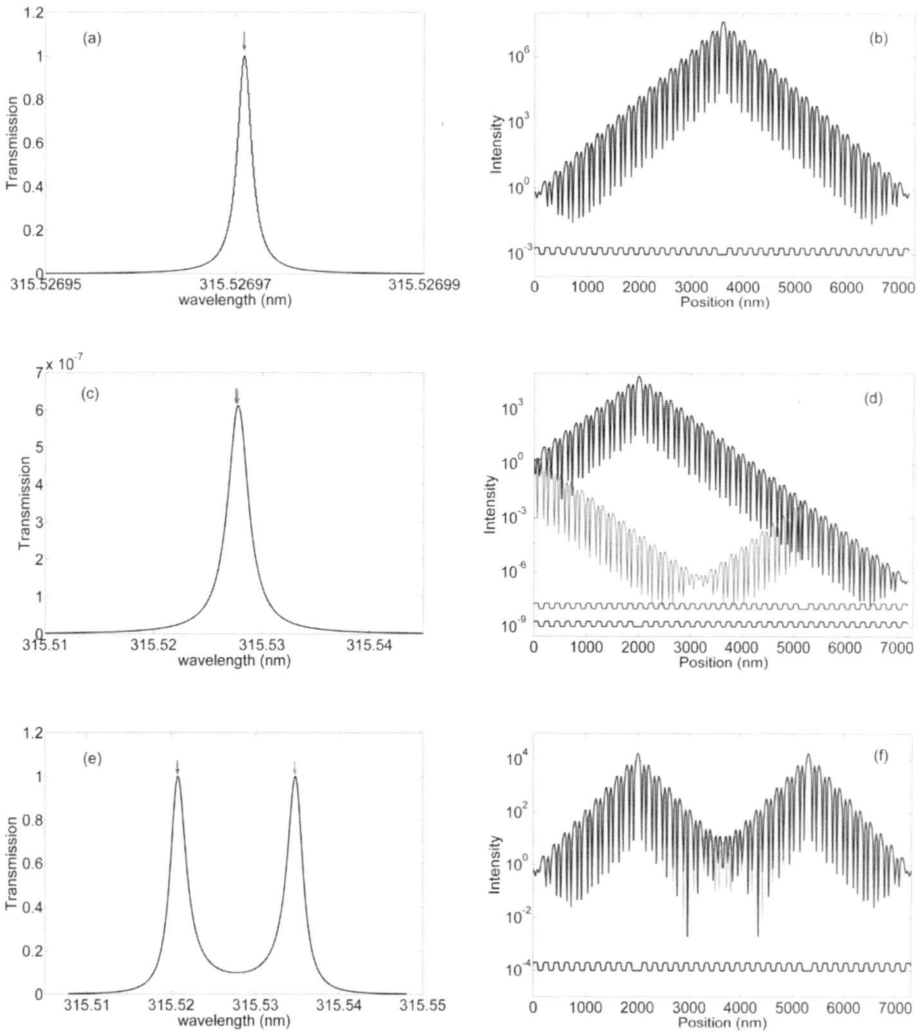

Figure 9.6 Simulation of propagation in 1D samples with defects in a periodic structure, in which the refractive index alternates between 1 and 2 in segments of length 100 nm and period of 200 nm. (a) Spectrum and (b) intensity distribution for sample with single defects at the center. (c) Spectrum and (d) intensity distribution for samples with single defects equally displaced to the left or right of the center. (e) Spectrum and (f) intensity distribution for sample with two defects symmetrically displaced from the center. The defect is an additional 100 nm with refractive index 1.

when the defect is in the second part of the sample, the incident wave couples to the defect state through an evanescent wave. The intensity first falls exponentially until the point at which its value matches the value of the intensity of the mode that is exponentially peaked at the defect. When the defect is at the center of the sample, the distances the exponential rise and fall are equal and the transmission coefficient is unity. This gives the largest integrated intensity within the sample, relative to the intensity value at the boundaries of the sample, and results in a small leakage rate of energy from the sample—corresponding to a long lifetime and narrow linewidth. The peak intensity within the sample, as well as the lifetime of the state, fall exponentially with displacement of the defect from the center of the sample, which is simply related to the displacement from the nearest boundary.

When two defects are positioned symmetrically with respect to the center of the sample, the intensity at the first defect rises to the same level as in the sample with a single defect at this position; however, the transmission is higher because the intensity rises exponentially as the second peak is approached. Because the intensity at the second defect is the same as at the first, and the distance to the output from the second defect equals the distance to the input of the first defect, the transmission coefficient is unity. In the case of two defects, two hybridized modes with slightly shifted central frequencies are produced as a result of coupling between the modes. In the sample with symmetrically positioned defects, both modes have the same intensity distribution and a transmission coefficient of unity. Spectrally isolated localized modes in random samples are similarly exponentially peaked, while spectrally overlapping modes are coupled to form multiply peaked quasimodes. Since both the peak transmissions and the linewidths of the overlapping modes are typically much larger than those of isolated modes, the overall transmission is then dominated by these multiply peaked modes. Propagation within the sample can also be understood as the coupling of a wave through single or multiple wells separated by barriers in which the intensity falls exponentially.[102] Measurements of quasimode transformation as the sample is continuously modified show an anticrossing behavior in which the closest approach of quasimode frequencies is determined by the strength of coupling between localization centers.[81]

9.3.3 Statistical characterization of localization

A dimensionless parameter that characterizes the statistical spatial and spectral properties of modes of the medium is the degree of level overlap. This is the ratio of the level width to the level spacing $\delta = \delta\nu/\Delta\nu$.[27,103] This ratio is the inverse of the average finesse of the spectrum and is closely related to the Thouless number, which is a measure of the sensitivity of the mode frequency to a change of boundary conditions.[27] Here,

$$\delta\nu = \frac{<\Gamma>}{2\pi} \tag{9.23}$$

is the average frequency width of modes. This corresponds to the width of the field correlation function with frequency shift.[103] The level spacing is the inverse of the density of states of the sample as a whole, and may be written as

$$\Delta\nu = \frac{1}{n(\nu)AL}, \tag{9.24}$$

where $n(\nu)$ is the density of states per unit volume per unit frequency, and AL is the sample volume given by the product of the area and thickness of the sample. δ is an indicator of localization, since when $\delta\nu < \Delta\nu$, modes generally do not overlap spectrally and are exponentially peaked within the sample. The linewidth is narrow since the leakage rate from the sample, which is proportional to the ratio of the energy density near the boundaries to the integrated energy within the sample, is small when the intensity is strongly peaked within the sample. On the other hand, when $\delta\nu > \Delta\nu$, modes overlap spectrally, and the wave is extended throughout the sample. Thus, $\delta = 1$ corresponds to the localization threshold.[27,28]

The critical dimension for localization can be found by considering the scaling of δ. For diffusive samples $\delta\nu \sim D/L^2$, where D is the diffusion coefficient controlling the spread of intensity or, correspondingly, the migration of particles in random systems and consistent with the diffusion relation[18,27,104]

$$\left\langle |\boldsymbol{r}(t) - \boldsymbol{r}(0)|^2 \right\rangle = 6Dt. \tag{9.25}$$

The level spacing is inversely proportional to the volume $\Delta\nu \sim 1/L^d$ for samples with $A = L^{d-1}$. In this case, δ scales as $\delta \sim L^{d-2}$, and δ decreases with increasing L for $d < 2$. On the other hand, δ increases with L for $d > 2$ for waves that are diffusive on a scale not much larger than the mean free path. Thus, we find again that $d = 2$ is the critical dimension for localization.[28]

The relationship between the degree of mode overlap δ and transmission can be seen by drawing an analogy between transmission of classical waves and electronic conductance. In the absence of inelastic processes, the suppression of average conductance due to the enhanced return of the wave to a coherence volume is connected to the average value of the dimensionless conductance $g = G/(e^2/h)$. Landauer showed that the dimensionless conductance may be expressed in terms of scattering coefficients.[105] When measured between points within perfect leads at a distance from the sample on the order of a few wavelengths so that all evanescent waves have decayed, this relation is

$$g = \langle T \rangle = \sum_{ab} \langle T_{ab} \rangle, \tag{9.26}$$

where the transmittance T is the sum of transmission coefficients T_{ab} over all of the N input and outgoing transverse modes a and b, respectively.[29,105–108] Thus, g

describes classical transmission as well as electronic conductance. Using the Einstein relation for the conductivity

$$\sigma = \left(\frac{e^2}{h}\right) D\,n(\nu), \tag{9.27}$$

the conductance can be written as

$$G = \frac{\sigma A}{L} = \left(\frac{e^2}{h}\right)\frac{D\,n(\nu) A}{L}. \tag{9.28}$$

We then find that[28]

$$g = \frac{G}{(e^2/h)} = \frac{D\,n(\nu)A}{L} = \frac{\delta\nu}{\Delta\nu = \delta}. \tag{9.29}$$

However, we have seen that the degree of mode overlap δ is a measure of localization. Thus g, which is a measure of average electronic transport, is also a measure of localization, which occurs at a threshold $g = \delta = 1$.[28]

In quasi-1D samples, which have reflecting sides and are much longer than their transverse dimensions $L >> \sqrt{A}$, $g = \langle T \rangle \sim N\langle T_a \rangle$, where $N \sim 2\pi A/\lambda^2$ for polarized radiation, and

$$T_a = \sum_b T_{ab} \tag{9.30}$$

is the coefficient of total transmission for a phase coherent incident wave. This may be a single incident transverse mode a, such as the plane wave or a propagating mode of a cavity, or a local coherent source placed near the sample. In the diffusive limit $g >> 1$, $\langle T_a \rangle \approx \ell / L$ and can be approximated by $g \approx N\langle T_a \rangle \approx N\ell / L$. If the cross-sectional area A is fixed as the length increases, as is the case in a long wire or microwave waveguide, g will always fall below unity. Thus, waves will always be localized in sufficiently long quasi-1D static samples, and electrons will always be localized in sufficiently long wires when the temperature is low enough that dephasing can be neglected.[27] For $L \approx N\ell$, $g \approx 1$, suggesting that the localization length in quasi-1D samples is $\xi \approx N\ell$.

The connection of intensity correlation to spatial localization can be seen by noting that the degree of correlation is related to the finesse $1/\delta$. For monochromatic excitation of waves in a random medium in which the ensemble average of transport is described by the diffusion equation, the excitation frequency falls within the half-width of δ modes on average, since approximately $\delta\nu / 2\Delta\nu$ modes just below the excitation frequency and a similar number just above this frequency overlap the excitation frequency. This relationship is shown schematically in Fig. 9.7(a). The field within the medium can be roughly represented by the superposition of these resonantly excited modes. Since the wave within the sample is approximately specified by the amplitudes of these δ modes, it may be specified roughly by δ parameters. As a result, the degree of correlation of intensity is approximately $1/\delta$, or equivalently, $1/g$. For localized waves, the excitation frequency generally either resonantly excites a single mode or falls between modes, as seen in Fig. 9.7(b). The fractional correlation of intensity or total transmission is then dominated by the on-resonance contribution. Because transmission is appreciable only on resonance, the typical value of transmission on resonance as compared to the average value for all frequencies is

$$\frac{\Delta\nu}{\delta\nu} = \frac{1}{\delta}. \tag{9.31}$$

Since the wave is only on resonance for a fraction $\delta\nu/\Delta\nu = \delta$ of the entire spectrum for localized waves, and in this case,

$$\frac{T_a}{\langle T_a \rangle} >> 1,\ \delta s_a = s_a - \langle s_a \rangle = \frac{T_a}{\langle T_a \rangle} - 1 \sim s_a, \tag{9.32}$$

we find that $\mathrm{var}(s_a = T_a/\langle T_a \rangle) = \langle(\delta s_a)^2\rangle \sim (1/\delta)^2\delta \sim 1/\delta$.

The degree of correlation at the output of the sample between two points for which the field correlation function vanishes or between orthogonal propagation modes of the region outside the sample b and b', $\kappa = \langle \delta s_{ab}\delta s_{ab'}\rangle$, can be shown to equal the variance of total transmission normalized to its ensemble average, $\kappa = \mathrm{var}(s_a = T_a/\langle T_a \rangle)$.[22]

The degree of correlation is also inversely proportional to the dimensionless conductance $\kappa \sim 1/g$.[6–22,43,72,73] We conclude that $\kappa = \mathrm{var}(s_a) \sim 1/\delta$ for localized as well as for diffusive waves. In the diffusive limit for nonabsorbing samples, $\kappa = 2/3g = 2/3\delta$.

We have seen above that the parameters δ, $\mathrm{var}(s_a)$, κ, and g are intimately connected. The presence of spectrally isolated modes for small values of δ signals the spatial localization of modes of the sample and corresponds to large fluctuations of transmission from the average value. $\mathrm{Var}(s_a)$ is a measure of the size of these fluctuations. However, when the speckle pattern is normalized to the total transmission, the field is a Gaussian random variable. This has been demonstrated in measurements of field statistics[22] and is a fundamental assumption of random matrix theory.[14–17,70–72] Since the intensity for Gaussian waves is not correlated on scales greater than the wavelength, fluctuations in the speckle pattern of the normalized field self-average to give a contribution to var(s_a) of ~ $1/N$. The large fluctuations observed in transmission are rather due to the value of long-range intensity correlation κ. Thus, the output field glows brightly or dimly as a whole as the frequency is tuned. Large fluctuations in transmission are associated with random ensembles with small values of g because the incident wave is then primarily off resonance with modes of the structure and decays exponentially within the sample. Transmission may be large, however, when the incident wave is resonant with modes of the sample. Azbel showed that the transmission coefficient on resonance approaches unity when the wave is localized near the center of the sample, but falls exponentially with separation from the center of the sample.[31] The four localization parameters discussed above are related in the absence of absorption as follows: $\delta = g = 2/3\text{var}(s_a) = 2/3\kappa$. The equality $\kappa = \text{var}(s_a)$ is maintained even in the face of absorption, while the equality $\delta = g$ is not. In absorbing samples, δ increases because linewidths are broadened by dissipation, while the density of states and its inverse, the average level spacing, do not change. However, $g = \langle T \rangle$ decreases when absorption is present. Thus the ratio δ/g increases with absorption.

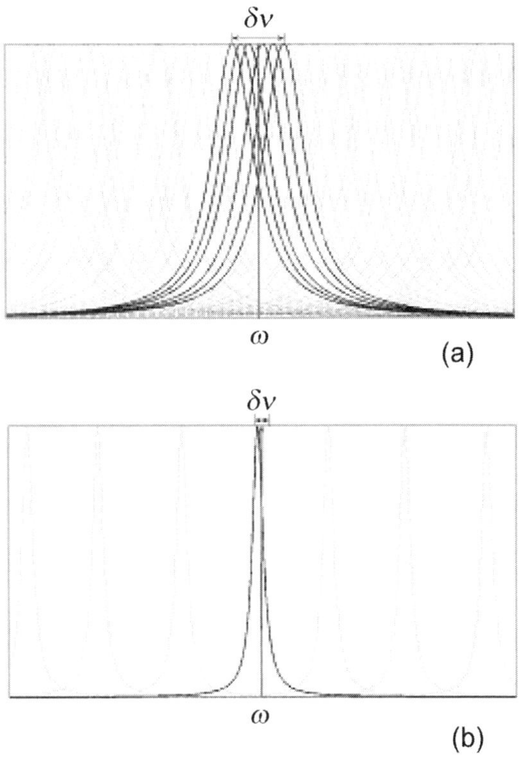

Figure 9.7 Schematic diagram of normalized spectra for (a) spectrally overlapping and (b) spectrally isolated modes.

In order to experimentally investigate the localization transition, it is useful to find a measurable localization parameter that scales exponentially with thickness and is not drastically affected by absorption. Localization parameters should exhibit a weak dependence on absorption, since the spatial distribution of quasimodes is hardly affected by moderate absorption. A parameter that measures localization should also reflect the statistical character of wave transport. We start by considering the distribution of normalized total transmission $P(s_a)$. An expression for $P(s_a)$ as a function of g was found in the diffusive limit without absorption by Van Rossum and Nieuwenhuizen[15,72] using diagrammatic calculations and by Kogan and Kaveh[16] using random matrix theory. The distribution $P(s_a)$ is found to be a function of a single parameter g with $\mathrm{var}(s_a) = 2/3g$, $P(s_a)$. It is therefore possible to express $P(s_a)$ as a function of $\mathrm{var}(s_a)$. The functional form of the distribution for $x = s_a$ in terms of its variance $\beta = \mathrm{var}(x)$ is[15,72]

$$\begin{aligned} P(x) &= \frac{1}{2\pi i}\int_{-i\infty}^{i\infty} \exp(qx) F\left(\frac{3q\beta}{2}\right) dq, \\ F(q) &= \exp\left[-\frac{2\ln^2(\sqrt{1+q}+\sqrt{q})}{3\beta}\right]. \end{aligned} \tag{9.33}$$

Though the direct link between $g = \langle T \rangle$ and localization breaks down in the presence of absorption so that $P(s_a)$ is no longer a function of g, $P(s_a)$ may still be expressed in terms of the variance of the distribution by Eq. (9.33).[74]

Detailed statistical measurements of microwave transmission are carried out in ensembles of statistically equivalent random quasi-1D samples contained in a copper tube. The samples are generally dielectric spheres with diameter of ~ 1 cm, which is of the order of one-half of the tunable microwave wavelength. The mean free path ℓ is determined by the local scattering strength, which depends on the dielectric constant of the sphere ε, the ratio of the sphere diameter to the wavelength, and the sphere density, which can be varied by placing the spheres in Styrofoam shells. Frequently studied samples are Polystyrene with index of refraction $n = 1.59$ and alumina with $n = 3.14$, where $\varepsilon = n^2$. The dimensionless conductance $g = N\ell / L$ is determined by the local scattering strength and the sample cross section and length, as well as the wavelength. Spectra of the polarized field transmission coefficient at points on the output are obtained by using a vector network analyzer from the ratio of the transmitted to the incident field detected with a short wire antenna. The amplitude and phase of the field are obtained from the measurement of the in-phase and out-of-phase components of the field. Field spectra in the frequency domain can be transformed to yield the temporal response to pulsed excitation. The source is a horn or wire antenna. Measurements can be taken at a single point or over a grid of points with spacing sufficiently tight that the 2D sampling theorem[109] can be used to obtain the full speckle pattern. After each spectrum is taken, the sample is briefly rotated to create a new random configuration.

The intensity is obtained by squaring the field amplitude, while the total transmission can be obtained either from the sum of intensity in the transmitted pattern or more rapidly by using an integrating sphere. The integrating sphere is comprised of a shell (with diameter much greater than the diameter of the sample tube) filled with movable scatterers and with a reflecting outer sphere.[74] The integrating sphere is rotated so that the speckle pattern within the shell is scrambled. The time average intensity detected by a Schottky diode in the center of the shell at a particular frequency is thus proportional to the total transmission.

Measurements of $P(s_a)$ for microwave radiation transmitted through samples of randomly mixed Polystyrene spheres contained in copper tubes of different length and diameter are shown in Fig. 9.8(a). These measurements are seen to be in agreement for strongly correlated absorbing samples with values of var(s_a) much smaller than and much greater than unity for weak and strong absorption. Thus, Eq. (9.33) describes the statistics of transmission far from the regime in which it was derived. It is perhaps even more surprising, as will be seen below, that Eq. (9.33) also describes the statistics of the speckle pattern evolution when the incident frequency is changed.

Long-range correlation of intensity across the output face of quasi-1D samples reflects fluctuations of total transmission. The field in individual transmitted speckle patterns is a Gaussian random variable and is correlated on a short length scale of $\lambda/2$. The Gaussian nature of the field in single speckle patterns is a central assumption of random matrix theory. This assumption is confirmed in the probability distribution measurements of the field in random ensembles[22,73] and in observations within individual speckle patterns. Because the intensity is the square of the field, the circular Gaussian distribution of the field

$$P(r,i) = \frac{1}{\sqrt{2\pi\sigma}} \exp\left[-\frac{(r^2+i^2)}{2\sigma^2} \right], \tag{9.34}$$

where r and i are the in-phase and out-of-phase components of a single polarization component of the field, respectively, leads to a negative exponential distribution for the normalized polarized intensity[1,2,4]

$$s_{ab} = \frac{T_{ab}}{\langle T_{ab} \rangle} \frac{1}{s_a} \exp\left(\frac{-s_{ab}}{s_a} \right) \tag{9.35}$$

with

$$\mathrm{var}\left(\frac{s_{ab}}{s_a} \right) = \left\langle \frac{s_{ab}}{s_a} \right\rangle = 1\,. \tag{9.36}$$

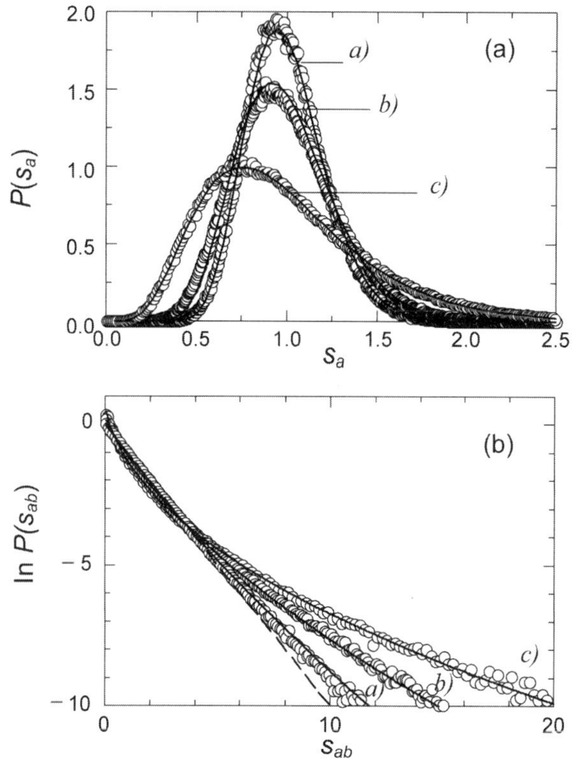

Figure 9.8 Probability distribution functions of (a) normalized total transmission $P(s_a)$ and (b) normalized transmitted intensity $P(s_{ab})$, respectively, for three Polystyrene samples with dimensions: *a)* d = 7.5 cm, L = 66.7 cm; *b)* d = 5.0 cm, L = 50 cm; *c)* d = 5 cm, L = 200 cm. Solid lines are given by Eqs. (9.33) and (9.37) using measured values of var(s_a) of 0.50, 0.65, and 0.22 for samples *a)*, *b)*, and *c)*, respectively. The dashed line in (b) is a plot of the Rayleigh distribution $P(s_{ab}) = \exp(-ss_{ab})$. [Reprinted from Ref. 74. © (2007) by the American Physical Society.]

The distribution of transmitted intensity normalized by its ensemble average value $P(s_{ab})$ is therefore a mixture of a negative exponential distribution and $P(s_a)$:[16]

$$P(s_{ab}) = \int_0^\infty \frac{ds_a}{s_a} P(s_a) \exp\left(\frac{-s_{ab}}{s_a}\right). \tag{9.37}$$

Measurements of $P(s_{ab})$ for the same samples for which measurements of $P(s_a)$ are shown in Fig. 9.8(a) are shown in Fig. 9.8(b) to be in agreement with plots of Eq. (9.37).

Thus, the distributions of intensity and total transmission are fully described by var(s_a). We next consider the scaling of var(s_a) in an absorbing quasi-1D sample over a scale of lengths in which the wave makes a crossover from photon diffusion to localization. This sample is composed of alumina spheres with diameter 0.95 cm and index of refraction 3.14 embedded within Styrofoam shells to produce a sample with alumina volume fraction 0.068.[104] The low sphere density in this sample ensures that distinct sphere resonances are not washed out. The sample is contained within a copper tube with diameter 7.3 cm and plastic end pieces. Measurements are made on large numbers of configurations by briefly rotating the tube between measurements. The variance spectrum of the transmitted intensity normalized by its ensemble average var(s_{ab}) for a 7.3-cm-diameter copper tube of length L = 80 shows a window of localization between 9.9 and 11 GHz[104] (see Fig. 9.9). The window is considerably narrower in shorter samples. As a consequence of Eq. (9.37), the moments of s_a and s_{ab} are related by[16]

$$\left\langle s_{ab}^{n} \right\rangle = n!\left\langle s_{a}^{n} \right\rangle.^{16} \tag{9.38}$$

This gives

$$\mathrm{var}(s_{ab}) = 2\,\mathrm{var}(s_{a}) + 1\,, \tag{9.39}$$

so that the localization threshold given by var(s_a) = 2/3 occurs at var(s_{ab}) = 7/3. Var(s_{ab}) rises above the localization threshold just above the first Mie resonance of the spheres with n = 3.14, which is located around frequency $\nu \approx 2\pi a/c$, where a is the radius of the sphere.[111] The average transit time is peaked,[110] while the transport velocity[111] exhibits a dip on resonance.[73] Localization is not achieved on resonance in this sample even though $\delta\nu$ is small as a consequence of the lengthened transit time in the sample[110] because the density of states is also peaked on resonance. This leads to a dip in the level spacing and a level overlap parameter exceeding unity $\delta = \delta\nu/\Delta\nu > 1$. The strongest scattering in this medium is achieved at 10.0–10.2 GHz. In this range, $k\ell \sim 2$. Somewhat higher values of $k\ell$ are reached at greater sphere density.

The scaling of var(s_a) in this sample attests to its utility as a localization parameter. This quantity scales linearly for values of var(s_a) < 2/3, but scales exponentially for larger values.[43] Measurements of the scaling of var(s_a) are shown in Fig. 9.10.

9.3.4 Time domain

The statistics of fluctuations and correlation can be studied in the time domain as well as in the frequency domain. The linewidth Γ_n represents the decay rate of the mode in the time domain due to both leakage in an open system and absorption. The statistics of Γ_n and of the delay time have been extensively studied based on the statistics of the scattering matrix **S**, since the complex eigenfrequencies of the quasimodes $\omega_n - \frac{i}{2}\Gamma_n$ correspond to the poles of the **S** matrix, and the Wigner delay time

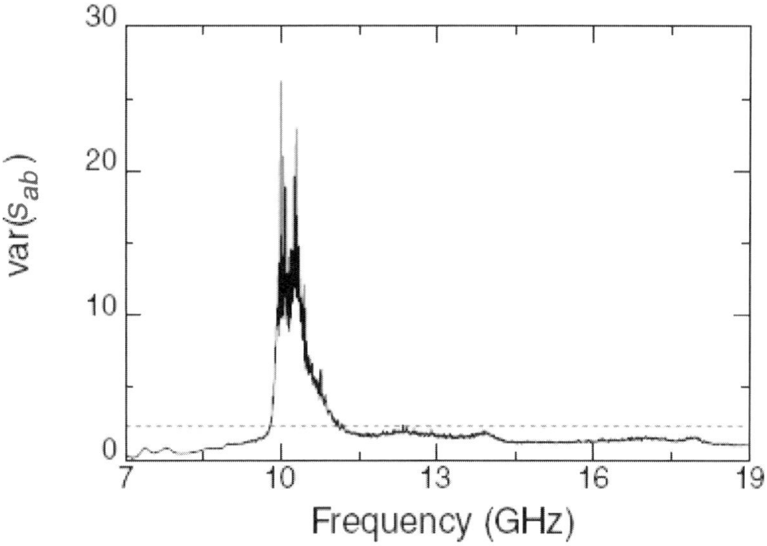

Figure 9.9 Spectrum of var(s_{ab}) in sample with length L = 80 cm. The dashed line indicates the localization threshold. A window of localization in which var(s_{ab}) > 7/3, corresponding to var(s_a) > 2/3, is found just above the first Mie resonance of the alumina spheres comprising the sample. [Reprinted from Ref. 110. © (2001) by the American Physical Society.]

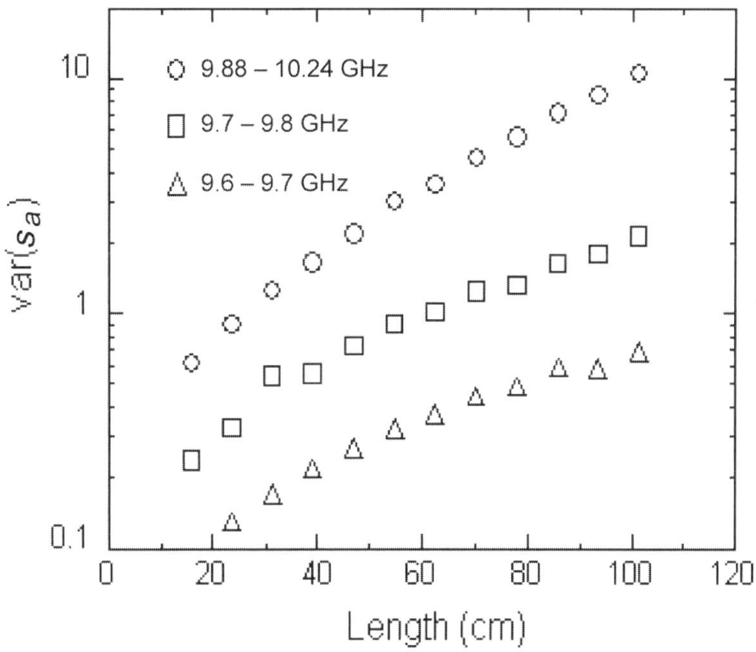

Figure 9.10 Scaling of var(s_a) in samples of alumina spheres. Above a value of the order of unity, var(s_a) increases exponentially. [Reprinted from Ref. 43. © (2000) by Nature Publishing Group.]

$$\tau_W(\omega) = \frac{1}{N} Tr\left(-iS^* dS / d\omega\right) \tag{9.40}$$

effectively characterizes the energy decay rate. For a review on this issue, see Ref. 112 and references therein.

Experimental investigation of the statistics of dynamics can take place directly via measurements following pulsed excitation or by Fourier transforming the product of transmitted field spectra and the spectrum of the incident pulse.[18,113–123] Measurements in the time domain make it possible to unravel the effects of absorption. Absorption in homogeneously absorbing samples simply introduces a multiplicative exponential factor in time.[40,43] All paths emerging from the sample at a given delay time t have the same length and have suffered the same diminution due to absorption. Because the weight of the path distribution is not changed by absorption at a given t, and since all partial waves arriving at a fixed time are equally suppressed by absorption, weak localization is not affected by absorption in the time domain.[40,43,117,122] The influence of localization can be seen in a reduction of the decay rate with increasing time delay[117,122] and in the increase of $\kappa = \text{var}(s_a)$ with time.[119]

In pulsed microwave measurements through diffusive samples, transmission is expected to fall exponentially at a rate equal to that of the lowest diffusion mode

$$1/\tau_1 = \pi^2 D / (L + 2z_0)^2 \tag{9.41}$$

after a time τ_1 in which higher-order modes with decay rates

$$1/\tau_n = n^2\pi^2 D/(L + 2z_0)^2 \tag{9.42}$$

have largely decayed. A breakdown of the diffusion model was found in quasi-1D random dielectric media composed of random mixtures of low-density alumina spheres at frequencies far from the localization window for which $\kappa =$ 0.09, 0.13, 0.25, and 0.125, respectively, in samples A–D.[117] The decay rate for the transmitted pulses shown in Fig. 9.11(a) for different random ensembles is seen in Fig. 9.11(b) to fall in time at a nearly constant rate. A linear decay of the decay rate would be associated with a Gaussian distribution of decay rates for quasimodes of the medium.[124] A slightly more rapid decrease of the decay rate is associated with a slower-than-Gaussian fall-off of the distribution of mode decay rates. The pulsed transmission can be regarded as a distribution of decaying modes. The slowing down of the decay rate at long times reflects the survival of more slowly decaying modes at long times. The decay rate distribution can then be found by taking the Laplace transform of the transmitted pulse intensity.[117,124]

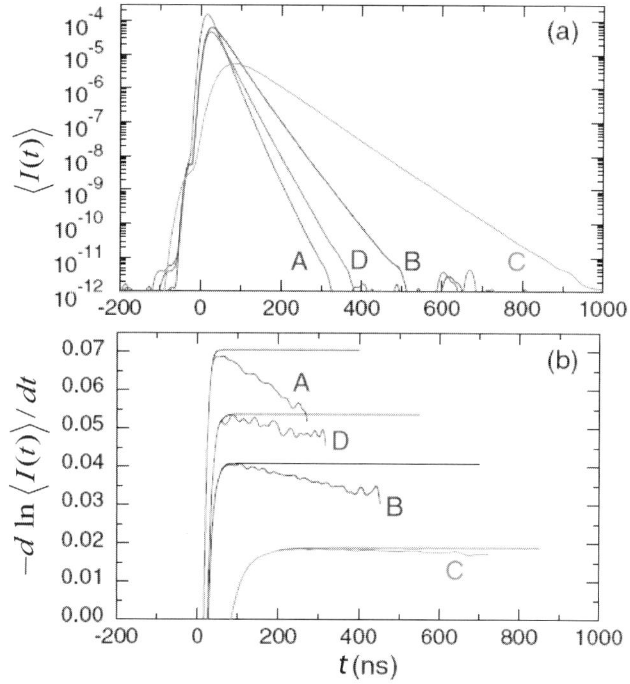

Figure 9.11 (a) Average pulsed transmitted intensity in alumina samples of L = 61 cm (A), 90 cm (B and D), and 183 cm (C). Sample D is the same as sample B except for the increased absorption by titanium foil inserted along the length of the sample tube D. (b) Temporal derivative of the intensity logarithm gives the rate γ of the intensity decay due to both leakage out of the sample and absorption.

The temporal variation of transmission can also be described in terms of the growing impact of weak localization on the dynamic behavior of waves, which can be expressed via the renormalization of a time-dependent diffusion constant or mean free path.[120] The decreasing decay rate has also been explained using a self-consistent theory of localization,[125–129] which incorporates an effective diffusion coefficient that varies with depth inside the sample[126–128] and with the frequency of modulation of the incident intensity. Though the sample is diffusive, significant suppression of the leakage rate is observed. At the same time, the extent of intensity fluctuations and correlation increases with time delay.[119] However, the field correlation function with displacement and polarization rotation is the same for steady state and pulsed transmission. This reflects the Gaussian statistics within the speckle pattern of a given sample configuration. The intensity correlation function at any time delay has the same form as in steady state, but exhibits a time-varying degree of correlation $\kappa_\sigma(t)$[119] that depends on the spectral bandwidth of the pulse σ. The total transmission and intensity distributions at various delay times are given using Eqs. (9.36) and (9.37) with the value of

$$\beta(t) = \mathrm{var}(s_a)(t) = \kappa_\sigma(t). \tag{9.43}$$

The increasing value of $\kappa_\delta(t)$ reflects the sharpened spectrum of transmission at the given time delay. This reflects the greater prominence of longer-lived modes that have narrower linewidths.

Microwave measurements of pulsed transmission at frequencies within the localization window show an evolution of the nature of propagation with increasing time delay reflecting the reduced role of short-lived overlapping states and the growing weight of long-lived spectrally isolated modes with increasing delay from the incident pulse. At short times, propagation is diffusive; at intermediate times, transport can be described in terms of a position- and frequency-dependent renormalized diffusion coefficient;[128,129] while at later times, energy flows largely from isolated localized modes and can be described via a distribution of localized modes in accord with the single parameter scaling theory of localization.[29,130] Different pulse dynamics for modes with different degrees of spatial and spectral overlap can be directly observed in measurements in single-mode waveguides.[131] Results are consistent with a change from localization to diffusion-like dynamics with an increasing degree of overlap in particular configurations.[131] The distinction between isolated and overlapping modes is also seen in pulsed optical spectra in random porous silicon slabs.[24] A full theory of dynamics could be obtained from the statistics of mode spacing and overlap as a function of δ.

Störzer et al.[132] also recently observed a falling decay rate in optical diffusion through a slab of particles of TiO_2 in the rutile phase with a refractive index of 2.8. The mean free path is determined from measurements of the coherent backscattering peak, while the pulse transmission profile is obtained from a histogram of time delays for single photons traversing the slab relative to the incident pulse. Substantial reductions in transmission relative to that predicted by a diffusion model are observed. The results for $k\ell$ with values of 4.3 and 2.5 are shown in Fig. 9.12.

9.3.5 Speckle

We have seen that fluctuations, correlation, and localization are closely related in quasi-1D samples. These characteristics of propagation can be assessed only in the context of an ensemble of random samples. Indeed, the statistics within an individual speckle pattern are independent of the nature of transport. For an ensemble of speckle patterns, the probability distribution of the field is Gaussian, once their average intensities are normalized.[16,22,23] Correspondingly, only short-range intensity correlation is observed and the cumulant intensity correlation function equals the square of the field correlation function. Typical speckle patterns for diffusive and localized waves are shown in Figs. 9.13(a) and (b). It is seen that the spatial variation of intensity is correlated with the variation in phase. Near an intensity peak, phase change is small, while in low-intensity regions, the phase change is rapid. Most importantly, phase singularities,[133–144] at which all equiphase lines cross, appear at points of vanishing intensity where the phase is not defined. In a Gaussian random wave field, phase singularities are generic, and the density of phase singularity is about twice that of intensity maxima.[3,139–141] Although the sizes of speckle spots differ in the two patterns due to the different wavelengths, the statistics of the structure of intensity and phase distributions for the normalized patterns for an ensemble of random samples are essentially the same. This is reflected in the statistics of the field near the core of phase singularities. The phase and intensity variation near a phase singularity has a simple geometric structure,[141,142,144] which is illustrated in Fig. 9.14(a). The contours of constant intensity and constant current magnitude are ellipses and circles, respectively. The orientation and eccentricity ε of the ellipses determine the phase variations in the vortex core.[144] The magnitude of the current increases linearly with the distance from the core, with a slope known as the vorticity, and

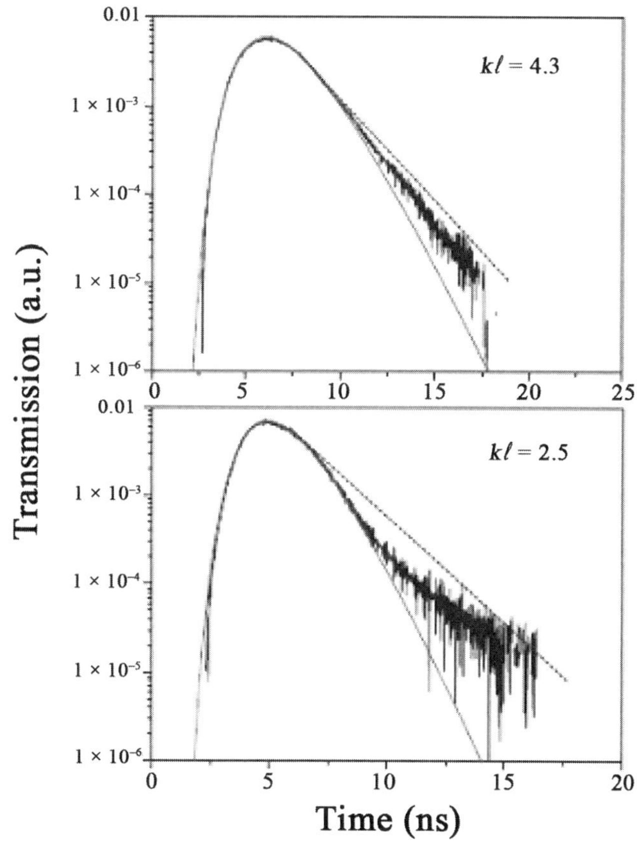

Figure 9.12 Time of flight distributions in two samples of TiO_2 particles with different scattering strengths. The experimental results are compared to an exponential decay at long times derived from diffusion theory including absorption (solid lines). The values of the absorption length are indicated by the slope of the dashed lines. Increasing deviations from diffusive propagation are observed with decreasing values of $k\ell$. [Reprinted with permission from Ref. 122. © (2006) by the American Physical Society.]

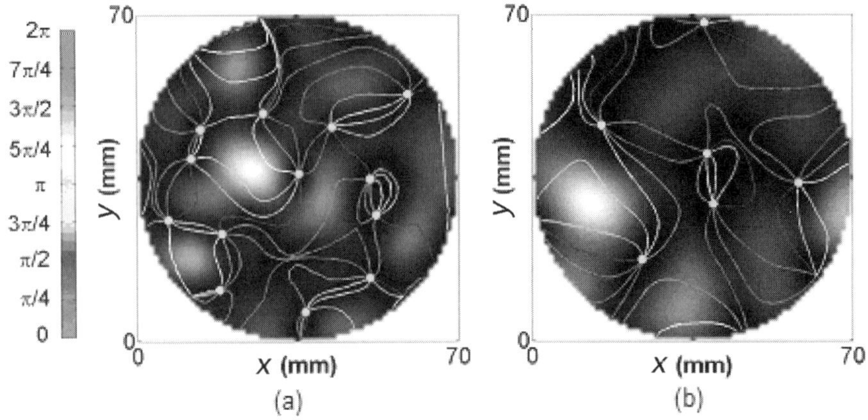

Figure 9.13 Examples of speckle patterns at the output of a quasi-1D sample for (a) diffusive and (b) localized waves. The gray scale shows the intensity variation, and the colored lines are equiphase lines. Green dots represent phase singularities. [Reprinted from Ref. 23. © (2007) by the American Physical Society.] (See color plate section.)

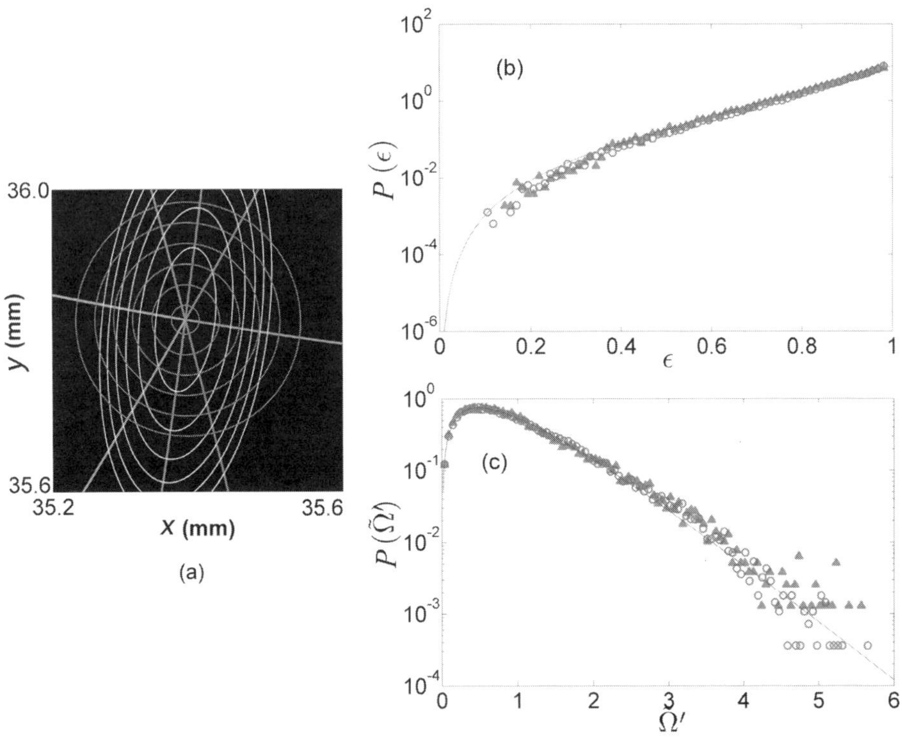

Figure 9.14 (a) Core structure of a phase singularity. The straight lines are equiphase lines with phase values shown in Fig. 9.10. Circles (green) are current contours, while ellipses (white) are intensity contours. (b) Probability distribution of ε. (c) Probability distribution of $\tilde{\Omega}'$. The solid line is a derivation from Gaussian statistics of random fields. Green triangles and red circles are experimental data for diffusive and localized waves, respectively. [Reprinted from Ref. 144. © (2007) by the American Physical Society.] (See color plate section.)

denoted as Ω. Since the average intensity across the entire speckle pattern is proportional to the total transmission s_a, we define a normalized vorticity

$$\Omega' = \Omega / s_a \tag{9.44}$$

to examine the relative intensity variation within one speckle pattern. The probability distributions of ε and

$$\tilde{\Omega}' = \Omega' / \langle \Omega' \rangle , \tag{9.45}$$

shown in Figs. 9.14(b) and (c), demonstrate that statistical distributions of phase and intensity in speckle patterns for diffusive and localized waves are the same. Both $P(\varepsilon)$ and $P(\tilde{\Omega}')$ can be derived from the Gaussian statistics of the random field.[144] The variation of vorticity with the var(s_a) is linear

$$\mathrm{var}(\tilde{\Omega}) = [3\,\mathrm{var}(s_a) + 1]/2 , \tag{9.46}$$

so that $\mathrm{var}(\tilde{\Omega})$ is a direct measure of localization.

Speckle patterns depend on the spatial configurations of all scatters in the sample and thus provide a fingerprint of the sample. Transmission for a given sample configuration can be fully characterized by the electromagnetic modes of the medium. Each mode is associated with a distinctive speckle pattern. In an open or dissipative sample, each mode has a finite linewidth. When δ is not too large, the modes that are superposed to produce the speckle pattern at a specific frequency can be determined. This superposition of fields can be expressed as

$$E_j(x, y, \omega) = \sum_n a_{n,j}(x, y) \frac{\Gamma_n / 2}{\Gamma_n / 2 + i(\omega - \omega_n)} , \tag{9.47}$$

where j is the polarization index and $a_{n,j}(x, y)$ is the spatial distribution of the field for the nth mode. When several modes overlap, the speckle pattern varies with frequency because the complex amplitudes of different modes are determined by a frequency-dependent Lorentzian factor. For diffusive waves, the incident wave always falls within the linewidth of several electromagnetic modes. Speckle patterns therefore change relatively uniformly with frequency since the amplitudes of a large number of modes contributing to the speckle pattern change. In contrast, when the wave is localized, the modes do not overlap and transmission increases dramatically as the source is tuned on resonance. When the incident frequency is on resonance with a single mode, transmission is dominated by the interaction with that mode. The speckle pattern then essentially corresponds to the distinctive pattern of the resonantly excited mode. Though the magnitude of transmission will vary appreciably as the frequency of the wave is tuned through resonance, the change in the spatial distribution of the field is small since the pattern is determined by the interaction with a single mode. However, once the frequency is tuned sufficiently far off resonance, the spatial distribution of the wave will change since the contribution of the initially resonant mode may be reduced to the level of the contributions from other modes. So, in contrast to the continual variation of the speckle pattern with frequency shift in the diffusive case, the speckle pattern may change abruptly as the source frequency is tuned for localized waves. Therefore, the statistics of speckle evolution with frequency shift reflects the degree of overlap of modes and can thus distinguish between diffusive and localized waves.

The speckle pattern changes when the internal structure of the sample changes, as may occur in colloidal samples[145–146] or when the frequency is tuned.[99] The difference in the evolution of speckle patterns for diffusive and localized waves may be quantified by following any of a number of the features within the speckle pattern or quantities computed for the speckle pattern as a whole. Below, we define three quantities that characterize the evolution of the speckle pattern. First, since the speckle pattern can be sketched by a number of critical points, e.g., maxima, minima, and saddle points of intensity or phase, the change in speckle patterns are roughly proportional to the displacement of these points. Here, we consider only the phase singularities and define $\tilde{\mathrm{R}}$ to be the average displacement of phase singularities. We also consider measures of the overall changes of phase and intensity $\tilde{\sigma}_{\Delta\phi}$, the standard deviation of phase changes

$$\Delta j(x,y) = j\left(x,y;\omega+\Delta\omega\right) - j\left(x,y;\omega\right), \tag{9.48}$$

and $\tilde{\sigma}_{\Delta I'}$, the standard deviation of fractional intensity change

$$\Delta I'(x,y) = \frac{2\left[I\left(x,y;\omega+\Delta\omega\right) - I\left(x,y;\omega\right)\right]}{I\left(x,y;\omega+\Delta\omega\right) + I\left(x,y;\omega\right)}. \tag{9.49}$$

The tilde on these variables indicates that the variables are normalized by their ensemble averages. All of these quantities characterize a relative change of speckle pattern but do not characterize the absolute changes of the field. When

$$E_j\left(x,y;\omega+\Delta\omega\right) = c\cdot E_j\left(x,y;\omega\right), \tag{9.50}$$

where c is a complex constant, the change of phase $\Delta\phi$ and intensity $\Delta I'$ are both uniform over the full pattern, and then

$$\tilde{R} = \tilde{\sigma}_{\Delta\phi} = \tilde{\sigma}_{\Delta r} = 0\,, \tag{9.51}$$

and the speckle pattern is unchanged. This occurs when the incident frequency is near resonance with a single mode. We find that for a small change in the speckle pattern, these three quantities are proportional on average. A linear relation between the average value of $\tilde{R}$ for given $\tilde{\sigma}_{\Delta\phi}$ (or $\tilde{\sigma}_{\Delta r}$) and $\tilde{\sigma}_{\Delta\phi}$ (or $\tilde{\sigma}_{\Delta r}$) is shown in Fig. 9.15.

The statistical properties of speckle evolution can be characterized by the probability distributions of $\tilde{R}$, $\tilde{\sigma}_{\Delta\phi}$, and $\tilde{\sigma}_{\Delta r'}$, which are shown in Fig. 9.16.[23] Fluctuations in the change of speckle patterns are much larger for localized waves than for diffusive waves. Such enhanced fluctuations of $\tilde{R}$, $\tilde{\sigma}_{\Delta\phi}$, and $\tilde{\sigma}_{\Delta r}$ for localized waves can also be seen in their spectra in a single configuration (Fig. 9.17). The spectrum of changes in the speckle pattern is continuous for diffusive waves and abrupt for localized waves. A comparison with spectra of s_a in Fig. 9.15(b) further shows that the change of speckle pattern is small on resonance and greatly enhanced off resonance. Both Figs. 9.16 and 9.17 show that fluctuation of speckle changes is greatly enhanced in the localization transition. The distributions $P(\tilde{R})$, $P(\tilde{\sigma}_{\Delta\phi})$, and $P(\tilde{\sigma}_{\Delta r})$ may all be described in terms of the corresponding variances utilized by Eq. (9.33), which described $P(s_a)$. The fit of Eq. (9.33) to the data is shown as the solid curves in Fig. 9.16.

The reason for the similarity between the statistics of changes of the speckle pattern and total transmission can be appreciated by comparing first- and second-order statistics of corresponding local quantities, such as the velocity of a single phase singularity and the intensity at a point.[147] The degree of long-range correlation of the velocity of phase singularities is nearly equal to var($\tilde{\sigma}_{\Delta r'}$); a similar relation exists between the degree of long-range intensity correlation κ and var(s_a). Once the velocity is normalized by $\tilde{\sigma}_{\Delta r'}$, the probability distribution and spatial correlation function of singularity velocity approach those for Gaussian random wave fields. Similarly, when the intensity is normalized by the average intensity in the speckle pattern, the probability distribution of intensity and its spatial correlation function also become coincident with results for Gaussian random waves.

We have so far considered the statistics of locally 3D samples within a quasi-1D geometry, which have a fixed cross section and length much greater than the transverse dimensions. An important unknown is the statistics in the 3D slab geometry. In this geometry, there is no well-defined area over which the total transmission s_a can be defined. Nonetheless, the intensity distribution given by Eqs. (9.33) and (9.37) coincided with that for transmitted ultrasound radiation in a slab of randomly positioned aluminum beads sintered to form an elastic network[148] and for light transmitted through thick glass stacks.[149] In the ultrasound measurements, the distribution corresponded to the quasi-1D distribution just beyond the localization threshold in one frequency range and otherwise to the distribution for diffusive waves. In thin glass stacks in which light does not spread beyond a single coherence area, the intensity probability distribution at the output plane $P(s_{ab})$ is in accord with 1D simulations and falls sharply to zero as $s_{ab} \rightarrow 0$. As the sample thickness increases, the spatial intensity pattern undergoes a topological transformation in which phase singularities are formed at intensity nulls as the lateral spread of the wave exceeds the field correlation length. $P(s_{ab})$ changes progressively to the distribution given in Eqs. (9.33) and (9.37). This distribution is a mixture of a mesoscopic distribution of s_a, $P(s_a)$, and a distribution of intensity for Gaussian waves, even though there is

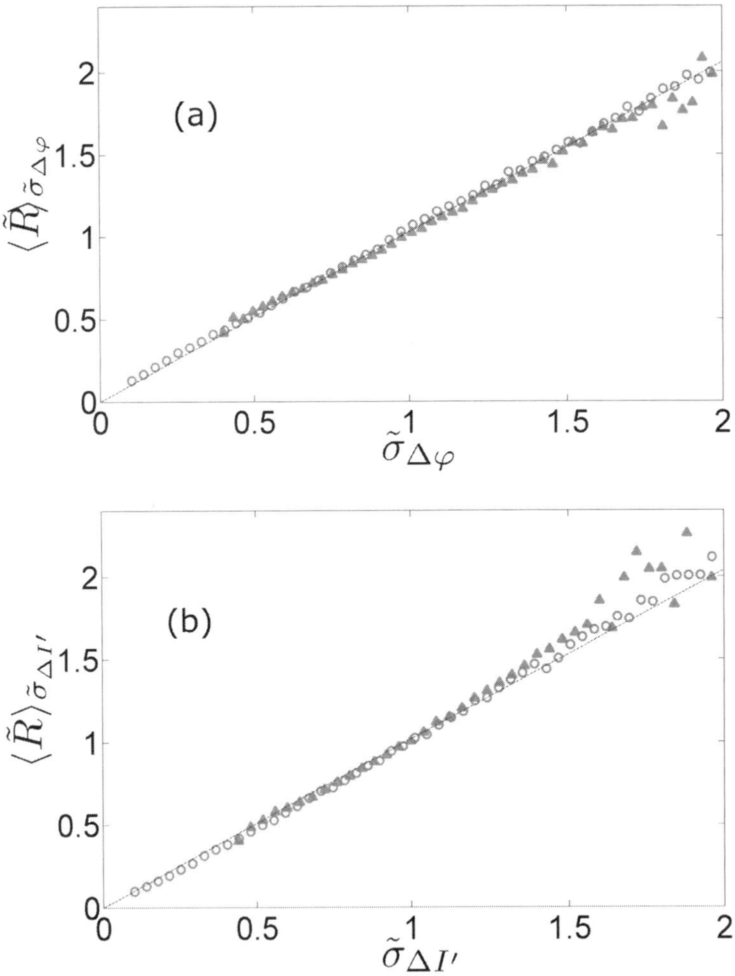

Figure 9.15 Average of (a) $\tilde{R}$ versus $\tilde{\sigma}_{\Delta\phi}$ for given $\tilde{\sigma}_{\Delta\phi}$ and of (b) $\tilde{R}$ versus $\tilde{\sigma}_{\Delta I'}$ for given $\tilde{\sigma}_{\Delta I'}$. Results are plotted in triangles for diffusive waves and in circles for localized waves. [Reprinted from Ref. 23. © (2007) by the American Physical Society.]

no area for which the distribution of integrated intensity is given by the quasi-1D expression for $P(s_a)$. These measurements show that mesoscopic intensity statistics have a universal form beyond one dimension and provide a measure of wave localization within the sample.

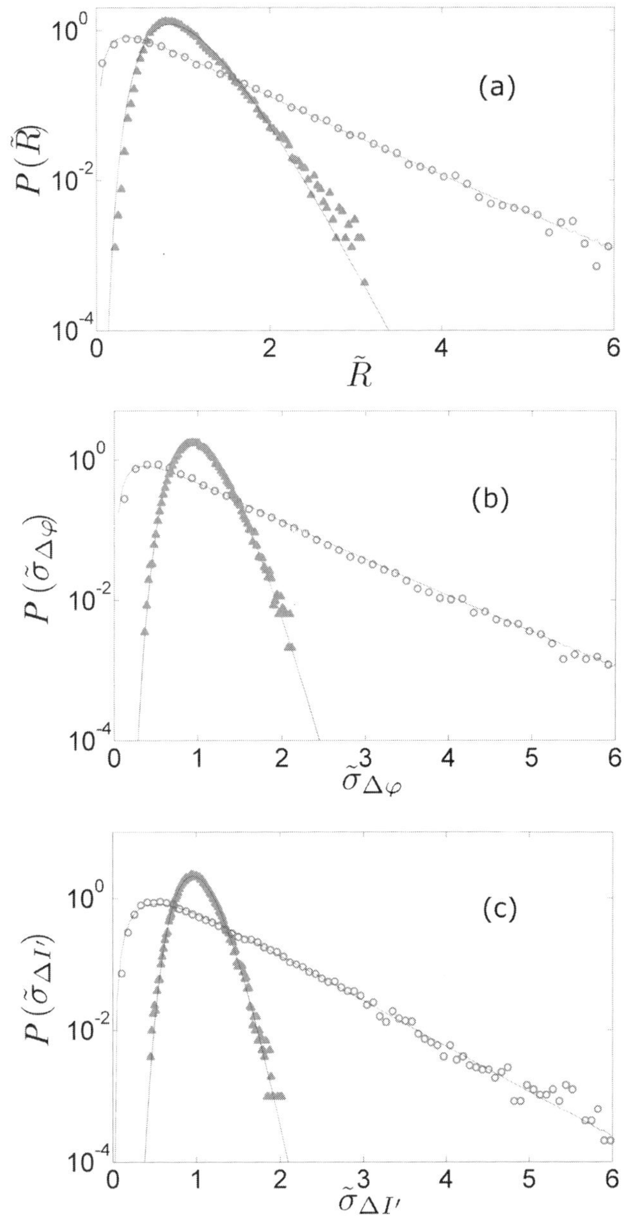

Figure 9.16 Probability distributions of (a) s_a, (b) $\tilde{R}$, (c) $\tilde{\sigma}_{\Delta\varphi}$, and (d) $\tilde{\sigma}_{\Delta I'}$ for diffusive and localized waves. Results are plotted in triangles for diffusive waves and in circles for localized waves. [Reprinted from Ref. 23. © (2007) by the American Physical Society.]

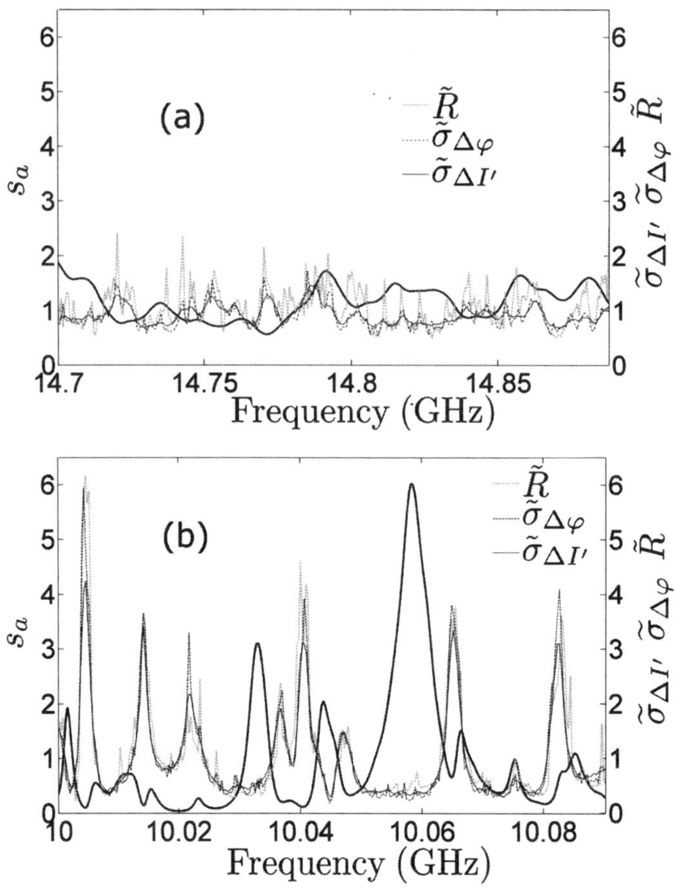

Figure 9.17 Spectra of s_a (black lines), $\tilde{R}$, $\tilde{\sigma}_{\Delta\phi}$, and $\tilde{\sigma}_{\Delta I'}$ for (a) diffusive and (b) localized waves.

9.4 Conclusions

The striking similarities between the statistics of different quantities computed over the entire speckle pattern arise from the common excitation characteristics of the underlying electromagnetic modes of the medium. The statistics of modes depends on the degree of spectral overlap of modes. This may be expressed directly via the ratio of the width to the spacing of levels, and indirectly, but no less precisely, in terms of its effect on the variance of normalized total transmission var(s_a), which equals the degree of intensity correlation κ and is essentially equal to the inverse of the dimensionless conductance g.

Wave propagation may also be seen as a consequence of the crossing of partial waves within the sample. The interference of partial waves produces a volume speckle pattern. The return of partial waves to a coherence volume leads to weak localization and to the suppression of transport. Local fluctuations produced by the interference of partial waves propagate throughout the sample.

This leads to intensity correlation and enhanced fluctuations of intensity, total transmission, and conductance.[11] These phenomena occur only when the wave is temporally coherent throughout the sample. They are referred to as mesoscopic because in the electronic context, they are observed only in samples at low temperatures and in samples that are intermediate in size between the atomic or microscopic scale and the macroscopic scale. For classical waves, scattering elements within a sample are generally correlated over scales much larger than the atomic size, so that the sample is essentially static, and such coherence phenomena are observed at room temperature. All of the above interference phenomena may be readily interpreted in terms of modes. The speckle pattern is the superposition of field patterns for the modes of the medium, while the probability of return to a coherence volume, which is directly related to localization and mesoscopic fluctuations, is essentially equal to $1/\delta$. This may be seen by noting that $\delta = \delta\nu/\Delta\nu$ may be expressed as the ratio of the Heisenberg time $\tau_H = 1/\Delta\nu$, which is the minimum time required for the wave to visit each coherence volume of the sample, and the Thouless time $\tau_{Th} = 1/\delta\nu$, which is proportional to the dwell time in the sample. Thus, $\tau_{Th}/\tau_H = 1/\delta$ is the average number of times the wave returns to a coherence volume. This demonstrates the close connection between the wave interference and mode pictures, each of which provides valuable insights into the nature of wave propagation and localization.

Acknowledgments

We thank Bing Hu, Patrick Sebbah, Andrey Chabanov, Jing Wang, Zhao-Qing Zhang, Jerome Klosner, and Shaolin Liao for valued interactions. This work was supported by the NSF under Grant No. DMR-0907285.

References

1. J. C. Dainty, *Laser Speckle and Related Phenomena*, Springer-Verlag, Heidelberg (1975).

2. J. W. Goodman, *Statistical Optics,* Wiley, New York (1985).

3. I. Freund, "'1001' correlations in random wave fields," *Waves in Random Media* **8,** 119–158 (1998).

4. J. W. Goodman, *Speckle Phenomena in Optics: Theory and Applications*, Roberts & Co. Publishers, Greenwood Village, CO (2007).

5. B. Shapiro, "Large intensity fluctuations for wave propagation in random media," *Phys. Rev. Lett.* **57**, 2168–2171 (1986).

6. M. J. Stephen and G. Cwilich, "Intensity correlation functions and fluctuations in light scattered from a random medium," *Phys. Rev. Lett.* **59**, 285–288 (1987).

7. P. A. Mello, E. Akkermans, and B. Shapiro, "Macroscopic approach to correlations in the electronic transmission and reflection from disordered conductors," *Phys. Rev. Lett.* **61**, 459–462 (1988).

8. S. Feng, C. Kane, P. A. Lee, and A. D. Stone, "Correlations and fluctuations of coherent wave transmission through disordered media," *Phys. Rev. Lett.* **61**, 834–837 (1988).

9. I. Freund, M. Rosenbluh, and S. Feng, "Memory effects in propagation of optical waves through disordered media," *Phys. Rev. Lett.* **61**, 2328–2331 (1988).

10. N. Garcia and A. Z. Genack, "Crossover to strong intensity correlation for microwave radiation in random media," *Phys. Rev. Lett.* **63**, 1678 (1989).

11. R. Pnini and B. Shapiro, "Fluctuations in transmission of waves through disordered slabs," *Phys. Rev. B* **39**, 6986–6994 (1989).

12. J. F. de Boer, M. P. van Albada, and A. Lagendijk, "Observation of long-range intensity correlation in the transport of coherent light through a random medium," *Phys. Rev. Lett.* **64**, 2787 (1990).

13. A. Z. Genack, N. Garcia, and W. Polkosnik, "Long-range intensity correlation in random media," *Phys. Rev. Lett.* **65**, 2129–2132 (1990).

14. E. Kogan and M. Kaveh, "Effect of absorption on long-range correlations in random media," *Phys. Rev. B* **45**, 1049–1052 (1992).

15. Th. M. Nieuwenhuizen and M. C. van Rossum, "Intensity distributions of waves transmitted through a multiple scattering medium," *Phys. Rev. Lett.* **74**, 2674–2677 (1995).

16. E. Kogan and M. Kaveh, "Random-matrix-theory approach to the intensity distributions of waves propagating in a random medium," *Phys. Rev. B* **52**, R3813–R3815 (1995).

17. P. W. Brouwer, "Transmission through a many-channel random waveguide with absorption," *Phys. Rev. B* **57**, 10526–10536 (1998).

18. P. Sebbah, R. Pnini, and A. Z. Genack, "Field and intensity correlation in random media," *Phys. Rev. E* **62**, 7348 (2000).

19. P. Sebbah, B. Hu, A. Z. Genack, R. Pnini, and B. Shapiro, "Spatial-field correlation: the building block of mesoscopic fluctuations," *Phys. Rev. Lett.* **88**, 123901 (2002).

20. A. A. Chabanov, A. Z. Genack, N. Tregoures, and B. A. van Tiggelen, "Polarization correlation in random media," *Phys. Rev. Lett.* **92**, 173901 (2004).

21. S.-H. Chang, A. Taflove, A. Yamilov, A. Burin, and H. Cao, "Numerical study of light correlations in a random medium close to Anderson localization threshold," *Opt. Lett.* **29**, 917–919 (2004).

22. A. A. Chabanov and A. Z. Genack, "Statistics of the mesoscopic field," *Phys. Rev. E* **72**, 055602 (2005).

23. S. Zhang, B. Hu, P. Sebbah, and A. Z. Genack, "Speckle evolution of diffusive and localized waves," *Phys. Rev. Lett.* **99**, 063902 (2007). http://link.aps.org/doi/10.1103/PhysRevLett.99.063902 (See EPAPS Document No. E-PRLTAO-99-005733 for movie clips.)

24. P. W. Anderson, "Absence of diffusion in certain random lattices," *Phys. Rev.* **109**, 1492–1505 (1958).

25. M. E. Gertsenshtein and V. B. Vasil'ev, "Waveguides with random inhomogeneities and Brownian motion in the Lobachevsky plane," *Theor. Probab. Appl.* **4**, 391–398 (1959).

26. A. Ioffe and A. R. Regel, "Non-crystalline, amorphous and liquid electronic semiconductors," *Prog. Semicond.* **4**, 237–291 (1960).

27. D. J. Thouless, "Maximum metallic resistance in thin wires," *Phys. Rev. Lett.* **39**, 1167–1169 (1977).

28. E. Abrahams, P. W. Anderson, D. C. Licciardello, and T. V. Ramakrishnan, "Scaling theory of localization: absence of quantum diffusion in two dimensions," *Phys. Rev. Lett.* **42**, 673–676 (1979).

29. P. W. Anderson, D. J. Thouless, E. Abrahams, and D. S. Fisher, "New method for a scaling theory of localization," *Phys. Rev. B* **22**, 3519–3526 (1980).

30. G. Papanicolaou, "Wave propagation in a one-dimensional random medium," *J. Appl. Math.* **21**, 13–18 (1971).

31. M. Y. Azbel and P. Soven, "Transmission resonances and the localization length in one-dimensional disordered systems," *Phys. Rev. B* **27**, 831–835 (1983).

32. S. John, "Electromagnetic absorption in a disordered medium near a photon mobility edge," *Phys. Rev. Lett.* **53**, 2169–2172 (1984).

33. S. John, "Localization of light," *Physics Today* **44**, 32–40 (1991).

34. S. John, "Strong localization of photons in certain disordered dielectric superlattices," *Phys. Rev. Lett.* **58**, 2486–2489 (1987).

35. G. Bergmann, Y. Bruynserade, and B. Kramer, *Localization, Interaction, and Transport Phenomena in Impure Metals*, Springer-Verlag, New York (1985).

36. S. He and J. D. Maynard, "Detailed measurements of inelastic scattering in Anderson localization," *Phys. Rev. Lett.* **57**, 3171–3174 (1986).

37. P. Sheng, Ed., *Scattering and Localization of Classical Waves in Random Media*, World Scientific, Singapore (1990).

38. N. Garcia and A. Z. Genack, "Anomalous photon diffusion at the threshold of the Anderson localization transition," *Phys. Rev. Lett.* **66**, 1850–1853 (1991).

39. A. Z. Genack and N. Garcia, "Observation of photon localization in a three-dimensional disordered system," *Phys. Rev. Lett.* **66**, 2064–2067 (1991).

40. R. Weaver, "Anomalous diffusivity and localization of classical waves in disordered media: the effect of dissipation," *Phys. Rev. B* **47**, 1077–1080 (1993).

41. D. S. Wiersma, P. Bartolini, A. Lagendijk, and R. Righini, "Localization of light in a disordered medium," *Nature* **390**, 671–673 (1997).

42. M. V. Berry and S. Klein, "Transparent mirrors: rays, waves and localization," *Eur. J. Phys.* **18**, 222–228 (1997).

43. A. A. Chabanov, M. Stoytchev, and A. Z. Genack, "Statistical signatures of photon localization," *Nature* **404**, 850–853 (2000).

44. V. Milner and A. Z. Genack, "Photon localization laser: low-threshold lasing in a random amplifying layered medium via wave localization," *Phys. Rev. Lett.* **94**, 073901 (2005).

45. P. Sebbah, B. Hu, J. Klosner, and A. Z. Genack, "Extended quasimodes within nominally localized random waveguides," *Phys. Rev. Lett.* **96**, 183902 (2006). http://link.aps.org/doi/10.1103/PhysRevLett.96.183902

46. J. Topolancik, B. Ilic, and F. Vollmer, "Experimental observation of strong photon localization in disordered photonic crystal waveguides," *Phys. Rev. Lett.* **99**, 253901 (2007). http://link.aps.org/doi/10.1103/PhysRevLett.99.253901

47. J. B. Pendry, "Quasi-extended electron states in strongly disordered systems," *J. Phys. C* **20**, 733–742 (1987).

48. R. A. Webb, S. Washburn, C. P. Umbach, and R. B. Laibowitz, "Observation of the Aharonov-Bohm oscillations in normal-metal rings," *Phys. Rev. Lett.* **54**, 2696–2699 (1985).

49. B. L. Altshuler and D. E. Khmelnitskii, "Fluctuation properties of small conductors," *JETP Lett.* **42**, 359–363 (1985).

50. P. A. Lee and A. D. Stone, "Universal conductance fluctuations in metals," *Phys. Rev. Lett.* **55**, 1622–1625 (1985).

51. B. L. Altshuler, P. A. Lee, and R. A. Webb, *Mesoscopic Phenomena in Solids*, Elsevier Science (1991).

52. M. P. van Albada and A. Lagendijk, "Observation of weak localization of light in a random medium," *Phys. Rev. Lett.* **55**, 2692–2695 (1985).

53. P. E. Wolf and G. Maret, "Weak localization and coherent backscattering of photons in disordered media," *Phys. Rev. Lett.* **55**, 2696–2699 (1985).

54. E. Akkermans, P. E. Wolf, and R. Maynard, "Coherent backscattering of light by disordered media: analysis of the peak line shape," *Phys. Rev. Lett.* **56**, 1471–1474 (1986).

55. D. S. Wiersma, M. P. van Albada, B. A. van Tiggelen, and A. Lagendijk, "Experimental evidence for recurrent multiple scattering events of light in disordered media," *Phys. Rev. Lett.* **74**, 4193–4196 (1995). http://link.aps.org/doi/10.1103/PhysRevLett.74.4193

56. P. Sheng, *Introduction to Wave Scattering, Localization and Mesoscopic Phenomena,* Springer-Verlag, New York (2006).

57. E. Akkermans and G. Montambaux, *Mesoscopic Physics of Electrons and Photons*, Cambridge University Press, Cambridge (2006).

58. W. H. Carter and E. Wolf, "Coherence and radiometry with quasi-homogeneous planar sources," *J. Opt. Soc. Am.* **67**, 785–796 (1977).

59. A. Z. Genack, "Fluctuations, Correlation and Average Transport of Electromagnetic Radiation in Random Media," Chap. 5 in *Scattering and Localization of Classical Waves in Random Media*, P. Sheng, Ed., World Scientific, Singapore (1990).

60. E. Akkermans, P. E. Wolf, R. Maynard, and G. Maret, "Theoretical study of the coherent backscattering of light by disordered media," *J. de Physique* **49**, 77–98 (1988).

61. S. Fiebig, C. M. Aegerter, W. Bührer, M. Störzer, E. Akkermans, G. Montambaux, and G. Maret, "Conservation of energy in coherent backscattering of light," *arXiv*:0706.0381v1 (2007).

62. G. Barton, *Elements of Green's Functions and Propagation*, Clarendon Press, Oxford (1989).

63. I. Freund and D. Eliyahu, "Surface correlations in multiple-scattering media," *Phys. Rev, A* **45**, 6133–6134 (1992).

64. F. J. Wegner, "The mobility edge problem: continuous symmetry and a conjecture," *Zeitschrift für Physik B* **35**, 207–210 (1979).

65. S. Hikami, "Anderson localization in a nonlinear-σ-model representation," *Phys. Rev. B* **24,** 2671–2679 (1981).

66. B. Kramer and A. MacKinnon, "Localization: theory and experiment," *Rep. Prog. Phys.* **56**, 1469–1564 (1993).

67. S. Chandrasekhar, *Radiative Transfer*, Dover Publications, New York (1960).

68. J. Lenoble, *Radiative Transfer in Scattering and Absorbing Atmospheres: Standard Computational Procedures*, A. Deepak Publishing, Hampton, VA, 583 (1985).

69. M. L. Mehta, *Random Matrices*, Academic Press, Boston (1991).

70. C. W. Beenakker, "Random-matrix theory of quantum transport," *Rev. Mod. Phys.* **69**, 731–808 (1997).

71. P. A. Mello, *Quantum Transport in Mesoscopic Systems: Complexity and Statistical*, Oxford University Press, Oxford (2004).

72. M. C. W. van Rossum and Th. M. Nieuwenhuizen, "Multiple scattering of classical waves: microscopy, mesoscopy, and diffusion," *Rev. Mod. Phys.* **71**, 313–371 (1999).

73. A. Z. Genack and A. A. Chabanov, "Signatures of photon localization," *J. Phys. A: Math. Gen.* **38,** 10465–10488 (2005).

74. M. Stoytchev and A. Z. Genack, "Measurement of the probability distribution of total transmission in random waveguides," *Phys. Rev. Lett.* **79**, 309–312 (1997). http://link.aps.org/doi/10.1103/PhysRevLett.79.309

75. P. T. Leung, S. Y. Liu, and K. Young, "Completeness and orthogonality of quasinormal modes in leaky optical cavities," *Phys. Rev. A* **49**, 3057 (1994).

76. E. S. C. Ching, P. T. Leung, A. Maassen van den Brink, W. M. Suen, S. S. Tong, and K. Young, "Quasinormal-mode expansion for waves in open systems," *Rev. Mod. Phys.* **70**, 1545–1554 (1998).

77. U. Kuhl, H. J. Stöckmann and R. Weaver, "Classical wave experiments on chaotic scattering," *J. Phys. A: Math. Gen.* **38**, 10433–10463 (2005).

78. H. E. Türeci, A. D. Stone, and B. Collier "Self-consistent multimode lasing theory for complex or random lasing media," *Phys. Rev. A* **74**, 043822 (2006).

79. H. E. Türeci, A. D. Stone, and L. Ge, "Theory of the spatial structure of nonlinear lasing modes," *Phys. Rev. A* **76**, 013813 (2007).

80. J. B. Pendry, "Symmetry and transport of waves in one-dimensional disordered systems," *Adv. Phys.* **43**, 461–542 (1994).

81. K. Yu. Bliokh, Yu. P. Bliokh, V. Freilikher, A. Z. Genack, and P. Sebbah, "Coupling and level repulsion in the localized regime: transition from isolated to quasi-extended modes," *arXiv:*0806.0704v1 (2008).

82. N. F. Mott, "Conduction in non-crystalline systems. IV. Anderson localization in a disordered lattice," *Philos. Mag.* **22**, 7–29 (1970).

83. I. M. Lifshits and V. Ya. Kirpichenkov, "On tunnel transparency of disordered systems," *Zh. Eksp. Teor. Fiz.* **77**, 989 (1979). [*Sov. Phys. JETP* **50**, 499 (1979).]

84. J. Bertolotti, M. Galli, R. Sapienza, M. Ghulinyan, S. Gottardo, L. C. Andreani, L. Pavesi, and D. S. Wiersma, "Wave transport in random systems: Multiple resonance character of necklace modes and their statistical behavior," *Phys. Rev. E* **74**, 035602 (2006).

85. V. Freilikher, M. Pustilnik, and I. Yurkevich, "Effect of absorption on the wave transport in the strong localization regime," *Phys. Rev. Lett.* **73**, 810–813 (1994).

86. K. Yu. Bliokh, Yu. P. Bliokh, V. Freilikher, A. Z. Genack, B. Hu, and P. Sebbah, "Localized modes in open one-dimensional dissipative random systems," *Phys. Rev. Lett.* **97**, 243904 (2006).

87. H. Cao, Y. Zhao, S. T. Ho, E. W. Seelig, Q. H. Wang, R. P. H. Chang, "Random laser action in semiconductor powder," *Phys. Rev. Lett.* **82**, 2278–2281 (1999).

88. P. Sebbah and C. Vanneste, "Random laser in localized regime," *Phys. Rev. B* **66**, 144202 (2002).

89. K. L. van der Molen, R. W. Tjerkstra, A. P. Mosk, and A. Lagendijk, "Spatial extent of random laser modes," *Phys. Rev. Lett.* **98**, 143901 (2007).

90. C. Vanneste, P. Sebbah, and. H. Cao, "Lasing with resonant feedback in weakly scattering random systems," *Phys. Rev. Lett.* **98**, 143902 (2007).

91. S. Mujumdar, V. Türck, R. Torre, and D. S. Wiersma, "Chaotic behavior of a random laser with static disorder," *Phys. Rev. A* **76**, 033807 (2007).

92. H. Cao, "Lasing in Random Media," Chapter 11 in *Tutorials in Complex Photonics Media*, M. A. Noginov, et al., Eds., SPIE Press, Bellingham, WA, 301–358 (2009).

93. D. S. Wiersma, S. Mujumdar, S. Cavalieri, R. Torre, G.-L. Oppo, S. Lepri, "Chaotic Behavior of Random Lasers," Chapter 10 in *Tutorials in Complex Photonics Media*, M. A. Noginov, et al., Eds., SPIE Press, Bellingham, WA, 277–299 (2009).

94. M. A. Noginov, "Feedback in Random Lasers," Chapter 12 in *Tutorials in Complex Photonics Media*, M. A. Noginov, et al., Eds., SPIE Press, Bellingham, WA 359–396 (2009).

95. D. Laurent, O. Legrand, P. Sebbah, C. Vanneste, and F. Mortessagne, "Localized modes in a finite-size open disordered microwave cavity," *Phys. Rev. Lett.* **99**, 253902 (2007).

96. O. Shapira and B. Fischer, "Localization of light in a random grating array in a single mode fiber," *J. Opt. Soc. of Am. B* **22**, 2542–2552 (2005).

97. V. I. Kopp, B. Fan, H. K. M. Vithana, and A. Z. Genack, "Low-threshold lasing at the edge of a photonic stop band in cholesteric liquid crystals," *Opt. Lett.* **23**, 1707–1709 (1998).

98. H. D. Raedt, A. Lagendijk, P. d. Vries, "Transverse localization of light," *Phys. Rev. Lett.* **62**, 47–50 (1989).

99. T. Schwartz, G. Bartal, S. Fishman and M. Segev, "Transport and Anderson localization in disordered two-dimentional photonic crystals," *Nature* **446**, 52–55 (2007).

100. Y. Lahini, A. Avidan, F. Pozzi, M. Sorel, R. Morandotti, D. N. Christodoulides, and Y. Silberberg, "Anderson localization and nonlinearity in one-dimensional disordered photonic lattices," *Phys. Rev. Lett.* **100**, 013906 (2008).

101. M. Ya. Azbel, "Resonance tunnelling and localization spectroscopy," *Solid State Commun.* **45**, 527 (1983).

102. K. Yu. Bliokh, Yu. P. Bliokh, V. Frielikher, S. Savelev, and F. Nori, "Unusual resonators: plasmonics, metamaterials, and random media," *Rev. Mod. Phys.* **80**, 1201 (2008).

103. A. Z. Genack, "Universality of wave propagation in random media," *Europhys. Lett.* **11**, 733–738 (1990).

104. A. Z. Genack, "Optical transmission in disordered media," *Phys. Rev. Lett.* **58**, 2043–2046 (1987).

105. R. Landauer, "Electrical resistance of disordered one-dimensional lattices," *Philos. Mag.* **21**, 863–876 (1970).

106. E. Economu and C. M. Soukoulis, "Static conductance and scaling theory of localization in one dimension," *Phys. Rev. Lett.* **46**, 618–621 (1981).

107. D. S. Fisher and P.A. Lee, "Relation between conductivity and transmission matrix," *Phys. Rev. B* **23**, 6851–6854 (1981).

108. A. S. Stone and A. Szafer, "What is measured when you measure a resistance? The Landauer formula revisited," *IBM J. Res. Dev.* **32**, 384–413 (1988).

109. J. W. Goodman, *Introduction to Fourier Optics*, McGraw-Hill, Columbus (1968).

110. A. A. Chabanov and A. Z. Genack, "Photon localization in resonant media," *Phys. Rev. Lett.* **87**, 153901 (2001). http://link.aps.org/ doi/10.1103/ PhysRevLett.87.153901

111. M. P. van Albada, B. A. van Tiggelen, A. Lagendijk, and A. Tip, "Speed of propagation of classical waves in strongly scattering media," *Phys. Rev. Lett.* **66**, 3132–3135 (1996).

112. T. Kottos, "Statistics of resonances and delay times in random media: beyond random matrix theory," *J. Phys. A: Math. Gen.* **38**, 10761–10768 (2005).

113. G. H. Watson, Jr., P. A. Fleury, and S. L. McCall, "Searching for photon localization in the time domain," *Phys. Rev. Lett.* **58**, 945–948 (1987).

114. J. M. Drake and A. Z. Genack, "Observation of nonclassical optical diffusion," *Phys. Rev. Lett.* **63**, 259–262 (1989).

115. K. M. Yoo, F. Liu, and R. R. Alfano, "When does the diffusion approximation fail to describe photon transport in random media?" *Phys. Rev. Lett.* **65**, 2210–2211 (1990).

116. A. Z. Genack and J. M. Drake, "Relationship between optical intensity fluctuations and pulse propagation in random media," *Europhys. Lett.* **11**, 331–336 (1990).

117. A. A. Chabanov, A. Z. Genack, and Z. Q. Zhang, "Breakdown of diffusion in dynamics of extended waves in mesoscopic media," *Phys. Rev. Lett.* **90**, 203903 (2003).

118. P. M. Johnson, A. Imhof, B. P. Bret, J. G. Rivas, and A. Lagendijk, "Time-resolved pulse propagation in a strongly scattering material," *Phys. Rev. E* **68**, 016604 (2003).

119. A. A. Chabanov, B. Hu, and A. Z. Genack, "Dynamic correlation in wave propagation in random media," *Phys. Rev. Lett.* **93**, 123901 (2004).

120. S. K. Cheung, X. Zhang, Z. Q. Zhang, A. A. Chabanov, and A. Z. Genack, "Impact of weak localization in the time domain," *Phys. Rev. Lett.* **92**, 173902 (2004).

121. J. Bertolotti, S. Gottardo, D. S. Wiersma, M. Ghulinyan, and L. Pavesi, "Optical necklace states in Anderson localized 1D systems," *Phys. Rev. Lett.* **94**, 113903 (2005).

122. M. Störzer, P. Gross, C. M. Aegerter, and G. Maret, "Observation of the critical regime near Anderson localization of light," *Phys. Rev. Lett.* **96**, 063904 (2006). http://link.aps.org/doi/10.1103/PhysRevLett.96.063904

123. M. Störzer, C. M. Aegerter, and G. Maret, "Reduced transport velocity of multiply scattered light due to resonant scattering," *Phys. Rev. E* **73**, 065602 (2006).

124. A. D. Mirlin, "Statistics of energy levels and eigenfunctions in disordered systems," *Phys. Rep.* **326**, 259–382 (2000).

125. D. Vollhardt and P. Wölfle, "Diagrammatic, self-consistent treatment of the Anderson localization problem in $d \leq 2$ dimensions," *Phys. Rev. B* **22**, 4666–4679 (1980).

126. B. A. van Tiggelen, "Localization of waves," Chapter 1 in *Diffusive Waves in Complex Media*, J. P. Fouque, Ed., Kluwer Academic Publisher, 1–60 (1999).

127. B. A. van Tiggelen, A. Lagendijk, and D. Wiersma, "Reflection and transmission of waves near the localization threshold," *Phys. Rev. Lett.* **84**, 4333–4336 (2000).

128. S. E. Skipetrov and B. A. van Tiggelen, "Dynamics of weakly localized waves," *Phys. Rev. Lett.* **92**, 113901 (2004).

129. S. E. Skipetrov and B. A. van Tiggelen, "Dynamics of Anderson localization in open 3D media," *Phys. Rev. Lett.* **96**, 043902 (2006).

130. Z. Q. Zhang, A. A. Chabanov, S. K. Cheung, C. H. Wong, and A. Z. Genack, "Dynamics of localized waves: Pulsed microwave transmissions in quasi-one-dimensional media," *Phys Rev. B* **79,** 144203 (2009).

131. P. Sebbah and A. Z. Genack, private communication (2008).

132. M. Störzer, P. Gross, C. M. Aegerter and G. Meret, "Observation of the critical regime near Anderson localization of light," *Phys. Rev. Lett.* **96**, 063904 (2006).

133. J. F. Nye and M. V. Berry, "Dislocations in wave trains," *Proc. R. Soc. Am.* **336,** 165–190 (1974).

134. M. V. Berry, "Disruption of wavefronts: statistics of dislocations in incoherent Gaussian random waves," *J. Phys. A: Math. Gen.* **11,** 27–37 (1978).

135. M. V. Berry, "Singularities in waves and rays," in *Physics of Defects*, Eds. R. Balian et al., North-Holland, Amsterdam, 453–543 (1981).

136. N. B. Baranova, A. V. Mamaev, N. F. Pilipetskii, V. V. Shukov, and B. Y. Zel'dovich, "Wave-front dislocations: topological limitations for adaptive systems with phase conjugation," *J. Opt. Soc. Am.* **73,** 525–528 (1983).

137. J. V. Hajnal, "Observation of singularities in electric and magnetic fields of freely propagating microwaves," *Proc. R. Soc. A* **430**, 413–421 (1990).

138. I. Freund, N. Shvartsman, and V. Freilikher, "Optical dislocation networks in highly random media," *Opt. Commun.* **101,** 247–264 (1993).

139. I. Freund, "Optical vortices in Gaussian random wave fields: statistical probability densities," *J. Opt. Soc. Am. A* **11,** 1644–1652 (1994).

140. N. Shvartsman and I. Freund, "Vortices in random wave fields: nearest neighbor anticorrelations," *Phys. Rev. Lett.* **72,** 1008 (1994).

141. M. V. Berry, and M. R. Dennis, "Phase singularities in isotropic random waves," *Proc. R. Soc. A.* **456**, 2059–2079 (2000).

142. W. Wang, S. G. Hanson, Y. Miyamoto, and M. Takeda, "Experimental investigation of local properties and statistics of optical vortices in random wave fields," *Phys. Rev. Lett.* **94**, 103902 (2005).

143. B. A. van Tiggelen, D. Anache, and A. Ghysels, "Role of mean free path in spatial phase correlation and nodal screening," *Europhys. Lett.* **74**, 999–1005 (2006).

144. S. Zhang and A. Z. Genack, "Statistics of diffusive and localized fields in the vortex core," *Phys. Rev. Lett.* **99**, 203901 (2007). http://link.aps.org/doi/10.1103/PhysRevLett.99.203901

145. G. Maret and P. E. Wolf, "Multiple light scattering from disordered media. The effect of Brownian motion of scatterers," *Zeitschrift für Physik B* **65**, 409–413 (1987).

146. D. J. Pine, D. A. Weitz, P. M. Chaikin, and E. Herbolzheimer, "Diffusing wave spectroscopy," *Phys. Rev. Lett.* **60**, 1134–1137 (1988).

147. S. Zhang, Y. Lockerman, and A. Z. Genack, "Mesoscopic speckle" *arXiv*: 0904.1908.v1.

148. H. Hu, A. Strybulevych, J. H. Page, S. E. Skipetrov, and B. A. van Tiggelen, "Localization of ultrasound in a three-dimensional elastic network," *Nature Physics* **4**, 945 (2008).

149. J. Park, S. Zhang, and A. Z. Genack, "Universal mesoscopic statistics and the localization of light," *arXiv*:0905.4111.v2.

Biographies

Azriel Genack is Distinguished Professor of Physics at Queens College of the City University of New York. He received his B.A. and Ph.D. degrees from Columbia University. Genack served as a postdoctoral researcher at the City College of CUNY and at the IBM Research Laboratory in San Jose. He joined the staff of the Exxon Research and Engineering Co. in 1977 and served there until 1984 when he began teaching at Queens College. Genack cofounded Chiral Photonics, Inc. in 1999 and has advised the company since its founding. He has published in the areas of microwave and optical propagation, localization and lasing in random and periodic media, band-edge lasing and photonics of planar and fiber chiral structures, acousto-optic tomography, surface-enhanced Raman scattering, coherent transient spectroscopy, photochemical hole burning in molecular solids, excitons in semiconductors, and nuclear spin diffusion in superconductors.

Sheng Zhang has been a senior scientist at Chiral Photonics, Inc. since 2008. He received his B.Eng. and M.S. degrees in optoelectronics from Zhejiang University and Huazhong University of Science and Technology, China, and his Ph.D. degree in quantum optics from the University of Bradford, UK. From 2005 until 2008, he worked as a postdoctoral research associate at Queens College of CUNY. His expertise includes wave propagation in random and nearly periodic media, speckle and singular optics, phase space methods in quantum optics, and the design and development of fiber optical devices.

Chapter 10
Chaotic Behavior of Random Lasers

Diederik S. Wiersma
LENS—European Laboratory for Non-Linear Spectroscopy, BEC-INFM, Florence, Italy

Sushil Mujumdar
Tata Institute of Fundamental Research, Mumbai, India

Stefano Cavalieri, Renato Torre
LENS—European Laboratory for Non-Linear Spectroscopy and Department of Physics, University of Florence, Italy

Gian-Luca Oppo
Department of Physics, University of Strathclyde, Glasgow, UK

Stefano Lepri
Institute of Complex Systems, CNR, Sesto Fiorentino, Italy

10.1 Introduction

10.1.1 Multiple scattering and random lasing

Light transport in disordered systems can be described as a multiple scattering process in which light waves are multiply scattered by random variations of the refractive index of the material. This has close analogies with other transport phenomena, such as electrons in a conducting material, sounds waves in random structures, and even matter waves.[1] A particularly interesting situation in light transport arises when optical amplification is included in the multiple scattering process. In this case, the intensity of the light waves increases during the multiple scattering. This can lead to an unstable situation wherein the overall gain exceeds the total losses. Multiple scattering and amplification form the two main ingredients to create what is called a random laser.[2–13] This is a system in which a threshold exists above which gain becomes larger than the losses due to the multiple scattering process.

One can wonder how it is possible to speak of laser action in a material that is random and therefore, obviously has no structure that resembles a laser cavity. To that end, one has to keep in mind two important points. The first point is that light waves that propagate in a dielectric material, even if that material has a disordered structure, will not lose memory of their phase. The disorder will scatter the light in a complicated way, thereby creating very complex amplitude and phase fields. However, this process is well defined, and a specific configuration of the disorder will lead to a unique phase and intensity distribution that can be predicted and measured. One can even define optical modes of such random structures as the (complex) field patterns that are formed when the samples are illuminated with monochromatic light.

The second point to keep in mind is that a laser cavity, as such, is not essential to obtain coherent emission. The fundamental property of a laser that leads to coherent emission is not the cavity itself, but the gain saturation that it creates. The cavity creates a situation in which gain is larger than loss, which leads to a rapid growth of the intensity until the gain medium is depleted. It is this saturation that leads to second-order coherence by suppressing fluctuations of the intensity. [In terms of photon statistics, second-order coherence reflects a certain level of 'antibunching' of the photons induced by the gain saturation. The photons from a chaotic source are 'bunched,' meaning that they arrive in clusters. In a coherent beam, the photons are more smeared-out (antibunched) in time, meaning that the intensity fluctuations are reduced.]

It is therefore not necessary to have 'coherent' feedback in order to obtain coherent laser emission. In the current literature, a distinction is sometimes made between 'coherent' and 'incoherent' random lasers. This distinction is confusing, since it would suggest that the multiple scattering process in certain materials can be 'incoherent,' which is not the case for dielectric materials.

The confusion came into existence because of the different models used to describe random lasing. In the initial extremely simplified picture of random lasing, the light transport was described as a diffusion process.[2,4,7,8] This model neglects all interference effects and local intensity and phase patterns but can be used to understand that random amplifying systems should exhibit a threshold behavior. In particular, by solving a diffusion equation with gain, one can see that there exists a critical volume above which the total gain (which is proportional to the volume) becomes larger than the total amount of light that leaves the sample (proportional to the sample surface). The intensity in a block of amplifying random dielectric material that is larger than this critical volume will diverge until the gain depletes and gain saturation sets it.

This simple diffusive model, as first discussed by Letokhov,[2] has proven to be very useful since it can explain several important aspects of a random laser, including an overall narrowing of the emission spectrum above threshold,[14–23] laser spiking, and relaxation oscillations.[7,14] It is also possible to describe the relatively high coherence of the output of a random laser by simply considering a diffusive picture with a gain term instead of an absorption term.[12]

However, a diffusion with gain model cannot explain certain observations, in particular, those of narrow spikes in emission spectra as first observed by Cao et al.[10,24] To explain these detailed spectral properties, it is necessary to take into account the details of the multiple scattering process and calculate the mode structure of the random system, for instance by finite-difference time-domain calculations. This does not mean, however, that certain materials are coherent and others are incoherent. The specific experimental configuration (e.g., single-shot experiments versus averaged spectra) and parameters (such as excitation energy and pulse duration of the excitation) determine if one observes only the overall narrowing of the spectrum or if one also observes the more detailed narrow spikes.

10.1.2 Mode coupling

To better understand the detailed mechanisms that play a role in a random laser, it is important to consider the optical modes of the materials. If one excites the modes of a random dielectric structure either with an incoming plane wave at a specific frequency or by a local narrow source inside the sample, one will find the specific intensity and phase distribution of the excited mode. In most random materials, these modes will have an extended nature, in the sense that they cover a large volume of the sample.

In very strongly scattering random systems, it is also possible to find modes that are spatially confined due to a phenomenon called optical Anderson localization. This phenomenon is the optical counterpart of Anderson localization of electrons[25] and is very difficult to obtain in dielectric materials. If it occurs, its effect is spectacular, however, since it leads to an inhibition of diffusive transport.[26,27] When Anderson localization occurs, the modes of the random system become strongly confined (localized) in a random fashion. That is, the modes have exponentially decaying tails in the system and have no spatial overlap. This inhibits transport and can occur when the scattering is extremely strong.

Anderson localization has been proposed and studied theoretically as a mechanism behind random lasing in spatially confined modes[28] and was initially linked to the observation of narrow emission spikes from zinc oxide (ZnO) powders.[29,30] By analyzing weakly scattering materials, it later became clear that Anderson localization was not essential to observing narrow emission spikes.[31] An alternative explanation was put forward that does not require localization but rather describes the effect in terms of long-lived extended modes.[31] Recent numerical calculations have shown that such extended modes indeed exist and can give rise to random lasing and narrow emission spikes.[32]

It was suggested that anomalously localized modes could exist in materials that are otherwise not Anderson localized.[33,34] Such modes would have a strong spatial confinement but are extraordinary given the fact that the majority of the modes in the system are extended. They would, therefore, require the co-existence of both localized and extended modes in the same material and at the same wavelengths. An alternative explanation was put forward that does not require localization but rather describes the effect in terms of long-lived extended modes[31,35] that were also observed later in finite-difference time-domain calculations.[32] Spatial confinement is in any case not required to obtain coherent laser emission, as was previously explained, so that both localized and extended modes can lead to coherent random laser emission.

Random lasers with either extended or localized modes share an important concept, namely that the number of modes is large and that these modes can couple either directly or via the gain mechanism. This coupling is even stronger for spatially extended modes, since they partially cover the same volume of the sample (and thereby compete for the same gain particles). This can lead to a strongly chaotic behavior. The threshold of ZnO-based random lasers was found, for instance, to show strong fluctuations, depending on the pulse duration of the excitation laser.[36] This inspired a systematic study on fluctuations in random lasers with dynamic disorder (e.g., particles in solution).[37,38]

In this chapter, we will look at experimental data on the emission from random lasers, and in particular, the reproducibility of these complex emission spectra over several excitation events. We will compare this with measurements of the speckle generated by diffusion of the excitation laser and show that in a certain range of parameters, the spectra show a chaotic behavior and are irreproducible in successive excitation events. Then, we will look at the results of Monte Carlo simulations on multiple scattering with gain, which exhibit chaotic behavior due to mode coupling. In particular, we can identify a region around the laser threshold in which the statistics of the emission intensity change from Gaussian into Lévy-type statistics.

10.2 Experiments on Emission Spectra

10.2.1 Sample preparation and setup

In these experiments, we wished to investigate the behavior of emission spectra from a random laser with static disorder. That is, we wished to examine samples in which the specific configuration of the scattering elements was stable over time. This can be done in practice by using, for instance, glass powder that is sintered in order to create physical connections between the glass particles. (See also Ref. 39.) We used SK11 glass obtained from Schott AG. A fine powder of the glass was obtained by grinding in a planetary micromill. The powder was chemically cleaned and sintered under a pressure of 1.2 GPa to obtain robust porous disks of 1-cm diameter and 6-mm thickness. The porosity of the sample enabled us to infiltrate it with a solution of rhodamine 6G in methanol (molar concentration 5×10^{-3} M/l). The mean free path of the sample, indicated as ℓ in the subsequent text, was found to be 50 μm, as determined from the measurement of the diffusion constant.

To determine the amount of disorder in a random material, the parameter $k\ell$ is commonly used, with k being the magnitude of the wave vector. This parameter reflects the comparison between the mean free path and the wavelength. Note that strong scattering corresponds to small values of $k\ell$. Anderson localization of optical waves, a strong interference phenomenon that leads to exponentially confined modes, occurs when $k\ell \leq 1$.[26] For our samples, we have $k\ell = 5.4 \times 10^2$ for the central wavelength of the emission band. The samples are therefore diffusively scattering and clearly far out of the Anderson localization regime.

Optical excitation was provided by the frequency-doubled output (λ = 532 nm) of a mode-locked Nd:YAG laser (2.5-Hz repetition rate, 25-ps pulse width) focused on the sample surface in a spot of 100-μm diameter and at quasi-perpendicular incidence. From previous calculations,[7] we know that the pump intensity is expected to fall off nearly linearly inside the sample, with the mean free path (in our case, 50 μm) as the characteristic length scale. A Peltier-cooled CCD array coupled to a monochromator was synchronized with the laser pulses to record single-shot emission spectra. The diffuse emission from the illuminated sample surface was collected by a collection lens with a focal length of 6 cm and numerical aperture 0.39, placed at an angle of 30 deg from the sample normal. The spectral resolution of the set-up was measured using a calibrated mercury lamp and was found to be 0.2 nm.

10.2.2 Emission spectra

Fig. 10.1 shows experimentally observed single-shot emission spectra from the sample for three successive pulses of the excitation laser. The excitation energy was 1.2 μJ, corresponding to a peak intensity of 6.1×10^8 W/cm^2. In each of the spectra, a broadband profile (pedestal) is observed. High-intensity ultranarrow lasing spikes are seen imposed on the broad pedestal. The bandwidth of the narrowest spike is < 0.2 nm. Clearly, the locations of the high-intensity lasing spikes in each spectrum are different from shot to shot, and the modes are not equispaced.

We find that the occurrence of the emission spikes is strongly nonlinear with excitation energy and exhibits a threshold behavior. In Fig. 10.2, the average height of the sharp emission peaks is plotted versus excitation energy. At every pump energy, a minimum of 100 single-shot emission spectra were collected. For every spectrum, the intensity of each spike was measured relative to the broad pedestal, and the average peak height was calculated. This was done for all of the spectra in the set for a given pump energy. The total average was found and is plotted in Fig. 10.2. Below an excitation energy of 1.2 μJ, no sharp peaks were observed, hence, the peak height is zero for excitation energies below 1.2 μJ. The bars on the points give the standard deviation at each pump pulse. As can be seen, the fluctuations in the random laser emission also increase as the emission intensity increases.

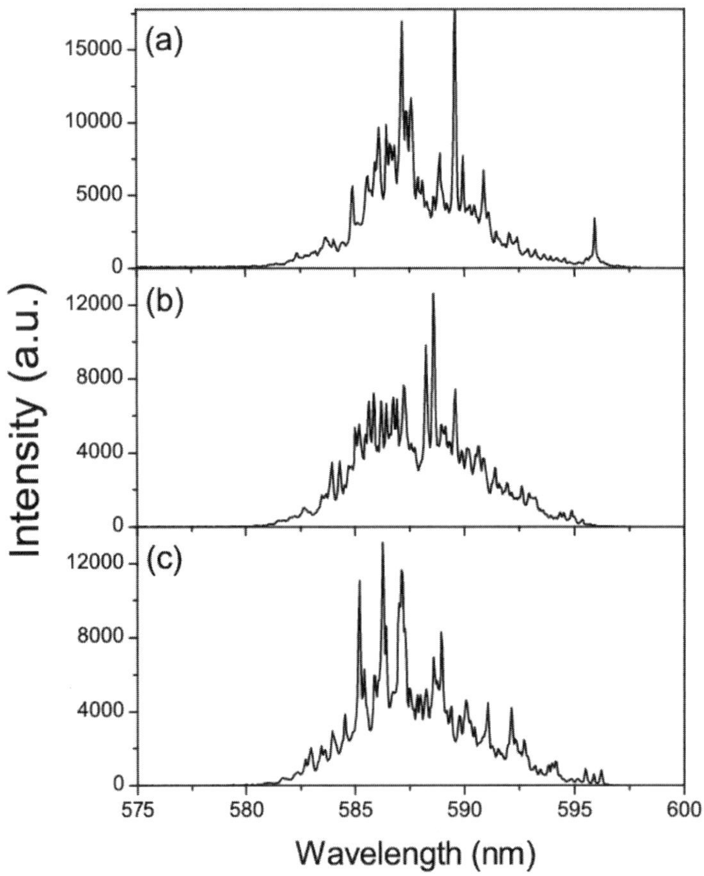

Figure 10.1 Experimentally observed single-shot emission spectra showing ultranarrow lasing modes from a porous glass disk doped with rhodamine 6G. The three spectra shown are taken at three successive excitation pulses. The narrow-banded lasing spikes occur at different wavelengths from pulse to pulse. [Reprinted from Ref. 39. © (2007) by the American Physical Society.]

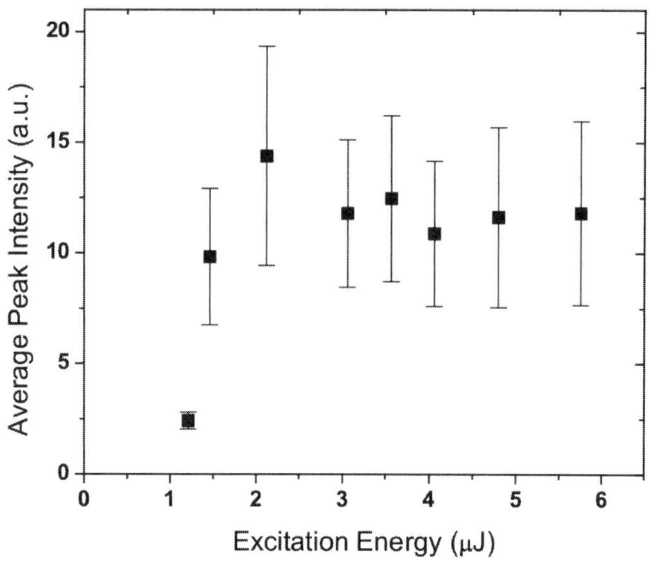

Figure 10.2 Average intensity (in arbitrary units) of the observed narrow emission spikes versus excitation energy (in μJ). Below an excitation energy of 1.2 μJ, no narrow spikes were observed, hence, their height is zero. [Reprinted from Ref. 39. © (2007) by the American Physical Society.]

10.3 Experiments on Speckle Patterns

The fact that the observed emission spectra are uncorrelated from one excitation shot to the next indicates a strong sensitivity of the system to its boundary condition given the static nature of the sample itself. For such an interpretation to be correct it is important, however, to assure that the positions of the scattering elements are indeed constant and that the sample is stable. Otherwise, the differences from one shot to another could simply be due to a rearrangement of the scattering elements.

An excellent test for sample stability is to study the speckle pattern that is generated by the multiple scattering through the sample. Since the speckle pattern is a direct consequence of the phases of waves that are multiply scattered throughout the random medium, it is extremely sensitive to any changes in the scatterer configuration. This sensitivity is applied in a technique called diffusing-wave spectroscopy to study nanometer-scale movements of particles with visible light.[40] A high sensitivity for the particle positions is obtained due to the multiple scattering process. The speckle pattern changes when the *cumulative* phase shift induced by movements of the particles becomes comparable to the wavelength λ. A static speckle pattern can therefore be used to demonstrate that the configuration of the disorder in the sample is constant on length scales that are relevant for the optical properties of the sample. On the same grounds, one can expect a similar sensitivity to any fluctuations of the phase of the incident coherent laser pulse. The method is, therefore, also suitable to verify the stability of the excitation laser and the experimental setup.

We monitored the speckle pattern generated by the excitation laser to assess any variation in the disorder configuration. Note that in these measurements we did not monitor the speckle generated by the emission of the material (the random lasing) but the speckle generated by multiple scattering of the excitation laser. A CCD camera was used in imaging mode to record the speckle pattern generated by individual laser shots. Two such speckle patterns are shown in Fig. 10.3. Clearly, the two speckle patterns are nearly identical, demonstrating that the realization of the disorder is not changing. Minor visible differences between the two patterns arise from CCD noise, which is unavoidable. These patterns are spaced by about 2 sec. Typically, we found that the correlation in the speckle pattern persisted for much longer times and over many excitation shots. The strong correlation between individual speckle patterns rules out any variation in the scatterer configuration on the length scale that is relevant for multiple scattering of light.

We performed a series of measurements on both the speckle pattern generated by these samples and their emission spectra and calculated the correlation in both cases. After subtracting the broadband pedestal, the correlation coefficient of the spectra was calculated using the standard technique by summing the product of the corresponding points of two spectral curves and normalizing to one. For the speckle patterns, the correlation coefficient was calculated as the normalized sum of the product of corresponding pixel values of successive CCD images. The results are plotted in Fig. 10.4. We can clearly see that while the speckle remains correlated over many shots, the emission spectra are completely uncorrelated.

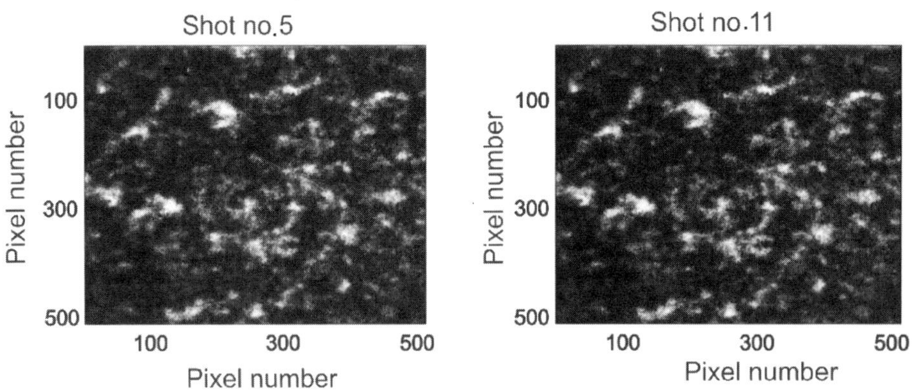

Figure 10.3 Far-field speckle patterns observed from two distinct laser pulses (fifth and eleventh in a series of pulses). The numbers on the axes indicate pixels on the CCD array. The highly correlated intensity distribution confirms that the configuration of scatterers does not change over subwavelength length scales, even over several laser pulses. Hence, the spatial mode distribution remains unchanged over several laser shots. [Reprinted from Ref. 39. © (2007) by the American Physical Society.]

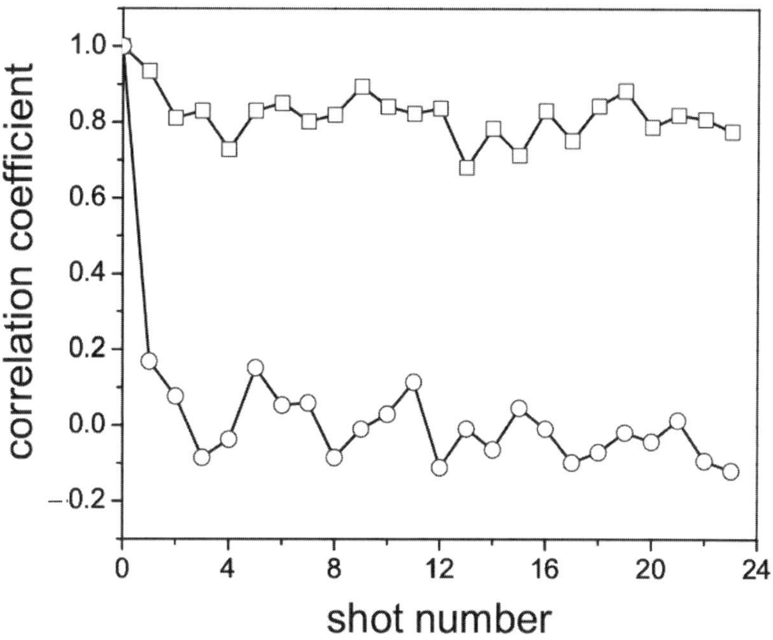

Figure 10.4 Spectral (circular markers) and speckle (square markers) correlation coefficient over several laser pulses. Speckle patterns are highly correlated over several laser pulses, whereas the emission spectra are uncorrelated. [Reprinted from Ref. 39. © (2007) by the American Physical Society.]

10.4 Modeling

10.4.1 Monte Carlo simulations

In order to gain more insight into the physical processes behind the chaotic behavior observed in random lasers, we use a general, yet easy to simulate, model of random lasing. We consider a sample partitioned into cells of linear size ℓ. Specifically, we consider a portion of a two-dimensional square lattice. Thus the center of each cell is identified by the vector index $\mathbf{r} = (x, y)$, with x, y integers. In the following discussion, we will consider a sample with a slab geometry, i.e., $1 \leq x \leq L$ and $1 \leq y \leq RL$. The total number of lattice sites is thus RL^2, where R defines the slab aspect ratio. Periodic boundary conditions in the y direction are assumed.

Within each cell we have the population $N(\mathbf{r}, t)$ of excitations. We consider a hypothetical three-level system with fast decay from the lowest laser level. If the population in the latter can be neglected, we can identify N as the number of atoms in the excited state of the lasing transition.

Isotropic diffusion of light is modelled as a standard random walk along the lattice sites. The natural time unit for the dynamics is thus given by $\Delta t = \ell / v$, with v as the speed of light in the medium. We choose to describe the photon dynamics in terms of a set of M walkers, each carrying a given number of photons n_1, ...n_M. Each of their intensities changes in time, due to processes of stimulated and spontaneous emission.

A basic description of these phenomena can be given in terms of a suitable master equation[12,41] that would require taking into account the discrete nature of the variables. To further simplify the model, we consider that the population and number of photons within each cell are so large for the evolution to be well approximated by the deterministic equation for their averages. In other words, the rate of radiative processes is much larger than that of the diffusive processes, and a huge number of emissions occur within the time scale Δt.[42] With these simplifications, N and n can be treated as continuous variables. Altogether, the model is formulated by the following discrete-time dynamics:

Step 0: pumping. The active medium is homogeneously excited at the initial time, i.e., $N(\mathbf{r}, 0) = N_0$. The value N_0 represents the pumping level due to some external field. The initial number of walkers is set to $M = 0$.

Step 1: spontaneous emission. At each time step and for every lattice site, a spontaneous emission event randomly occurs with probability $\gamma N \Delta t$, where γ denotes the spontaneous emission rate of the single atom. The local population is decreased by one

$$N \longrightarrow N-1, \tag{10.1}$$

and a new walker is started from the corresponding site with initial photon number $n=1$. The number of walkers M is increased by one accordingly.

Step 2: diffusion. Parallel and asynchronous update of the photons' positions is performed. Each walker moves with equal probability to one of its four nearest neighbours. If the boundaries $x = 1, L$ of the system are reached, the walker is emitted and its photon number n_{out} recorded in the output. The walker is then removed from the simulation and M is diminished by one.

Step 3: stimulated emission. At each step, the photon numbers of each walker and the population are deterministically updated according to the following rules:

$$\begin{aligned} n_i &\longrightarrow (1+\gamma \Delta t\, N)\, n_i \\ N &\longrightarrow (1-\gamma \Delta t\, n_i)\, N, \end{aligned} \tag{10.2}$$

where N is the population at the lattice site on which the ith walker resides.

Stochasticity is thus introduced in the model by both the randomness of spontaneous emission events (Step 1) and the diffusive process (Step 2). Note that the model in the above formulation does not include nonradiative decay mechanisms of the population. Furthermore, no dependence on the wavelength is, at present, accounted for, and in general, $\gamma = \gamma(\lambda)$.

The initialization described in Step 0 is a crude modelling of the pulsed pumping employed experimentally. It amounts to considering an infinitely short excitation during which the sample absorbs N_0 photons from the pump beam. As a further simplification, we also assumed that the excitation is homogeneous throughout the sample. More realistic pumping mechanisms can be easily included in this type of modelling.[49] More importantly, as we wish to study the time dependence of the emission, this type of scheme applies to the case in which the time separation between subsequent pump pulses is much larger than the duration of the emitted pulse (i.e., no repumping effects are present).

Steps 1–3 are repeated up to a preassigned maximum number of iterations. The sum of all of the photon numbers of walkers flowing out of the medium at each time step is recorded. The resulting time series is binned on a time window of duration T_W to reconstruct the output pulse as it would be measured by an external photocounter. This ensures that each point is a sum over a large number of events and allows a comparison with ensemble-averaged results, as we will see in the next section.

It should be emphasized that, although each walker evolves independently from all of the others, they all interact with the same population distribution, which, in turn, determines the photon number distribution. In spite of its simplicity, the model describes these two quantities in a self-consistent way. For convenience, we choose to work henceforth in dimensionless units such that $v = 1$ and $\ell = 1$ (and thus, $\Delta t = 1$). The only independent parameters are then γ, the initial population N_0 (i.e., the pumping level), and the slab sizes L and RL.

10.4.2 Results and interpretation

Preliminary runs of the Monte Carlo code were performed to check that lasing thresholds exist after increasing either the pumping parameter N_0 or the slab width L.[43] The values are in agreement with the theoretical analysis presented above. We monitored the outgoing flux (per unit length) ϕ as a function of time. Here, ϕ is defined from the discrete continuity equation to be

$$\phi = \frac{D}{2}\left[I(1,t) + I(L,t)\right]. \tag{10.3}$$

$I(x\,,t)$ is the number of photons in each cell. The factor 2 takes into account the contribution from the two boundaries $x = 0$, L of the lattice.

The results of Monte Carlo simulation for a lattice with $L = 30$, $R = 20$ (18,000 sites), and $\gamma = 10^{-12}$ (yielding $N_c = 2.5673 \times 10^9$) are reported in Fig. 10.5. The three chosen values of N_0 are representative of the three relevant statistical regions depicted in Fig. 10.9. They correspond to the values of the gain length $\ell_G = 500$ (just below threshold), $\ell_G = 200$ (just above threshold), and $\ell_G = 20$ (far above threshold), respectively. At the threshold, the total number of generated photons is equal to the number of photons that leave the sample through the boundaries. In the first two cases, the total emission is highly irregular with huge departures from the expected mean-field behavior. Above the lasing threshold [Fig. 10.5(b)] single events (corresponding to rare events with very long path length and huge amplification) may carry values of n_i up to 10^{10}. The resulting time series are quite sensitive to initialization of the random number generator used in the simulation. On the contrary, in the third case [Fig. 10.5(c)] the pulse is quite smooth and reproducible, except perhaps for its tails that have a much smaller relative intensity.

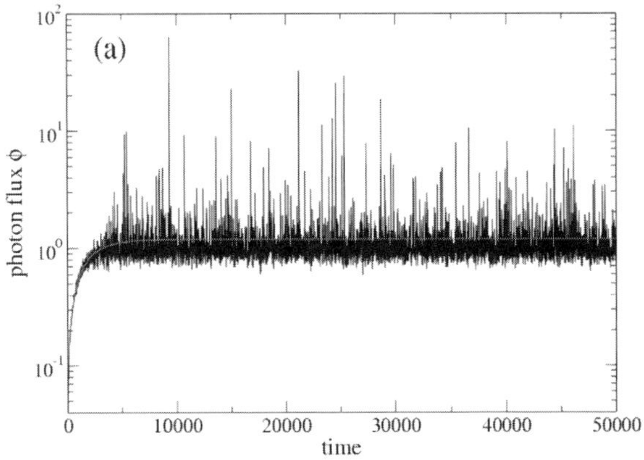

Figure 10.5 (a) The photon flux (per unit length) as a function of time for a single shot and for $N_0 = 2 \times 10^9$. The data have been binned over consecutive time windows of duration $T_W = 10$. Note the difference in the vertical-axis scales. Smooth lines are the mean-field limit for large population and photon number. In this limit, the dynamics is described by the rate equations for the macroscopic averages. [Reprinted from Ref. 43. © (2007) by the American Physical Society.]

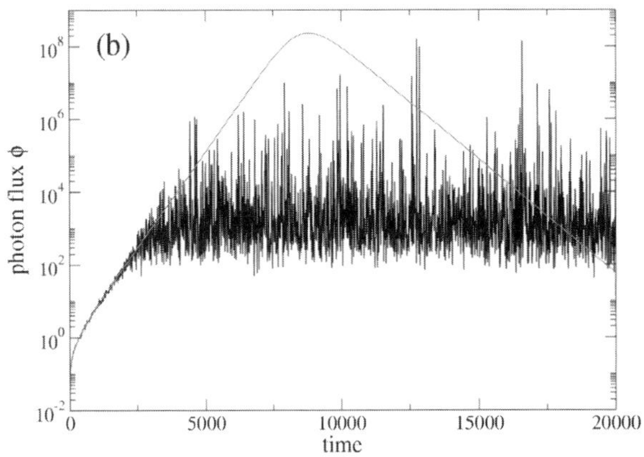

Figure 10.5 (b) The photon flux (per unit length) as a function of time for a single shot and for $N_0 = 5 \times 10^9$. The data have been binned over consecutive time windows of duration $T_W = 10$. Note the difference in the vertical-axis scales. Smooth lines are the mean-field limit for large population and photon number. In this limit, the dynamics is described by the rate equations for the macroscopic averages. [Reprinted from Ref. 43. © (2007) by the American Physical Society.]

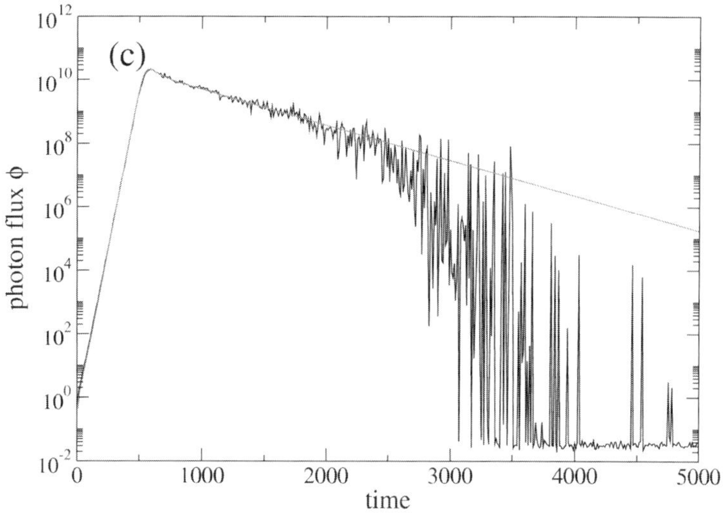

Figure 10.5 (c) The photon flux (per unit length) as a function of time for a single shot and for $N_0 = 50 \times 10.^9$ The data have been binned over consecutive time windows of duration $T_W = 10$. Note the difference in the vertical-axis scales. Smooth lines are the mean-field limit for large population and photon number. In this limit, the dynamics is described by the rate equations for the macroscopic averages. [Reprinted from Ref. 43. © (2007) by the American Physical Society.]

The evolution of the population N displays similar features. For a better comparison we have chosen to monitor the volume-averaged population

$$\frac{1}{RL^2}\sum_{\mathbf{r}} N(\mathbf{r},t) \tag{10.4}$$

normalized to its initial value. Fig. 10.6 shows the corresponding time series for the same runs of Fig. 10.5. Again, large deviations from mean-field theory appear for the first two values of N_0. The inset shows that, in correspondence with large-amplitude events, the population abruptly decreases, yielding a distinctive stepwise decay.

The nonsmooth time decay is accompanied by irregular evolution in space. Indeed, a snapshot of $N(\mathbf{r}, t)$ reveals a highly inhomogeneous profile (see Fig. 10.7). Light regions are traces of high-energy events that locally deplete the population before exiting the sample. For the case far above threshold, similar considerations as those made for the corresponding pulse apply. Note that now the population level decays extremely fast. It reaches 10% of its initial value at $t \approx 600$, which is within a factor of 2 of the average residence time in the sample. This means that photons emitted after a few hundred time steps have a very slight chance of being significantly amplified (i.e., ℓ_G has become too large).

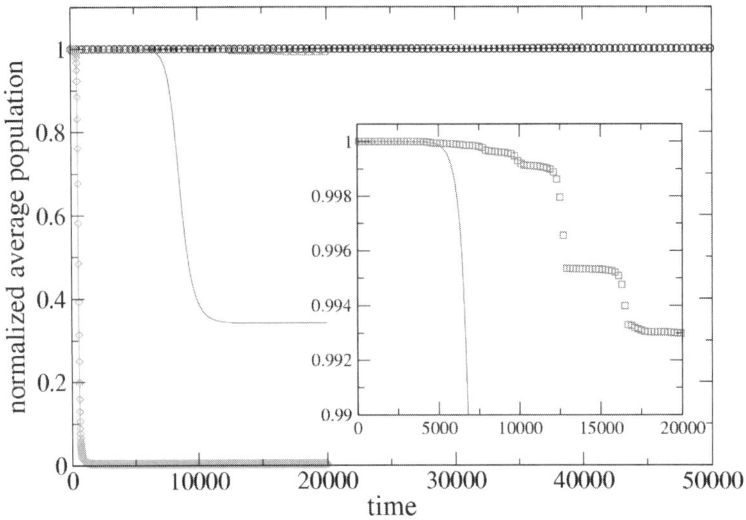

Figure 10.6 The normalized volume-averaged atomic population as a function of time for a single shot and for the same values of N_0 as in Fig. 10.5. Solid lines are the mean-field results. Top curve: below threshold; middle curve: around threshold; lower curve: far above threshold, corresponding to the cases (a), (b), and (c) of Fig. 10.5. The inset shows a magnification of the middle curves [where $N_0 = 5 \times 10^9$ of Fig. 10.5(b)]. The abrupt drops in the population correspond to strong emission spikes. [Reprinted from Ref. 43. © (2007) by the American Physical Society.]

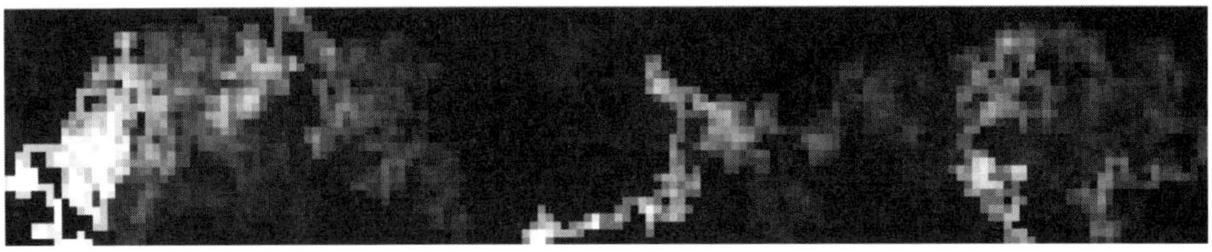

Figure 10.7 A grayscale plot of the atomic population distribution along a portion of the lattice for $t = 10^5$, $N_0 = 5 \times 10^9$. White regions correspond to small values of N. [Reprinted from Ref. 43. © (2007) by the American Physical Society.]

We have plotted the histograms of the photon number n_{out} for each and every collected event during the whole simulation run. The results are given in Fig. 10.8 for three values of N_0 around the laser threshold. A clear power-law tail extending over several decades is observed, which indicates that the statistics of the intensity are of the Lévy type. We observed Lévy statistics in the region just above and below the lasing threshold and Gaussian statistics elsewhere.

To check our results, we also performed a series of simulations increasing the number of lattice sites. For comparison, we kept $L = 30$ fixed and increased the aspect ratio R up to a factor 4. In this way, we increased the number of walkers, accordingly. For the case far below and above threshold where we expect regular Gaussian statistics, we do observe the expected reduction of fluctuations around the mean-field solution. On the contrary, the wild fluctuations of the Lévy case are hardly affected.

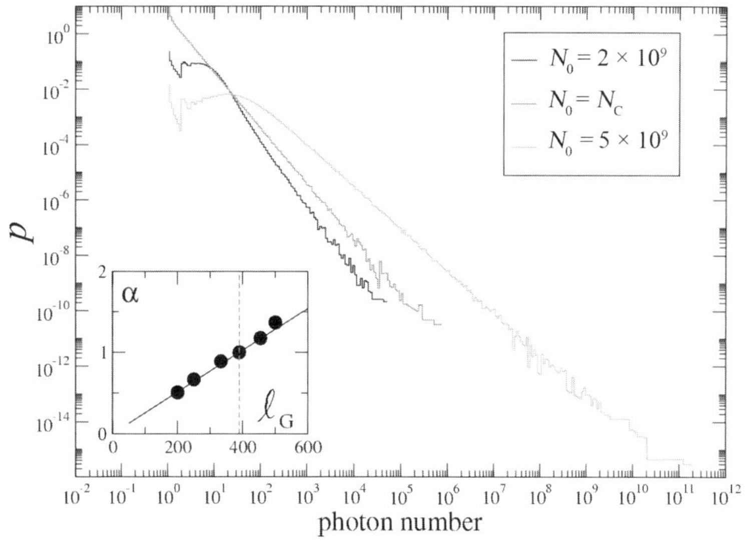

Figure 10.8 Histogram $p(n_{out})$of the emitted photon numbers n_{out} for the two values of N_0 of Figs. 10.5(a) and (b) (lower and upper curves) and $N_0 = N_c$ (middle). This middle curve has been vertically shifted for clarity. The inset reports the values of α obtained by a power-law fit of the histograms as a function of the gain length ℓ_G. The thin solid line is the theoretical curve as computed from Eqs. (10.7) and (10.8). The dashed vertical line marks the lasing threshold. [Reprinted from Ref. 43. © (2007) by the American Physical Society.]

10.5 Lévy Statistics in Random Laser Emission

The origin of the Lévy statistics can be understood by means of the following reasoning.[43] Spontaneously emitted photons are amplified within the active medium due to stimulated emission. Their emission energy is exponentially large in the path length l, i.e.,

$$I(l) = I_0 \exp(l / \ell_G), \tag{10.5}$$

where ℓ_G is the gain length. On the other hand, the path length in a diffusing medium is a random variable with exponential probability distribution

$$p(l) = \frac{\exp(-l / \langle l \rangle)}{\langle l \rangle}, \tag{10.6}$$

where $\langle l \rangle$ is the length of the photon path within the sample. The path length depends on both the geometry of the sample and the photon diffusion constant D. A simple estimate of $\langle l \rangle$ can be provided by noting that for a diffusive process with diffusion coefficient D, $\langle l \rangle$ is proportional to the mean first-passage time, yielding[44]

$$\langle l \rangle = \frac{v}{D\Lambda}, \tag{10.7}$$

where v is the speed of light in the medium as previously defined, and Λ is the smallest eigenvalue of the Laplacian in the active domain (with absorbing boundary conditions). For instance, $\Lambda = q^2$ with $q = \pi / L$ for an infinite slab or a sphere with L being the thickness or the radius, respectively. The combination of Eqs. (10.5) and (10.6) immediately shows that the probability distribution of the emitted intensity follows a power law

$$p(I) = \frac{\ell_G}{\langle l \rangle} I^{-(1+\alpha)} \quad , \qquad \alpha = \frac{\ell_G}{\langle l \rangle}. \tag{10.8}$$

Obviously the heavy-tailed Eq. (10.8) holds asymptotically, and the distribution should be cut off at small I. The properties of the Lévy distribution (more properly termed Lévy-stable) are well known.[45] The most striking one is that for $0 < \alpha \leq 2$ the average $\langle I \rangle$ exists, but the variance (and all higher order moments) diverges. This has important consequences on the statistics of experimental measurements, yielding highly irreproducible data with a lack of self-averaging of sample-to-sample fluctuations. On the contrary, for $\alpha > 2$, the standard central-limit theorem holds, and fluctuations are Gaussian.

The gain length ℓ_G is basically controlled by the pumping energy, i.e., by the population of the medium excitations. Increasing the number of excitations decreases both ℓ_G and the exponent α, thereby enhancing the fluctuations. At first glance, one can infer that the larger the pumping the stronger the effect. On the other hand, ℓ_G is a time-dependent quantity that should be determined self-consistently from the dynamics. Indeed, above threshold the release of a huge number of photons may lead to such a sizeable depletion of the population itself that it forces ℓ_G to increase. It can then be argued that when the depletion is large enough, the Lévy fluctuation may be barely detectable.

To put the above statements on a more quantitative ground, we need to estimate the lifetime of the population as created by the pumping process. Following Ref. 3, we write the threshold condition as

$$r = v / \ell_G - D\Lambda > 0, \tag{10.9}$$

which is interpreted as "gain larger than losses," the latter being caused by the diffusive escape of light from the sample. Note that the condition $r = 0$ along with Eqs. (10.7) and (10.8) imply that $\alpha = 1$ at the laser threshold.

For short pump pulses the time necessary for the intensity to become large is of the order of the inverse of the growth rate r. When this time is smaller than the average path duration within the sample $\langle l \rangle / v$, a sizeable amplification occurs on average for each spontaneously emitted photon, leading to a strong depletion of the population. In this case, we expect a Gaussian regime where a mean-field description is valid. The conditions for the Lévy regime are $1/r > \langle l \rangle / v$ and $\alpha \leq 2$ and can be written as

$$\frac{1}{2}\frac{v}{D\Lambda} < \ell_G < 2\frac{v}{D\Lambda}. \tag{10.10}$$

Note that the lower bound of the above inequalities corresponds to $\alpha = 1/2$.

Without losing generality and for later convenience, let us focus on the case of a two-dimensional infinite slab of thickness L. In Fig. 10.9 we graphically summarize Eq. (10.10) by drawing a diagram in the (L, ℓ_G) plane. This representation allows identification of three different regions corresponding to different statistics. For convenience, the line corresponding to the threshold $\alpha = 1$ is also drawn. The three regions of statistical interest are:

Subthreshold Lévy: weak emission with Lévy statistics, where $1 < \alpha < 2$ (shaded region in Fig. 10.9 above the laser threshold line);

Suprathreshold Lévy: strong emission with Lévy statistics, where $1/2 < \alpha \leq 1$ (shaded region in Fig. 10.9 below the laser threshold line);

Gaussian: strong emission with Gaussian statistics where $\alpha < 1/2$ and weak emission with Gaussian fluctuations, where $\alpha > 2$.

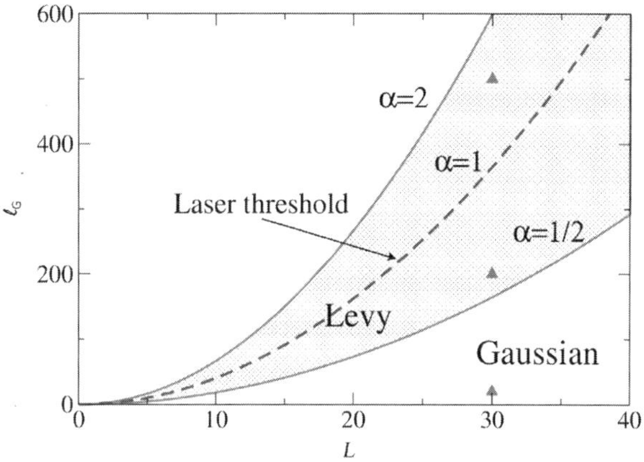

Figure 10.9 Different statistical regimes of fluctuations of a random laser with a two-dimensional slab geometry of thickness L. The dashed line corresponds to the threshold, and the region below this line is the region above threshold. For comparison with the simulations reported below, all quantities are expressed in dimensionless units $\ell = 1$, $v = 1$. The symbols correspond to the parameter of the previous figure. [Reprinted from Ref. 43. © (2007) by the American Physical Society.]

Note that the first region corresponds to a nonlasing state that should display anomalous fluctuations as “precursors” to the transition. It should also be emphasized that the boundary between Lévy and Gaussian statistics is not expected to correspond to a sharp transition (as displayed in the figure) but rather to a crossover region.

10.6 Discussion

Based on heuristic arguments, we have shown that, depending on the value of the dimensionless parameter $D\Lambda\ell_G / v$, the fluctuations in the emission of a random laser subject to short pump pulses can be drastically different. In a parameter region extending both above and below threshold, the photon number fluctuations follow a Lévy distribution, thus displaying wild fluctuations and huge differences in the emission from pulse to pulse. The latter is what we observe in the experiments described in the previous section. Lévy statistics were also observed in the suprathreshold case in recent experiments.[46,47]

The exponent α of the Lévy distribution can be tuned after changing the pumping level but must be somehow bounded from below ($\alpha \geq 1/2$), since a further crossover to Gaussian statistics is attained there. Indeed, far above threshold when the gain length is very small, a large and fast depletion of the population occurs (saturation). This hinders the possibility of huge amplification of individual events. In this case, all photons behave in a statistically similar way. As a consequence, the statistics is Gaussian and a mean-field description applies. The above considerations have been substantiated by comparison with a simple stochastic model that includes population dynamics in a self-consistent manner. In the Lévy regions, the simulation data strongly depart from the predictions of the mean-field approximation due to the overwhelming role of individual rare events. As a consequence, the evolution of the population displays abrupt changes in time and is highly inhomogeneous in space.

To conclude this general discussion, we remark that the width of the Lévy region as defined by inequalities (Eq. 10.6) and, as depicted in Fig. 10.9, is of the order of L^2. Since in our simple model ℓ_G is inversely proportional to the pump parameter, the range of N_0 values for which the Lévy fluctuations occur shrinks as $1/L^2$. Therefore, the larger the lattice, the closer to threshold one must be to observe them.

The existence of different statistical regimes, their crossovers, and their dependence on various external parameters all enrich the possible experimental scenarios. The emission statistics of random amplifying media has diverging moments in a *finite* region of parameters extending across the threshold curve. Our work has shown that, depending on size, geometry, pumping protocols, etc., the emission of random lasers may change considerably. This general conclusion should be a useful guide in understanding past and future experiments on random amplifying media.

Acknowledgments

We wish to thank Roberto Livi, Antonio Politi, and the entire micro- and nanophotonics group at LENS for stimulating discussions. This work was financially supported by PRIN2004/5 projects *Transport properties of classical and quantum systems* and *Silicon based photonic crystals*, and funded by MIUR-Italy, by LENS under EC contract RII3-CT-2003-506350, and by the European Network of Excellence Phoremost (IST-2-511616-NOE).

References

1. P. Sheng, *Introduction to Wave Scattering, Localization, and Mesoscopic Phenomena*, Academic Press, San Diego (1995).
2. V. S. Letokhov, "Generation of light by a scattering medium with negative resonance absorption," *Zh. 'Eksp.Teor. Fiz.* **53**, 1442 (1967) [*Sov. Phys. JETP* **26**, 835 (1968)].
3. V. M. Markushev, V. F. Zolin, and Ch. M. Briskina, "Powder laser," *Zh. Prikl. Spektrosk.* **45**, 847 (1986).
4. A. Yu. Zyuzin, "Weak localization in backscattering from an amplifying medium," *Europhys. Lett.* **26**, 517–520 (1994).
5. N. M. Lawandy, R. M. Balachandran, A. S. L. Gomes, and E. Sauvain, "Laser action in strongly scattering media," *Nature* **368**, 436 (1994).
6. D. S. Wiersma, M. P. van Albada, and A. Lagendijk, "Random laser?," *Nature* **373**, 203–204 (1995); N. M. Lawandy and R. M. Balachandran, "Random laser? – Reply," *Nature* **373**, 204 (1995).
7. D. S. Wiersma and A. Lagendijk, "Light diffusion with gain and random lasers," *Phys. Rev. E* **54**, 4256 (1996); "Light in Strongly Scattering and Amplifying Random Systems," D. S. Wiersma, Ph.D. thesis, University of Amsterdam (1995).
8. S. John and G. Pang, "Theory of lasing in a multiple-scattering medium," *Phys. Rev. A* **54**, 3642 (1996).
9. C. W. J. Beenakker, "Thermal radiation and amplified spontaneous emission from a random medium," *Phys. Rev. Lett.* **81**, 1829 (1998).
10. H. Cao, Y. G. Zhao, S. T. Ho, E. W. Seelig, Q. H. Wang, and R. P. H. Chang, "Random laser action in semiconductor powder," *Phys. Rev. Lett.* **82**, 2278 (1999).
11. Gijs van Soest, "Experiments on Random Lasers," Ph.D. thesis, University of Amsterdam (2001).
12. L. Florescu and S. John, "Photon statistics and coherence in light emission from a random laser," *Phys. Rev. Lett.* **93**, 013602 (2004).

13. V. Milner and A. Z. Genack, "Photon localization laser: low-threshold lasing in a random amplifying layered medium via wave localization," *Phys. Rev. Lett.* **94,** 073901 (2005).

14. C. Gouedard, D. Husson, C. Sauteret, F. Auzel, and A. Migus, "Generation of spatially incoherent short pulses in laser-pumped neodymiun stoichiometric crystals and powders," *J. Opt. Soc. Am. B* **10**, 2362 (1993).

15. J. Martorell, R. M. Balachandran, and N. M. Lawandy, "Radiative coupling between photonic paint layers," *Opt. Lett.* **21**, 239 (1996).

16. W. L. Sha, C. H. Liu, and R. R. Alfano, "Spectral and temporal measurements of laser action of Rhodamine-640 dye in strongly scattering media," *Opt. Lett.* **19**, 1922 (1994).

17. M. Bahoura, K. J. Morris, and M. A. Noginov, "Threshold and slope efficiency of Nd La Al (BO) ceramic random laser: effect of the pumped spot size," *Opt. Comm.* **201**, 405 (2002).

18. D. S. Wiersma and S. Cavalieri, "A temperature tunable random laser," *Nature* **414**, 708 (2001).

19. R. C. Polson, A. Chipouline, and Z. V. Vardeny, "Random lasing in pi-conjugated films and infiltrated opals," *Adv. Materials* **13**, 760–764 (2001).

20. M. A. Noginov, *Solid-State Random Lasers*, Springer, Berlin (2005).

21. S. Gottardo, S. Cavalieri, O. Yaroschuck, and D. S. Wiersma, "Quasi 2D random laser action," *Phys. Rev. Lett.* **93**, 263901 (2004).

22. R. M. Laine, S. C. Rand, T. Hinklin, and G. R. Williams, "Ultrafine powders and their use as lasing media," U.S. Patent No. 6656588 (2 December 2003).

23. J. Dubois and S. La Rochelle, "Active cooperative tuned identification friend or foe (ACTIFF)," U.S. Patent No. 5966227 (12 October 1999).

24. H. Cao, Y. Ling, J. Y. Xu, C. Q. Cao, and P. Kumar, "Photon statistics of random lasers with resonant feedback, *Phys. Rev. Lett.* **86**, 4524–4527 (2001).

25. P. A. Lee and T. V. Ramakrishnan, "Disordered electronic systems," *Rev. Mod. Phys*. **57**, 287–337 (1985).

26. S. John, "Electromagnetic absorption in a disordered medium near a photon mobility edge," *Phys. Rev. Lett.* **53**, 2169 (1984); P. W. Anderson, "The question of classical localization: a theory of white paint?," *Philos. Mag. B* **52**, 505 (1985); A. Lagendijk, M. P. van Albada, and M. B. van der Mark, "Localization of light: the quest for the white hole," *Physica A* **140**, 183 (1986).

27. D. S. Wiersma et al., *Nature* (London) **390**, 671 (1997); A. A. Chabanov and A. Z. Genack, *Phys. Rev. Lett.* **87**, 153901 (2001); M. Stoerzer et al., *Phys. Rev. Lett.* **96**, 063904 (2006).

28. P. Pradhan and N. Kumar, "Localization of light in coherently amplifying random media," *Phys. Rev. B* **50**, 9644 (1994).

29. X. Y. Jiang and C. M. Soukoulis, "Time-dependent theory for random lasers," *Phys. Rev. Lett.* **85**, 70–73 (2000).

30. H. Cao, J. Y. Xu, D. Z. Zhang, S.-H. Chang, S. T. Ho, E. W. Seelig, X. Liu, and R. P. H. Chang, "Spatial confinement of laser light in active random media," *Phys. Rev. Lett.* **84**, 5584–5587 (2000).

31. S. Mujumdar, M. Ricci, R. Torre, and D. S. Wiersma, "Amplified extended modes in random lasers," *Phys. Rev. Lett.* **93**, 053903 (2004).

32. C. Vanneste, P. Sebbah, and H. Cao, "Lasing with resonant feedback in weakly scattering random systems." *Phys. Rev. Lett.* **98**, 143902 (2007).

33. V. M. Apalkov, M. E, Raikh, and B. Shapiro, "Random resonators and prelocalized modes in disordered dielectric films," *Phys. Rev. Lett.* **89**, 016802 (2002).

34. S. E. Skipetrov and B. A. van Tiggelen, "Dynamics of weakly localized waves," *Phys. Rev. Lett.* **92**, 113901 (2004).

35. A. A. Chabanov, Z. Q. Zhang, and A. Z. Genack, "Breakdown of diffusion in dynamics of extended waves in mesoscopic media," *Phys. Rev. Lett.* **90**, 203903 (2003).

36. D. Anglos, A. Stassinopoulos, R. N. Das, G. Zacharakis, M. Psyllaki, R. Jakubiak, R. A. Vaia, E. P. Giannelis, and S. H. Anastasiadis, "Random laser action in organic-inorganic nanocomposites," *J. Opt. Soc. Am. B* **21**, 208 (2004).

37. K. van der Molen, A. P. Mosk, and A. Lagendijk, "Intrinsic intensity fluctuations in random lasers," *Phys. Rev. A* **74**, 053808 (2006).

38. K. L. van der Molen, R. W. Tjerkstra, A. P. Mosk, and A. Lagendijk, "Spatial extent of random laser modes, *Phys. Rev. Lett.* **98**, 143901 (2007).

39. S. Mujumdar, V. Tuerck, R. Torre, and D. S. Wiersma, "Chaotic behavior of a random laser with static disorder," *Phys. Rev. A* **76**, 033807 (2007). http://link.aps.org/doi/10.1103/PhysRevA.76.033807

40. G. Maret and P. E. Wolf, *Z. Phys. B* **65**, 409 (1987); D. J. Pine, D. A. Weitz, P. M. Chaikin, and E. Herbolzheimer, *Phys. Rev. Lett.* **60**, 1134 (1988).

41. H. J. Carmichael, *Statistical Methods in Quantum Optics 1*, Springer-Verlag, Berlin (1999).

42. In our units, this corresponds to the condition $\gamma N n << 1$. For the parameters in the simulations, $\gamma N \approx 10^{-3}$; i.e., the condition may be violated for short times. On the other hand, this initial regime is irrelevant for the effects we are interested in.

43. S. Lepri, S. Cavalieri, G.-L. Oppo, and D. S. Wiersma, "Statistical regimes of random laser fluctuations," *Phys. Rev. A* **75**, 063820 (2007). http://link.aps.org/doi/10.1103/PhysRevA.75.063820

44. S. Redner, *A Guide to First-Passage Processes*, Cambridge University Press, Cambridge (2001).

45. J. P. Bouchaud and A. Georges, "Anomalous diffusion in disordered media: Statistical mechanisms, models and physical applications," *Phys. Rep.* **195**, 127 (1990).

46. D. Sharma, H. Ramachandran, and N. Kumar, "Lévy statistics of emission from a novel random amplifying medium: an optical realization of the Arrhenius cascade," *Optics Lett.* **31**, 1806 (2006).

47. D. Sharma, H. Ramachandran, and N. Kumar, "Lévy statistical fluctuations from a random amplifying medium," *Fluct. Noise Lett.* **6** L95–L101 (2006).

Biographies

Diederik S. Wiersma is research director at both the European Laboratory for Non-linear Spectroscopy (LENS), University of Florence, and the National Institute for the Physics of Matter (INFM-CNR), Italy. He leads a research group that deals with photonic materials on micro- and nanometer length scales. His group has played a leading role in the understanding of light transport in periodic, quasi-crystalline, and disordered structures, including the observation of phenomena such as light localization, Bloch oscillations of light, and random lasers. The group has also developed a patented technology to realize rewritable photonic circuits with a unique liquid infiltration technique. Recently, the group was the first to observe the optical analogy of Lévy flights and superdiffusion of photons in a newly developed material called Lévy glass.

Sushil Mujumdar received his Ph.D. in 2002 at the Raman Research Institute, Bangalore, India. He carried out postdoctoral research first at the European Laboratory for Nonlinear Spectroscopy (LENS), Italy, then at the University of Alberta, Canada, and finally at the ETH, Zurich. His research interests cover several aspects of light propagation in disordered media, including random lasers and biomedical imaging. He also has expertise in near-field scanning optical microscopy of structured materials such as photonic crystals. Currently, he is setting up his own research program in nano- and mesoscopic optics at the Tata Institute of Fundamental Research, Mumbai, India.

Stefano Cavalieri is Associate Professor of Physics of Matter at the University of Florence and a member of LENS. His main recent research activities have been devoted to the study of high-order harmonics and their applications to spectroscopy, optical properties of complex systems, coherent laser-atom interactions in the discrete and ionization continuum spectrum, and coherent control of photo-transitions and field propagation characteristics in atomic media. His research, which is mainly experimental, has produced more than 80 research publications.

Renato Torre received his Ph.D. in Physics in 1991 from the University of Florence. He was a visiting scholar at Stanford University in 1992. Since 2001, he has been a researcher in the Department of Physics at the University of Florence. Since 2002, he has coordinated the Structured Fluids and Glasses Group at the European Laboratory for Nonlinear Spectroscopy (LENS) at the University of Florence. His research interests range from nonlinear spectroscopy to transport phenomena in complex matter. He has (co)authored about 72 papers in international peer-reviewed physics journals.

Gian-Luca Oppo has been a professor of Computational and Nonlinear Physics in the Department of Physics of the University of Strathclyde in Glasgow, UK since 1998 and is the director of the Institute of Complex Systems (ICS-CNR) in Florence. His research group was flagged as top mark 5* in the UK Research Assessment Exercise of 2001. He has more than 130 research publications on topics that include self-organizing structures in complex systems, such as patterns, spatial solitons, domain fronts, vortices and defects, as well as in the dynamics and control of discrete lattices and stochastic systems.

Stefano Lepri received his Ph.D. in Theoretical Physics in 1996 from the University of Bologna. Since 2005, he has been a researcher at the Institute of Complex Systems (ISC-CNR) in Florence. His research interests range from the theory of nonlinear systems to the statistical mechanics of out-of-equilibrium processes and, in particular, include anomalous transport and relaxation. He has (co)authored about 40 papers in international peer-reviewed physics journals.

Chapter 11
Lasing in Random Media

Hui Cao
Departments of Applied Physics and Physics
Yale University, New Haven, CT, USA

11.1 Introduction

11.1.1 "LASER" versus "LOSER"

A photon, unlike an electron, can stimulate an excited atom to emit a second photon into the same electromagnetic mode. This stimulated emission process is the foundation for light amplification and oscillation (i.e., self-generation). Initially, the term LASER referred to Light Amplification by Stimulated Emission of Radiation. Nowadays laser often means Light *Oscillation* by Stimulated Emission, which should literally be called "LOSER" instead of "LASER." To distinguish the above two devices, the former is called a laser amplifier, the latter a laser oscillator (Siegman 1986).

In a laser amplifier, input light is amplified when the net gain coefficient $g_{eff} = g_a - \alpha_r > 0$, where g_a and α_r represent gain and absorption coefficients, respectively. In the absence of input light, photons spontaneously emitted by excited atoms are multiplied, giving amplified spontaneous emission (ASE).

Laser oscillation occurs when the photon generation rate exceeds the photon loss rate in a system. If gain saturation were absent, the photon number in a laser oscillator would diverge in time. In other words, the rate equation for the photon number would acquire an unstable solution above the oscillation threshold. In reality, gain saturation reduces the photon generation rate to the photon loss rate so that the number of photons in the oscillator remains at a finite value.

11.1.2 Random lasers

For a long time, optical scattering was considered to be detrimental to lasing because such scattering removes photons from the lasing modes of a conventional laser cavity. However, in a disordered medium with gain, light scattering plays a positive role in both laser amplification and laser oscillation. Multiple scattering increases the path length or dwell time of light in an active medium, thereby enhancing laser amplification. In addition, strong scattering increases the chance of light (of wavelength λ) returning to a coherence volume ($\sim \lambda^3$) it has visited previously, providing feedback for laser oscillation.

Since the pioneering work of Letokhov and coworkers (Ambartsumyan et al. 1966), lasing in disordered media has been a subject of intense theoretical and experimental studies. It represents the process of light amplification by stimulated emission with feedback mediated by random spatial fluctuations of the dielectric constant. There are two kinds of feedback: one is *intensity* or *energy* feedback, the other is *field* or *amplitude* feedback (Cao 2003). Field feedback is phase sensitive (i.e., coherent) and therefore frequency dependent (i.e., resonant). It requires light scattering to be elastic and the spatial distribution of the dielectric constant to be time-invariant. The intensity feedback is phase insensitive (i.e., incoherent) and frequency independent (i.e., nonresonant). This can occur, e.g., in the presence of inelastic scattering, mobile scatterers, dephasing, and nonlinearity. Based on the feedback mechanisms, random lasers are classified into two categories: (i) random lasers with incoherent and nonresonant feedback, (ii) random lasers with coherent and resonant feedback (Cao et al. 2000a; Cao et al. 2003a).

Random lasers have been realized primarily in disordered dielectric media of finite size. They differ from the chaotic cavity lasers which have been a focus of many theoretical studies (Beenakker 1999; Türeci et al. 2005). In the latter, the cavity is nearly totally enclosed by a metallic boundary. Its dimension is much larger than the wavelength λ of radiation. There are a few openings at the boundary but their size is smaller than λ. Owing to the irregular shape of the boundary and/or scatterers placed at random positions inside the cavity, the intracavity ray dynamics is chaotic. The small leakage rate allows light to ergodically explore the entire cavity volume. In contrast, the random lasers considered in this chapter have completely open boundaries, and light can escape from the dielectric random media via any point on the boundaries. Hence, they are open systems with strong coupling to the environment.

11.1.3 Characteristic length scales for the random laser

Correlation radius R_c. In a disordered dielectric medium, the dielectric constant $\epsilon(\mathbf{r})$ fluctuates randomly in space. The spatial variation of $\epsilon(\mathbf{r})$ can be characterized statistically by the correlator $K(\Delta\mathbf{r}) \equiv \langle\epsilon(\mathbf{r})\epsilon(\mathbf{r}+\Delta\mathbf{r})\rangle$, where $\langle ... \rangle$ represents an ensemble average. When the random medium is isotropic, the width of $K(\Delta r)$ is called the correlation radius R_c. It reflects the length scale of spatial fluctuation of the dielectric constant. If $R_c \gg \lambda$, light is deflected by long-range disorder. When R_c is comparable to or less than λ, light is scattered by short-range disorder.

Scattering mean free path l_s and transport mean free path l_t. The relevant length scales that describe the light scattering process are the scattering mean free path l_s and the transport mean free path l_t. The scattering mean free path l_s is defined as the average distance that light travels between two consecutive scattering events. The transport mean free path l_t is defined as the average distance a wave travels before its direction of propagation is randomized. These two length scales are related:

$$l_t = \frac{l_s}{1 - \langle\cos\theta\rangle} . \tag{11.1}$$

$\langle\cos\theta\rangle$ is the average cosine of the scattering angle, which can be found from the differential scattering cross-section. Rayleigh scattering is an example of $\langle\cos\theta\rangle = 0$ and $l_t = l_s$, while Mie scattering may have $\langle\cos\theta\rangle \approx 0.5$ and $l_t \approx 2l_s$.

Gain length l_g and amplification length l_{amp}. Light amplification by stimulated emission in a random medium is described by the gain length l_g and the amplification length l_{amp}. The gain length l_g is defined as the path length over which light intensity is amplified by a factor e. The amplification length l_{amp} is defined as the (rms) average distance between the beginning and ending point for paths of length l_g. In a homogeneous medium, light travels in a straight line, thus

$l_{amp} = l_g$. In a diffusive sample, $l_{amp} = \sqrt{D\,\tau_{amp}}$, where D is the diffusion coefficient, $\tau_{amp} = l_g/v$, and v the speed of light. In a three-dimensional (3D) system, $D = v\,l_t/3$, thus

$$l_{amp} = \sqrt{\frac{l_t l_g}{3}}\,. \tag{11.2}$$

The gain length l_g is the analogue of the inelastic length l_i, defined as the travel length over which light intensity is reduced to $1/e$ by absorption. Hence, the amplification length l_{amp} is analogous to the absorption length $l_{abs} = \sqrt{l_t\,l_i/3}$.

Dimensionality d *and size* L. Light transport in a random medium depends on its dimensionality d and size L. For a random medium of $d > 1$, L refers to its smallest dimension. The average trapping time of photons in a diffusive random medium $\tau_d = L^2/D$. In an active random medium, the gain volume may be smaller than the volume of the entire random medium, e.g., when only part of the disordered medium is pumped. The gain volume is characterized by its dimension L_g, and $L_g \leq L$.

11.1.4 Light localization

There are three regimes for light transport in a 3D random medium: (i) ballistic regime, $L \leq l_t$; (ii) diffusive regime, $L \gg l_t \gg \lambda$; (iii) localization regime, $k\,l_t \simeq 1$ (k is the wave vector in the random medium) (John 1991).

Light localization can also be understood in the mode picture. Quasimodes, also called quasi-states, are the eigenmodes of Maxwell's equations in a passive random medium. Due to the finite size of a dielectric medium and its open boundary, the frequency of a quasimode is a complex number: $\Omega = \omega_r + i\gamma$. γ is the decay rate of a quasimode as a result of its coupling to the environment. It also represents the linewidth of the quasimode in frequency. The Thouless number δ is defined as the ratio of average linewidth $\delta\nu = \langle\gamma\rangle$ to average frequency spacing $d\nu$ of adjacent quasimodes, $\delta \equiv \delta\nu/d\nu$. In the delocalization regime, the quasimodes are spatially extended over the entire random system. They have large decay rates and overlap in frequency, $\delta > 1$. In the localization regime, quasimodes are localized inside the system and have small decay rates. They do not overlap spectrally, thus $\delta < 1$. The localization threshold is set at $\delta = 1$; this is the Thouless criterion. Thus, the localization transition corresponds to a transition from overlapping modes to nonoverlapping modes.

The openness of a random laser, namely, its coupling to the environment, can also be characterized by δ. Note that the value of δ is obtained in the absence of gain or absorption so that it describes solely light leakage from the random system. Typical chaotic cavity lasers and photon localization lasers have nonoverlapping modes, $\delta < 1$; whereas in diffusive random lasers most quasimodes overlap, $\delta > 1$.

This chapter is a compendium of a wide range of experimental studies on light amplification and lasing in random media. Brief descriptions of some theoretical models and numerical calculations are given for qualitative explanations of experimental phenomena.

11.2 Random Lasers with Incoherent Feedback

11.2.1 Lasers with a scattering reflector

The two essential components of a laser are a gain medium and a cavity. The gain medium amplifies light through stimulated emission, and the cavity provides positive feedback. A simple laser cavity is a Fabry-Perot cavity shown in Fig. 11.1(a). It is made with two parallel mirrors. We assume that the gain medium is uniformly distributed between the two mirrors. After traveling one round trip between the mirrors, light returns to its original position. The requirement of constructive inference determines the resonant frequencies, namely

$$2kL_c + \phi_1 + \phi_2 = 2\pi m \, , \tag{11.3}$$

where k is the wave vector, L_c is the cavity length, ϕ_1 and ϕ_2 represent the phases of the reflection coefficients of the two mirrors, and m is an integer. Only light at the resonant frequencies experiences minimum loss and spends a long time in the cavity. The long dwell time in the cavity facilitates light amplification. When the optical gain balances the loss of a resonant mode, lasing oscillation occurs in this mode. The threshold condition is

$$R_1 \, R_2 \, e^{2g_{eff}L_c} = 1 \, , \tag{11.4}$$

where R_1 and R_2 represent the reflectivities of the two mirrors, and g_{eff} is the net gain coefficient.

In 1966, Ambartsumyan et al. (1966) realized a different type of laser cavity that provides nonresonant feedback. They replaced one mirror of the Fabry-Perot cavity with a scattering surface, as shown in Fig. 11.1(b). Light in the cavity suffers multiple scattering; its direction is changed every time it is scattered. Thus, light does not return to its original position after one round trip. The spatial resonances for the electromagnetic field are absent in such a cavity. The dwell time of light is not sensitive to its frequency. The feedback in such a laser is used merely to

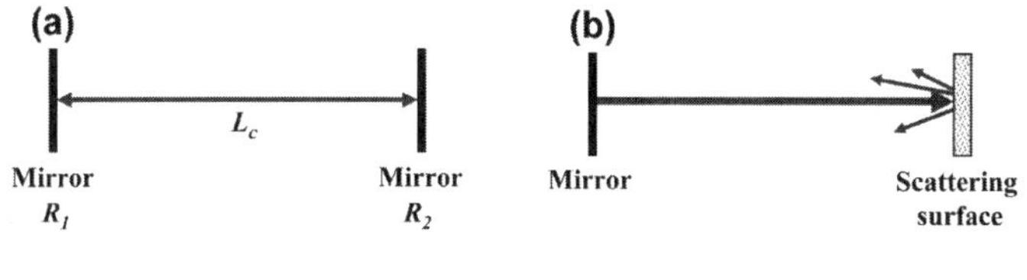

Figure 11.1 (a) Schematic of a Fabry-Perot cavity made of two parallel mirrors with reflectivity R_1 and R_2. The cavity length (distance between the two mirrors) is L_c. (b) One of the mirrors is replaced by a scattering surface that scatters light instead of reflecting it.

return part of the energy or photons to the gain medium, i.e., it depends on energy or intensity feedback.

The nonresonant feedback can also be interpreted in terms of "modes." When one end mirror of a Fabry-Perot cavity is replaced by a scattering surface, escape of light from the cavity by scattering becomes the predominant loss mechanism for all modes. Instead of individual high-Q resonances, there appear a large number of low-Q resonances that spectrally overlap and form a continuous spectrum. This corresponds to nonresonant feedback. The absence of resonant feedback means that the cavity spectrum tends to be continuous, i.e., it does not contain discrete components at selected resonant frequencies. The only resonant element left in this kind of laser is the amplification line of the active medium. With an increase of pumping intensity, the emission spectrum narrows continuously toward the center of the amplification line. However, the process of spectral narrowing is much slower than in ordinary lasers (Ambartsumyan et al. 1967a). Since many modes in a laser cavity with nonresonant feedback interact with the active medium as a whole, the statistical properties of laser emission are quite different from those of an ordinary laser. As shown by Ambartsumyan et al. (1968), the statistical properties of the emission of a laser with nonresonant feedback are very close to those of the emission from an extremely bright "blackbody" in a narrow range of the spectrum. The emission of such a laser has no spatial coherence and is not stable in phase.

Because the only resonant element in a laser with nonresonant feedback is the amplification line of the gain medium, the mean frequency of laser emission does not depend on the dimensions of the laser but is determined only by the center frequency of the amplification line. If this frequency is sufficiently stable, the emission of this kind of laser has a stable mean frequency. Ambartsumyan et al. (1967b) proposed using the method of nonresonant feedback to produce an optical standard for frequency. To realize it, they built continuous gas lasers with nonresonant feedback based on the scattering surface (Ambartsumyan et al. 1970).

11.2.2 The photonic bomb

In 1968, Letokhov (1968) took one step further and proposed self-generation of light in an active medium filled with scatterers. When the photon mean free path is much smaller than the dimensions of the scattering medium, the motion of photons is diffusive. Letokhov solved the diffusion equation for the photon energy density $W(\vec{r},t)$ in the presence of a uniform and linear gain:

$$\frac{\partial W(\vec{r},t)}{\partial t} = D\nabla^2 W(\vec{r},t) + \frac{v}{l_g} W(\vec{r},t) \; , \tag{11.5}$$

where v is the transport velocity of light inside the scattering medium, l_g is the gain length, and D is the diffusion constant given by

$$D = \frac{v\, l_t}{3} \; . \tag{11.6}$$

The general solution to Eq. (11.5) can be written as

$$W(\vec{r},t) = \sum_n a_n \Psi_n(\vec{r}) e^{-(DB_n^2 - v/l_g)t} \,, \tag{11.7}$$

where $\Psi_n(\vec{r})$ and B_n are the eigenfunctions and eigenvalues of the following equation:

$$\nabla^2 \Psi_n(\vec{r}) + B_n^2 \Psi_n(\vec{r}) = 0, \tag{11.8}$$

with the boundary condition that $\Psi_n = 0$ at a distance z_e from the boundary. z_e is the extrapolation length. Usually z_e is much smaller than any dimension of the scattering medium and can be neglected. Hence, the boundary condition becomes that $\Psi_n = 0$ at the boundary of the random medium.

The solution for $W(\vec{r},t)$ in Eq. (11.7) changes from exponential decay to exponential increase in time when crossing the threshold

$$DB_1^2 - \frac{v}{l_g} = 0 \,, \tag{11.9}$$

where B_1 is the lowest eigenvalue of Eq. (11.8). If the scattering medium has the shape of a sphere of diameter L, $B_n = 2\pi n/L$, and the smallest eigenvalue $B_1 = 2\pi/L$. If the scattering medium is a cube whose side length is L, the smallest eigenvalue $B_1 = \sqrt{3}\pi/L$. Regardless of the shape of the scattering medium, the lowest eigenvalue B_1 is on the order of $1/L$. Substituting $B_1 \approx 1/L$ into Eq. (11.9), the threshold condition predicts a critical volume

$$V_{cr} \approx L^3 \approx \left(\frac{l_t l_g}{3}\right)^{3/2} \,. \tag{11.10}$$

With fixed gain length l_g and transport mean free path l_t, once the volume of the scattering medium V exceeds the critical volume V_{cr}, $W(\vec{r},t)$ increases exponentially with t.

This can be understood intuitively in terms of two characteristic length scales. One is the generation length L_{gen}, which represents the average distance a photon travels before generating a second photon by stimulated emission. L_{gen} can be approximated by the gain length l_g. The other is the mean path length L_{pat} that a photon travels in the scattering medium before escaping through its boundary. $L_{pat} \sim vL^2/D$. When $V \geq V_{cr}$, $L_{pat} \geq L_{gen}$. This means that on average every photon generates another photon before escaping the medium, thus triggering a "chain reaction," i.e., one photon generates two photons, and two photons generate four photons, etc. Hence the photon number increases with time; this is the onset of photon self-generation. Because this process of photon generation is analogous to the multiplication of neutrons in an atomic bomb, this device is sometimes called a photonic bomb.

In reality the light intensity will not diverge (there is no "explosion") because gain depletion quickly sets in and l_g increases. Taking into account gain saturation,

Letokhov calculated the emission linewidth and the generation dynamics. If the scattering centers are stationary, the limiting width of the generation spectrum is determined by the spontaneous emission. Otherwise, the Brownian motion of the scattering particles leads to a random variation (wandering) of the photon frequency as a result of the Doppler effect on the scattering particles. Letokhov also predicted damped oscillation (pulsation) in the transient process of generation.

11.2.3 The powder laser

In 1986 Markushev et al. (1986) reported intense stimulated radiation from a sample of $Na_5La_{1-x}Nd_x(MoO_4)_4$ powder under resonant pumping at low temperature (77 K). When the pumping intensity exceeded a threshold, the Nd^{3+} emission spectrum was narrowed to a single line, and the emission pulse duration was shortened by approximately four orders of magnitude. Later on, they reported similar phenomena in a wide range of Nd^{3+}-activated scattering materials, including La_2O_3, La_2O_2S, $Na_5La(MoO_4)_4$, La_3NbO_7, and $SrLa_2WO_7$ (Markushev et al. 1990). The powder was pumped by a 20-ns Q-switched laser pulse. When the pump energy reached a threshold (in the range 0.05–0.1 J/cm^2), a single emission pulse of 1- to 3-ns duration was observed. With a further increase of pump energy, the number of emission pulses increased to three or four. At a constant pumping intensity, the number of pulses, their duration, and the interval between them were governed by the properties of the materials. The emission spectrum above threshold was related to the particle shape. In a powder of particles of various shapes, there was only one narrow emission line at the center of the luminescence band, while in a powder of particles of one specific shape, the emission spectrum consisted of several lines in the range of the luminescence band (Ter-Gabrielyan et al. 1991a). In all cases, the spectral width of the emission lines above threshold is on the order of 0.1 nm. The observed emission was very much like laser emission. Because the particle size ($\sim$ 10 μm) was much larger than the emission wavelength, Markushev et al. (1990) speculated that individual particles served as effective resonators and lasing occurred in the modes formed by total internal reflection at the surface of a particle. However, there might be some weak coupling between the neighboring particles. In a mixture of two powders with slightly shifted luminescence bands [e.g., $Na_5La_{1-z}Nd_z(MoO_4)_4$ powder with significantly different Nd^{3+} concentrations], the emission wavelength depended on the relative concentrations of components in the mixture and the excitation wavelength (Ter-Gabrielyan et al. 1991b). Briskina et al. set up a model of coupled microcavities to interpret the experimental result (Briskina et al. 1996; Briskina and Li 2002). They treated the powder as an aggregate of active optically coupled microcavities and calculated the modes formed by total internal reflection (an analog of the whispering-gallery modes). They found the quality factor of a coupled-particle cavity in the compact powder could be higher than that of a single-particle cavity due to optical coupling. To confirm their model, they measured the spot size of laser-like radiation from a powder of $Al_3Nd(BO_3)_4$ and NdP_5O_{14} (Lichmanov et al. 1998). The minimum spot size was 20–30 μm. Since the particle size was 20 μm, the laser-like radiation was from a single particle or a few particles.

Later, the powder laser was realized with nonresonant pumping at room temperature (Gouedard et al. 1993; Noginov et al. 1996). The gain materials were extended from Nd^{3+}-doped powder to Ti:Sapphire powder (Noginov et al. 1998a), Pr^{3+}-doped powder (Zolin 2000), and pulverized LiF with color centers (Noginov et al. 1997). Despite the difference in material systems, the observed phenomena are similar: (i) drastic shortening of the emission pulse and spectral narrowing of the emission line above a pumping threshold, (ii) damped oscillation of the emission intensity under pulsed excitation, and (iii) drifting of the stimulated emission frequency and hopping of emission line from one discrete frequency to another within the same series of pulses. Gouedard et al. (1993) analyzed the spatial and temporal coherence of the powder laser. From the contrast of the near-field speckle pattern, they concluded that the powder emission above threshold is spatially incoherent. This result was explained by the incoherent superposition of uncorrelated speckle patterns. Their time-resolved measurement showed that the speckle pattern changed rapidly in time. The estimated coherence time of $\sim$ 10 ps indicated low temporal coherence of the powder emission. Noginov et al. (1999) also performed quantitative measurement of the longitudinal and transverse coherence with interferometric techniques. Using a Michelson (Twyman-Green) interferometer, they found that the longitudinal coherence time of $Nd_{0.5}La_{0.5}Al_3(BO_3)_4$ powder (ceramic) emission was 56 ps at a pump energy of twice the threshold. This value corresponded to a 0.07-nm linewidth, in agreement with the results of direct spectroscopic linewidth measurements. They also examined the transverse spatial coherence using Young's double-slit interferometric scheme. The transverse coherence was not noticeable when the distance between two points on the emitting surface was approximately 85 μm or larger.

Despite the detailed experimental study of the powder laser, the underlying mechanism was not fully understood. Gouedard et al. (1993) conjectured that the grains of the powder emitted collectively in a subnanosecond pulse with a kind of distributed feedback provided by multiple scattering. Auzel and Goldner identified two processes of coherent light generation in a powder: (i) amplification of spontaneous emission by stimulated emission and (ii) synchronized spontaneous emission, namely superradiance and superfluorescence (Auzel and Goldner 2000; Zyuzin 1998; Zyuzin 1999). Noginov et al. (1996) noted the essential role played by photon diffusion in stimulated emission when comparing the powder laser with the single crystal laser. The diffusive motion of photons led to a long path length for light emitted from the powder, resulting in a threshold. Wiersma and Lagendijk (1996) proposed a model based on light diffusion with gain. They considered a pump pulse and a probe pulse incident on a powder slab. The active material was approximated as a four-level (2,1,0′,0) system with the radiative transition from level 1 to 0′ and the pumping from level 0 to 2. Fast relaxation from level 2 to 1 and from level 0′ to 0 made both level 2 and level 0′ nearly unpopulated, thus the population of level 1 can be described by one rate equation. The whole system was described by three diffusion equations for the energy densities of the pump light $W_G(\vec{r}, t)$, the probe light $W_R(\vec{r}, t)$, the (amplified) spontaneous emission $W_A(\vec{r}, t)$, and one rate equation for the population density $N_1(\vec{r}, t)$ of level 1. These diffusion and rate equations are:

$$\frac{\partial W_G(\vec{r},t)}{\partial t} = D\nabla^2 W_G(\vec{r},t) - \sigma_{abs} v[N_t - N_1(\vec{r},t)]W_G(\vec{r},t) + \frac{1}{l_G} I_G(\vec{r},t) , \quad (11.11)$$

$$\frac{\partial W_R(\vec{r},t)}{\partial t} = D\nabla^2 W_R(\vec{r},t) + \sigma_{em} v N_1(\vec{r},t) W_R(\vec{r},t) + \frac{1}{l_R} I_R(\vec{r},t) , \quad (11.12)$$

$$\frac{\partial W_A(\vec{r},t)}{\partial t} = D\nabla^2 W_A(\vec{r},t) + \sigma_{em} v N_1(\vec{r},t) W_A(\vec{r},t) + \frac{1}{\tau_e} N_1(\vec{r},t) , \quad (11.13)$$

$$\frac{\partial N_1(\vec{r},t)}{\partial t} = \sigma_{abs} v[N_t - N_1(\vec{r},t)]W_G(\vec{r},t) - \sigma_{em} v N_1(\vec{r},t)[W_R(\vec{r},t) + W_A(\vec{r},t)] - \frac{1}{\tau_e} N_1(\vec{r},t) , \quad (11.14)$$

where σ_{abs} and σ_{em} are the absorption and emission cross sections, τ_e is the lifetime of level 1, $I_G(\vec{r}, t)$ and $I_R(\vec{r}, t)$ are the intensities of the incoming pump and probe pulses, l_G and l_R are the transport mean free paths at the pump and probe frequencies, and N_t is the total concentration of four-level atoms. Wiersma and Lagendijk (1996) numerically solved the above coupled nonlinear differential equations. Their simulation result reproduced the experimental observation of transient oscillation (spiking) of the emission intensity under pulsed excitation.

In the slab geometry, the critical volume predicted by Letokhov is reduced to the critical thickness $L_{cr} = \pi\sqrt{l_t l_g/3} = \pi l_{amp}$. For the fixed slab thickness L, there exists a critical amplification length $l_{cr} = L/\pi$. At the beginning of the pump pulse, the average amplification length l_{amp} decreases due to an increasing excitation level. Once l_{amp} crosses l_{cr}, the gain in the sample becomes larger than the loss through the boundaries, and the system becomes unstable. This leads to a large increase of the amplified spontaneous emission (ASE) energy density. The characteristic time scale corresponding to the buildup of ASE is l_g/v. The large ASE energy density de-excites the system again, which leads to an increase of l_{amp}. This de-excitation continues as long as the large ASE energy density is present. The characteristic time scale over which the ASE energy density diffuses out of the medium through the front or rear surface is given by L^2/D. On one hand, an overshoot of the excitation takes place because the de-excitation mechanism needs some time to set in. On the other hand, once the ASE has built up considerably, the ASE energy density can disappear only slowly due to the presence of multiple scattering, which leads to an undershoot below the threshold. These two processes result in transient oscillations of the outgoing ASE flux. The oscillations are damped because the increase of l_{amp} during the de-excitation is opposed by re-excitation owing to the presence of pump light. Therefore the system reaches the equilibrium situation $l_{amp} = l_{cr} = L/\pi$ after a few oscillations.

Both models based on light diffusion and intraparticle resonances reproduced the experimental phenomena. It is difficult to determine whether feedback in the powder laser is provided by multiple scattering or total internal reflection because the gain medium and scattering elements are not separated in the powder. Lawandy et al. (1994) separated the scattering and amplifying media in liquid solutions. This separation allowed the scattering strength to be varied independently of the gain coefficient and facilitated a systematic study of the scattering effect on feedback.

11.2.4 Laser paint

In 1994, Lawandy et al. (1994) observed laser-like emission from a methanol solution of rhodamine 640 perchlorate dye and TiO_2 microparticles. The dye molecules were optically excited by laser pulses and provided optical gain. The TiO_2 particles, with a mean diameter of 250 nm, were scattering centers. As shown in Fig. 11.2(a), the (input-output) plot of the peak emission intensity versus the pump energy exhib-

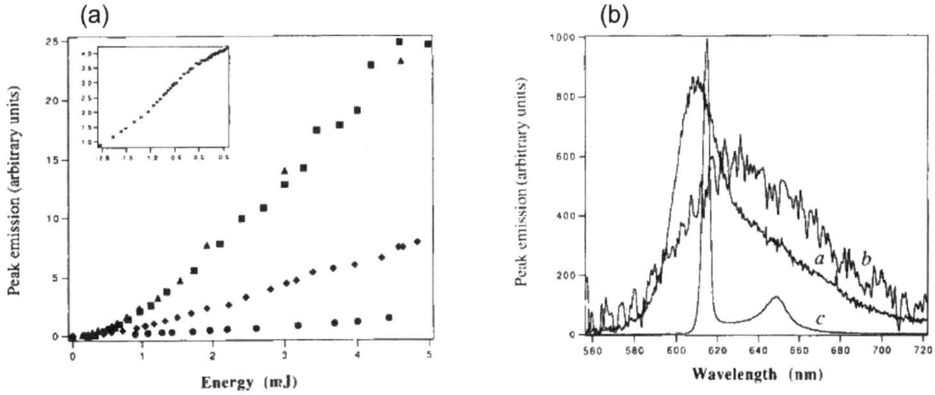

Figure 11.2 (a) Peak emission intensity as a function of the pump pulse energy for four TiO_2 nanoparticle colloidal dye solutions. The TiO_2 particle densities were 1.4×10^9 cm^{-3} (circles), 7.2×10^9 cm^{-3} (diamonds), 2.8×10^{10} cm^{-3} (squares), and 8.6×10^{11} cm^{-3} (triangles). (b) Curve a is the emission spectrum of a 2.5×10^{-3}M solution of rhodamine 640 perchlorate in methanol pumped by 3-mJ (7-ns) pulses at 532 nm. Curves b and c are emission spectra of the titanium dioxide (TiO_2) nanoparticle (2.8×10^{10} cm^{-3}) colloidal dye solution pumped by 2.2-μJ and 3-mJ (7-ns) pulses. The amplitude of spectrum b is scaled up by a factor of 10, whereas that of spectrum c is scaled down by a factor of 20. [Reprinted with permission from Lawandy et al. 1994. © (1994) by Nature Publishing Group.]

ited a well-defined pumping threshold for the slope change. At the same threshold, the emission linewidth (FWHM: full width at half maximum) collapsed rapidly from 70 to 4 nm [Fig. 11.2(b)], and the duration of emission pulses was shortened dramatically from 4 ns to 100 ps. This threshold behavior suggested the existence of feedback. The relatively broad and featureless emission spectrum above threshold indicated that the feedback was frequency insensitive (nonresonant). In the solution, a feedback mechanism based on morphology-dependent resonance can be ruled out because the gain was outside the scatterer, and individual scatterers were too small to serve as morphology-dependent resonators. It was found experimentally that the threshold was reduced by more than 2 orders of magnitude when the density of scattering particles was increased from 5×10^9 to 2.5×10^{12} cm^{-3} at the fixed dye concentration of 2.5×10^{-3} M (Sha et al. 1994). The strong dependence of the threshold on the transport mean free path revealed that the feedback was related to scattering (Lawandy and Balachandran 1995; Zhang et al. 1995a; Balachandran and Lawandy 1995).

However, light diffusion is negligible unless the smallest dimension of the scattering medium is much larger than the transport mean free path. Experimentally, when a spatially broad pump pulse is incident on a dye cell, a disk-shaped amplifying region is formed near the front window (Wiersma et al. 1995). The thickness of the disk is determined by the penetration depth L_{pen} of the pump light. In Lawandy's experiment, L_{pen} was close to the transport mean free path. However, the actual sample thickness (i.e., the thickness of the entire suspension) was much larger than the transport mean free path. Hence, light transport in the suspension was diffusive. Nevertheless, the emitted photons could easily escape from the thin amplifying region. Some of them escaped through the front surface into the air, while the rest went deeper into the unpumped region of the suspension. After multiple scattering (or random walk), some of these photons returned to the active volume for more amplification. This return process provided energy feedback. When scattering is stronger, the return probability is higher, thus the feedback is stronger. However, incomplete feedback (less than 100% return probability) gives rise to loss. The lasing threshold is set by the condition that the photon loss rate is balanced by the photon generation rate in the amplifying region. On one hand, the total amount of gain or amplification is the product of the amplification per unit path length and the path length traveled through the amplifying volume. The frequency dependence of the amplification per unit path length gives the highest photon generation rate at the peak of the gain spectrum. On the other hand, owing to the weak frequency dependence of the transport mean free path, the feedback is nearly frequency independent within the gain spectrum, as is the loss rate for photons. Therefore, with an increase of the pumping rate, the photon generation rate in the spectral region of maximum gain balances the photon loss rate, while outside this frequency region the photon generation rate is still below the loss rate, in which case the photon density around the frequency of the gain maximum builds up quickly. The sudden increase of photon density near the peak of the gain spectrum results in the collapse of the emission linewidth.

A model based on the ring laser in the random phase limit was proposed by Balachandran and Lawandy (1997) to quantitatively explain the experimental data. The amplifying volume is approximated as a sharply bounded disk with a homogeneous gain coefficient. In a Monte Carlo simulation of the random walk of photons, they calculated the return probabilities R_{t1} and R_{t2} of photons to the disk after being launched from the disk bottom (either toward the disk interior or away from it) and the average total path length L_{pat}. The threshold gain g_{th} is determined by the steady state condition

$$R_{t1}\, R_{t2}\, e^{g_{th}\, L_{pat}} = 1. \tag{11.15}$$

This condition is analogous to the threshold condition for a ring laser. Note that a typical ring laser has a second condition on the round-trip phase shift: $kL_{pat} = 2\pi m$, which determines the lasing frequencies. In the scattering medium the phase condition can be ignored because the diffusive feedback is nonresonant, i.e., it requires that light return only to the gain volume instead of to its original position. Therefore, this kind of laser is a random laser with nonresonant or incoherent feedback.

11.2.5 Further developments

The discovery of Lawandy et al. triggered many experimental and theoretical studies that are briefly summarized here.

Lasing threshold. The dependence of the lasing threshold on the dye concentration and the gain length was investigated (Zhang et al. 1995a). Usually the threshold was reached at the point at which the pump transition was bleached. Such bleaching increased the penetration depth of the pump and consequently led to longer path lengths for the emitted light within the gain region, which resulted in a reduced threshold (Siddique et al. 1996). The influence of the excitation spot diameter on the threshold was also examined (van Soest et al. 1999). In a suspension of TiO_2 scatterers in sulforhodamine B dye, the threshold pump intensity increased by a factor of 70 when the excitation beam diameter got close to the mean free path. This is because a large pump beam spot produced a large amplifying volume. The emitted light could travel a long path inside the active region and experienced more amplification before escaping. After the light went into the passive (unexcited) region, there was a large probability that it would return to the amplifying region because of the large pumped area. For a small excitation beam diameter, the emitted light would very likely leave the active volume after a short time, with a small chance of returning. This gave a larger photon loss rate and higher threshold. The amplification by stimulated emission was found to be strongest when the absorption length of the pump light and the transport mean free path had approximately the same magnitude (Beckering et al. 1997). A critical transport mean free path was identified for each beam diameter below which the threshold was almost independent of the mean free path (Totsuka et al. 2000). These results can be explained in terms of the spatial overlap of the gain volume and the diffusion volume. By solving the coupled rate and diffusion equations, Totsuka et al. calculated the spatial distribution of the excited state population (gain volume) and the spatial spreading of the trajectories for the luminescence light (diffusion volume). When the gain volume was smaller than the diffusion volume of the luminescence light, the amplification was not efficient, as the light propagated primarily through the gainless region. If the gain volume was larger than the diffusion volume of the luminescence light, the excitation pulse energy was not used efficiently for amplification. There existed an optimized condition under which the pulse energy was used most efficiently for stimulated emission.

Emission spectra. The stimulated emission spectrum was shifted with respect to the luminescence spectrum. This spectral shift was explained by a simple ASE model accounting for absorption and emission at the transition between the ground and first singlet excited states of the dye (Noginov et al. 1995). Bichromatic emission was produced in a binary dye mixture in the presence of scatterers (Zhang et al. 1995b). The dye molecules were of the donor-acceptor type, and the energy transfer between them gave double emission bands. The relative intensity of stimulated emission of the donor and the acceptor depended on the scatterer density, in addition to the pumping intensity and the concentration of the dyes. The narrow-linewidth bichromatic emission was also observed in a single dye solution with scatterers at large pumping intensity or high dye concentration (Sha et al. 1996; Balachandran and Lawandy 1996a). John and Pang (1996) explained the bichromatic emission in terms of dye molecules' singlet and triplet transitions. Using physically reasonable estimates for the absorption and emission cross-section for the single and triplet manifolds and the singlet-triplet intersystem crossing rate, they solved the nonlinear rate equations for the dye molecules. This led to a diffusion equation for the light intensity in the scattering medium with a nonlinear intensity-dependent gain coefficient. Their model could account for most experimental observations, e.g., the collapse of emission linewidth at a specific threshold pump intensity, the variation of the threshold intensity with the transport mean free path, and the dependence of peak emission intensity on the transport mean free path, the dye concentration, and the pump intensity.

Dynamics. One surprising result concerning the dynamics of stimulated emission from colloidal dye solutions was that the emission pulses can be much shorter than the pump pulses when the pumping rate is well above threshold. For instance, 50-ps pulses of stimulated emission were obtained from the colloidal solution excited by 3-ns pulses (Sha et al. 1994). The shortest emission pulses were $\sim$ 20-ps long and produced by 10-ps pump pulses (Siddique et al. 1996). Berger et al. (1997) modeled the dynamics of stimulated emission from random media using a Monte Carlo simulation of the random walk of pump and emitted photons. They tracked the temporal and spectral evolution of emission by following the migration of photons and molecular excitation as determined purely by local probabilities. Their simulation results revealed a sharp transition to ultrafast, narrow line-width emission for a 10-ps incident pump pulse and a rapid approach to steady state for longer pump pulses. Using a different approach van Soest et al. (2001) also studied the dynamics of stimulated emission. They numerically solved the coupled diffusion equations for the pump light and the emitted light and the rate equation for the excited population. Their simulation result illustrated that the slow response rate of the population inversion density, compared to the light transport time, started a relaxation oscillation at the threshold crossing.

β factor. In analogy with traditional laser theory, the spontaneous emission coupling factor β was used to characterize the random laser (van Soest and Lagendijk 2002). In a conventional laser, β is defined as the ratio of the rate of spontaneous emission into the lasing modes to the total rate of spontaneous emission. Its value is determined by the overlap in the wave-vector space between the spontaneous emission and laser field. In conventional macroscopic lasers, the spontaneous emission is isotropic, while the cavity modes occupy small solid angles. The directional mismatch contributes to a small β value (less than 10^{-5}). In a scattering medium the diffusive feedback is nondirectional, thus the spatial distinction between lasing and nonlasing modes vanishes, and the only relevant criterion is the spectral overlap of the spontaneous emission spectrum with the lasing spectrum. This yields a large β value (~ 0.1).

Control and optimization. Liquid solutions are awkward to handle; e.g., the sedimentation of scattering particles in the solvent causes instability. Thus liquid solvents were replaced by polymers as host materials (Balachandran and Lawandy 1996a). The polymer sheets containing laser dyes and TiO_2 nanoparticle scatterers were made with the cell-casting technique. The lasing phenomenon in solid dye solutions is similar to that in liquid dye solutions, although the different embedding environments affect the fluorescence characteristics of the dye (Zacharakis et al. 1999). This kind of random laser is called a painted-on laser, or photonic paint, as the polymer film can be deposited on any substrate (Lawandy 1994; Wiersma and Lagendijk 1997). Many techniques developed for traditional lasers were exploited to optimize and control random lasers. For example, external feedback was introduced to control the lasing threshold. De Oliveira et al. (1997) placed a mirror close to the high-gain scattering medium and measured the spectral line shapes of the emitted light as a function of the distance between them. The main effect of the feedback from the mirror was to increase the lifetime of the photons inside the pump region, resulting in a reduction of the threshold pump energy. The injection-locking technique was also utilized to control the emission wavelength. Introducing a seed into the optically pumped scattering gain medium resulted in an intense isotropic emission whose wavelength and linewidth were locked to the seeding beam properties (Balachandran et al. 1996b). Moreover, multiple narrow-linewidth emission was obtained by pumping one laser paint with the output from another laser paint (Martorell et al. 1996). Later, temperature tuning was employed to turn random lasers on and off. A liquid crystal was infiltrated into macroporous glass, and the diffusive feedback was controlled through a change of the refractive index of the liquid crystal with temperature (Wiersma and Cavalieri 2001). In a different approach, a mixture with a lower critical temperature, which could be reversibly transformed between a transparent state and a highly scattering colloid with a small temperature change, was used to tune the lasing threshold with temperature (Lee and Lawandy 2002). Finally, an external electric field was used to switch random lasing in dye-doped polymer dispersed liquid crystals from a 3D random walk to a quasi-two-dimensional (2D) type of transport (Gottardo et al. 2004). The laser emission was anisotropic, and the polarization was enhanced due to strong scattering anisotropies.

Solid state random lasers. Recently there has been much progress in the development of solid state random lasers (Noginov et al. 2004a; Noginov et al. 2004b; Noginov et al. 2004d; Bahoura et al. 2005a; Noginov et al. 2006). For example, a low-threshold GaAs powder laser was realized (Noginov et al. 2004c; Noginov et al. 2005b). Both the lasing threshold and the slope efficiency were significantly improved when the pulverized gain medium was mixed with a powder of optically inert material (without any absorption or emission). The employment of a fiber-coupling scheme significantly improved the performance of a random laser (Noginov et al. 2005). When the tip of a fiber is relocated from the surface to deep inside the powder's volume, the lasing threshold is reduced twofold and the slope efficiency is increased fivefold. High absorption efficiency (85%) and high conversion efficiency of population inversion to stimulated emission (90%) make the fiber-coupled random laser a promising laser source.

11.3 Random Lasers with Coherent Feedback

11.3.1 "Classical" versus "quantum" random lasers

Random lasers with coherent feedback can operate either in the classical optics regime or wave optics regime (Polson et al. 2002). In the former, $R_c \gg \lambda$, whereas in the latter, $R_c \leq \lambda$. These regimes have analogues in chaotic cavity lasers whose cavity shape is irregular. In the classical regime the spatial variation of the chaotic cavity shape is much larger than the wavelength of radiation, whereas in the quantum/wave regime the spatial variation is comparable to or smaller than the wavelength.

When the dielectric constant in a random medium varies over length scales much larger than λ, geometrical optics can be applied to describe light propagation in terms of ray trajectories. The majority of the ray trajectories are chaotic and open, yet unstable periodic orbits exist when the sample size is large enough. In an open system, unstable periodic orbits might trap light for a longer time than chaotic trajectories. Thus lower optical gain is needed to realize lasing oscillation in certain "scar" modes that concentrate about some unstable periodic orbits.

When the dielectric constant fluctuates over length scales comparable to or even smaller than λ, ray optics no longer holds. It is replaced by wave optics that not only describes light scattering by short-range disorder but also takes into account interference of scattered waves. The interference effect is crucial to light localization in a random medium, which is analogous to the (quantum) Anderson localization of electrons in a short-range potential. Even when $l_t > \lambda$, light may still be partially trapped in a random medium via the process of multiple scattering and wave interference. Incomplete confinement can be compensated for by photon amplification when optical gain is introduced into a random medium, leading to lasing oscillation.

11.3.2 Classical random lasers with coherent feedback

The classical type of random lasers with coherent feedback was first demonstrated by Vardeny and coworkers in weakly disordered media such as π-conjugated polymer films (Frolov et al. 1999a; Polson et al. 2001a), organic dye-doped gel films (Frolov et al. 1999b), and synthetic opals infiltrated with π-conjugated polymers and dyes (Frolov et al. 1999b; Yoshino et al. 1999; Polson et al. 2001b). The long-range fluctuations of refractive indices in their polymer films were most likely caused by inhomogeneity of the film thickness. Since light is confined within a film due to waveguiding, larger thickness leads to a higher effective index of refraction.

The samples were excited by short laser pulses and emitted broad-band luminescence at low pumping. With increasing pump intensity the photoluminescence band narrowed drastically. As the excitation intensity increased even further, the emission spectrum transformed into a fine structure that consisted of a number of sharp peaks [Fig. 11.3(a)]. The spectral width of these peaks was less than 1 nm. When the pump light excited a different sample or a different part of the same sample, the narrow peaks changed completely. However, when the same part of the sample was excited repeatedly by different pump pulses, the spectral peaks were reproducible [Fig. 11.3(b)].

Polson et al. (2002) suggest that the lasing modes in the polymer films are formed by total internal reflection at the boundaries of high refractive index regions. The long-range fluctuation of the refractive index is similar to a ring resonator, in the sense that it gives rise to a number of resonant modes having close frequencies and quality factors. These modes are revealed in the emission spectrum, and their frequency spacings are determined by the cavity lengths. Hence the size L_r of the underlying resonators can be estimated from the power Fourier transform (PFT) of lasing spectra. They indeed found that $L_r \gg \lambda$ in their samples.

Along with long-range fluctuations of refractive index, short-range disorder is also present in polymer films. It is conjectured that the ability of most random resonators to trap light by total internal reflection is suppressed by short-range disorder. The dramatic consequence of this suppression is that the resonators that "survive" the short-range disorder are sparse, and consequently almost identical.

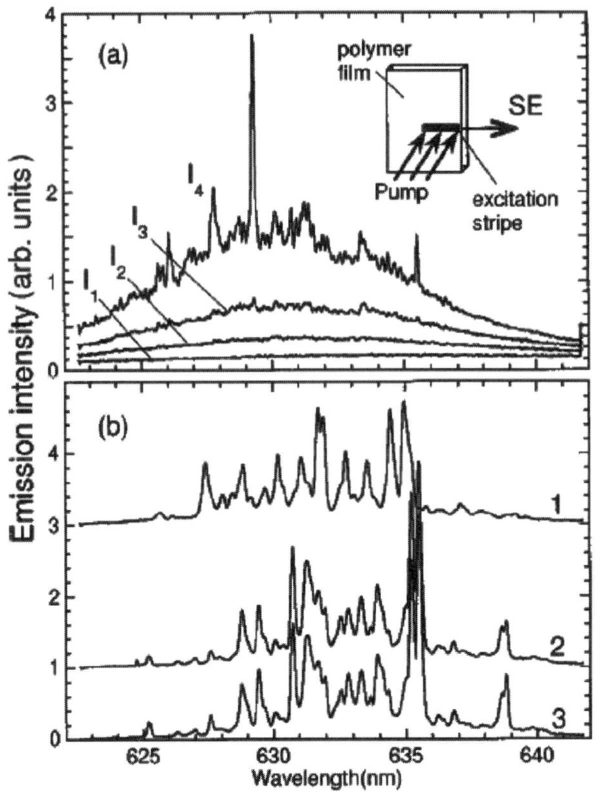

Figure 11.3 Stimulated emission (SE) spectra of a DOO-PPV film obtained using a stripelike excitation area with length $L = 51$ mm and width $a = 530$ mm: (a) SE at different excitation intensities I: $I_1 = I_A = 1$ MW/cm^2, $I_2 = 1.25 I_A$, $I_3 = 1.4 I_A$, $I_4 = 1.6 I_A$; the inset schematically shows the excitation geometry. (b) SE spectra (offset for clarity) measured sequentially at $I \simeq 2 I_B$ from the same DOO-PPV film: line 2 was obtained from a different excited area vertically shifted by 0.3 mm from that of line 1; line 3 was obtained after a 3-min delay from the same area as line 2. [Reprinted with permission from Frolov et al. 1999a. © (1999) by the American Physical Society.]

Experimentally, despite the PFT of individual random lasing spectra exhibiting position-specific multipeak structures, averaging the PFTs over various positions on the sample does not smear these features; on the contrary, averaging yields a series of distinct transform peaks. Moreover, the shape of the averaged PFT is universal, i.e., increasing the disorder and correspondingly reducing l_t does not change its shape: the average of the PFT spectra at different l_t scales with l_t to a universal curve (Polson et al. 2002).

To understand the classical type of random laser with coherent feedback, Apalkov, Raikh, and Shapiro (2003) conducted a theoretical analysis. They believe the lasing modes are the almost-localized states in the passive medium. Such states are formed from rare disorder configurations that can trap light for a long time in a sub-mean-free-path region in space. In the case of a continuous random potential, the almost-localized states are confined to small rings of a sub-mean-free-path size. The almost-localized states are nonuniversal, i.e., their character and formation probability depend not only on the average strength of the disorder, but also on the microscopic details of the disorder (Apalkov et al. 2004a). Apalkov, Raikh, and Shapiro (2002) calculated the areal density of the almost-localized states in a film with fluctuating refractive index. The rings formed by disorder can be viewed as waveguides that support the whispering-gallery type modes. Because of azimuthal symmetry these modes are characterized by the angular momentum quantum number m. The areal density of ring resonators with quality factor Q can be expressed as $N_m(k\,l_t, Q) = N_0 \exp[-S_m(k\,l_t, Q)]$ for $k\,l_t > 1$. In the case of smooth disorder, $kR_c \gg 1$ and $S_m \sim l_t(\ln Q)^{4/3}/(kR_c^2 m^{1/3})$. In the opposite limit of short-range disorder, $S_m \sim k\,l_t \ln Q$. Therefore, when on average the light propagation is diffusive, the likelihood for finding an almost-localized state increases sharply with the disorder correlation radius R_c for a given $k\,l_t$. Note that this conclusion applies only to a continuous (Gaussian) random potential. In the presence of a discrete lattice (the Anderson model), a new type of almost-localized state is formed; its formation probability is reduced by correlation in disorder (Patra 2003b; Apalkov et al. 2004b).

Apalkov and Raikh (2005) also investigated the fluctuation of the random lasing threshold. The distribution of the threshold gain over the ensemble of statistically independent finite-size samples is found to be universal. This universality stems from two results; (i) the lasing threshold in a given sample is determined by the highest-quality mode of all the random resonators present in the sample, and (ii) the areal density of the random resonators decays sharply with the quality factor of the mode that they trap. In a 2D sample of area S, the distribution function of the threshold excitation intensity I_{th} is:

$$F_S(I_{th}) = \frac{\beta_S}{I_{th}} \left(\frac{I_{th}}{I_S}\right)^{-\beta_S} \exp\left[\left(\frac{I_{th}}{I_S}\right)^{-\beta_S}\right]. \tag{11.16}$$

The typical value of I_S is related to the sample area S:

$$I_S \propto \exp\left\{-\left[\frac{\ln(S/S_0)}{G}\right]^{1/\alpha}\right\}, \tag{11.17}$$

where S_0 is the typical area of a random resonator. The parameters α and G are determined by the intrinsic properties of the disordered medium and are independent of S. These two parameters play different roles: α is determined exclusively by the shape of the disorder correlator, while G is a measure of the disorder strength. $\beta_S \propto [\ln(S/S_0)]^{(\alpha-1)/\alpha}$. The parameter $\beta_S \propto [\ln(S/S_0)]^{(\alpha-1)/\alpha}$ decreases with $k\,l_t$ as a power law, and the exponent depends on the microscopic properties of the disorder. For a weakly scattering medium, $\beta_S \gg 1$, and $F_S(I_{th})$ is close to a Gaussian distribution: $F_S(I_{th}) \propto \exp[-(\beta_S^2/2)(I_{th}/I_S - 1)^2]$. When β_S is small, the distribution $F_S(I_{th})$ is broad and strongly asymmetric. It has a long tail toward high thresholds and falls off abruptly toward low thresholds.

11.3.3 Quantum random lasers with coherent feedback

In the classical type of random laser with coherent feedback, formation of "closed periodic orbits" with small leakage results in light confinement. Interference plays a secondary role as it determines only the resonant frequencies of the periodic orbits. However, in the quantum/wave type of random lasers with coherent feedback, random media have discrete scatterers and strong short-range disorder, thus the interference of scattered waves is essential to light trapping in a random medium. The active random media used for the quantum/wave type of random laser can be divided into two categories: (i) aggregation of active scatterers such as ZnO powder and (ii) passive scatterers in continuous gain media, e.g., TiO_2 particles in a dye solution. Both have their advantages and disadvantages. In (ii), gain media and scattering centers are separated, allowing independent variation of the amount of scattering and gain. However, the scattering strength in (i) is usually higher than that in (ii), owing to larger contrast of refractive index and higher density of scatterers. In the following subsections, lasing in both types of random media are discussed with some examples.

11.3.3.1 Lasing oscillation in semiconductor nanostructures

Figure 11.4 shows the scanning electron microscope (SEM) images of some semiconductor nanostructures that were used for random laser experiments. The ZnO nanorods in Fig. 11.4(a) were grown on a sapphire substrate by metalorganic chemical vapor deposition (MOCVD) (Liu et al. 2004). The rods are uniform in diameter and height, but randomly located on the substrate. The average rod diameter

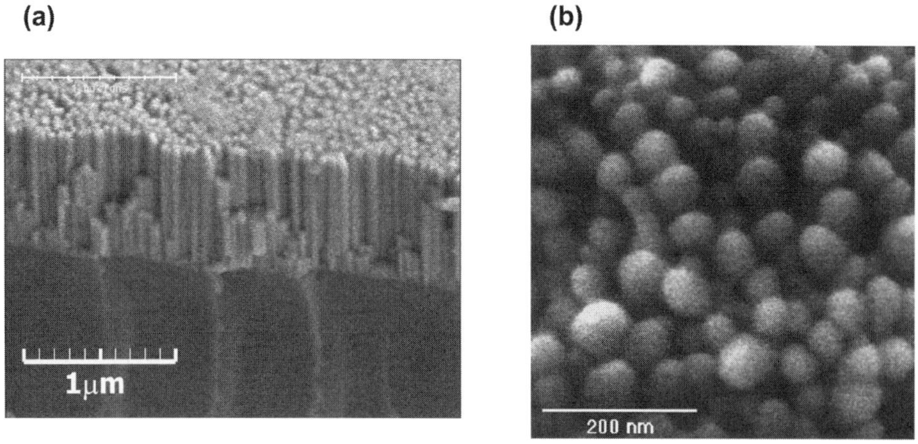

Figure 11.4 SEM images of (a) ZnO nanorods on a sapphire substrate, (b) closely packed ZnO nanoparticles. (Reprinted from Cao 2005.)

is about 50 nm. The ZnO nanorod array is a 2D scattering system, as light is scattered by the nanorods in the plane perpendicular to the rods. When the layer of ZnO nanorods has a larger refractive index than the substrate, index guiding leads to light confinement in the third direction parallel to the rods.

Figure 11.4(b) shows the ZnO nanoparticles synthesized in a wet chemical reaction (Seelig et al. 2003). The particles are polydisperse with an average size of $\sim$ 100 nm. They are randomly close packed with a filling fraction of $\sim 50\%$. Since light is scattered by ZnO nanoparticles in all directions, the ZnO powder represented a 3D scattering system. The above two examples of random media have discrete scatterers of subwavelength size. Short-range disorder results in strong light scattering. In ZnO powder, the transport mean free path $l_t \sim \lambda$. To introduce optical gain, the ZnO samples were optically pumped by the frequency-tripled output (λ = 355 nm) of a mode-locked Nd:YAG laser (10 Hz repetition rate, 20 ps pulse width). The ZnO nanorods and nanoparticles were active scatterers.

Lasing with coherent feedback has been observed in both ZnO nanorods and nanoparticles. Since their lasing behavior is similar, only the measurement results of ZnO powder are presented next. Figure 11.5 shows the measured spectra and spatial distribution of emission in a ZnO powder film at two pumping intensities

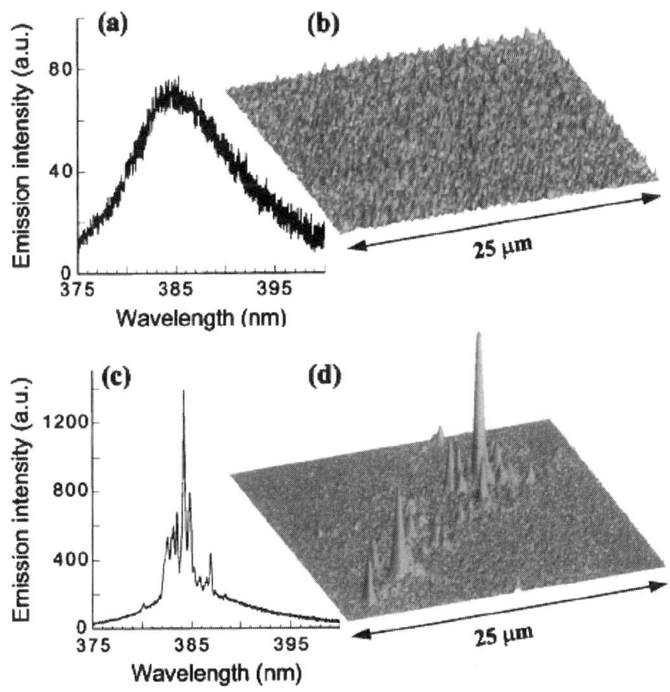

Figure 11.5 (a) and (c) are the measured spectra of emission from a ZnO powder sample. (b) and (d) are the measured spatial distribution of emission intensity on the sample surface. The incident pump pulse energy was 5.2 nJ for (a) and (b), and 12.5 nJ for (c) and (d). (Reprinted from Cao et al. 2000b.)

(Cao et al. 1999a). At a low pumping level the spectrum consisted of a single broad spontaneous emission band. Its full width at half maximum (FWHM) is about 12 nm [Fig. 11.5(a)]. In Fig. 11.5(b), the spatial distribution of the spontaneous emission intensity is smooth across the excitation area. Due to pump intensity variation across the excitation spot, the spontaneous emission at the center of the excitation spot was stronger. When the pump intensity exceeded a threshold, discrete narrow peaks emerged in the emission spectrum [Fig. 11.5(c)]. The FWHM of these peaks is about 0.2 nm. Simultaneously, tiny bright spots appeared in the image of the emitted light from the film [Fig. 11.5(d)]. The size of the bright spots shown is between 0.3 and 0.7 μm. When the pump intensity was increased further, additional sharp peaks emerged in the emission spectrum, and more bright spots appeared in the image of the emitted light distribution. Above threshold the total emission intensity increased much more rapidly with the excitation intensity.

The frequencies of the sharp peaks depended on the sample position. As the excitation spot was moved across the sample, the frequencies of these peaks changed significantly. However, at a fixed sample position, the peak frequencies remained the same, while the peak heights fluctuated from shot to shot. These phenomena suggest that the discrete spectral peaks result from spatial resonances for light in the ZnO powder, and that such resonances depend on the local configurations of nanoparticles. To find the spatial size of each lasing mode, spectrally resolved speckle analysis was employed to map the spatial profile of individual lasing modes at the sample surface (Cao et al. 2002). The far-field speckle pattern of one lasing mode was recorded, then Fourier-transformed to generate the spatial field correlation function in the near-field zone. Once above the lasing threshold, spatial coherence was established across the entire lasing mode. Hence, the spatial extent of the field correlation function directly reflects the mode size. The experimental data reveal that individual lasing modes have dimensions of a few microns. This can be understood as follows. Due to local variations in particle density and spatial configuration, there exist small regions of stronger scattering. Light can be trapped in these regions through the process of multiple scattering and wave interference. For a particular configuration of particles, only light at certain frequencies can be confined, because the interference effect is frequency sensitive. In a different part of the sample, the particle configuration is different, thus light is confined at different frequencies. However, the confinement is incomplete, as light can escape through the sample surface. When the photon generation rate reaches the photon escape rate, lasing oscillation occurs at the local resonant frequencies, resulting in discrete lasing peaks in the emission spectrum.

The dependence of random lasing on the pump area A_p was investigated by Cao et al. (1999a). The lasing threshold decreases with increasing A_p, as it is more likely to find a stronger trapping site for light within a larger gain volume. At a fixed pump intensity, more lasing peaks appear when A_p increases. This is simply because there are more trapping sites for light. Eventually, at a very large pump area, the lasing peaks are so close to each other in frequency that they can no longer be resolved. However, when A_p is reduced below a critical value, lasing oscillation stops. The critical pump area decreases with increasing pumping intensity.

The temporal evolution of emission was measured by a streak camera (Soukoulis et al. 2002). Below the lasing threshold, the decay time of the emission was 167 ps. When the pump intensity exceeded the threshold, the emission pulse was shortened dramatically. The initial decay of emission intensity was very fast, being just 27 ps. After about 50 ps, the decay slowed to 167 ps, which was also the decay time below threshold. The initial fast decay was caused by stimulated emission, and the later slow decay resulted from spontaneous emission and nonradiative recombination. As the pump intensity increased further, the initial stimulated emission became much stronger than the later spontaneous emission. The dynamics of individual lasing modes were also measured by combining a spectrometer with a streak camera. The time traces of individual lasing modes revealed that lasing in different modes was not synchronized. Just above the lasing threshold, relaxation oscillation was observed for some of the lasing modes. Since the pump pulse was shorter than both radiative and nonradiative recombination times for ZnO particles, lasing was in the transient regime. Recently, lasing in ZnO powder was realized with 10-ns pump pulses (Markushev et al. 2005). Since the pumping time was much longer than all of the characteristic time scales in the system, the lasing oscillation could be regarded as quasi-continuous.

The quantum-statistical property of laser emission from the ZnO powder was also probed in a photon-counting experiment (Cao et al. 2001). The photon number distribution for coherent light in a single electromagnetic mode satisfies the Poisson distribution, whereas the photon number distribution for chaotic light in a single mode meets the Bose-Einstein distribution (Mandel and Wolf 1995). However, the photon number distribution for chaotic light in multiple modes approaches the Poisson distribution as the number of modes increases to infinity. Hence, it is difficult to distinguish coherent light from chaotic light in a measurement of photon statistics of multimode lasing. Experimentally, the number of photons in a single electromagnetic mode is counted. The counting time is shorter than the inverse of the frequency bandwidth of each lasing mode, and the collection angle in the far-field zone is less than one speckle's angular width. The photon number distribution in a single mode changes continuously from a Bose-Einstein distribution near the threshold to a Poisson distribution well above threshold. The second-order correlation coefficient G_2 decreases gradually from 2 to 1. It is well known that for single-mode chaotic light, $G_2 = 2$, while for single-mode coherent light, $G_2 = 1$ (Mandel and Wolf 1995). Hence, coherent light is indeed generated in the highly disordered ZnO powder.

Note that the photon statistics of a random laser with resonant feedback is very different from that of a random laser with nonresonant feedback. For a random laser with nonresonant feedback, lasing occurs simultaneously in a large number of modes that are spatially and/or spectrally overlapped. Mode coupling prevents quenching of photon number fluctuations in a single mode (Ambartsumyan et al. 1967c; Ambartsumyan et al. 1968; Ambartsumyan et al. 1970), resulting in excess photon noise (Beenakker 1998; Misirpashaev and Beenakker 1998; Patra and Beenakker 1999; Mishchenko et al. 2001; Patra 2002). Above the lasing threshold, the fluctuation of laser emission intensity is quenched by gain saturation (Florescu and John 2004a; Florescu and John 2004b). The total number of photons in all lasing modes exhibits a fluctuation much smaller than that of the blackbody radiation in the same number of modes. However, the number of photons in a single lasing mode is subject to large fluctuations as a result of mode coupling. Well above the lasing threshold, the amount of photon number fluctuation in each lasing mode is increased above the Poissonian value by an amount that depends on the number of lasing modes. In contrast, a random laser with resonant feedback has lasing modes well separated in frequency. Due to little mode coupling, the photon number fluctuation in each lasing mode can be quenched efficiently by gain saturation at the mode frequency.

11.3.3.2 Random microlasers

The small size of lasing modes in closely packed ZnO nanoparticles indicates strong optical scattering, which not only supplies coherent feedback for lasing but also leads to spatial confinement of laser light in a micron-size volume (Cao et al. 2000b). This makes it possible to realize a new type of microlaser made from a disordered medium (Cao et al. 2000c).

Figure 11.6(a) is a SEM image of a microcluster of ZnO nanoparticles. The size of the cluster was about 1.7 μm. It contained roughly 20,000 nanoparticles. A single cluster was optically pumped by the third harmonic of a pulsed Nd:YAG laser. At low pumping level, the emission spectrum consisted of a single broad spontaneous emission peak. Its FWHM was 12 nm. The spatial distribution of the spontaneous emission intensity was uniform across the cluster. When the pump intensity exceeded a threshold, a sharp peak emerged in the emission spectrum [Fig. 11.6(b)]. Its FWHM was 0.2 nm. Simultaneously, a few bright spots appeared in the image of the emitted light distribution in the cluster [Fig. 11.6(c)]. When the pump intensity increased further, a second sharp peak emerged in the emission spectrum. Correspondingly, additional bright spots appeared in the image of the emitted light distribution. Note that the frequencies of the sharp peaks and the positions of the bright spots did not change from pump pulse to pulse (from shot to shot).

The total emission intensity is plotted against the pump intensity in Fig. 11.6(d). The curve exhibits a distinct change in slope at the threshold where sharp spectral peaks and bright spots appeared. Well above the threshold the total emission intensity increased almost linearly with the pump intensity. These data revealed lasing oscillation in micron-size clusters.

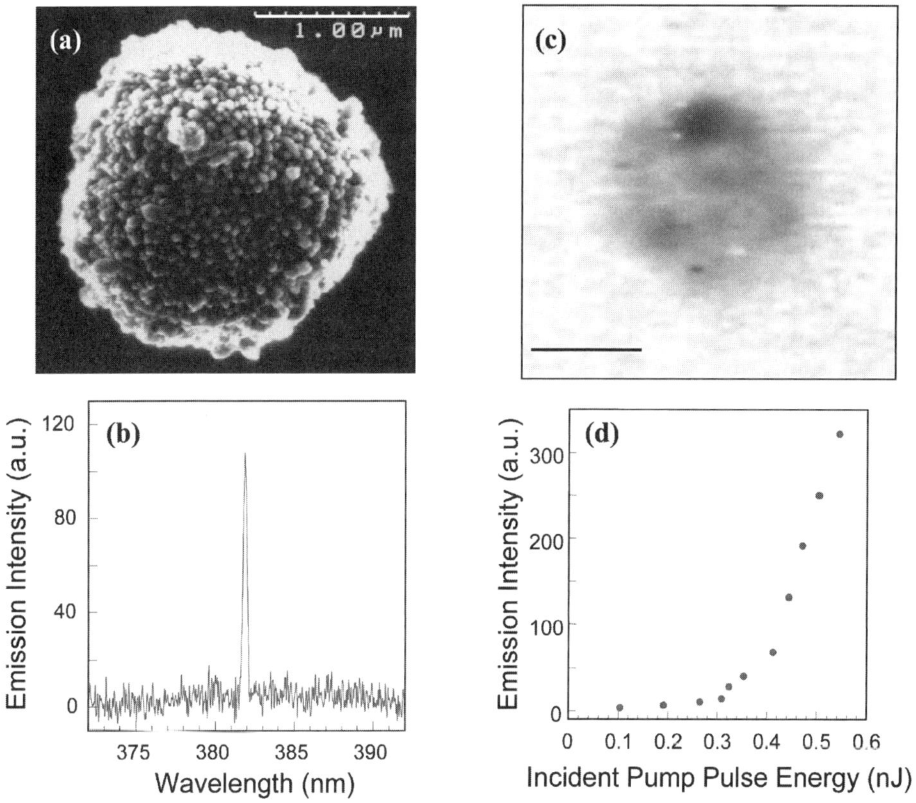

Figure 11.6 (a) Scanning electron micrograph of a microcluster of ZnO nanocrystallites. (b) The spectrum of emission from the cluster at an incident pump pulse energy of 0.35 nJ. (c) Optical image of the emitted light distribution across the cluster. The incident pump pulse energy was 0.35 nJ. The scale bar represents 1 μm. (d) Spectrally integrated emission intensity as a function of the incident pump pulse energy. (Reprinted from Cao et al. 2000c.)

Since clusters are very small, optical reflection from the boundary of a cluster may have some contribution to light confinement in the cluster (Wiersma 2000; Kretschmann and Maradudin 2004). However, the laser cavity is not formed by total internal reflection at the boundary. Otherwise, the spatial pattern of laser light would be a bright ring near the edge of the cluster (Taniguchi et al. 1996). The 3D optical confinement in a micron-size cluster is realized through multiple scattering and interference. Since the interference effect is wavelength sensitive, only light at certain wavelengths can be confined in a cluster. In another cluster of different particle configuration, light at different wavelengths is confined. Therefore, the lasing frequencies are fingerprints of individual random clusters.

11.3.3.3 Collective modes of resonant scatterers

A simplified way of simulating the closely packed ZnO nanoparticles or nanorods is to approximate individual particles or rods as dipolar oscillators (Burin et al.

2001). Burin et al. calculated the quasimodes in a random ensemble of point dipoles. The k-th dipolar oscillator was represented by its resonant frequency ω_k and transition dipole moments $\mathbf{d}_k$, where $k = 1, 2, ...N$. N was the total number of scatterers. Gain was introduced into each scatterer by adding an imaginary term $i\tilde{g}$ to its resonant frequency. A quasimode of this system represented a collective excitation of the coupled dipoles, thus it can also be called a collective mode. The equation of motion for the k-th oscillator's polarization component $\mathbf{p}_k = p_k \mathbf{d}_k / d_k$ can be written as:

$$-\Omega^2 p_k = -(\omega_k - i\tilde{g})^2 p_k + 2\omega_k d_k (\mathbf{d}_k \cdot \mathbf{E}_k). \tag{11.18}$$

$\mathbf{E}_k$ is the local electric field,

$$\mathbf{E}_k = \sum_{j \neq k} \mathbf{E}_{kj} + i\frac{2}{3}q^3 \mathbf{p}_k, \tag{11.19}$$

where $q = \Omega/c$, and $\mathbf{E}_{kj}$ represents the electric field generated by the j-th dipole at the location of the k-th dipole. The solution to Maxwell's equations for the electric field of a single dipole gives:

$$\mathbf{E}_{kj} = e^{iqR_{kj}} \frac{\mathbf{p}_j - 3\mathbf{n}_{kj}(\mathbf{n}_{kj} \cdot \mathbf{p}_j)}{R_{kj}^3}(1 - iqR_{kj}) + q^2 e^{iqR_{kj}} \frac{\mathbf{p}_j - \mathbf{n}_{kj}(\mathbf{n}_{kj} \cdot \mathbf{p}_j)}{R_{kj}}, \tag{11.20}$$

where $\mathbf{n}_{kj} = \mathbf{R}_{kj}/R_{kj}$, and $\mathbf{R}_{kj}$ is the vector from the j-th dipole to the k-th dipole.

There are N solutions to the above equations for N coupled dipolar oscillators. Hence, there are N collective modes, and each is characterized by a complex frequency Ω_α ($\alpha = 1, 2, ...N$). The imaginary part of Ω_α, γ_α, represents the decay rate of a collective mode caused by light leakage out of the system. In the absence of gain ($\tilde{g} = 0$), the system is lossy and all of the decay rates are positive. An increase of gain leads to a decrease of γ_α. At some finite value of gain $\tilde{g}_{th}$, the decay rate for one collective mode vanishes. This corresponds to the onset of the lasing instability. The collective mode with the smallest decay rate in the passive system ($\tilde{g} = 0$) turns out to be the first lasing mode.

Burin et al. (2001) numerically calculated the threshold gain $\tilde{g}_{th}$ in 2D random arrays of dipolar oscillators with N up to 1000. All of the dipoles were assumed to have the same resonant frequency: $\omega_k = \omega_0$. They were positioned randomly within a circle of radius R_0. The average interdipole distance, normalized by the resonant wavelength $\lambda_0 = 2\pi c/\omega_0$, was described by $\eta = 2\pi R_0/\sqrt{N}$. The ensemble-averaged lasing threshold decreases with increasing N as $\langle g_{th} \rangle \propto N^{-\beta}$, where the exponent β is a function of η. It is equivalent to a $1/A^\beta$ dependence on the sample area A for a fixed particle density with $\beta = 0.52$ for $\eta = 0.3$, $\beta = 0.51$ for $\eta = 1$, and $\beta = 0.335$ for $\eta = 3$. The size of the lasing modes is usually smaller than that of the entire system. Hence, the decrease of $\tilde{g}_{th}$ with the system size results from the increase in the probability of finding optimum particle configurations for minimum γ_α, rather than the formation of larger modes. The calculation also included dispersion (random variation) of the resonant frequencies ω_k of the dipolar oscillators. As the dispersion increased, the efficiency of forming collective modes with small γ_α decreased. The threshold gain $\tilde{g}_{th}$ became nearly size independent when the ω_k deviated from one another by more than the near-neighbor dipolar coupling constant.

Collective excitations can be formed most efficiently when all scatterers have identical resonant frequencies (Burin et al. 2001). This condition was realized experimentally with ZnO particles of uniform shape and size (Wu et al. 2004a). Seelig et al. (2003) developed a two-stage chemical reaction process to synthesize monodisperse ZnO nanospheres. The mean diameter of ZnO spheres varied from 85 to 617 nm. The dispersion of the sphere diameter was 5–8%. The ZnO spheres were closely packed with the volume fraction of $\sim 58\%$. In the lasing experiment, the threshold pump intensity I_{th} decreased drastically with increasing sphere diameter from 85 to 137 nm. This rapid drop was replaced by a slow decrease as d_s increased from 137 to 355 nm. When d_s increased further to 617 nm, I_{th} increased slightly. The variation of I_{th} with d_s followed roughly the trend of the scattering cross-section σ_{sc} for ZnO nanospheres. The range of d_s covered the first few Mie resonances at the ZnO emission wavelength. σ_{sc} exhibited a drastic increase with d_s before reaching the first Mie resonance at $d_s \simeq 200$ nm. The value of σ_{sc} reached its maximum at $d_s \sim 370$ nm. Then it decreased with a further increase of d_s to 617 nm. At the Mie resonances, the photon dwell time within individual scatterers was drastically increased. This led to a significant enhancement of light amplification because optical gain existed inside the particles. Of course, in such a densely packed system, scattering particles cannot be considered independent; the resonances of individual scatterers are significantly modified by the interactions among them. Strong coupling of resonant scatterers leads to formation of collective modes with small to large decay rates (Ripoll et al. 2004; Vanneste and Sebbah 2005). The former serve as the lasing modes. The experimental data for I_{th} can be explained qualitatively in terms of σ_{sc}. The larger the scattering cross-section of individual spheres, the stronger the coupling among them and the higher the chance of forming collective modes with smaller decay rates. Hence, the lasing threshold is lower.

11.3.3.4 Time-dependent theory of the random laser

Simulation of random lasers above threshold requires a time-dependent model that takes into account gain saturation. Jiang and Soukoulis (2000) have developed a time-dependent theory for random lasers that couples Maxwell's equations to the rate equations for the electronic population density. The gain medium has four electronic levels. Electrons are pumped from level 0 to level 3, then relax quickly (with time constant τ_{32}) to level 2. Level 2 and level 1 are the upper and lower levels of the lasing transition at frequency ω_a. After radiative decay (with time constant

τ_{21}) from level 2 to 1, electrons relax rapidly (with time constant τ_{10}) from level 1 back to level 0. If the dephasing time is much shorter than all other time scales in the system, the populations in four levels (N_3, N_2, N_1, N_0) satisfy the following rate equations:

$$\begin{aligned}
\frac{dN_3(\mathbf{r},t)}{dt} &= P_r(t)\,N_0(\mathbf{r},t) - \frac{N_3(\mathbf{r},t)}{\tau_{32}}\,, \\
\frac{dN_2(\mathbf{r},t)}{dt} &= \frac{N_3(\mathbf{r},t)}{\tau_{32}} + \frac{\mathbf{E}(\mathbf{r},t)}{\hbar\omega_a}\cdot\frac{d\mathbf{P}(\mathbf{r},t)}{dt} - \frac{N_2(\mathbf{r},t)}{\tau_{21}}\,, \\
\frac{dN_1(\mathbf{r},t)}{dt} &= \frac{N_2(\mathbf{r},t)}{\tau_{21}} - \frac{\mathbf{E}(\mathbf{r},t)}{\hbar\omega_a}\cdot\frac{d\mathbf{P}(\mathbf{r},t)}{dt} - \frac{N_1(\mathbf{r},t)}{\tau_{10}}\,, \\
\frac{dN_0(\mathbf{r},t)}{dt} &= \frac{N_1(\mathbf{r},t)}{\tau_{10}} - P_r(t)\,N_0(\mathbf{r},t)\,.
\end{aligned} \tag{11.21}$$

$P_r(t)$ represents the external pumping rate. $\mathbf{P}(\mathbf{r},t)$ is the polarization density that obeys the equation:

$$\frac{d^2\mathbf{P}(\mathbf{r},t)}{dt^2} + \Delta\omega_a\frac{d\mathbf{P}(\mathbf{r},t)}{dt} + \omega_a^2\mathbf{P}(\mathbf{r},t) = \frac{\Gamma_r}{\Gamma_c}\frac{e^2}{m}\left[N_1(\mathbf{r},t) - N_2(\mathbf{r},t)\right]\mathbf{E}(\mathbf{r},t), \tag{11.22}$$

where ω_a and $\Delta\omega_a$ represent the center frequency and linewidth of the atomic transition from level 2 to level 1. $\Gamma_r = 1/\tau_{21}$ and $\Gamma_c = e^2\omega_a^2/6\pi\varepsilon_0 mc^3$, where e and m are electron charge and mass. $\mathbf{P}(\mathbf{r},t)$ introduces gain into Maxwell's equations:

$$\nabla\times\mathbf{E}(\mathbf{r},t) = -\frac{\partial\mathbf{B}(\mathbf{r},t)}{\partial t}\,, \tag{11.23}$$

$$\nabla\times\mathbf{H}(\mathbf{r},t) = \varepsilon(\mathbf{r})\frac{\partial\mathbf{E}(\mathbf{r},t)}{\partial t} + \frac{\partial\mathbf{P}(\mathbf{r},t)}{\partial t}\,, \tag{11.24}$$

where $\mathbf{B}(\mathbf{r},t) = \mu\,\mathbf{H}(\mathbf{r},t)$. Disorder is described by the spatial fluctuation of the dielectric constant $\varepsilon(\mathbf{r})$. Equations 11.22–11.24 were solved with the finite-difference time-domain (FDTD) method (Taflove and Hagness 2000) to obtain the electromagnetic field distribution in the random medium. Fourier transforming $\mathbf{E}(\mathbf{r},t)$ gave the local emission spectrum. To simulate an open system, the random medium had a finite size and was surrounded by air. The air was then surrounded by strongly absorbing layers, i.e., by uniaxial perfectly matched layers that absorb all of the light escaping through the boundary of the random medium. Within a semiclassical framework, spontaneous emission can be included in Maxwell's equations as a noise current.

Jiang and Soukoulis (2000) simulated the lasing phenomenon in a one-dimensional (1D) random system with a time-dependent theory. A critical pumping rate exists for the appearance of lasing peaks in the spectrum. The number of lasing modes increases with the pumping rate and the length of the system. When the pumping rate increases even further, the number of lasing modes does not increase any more, but saturates at a constant value that is proportional to the system size for a given randomness. This saturation is caused by spatial repulsion of lasing modes that results from gain competition and spatial localization of the lasing modes. This prediction was later confirmed experimentally (Ling et al. 2001; Anni et al. 2004). The time-dependent theory is especially suitable for the simulation of laser dynamics. Soukoulis et al. (2002) simulated the dynamic response and relaxation oscillation in random lasers. The simulation results reproduced experimental observations and provided an understanding of the dynamic response of random lasers.

Vanneste and Sebbah (2001; 2002) calculated the spatial profile of lasing modes in 2D random media with the same method. They compared the passive modes of a 2D localized system with lasing modes having gain and found that they were identical. When the external pump light focused, the lasing modes changed with the location of the pump, in agreement with experimental observations (Cao et al. 1999a; van der Molen et al. 2007). Therefore, local pumping of the system allows selective excitation of individual localized modes. Jiang and Soukoulis (2002) also showed that a knowledge of the density of states and the eigenstates of a random system without gain, in conjunction with the frequency profile of gain, are sufficient to accurately predict the mode that lases first when optical gain is added.

The advantage of the time-dependent theory is that it can simulate lasing in a real random structure. The numerical simulation gives the lasing spectra, the spatial distribution of lasing modes, and the dynamic response that can be compared directly with experimental measurements. The problem is that simulation of large samples requires much computing power, and the computing time is quite long. So far, numerical simulations have been carried out only in 1D and 2D systems, even though the method can be applied to 3D systems. Furthermore, the simulation must be done for thousands of samples with different configurations before any statistical conclusion can be drawn (Li et al. 2001).

11.3.3.5 Lasing modes in diffusive samples

The above experimental and theoretical studies were focused on random systems near or in the localization regime where $l_t \sim \lambda$. The coherent feedback for lasing supplied by multiple scattering is expected to be strong. However, lasing with coherent feedback is also observed in diffusive samples with $l_t \gg \lambda$, e.g., in polymers doped with dye molecules and microparticles (Ling et al. 2001). In such samples, the microparticles are the scattering centers, and the excited dye molecules provide optical gain. By varying the particle density, the scattering strength changes from $l_t \sim \lambda$ to $l_t \gg \lambda$. As l_t increases, the lasing threshold I_{th} rises quickly. The strong dependence of the lasing threshold on the transport mean free path confirms the essential contribution of scattering to lasing oscillation. With a decrease in optical scattering, the feedback provided by scattering becomes weaker, and the lasing threshold is increased. Experimental data imply that $I_{th} \propto {l_t}^{0.52}$. At a fixed pump intensity, the number of lasing modes increases with decreasing l_t. When l_t approaches λ, the lasing threshold pump intensity drops rapidly and the number of lasing modes increases dramatically. This result agrees with John and Pang's prediction of a dramatic threshold reduction in the regime $l_t \rightarrow \lambda$ of incipient photon localization (John and Pang 1996).

In the diffusion regime $l_t \gg \lambda$, coherent feedback supplied by scattering is weak, and light leakage from the sample is large. Nevertheless, lasing oscillation can still occur as long as the optical gain is high enough to compensate for leakage. One important question is whether the lasing modes are still the quasimodes of the passive system as in the localization regime. In the absence of gain, the quasimodes of a diffusive system strongly overlap in space and frequency. The spectral width of individual quasimodes is much larger than the spacing of adjacent modes. Even monochromatic light impinging on the sample, with a frequency equal to that of a particular quasimode, cannot excite only that mode. Instead it excites several modes that overlap with the incident frequency. These modes are excited with a constant phase relation, leading to the formation of a speckle pattern inside the system. In contrast, the lasing modes in a diffusive system can be well separated in frequency. To check the relation between the lasing modes and the quasimodes of weakly scattering systems, Vanneste et al. (2007) simulated lasing in 2D random systems under uniform pumping. They carefully adjusted the pumping rate so that only one mode lased. (A slight increase of pumping rate would lead to multimode lasing.) Their calculations showed that the first lasing mode above threshold corresponded to the quasimode of the passive system. The reason that lasing can occur in a single quasimode is that optical amplification greatly enhances the interference of scattered light. In a naïve picture, a quasimode is formed by the constructive interference of scattered waves returning to the same coherence volume via different closed paths (loops). The longer the path, the lower the amplitude of the returning field. In a weakly scattering system, the amplitude of the scattered field returning via a long loop is much lower than that via a short loop. This significant amplitude difference weakens the interference effect. In the presence of gain, light traveling along a long path is amplified more than it is amplified along a short path. The preferable amplification of the scattered field over a long loop makes its amplitude approach that of light traversing a short loop; this greatly enhances the interference effect. Consequently, the spectral width of these modes decreases dramatically. At the lasing threshold the calculated mode width is reduced to zero provided spontaneous emission is neglected.

11.3.3.6 Spatial confinement of lasing modes by absorption

In a diffusive sample, the decay rate distribution of quasimodes is narrow, thus many modes have a similar lasing threshold (Chabanov et al. 2003). When optical gain is introduced into the entire sample, many modes can lase simultaneously and are closely packed in frequency. If the spectral width of individual lasing modes is larger than the frequency spacing of adjacent lasing modes, the lasing modes cannot be resolved; instead, they merge to a single broad peak. In order to resolve the lasing modes, local pumping is often used. The pump beam is focused to a tiny spot on the sample so that only a small part of the random system has gain. The number of lasing modes is usually reduced, leading to an increase of mode spacing. In contrast, this reduction in the number of lasing modes is not expected for a diffusive system having quasimodes that are usually spread over the entire volume. Under local pumping, all modes with frequencies within the gain spectrum experience similar gain. Shrinking the gain volume should not decrease the number of potential lasing modes, but only increase their lasing threshold. Imaging the laser light leaving the sample surface reveals that the lasing modes are not extended over the entire random medium; instead, they are localized in the vicinity of excitation region.

The initial explanation for the above experimental observation was that the lasing modes are anomalously localized states (Cao et al. 2002). They are analogous to the prelocalized electronic states in diffusive conductors that are responsible for the long-time asymptotics for current relaxation (Altshuler et al. 1991; Mirlin 2000). Such states exhibit an anomalous buildup of intensity in a region of space. If they are located inside the excitation volume, they tend to be the lasing modes because they experience more amplification and less leakage. The anomalously localized states should be rare in diffusive samples (Mirlin 2000). Experimentally, no matter where on the sample the pump beam is focused, the lasing modes are always confined in thc pumped region. Moreover, the lasing threshold does not fluctuate much as the pump spot is moved across the random medium. These experimental observations contradict the theory of anomalously localized states. The discrepancy originates with the assumption that the lasing modes are equal to the eigenmodes of the passive random system. As first illustrated numerically by Yamilov et al. (2005) and later confirmed experimentally by Wu et al. (2006), this assumption no longer holds if absorption at the emission wavelength is significant outside the gain volume.

Many laser dyes used in random lasing experiments have significant overlap between the absorption band and the emission band. Therefore, the photons that are emitted by the excited dye molecules inside the pumped region may diffuse into the surrounding unpumped region and be absorbed by the unexcited dye molecules there. This absorption reduces the probability of light returning to the pumped region, thus suppressing the feedback from the unpumped region. Hence, the lasing modes differ from the quasimodes of the total random system (Yamilov et al. 2005; Wu et al. 2006). Even if all of the quasimodes are extended across the entire random system, the lasing modes are confined to the gain volume with an exponential tail outside it. This is because absorption in the unpumped region effectively reduces the system size to the size of the pumped region plus the absorption length l_{abs}.

In a 3D random medium, the reduction of the effective system size L_{eff} leads to a decrease of the Thouless number $\delta \propto L_{eff}$, as $\delta\nu \propto L_{eff}^{-2}$ and $d\nu \propto L_{eff}^{-3}$.

The smaller the value of δ, the larger the fluctuation of the decay rate γ of the quasimodes (Chabanov et al. 2003). The variance of the decay rate is $\sigma_\gamma^2 = \langle\gamma\rangle^2/\delta$ (Mirlin 2000), thus $\sigma_\gamma/\langle\gamma\rangle \propto L_{eff}^{-1}$. The broadening of the decay rate distribution $P(\gamma)$ along with the reduction of the density of states is responsible for the observation of discrete lasing peaks in the tight focusing condition. Despite the reduction in the effective Thouless number, it is still much larger than 1 because of weak scattering. As a result, the lasing modes are the extended states within the effective volume. Because $\sigma_\gamma/\langle\gamma\rangle \ll 1$, the minimum decay rate γ_{min} is still close to $\langle\gamma\rangle$, leading to relatively small fluctuations in the lasing threshold. The threshold gain can be estimated as $g_{th} = \gamma_{min} \approx \langle\gamma\rangle \sim D/L_{eff}^2$. This estimate gives a threshold equal to that predicted by Letokhov for random lasing with nonresonant feedback (Letokhov 1968), $l_{amp} \sim L_{eff}$ or $l_g \sim L_{eff}^2/l_t$. Using this approximation and taking into account the saturation of absorption of pump light, Burin et al. (2003a) reproduced the experimentally measured dependence of the lasing threshold pump intensity I_{th} on l_t and the pump area A_p, namely, $I_{th} \propto l_t^{1/2}/A_p^q$, where $0.5 \le q \le 1$.

Note that the above theory does not exclude the possibility of anomalously localized states in the sense that it does not eliminate the possibility of rare events. If they happen to be *within* the pumped volume, the anomalously localized states could serve as low-threshold lasing modes. However, even in the absence of such rare events, the nonuniform distribution of gain and absorption could result in spatial localization of lasing modes in the pumped region. In other words, local pumping in an absorbing medium creates a "trapping" site for lasing modes.

11.3.3.7 Effect of local gain on random lasing modes

Even without absorption, local pumping may also modify the lasing modes, especially in a weakly scattering system. Recently Wu et al. (2007) developed a numerical method based on the transfer matrix to calculate the quasimodes of 1D random systems, as well as the lasing modes with arbitrary spatial profile of pumping. The boundary condition is that there are only outgoing waves through the system boundary. In a passive system such a boundary condition gives the frequency and decay rate of every quasimode, while in an active system it determines the frequency and threshold gain of each lasing mode. This method is valid for linear gain up to the lasing threshold for which both gain saturation and mode competition for gain can be neglected. This simplification allows the identification of all *potential* lasing modes, regardless of the material-specific nonlinear gain.

The quasimodes, as well as the lasing modes, are formed by distributed feedback in a random system. The conventional distributed feedback (DFB) laser, made of periodic structures, operates either in the over-coupling regime or the under-coupling regime (Kogelnik and Shah 1971). The random laser, which can be considered as a random DFB laser, also has these two regimes of operation. In the under-coupling regime, the system size L is much less than the localization length

ξ, while in the over-coupling regime, $L > \xi$. In the under-coupling regime, the electric field of a quasimode grows exponentially toward the system boundaries, while in the over-coupling regime the field maxima are located inside the random system. The frequency spacing of adjacent modes is more regular in the under-coupling regime, and there is less fluctuation in the mode decay rate. The distinct characteristics of the quasimodes in these two regimes result from the different mechanisms of mode formation. In an over-coupling system, the quasimodes are formed mainly through the interference of multiply scattered waves by the particles in the interior of the random system. In contrast, feedback from the system boundaries becomes important in the formation of quasimodes in an under-coupling system. Contributions from the scatterers in the interior of the random system to the mode formation are relatively weak but not negligible. They induce small fluctuations in mode spacing and decay rate. As the scattering strength is increased, feedback from the scatterers in the interior of the system becomes stronger, and the frequency spacing of the quasimodes becomes more irregular.

When gain is uniformly distributed across a random system, the lasing modes (at threshold) have a one-to-one correspondence with the quasimodes in both the over-coupling and under-coupling regimes. However, the lasing modes may differ slightly from the corresponding quasimodes in frequency and spatial profile, especially in an under-coupled system. This is because the introduction of uniform gain removes the feedback caused by spatial inhomogeneity of the imaginary part of the wave vector within the random system and creates additional feedback by the discontinuity of the imaginary part of the wave vector at the system boundaries. As long as the scattering is not too weak, the quasimodes are only slightly modified by the introduction of uniform gain in a random system and serve as the lasing modes. Because of the correspondence between the lasing modes and the quasimodes, the frequency spacing of adjacent lasing modes is more regular in the under-coupled systems with smaller mode-to-mode variations in the lasing threshold.

When optical gain is introduced to a local region of a random system (with no absorption outside the gain region), some quasimodes fail to lase no matter how high the gain is. Other modes can lase but their spatial profiles may be significantly modified. Such modifications originate in the strong enhancement of feedback from the scatterers within the pumped region. This increases the weight of a lasing mode within the gain region. Moreover, the spatial variation in the imaginary part of the refractive index generates additional feedback for lasing. As the size of the gain region L_g decreases, the number of lasing modes N_l is reduced, and the frequency spacing of lasing modes is increased. The sublinear decrease of N_l with L_g indicates that feedback from the scatterers outside the pumped region is not negligible. In an under-coupled system, the regularity in the lasing mode spacing remains under local excitation. Note that the increase of mode concentration in the vicinity of a region in which local pumping leads to gain is due to a physical mechanism distinct from the absorption-induced localization of lasing modes in the pumped region. The former is based on selective enhancement of feedback within the gain region, while the latter is based on suppression of the feedback outside the pumped region by absorption.

11.3.3.8 The 1D photon localization laser

One problem of 3D random lasers is that multiple scattering of the pump light restricts the excitation to the proximity of a sample's surface. The emitted photons readily escape through the sample surface, giving a high lasing threshold. This problem is less serious in 2D random lasers, as the pump light incident from the third dimension does not experience scattering (Stassinopoulos et al. 2005). The pump light is not confined inside a 2D random system. Recently, Milner and Genack (2005) realized a photon localization laser in which the pump light was localized deep inside a 1D random structure.

Their 1D sample was a stack of partially reflecting glass slides of random thickness between 80 and 120 μm with interspersed dye films. The stack was illuminated at normal incidence by the second harmonics of a pulsed Nd:YAG laser. In this 1D structure, light is localized by multiple scattering from the parallel layers that returns the wave upon itself. The average transmittance through a stack of glass slides without intervening dye solution decayed exponentially with the number of slides. Even though the average transport of light was suppressed, resonant tunneling through localized states gave spectrally narrow transmission peaks. A large number of narrow peaks were observed in the transmission spectra. When the pump wavelength was tuned into one of the narrow transmission lines, the pump light penetrated deeply into the sample's interior via resonant excitation of a long-lived spatially localized mode. Energy absorbed from this mode was subsequently emitted into long-lived localized modes that fell within the dye emission spectrum. Stimulated emission was enhanced when the spatial energy distributions at both the excitation and emission wavelengths overlapped. The deposition of pump energy deep within the sample and its efficient coupling to long-lived emission modes removed a major barrier to achieving low-threshold lasing in the presence of disorder. The threshold was sufficiently low for lasing that it could be induced with quasi-continuous illumination by a 3-W argon-ion laser.

Another feature of the 1D photon localization laser is the large fluctuation in lasing power when the pump beam is focused onto different parts of the sample that correspond to different realizations of disorder. The pump transmission is strongly correlated with the output lasing power. High pump transmission results from resonant excitation of a localized mode that is spatially peaked near the center of the sample. Therefore, the pump energy is exponentially enhanced within the sample and is efficiently transferred to the gain medium. Since excitation in the center of the sample is likely to escape via emission into localized modes having a long lifetime, the opportunity for stimulated emission is enhanced, and the laser output is high.

In the localization regime, the sample length L is much larger than the localization length ξ. If the gain is uniform throughout the 1D structure, the first lasing mode is usually the most localized mode in the middle of the random structure. Let us consider such a mode located at $x_0 \sim L/2$. The random media on its left and right sides are characterized by the reflection coefficients (r_l, r_r) and transmission coefficients (t_l, t_r). The threshold gain g_{th} is related to the reflection coefficients (Burin et al. 2002):

$$|r_l r_r| \exp[g_{th}(d\Phi_l/d\omega + d\Phi_r/d\omega)/2] = 1, \tag{11.25}$$

where $r = |r|e^{i\Phi}$. The frequency dependence of $|r|$ can be neglected because $L \gg \xi$ and $1 - |r| \ll 1$. The factor in the exponent of Eq. (11.25) represents a product of photon amplification rate and the trapping time of photons inside the system $\tau_0 = d\Phi_l/d\omega + d\Phi_r/d\omega$. Since $|r_l|$ and $|r_r|$ are very close to 1, g_{th} can be expressed in terms of the transmission coefficients through a linear expansion leading to:

$$g_{th} \approx \frac{|t_l|^2 + |t_r|^2}{d\Phi_l/d\omega + d\Phi_r/d\omega}. \tag{11.26}$$

In the localization regime, $t_l \sim \exp(-x_0/\xi)$ and $t_r \sim \exp[-(L - x_0)/\xi]$. Hence, $g_{th} \sim \{\exp(-2x_0/\xi) + \exp[-2(L - x_0)/\xi]\}/\tau_0$. Since $x_0 \sim L/2$ and $g_{th} \propto \exp(-L/\xi)$, the lasing threshold depends exponentially on the system length L. The photons emitted inside the random system need an exponentially long time to escape from it, due to the exponentially small transmission in the localization regime. Therefore, an exponentially small gain is enough to initiate lasing oscillation.

11.3.4 Amplified spontaneous emission (ASE) spikes versus lasing peaks

One fundamental difference between a random laser with resonant feedback and a random laser with nonresonant feedback is that the lasing frequency of the former is determined by the spatial resonance of a random structure, and the latter by the maximum of a gain spectrum. The emergence of discrete narrow peaks in the emission spectrum, with frequencies dependent on the spatial distribution of refractive index, is a distinct feature of a random laser with resonant and coherent feedback. In addition to the lasing peaks, stochastic spikes are also observed in the single-shot spectra of ASE from colloidal solutions over a wide range of scattering strength (Mujumdar et al. 2004b). They are attributed to single spontaneous emission events that happen to take long open paths inside the amplifying random medium and pick up large gain. One distinction between the ASE spikes and lasing peaks is that even when the random structure is fixed the ASE spikes change completely from pulse to pulse, while the lasing peaks remain constant in frequency with only amplitude fluctuation. The next section describes how the spectral correlation and intensity statistics of random lasing peaks are very different from those of the ASE spikes, attributes related to their distinct physical mechanisms.

The experiment was performed on diethylene glycol solutions of stilbene 420 dye and TiO_2 microparticles (Wu and Cao 2007). The motion of particles in solution provided different random configurations for each pump pulse, allowing for an ensemble measurement under identical conditions. The dye molecules were excited by the third harmonic of a pulsed Nd:YAG laser that was focused into the solution. Stilbene 420 has well-separated absorption and emission bands, thus the absorption of emitted light outside the pumped region was negligible. At a particle density $\rho = 3 \times 10^9 \text{cm}^{-3}$, the scattering mean free path l_s was on the order of 1 mm, exceeding the penetration depth of the pump light. The excitation volume had a cone shape with a length of a few hundred microns and a base diameter of 30 µm. Because its length is smaller than l_s, the excitation cone is almost identical to that in the neat dye solution. On one hand, the transport of emitted light is diffusive in the entire colloidal solution with dimension much larger than l_s. On the other hand, light amplification occurred only in a pumped region of a size smaller than the mean free path.

The single-shot emission spectra from the colloidal solution are shown in Figs. 11.7(a)–(c) with increasing pump pulse energy E_p. At E_p = 0.05 µJ [Fig. 11.7(a)], the spectrum exhibits sharp spikes on top of a broad ASE band. The spikes change completely from shot to shot. The typical linewidth of the spikes is about 0.07 nm. The neighboring spikes often overlap partially. As pump power increases, the spikes grow in intensity. When E_p exceeds a threshold, a different type of peak emerges in the emission spectrum [Fig. 11.7(b)]. These peaks grow much faster with pumping than the spikes, and dominate the emission spectrum at $E_p = 0.13\,\mu\text{J}$ [Fig. 11.7(c)]. These peaks, with a typical width of 0.13 nm, are also notably broader than the spikes. Unlike the spikes, the spectral spacing of adjacent peaks is more or less regular. The above experiment was repeated with the neat dye solution of the same M [Fig. 11.7(d)–(f)]. Although they are similar at $E_p = 0.05\,\mu\text{J}$, the emission spectra with and without particles are dramatically different at $E_p = 0.13\,\mu\text{J}$. Under intense pumping, the emission spectrum of the neat dye solution has only random and closely spaced spikes but no strong and regularly spaced peaks [Fig. 11.7(f)]. The maximum spike intensity is about $50\times$ lower than the maximum peak intensity from the colloidal solution at the same pumping rate [Fig. 11.7(c)]. While the pump threshold for the appearance of peaks depends on ρ, the threshold for the emergence of spikes in solutions with low ρ is similar to that with $\rho = 0$.

The large peaks represent the lasing modes formed by distributed feedback in the colloidal solution. Although the feedback is weak at low ρ, the intense pumping strongly amplifies the backscattered light and greatly enhances the feedback. In contrast, feedback from particles is not necessary for the spikes that also exist in the neat dye solution. Thus, the spikes are attributed to amplified spontaneous emission.

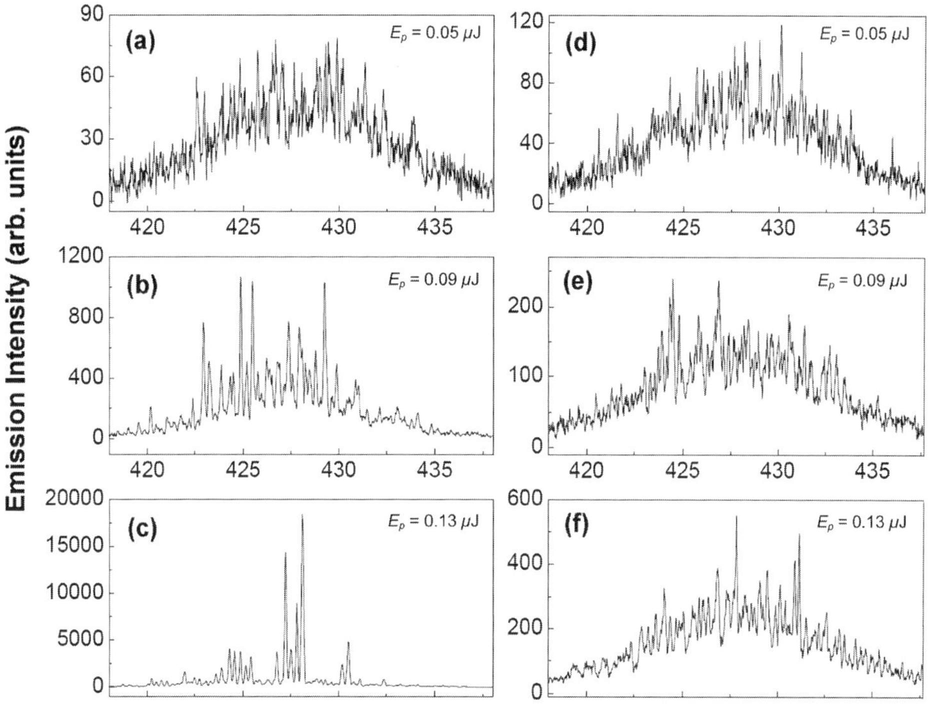

Figure 11.7 Single-shot spectra of emission from the 8.5-mM stilbene 420 dye solutions with [(a)–(c)] and without TiO_2 particles. The particle density $\rho = 3 \times 10^9 \text{cm}^{-3}$ in (a)–(c). The pump pulse energy $E_p = 0.05$ μJ for (a) & (d), 0.09 μJ for (b) & (e), and 0.13 μJ for (c) & (f). (Reprinted from Wu and Cao 2007.)

The ensemble-averaged spectral correlation function $C(\Delta\lambda) = \langle I(\lambda)I(\lambda + \Delta\lambda)\rangle / \langle I(\lambda)\rangle\langle I(\lambda + \Delta\lambda)\rangle - 1$ was obtained from 200 single-shot emission spectra over the wavelength range 425–431 nm within which the gain coefficient had only slight variation. For $\rho = 3 \times 10^9 \text{cm}^{-3}$, $C(\Delta\lambda)$ changed dramatically with pumping [Fig. 11.8(a)]. Below the lasing threshold, it started with a small value at $\Delta\lambda = 0$ and decayed quickly to zero as $\Delta\lambda$ increased. Above the lasing threshold, the amplitude of $C(\Delta\lambda)$ grew rapidly, and regular oscillations with $\Delta\lambda$ developed. The oscillation period was about 0.27 nm, corresponding to the average spacing of lasing peaks. Despite the change of lasing peaks from shot to shot, the oscillations survived the ensemble average. This result confirmed that the lasing peaks in a single-shot spectrum are more or less regularly spaced, and the average peak spacing was nearly the same for different shots. The periodicity of lasing peaks in similar random samples was also revealed by Polson and Vardeny (2005) using the power Fourier transform technique. $C(\Delta\lambda)$ for the ASE spikes at $\rho = 0$ barely changes with pumping [Fig. 11.8(b)]. It is similar to that of the colloidal solution below the lasing threshold where the spectrum has only ASE spikes. Although ASE spikes produce irregular oscillations in the spectral correlation function of in-

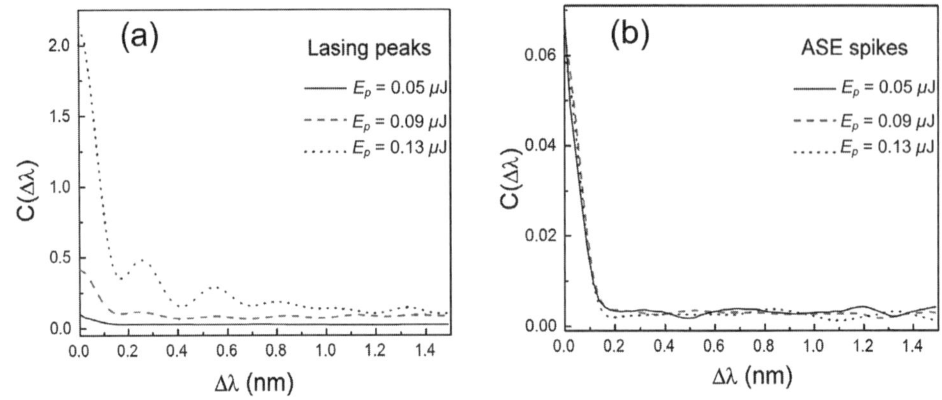

Figure 11.8 Ensemble-averaged spectral correlation function $C(\Delta\lambda)$ of single-shot emission spectra. $M = 8.5$ mM. $\rho = 3 \times 10^9\text{cm}^{-3}$ in (a) and 0 in (b). (Reprinted from Wu and Cao 2008.)

dividual single shot spectra, such oscillations are smeared out after averaging over many shots. This result reflects the stochastic nature of the ASE spikes.

The statistical distribution $P(\delta\lambda)$ of wavelength spacing $\delta\lambda$ between adjacent lasing peaks is distinct from that of ASE spikes. $P(\delta\lambda)$ for the ASE spikes decays exponentially as $\delta\lambda$ increases from zero. This suggests that the spectral spacing of ASE spikes satisfies Poisson statistics, which means the frequencies of individual ASE spikes are uncorrelated. $P(\delta\lambda)$ for the lasing peaks had a value close to zero at $\delta\lambda \sim 0$. It increased with $\delta\lambda$ and reached a maximum at $\delta\lambda = 0.27$ nm, which coincided with the average lasing peak spacing obtained from the oscillation period of $C(\Delta\lambda)$. This distribution reflects the spectral repulsion of lasing peaks.

In the study of the statistics of emission intensity, the average intensity $\langle I(\lambda)\rangle$ was first computed from 200 single-shot emission spectra, then the statistical distribution of the normalized emission intensity $I(\lambda)/\langle I(\lambda)\rangle$ was computed. In Fig. 11.9(a), the log-log plot of $P(I/\langle I\rangle)$ for $\rho = 3 \times 10^9\text{cm}^{-3}$ clearly reveals a power-law decay at large I above the lasing threshold. The solid lines represent the fit of $(I/\langle I\rangle)^{-b}$ to $P(I/\langle I\rangle)$, with $b = 3.3$ and 2.5 for $E_p = 0.09$ and 0.13 μJ, respectively. Since only the high lasing peaks contribute to the tail of $P(I/\langle I\rangle)$, the power-law decay reflects the intensity statistics of lasing peaks. Below the lasing threshold, $P(I/\langle I\rangle)$ is similar to that of the neat dye solution, which exhibits an exponential tail. As shown in the log-linear plot of Fig. 11.9(b), the exponential decay rate is almost the same for different pumping levels at $\rho = 0$. The solid line is an exponential fit $P(I/\langle I\rangle) \sim \exp(-a\, I/\langle I\rangle)$ with $a = 5.1$. In the absence of lasing peaks, the ASE spikes contribute to $P(I/\langle I\rangle)$ at large I. Therefore the exponential decay describes the intensity statistics of ASE spikes.

The above experimental results demonstrate the fundamental difference between ASE spikes and lasing peaks. The stochastic structures of the pulsed ASE

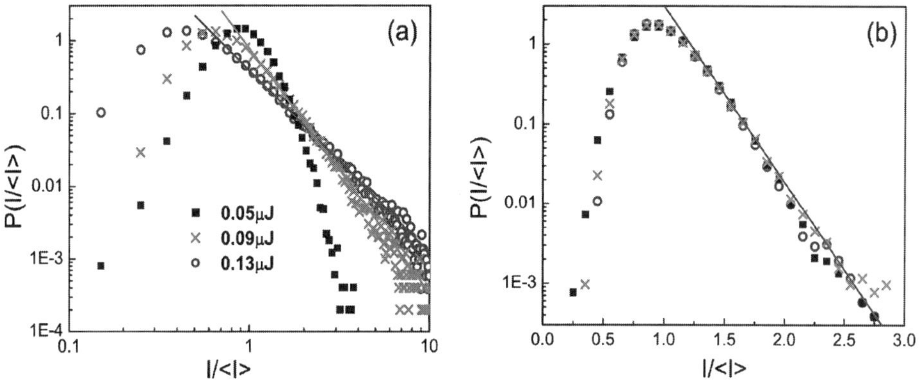

Figure 11.9 Statistical distributions of normalized emission intensities $I(\lambda)/\langle I(\lambda)\rangle$ for 425 nm $< \lambda <$ 431 nm. $\rho = 3 \times 10^9 \text{cm}^{-3}$ in (a) and 0 in (b). $E_p = 0.05$ μJ (squares), 0.09 μJ (crosses), and 0.13 μJ (circles). The solid lines represent fits to the data. In (a) $P(I/\langle I\rangle) = 0.77(I/\langle I\rangle)^{-3.3}$ (for crosses), and $P(I/\langle I\rangle) = 0.38(I/\langle I\rangle)^{-2.5}$ (for circles). In (b), $P(I/\langle I\rangle) = 467\exp(-5.1\, I/\langle I\rangle)$. (Reprinted from Wu and Cao 2007.)

spectra of neat dye solutions were observed long ago (Sperber et al. 1988). In the above experiment, the observed ASE spikes originate from photons spontaneously emitted near the excitation cone tip in the direction toward the cone base. As they propagate along the cone, these photons experience the largest amplification due to their longest path length inside the gain volume. The ASE at the frequencies of these photons is the strongest, leading to the spikes in the emission spectrum. Although the spontaneous emission time is a few nanoseconds, the 25-ps pump pulse creates the transient gain, and only the initial part of the spontaneous emission pulse is strongly amplified. Hence, the ASE pulse is a few tens of picoseconds long, followed by a spontaneous emission tail. The spectral width of the ASE spikes is determined by the ASE pulse duration. Since different ASE spikes originate from independent spontaneous emission events, their frequencies are uncorrelated. This leads to Poisson statistics for the frequency spacing of neighboring ASE spikes. Although the occurrence of ASE spikes does not rely on scattering, multiple scattering can elongate the path lengths of spontaneously emitted photons inside the gain volume and increase the amplitudes of some spikes. In the above experiment, however, the size of the gain volume was less than l_s, thus the effect of scattering on the ASE spikes was negligibly small.

In the colloidal solution of low ρ, the large aspect ratio of the excitation cone results in lasing along the cone, confirmed by the directionality of the lasing output. The large gain inside the cone greatly amplifies the feedback from the scatterers within the cone as compared to that from outside the cone. Thus, the lasing modes deviate from the quasimodes of the passive system. This explains why the statistical distribution of the lasing peak spacing does not satisfy the Wigner-Dyson distribution, which holds for the statistical distribution of quasimode spacing in a diffusive colloidal solution. It also indicates that the statistical distribution of decay rate of the quasimodes cannot be applied directly to the calculation of $P(I/\langle I\rangle)$ (van der Molen et al. 2006). Moreover, mode competition and gain saturation, as well as the initial spontaneous emission into individual modes, must be taken into account to reproduce $P(I/\langle I\rangle)$. The rapid variation of gain in time and space makes the calculation of intensity statistics more difficult. An extensive theoretical study is needed to quantitatively understand the statistical distribution of peak intensity in random lasing.

11.3.5 Recent developments

Over the past few years, random lasers with coherent feedback were realized in many material systems such as semiconductor nanostructrues (Cao et al. 1998; Mitra and Thareja 1999; Thareja and Mitra 2000; Sun et al. 2003; Yu et al. 2004a; Yu et al. 2004b; Leong et al. 2004; Hsu et al. 2005; Yuen et al. 2005a; Lau et al. 2005), organic films and nanofibers (Anni et al. 2003; Quochi et al. 2004; Quochi et al. 2005; Klein et al. 2005; Sharma et al. 2006), and hybrid organic-inorganic composites (Yokoyama and Mashiko 2003; Anglos et al. 2004; Song et al. 2005). Various schemes have been proposed to improve the performance of random lasers, e.g., application of external feedback to reduce the lasing threshold and control the output direction of laser emission (Cao et al. 1999b), optimal tuning of random lasing modes through collective particle resonances (Ripoll et al. 2004), coupled-cavity ZnO thin-film random lasers for high-power one-mode operation (Yu and Leong 2004), one-mirror random lasers for quasi-continuous operation (Feng and Ueda 2003; Feng et al. 2004), and waveguide random lasers for directional output (Yuen et al. 2005b; Watanabe et al. 2005). The progress is so rapid that it is impossible to detail all of the advances. Only a few examples are mentioned briefly here.

Partially ordered random laser. One way of reducing the random laser threshold is to incorporate some degree of order into an active random medium (Chang et al. 2003; Yamilov and Cao 2004; Burin et al. 2004). Shkunov et al. (2001) have observed both photonic lasing and random lasing in dye-infiltrated opals. However, random lasing has a higher threshold than photonic lasing. We can numerically simulate lasing in a random system with a variable degree of order. When disorder is introduced to a perfectly ordered system, the lasing threshold is reduced. At a certain degree of disorder, the lasing threshold reaches a minimum. Then it rises with a further increase of disorder. Therefore, there exists an optimum degree of order for a minimum lasing threshold. We map out the transition from full order to complete disorder, and identify five scaling regimes for the mean lasing threshold versus the system size L. For increasing degree of disorder, the five regimes are (a) photonic band-edge $1/L^3$, (b) transitional super-exponential, (c) bandgap-related exponential, (d) diffusive $1/L^2$, and (e) disorder-induced exponential. Experimentally, disordered photonic crystal lasers have been fabricated (Wu et al. 2004b). The most efficient lasing modes are localized defect states near the edge of a photonic band gap. Such defect states are formed by structural disorder in a 2D triangular lattice. Another advantage of the partially ordered random laser is efficient pumping. For example, in a 1D random stack of resonant dielectric layers, the pump wavelength can be tuned to a pass band while the emission wavelength stays in a stop band (Feng and Ueda 2004). Then the pump light penetrates into the sample, while the emission is confined inside the system. As a result, the lasing threshold can be significantly reduced.

Mode interaction. The interaction of lasing modes in a random medium is interesting but complicated. Gain competition may lead to mode repulsion in real space for a homogeneously broadened gain spectrum or in the frequency domain for an inhomogeneously broadened gain spectrum (Cao et al. 2003b; Jiang et al. 2004). In a diffusive random medium, the finesse, defined as the ratio of the quasimode spacing to the mode width, is much less than unity. Strong modal interactions throughout the gain medium lead to a uniform spacing of lasing peaks (Türeci et al. 2006; Türeci et al. 2007; Türeci et al. 2008). In addition, the inhomogeneity of dielectric constant $\epsilon(\mathbf{r})$ modifies the ortho-normalization condition for the quasimodes and introduces a linear coupling between the quasimodes mediated by the polarization of the gain medium (Deych 2005). Finally, the overlapping quasimodes may couple via an external bath that generates excess noise and broadens the lasing linewidth (Patra et al. 2000; Frahm et al. 2000; Schomerus et al. 2000). The openness of random laser cavities strongly affects mode interaction, which determines the number of lasing modes and their spacing statistics (Hackenbroich 2005; Zaitsev 2006; Zaitsev 2007).

Nonlinear random laser. Random lasers offer an opportunity to study the interplay between nonlinearity and localization. Nonlinearity is strong in a random laser because the nonlinear coefficient is resonantly enhanced at the lasing frequency; the light intensity is high due to spatial confinement in random media. Noginov et al. (1998b) demonstrated second-harmonic generation in a mixture of powders of laser and frequency-doubling materials. Our recent study on the dynamic nonlinear effect in a random laser illustrates that the third-order nonlinearity not only changes the frequency and size of the lasing modes, but also modifies the laser emission intensity and laser pulse width (Liu et al. 2003). How nonlinearity affects the random lasing process depends on the speed of the nonlinear response. We find two regimes that are differentiated by the relative values of two time scales; one is the nonlinear response time, the other is the lifetime of the lasing state. For slow nonlinear response, collective scattering of many particles determines the buildup of a lasing mode. Nonlinearity changes the lasing output through modification of the spatial size of the lasing mode. However, when the nonlinear response is faster than the buildup of a lasing mode, the lasing mode cannot respond fast enough to the nonlinear refractive index change. Rapid change of the phase of scattered waves undermines the interference effect of multiple scattering. Instead, the nonlinear effect of single-particle scattering becomes dominant. Strong nonlinearity could lead to temporal instability. One application of optical nonlinearity is up-conversion lasing in random media via two-photon or multiphoton pumping (Zacharakis et al. 2002; Fujiwara and Sasaki 2004). The small two-photon/multiphoton absorption coefficient and weak scattering at long pumping wavelength allow the pump light to penetrate deeply into a 3D random medium and excite resonant modes far from the surface. These have a better spatial confinement and a lower lasing threshold (Burin et al. 2003b).

The latest theoretical and experimental studies provide insight into the physical mechanisms for lasing in random media (Patra 2003a; Florescu and John 2004c; Kretschmann and Maradudin 2004; Noginov et al. 2004a; Noginov et al. 2004b; Noginov et al. 2004c; Polson and Vardeny 2004; Mujumdar et al. 2004a; Li et al. 2005; Lubatsch et al. 2005; Vasa et al. 2005). However, our understanding of random lasers is far from complete. New ideas and surprises arise frequently, keeping up the momentum of random laser study. For example, Rand and coworkers investigated the electrical generation of stationary light (an evanescent wave) in ultrafine laser crystal powder (Redmond et al. 2004). Dice et al. (2005) reported the surface-plasmon-enhanced random laser emission from a suspension of silver nanoparticles in a laser dye.

11.4 Potential Applications of Random Lasers

A random laser is a nonconventional laser whose feedback mechanism is based on light scattering, as opposed to mirror reflection in a conventional laser. This alternative feedback mechanism has important applications in the fabrication of lasers in the spectral regimes where efficient reflective elements are not available, e.g., x-ray laser, γ-ray laser. Furthermore, the low fabrication cost, sample-specific wavelength of operation, small size, flexible shape, and substrate compatibility of random lasers lead to many potential applications (Lawandy 1994; Wiersma 2000; Rand 2003; Cao 2005).

One application is document encoding and material labeling, as the lasing frequencies represent the "signature" of a random structure. The microrandom laser described in the last section may be used as an optical tag in biological and medical studies. When nanoparticle clusters are attached to biological targets, the positions of the targets can be traced by detecting the lasing emission from the clusters. Each nanoparticle cluster has its unique set of lasing frequencies, which allows us to differentiate the targets.

The multidirectional output of a random laser makes it suitable for display applications. For example, a thin layer of a random medium doped with emitters can be coated on an arbitrarily shaped display panel. Compared to light-emitting diodes that are often used for display, random lasers have a much shorter turn-on/off time and would be useful for high-speed displays. A multicolor display may be made by incorporating emitters of different frequencies into a single random medium. The shape flexibility and substrate compatibility of random lasers allow them to be used for machine vision of manufactured parts in assembly lines, search and rescue of downed aircraft or satellites, etc.

So far most random lasers are pumped optically. Some applications such as flat-panel, automotive, and cockpit displays require electrical pumping. Recently electrically pumped continuous-wave laser action was reported in rare-earth-metal-doped dielectric nanophosphors (Williams et al. 2001; Li et al. 2002). Electrical pumping is much more efficient than optical pumping, because in the latter case, most of the pump light is scattered instead of being absorbed by the random medium.

In the medical arena, random lasers may be used for tumor detection and photodynamic therapy. Polson and Vardeny (2000) have shown that human tissue has strong scattering and can support random lasing when infiltrated with a concentrated laser dye solution. Since cancerous cells grow faster than normal cells, they generate more fragments or waste. The additional disorder in the malignant tissue leads to stronger scattering and more efficient lasing. One could imagine a novel method of probing tumors by scanning a focused laser beam across the human tissue. The laser light pumps the infiltrated dye molecules in a local region. When the pumping is not very strong, lasing can occur only at the location of a tumor where scattering is stronger. Since lasing may happen in a region much smaller than 1 mm, it is possible to detect a tumor at a very early stage of its development.

Random lasers could also serve as active elements in photonic devices and circuits. For example, a microrandom laser may play a crucial role as a miniature light source in a photonic crystal. The temperature-tunable random laser is expected to find applications in photonics, temperature-sensitive displays and screens, and remote temperature sensing. Finally, the study of random lasers could help us understand galactic masers and interstellar lasers in which feedback is also caused by scattering (Letokhov 1972; Letokhov 1996).

Acknowledgments

I wish to thank my coworkers on the study of random lasers: J. Y. Xu, Y. Ling, X. Wu, Y. G. Zhao, and Prof. Prem Kumar for the experimental work on random lasers; Drs. A. Yamilov, A. L. Burin, B. Liu, S.-H. Chang, Profs. S. T. Ho, A. Taflove, M. A. Ratner, and G. C. Schatz for the theoretical investigations of random lasers. Prof. R. P. H. Chang and his students E. W. Seelig and X. Liu fabricated ZnO nanorods and nanoparticles. We enjoyed a fruitful collaboration with Prof. C. M. Soukoulis and Dr. Xunya Jiang on the simulation of random lasers. Stimulating discussions are acknowledged with Drs. A. A. Asatryan, A. A. Chabanov, Ch. M. Brikina, L. Deych, M. Giudici, B. Grémaud, F. Haake, G. Hackenbroich, A. Z. Genack, S. John, R. Kaiser, T. Kottos, V. M. Letokhov, C. Miniatura, M. A. Noginov, M. Patra, M. E. Raikh, S. C. Rand, P. Sebbah, B. Shapiro, C. M. de

Sterke, A. D. Stone, J. Tredicce, H. E. Türeci, C. Vanneste, Z. V. Vardeny, C. Viviescas, T. Wellens, and D. S. Wiersma. Our research program is partly sponsored by the National Science Foundation through the grants ECS-9877113, DMR-0093949, and ECS-0244457, and by the David and Lucille Packard Foundation, the Alfred P. Sloan Foundation, and the Northwestern University Materials Research Center.

References

Altshuler, B. L., Kravtsov, V. E., and Lerner, I. V., 1991. "Distribution of mesoscopic fluctuations and relaxation properties in disordered conductors," in *Mesoscopic Phenomena in Solids,* B. L. Altshuler, P. A. Lee, and R. A. Wcbb, Eds., North-Holland, Amsterdam, 449–521.

Anglos D., Stassinopoulos A., Das, R. N., Zacharakis, G., Psyllaki, M., Jakubiak, R., Vaia, R. A., Giannelis, E. P., Anastasiadis, S. H., 2004. "Random laser action in organic–inorganic nanocomposites," *J. Opt. Soc. Am. B* **21**, 208.

Anni, M., Lattante, S., Cingolani, R., Gigli, G., Barbarella, G., and Favaretto, L., 2003. "Far-field emission and feedback origin of random lasing in oligothiophene dioxide neat films," *Appl. Phys. Lett.* **83**, 2754.

Anni, M., Lattante, S., Stomeo, T., Cingolani, R., Gigli, G., Barbarella, G., and Favaretto, L., 2004. "Modes interaction and light transport in bidimensional organic random lasers in the weak scattering limit," *Phys. Rev. B* **70**, 195216.

Ambartsumyan, R. V., Basov, N. G., Kryukov, P. G., and Letokhov, V. S., 1966. "A laser with a non-resonant feedback," *IEEE J. Quant. Electron.* **QE-2**, 442.

Ambartsumyan, R. V., Kryukov, P. G., and Letokhov, V. S., 1967a. "Dynamics of emission line narrowing for a laser with nonresonant feedback," *Sov. Phys. JETP* **24**, 1129.

Ambartsumyan, R. V., Basov, N. G., Kryukov, P. G., and Letokhov, V. S., 1967b. "A laser with nonresonant feedback," *Sov. Phys. JETP* **24**, 481.

Ambartsumyan, R. V., Kryukov, P. G., Letokhov, V. S., and Matveets, Yu. A., 1967c. "Emission statistics of a laser with nonresonant feedback," *JETP Lett.* **5**, 378.

Ambartsumyan, R. V., Basov, N. G., and Letokhov, V. S., 1968. "Statistical emission properties of a nonresonant feedback laser," *Sov. Phys. JETP* **26**, 1109.

Ambartsumyan, R. V., Bazhulin, S. P., Basov, N. G., and Letokhov, V. S., 1970. "Emission spectrum of an He-Xe laser with nonresonant feedback," *Sov. Phys. JETP* **31**, 234.

Apalkov, V. M., Raikh, M. E., and Shapiro, B., 2002. "Random resonators and prelocalized modes in disordered dielectric films," *Phys. Rev. Lett.* **89**, 016802.

Apalkov, V. M., Raikh, M. E., and Shapiro, B., 2003. In *The Anderson Transition and its Ramifications: Localization, Quantum Interference, and Interactions.* Vol. 630 of *Lecture Notes in Physics*, T. Brandes and S. Ketermann, Eds., Springer-Verlag, Berlin, 119.

Apalkov, V. M., Raikh, M. E., and Shapiro, B., 2004a. "Almost localized photon modes in continuous and discrete models of disordered media," *J. Opt. Soc. Am. B* **21**(1), 132.

Apalkov, V. M., Raikh, M. E., and Shapiro, B., 2004b. "Anomalously localized states in the Anderson Model," *Phys. Rev. Lett.* **92**, 066601.

Apalkov, V. M. and Raikh, M. E., 2005. "Universal fluctuations of the random lasing threshold in a sample of a finite area," *Phys. Rev. B* **71**, 054203.

Auzel, F. and Goldner, P., 2000. "Coherent light sources with powder: stimulated amplification versus super-radiance," *J. Alloy. Compd.* **300**, 11.

Bahoura, M., Morris, K. J., Zhu, G., and Noginov, M. A., 2005. "Dependence of the neodymium random laser threshold on the diameter of the pumped spot," *IEEE J. Quantum Electron.* **41**, 677.

Balachandran, R. M. and Lawandy, N. M., 1995. "Interface reflection effects in photonic paint," *Opt. Lett.* **20**(11), 1271.

Balachandran, R. M., Perkins, A. E., and Lawandy, N. M., 1996a. "Understanding bichromatic emission from scattering gain media," *Opt. Lett.* **21**, 1603.

Balachandran, R. M., Perkins, A. E., and Lawandy, N. M., 1996b. "Injection locking of photonic paint," *Opt. Lett.* **21**, 650.

Balachandran, R. M. and Lawandy, N. M., 1997. "Theory of laser action in scattering gain media," *Opt. Lett.* **22**, 319.

Beckering, G., Zilker, S. J., and Haarer, D., 1997. "Spectral measurements of the emission from highly scattering gain media," *Opt. Lett.* **22**, 1427.

Beenakker, C. W. J., 1998. "Thermal radiation and amplified spontaneous emission from a random medium," *Phys. Rev. Lett.* **81**, 1829.

Beenakker, C. W. J., 1999. "Photon statistics of a random laser," in *Diffusive Waves in Complex Media* J.-P. Fouquet, Ed., Vol. 531 of *NATO Advanced Studies Institute, Series C: Mathematical and Physical Sciences,* Kluwer, Dordrecht, 137–164.

Berger, G. A., Kempe, M., and Genack, A. Z., 1997. "Dynamics of stimulated emission from random media," *Phys. Rev. E* **56**, 6118.

Briskina, Ch. M., Markushev, V. M., and Ter-Gabrielyan, N. E., 1996. "Use of a model of coupled microcavities in the interpretation of experiments on powder lasers," *Quant. Electron.* **26**, 923.

Briskina, Ch. M. and Li, L. F., 2002. "Analysis of plausible mechanisms of lasing of powdered zinc oxide," *Laser Physics* **12**, 724.

Burin, A. L., Ratner, M. A., Cao, H., and Chang R. P. H., 2001. "Model for a random laser," *Phys. Rev. Lett.* **87**, 215503.

Burin, A. L., Ratner, A. M., Cao, H., and Chang, S.-H., 2002. "Random laser in one dimension," *Phys. Rev. Lett.* **88**, 093904.

Burin, A. L., Cao, H., and Ratner, M. A., 2003a. "Understanding and control of random lasing," *Physica B* **338**, 212.

Burin, A. L., Cao, H., and Ratner, M. A., 2003b. "Two-photon pumping of a random laser," *IEEE J. Select. Top. Quant. Electron.* **9**, 124.

Burin, A. L., Cao, H., Schatz, G. C., and Ratner, M. A., 2004. "High-quality optical modes in low-dimensional arrays of nanoparticles: application to random lasers," *J. Opt. Soc. Am. B* **21**(1), 121.

Cao, H., Zhao, Y. G., Ong, H. C., Ho, S. T., Dai, J. Y., Wu, J. Y., and Chang, R. P. H., 1998. "Ultraviolet lasing in resonators formed by scattering in semiconductor polycrystalline films," *Appl. Phys. Lett.* **73**, 3656.

Cao, H., Zhao, Y. G., Ong, H. C., Ho, S. T., Seelig, E. W., Wang, Q. H., and Chang, R. P. H., 1999a. "Random laser action in semiconductor powder," *Phys. Rev. Lett.* **82**, 2278.

Cao, H., Zhao, Y. G., Liu, X., Seelig, E., W., and Chang, R. P. H., 1999b. "Effect of external feedback on lasing in random media," *Appl. Phys. Lett.* **75**, 1213.

Cao, H., Xu, J. Y., Chang, S.-H., and Ho, S. T., 2000a. "Transition from amplified spontaneous emission to laser action in strongly scattering media," *Phys. Rev. E* **61**, 1985.

Cao, H., Xu, J. Y., Chang, S.-H., Ho, S. T., Seelig, E. W., Liu, X., and Chang, R. P. H., 2000b. "Spatial confinement of laser light in active random media," *Phys. Rev. Lett.* **84**, 5584.

Cao, H., Xu J. Y., Seelig, E. W., and Chang, R. P. H., 2000c. "Microlaser made of disordered media," *Appl. Phys. Lett.* **76**, 2997.

Cao, H., Ling, Y., Xu, J. Y., Cao, C. Q., and Kumar, P., 2001. "Photon statistics of random lasers with resonant feedback," *Phys. Rev. Lett.* **86**, 4524.

Cao, H., Ling Y., Xu, J. Y., and Burin, A. L., 2002. "Probing localized states with spectrally resolved speckle techniques," *Phys. Rev. E* **66**, 025601.

Cao, H., 2003. "Lasing in random media," *Waves Random Media* **13**, R1.

Cao, H., 2005. "Random lasers: development, features and applications," *Opt. Photon. News* **16**, 24.

Cao, H., Xu J. Y., Ling Y., Burin, A. L., Seelig, E. W., Liu, X., and Chang, R. P. H., 2003a. "Random lasers with coherent feedback," *IEEE J. Select. Top. Quant. Electron.* **9**, 111.

Cao, H., Jiang, X., Ling, Y., Xu J. Y., and Soukoulis, C. M., 2003b. "Mode repulsion and mode coupling in random lasers," *Phys. Rev. B* **67**, 161101.

Chabanov, A. A., Zhang, Z. Q., and Genack, A. Z., 2003. "Breakdown of diffusion in dynamics of extended waves in mesoscopic media," *Phys. Rev. Lett.* **90**, 203903.

Chang, S.-H., Cao, H., and Ho, S. T., 2003. "Cavity formation and light propagation in partially ordered and completely random one-dimensional systems," *IEEE J. Quantum Electron.* **39**, 364.

de Oliveira, P. C., McGreevy, J. A., and Lawandy, N. M., 1997. "External-feedback effects in high-gain scattering media," *Opt. Lett.* **22**(12), 895.

Deych, L. I., 2005. "Effects of spatial nonuniformity on laser dynamics," *Phys. Rev. Lett.* **95**, 043902.

Dice, G. D., Mujumdar, S., and Elezzabi, A. Y., 2005. "Plasmonically enhanced diffusive and subdiffusive metal nanoparticle-dye random laser," *App. Phys. Lett.* **86**, 131105.

Feng, Y. and Ueda, K., 2003. "One-mirror random laser," *Phys. Rev. A* **68**, 025803.

Feng Y. and Ueda, K., 2004. "Random stack of resonant dielectric layers as a laser system," *Opt. Express* **12**(15), 3307.

Feng, Y., Bisson, J.-F., Lu, J., Huang, S., Takaichi, K., Shirakawa, A., Musha, M., and Ueda K., 2004. "Thermal effects in quasi-continuous-wave Nd^{3+}:$Y_3Al_5O_{12}$ nanocrystalline-powder random laser," *Appl. Phys. Lett.* **84**, 1040–1042.

Florescu, L. and John, S., 2004a. "Photon statistics and coherence in light emission from a random laser," *Phys. Rev. Lett.* **93**, 013602.

Florescu, L. and John, S., 2004b. "Theory of photon statistics and optical coherence in a multiple-scattering random-laser medium," *Phys. Rev. E* **69**, 046603.

Florescu, L. and John, S., 2004c. "Lasing in a random amplifying medium: Spatiotemporal characteristics and nonadiabatic atomic dynamics," *Phys. Rev. E* **70**, 036607.

Frahm, K. M., Schomerus, H., Patra, M., and Beenakker, C. W. J., 2000. "Large Petermann factor in chaotic cavities with many scattering channels," *Europhys. Lett.* **49**, 48.

Frolov, S. V., Vardeny, Z. V., Yoshino, K., Zakhidov, A., and Baughman, R. H., 1999a. "Stimulated emission in high-gain organic media," *Phys. Rev. B* **59**, R5284. http://link.aps.org/doi/10.1103/PhysRevB.59.R5284

Frolov, S. V., Vardeny, Z. V., Yoshino, K., Zakhidov, A., and Baughman, R. H., 1999b. "Laser-like emission in opal photonic crystals," *Opt. Commun.* **162**, 241.

Fujiwara, H. and Sasaki, K., 2004. "Observation of upconversion lasing within a thulium-ion-doped glass powder film containing titanium dioxide particles," *Jpn. J. App. Phys.* **43**, L1337.

Genack, A. Z. and Drake, J. M., 1994. "Scattering for super-radiation," *Nature* **368**, 400.

Gottardo, S., Cavalieri, S., Yaroshchuk, O., and Wiersma, D. S., 2004. "Quasi-two-dimensional diffusive random laser action," *Phys. Rev. Lett.* **93**, 263901.

Gouedard, C., Husson, D., Sauteret, C., Auzel, F., and Migus, A., 1993. "Generation of spatially incoherent short pulses in laser-pumped neodymium stoichiometric crystals and powders," *J. Opt. Soc. Am. B* **10**(12), 2358.

Hackenbroich, G., Viviescas, C., and Haake, F., 2002. "Field quantization for chaotic resonators with overlapping modes," *Phys. Rev. Lett.* **89**, 083902.

Hackenbroich, G., Viviescas, C., and Haake, F., 2003. "Quantum statistics of overlapping modes in open resonators," *Phys. Rev. A* **68**, 063805.

Hackenbroich, G., 2005. "Statistical theory of multimode random lasers," *J. Phys. A: Math. Gen.* **38**, 10537.

Herrmann, J. and Wilhelmi, B., 1998. "Mirrorless laser action by randomly distributed feedback in amplifying disordered media with scattering centers," *Appl. Phys. B* **66**, 305.

Hsu, H.-C., Wu, C.-Y., and Hsieh, W.-F., 2005. "Stimulated emission and lasing of random-growth oriented ZnO nanowires," *J. Appl. Phys.* **97**, 064315.

Jiang, X. and Soukoulis, C. M., 2000. "Time dependent theory for random lasers," *Phys. Rev. Lett.* **85**, 70.

Jiang, X. and Soukoulis, C. M., 2002. "Localized random lasing modes and a path for observing localization," *Phys. Rev. E* **65**, 025601(R).

Jiang, X., Song, F., Soukoulis, C. M., Zi, J., Joannopoulos, J. D., and Cao, H., 2004. "Coupling, competition, and stability of modes in random lasers," *Phys. Rev. B* **69**, 104202.

John, S., 1991. "Localization of light," *Phys. Today* **44**(5), 32.

John, S. and Pang, G., 1996. "Theory of lasing in a multiple-scattering medium," *Phys. Rev. A* **54**, 3642.

Klein, S., Crégut, O., Gindre, D., Boeglin, A., and Dorkenoo, K. D., 2005. "Random laser action in organic film during the photopolymerization process," *Opt. Express* **13**(14), 5387.

Kogelnik, H. and Shank, C. V., 1972. "Coupled-wave theory of distributed feedback lasers," *J. Appl. Phys.* **43**, 2327.

Kretschmann, M. and Maradudin, A. A., 2004. "Lasing action in waveguide systems and the influence of rough walls," *J. Opt. Soc. Am. B* **21**(1), 150.

Lau, S. P., Yang, H. Y., Yu, S. F., Li, H. D., Tanermura, M., Okita, T., Hatano, H., and Hng, H. H., 2005. "Laser action in ZnO nanoneedles selectively grown on silicon and plastic substrates," *Appl. Phys. Lett.* **87**, 013104.

Lawandy N. M., Balachandran, R. M., Gomes, A. S. L., and Sauvain, E., 1994. "Laser action in strongly scattering media," *Nature* **368**, 436. http://www.nature.com/nature/index.html

Lawandy, N. M., 1994. "Paint-on lasers light up the way to new technologies," *Photon. Spectra* **28**(7), 119.

Lawandy, N. M. and Balachandran, R. M., 1995. "Random laser, Lawandy and Balachandran reply," *Nature* **373**, 204.

Lee, K. and Lawandy, N. M., 2002. "Laser action in temperature-controlled scattering media," *Opt. Commun.* **203**, 169.

Leong, E., Chong, M. K., Yu, S. F., and Pita, K., 2004. "Sol-gel ZnO-SiO_2 composite waveguide ultraviolet lasers," *IEEE Photon. Tech. Lett.* **16**(11), 2418.

Letokhov, V. S., 1968. "Quantum statistics of multi-mode radiation from an ensemble of atoms," *Sov. Phys. JETP* **26**, 1246.

Letokhov, V. S., 1972. "Laser action in stellar atmospheres," *IEEE J. Quant. Electron.* **QE-8**(8), 615.

Letokhov, V. S., 1996. "Noncoherent feedback in space masers and stellar lasers," In *Amazing Light: A Volume Dedicated to Charles Hard Townes on his 80th birthday.* R. Y. Chiao, Ed., Berlin: Springer. 409.

Li B., Williams, G., Hinklin, T., and Laine, R. M., 2002. "Continuous-wave ultraviolet laser action in strongly scattering Nd-doped alumina," *Opt. Lett.* **27**(6), 394.

Li, Q., Ho, K. M. and Soukoulis, C. M., 2001. "Mode distribution in coherently amplifying random media," *Physica B* **296**, 78.

Li, S., Wang, Z.-J., Chen, L.-S., Sun, X., and George, T. F., 2005. "Collective behavior and disorder-induced resonator of random lasers," *Appl. Phys. Lett.* **86**, 171109.

Lichmanov, A. A., Briskina, Ch. M., Markushev, V. M., Lichmanova, V. N., and Soshchin, N. P., 1998. "Degree of coherence and dimensions of the generation region of powder lasers," *J. Appl. Spectroscopy* **65**(6), 818.

Ling, Y., Cao, H., Burin, A. L., Ratner, M. A., Liu, X., and Chang, R. P. H., 2001. "Investigation of random lasers with resonant feedback," *Phys. Rev. A* **64**, 063808.

Liu, B., Yamilov, A., Ling, Y., Xu, J. Y., and Cao, H., 2003. "Dynamic nonlinear effect on lasing in a random medium," *Phys. Rev. Lett.* **91**, 063903.

Liu, X., Wu, X., Cao, H., and Chang, R. P. H., 2004. "Growth mechanism and properties of ZnO nanorods synthesized by plasma-enhanced chemical vapor deposition," *J. Appl. Phys.* **95**, 3141.

Lubatsch, A., Kroha, J., and Busch, K., 2005. "Theory of light diffusion in disordered media with linear absorption or gain," *Phys. Rev. B* **71**, 184201.

Mandel, L. and Wolf, E. 1995. *Optical Coherence and Quantum Optics.* Cambridge University Press, Cambridge.

Markushev, V. M., Zolin, V. F., and Briskina, Ch. M., 1986. "Luminescence and stimulated emission of neodymium in sodium lanthanum molybdate powders," *Sov. J. Quantum Electron.* **16**, 281.

Markushev, V. M., Ter-Gabrielyan, N. E., Briskina, Ch. M., Belan, V. R., and Zolin, V. F., 1990. "Stimulated emission kinetics of neodymium powder lasers," *Sov. J. Quantum Electron.* **20**, 773.

Markushev, V. M., Ryzhkov, V. M., Briskina, Ch. M., Cao, H., Zadorozhnayac, L. A., Li, L. E., Gevargizovc, E. I., and Demianets, L. N. 2005. "UV radiation of powdered ZnO pumped by nanosecond pulses," *Laser Physics* **15**, 1611.

Martorell, J., Balachandran, R. M., and Lawandy, N. M., 1996. "Radiative coupling between photonic paint layers," *Opt. Lett.* **21**(4), 239.

Milner, V. and Genack, A. Z., 2005. "Photon localization laser: low-threshold lasing in a random amplifying layered medium via wave localization," *Phys. Rev. Lett.* **94**, 073901.

Mirlin, A. D., 2000. "Statistics of energy levels and eigenfunctions in disordered systems," *Phys. Rep.* **326**, 259.

Mishchenko, E. G., Patra, M., and Beenakker C. W. J., 2001. "Frequency dependence of the photonic noise spectrum in an absorbing or amplifying diffusive medium," *Euro. Phys. J. D* **13**, 289.

Misirpashaev, T. Sh. and Beenakker, C. W. J., 1998. "Lasing threshold and mode competition in chaotic cavities," *Phys. Rev. A* **57**, 2041.

Mitra, A. and Thareja, R. K., 1999. "Ultraviolet laser emission from optically pumped ZnO," *Mod. Phys. Lett. B* **13**(29/30), 1075.

Mujumdar, S., Cavalieri, S., and Wiersma, D. S., 2004a. "Temperature-tunable random lasing: numerical calculations and experiments," *J. Opt. Soc. Am. B* **21**(1), 201.

Mujumdar, S., Ricci, M., Torre, R., and Wiersma, D. S., 2004b. "Amplified extended modes in random lasers," *Phys. Rev. Lett.* **93**, 053903.

Noginov, M. A., Egarievwe, S. U., Noginova, N. E., Wang, J. C., and Caulfield, H. J., 1998. "Demonstration of a second-harmonic powder laser," *J. Opt. Soc. Am. B* **15**(12), 2854.

Noginov, M. A., Caulfield, H. J., Noginova, N. E., and Venkateswarlu, P., 1995. "Line narrowing in the dye solution with scattering centers," *Opt. Commun.* **118**, 430.

Noginov, M. A., Noginova, N. E., Egarievwe, S. U., Caulfield, H. J., Venkateswarlu, P., Thompson T., Mahdi M., and Ostroumov, V., 1996. "Short-pulsed stimulated emission in the powders of $NdAl_3(BO_3)_4$, $NdSc_3(BO_3)_4$, and Nd:$Sr_5(PO_4)_3F$ laser crystals," *J. Opt. Soc. Am. B* **13**(9), 2024.

Noginov, M. A., Noginova, N. E., Egarievwe, S. U., Caulfield, H. J., Venkateswarlu, P., Williams, A., and Mirov, S. B., 1997. "Color-center powder laser: the effect of pulverization on color-center characteristics," *J. Opt. Soc. Am. B* **14**(8), 2153.

Noginov, M. A., Noginova, N. E., Egarievwe, S. U., Caulfield, H. J., Cochrane, C., Wang, J. C., Kokta, M. R., and Paitz J., 1998a. "Study of the pumping regimes in Ti-sapphire and $Nd_{0.5}La_{0.5}Al_3(BO_3)_4$ powders," *Opt. Mater.* **10**(4), 297.

Noginov, M. A., Egarievwe, S. U., Noginova, N. E., Wang, J. C. and Caulfield, H. J., 1998b. "Demonstration of a second-harmonic powder laser," *J. Opt. Soc. Am. B* **15**(12), 2854.

Noginov, M. A., Egarievwe S. U., Noginova, N. E., Caulfield H. J., and Wang, J. C., 1999. "Interferometric studies in a powder laser," *Opt. Mater.* **12**(1), 127.

Noginov, M. A., Fowlkes, I. N., Zhu, G., Novak J., 2004a. "Random laser thresholds in cw and pulsed regimes," *Phys. Rev. A* **70**, 043811.

Noginov, M. A., Novak, J., and Williams, S., 2004b. "Modeling of photon density dynamics in random lasers," *Phys. Rev. A* **70**, 063810.

Noginov, M. A., Zhu, G., Fowlkes, I., and Bahoura, M., 2004c. "GaAs random laser," *Laser Phys. Lett.* **1**, 291.

Noginov, M. A., Zhu, G., Frantz, A. A., Novak, J., Williams, S. N., and Fowlkes, I., 2004d. "Dependence of $NdSc_3(BO_3)_4$ random laser parameters on particle size," *J. Opt. Soc. Am. B* **21**(1), 191.

Noginov, M. A., Fowlkes, I. N., and Zhu, G., 2005a. "Fiber-coupled random laser," *Appl. Phys. Lett.* **86**, 161105.

Noginov, M. A., Novak, J., Grigsby, D., Zhu, G., and Bahoura, M., 2005b. "Optimization of the transport mean free path and the absorption length in random lasers with non-resonant feedback," *Opt. Express* **13(22)**, 8829.

Noginov, M. A., Zhu, G., Small, C., and Novak, J., 2006. "Neodymium random laser with external mirrors," *Appl. Phys. B* **84**(1–2), 269.

Patra, M. and Beenakker, C. W. J., 1999. "Excess noise for coherent radiation propagating through amplifying random media," *Phys. Rev. A* **60**, 4059.

Patra, M., Schomerus, H., and Beenakker, C. W. J., 2000. "Quantum-limited linewidth of a chaotic laser cavity," *Phys. Rev. A* **61**, 023810.

Patra, M., 2002. "Theory for photon statistics of random lasers," *Phys. Rev. A* **65**, 043809.

Patra, M., 2003a. "Decay rate distributions of disordered slabs and application to random lasers," *Phys. Rev. E* **67**, 016603.

Patra, M., 2003b. "Influence of spatial correlations on the lasing threshold of random lasers," *Phys. Rev. E* **67**, 065603.

Polson, R. C. and Vardeny, Z. V., 2000. "Magnetic flux pinning enhancement in $HgBa_2CaCu_2O_{6+\delta}$ films on vicinal substrates," *Appl. Phys. Lett.* **85**, 70.

Polson, R. C., Huang, J. D., and Vardeny, Z. V., 2001a. "Random lasers in π-conjugated polymer films," *Syn. Metals* **119**(1–3), 7.

Polson, R. C., Chipouline, A., Vardeny, Z. V., 2001b. "Random lasing in π-conjugated films and infiltrated opals," *Adv. Mater.* **13**(10), 760.

Polson, R. C., Raikh, M. E., and Vardeny, Z. V., 2002. "Universality in unintentional laser resonators in π-conjugated polymer films," *Comptes Rendus de l'Academie des Sciences, Serie IV (Physique, Astrophysique)* **3**(4), 509.

Polson, R. and Vardeny, Z. V., 2004. "Random lasing in human tissues," *Appl. Phys. Lett.* **85**, 1289.

Polson, R. and Vardeny, Z. V., 2005. "Organic random lasers in the weak- scattering regime," *Phys. Rev. B* **71**, 045205.

Quochi, F., Cordella, F., Orrù, R., Communal, J. E., Verzeroli, P., Mura, A., Bongiovanni G., Andreev A., Sitter, H., and Sariciftci, N. S., 2004. "Random laser action in self-organized para-sexiphenyl nanofibers grown by hot-wall epitaxy," *Appl. Phys. Lett.* **84**, 4454.

Quochi, F., Andreev, A., Cordella, F., Orrù R., Mura, A., Bongiovanni, G., Hoppe, H., Sitter, H., and Sariciftci, N. S., 2005. "Low-threshold blue lasing in epitaxially grown para-sexiphenyl nanofibers," *J. Luminescence* **112**, 321.

Rand, S., 2003. "Bright storage of light," *Opt. Photon. News* **33**.

Redmond, S. M., Armstrong, G. L., Chan, H.-Y., Mattson, E., Mock, A., Li, B., Potts, J. R., Cui, M., Rand, S. C., Oliveira, S. L., Marchal, J., Hinklin, T., and Laine, R. M., 2004. "Electrical generation of stationary light in random scattering media," *J. Opt. Soc. Am. B* **21**(1), 214.

Ripoll, J., Soukoulis, C. M., and Economou, E. N., 2004. "Optimal tuning of lasing modes through collective particle resonance," *J. Opt. Soc. Am. B* **21**(1), 141.

Schomerus, H., Frahm, K. M., Patra, M., and Beenakker, C. W. J., 2000. "Quantum limit of the laser line width in chaotic cavities and statistics of residues of scattering matrix poles," *Physica A* **278**, 469.

Sebbah, P. and Vanneste, C., 2002. "Random laser in the localized regime," *Phys. Rev. B* **66**, 144202.

Seelig, E. W., Tang, B., Yamilov, A., Cao, H., Chang, R. P. H., 2003. "Self-assembled 3D photonic crystals from ZnO colloidal spheres," *Mat. Chem. Phys.* **80**, 257.

Sha, W. L., Liu, C.-H. and Alfano, R. R., 1994. "Spectral and temporal measurements of laser action of Rhodamine 640 dye in strongly scattering media," *Opt. Lett.* **19**(23), 1922.

Sha, W. L., Liu, C.-H,. Liu, F., and Alfano, R. R., 1996. "Competition between two lasing modes of Sulforhodamine 640 in highly scattering media," *Opt. Lett.* **21**(16), 1277.

Sharma, D., Ramachandran, H., and Kumar, N., 2006. "Lévy statistics of emission from a novel random amplifying medium: an optical realization of the Arrhenius cascade," *Opt. Lett.* **31**(12), 1806.

Shkunov, M. N., DeLong, M. C., Raikh, M. E., Vardeny, Z. V., Zakhidov, A. A., and Baughman, R. H., 2001. "Photonic versus random lasing in opal single crystals," *Syn. Metals* **116**, 485.

Siddique, M., Alfano, R. R., Berger, G. A., Kempe, M., and Genack, A. Z., 1996. "Time-resolved studies of stimulated emission from colloidal dye solutions," *Opt. Lett.* **21**(7), 450.

Siegman, A., 1986. *Lasers*. University Science Books, Mill Valley.

Song, Q., Wang, L., Xiao, S., Zhou, X., Liu, L., and Xu, L., 2005. "Random laser emission from a surface-corrugated waveguide," *Phys. Rev. B* **72**, 035424.

Soukoulis, C. M., Jiang, X., Xu, J. Y., and Cao, H., 2002. "Dynamic response and relaxation oscillations in random lasers," *Phys. Rev. B* **65**, 041103.

Sperber, P., Spangler, W., Meier, B., and Penzkofer, A., 1988. "Experimental and theoretical investigation of tunable picosecond pulse generation in longitudinally pumped dye laser generators and amplifiers," *Opt. Quantum Electron.* **20**, 395.

Stassinopoulos, A., Das, R. N., Giannelis, E. P., Anastasiadis, S. H., and Anglos, D., 2005. "Random lasing from surface modified films of zinc oxide nanoparticles," *App. Surf. Sci.* **247**, 18.

Sun, B. Q., Gal, M., Gao, Q., Tan H. H., Jagadish, C., Puzzer, T., Ouyang, L., and Zou, J., 2003. "Epitaxially grown GaAsN random laser," *J. Appl. Phys.* **93**, 5855.

Taniguchi, H., Tanosaki, S., Tsujita, K., and Inaba, H., 1996. "Biological laser emission achieved by mixed highly scattering intralipid-10 in microdroplets including pigment extracted from green algae," *IEEE J. Quant. Electron.* **32**, 1864.

Taflove, A. and Hagness, S. C., 2000. *Computational Electrodynamics*. Artech House, Boston.

Ter-Gabrielyan, N. E., Markushev, V. M., Belan, V. R., Briskina, Ch. M., Dimitrova, O. V., Zolin, V. F., and Lavrov, A. V., 1991a. "Stimulated radiation emitted by lithium neodymium tetraphosphate $LiNd(PO_3)_4$ and neodymium pentaphosphate NdP_5O_{14} powders," *Sov. J. Quantum Electron.* **21**, 840.

Ter-Gabrielyan, N. E., Markushev, V. M., Belan, V. R., Briskina, Ch. M., and Zolin, V. F., 1991b. "Stimulated emission spectra of powders of double sodium and lanthanum tetramolybdate," *Sov. J. Quantum Electron.* **21**, 32.

Thareja, R. K. and Mitra, A., 2000. "Random laser action in ZnO," *Appl. Phys. B* **71**(2), 181.

Totsuka, K., van Soest, G., Ito, T., Lagendijk, A., and Tomita, M., 2000. "Amplification and diffusion of spontaneous emission in strongly scattering medium," *J. Appl. Phys.* **87**, 7623.

Türeci, H. E., Schwefel, H. G. L., Jacquod, P., and Stone, A. D., 2005. In *Progress in Optics*. E. Wolf, Ed., North Holland, Amsterdam.

Türeci, H. E., Stone, A. D., and Collier, B., 2006. "Self-consistent multimode lasing theory for complex or random lasing media," *Phys. Rev. A* **74**, 043822.

Türeci, H. E., Stone, A. D., and Li, G., 2007. "Theory of the spatial structure of nonlinear lasing modes," *Phys. Rev. A* **76**, 013813.

Türeci, H. E., Stone, A. D., Rotter, S., and Li, G., 2008. "Strong interactions in multimode random lasers," *Science* **320**, 643.

van der Molen, K. L., Mosk, A. P., and Lagendijk, A., 2006. "Intrinsic intensity fluctuations in random lasers," *Phys. Rev. A* **74**, 053808.

van der Molen, K. L., Tjerkstra, R. W., Mosk, A. P., and Lagendijk, A., 2007. "Spatial extent of random laser modes," *Phys. Rev. Lett.* **98**, 143901.

van Soest, G., Tomita, M., Sprik, R., and Lagendijk, A., 1999. "Amplifying volume in scattering media," *Opt. Lett.* **24**(5), 306.

van Soest, G., Poelwijk, F. J., Sprik, R., and Lagendijk, A., 2001. "Dynamics of a random laser above threshold," *Phys. Rev. Lett.* **86**, 1522.

van Soest, G. and Lagendijk, A., 2002. "β factor in a random laser," *Phys. Rev. E* **65**, 047601.

Vanneste, C. and Sebbah, P., 2001. "Selective excitation of localized modes in active random media," *Phys. Rev. Lett.* **87**, 183903.

Vanneste, C. and Sebbah, P., 2005. "Localized modes in random arrays of cylinders," *Phys. Rev. E* **71**, 026612.

Vanneste, C., Sebbah, P., and Cao, H., 2007. "Lasing with resonant feedback in weakly scattering random systems," *Phys. Rev. Lett.* **98**, 143902.

Vasa, P., Singh, B. P., and Ayyub, P. J., 2005. "Coherence properties of the photoluminescence from CdS-ZnO nanocomposite thin films," *Phys. Condens. Matter* **17**, 189.

Watanabe, H., Oki, Y., Maeda, M., and Omatsu, T., 2005. "Waveguide dye laser including a SiO2 nanoparticle-dispersed random scattering active layer," *Appl. Phys. Lett.* **86**, 151123.

Wiersma, D., van Albada, M. P., and Lagendijk, A., 1995. "Random laser?" *Nature* **373**, 203.

Wiersma, D. S. and Lagendijk, A., 1996. "Light diffusion with gain and random lasers," *Phys. Rev. E* **54**, 4256.

Wiersma, D. S. and Lagendijk, A., 1997. "Laser action in very white paint," *Physics World* **10**, 33.

Wiersma, D., 2000. "Laser physics: The smallest random laser," *Nature* **406**, 132.

Wiersma, D. and Cavalieri, S., 2001. "Light emission: A temperature-tunable random laser," *Nature* **414**, 708.

Williams, G. R., Bayram, S. B., Rand, S. C., Hinklin, T., and Laine, R. M., 2001. "Laser action in strongly scattering rare-earth-metal-doped dielectric nanophosphors," *Phys. Rev. A* **65**, 013807.

Wu, X., Yamilov, A., Noh, H., Cao, H., Seelig, E. W., and Chang, R. P. H., 2004a. "Random lasing in closely packed resonant scatterers," *J. Opt. Soc. Am. B* **21**(1), 159.

Wu, X., Yamilov, A., Liu, X., Li, S., Dravid, V. P., Chang, R. P. H., and Cao, H., 2004b. "Ultraviolet photonic crystal laser," *App. Phys. Lett.* **85**, 3657.

Wu, X., Fang, W., Yamilov, A., Chabanov, A. A., Asatryan, A. A., Botten, L. C., and Cao, H., 2006. "Random lasing in weakly scattering systems. *Phys. Rev. A* **74**, 053812.

Wu, X. and Cao, H., 2007. "Statistics of random lasing modes and amplified spontaneous emission spikes in weakly scattering systems," *Opt. Lett.* **32**, 3089.

Wu, X. and Cao, H., 2008. "Statistical studies of random-lasing modes and amplified spontaneous-emission spikes in weakly scattering systems," *Phys. Rev. A* **77**, 013832.

Yamilov, A. and Cao, H., 2004. "Highest-quality modes in disordered photonic crystals," *Phys. Rev. A* **69**, 031803.

Yamilov, A., Wu, X., Cao, H., and Burin, A. L., 2005. "Absorption-induced confinement of lasing modes in diffusive random media," *Opt. Lett.* **30**(18), 2430.

Yokoyama, S. and Mashiko, S., 2003. "Tuning of laser frequency in random media of dye-doped polymer and glass particle hybrid," *Jpn. J. Appl. Phys.* **42**, L970.

Yoshino, K., Tatsuhara, S., Kawagishi, Y., and Ozaki, M., 1999. "Amplified spontaneous emission and lasing in conducting polymers and fluorescent dyes in opals as photonic crystals," *Appl. Phys. Lett.* **74**, 2590.

Yu, S. F. and Leong, E. S. P., 2004. "High-power single-mode ZnO thin-film random lasers," *IEEE J. Quant. Electron.* **40**(9), 1186.

Yu, S. F., Yuen, C., Lau, S. P., Park, W. I., and Yi, G.-C., 2004a. "Random laser action in ZnO nanorod arrays embedded in ZnO epilayers," *Appl. Phys. Lett.* **84**, 3241.

Yu, S. F., Yuen, C., Lau, S. P., and Lee, H. W., 2004b. "Zinc oxide thin-film random lasers on silicon substrate," *Appl. Phys. Lett.* **84**, 3244.

Yuen, C., Yu, S. F., Leong, E. S. P., Yang, H. Y., and Hng, H. H., 2005a. "Formation conditions of random laser cavities in annealed ZnO epilayers," *IEEE J. Quant. Electron.* **41**, 970.

Yuen, C., Yu, S. F., Leong, E. S. P., Yang, H. Y., Lau, S. P., Chen, N. S., and Hng, H. H., 2005b. "Low-loss and directional output ZnO thin-film ridge waveguide random lasers with MgO capped layer," *Appl. Phys. Lett.* **86**, 031112.

Zacharakis, G., Heliotis, G., Filippidis, G., Anglos, D., and Papazoglou, T. G., 1999. "Investigation of the laserlike behavior of polymeric scattering gain media under subpicosecond laser excitation," *Appl. Opt.* **38**(28), 6087.

Zacharakis, G., Papadogiannis, N. A., and Papazoglou, T. G., 2002. "Random lasing following two-photon excitation of highly scattering gain media," *Appl. Phys. Lett.* **81**, 2511.

Zaitsev, O., 2006. "Mode statistics in random lasers," *Phys. Rev. A* **74**, 063803.

Zaitsev, O., 2007. "Spacing statistics in two-mode random lasing," *arXiv:cond-mat/0703783*.

Zhang, W., Cue, N., and Yoo, K. M., 1995a. "Emission linewidth of laser action in random gain media," *Opt. Lett.* **20**(9), 961.

Zhang, W., Cue, N., and Yoo, K. M., 1995b. "Effect of random multiple light scattering on the laser action in a binary-dye mixture," *Opt. Lett.* **20**(9), 1023.

Zolin, V. F., 2000. "The nature of plaser-powdered laser," *J. Alloy. Compd.* **300**, 214.

Zyuzin, A. Yu., 1998. "Superfluorescence of photonic paint," *JETP* **86**, 445.

Zyuzin, A. Yu., 1999. "Superfluorescent decay in quasiballistic disordered systems," *Europhys. Lett.* **46**(2), 160.

Biography

Hui Cao is a Professor of Applied Physics and Physics at Yale University, New Haven, Connecticut. Her main research interests lie in nanophotonics and quantum optics. She received a bachelor degree in Physics from Peking University (China) in 1990, a masters degree from the Mechanical and Aerospace Engineering Department of Princeton University in 1992, and a Ph.D. in Applied Physics (major) and Electrical Engineering (minor) from Stanford University in 1997. Her Ph.D. thesis on semiconductor cavity quantum electrodynamics was the foundation for a research monograph published by Springer-Verlag in 2000. She was a faculty member in the Department of Physics and Astronomy of Northwestern University from 1997 to 2007, where she pioneered the experimental studies of random laser and light localization in disordered nanomaterials. In 2004 she took a sabbatical with the Max-Planck Research Group on Optics, Information and Photonics in Erlangen Germany. In January 2008 she moved to Yale University with a primary appointment in the Applied Physics Department and a joint appointment in the Physics Department. Her current research focuses on understanding and controlling quantum optical processes in nanostructures, both for fundamental physical studies and for device applications.

Dr. Cao has coauthored one book and has published three invited book chapters, four review articles, and over 100 research papers. She became a David and Lucille Packard Fellow in 1999 and an Alfred P. Sloan Fellow in 2000. She won the Early CAREER Award from the U.S. National Science Foundation in 2001. In 2004 she won the Friedrich Wilhelm Bessel Research Award from the Alexander von Humboldt Foundation, and the outstanding Young Researcher Award from the Overseas Chinese Physics Association. She was the recipient of the 2006 Maria Goeppert-Mayer Award from the American Physical Society. In 2007 she became a fellow of the Optical Society of America and of the American Physical Society. From 2006 to 2008, Cao chaired the OSA Technical Group "Photonic Metamaterials," and served as vice chair of the OSA Technical Group "Waves in Random and Periodic Media" from 2004 to 2006. She has served on and chaired many program committees and subcommittees of international optics conferences including CLEO, QLES, IQEC, and FiO. Dr. Cao will be the program chair for 2010 QELS, and general chair for 2012 QELS.

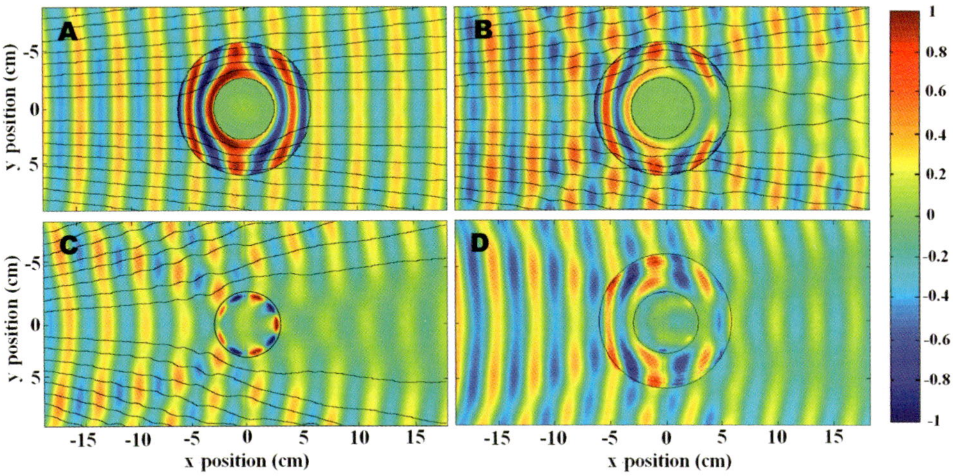

Plate 1.16 Theoretical simulation and experimental realization of the electromagnetic cloak. The instantaneous value of the electric field (color bar) is shown for (A) calculations using the exact material properties prescribed by Eqs. (1.58)–(1.60), (B) calculations using the reduced material properties prescribed by Eqs. (1.61)–(1.63), (C) measurements with a copper disc without the surrounding cloak and (D) measurements with a copper disc surrounded by the metamaterial cloak of Fig. 1.15. Note that the wavefront kinking in the geometric shadow of the object in (C) is largely absent in (D). [Reprinted with permission from Ref. 32. © (2006) by the American Association for the Advancement of Science.]

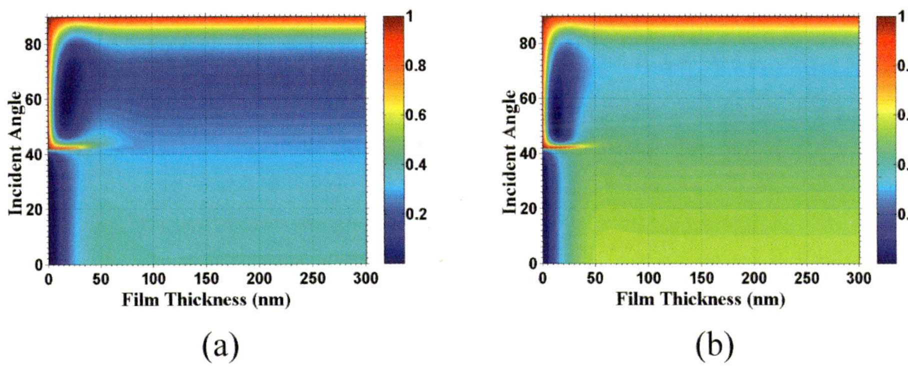

Plate 3.8 Variation of the *p*-polarised reflectivity of the ATR system over the plane defined by the incident angle and the film thickness. (a) MnBi: λ = 578 nm, n = 2.44 + 2.92i, Q = –0.089 + 0.034i. (b) Ni: λ = 600 nm, n = 2.09 + 3.89i, Q = 0.01 – 0.0052i, where λ is the wavelength of an incident light beam. The refractive index of the glass prism is 1.55.

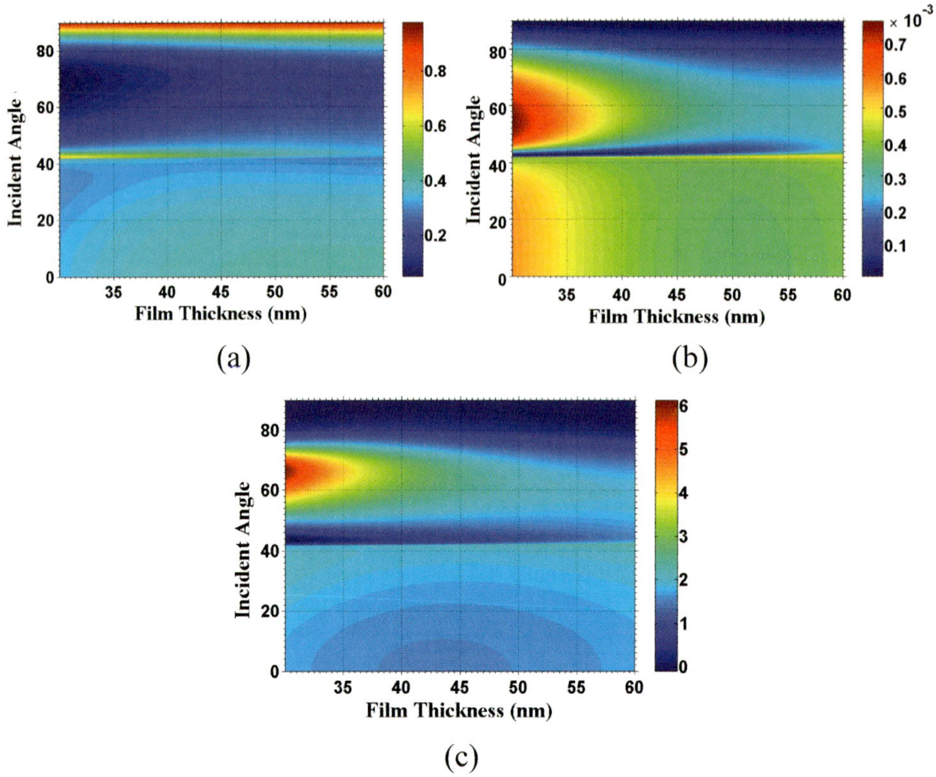

Plate 3.10 Incident angle–film thickness variation of (a) $|R_p|^2$, (b) $|R_s|^2$, and (c) Θ_K, for MnBi. Data is the same as is presented in Figs. 3.8 and 3.9. Note the difference in the color scale for (b).

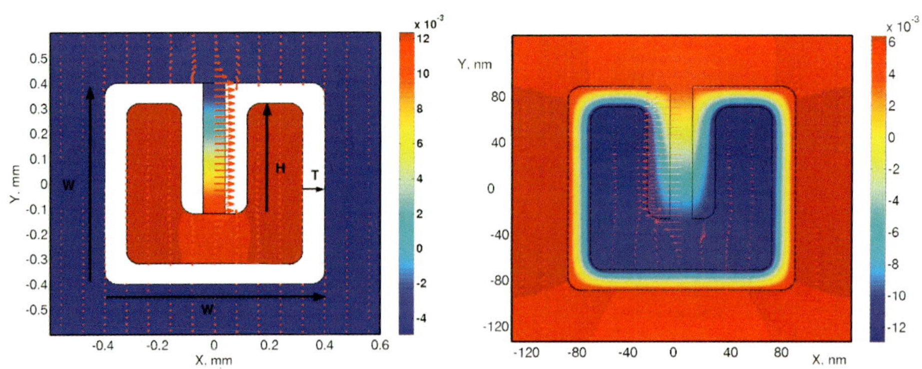

Plate 5.1 Magnetic field distribution inside a lattice of split-ring resonators. Color: magnetic field H_z. Arrows: electric field vectors. Left: SRR made of a perfectly conducting (PEC) material. The gap is filled with a high-permittivity ($\varepsilon_d = 4$) dielectric. Resonator parameters: periods $a_x = a_y = 1.2$ mm, ring size $W = 0.8$ mm, ring thickness $T = 80$ μm, gap height $H = 0.44$ mm, gap width $G = 80$ μm. The gap is filled with a high-permittivity dielectric $\varepsilon_d = 4$. At the magnetic cutoff ($\mu_{eff} = 0$) shown here, the vacuum wavelength $\lambda = 1.57$ cm. Right: the scaled-down by a factor of ≈ 4400 plasmonic version of the silver SRR operating at $\lambda = 3.44$ μm. (Reprinted from Ref. 29.)

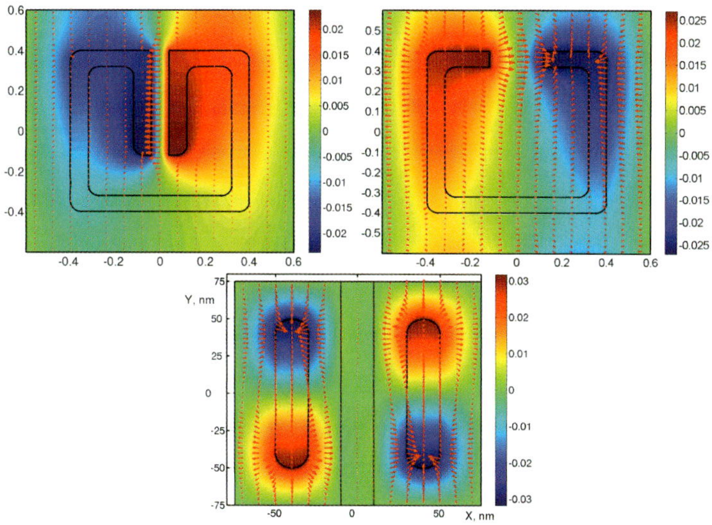

Plate 5.2 Potential distribution φ (color-coded) inside a lattice of (a) split-ring resonators (SRRs), (b) split rings (SRs), and (c) metal strips separated by a metal film corresponding to electrostatic resonances responsible for the magnetic response. Arrows: $\vec{E} = -\nabla\phi$. Electrostatic resonances occur at (a) $\varepsilon_{res} \equiv -330$ (λ = 3 μm) for SRR, (b) $\varepsilon_{res} \equiv -82$ (λ = 1.5 μm) for SR, and (c) $\varepsilon_{res} \equiv -8.8$ (λ = 0.5 μm), assuming that the plasmonic material is silver. (Reprinted from Refs. 29 and 31.)

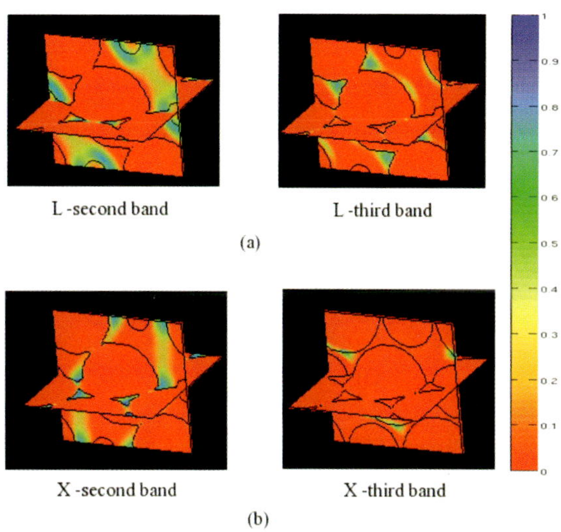

Plate 8.16 Electric field energy density distributions of the second and third bands at the points (a) L, and (b) X in an inverse opal of close-packed air spheres ($R_2 = 0.35a$) in titania (model 1). The black contours indicate horizontal and vertical cross-sections of the dielectric structures. The electric field energy density distributions are shown in real space for the modes corresponding to the X- and L-points of the reciprocal space. The horizontal and vertical cross sections shown are perpendicular to the (111) and (100) crystallographic directions. In each plot, the maximum occurring electric field energy density distribution was normalized to a value of 1 (dark blue). The corresponding reciprocal space band structure is shown in Fig. 8.15(a), and the real space structure is shown in Figs. 8.13(a) and 8.14.

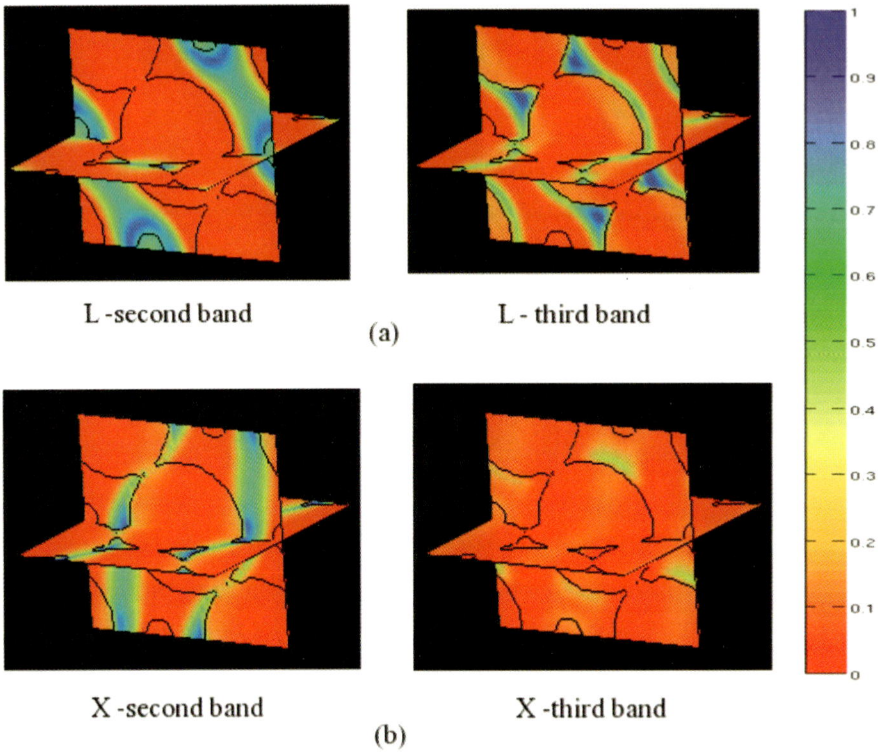

Plate 8.17 As in Fig. 8.16, except that n_s = 1.5 (i.e., silica spheres), the corresponding band structure is that of Fig. 8.15(b). In the right-hand panel of (b), the horizontal and vertical cross sections do not retain the points of maximum electric field energy distributions; they were instead chosen to clearly depict the structure's dielectric distribution contours.

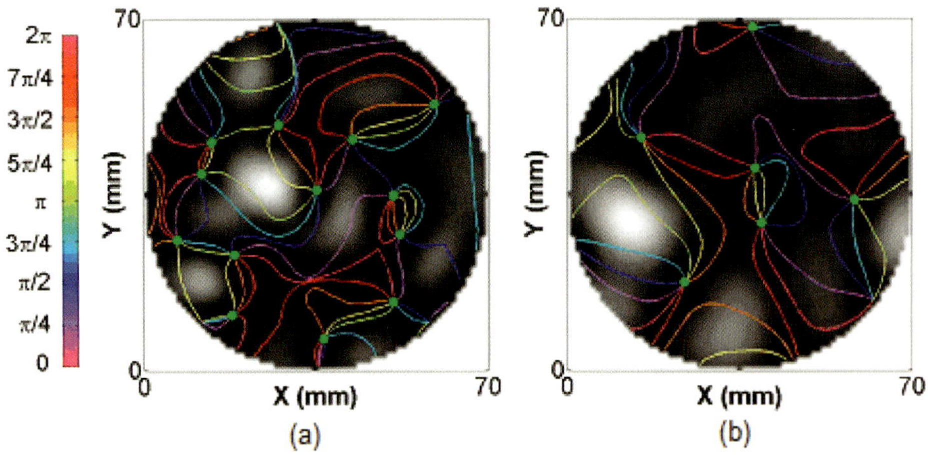

Plate 9.13 Examples of speckle patterns at the output of a quasi-1D sample for (a) diffusive and (b) localized waves. The gray scale shows the intensity variation, and the colored lines are equiphase lines. Green dots represent phase singularities. [Reprinted from Ref. 23. © (2007) by the American Physical Society.]

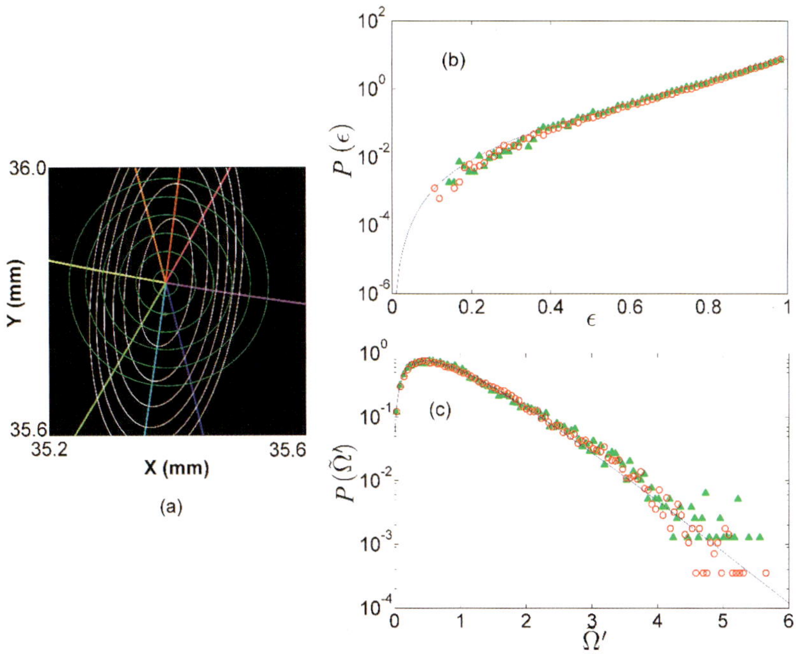

Plate 9.14 (a) Core structure of a phase singularity. The straight lines are equiphase lines with phase values shown in Fig. 9.10. Circles (green) are current contours, while ellipses (white) are intensity contours. (b) Probability distribution of ε. (c) Probability distribution of $\tilde{\Omega}'$. The solid line is a derivation from Gaussian statistics of random fields. Green triangles and red circles are experimental data for diffusive and localized waves, respectively. [Reprinted from Ref. 144. © (2007) by the American Physical Society.]

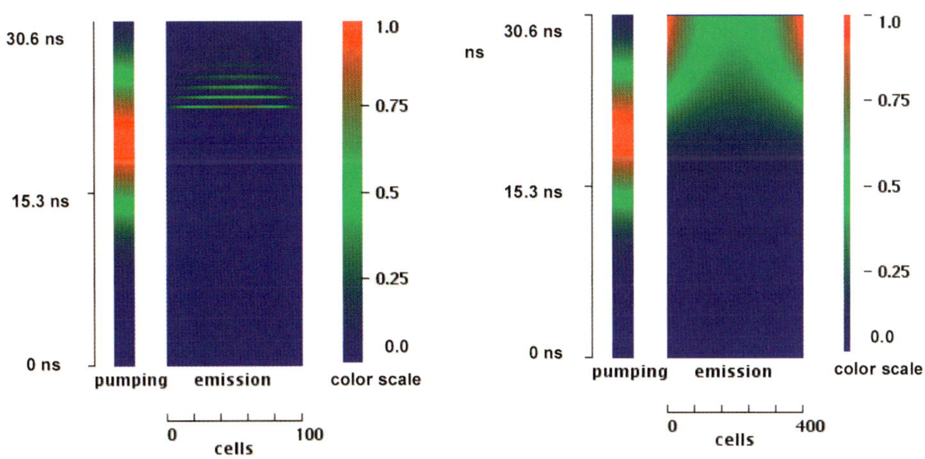

Plate 12.7 Calculated emission dynamics in (a) a strip of 100 cells; pumping intensity corresponds to 1 J/cm^2 of absorbed energy, r_{aver} = 8%, τ_{res} is 0.43 ps, and maximum emission intensity is 24,566 arb. units. (b) Strip of 400 cells pumped with the same intensity, with no reflection at the boundaries between cells; τ_{res} is 0.67 ps and the maximum emission intensity is 16 arb. units. [Reprinted from Ref. 35. © (2004) by the American Physical Society.]

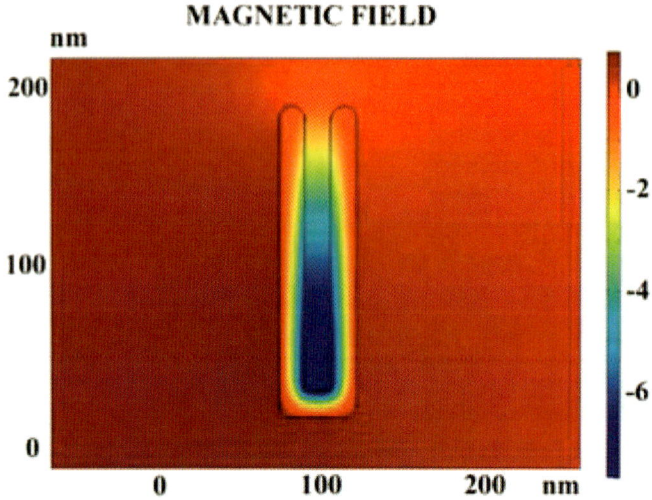

Plate 13.2 Magnetic plasmon resonance in a silver horseshoe nanoantenna placed in the maximal external magnetic field, directed perpendicular to the plan of the paper. The incident wavelength is λ = 1.5 μm; the silver permittivity is estimated from Eq. (13.19) and $\varepsilon_d = 2$. The magnetic field inside the horseshoe is opposite in direction to the external field. (Reprinted from Ref. 67.)

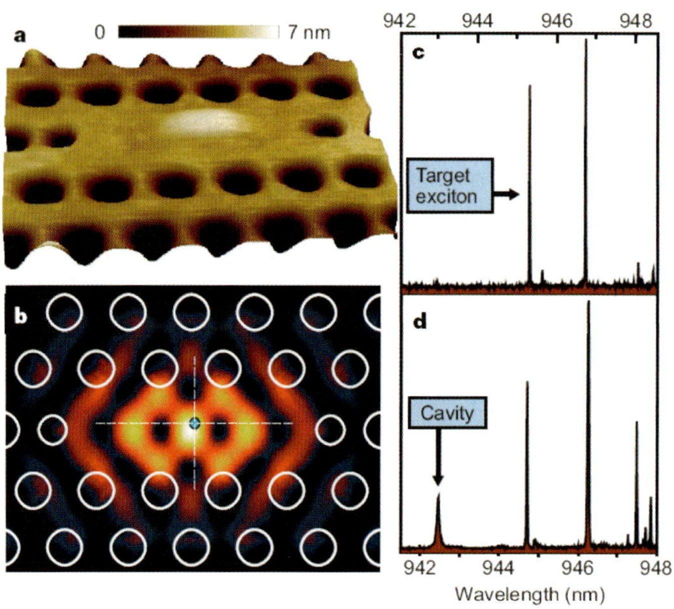

Plate 15.7 Positioning a photonic crystal cavity mode relative to a single buried QD. (a) AFM topography of photonic crystal nanocavity aligned to a hill from a single QD. Depth is depicted by the bar on the top. (b) Electric field intensity distribution of photonic cavity mode showing overlap field maximum with the QD. (c) Photoluminescence spectrum of a single QD (before cavity fabrication) showing bounded exciton excitations. (d) Photoluminescence spectrum after cavity fabrication showing emission from the cavity at 942.5 nm. [Reprinted with permission from Ref. 23. © (2007) by Nature Publishing Group.]

Chapter 12
Feedback in Random Lasers

Mikhail A. Noginov
Center for Materials Research, Norfolk State University, Norfolk, VA, USA

12.1 Introduction

The purpose of this chapter is to introduce the concept of a *random laser* and show how strongly the properties of random lasers depend on the strength of light scattering in a gain medium. Random lasers, theoretically predicted forty years ago[1,2] and experimentally demonstrated more than twenty years ago,[3,4] have developed into a large area of research. It is not the goal of this chapter to provide a comprehensive review of all of the results and phenomena known in the field, many of which are discussed in Refs. 5–7. Instead, using a limited number of examples, I attempt to identify the major regimes of operation of random lasers and relate them to the strength of scattering or character of feedback.

The chapter is organized as follows. The concept of regular lasers is briefly discussed in Sec. 12.2, and the idea of a laser with nonresonant feedback and a random laser are introduced in Sec. 12.3. Photon migration and localization in scattering media, their theoretical description, and applications to random lasers are discussed in Sec. 12.4. The experimental behavior of random lasers with nonresonant feedback, exemplified by neodymium random lasers, is described in detail in Sec. 12.5. Stimulated emission in ZnO random lasers with resonant feedback is presented in Sec. 12.6. Different strengths of scattering, regimes of the stimulated emission feedback, and their effect on the coherence of random laser operation are discussed in Sec. 12.7. Finally, the major regimes of the operation of random lasers are summarized in Sec. 12.8.

12.2 The Concept of a Laser

Many readers are most likely familiar with the basic concept of lasers,[8–11] which is briefly summarized below for completeness of the chapter. The major elements of conventional lasers are (*i*) a gain medium and (*ii*) a cavity formed by mirrors. Optical or any other type of pumping produces a *population inversion* in a gain medium, at which the population of the upper laser level $|3\rangle$ exceeds that of the lower laser level $|2\rangle$, as shown in Fig. 12.1. Photons, with energy that matches the energy gap between states $|3\rangle$ and $|2\rangle$, can induce *stimulated emission* accompanied by the electronic transition $|3\rangle \rightarrow |2\rangle$ (see Fig. 12.1). The energy of the emitted photon (color of light), its momentum (direction of light propagation), and its polarization correspond to those of the stimulating photon.

Excited atoms or molecules can also spontaneously relax to lower electronic states, without perturbation by another photon.

Many spontaneously emitted photons leave the cavity and do not participate in the laser action. However, some photons are emitted along the optical axis of the cavity, which in the simplest case is formed by two parallel mirrors—a Fabry-Perot resonator.[12,13]

As photons propagate through a gain medium possessing a population inversion, the number of photons increases due to stimulated emission. After reflection off the mirror (in the direction along the optical axis of the cavity) the flux of photons again passes through the gain medium, increasing its intensity. The second mirror sends photons back to the gain medium and the process repeats continuously, as seen in Fig. 12.2. As a rule, one of the mirrors (the back mirror) has 100% reflectivity at the stimulated emission wavelength and another mirror (the output mirror) is partially transparent. Leakage of photons through the output mirror constitutes the laser beam. As photons circulate in the laser cavity, on each round trip they experience (*i*) amplification in the gain medium, (*ii*) losses due to the leakage through the mirror(s), and (*iii*) other losses, which often include unwanted scattering in the laser medium. If light intensity at any cross section of the laser cavity at time t_0 is equal to I_0, then after one round trip in the laser cavity, the light intensity will become

$$I_1 = I_0 \exp(2n\sigma_e l) R_1 R_2 (1-\alpha), \tag{12.1}$$

where the product of the population inversion n and the emission *cross section* σ_e constitutes gain g in a laser medium, $2l$ is the double pass of photons through the laser medium, R_1 and R_2 are the reflectivities of the mirrors, and α is the net parasitic loss per round trip.

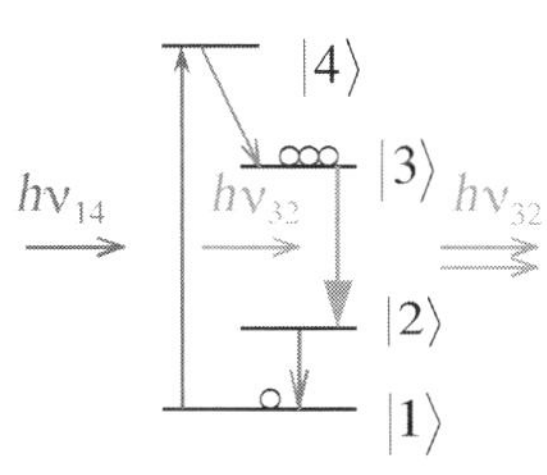

Figure 12.1 Excitation and relaxation processes in a *four-level laser scheme*. In this excitation scheme, optical pumping at the transition $|1\rangle \rightarrow |4\rangle$ (photon energy $h\nu_{14}$) is followed by fast radiationless relaxation of the excitation to the *metastable* state $|3\rangle$. This creates a *population inversion* between the metastable state $|3\rangle$ and the lower laser state $|2\rangle$. In a four-level laser scheme, the state $|2\rangle$ is practically empty because of fast radiationless relaxation of its excitation to the ground state $|2\rangle \rightarrow |1\rangle$. *Stimulated emission* occurs at the transition $|3\rangle \rightarrow |2\rangle$ (photon energy $h\nu_{32}$).

When I_1 is smaller than I_0, the photon flux decays with time, and no laser action occurs. With an increase of the pumping intensity, the population inversion increases as well. At the critical *threshold* population inversion n_{th}, I_1 becomes I_0. This implies that the photon flux does not decay as it circulates in the laser cavity. With a further increase in pumping, as I_1 becomes larger than I_0, photons spontaneously emitted along the laser cavity stimulate avalanche growth of emission intensity and produce a laser beam emerging from the output mirror. The laser output is nearly zero below the threshold and grows linearly with the increase of the pumping intensity above the threshold. The overall amplification, which is determined by the properties of the gain medium as well as the cavity, is typically not uniform over the spectrum. Above the threshold, the emission spectrum rapidly narrows to one or several closely spaced lines, corresponding to cavity modes, which have maximal gain.

The laser cavity supports a discrete set of *modes* (eigenstates of the wave equation with appropriate boundary conditions) corresponding to standing electromagnetic waves. The double cavity length $2L$ is an integer number m of supported wavelength[1]†

$$L = (\lambda/2)m. \tag{12.2}$$

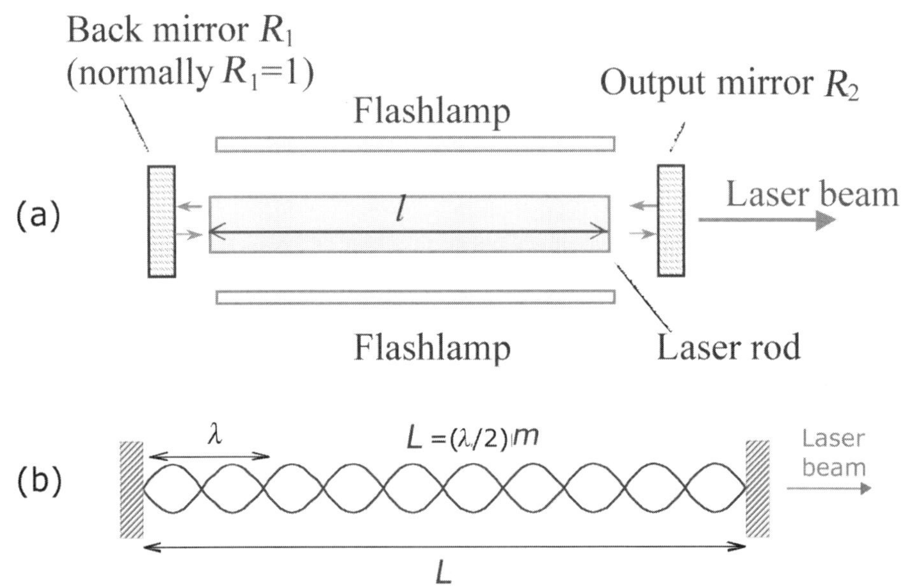

Figure 12.2 (a) Simplified laser scheme. Flashlamps provide pumping to the laser gain medium (laser rod). (b) Laser cavity supporting standing electromagnetic waves.

[1]†This is valid under the commonly satisfied assumption that the net phase shift due to reflection off both mirrors is an integral multiple of 2π.

Diffraction of light at the apertures of laser mirrors or the gain element determines a rich *transversal* structure of the laser modes, in addition to the *longitudinal* modes depicted in Fig. 12.2(b) and described by Eq. (12.2). The feedback in regular lasers is *resonant*, and a laser beam, as a rule, has a high degree of *longitudinal* and *transversal* coherence.[14–16] (Oscillation of the electromagnetic field in points A and B is said to be *coherent* when the phase difference between these two points remains constant. Longitudinal coherence is measured along the laser beam, and transversal coherence is measured across the laser beam.) Coherence of laser radiation is very useful for certain applications, such as metrology and holography.[14,15] However, because of the tendency of coherent light to form *speckle* and *fringe* patterns,[16,14] coherence presents a severe drawback if high spatial uniformity of illumination is desired. Equation (12.2) suggests that the laser wavelength (or frequency) depends on the cavity length L and can be affected by thermal expansion of the optical resonator or by vibrations of the mirrors. This hinders many applications of lasers.

12.3 Lasers with Nonresonant Feedback and Random Lasers

In order to overcome the disadvantages caused by resonant feedback and spatial coherence of a laser beam, Ambartsumyan et al. (in 1966) proposed a new type of laser in which *nonresonant* feedback occurred via diffuse reflection off a highly scattering medium used in place of a back laser mirror[17–20] [see Fig. 12.3(a)]. Alternative realizations of multimode cavities in which the conditions of nonresonant feedback can be fulfilled are (*i*) a cavity with rough reflecting inner walls and a small pinhole opening [Fig. 12.3(b)], and (*ii*) a quasi-concentric resonator formed by two concave mirrors [Fig. 12.3(c)].[20] A laser with nonresonant feedback is an extreme case of a multimode laser (a laser with a rich structure of longitudinal and transversal modes), which has very strong interactions between modes.[20] The central emission frequency of such a laser is determined by the resonant frequency of the gain medium rather than by the eigenmodes of the cavity.[21–24]

In 1967 and 1968, Letokhov took one step farther and theoretically predicted the possibility of generating laser-like light by scattering particles with gain (negative absorption) in the case where the mean free path of photons due to scattering was much smaller than the dimensions of the system, i.e., when the motion of photons was diffuse.[25,26] In the proposed system, the scattering material played the roles of both the active laser medium and the effective resonator, providing for nonresonant feedback [Fig. 12.3(d)].This was most likely the first report of what we now call a *random laser*. The proposed applications of nonresonant random lasers included highly stable optical frequency standards and express testing of laser materials that could not be easily produced in the form of large, homogeneous crystals.

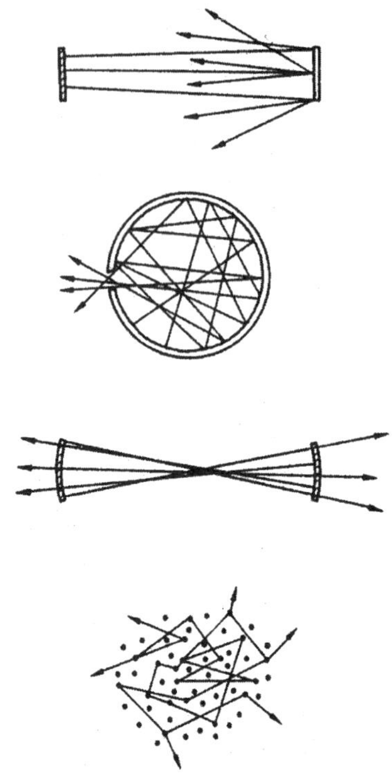

Figure 12.3 Examples of different configurations of multimode cavities in which nonresonant feedback conditions can be fulfilled: (a) scattering surface and mirror, (b) cavity with scattering walls and small outlet hole, (c) quasi-concentric resonator, and (d) combination of scattering particles in an amplifying medium. (Reprinted with permission from Ref. 20.)

12.4 Photon Migration and Localization in Scattering Media and Their Applications to Random Lasers

12.4.1 Diffusion

Rigorously defined, photon motion in a scattering medium is diffusive if $\langle x^2 \rangle$, the average square of the distance a photon moves from its original position in time *t*, is

$$\langle x^2 \rangle = Dt^z, \tag{12.3}$$

where D is the time-independent diffusion coefficient and $z = 1$. (Note that one can find in the literature discussions of diffusion with the time-dependent diffusion coefficient or $z \neq 1$.)

The diffusive character of photon motion is realized provided $\lambda << l^* << L$, where λ is the light wavelength, l^* is the mean scattering length or the distance between scattering events, and L is the linear dimension of the gain medium.[26]

The condition $l^* << L$ implies multiple scattering of light in the sample. On the other hand, the inequality $\lambda << l^*$ suggests that the light scattering occurs at a scale much larger than the wavelength. Correspondingly, wave interference phenomena can be neglected, and light scattering can be described in terms of classical photon trajectories.

Note that in this chapter, we do not distinguish between the mean scattering length l^* and the transport mean free path l_t defined as $l_t = 1 \big/ 1 - \langle \cos\theta \rangle l^*$, where $\langle \cos\theta \rangle$ is the average cosine of the scattering angle. The value $\langle \cos\theta \rangle$ is difficult to evaluate experimentally. Therefore, very often the measurement of l_t based on coherent backscattering is based on the assumption that $l_t = l^*$. (The concept of weak localization of light resulting in coherent backscattering is discussed in detail in Refs. 27–30.)

12.4.2 Prediction of stimulated emission in a random laser operating in the diffusion regime

Assuming that photon migration in a scattering medium is diffusive, Letohkov proposed the following equation describing the flux density $\Phi_\omega(r, t)$ of emitted photons in a pumped volume with a high concentration of scatterers:[25,26,20]

$$\frac{1}{c}\frac{\partial \Phi_\omega(r,t)}{\partial t} = D\Delta\Phi_\omega(r,t) + Q_\omega(r,t)n_0\Phi_\omega(r,t), \tag{12.4}$$

where D is the diffusion coefficient, n_0 is the concentration of particles with gain, Q_ω is the cross section of *negative* absorption of light (gain), c is the speed of light, r is the coordinate, ω is the frequency of light, and t is time. The general solution of Eq. (12.4) is given by

$$\Phi_\omega(r,t) = \sum_n a_n\psi_n(r)\exp\left[-\left(DB_n^2 - Q_\omega n_0\right)ct\right], \tag{12.5}$$

where $\psi_n(r)$ and B_n are the eigenfunctions and eigenvalues of the equation

$$\Delta\psi_n(r) + B_n^2\psi_n(r) = 0\,. \tag{12.6}$$

The threshold condition [corresponding to an avalanche growth of the flux density $\Phi_\omega(r, t)$] comes from the requirement[26]

$$g_{\text{th}} = n_0Q_0 = DB^2\,, \tag{12.7}$$

where B is the smallest eigenvalue of Eq. (12.6) and g_{th} is the threshold value of the gain. For a cylinder with radius r and height h (which is a typical geometry for a pumped volume in random laser experiments), the threshold gain g_{th} depends on h and r as

$$g_{\text{th}} \propto B = \left[\left(\frac{2.4}{r} \right)^2 + \left(\frac{\pi}{h} \right)^2 \right]^{1/2}. \tag{12.8}$$

If the gain in the system is assumed to be constant in time, then Eqs. (12.4)–(12.8) predict an infinite increase of the emission flux density above the threshold. This is qualitatively similar to an "atomic bomb" behavior for multiple scattering of photons in an amplifying medium at a critical volume and gain. Following Letokhov, this avalanche growth of emission flux is identified with the random laser threshold.

However, in a real experimental situation, gain [or the imaginary part of the permittivity $\varepsilon''(r, t)$] is not constant. According to Ref. 26, gain is described by the equation

$$\frac{\partial \varepsilon''(r,t)}{\partial t} + \frac{1}{T_1} \partial \varepsilon''(r,t) = -2\sigma_a \partial \varepsilon''(r,t) \int a(\omega) \Phi_\omega (r,t)\, d\omega + \frac{1}{T_1} \tilde{\varepsilon}''(r), \tag{12.9}$$

where $\sigma_0 = \sigma(\omega_0)$ is the emission cross section of active atoms, T_1 is the spontaneous decay-time of excitations, $a(\omega)$ is the spectral shape of the emission line normalized to unity, and $\tilde{\varepsilon}''(r)$ is the term proportional to the pumping power. Above the threshold, the exponential rise of emission intensity continues until the increased stimulated emission depopulates the upper laser level and drives $\varepsilon''(r, t)$ below its threshold level. As a result, the emission intensity falls sharply, but then, because of pumping $\tilde{\varepsilon}''(r)$, the threshold is exceeded again and pulsations are repeated. This causes *relaxation oscillations* of emission density and population inversion, which are routinely predicted and observed in regular lasers as well as in random lasers (see Fig. 12.4).

12.4.3 Modeling of stimulated emission dynamics in neodymium random lasers

Neodymium random lasers, characterized by the hierarchy of the length scales $\lambda << l^* << L$, operate in a diffusion regime with nonresonant feedback (i.e., their behavior can be adequately described in terms of the model, neglecting any coherence effects) and are discussed in detail in Sec. 12.5. In Ref. 31, the dynamics of stimulated emission in a neodymium random laser was calculated with a simpler version of Eqs. (12.4) and (12.9), accounting for population inversion n and energy density of emitted photons E:

$$\begin{aligned} \frac{dn}{dt} &= \frac{P(t)}{S l_p h\nu_{\text{pump}}} - \frac{n}{\tau} - \frac{E}{h\nu_{\text{em}}} c\sigma_{\text{em}} n \\ \frac{dE}{dt} &= -\frac{E}{\tau_{\text{res}}} + \zeta \frac{n}{\tau} h\nu_{\text{em}} + E c\sigma_{\text{em}} n. \end{aligned} \tag{12.10}$$

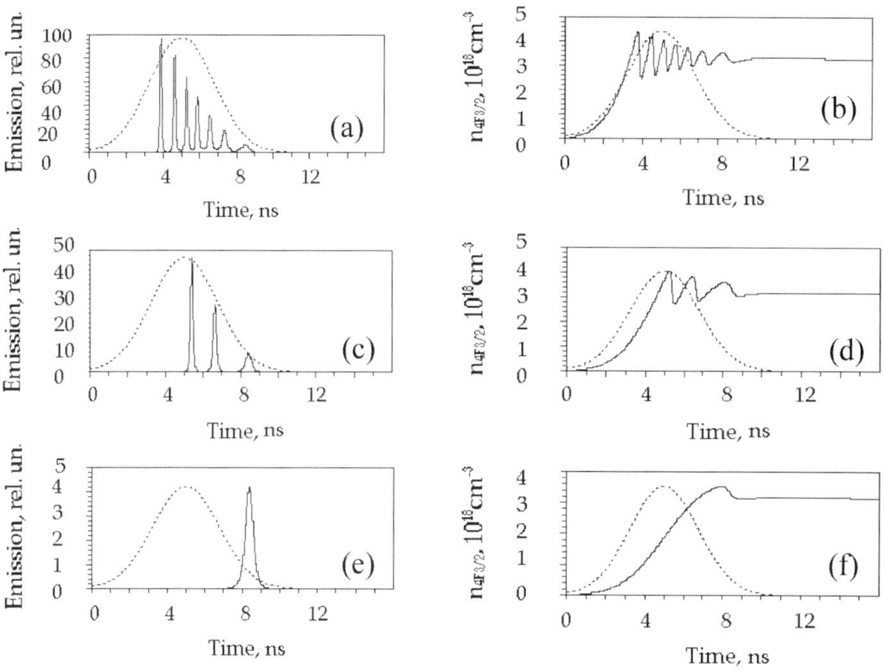

Figure 12.4 (a), (c), and (f) Calculated dynamics of stimulated emission. (b), (d), and (e) $^4F_{3/2}$ upper laser level concentration in a $NdAl_3(BO_3)_4$ random laser. The calculation according to Eq. (12.10) is explained in detail in Sec. 12.5.4. Pumping density: 1000 mJ/cm^2 in (a) and (b), 400 mJ/cm^2 in (c) and (d), and 200 mJ/cm^2 (the threshold) in (e) and (f). The dashed line is the pumping pulse (not plotted to scale). (Reprinted with permission from Ref. 31.)

Here $P(t)/S$ is the pumping power density; l_p is the penetration depth of pumping light; σ_{em} is the emission cross section; $h\nu_{pump}$ ($h\nu_{em}$) is the photon energy at the pumping (emission) wavelength; τ is the lifetime of the upper laser level $^4F_{3/2}$; ζ is the quantum yield of emission (at the transition $^4F_{3/2}\rightarrow{}^4I_{11/2}$) to the lasing mode(s); τ_{res} is the effective residence time of a photon in a pumped volume (in conventional lasers, a similar term represents the lifetime of a photon in the cavity); and c is the speed of light. For simplicity, spatial and spectral nonuniformity of emission intensity has been neglected in Eq. (12.10).

The calculated dynamics of the emission energy density and the population inversion are shown in Fig. 12.4. Calculations predict one short ($\approx$ 1 ns) high-intensity pulse to appear in the emission kinetics at the lasing threshold. More pulses (relaxation oscillations in a highly nonlinear regime) emerge in the emission kinetics at stronger pumping, and the time intervals between pulses shorten.

Spectroscopic parameters used in the numerical solution of Eq. (12.10) approximated those in $NdAl_3(BO_3)_4$ powder: $\sigma_{em} = 1 \times 10^{-18}$ cm^2, $l_p \approx N\bar{\sigma} = (5 \times 10^{21}\text{cm}^{-3}) \times (3 \times 10^{-21}\text{cm}^2) \approx 0.67$ mm, $h\nu_{abs} = 4 \times 10^{-19}$ J, $h\nu_{em} = 2 \times 10^{-19}$ J, τ = 20 μs, and $\zeta \approx 0.9$.[31] The only unknown value in Eq. (12.10) is τ_{res}, which was used as an adjustable parameter to fit the experimental energy threshold, 200 mJ/cm^2.[31] The best fit was obtained at τ_{res} = 10 ps, which corresponds to the effective photon path length l = 1.7 mm (at the index of refraction $n \approx 1.8$).[32] The estimated value of l appears to be reasonable for the experiment in which the characteristic pumped volume was of the order of $\approx$ 1 mm^3.[31]

No diffusion or photon scattering enters explicitly into Eq. 12.10. However, photon migration determines τ_{res}. Assuming a diffusive character of photon motion (since $\lambda << l^* << L$), τ_{res} can be calculated for any given geometry of the pumped volume.[29,30]

Predicted oscillations of emission energy density E and population inversion n (see Fig. 12.4), are known in laser theory as *relaxation oscillations*. Figure 12.5 depicts the calculated input-output curve of stimulated emission and thc dependence of the steady state population inversion (detected after the last stimulated emission pulse) on the pumping density. Both curves are typical of lasers operating according to the *four-level scheme*.

The calculated threshold pumping energy density was found to be proportional to the thickness of the pumped layer l_p and inversely proportional to the emission cross section σ_{em} and the photon residence time τ_{res}.

12.4.4 Stimulated emission in a one-dimensional array of coupled lasing volumes

The dynamics of stimulated emission in a one-dimensional strip of amplifying volumes separated by partially reflective walls (granules composing random laser material) was studied theoretically in Ref. 35 (see Fig. 12.6). The cell size was 1 μm. All cells were assumed to be uniformly pumped. Emission could propagate only along the strip, one photon flux moving to the right and one to the left. Reflection coefficients of the cell walls r were calculated with the help of a random function generator and were randomly distributed between $r = 0$ and $r = 2r_{aver}$, where r_{aver} was the average reflection coefficient.

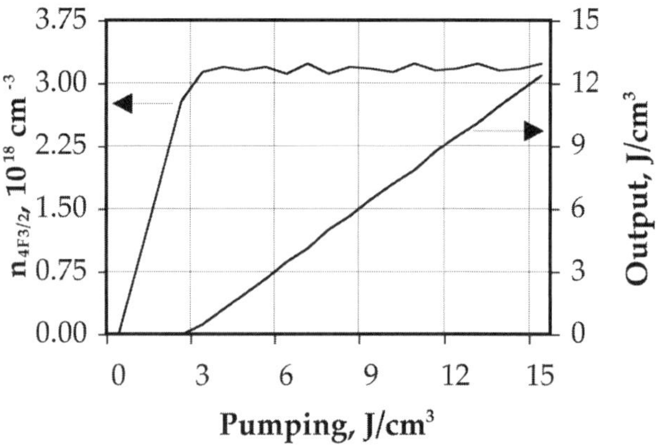

Figure 12.5 Calculated input-output curve (right vertical axis) and dependence of the $^4F_{3/2}$ excited state concentration (left vertical axis) on pumping density in $NdAl_3(BO_3)_4$ powder laser. (Reprinted with permission from Ref. 31.)

The model was based on the system of rate Eq. (12.10) that neglects any coherence effects. Random laser parameters were similar to those presented in Sec. 12.4.3. Instead of direct introduction of the residence time τ_{res}, neighboring cells were allowed to communicate with each other, exchanging photons in accordance with the reflection and transmission coefficients assigned to each intercell boundary. The bouncing back and forth of photons determined photon residence time in a strip.

The typical calculated two-dimensional (time/cell number) kinetics of stimulated emission characterized by relaxation oscillations is shown in Fig. 12.7(a). In this particular calculation, the strip consisted of 100 cells, r_{aver} was 8%, and the mean photon residence time τ_{res} was 0.43 ps. As follows from Fig. 12.7(a), in the system with feedback, the emission is localized (confined) in the central region of the pumped volume.

Figure 12.7(b) shows the dynamics of stimulated emission in a strip consisting of 400 cells without any reflection at the cell boundaries ($r = 0$). Because of the larger strip size, the mean photon residence time τ_{res} was 0.67 ps, longer than that in the strip of Fig. 12.7(a). The amount of pumping energy per cell in Fig. 12.7(b) was the same as that in Fig. 12.7(a). However, despite the longer residence time, no short high-intensity pulses of stimulated emission (*relaxation oscillations*) were calculated in the long strip. Furthermore, no spatial confinement of emission to the central region of the pumped volume is seen in Fig. 12.7(b). Instead, the emission has its maximum values at the ends of the strip, which is an expected behavior in the case of amplification in open paths. Thus, wc conclude that (*i*) relaxation oscillations and spatial confinement of emission accompany each other and serve as evidence of feedback in random lasers and (*ii*) stimulated emission in the presence of even small feedback is dramatically different from stimulated emission with no feedback.

Note that the concept of random laser operation without feedback was introduced in Ref. 36. It was proposed[36] that some 'lucky' photons in a random laser medium might gain extremely large amplification propagating down exponentially long, although very rare, paths. Such photons can be responsible for narrow lines observed in emission spectra of random lasers based on a liquid dye with scatterers. No emission kinetics in narrow spectral lines were studied in order to experimentally discriminate between the 'feedback' and 'no-feedback' stimulated emission regimes.

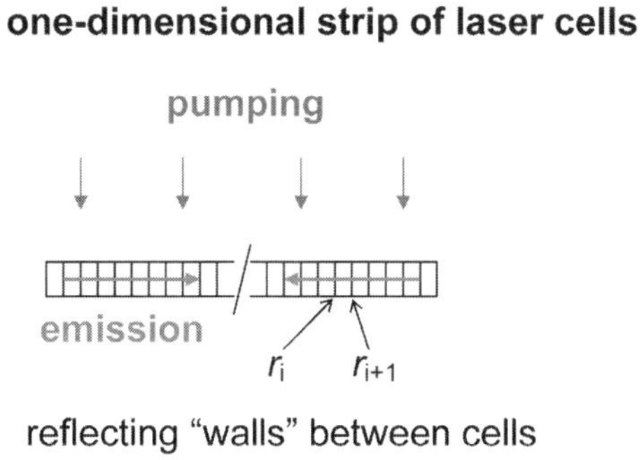

Figure 12.6 Schematic diagram of a one-dimensional strip of lasing volumes. [Reprinted with permission from Ref. 35. © (2004) by the American Physical Society.]

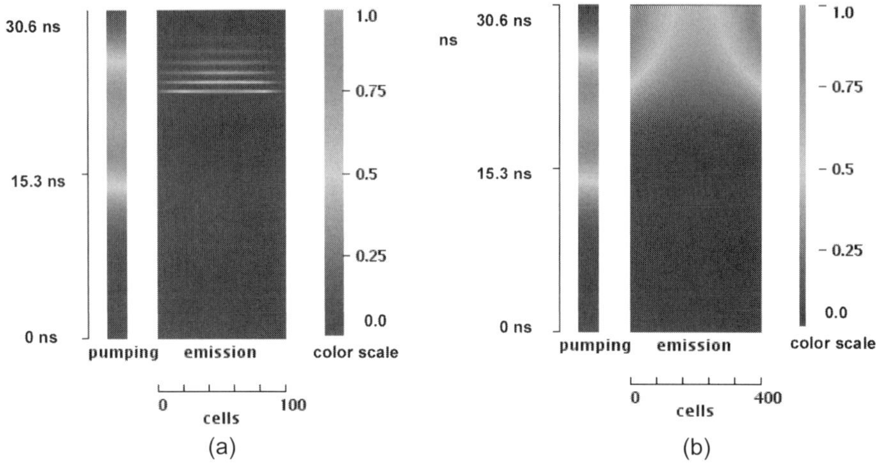

Figure 12.7 Calculated emission dynamics in (a) a strip of 100 cells; pumping intensity corresponds to 1 J/cm^2 of absorbed energy, r_{aver} = 8%, τ_{res} is 0.43 ps, and the maximum emission intensity is 24,566 arb. units. (b) Strip of 400 cells pumped with the same intensity as in (a), with no reflection at the boundaries between cells; τ_{res} is 0.67 ps, and the maximum emission intensity is 16 arb. units. [Reprinted with permission from Ref. 35. © (2004) by the American Physical Society.] (See color plate section.)

12.4.5 Random laser feedback in a weakly scattering regime: space masers and stellar lasers

For the case in which the photon mean free path l^* is larger than the characteristic size of the pumped medium L, only a small fraction of emitted radiation is scattered in the ensemble of particles, and lasing can occur only when the amplification g is sufficiently large. The threshold condition in this case can be easily derived for a simple configuration of an elongated amplifying cloud with two radiation fluxes, Φ_+ and Φ_-, propagating in opposite directions along the cloud (as shown in Fig. 12.8):[20,37]

$$\begin{aligned}\frac{1}{c}\frac{\partial \Phi_+}{\partial t}+\frac{\partial \Phi_+}{\partial x}&=g\Phi_+ +\chi\Phi_-\\ \frac{1}{c}\frac{\partial \Phi_-}{\partial t}-\frac{\partial \Phi_-}{\partial x}&=g\Phi_- +\chi\Phi_+.\end{aligned} \tag{12.11}$$

Here $\chi = 1/l^*$ is the backscattering coefficient per unit length (along the cloud), g is the gain coefficient corrected for parasitic absorption and parasitic scattering (in all directions), and it is assumed that $\chi << g, L^{-1}$. By solving this system of equations, one can find a condition for exponential growth of power with time (a threshold condition)

$$e^{gL}\frac{\chi}{g}>1. \tag{12.12}$$

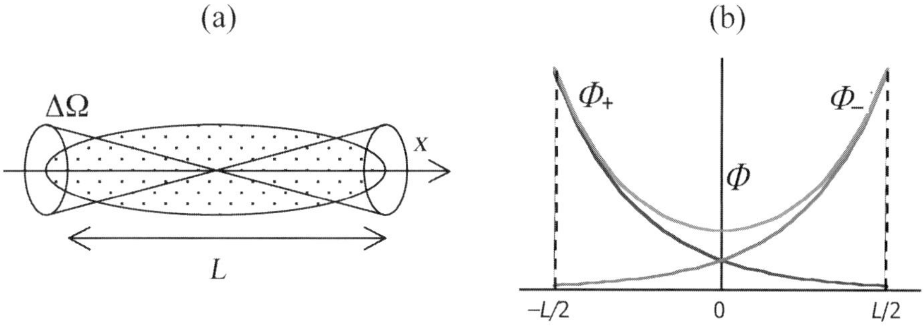

Figure 12.8 Elongated amplifying cloud with weak backscattering. (a) Angular distribution of radiation. (b) Intensity distribution along cloud: Φ_+, Φ_- and $\Phi = \Phi_+ + \Phi_-$. (Reprinted with permission from Ref. 20.)

This case is most likely realized in interstellar clouds of optically pumped HO radicals[38] with dust particles and electrons acting as scattering centers. Stimulated emission in such a natural maser ($\lambda = 18.5$ cm) was used to explain emission pulsations registered in radio astronomy experiments. Another type of space maser emitting at $\lambda = 1.35$ cm is based on rotational transitions of H_2O molecules.[39] The possibility of lasing effect in stellar atmospheres was predicted in Ref. 39, where OI-oxygen was considered to be an active laser medium and scattering (providing for nonresonant feedback) was assumed to be due to amplifying transitions of excited atoms.

Note that although it is weak, stimulated emission feedback plays an important role in cosmic sources of stimulated emission radiation. Thus, initial attempts to describe space masers in terms of a traveling-wave amplifier model without feedback were not highly successful.[38] The principal difference between stimulated emission with and without feedback is illustrated in Fig. 12.7. Studies of nonresonant feedback in space masers and stellar lasers (which were experimentally observed first in 1965[40]) as well as naturally occurring random lasers are summarized in Ref. 37.

12.4.6 Localization of light and random lasers

Disorder in crystals hinders electrical conductivity, and for some energies, electronic wavefunctions are spatially localized. This phenomenon, known as Anderson localization,[41] arises from interference of wavefunctions representing electrons. *Light localization* is an effect that arises from coherent multiple scattering and interference of light waves.[42,43]

In quantum mechanics, wave-like phenomena play a significant role when the characteristic scale of confinement or potential variation is comparable to the wavelength. In this case, wavefunctions describing quantum mechanical objects should be sought in the form of eigenfunction solutions of wave equations with appropriate boundary conditions. On the other hand, if characteristic dimensions of the system are macroscopic (much larger than the wavelength), the particle's motion can be treated as a classical trajectory, neglecting any wave interference effects.

In many studies of electromagnetic wave propagation in dielectrics, the scattering length l^* is longer than the correlation length a in spatial variation of the dielectric constant, which in turn, is much longer than the wavelength λ. This short-wavelength limit, in which interference corrections to the optical transport become less and less important, is described in terms of geometrical ray optics.

In the opposite long-wavelength limit $\lambda >> a$, the elastic scattering cross section decreases, and the scattering length increases as $\sim \lambda^4$ (Rayleigh scattering). Furthermore, the scattering term $(\omega/c)^2\, \varepsilon_{\text{fluct}}(x)E$ in a wave equation for light[43] is proportional to ω^2. (Here $\varepsilon_{\text{fluct}}(x)$ is the spatially fluctuating dielectric constant, E is the electric field, ω is the frequency, and c is the speed of light.) As $\omega \to 0$, the scattering strength approaches zero. Thus, in both high- and low-frequency limits, the electromagnetic field is not localized.

In disordered systems, wave interference effects lead to large spatial fluctuations in light intensity. At $l^* >> \lambda$, these fluctuations average out, resulting in essentially noninterfering, multiple scattering paths for electromagnetic transport.[43] However, when $l^* \to \lambda/2\pi$, interference between multiple scattering paths dramatically modifies the average transport properties, and a transition from extended to localized modes occurs.[42,43] According to the Ioffe-Regel criterion,[44] the condition of localization is $kl^* \approx 1$, where k is the wave number of a light wave. The dependence of the scattering length l^* on the light wavelength λ as well as the range of Anderson localization of photons are shown in Fig. 12.9.

Localized modes are the eigenfrequencies calculated as solutions of wave equations with appropriate boundary conditions. The widths of spectral lines $\delta\lambda$ are determined by the modes' losses. In the localization regime, $\delta\lambda$ is smaller than the average distance between modes $\Delta\lambda$. This condition, known as Thouless criterion $\delta\lambda/\Delta\lambda \leq 1$,[45,46] ensures that modes in neighboring subvolumes of a disordered medium do not spectrally overlap. Thus, energy transfer from one subvolume to another is prevented, providing for localization of light.[47]

Claims of the experimental demonstration of Anderson localization in microwave and optical spectral ranges have been made in Refs. 47–49. In spite of several publications available in the literature (e.g., Refs. 50 and 51), a comprehensive theory describing random laser emission in the strong localization regime has not yet been developed. Similarly to the case of photonic band crystals, Anderson localization should alter both spontaneous and stimulated emission in disordered systems. Thus, no true spontaneous emission is expected to occur. Instead, coupled eigenstates of the electronic degrees of freedom and electromagnetic modes of the dielectric are formed.[43] These photon–atom-bound states are optical analogs of electron impurity states in the gap of a semiconductor. When a collection of excited atoms is placed into a disordered system supporting light localization, excited atoms can transfer their bound photons to neighboring atoms by a resonant dipole–dipole interaction. The effective 'impurity' band defines a novel quantum many-body system in which the processes of spontaneous and stimulated emission of light are completely confined and mediated by photonic hopping conduction between atoms.[43]

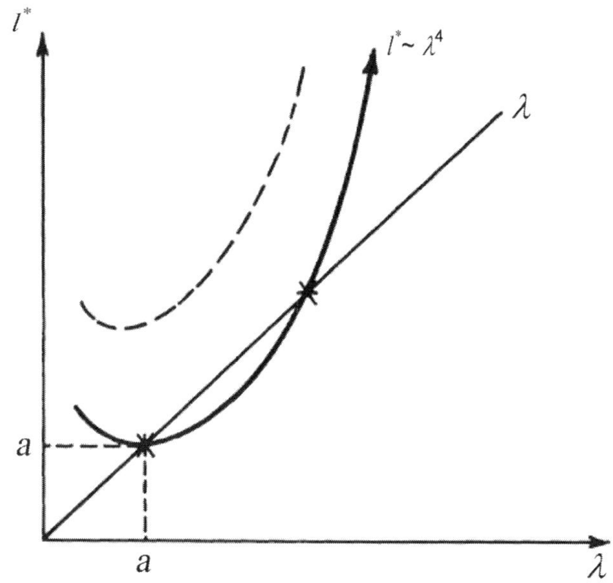

Figure 12.9 Scattering length l^* as a function of wavelength λ. In the long-wavelength Raleigh scattering limit, $l^* \sim \lambda^4$. In the short-wavelength range, $l^* \geq a$, where a is the correlation length. In a strongly scattering disordered medium (solid line), a localization range may exist for which $l^*/(2\pi\lambda) \approx 1$. Localization does not occur in a dilute system of scatterers (dashed line). [Reprinted with permission from Ref. 42. © (1984) by the American Physical Society.]

According to another set of arguments,[7,52,53] the photon scattering length l^* sets a distance for directional randomization of the optical wave vector. For $l^* < \lambda$, the coherence length for light is shorter than the wavelength because scattering events redirecting the wave also alter its phase. Light encountering such strong scattering cannot propagate even a single spatial period without being dephased and directionally randomized.[7] This forms a perfectly incoherent field with infinitesimally fine ($\ll \lambda$) spatial structure that corresponds to nonpropagating evanescent waves.[7,16,52,53] It has been argued[7,52,53] that for $l^* < \lambda$ and in the presence of gain, amplified spontaneous emission (ASE) with a coherence length shorter than a wavelength will be generated in subwavelength volumes. This suggests that laser action may be possible in high-quality subwavelength cavities enclosing a localizing random medium that cannot support constructive interference. In this case, no narrow lines corresponding to coherent random laser modes should be expected in the stimulated emission spectra. (As claimed by the authors of Refs. 52 and 53, the random laser discussed in Sec. 12.7.3 operates in this high-scattering regime. However, many researchers are skeptical about the possibility of this feedback mechanism.) A similar result (the absence of resolvable lasing modes) is expected in the regime of classical diffusion, but for a completely different reason. In the case of diffusion, the density of modes is so high that they cannot be resolved.

12.5 Neodymium Random Lasers with Nonresonant Feedback

12.5.1 First experimental observation of random lasers

In 1986, Markushev et al., conducting routine emission spectroscopy experiments with powders of neodymium-activated luminophosphors ($Nd{:}La_2O_3$,[54] $Nd{:}La_2O_2S$,[54] and $Na_5La_{1-x}Nd_x(MoO_4)_4$[55]), found that above a certain pumping energy threshold, the duration of the emission pulse shortened by approximately four orders of magnitude.

An approximately equally strong enhancement was found in the intensity of the strongest spectral component of the $^4F_{3/2}$–$^4I_{11/2}$ emission transition (≈ 1.06–1.08 μm), for which the linewidth narrowed significantly.[54–56] The energy level diagram of Nd^{3+} ions and relevant excitation and relaxation processes are shown in Fig. 12.10.

Only one narrow emission line was observed in the spectrum above the threshold, as shown in Fig. 12.11. The intensity of this emission line plotted versus pumping energy resembled the input-output dependence in conventional lasers (see Fig. 12.12). Since the experimental behavior of the observed radiation was characteristic of lasers, Markushev et al. explained the behavior in terms of stimulated emission from excited powders.[54,55]

Note that in 1981, Nikitenko et al. reported stimulated emission in ZnO powder.[57] However, the communication of Ref. 57 was very brief and lacked many important details; for this reason, the credit for the first experimental demonstration of a random laser should be given to Markushev et al.[54,55]

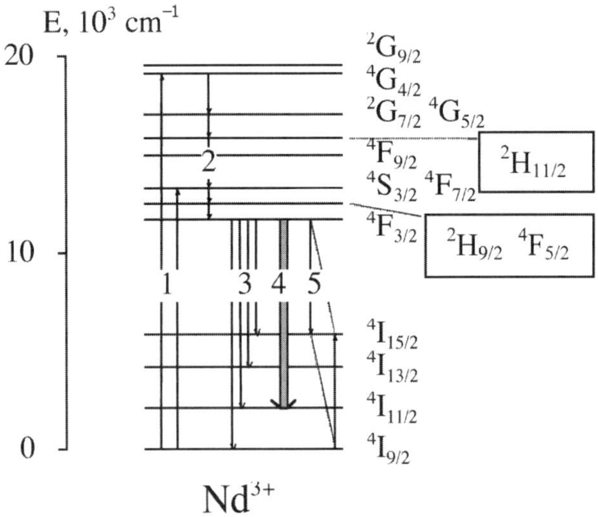

Figure 12.10 Nd^{3+} energy level diagram and the most important excitation and relaxation processes: (1) pumping, (2) multiphonon relaxation populating the metastable level $^4F_{3/2}$, (3) spontaneous radiation and multiphonon relaxation of the level $^4F_{3/2}$, (4) stimulated emission at the transition $^4F_{3/2}$–$^4I_{9/2}$, and (5) cross relaxation. (Reprinted with permission from Ref. 31.)

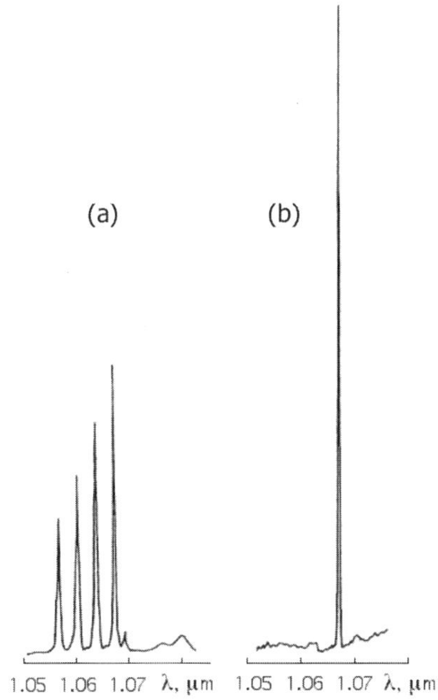

Figure 12.11 Emission spectrum of Nd^{3+} ($^4F_{3/2}$–$^4I_{11/2}$ transition, T = 77K) (a) below and (b) above the threshold. (Reprinted from Ref. 55.)

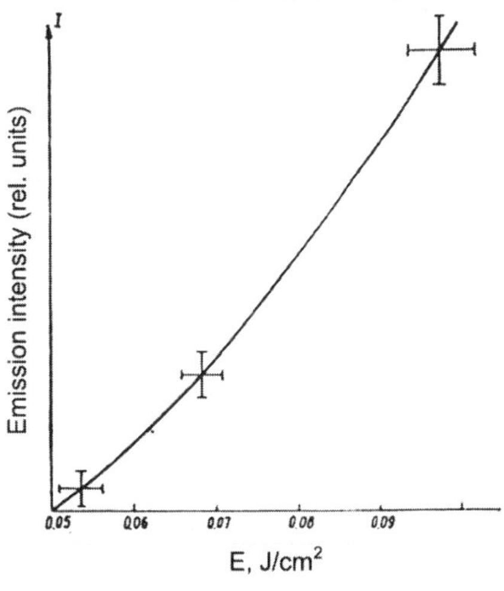

Figure 12.12 Input-output curve of stimulated emission in $Na_5La_{1-x}Nd_x(MoO_4)_4$ powder. (Reprinted from Ref. 55.)

12.5.2 Emission kinetics in neodymium random lasers

In a series of studies that followed the pioneering works,[54,55] the list of neodymium-activated random lasers was significantly extended. Of special interest is the emission kinetics in neodymium random lasers.

In different experiments, the duration of the pumping pulse varied between ~10 and ~30 ns.[31,55,58,59] At weak pumping, only spontaneous emission of neodymium ions can be observed. With an increase of the pumping energy, the regime of ASE predominates over regular spontaneous emission, causing (*i*) an increase in the peak emission intensity, (*ii*) a narrowing of the strongest emission spectral line, and (*iii*) a shortening of the emission decay kinetics. As the pumping energy is further increased to reach the critical threshold level, one short and very intense emission pulse appears close to the end of the pumping pulse.[31,59,60] The duration of this pulse varies in different experiments between ~0.3 and 10 ns.

At stronger pumping, a second emission pulse, separated from the first by several nanoseconds, emerges in the kinetics. As the pumping energy is further increased, the number of pulses increases, the pulses become shorter, and the time delay between the beginning of the pumping pulse and the first emission pulse becomes shorter.[31,58,59] The emission kinetics recorded in $NdAl_3(BO_3)_4$ powder at different pumping energies are depicted in Fig. 12.13.[31]

Figures 12.13 and 12.4 show a remarkable similarity between the experimental emission kinetics observed in a $NdAl_3(BO_3)_4$ random laser and the calculation using Eq. 12.10. This similarity, together with a fairly accurate fit of the lasing threshold (as described in Sec. 12.4.3) suggest that Eq. (12.10) adequately describes stimulated emission in neodymium random lasers, both qualitatively and quantitatively.

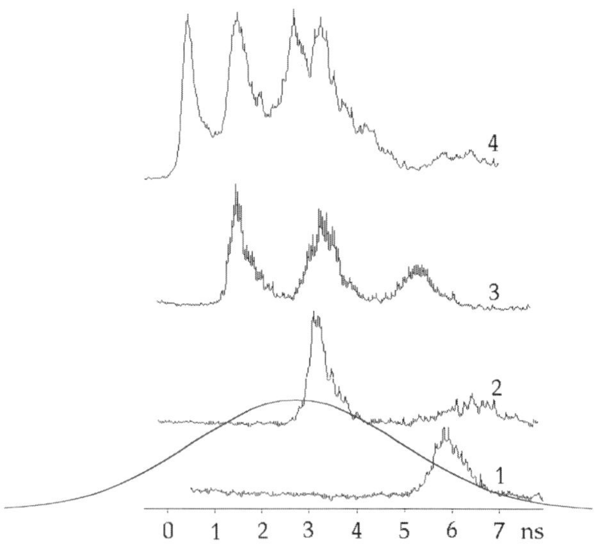

Figure 12.13 Pulses of stimulated emission in $NdAl_3(BO_3)_4$ powder (1) near the threshold (200 mJ/cm^2), (2) at x = 1.6 times the threshold energy, (3) x = 1.9, and (4) x = 3.9. The bell-shaped line in the bottom of the figure shows the approximate position and shape of the pumping pulse (λ_{pump} = 532 nm). The mean linear size of powder particles in this experiment was 3.6 μm. (Reprinted with permission from Ref. 31.)

12.5.3 Analysis of speckle pattern and coherence in neodymium random lasers

In most neodymium random lasers, the scattering length l^* (~ 6 μm) is longer than the wavelength λ (~ 1 μm) and much shorter than the size of the pumped volume L (several tens of micrometers to ~1 mm).[60,61] This defines the diffusive regime of operation in which nonresonant feedback supports low-coherence stimulated emission. Experimental studies of coherence in neodymium random lasers have been carried out in Refs. 59, 62, and 63.

The contrast of a speckle pattern is one of the important parameters characterizing coherence of a light source.[16] In order to evaluate the degree of coherence in $NdCl_3$:H_2O and $Nd_{0.75}La_{0.25}P_5O_{14}$ random lasers, their speckle patterns have been studied in Ref. 59. Figure 12.14(a) depicts a polarized speckle pattern of 1.06-μm Nd:YAG laser light scattered off a random laser medium. Its intensity distribution histogram $P(I)$ corresponds to that of a highly coherent light source with the contrast close to unity.[59] The corresponding degree of coherence was calculated to be $\sigma/\langle I\rangle = 0.98$. (Here $\langle I\rangle$ is the mean intensity and σ is the variance of the distribution.) The analogous speckle patterns from 0.3-ns emission pulses of a $NdCl_3$:$6H_2O$ random laser and 1-ns pulses of a $Nd_{0.75}La_{0.25}P_5O_{14}$ (Nd:LaPP) random laser had much smaller contrasts. The corresponding intensity histogram is shown in Figure 12.14(b). The degree of coherence of the neodymium random laser emission was calculated to be $\sigma/\langle I\rangle = 0.14$ in $NdCl_3$:$6H_2O$ and $\sigma/\langle I\rangle = 0.08$ in Nd:LaPP.

Qualitatively similar results were obtained in Ref. 63, where the speckle patterns of $Al_3Nd(BO_3)_4$ and NdP_5O_{14} random lasers were analyzed. In Ref. 62, the coherence of a $Nd_{0.5}La_{0.5}Al_3(BO_3)_4$ random laser was studied in a Young interferometer. In agreement with the studies in Refs. 59 and 63, it was found that two subareas on an emitting surface separated by only 80 μm lack any spatial coherence. These results are in line with the theoretical prediction of Letokhov,[2] who proposed that a 'stochastic resonator' in the form of a scattering medium constitutes a system with a large number of modes that are strongly coupled by scattering and have large radiation losses. If the number of interacting modes is sufficiently large, the feedback becomes nonresonant. Correspondingly, emission of such a "super-multimode" laser with photon exchange between modes has low coherence.

12.6 ZnO Random Lasers with Resonant Feedback

12.6.1 Narrow modes in emission spectra

Cao et al.[64] investigated 300- to 350-nm-thick ZnO films composed of 50- to 150-nm columnar grains. The in-plane pattern of deposited particles was highly disordered, resulting in strong optical scattering of the film. The scattering mean free path, measured using the coherent backscattering technique, was of the order of 0.4 μm.

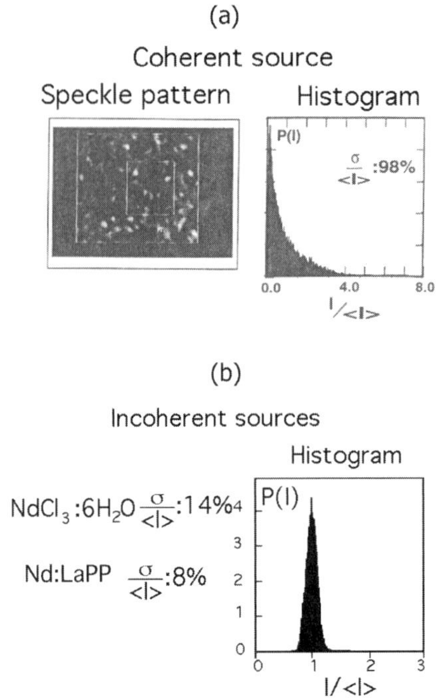

Figure 12.14 (a) Speckle pattern of a coherent layer pumping light scattered off a NdP_5O_{14} random laser sample. (b) Analogous distribution diagram of the NdP_5O_{14} random laser emission. (Reprinted with permission from Ref. 59.)

Samples were optically pumped with 15-ps pulses of a frequency-tripled Nd:YAG laser at 355 nm. At low pumping intensity, a spontaneous emission band corresponding to the transition from the conduction band to the valence band of ZnO with the maximum at ≈ 387 nm and the full width at half maximum (FWHM) equal to ≈ 10 nm was observed (see Fig. 12.15). Note that the emission wavelength was nearly equal to the scattering mean free path, $\lambda \approx l^*$.

As pumping power increased, the emission band narrowed due to amplification of spontaneous emission that was particularly strong in the vicinity of the maximum of the gain band. When the pumping energy exceeded the critical threshold level, narrow emission lines, with the FWHM smaller than 0.4 nm, appeared in the spectrum [Fig. 12.15(b)]. The number of narrow lines in the emission spectrum increased with a further increase in pumping energy [Fig. 12.15(c)]. The dependence of the integrated emission intensity on the pumping energy resembled a typical input-output curve for regular lasers [inset of Fig. 12.15(b)]. The emission spectrum varied with the observation angle or when the pumped spot was moved along the sample.

A qualitatively similar emission behavior has been observed in ZnO microlasers: ~1-μm spheres formed by 50-nm ZnO particles.[65,66] Of special interest was the study of a near-field pattern of random laser emission. Below the threshold, the spatial distribution of emission intensity in a pumped spot was nearly uniform, resembling that of the pumping.[65,66] At the lasing threshold, when the first narrow lines emerged in the emission spectrum, two or several bright spots appeared in the near-field emission pattern in Fig. 12.16. With a further increase in pumping energy, the number of bright spots in the near-field emission increased, as did the number of narrow emission spectral lines.

It has been inferred that narrow spectral lines and corresponding bright spots in the near-field pattern belong to *spatially confined modes* of stimulated emission. With an increase in the size of a pumped spot, the number of localized modes increases, resulting in an increase in the number of narrow emission lines in the spectrum.

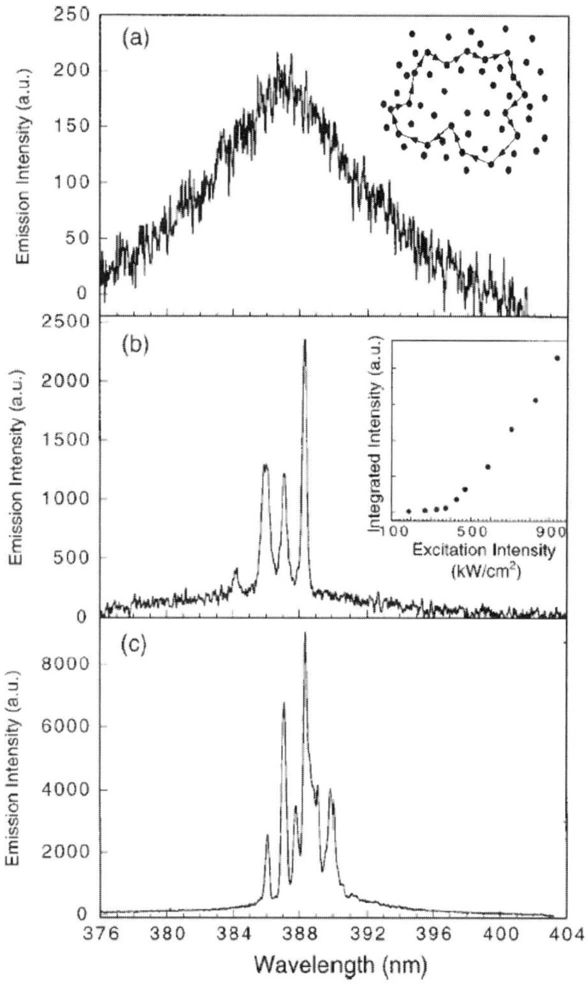

Figure 12.15 Emission spectra of a ZnO film at *absorbed* pumping intensities equal to (a) 330 kW/cm^2, (b) 380 kW/cm^2, and (c) 600 kW/cm^2. Inset in panel (a) is a schematic diagram showing the formation of a closed-loop path for light through multiple optical scattering in a random medium. Inset in panel (b) shows the integrated emission intensity as a function of pumping intensity. [Reprinted with permission from Ref. 64. © (1998) by the American Institute of Physics.]

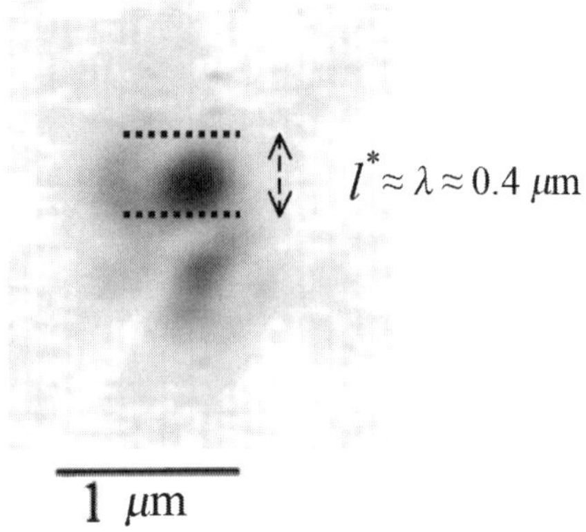

Figure 12.16 Spatial distribution of emission intensity above the threshold. The incident pump pulse energy is 0.25 nJ. The characteristic size of the lasing mode is of the order of l^* and λ. [Reprinted with permission from Ref. 65. © (2000) by the American Physical Society.]

12.6.2 Photon statistics in a ZnO random laser

The photon statistics of a ZnO random laser was studied in Ref. 67. The experimental sample was a cold-pressed pellet composed of 80-nm ZnO particles. The diameter of the focused laser spot on the sample was ≈ 15 μm, and the photon mean scattering path was ≈ 2.3 λ (0.9 μm).

Random laser emission was experimentally studied using a streak camera with 2-ps time resolution attached to the output port of a spectrometer. This setup allowed one to monitor emission dynamics in several laser modes (narrow spectral lines) simultaneously. A typical two-dimensional (time–wavelength) streak camera image of ZnO random laser emission is shown in Fig. 12.17.[68]

The photon statistics were measured in a rectangular area $\Delta t \Delta \nu$ of a streak camera image in the maximum of an emission pulse in one of the strongest lasing modes. The sampling time Δt was chosen to be shorter than the coherence time of the radiation field $1/\Delta \nu$. Thus, the area $\Delta t \Delta \nu < 1$ corresponded to a single electromagnetic mode.

For single-mode *coherent* light, the photon number distribution $P(n)$ is Poissonian (as in a simple harmonic oscillator):

$$P(n) = \frac{\langle n \rangle^n e^{-n}}{n!}, \tag{12.13}$$

where $\langle n \rangle$ is the average photon number.[16,69] In contrast, for single-mode *chaotic* light, $P(n)$ corresponds to the Bose-Einstein distribution

$$P(n) = \frac{\langle n \rangle^n}{\left(1 + \langle n \rangle\right)^{n+1}} . \tag{12.14}$$

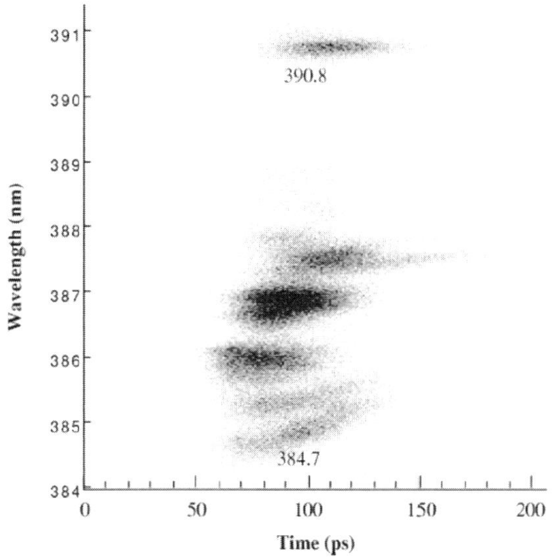

Figure 12.17 Typical two-dimensional streak camera image of ZnO random laser emission. [Reprinted with permission from Ref. 68. © (2003) by the American Physical Society.]

Thus, by analyzing the experimentally measured photon number distribution $P(n)$, one can determine the degree of coherence of laser emission.[68] The photon number distributions $P(n)$ measured in a ZnO random laser at different pumping energies are shown in Fig. 12.18.

One can see that the photon statistics changes continuously from Bose-Einstein statistics at the threshold to Poissonian statistics well above the threshold. The Poissonian photon statistics (typical of regular lasers and harmonic oscillators in general) that were observed in this experiment are evidence of both the coherent character of stimulated emission and resonant feedback in a ZnO random laser.

12.6.3 Modeling of a ZnO random laser

To explain the coherent character of stimulated emission in a ZnO random laser, it was proposed[66] that, due to strong scattering, emitted photons may return to the same positions where they were born or visited before, thereby forming closed-loop paths, as seen in the inset of Fig. 12.15(a). Such loops can serve as ring resonators for laser light, determining the frequencies of narrow stimulated emission lines. Since different resonators (localized modes) have different losses and support different oscillation frequencies, different stimulated emission lines appear in the spectrum at different thresholds.

To model the near-field emission pattern and the spectrum of stimulated emission in a ZnO random laser, the electromagnetic field distribution in a random medium was calculated by solving Maxwell's equations using a finite-difference time-domain (FDTD) method.[65] The gain was introduced into the system via negative conductance σ, and the randomness via spatially varying dielectric constant ε. The calculations show that when the optical gain exceeds the threshold value, the electromagnetic field oscillation starts to build up in the time domain. Using the Fourier transform of the time domain data, the emission spectrum was calculated.[65] In agreement with experiments, one narrow spectral line was predicted to appear in the emission spectrum above the threshold, accompanied by several bright spots emerging in the near-field emission pattern close to the center of the pumped area (compare Figs. 12.15, 12.16, and 12.19).

An independence of the near-field emission distribution and of the emission spectrum of the boundary conditions indicates that the lasing mode is formed deep inside the disordered medium and can be treated as a *localized* mode.[65] The question as to whether this mode confinement can be regarded as Anderson localization remains open. In a ZnO random laser, the Ioffe-Regel criterion[44] for

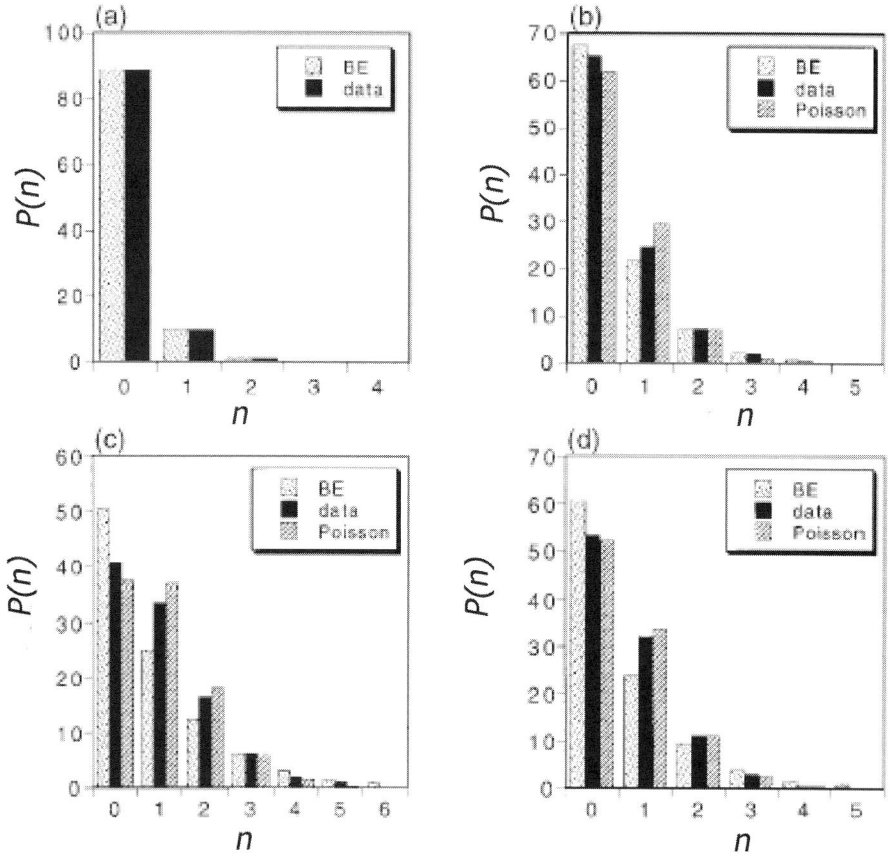

Figure 12.18 The solid columns are the measured photon count distribution of emission from a ZnO pellet. The dotted (dashed) columns are the Bose-Einstein (Poissonian) distribution of the same count mean. The incident pump intensity is (a) 1 ×, (b) 1.5 ×, (c) 3.0 ×, and (d) 5.6 × the threshold intensity where discrete spectral peaks appear. [Reprinted with permission from Ref. 67. © (2001) by the American Physical Society.]

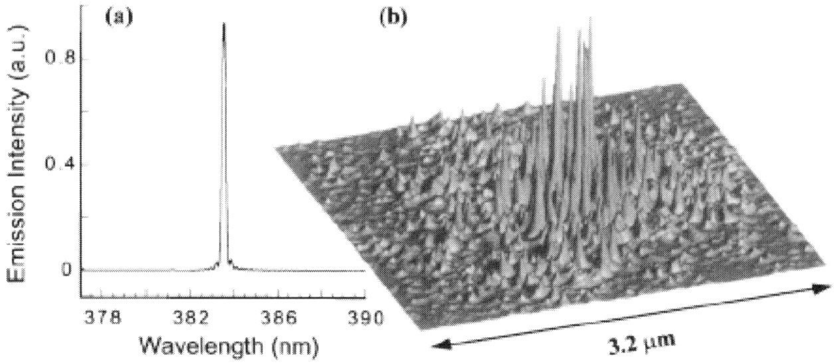

Figure 12.19 (a) Calculated emission spectrum. (b) Calculated spatial distribution of emission intensity in the random medium. [Reprinted with permission from Ref. 65. © (2000) by the American Physical Society.]

Anderson localization[41–43] is not satisfied ($kl^* \approx 3$), but the Thouless criterion[45] is satisfied ($\delta\lambda/\Delta\lambda \leq 1$).[65] In the literature, Anderson localization is defined and studied in passive systems. Research in random lasers may require its redefinition for the case of an active medium.

12.7 Stimulated Emission Feedback: From Nonresonant to Resonant and Back to Nonresonant

12.7.1 Mode density and character of stimulated emission feedback

An effective three-dimensional random laser cavity formed by scatterers can be treated in the first approximation as a closed (three-dimensional) resonator with losses. This suggests that in macroscopic volume V the number of modes N per frequency interval $\Delta\nu$ around the optical frequency ν can be very large. According to the Rayleigh-Jeans formula,

$$N = 4\pi\frac{V}{c^3}\nu^2\Delta\nu = 4\pi\frac{V}{\lambda^4}\Delta\lambda, \tag{12.15}$$

where $\lambda = c/\nu$ is the wavelength, $\Delta\lambda = \Delta\nu\lambda^2/c$ is the frequency interval, and c is the speed of light.

For the set of parameters characteristic of neodymium random lasers ($V = 0.1$ mm^3 and $\lambda = 1$ μm), the number of modes per $\Delta\lambda = 1$ nm is $N = 1.25 \times 10^6$ (very large!). Such modes cannot be resolved, since the mode width $\delta\nu$ significantly exceeds the spacing between modes $\Delta\nu$. A 'stochastic resonator' formed by scatterers constitutes a system with a very large number of modes that are strongly coupled by scattering and have large radiation losses.[2] Radiation losses and strong interaction between modes lead to a complete overlap of their frequency spectra. Here, the concept of 'mode' loses its usual meaning, and the spectrum becomes a continuum. If the number N of interacting 'modes' is sufficiently large, the feedback becomes nonresonant.

If one substitutes into Eq. (12.15) the volume $V \approx 5 \times 10^{-12}$ cm^3 and the wavelength $\lambda = 0.38$ μm of a ZnO random microlaser,[65,66] one obtains the mode density $N/\Delta\lambda \approx 3 \times 10^7$ cm^{-1}, corresponding to three modes per nanometer. If, instead, the volume of a localized mode is $V \approx 4/3\pi(0.4\ \mu\text{m})^3 \approx 2.6 \times 10^{13}$ cm^3 (Fig. 12.16), then $N/\Delta\lambda$ becomes $\approx 16.5 \times 10^5$ cm^{-1}, which corresponds to one mode per 6 nm. The experimentally observed mode density, $N/\Delta\lambda \approx$ one mode per nanometer (Fig. 12.15), perfectly fits the range of calculated parameters presented above. The mode spacing in a ZnO random laser exceeds the mode width, $\Delta\nu > \delta\nu$; likewise, the modes are well resolved and not mixed. Correspondingly, the feedback within each mode is resonant, and the simulated emission is coherent.

12.7.2 Transition from the nonresonant to the resonant regime of operation

Random lasers based on the suspension of ZnO nanoparticles (scatterers) in rhodamine 640 dye solution (gain medium) have been studied in Refs. 70 and 71. The concentration of dye molecules in the solution was 5×10^{-3} M. At a small concentration of scatterers (2.5×10^{11} cm^{-3}), a narrowing of the emission band to 5 nm (Fig. 12.20) and an increase in the slope of the input-output curve above the threshold were observed as the pumping energy increased. No narrow lines corresponding to individual coherent modes were seen in the spectrum. This behavior is typical of random lasers with nonresonant feedback.

The character of stimulated emission changed dramatically when the concentration of particles was quadrupled to 1×10^{12} cm^{-3}. In this case, discrete narrow lines ($\delta\lambda \sim 0.2$ nm) appeared on top of a much broader (~ 5 nm) emission band above the threshold, as shown in Fig. 12.21.

The stimulated emission at a high concentration of scatterers appeared qualitatively to be similar to that in a ZnO random laser (see Sec. 12.6.1). Correspondingly, the observed effect was explained in terms of coherent (resonant) feedback, where the interference of scattered waves created a quasi-stationary distribution of the electromagnetic field and determined the resonant lasing frequencies. (Note that because of continuous Brownian motion of scattering particles, the positions of narrow emission lines changed from pulse to pulse.)

The behavior of stimulated emission was particularly interesting at the intermediate concentration of scatterers: 6×10^{11} cm^{-3}. In this case, as the pumping intensity was increased, a narrowing of the emission band (similar to that in a low-scattering solution) occurred first. Then, at a higher pumping energy threshold, discrete narrow peaks (similar to those in highly scattering solution) emerged on the top of the narrowed emission band, as shown in Fig. 12.22.

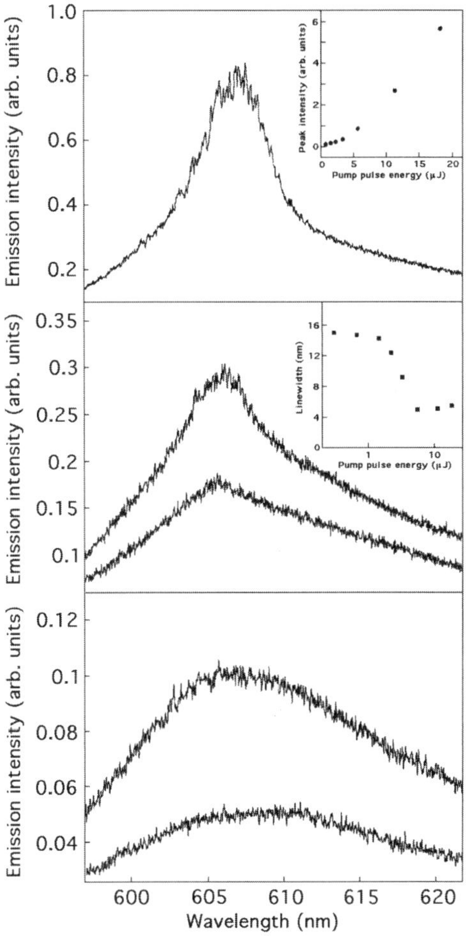

Figure 12.20 Emission spectra corresponding to pumping energies (from bottom to top): 0.74, 1.35, 1.7, 2.25, and 3.4 mJ. The ZnO particle density is ≈ 6×10^{11} cm^{-3}. Inset of (a): input-output dependence of emission. Inset of (b): dependence of emission linewidth on pumping intensity. (Reprinted from Ref. 71 with kind permission from Springer Science+Business Media.)

The experimental results presented above have been explained[71] in terms of quasi-state—eigenmode solutions of Maxwell's equations in a finite-sized random medium. It has been argued that losses in quasi-states are due to (*i*) losses at the medium–air boundaries and (*ii*) transfer of energy to other quasi-states. Energy coupling redistributes the number of photons between quasi-states but does not affect the loss in an ensemble of quasi-states as a whole. Thus, stimulated emission is first achieved in an ensemble of (more strongly or more weakly coupled) quasi-states that have relatively low loss, which corresponds to the regime of nonresonant feedback.[20] At the higher energy threshold, the laser action occurs in individual quasi-states that have higher loss.[71] This is the regime of coherent feedback.

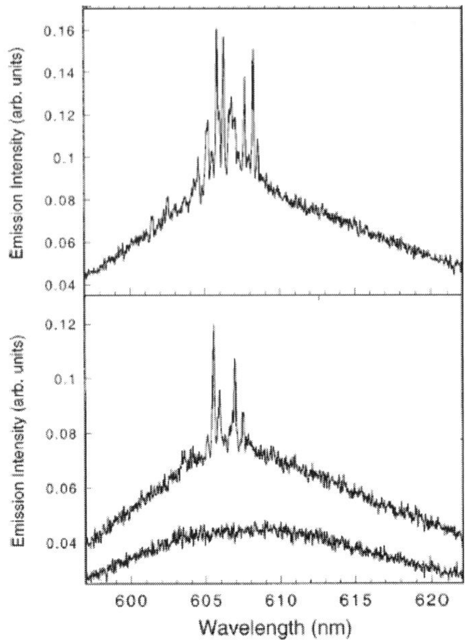

Figure 12.21 Emission spectra in a solution of rhodamine 640 laser dye (5×10^{-3} M) with a high concentration of ZnO scatterers (1×10^{12} cm^{-3}). The pumping pulse energy is 0.68 μJ (bottom trace),1.1 μJ (middle trace), and 1.3 μJ (upper trace). (Reprinted from Ref. 71 with kind permission from Springer Science+Business Media.)

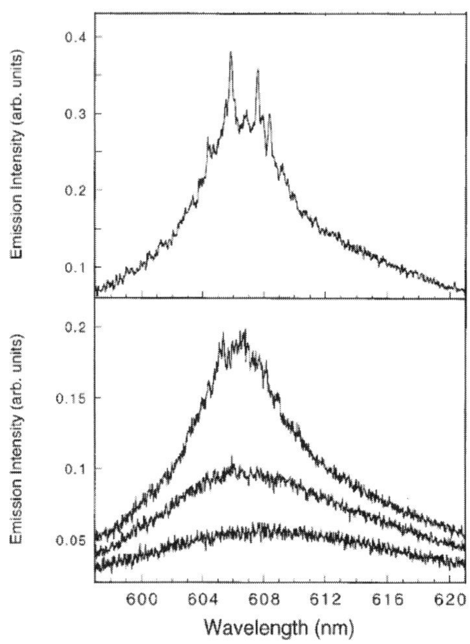

Figure 12.22 Emission spectra in a solution of rhodamine 640 laser dye (5×10^{-3} M) with a medium concentration of ZnO scatterers (6×10^{11} cm^{-3}). The pumping pulse energy is (from bottom trace to top trace) 0.74 μJ, 1.35 μJ, 1.7 μJ, and 2.25 μJ. (Reprinted from Ref. 71 with kind permission from Springer Science+Business Media.)

One should note that discrete (nondegenerate) laser modes such as those in Figs. 12.21 and 12.22 are states with different photon energies. Thus, a coupling of quasi-states is possible only if there is a mechanism by which the frequency of laser oscillation may change. Acoustic vibrations, Doppler shift due to moving scattering particles, and random frequency modulation in the active medium were discussed in Ref. 20 as possible mechanisms of mode mixing.

12.7.3 Nonresonant feedback in the regime of ultrastrong scattering: electron-beam-pumped random lasers

The majority of random lasers are optically pumped through the sample surface. In such a pumping geometry, scattering, which is necessary to provide stimulated emission feedback, also plays the negative role of reflecting pumping light and reducing pumping efficiency. The thickness of the pumped layer and, correspondingly, the quality factor of an effective resonator formed by scatterers can be increased if electron-beam pumping is used instead of optical pumping.[7,52,53,72,73] In this case, the penetration depth of pumping can be controlled by the energy of the electron beam.[52]

Stimulated emission in Ce^{3+}-doped δ-alumina (δ-Al_2O_3) was studied in Refs. 52, 53, and 74. In the experiment described in Ref. 53, the mean particle size was 20 nm, and the photon scattering path determined in the coherent backscattering experiment was $l^* = 114$ nm at the wavelength $\lambda = 363.8$ nm. Lightly compressed powder was placed in an ultrahigh-vacuum chamber and irradiated with an electron beam. The spectrum of cathodoluminescence consisted of two practically unresolved bands, corresponding to the transitions 5d–4f $^2F_{5/2}$ (higher energy) and 5d–4f$^2F_{7/2}$ (lower energy), as shown in Fig. 12.23.

At a small pumping current, the emission band had a maximum at approximately 2.6×10^4 cm^{-1} (≈ 0.385 μm) and a FWHM equal to 6.6×10^3 cm^{-1}. With an increase in the pumping current above the threshold, the emission band narrowed to 4.8×10^3 cm^{-1}, and its maximum shifted to approximately 2.8×10^4 cm^{-1} (≈ 0.36 μm). The dependence of the emission intensity (at $\lambda = 362$ nm) on the pumping current is a combination of two straight lines with different slopes, as shown in Fig. 12.24. The change in the slope occurs at exactly the same threshold at which the emission band starts narrowing (see the inset of Fig. 12.24.)

The experimental results presented above were explained by an onset of cw electron-beam-pumped stimulated emission at the Ce^{3+} 5d–4f $^2F_{5/2}$ transition. In fact, in the case of lasing, only the emission intensity at the laser transition continued to grow above the threshold, while the emission intensity at the other transition originating from the same upper laser level (5d–4f $^2F_{7/2}$) saturated at its threshold value. This explains the experimentally observed spectral changes. The fact that the input-output curve consisted of straight lines (as opposed to an exponential function) was invoked as evidence that the observed phenomenon was stimulated emission with feedback as opposed to amplified spontaneous emission in open paths.[53]

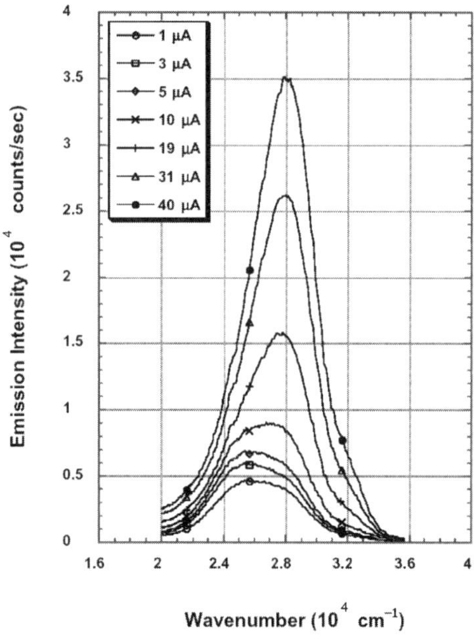

Figure 12.23 Cathodoluminescence spectra of Ce:δ-Al_2O_3 nanoparticles excited by various electron-beam current levels (4 keV, 2-mm beam diameter). [Reprinted with permission from Ref. 53. © (2001) by the American Physical Society.]

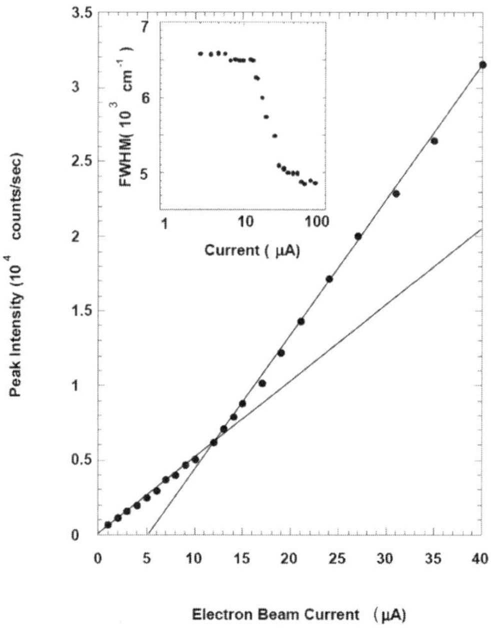

Figure 12.24 Input-output curve in Ce:δ-Al_2O_3 electron-beam-pumped random laser. Inset: dependence of the FWHM of the emission band on the pumping current. [Reprinted with permission from Ref. 53. © (2001) by the American Physical Society.]

The fact that the stimulated emission threshold is first achieved at the transition terminating at the ground state $4f\,^2F_{5/2}$ (three-level laser scheme) rather than at the empty excited state $4f\,^2F_{7/2}$ (four-level laser scheme) can be explained if one assumes that the transition terminating at $4f\,^2F_{5/2}$ has a higher cross section than the transition terminating at $4f\,^2F_{7/2}$ and that the ground state $4f\,^2F_{5/2}$ is significantly depopulated by the strong pumping.

The spectrum of Ce:δ-Al_2O_3-stimulated emission lacked any sharp lines that could be attributed to coherent laser modes (see Fig. 12.23). In addition, no speckle pattern was observed in the cerium random laser emission.[53] These two facts served as evidence of the low coherence of the Ce:δ-Al_2O_3 random laser. The low degree of coherence was explained by a very small (most likely the smallest known among all random lasers) ratio of the photon scattering path, $l^* =$ 114 nm, to the wavelength, ~ 385 nm. It has been argued[7,53] that if emitted electromagnetic energy is significantly attenuated by scattering at a distance shorter than a wavelength, the emitted light does not propagate or diffuse at all. Instead, light experiences three-dimensional evanescent attenuation, and no interference is required to produce frequency selectivity.[53] (This effect was discussed in more detail in Sec. 12.4.6.) Qualitatively similar results have been observed in electron-beam-pumped praseodymium[52,53,74] and neodymium[73] nanopowder random lasers.

12.8 Summary of Various Random Laser Operation Regimes

In this chapter, several representative classes of random lasers demonstrating strongly dissimilar properties have been discussed. The argument was made that the difference in operating regimes of random lasers is determined by the type of feedback, the strength of the photon scattering, and the experimental conditions. The predominant characteristics of feedback in random lasers are outlined in four subsections below. Note that the boundaries between different regimes are fuzzy and depend on conditions of the experiments as well as the particular random laser properties being studied. For example, following the arguments in Sec. 12.7.1, the same random laser medium, depending on its volume, can operate in the resonant regime or in the diffusive regime. In turn, the random laser volume and shape, as well as the losses of stimulated emission modes, can be controlled by, e.g., the penetration depth of pumping. Therefore, the following classification, although useful, is rather approximate.

12.8.1 Amplification in open paths: the regime of amplified stimulated emission without feedback

In the regime of amplified stimulated emission without feedback, the emission grows exponentially as it moves toward the ends of open paths; no relaxation oscillations are predicted in the emission kinetics. In Ref. 36, narrow lines in random laser spectra are explained by light amplification in very long and very rare open light paths.

12.8.2 Extremely weak feedback

In the case of extremely weak feedback, the mean scattering length l^* is longer than the size of the gain medium L and much longer than the wavelength λ. This type of feedback is realized in space masers and stellar lasers characterized by pulsating emission of very high spectral brightness.

The difference between this case and the previous case (stimulated emission without feedback) is fundamental. In the presence of feedback, the population inversion n in a small volume ΔV at time $t+\Delta t$ depends on the photon energy density E in the same volume at time t. Similarly, the energy density E in the volume ΔV at time $t+\Delta t$ depends on the population inversion n in the same volume at time t. This situation can be described by a system of coupled differential rate equations for n and E that predicts relaxation oscillations and confinement of stimulated emission to the interior of the random laser sample. One can argue that since spatial nonuniformity of pumping causes spatial nonuniformity of the refractive index, some feedback, although a very small amount, is always present in random lasers.

12.8.3 Medium-strength feedback: diffusion

In the diffusion regime, the photon scattering length l^* is much longer than the wavelength λ and much shorter than the size of the gain medium L. Stochastic resonators formed by scatterers support a very large number of leaky modes that interact with each other, with widths $\delta\nu$ significantly exceeding the intermode spacing $\Delta\nu$. This is the case of nonresonant feedback, characterized by low coherence of stimulated emission and Bose-Einstein (blackbody-like) photon statistics. The stimulated emission wavelength in this case is determined by the maximum of the gain spectrum rather than by resonances of any individual modes.

The threshold intensity I_{th} in random lasers is inversely proportional to the photon residence time τ_{res} in a gain volume. In the diffusion regime, the average square of a photon deviation from its original position $<\Delta x^2>$ is proportional to time t. Accordingly, for three-dimensional (sphere-like) random laser volumes, I_{th} is approximately $V^{-2/3}$, and for flat (pancake-like) random laser layers, I_{th} is approximately V^{-1}, where V is the volume of the uniformly pumped random laser sample.

12.8.4 The regime of strong scattering

Strong feedback realized for $\lambda \leq l^* << L$ facilitates formation of spatially confined stimulated emission modes radiating light in narrow spectral lines. Random laser emission in this regime exhibits Poissonian statistics and has the high-coherence characteristic of lasers with resonant feedback. Random laser modes are resolved spectrally and often spatially. Wave interference phenomena play an important role in the mode formation. Although the spatial extent of resonant modes can significantly exceed the emission wavelength, this regime of random laser operation may be considered as a precursor to Anderson localization of light in a gain medium. (In fact, in this case, the Thouless criterion for localization, $\Delta\nu > \delta\nu$, is satisfied, while the Ioffe-Regel criterion, $kl^* \leq 1$, is not.)

Random laser emission in the regime of Anderson localization, which relies on strong interference of scattered waves and is expected at $kl^* \leq 1$, has never been experimentally observed, and its spectroscopic and kinetics properties are unknown. Instead, a stimulated emission with nonresonant feedback has been observed in a random laser with $kl^* \leq 1.85$ (the smallest kl^* product reported in the literature).[53] It has been argued that an effective random laser cavity smaller than a half-wavelength cannot have any standing waves and because of this cannot support any coherent spectrally resolved modes. Instead, stimulated emission in this case is emitted into evanescent (not propagating) modes with highly nonresonant feedback.

Acknowledgments

This work was supported by the NSF PREM grant # DMR 0611430, the NSF CREST grant # HRD 0317722, the NSF NCN grant # EEC-0228390, and the NASA URC grant # NCC3–1035.

References

1. V. S. Letokhov, "Stimulated emission of an ensemble of scattering particles with negative absorption," *JETP Lett.* **5**, 212–215 (1967).

2. V. S. Letokhov, "Generation of light by a scattering medium with negative resonance absorption," *Sov. Phys. JETP* **26**, 835–840 (1968).

3. V. M. Markushev, V. F. Zolin, and Ch. M. Briskina, "Powder laser," *Zhurnal Prikladnoy Spektroscopii* **45**, 847–850 (1986). [in Russian]

4. V. M. Markushev, V. F. Zolin, and Ch. M. Briskina, "Luminescence and stimulated emission of neodymium in sodium lanthanum molybdate powders," *Sov. J. Quantum Electron.* **16**, 281–283 (1986).

5. M. A. Noginov, *Solid-State Random Lasers*, Springer, New York (2005).

6. H. Cao, "Review on latest developments in random lasers with coherent feedback," *J. Phys. A: Math. Gen.* **38**, 10497–10535 (2005).

7. S. C. Rand, "Bright storage of light," *Optics and Photonics News*, 32–37, May (2004).

8. O. Svelto, *Principles of Lasers*, 4th ed., translated and edited by D. C. Hanna, Plenum Press, New York (1998).

9. W. Koechner, *Solid-State Laser Engineering*, 5th ed., Springer, New York (1999).

10. A. Yariv, *Quantum Electronics*, 3rd ed., Wiley, New York (1998).

11. A. E. Siegman, *Lasers*, University Science Books, Herndon, VA (1986).

12. A. M. Prokhorov, "Molecular amplifier and generator for submillimeter waves," *Soviet Phys. JETP* **7**, 1140–1141 (1958).

13. A. L. Shawlow and C. H. Townes, "Infrared and optical masers," *Phys. Rev.* **112**, 1940–1949 (1958).

14. M. V. Klein and T. E. Furtak, *Optics*, Wiley, New York (1986).

15. W. Demtröder, *Laser Spectroscopy*, 3rd ed., Springer, New York, (2002).

16. J. W. Goodman, *Statistical Optics*, Wiley, New York (2000).

17. R. V. Ambartsumyan, N. G. Basov, P. G. Kryukov, and V. S. Letokhov, "Laser with nonresonant feedback," *JETP Lett.* **3**, 167–169 (1966).

18. R. V. Ambartsumyan, N. G. Basov, P. G. Kryukov, and V. S. Letokhov, "A laser with nonresonant feedback," *Sov. Phys. JETP* **24**, 481–485 (1967).

19. R. V. Ambartsumyan, N. G. Basov, P. G. Kryukov, and V. S. Letokhov, "A laser with a nonresonant feedback," *IEEE J. Quantum Electron.* **QE-2**, 442–446 (1966).

20. R. V. Ambartsumyan, N. G. Basov, P. G. Kryukov, and V. S. Letokhov, "Non-resonant feedback in lasers," in *Progress in Quantum Electronics,* Vol. 1, J. H. Sanders and K. W. H. Stevens, Eds., Pergamon Press, Oxford, 107–185 (1970).

21. R. V. Ambartsumyan, P. G. Kryukov, and V. S. Letokhov, "Dynamics of emission line narrowing for a laser with nonresonant feedback," *Sov. Phys. JETP* **24**, 1129–1134 (1967).

22. R. V. Ambartsumyan, N. G. Basov, and V. S. Letokhov, "Investigation of frequency characteristics of a He-Ne laser with a diffuse mirror," *JETP Lett.* **7**, 66–68 (1968).

23. R. V. Ambartsumyan, N. G. Basov, and V. S. Letokhov, "Frequency stability of a He-Ne laser with nonresonant feedback," *IEEE Trans. Instrum. Meas.* **IM-17**, 338–343 (1968).

24. R. V. Ambartsumyan, S. P. Bazhunin, N. G. Basov, and V. S. Letokhov, "Emission spectrum of a He-Ne laser with nonresonant feedback," *Sov. Phys. JETP* **31**, 234–241 (1970).

25. V. S. Letokhov, "Stimulated emission of an ensemble of scattering particles with negative absorption," *JETP Lett.* **5**, 212–215 (1967).

26. V. S. Letokhov, "Generation of light by a scattering medium with negative resonance absorption," *Sov. Phys. JETP* **26**, 835–840 (1968).

27. M. P. Van Albada and A. Lagendijk, "Observation of weak localization of light in random medium," *Phys. Rev. Lett* **55**, 2692–2695 (1985).

28. P. E. Wolf and G. Maret, "Weak localization and coherent backscattering of photons in disordered media," *Phys. Rev. Lett.* **55**, 2696–2695 (1985).

29. P. E. Wolf, G. Maret, E. Akkermans, and R. Maynard, "Optical coherent backscattering by random media: an experimental study," *J. Phys.* (Paris) **49**, 63–75 (1988).

30. M. B. Vad der Mark, M. P. van Albada, and A. Lagendijk, "Light scattering in strongly scattering medium: multiple scattering and weak localization," *Phys. Rev. B* **37**, 3575–3592 (1988).

31. M. A. Noginov, N. E. Noginova, H. J. Caulfield, P. Venkateswarlu, T. Thompson, M. Mahdi, and V. Ostroumov, "Short-pulsed stimulated emission in the powders of $NdAl_3(BO_3)_4$, $NdSc_3(BO_3)_4$, and Nd:$Sr_5(PO_4)_3F$ laser crystals," *J. Opt. Soc. Am. B* **13**, 2024–2033 (1996).

32. S. T. Durmanov, O. V. Kuzmin, G. M. Kuzmiheva, S. A. Kutovoi, A. A. Martynov, E. K. Nesynov, V. L. Panyutin, Yu. P. Rudnitsky, G. V. Smirnov, and V. I. Chizhikov, "Binary rare-earth scandium borates for diode-pumped lasers," *Opt. Mater.* **18**, 243–284 (2001).

33. M. A. Noginov, G. Zhu, A. A. Frantz, J. Novak, S. N. Williams, and I. Fowlkes, "Dependence of $NdSc_3(BO_3)_4$ random laser parameters on particle size," *J. Opt. Soc. Am. B* **21**, 191–200 (2004).

34. M. Bahoura, K. J. Morris, G. Zhu, and M. A. Noginov, "Dependence of the neodymium random laser threshold on the diameter of the pumped spot," *IEEE J. Quantum Electron.* **41**, 677–685 (2005).

35. M. A. Noginov, J. Novak, and S. Williams, "Modeling of photon density dynamics in random lasers," *Phys. Rev. A* **70**, 063810 (2004). http://link.aps.org/doi/10.1103/PhysRevA.70.063810

36. S. Mujumdar, M. Ricci, R. Torre, and D. S. Wiersma, "Amplified extended modes in random lasers," *Phys. Rev. Lett.* **93**, 053903 (2004).

37. V. S. Letokhov, "Noncoherent feedback in space masers and stellar lasers," in *Amazing Light: a Volume Dedicated to Charles Hard Townes on His 80th Birthday*, Raymond Y. Chiao, Ed., Springer-Verlag, New York (1996).

38. V. S. Letokhov, "Stimulated radio emission of the interstellar medium," *JETP Lett.* **4**, 321–323 (1966).

39. N. N. Lavrinovich and V. S. Letokhov, "The possibility of the laser effect in stellar atmospheres," *Sov. Phys. JETP* **40**, 800–805 (1975).

40. H. Weaver, D. R. W. Williams, N. H. Dieter, and W. T. Lum, "Observation of a strong unidentified microwave line and of emission from the OH molecule," *Nature* **208**, 29–31 (1965).

41. P. W. Anderson, "Absence of diffusion in certain random lattices," *Phys. Rev.* **109**, 1492–1505 (1958).

42. S. John, "Electromagnetic absorption in a disordered medium near a photon mobility edge," *Phys. Rev. Lett.* **53**, 2169–2172 (1984). http://link.aps.org/doi/10.1103/PhysRevLett.53.2169

43. S. John, "Localization of light," *Physics Today*, 32–40, May (1991).

44. A. F. Ioffe and A. R. Regel, "Non-crystalline, amorphous, and liquid electronic semiconductors," in *Progress in Semiconductors*, Vol. 4, A. F. Gibbson, gen. Ed., R. E. Burgess and F. A. Kröger, Eds., 237–291, Heywood & Company Ltd., London (1960).

45. D. J. Thouless, "Electrons in disordered systems and the theory of localization," *Phys. Rep.* **13**, 93–142 (1974).

46. D. J. Thouless, "Maximum metallic resistance in thin wires," *Phys. Rev. Lett.* **39**, 1167–1169 (1973).

47. A. A. Chabanov and A. Z. Genack, "Photon localization in resonant media," *Phys. Rev. Lett.* **87**, 153901 (2001).

48. D. S. Wiersma, P. Bartolini, A. Lagendijk, and R. Righini, "Localization of light in a disordered medium," *Nature* **390**, 671–673 (1997).

49. M. Störzer, P. Gross, C. M. Aegerter, and G. Maret, "Observation of the critical regime near Anderson localization of light," *Phys. Rev. Lett.* **96**, 063904 (2006).

50. A. L. Burin, M. A. Ratner, H. Cao, and S. H. Chang, "Random laser in one dimension," *Phys. Rev. Lett.* **88**, 093904 (2002).

51. P. Sebbah and C. Vanneste, "Random laser in the localized regime," *Phys. Rev. B* **66**, 144202 (2002).

52. S. C. Rand, "Strong localization of light and photonic atoms," *Can. J. Phys.* **78**, 625–637 (2000).

53. G. Williams, B. Bayram, S. C. Rand, T. Hinklin, and R. M. Laine, "Laser action in strongly scattering rare-earth-doped dielectric nanophosphors," *Phys. Rev. A* **65**, 013807 (2001).

54. V. M. Markushev, V. F. Zolin, and Ch. M. Briskina, "Powder laser," *Zhurnal Prikladnoy Spektroskopii* **45**, 847–850 (1986). [in Russian]

55. V. M. Markushev, V. F. Zolin, and Ch. M. Briskina, "Luminescence and stimulated emission of neodymium in sodium lanthanum molybdate powders," *Sov. J. Quantum Electron.* **16**, 281–283 (1986).

56. V. F. Zolin, "The nature of plaser-powdered laser," *J. Alloys and Compounds* **300–301**, 214–217 (2000).

57. V. A. Nikitenko, A. I. Tereschenko, I. P. Kuz'mina, and A. N. Lobachev, "Stimulated emission of ZnO at high level of single photon excitation," *Optika i Spektroskopiya* **50**, 605–607 (1981). [in Russian]

58. V. M. Markushev, N. È. Ter-Gabriélyan, Ch. M. Briskina, V. R. Belan, and V. F. Zolin, "Stimulated emission kinetics of neodymium powder lasers," *Sov. J. Quantum Electron.* **20**, 772–777 (1990).

59. C. Gouedard, D. Husson, C. Sauteret, F. Auzel, and A. Migus, "Generation of spatially incoherent short pulses in laser-pumped neodymium stoichiometic crystals and powders," *J. Opt. Soc. Am. B* **10**, 2358–2363 (1993).

60. M. Bahoura and M. A. Noginov, "Determination of the transport mean free path in a solid-state random laser," *J. Opt. Soc. Am. B* **20**, 2389–2394 (2003).

61. M. A. Noginov, N. Noginova, S. Egarievwe, J. C. Wang, M. R. Kokta, and J. Paitz, "Study of light propagation in scattering powder laser materials," *Opt. Mater.* **11**, 1–7 (1998).

62. M. A. Noginov, S. U. Egarievwe, N. Noginova, H. J. Caulfield, and J. C. Wang, "Interferometric studies in a powder laser," *Opt. Mater.* **12**, 127–134 (1999).

63. A. A. Lichmanov, Ch. M. Brickina, V. N. Lichmanova, N. P. Soshchin, and V. F. Zolin, "Experimental studies of the lanthanide doped lasing powders (plasers)," *Proc. Int. Conf. LASERS'98*, 725–731 (1999).

64. H. Cao, Y. G. Zhao, H. C. Ong, S. T. Ho, J. Y. Dai, J. Y. Wu, and R. P. H. Chang, "Ultraviolet lasing in resonators formed by scattering in semiconductor polycrystalline films," *Appl. Phys. Lett.* **73**, 3656–3658 (1998).

65. H. Cao, J. Y. Xu, D. Z. Zhang, S.-H. Chan, S. T. Ho, E. W. Seelig, X. Liu, and R. P. H. Chang, "Spatial confinement of laser light in active random media," *Phys. Rev. Lett.* **84**, 5584–5587 (2000). http://link.aps.org/doi/10.1103/PhysRevLett.84.5584

66. H. Cao, J. Y. Xu, E. W. Seeling, and R. P. Chang, "Microlaser made of disordered media," *Appl. Phys. Lett.* **76**, 2997–2999 (2000).

67. H. Cao, Y. Ling, and C. Q. Cao, "Photon statistics of random lasers with resonant feedback," *Phys. Rev. Lett.* **86**, 4524–4527 (2001).

68. B. Liu, A. Yamilov, Y. Ling, J. Y. Xu, and H. Cao, "Dynamic nonlinear effect on lasing in a random media," *Phys. Rev. Lett.* **91**, 063903 (2003).

69. R. Loudon, *The Quantum Theory of Light*, 2nd ed., Oxford University Press, Oxford (1983).

70. H. Cao, J. Y. Xu, S.-H. Chang, and S. T. Ho, "Transition from amplified spontaneous emission to laser action in strongly scattering medium," *Phys. Rev. E* **61**, 1985–1989 (2000).

71. H. Cao, "Random lasers with coherent feedback," in *Optical Properties of Nanostructured Random Media*, V. M. Shalaev, Ed., 303–328, Springer, New York (2002).

72. G. Williams, S. C. Rand, T. Hinklin, and R. M. Laine, "Ultraviolet laser action in strongly scattering Ce:alumina nanoparticles," in *Conf. Lasers and Electo-Optics*, OSA Technical Digest, p. 90, Optical Society of America, Washington, DC (1999).

73. B. Li, G. Williams, S. C. Rand, T. Hinklin, and R. M. Laine, "Continuous-wave ultraviolet laser action in strongly scattering Nd-doped alumina," *Opt. Lett.* **27**, 394–396 (2002).

74. R. M. Laine, T. Hinklin, G. Williams, and S. C. Rand, "Low-cost nano-powders for phosphor and laser applications by flame spray pyrolysis," *Mater. Sci. Forum.* **343**, 500–510 (2000).

Biography

Mikhail A. Noginov graduated from Moscow Institute for Physics and Technology (Moscow, Russia) in 1985 with an M.S. in Electronics Engineering. In 1990 he received a Ph.D. in Physical-Mathematical Sciences from General Physics Institute of the USSR Academy of Sciences (Moscow, Russia). Dr. Noginov's affiliations include: General Physics Institute of the USSR Academy of Sciences (Moscow, Russia, 1985–1991); Massachusetts Institute of Technology (Cambridge, MA, 1991–1993); Alabama A&M University (Huntsville, AL, 1993–1997); and Norfolk State University (NSU) (Norfolk, VA, 1997–present).

Dr. Noginov has published one book, four book chapters, over 100 papers in peer-reviewed journals, and over 100 publications in proceedings of professional societies and conference technical digests (more than 20 of them being invited papers). He is a member of Sigma Xi, Optical Society of America, SPIE, and the American Physical Society and has served as a chair and a committee member on several conferences of SPIE and OSA. Since 2003, Dr. Noginov has been a faculty advisor of the OSA student chapter at NSU. His research interests include metamaterials, nanoplasmonics, random lasers, solid state laser materials, and nonlinear optics.

Chapter 13
Optical Metamaterials with Zero Loss and Plasmonic Nanolasers

Andrey K. Sarychev
Institute of Theoretical and Applied Electrodynamics, Moscow, Russia

13.1 Introduction

We consider plasmonic nanoantennas immersed in an active host medium. Specifically shaped metal nanoantennas can exhibit strong magnetic properties in the optical spectral range due to excitation of the magnetic plasmon resonance. A case in which a metamaterial comprising such nanoantennas can demonstrate both left handedness and negative permeability in the optical range is discussed. We show that high losses predicted for optical left-handed materials can be compensated in the gain medium. Gain allows one to achieve local generation in magnetically active metamaterials. We propose a plasmonic nanolaser where the metal nanoantenna operates in a fashion similar to a resonator. The size of the proposed plasmonic laser is much smaller than the light wavelength. Therefore, it can serve as a very compact source of coherent electromagnetic radiation and can be incorporated in future plasmonic devices.

Extending the range of electromagnetic properties of naturally occurring materials motivates the development of artificial metamaterials. For example, metamaterials with artificial microwave magnetism were known since the beginning of the 1950s.[1] It has been demonstrated recently that metamaterials may exhibit such exotic properties as negative dielectric permittivity,[2–4] negative magnetic permeability,[5–7] and even both.[8–10] The double-negative case of $\mathrm{Re}\,\varepsilon < 0$ and $\mathrm{Re}\,\mu < 0$ is often referred to as a left-handed material (LHM). Situations in which a negative refractive index can be realized in practice are particularly interesting because of the possibility of a "perfect" lens with subwavelength spatial resolution.[10] In addition to the superresolution not being limited by classical diffraction, many unusual and sometimes counterintuitive properties of negative refraction index materials (NIMs) make them very promising for applications in resonators, waveguides, and other microwave and optical elements.[11–13] Negative refraction[9,11,12,14–18] and subwavelength imaging[19–24] have been demonstrated in the microwave and RF regimes.

For microwave NIMs, artificial magnetic elements providing $\mathrm{Re}\,\mu < 0$ are the resonators of the split-ring type or helix type. In the microwave spectral range, metals can be considered as almost perfect conductors because the skin depth is much smaller than the metallic feature size. The strong magnetic response is achieved by operating in the vicinity of the LC resonance of the split ring.[2,5,7,25–27] The same technique for obtaining $\mathrm{Re}\,\mu < 0$ using split rings was extended to the mid-IR[27] by scaling down the dimensions of the split rings. Therefore, the frequencies of the LC resonances are determined entirely by the split-ring geometry and size, not by the electromagnetic properties of the metal. In accordance with this statement, the ring response is resonantly enhanced at some particular ratio of the radiation wavelength and the structure size. Thus, we refer to the LC resonances of perfectly conducting metallic structures as geometric LC (GLC) resonances.

The situation drastically changes in the optical and near-IR part of the spectrum, where thin subwavelength metal components behave very differently when their size becomes less than the skin depth. For example, the electrical surface plasmon resonance (SPR) occurs in the optical and near-IR parts of the spectrum due to collective electron oscillations in metal structures. Many important plasmon-enhanced optical phenomena, including surface plasmon (SP) propagation, anomalous absorption, surface-enhanced Raman scattering, and extraordinary optical transmittance are based on the electrical SPR (for review see Refs. 28 and 29).

Near-field optical superresolution is also based on the plasmon excitation in a material with $\varepsilon = -1$.[10] The near-field superresolution follows from rather elementary consideration presented, for example, in Ref. 30 (problem #209). The superresolution was demonstrated for silver nanofilms in optical experiments[31,32] and for silicon carbide film in infrared experiments.[33]

Superresolution in the far zone can be achieved using a cylindrical or spherical plasmonic lens with a layered structure, where the permittivity changes its sign from layer to layer.[34–36] Far-field superresolution was indeed obtained in a three-dimensional cylindrical[37,38] and in a planar, circular hyperlens.[39,40] In the case of a planar hyperlens, the propagation of a surface plasmon-polariton over a gold surface is used to form a magnified image of subwavelength objects. Negative refraction and image magnification can be also achieved in a system of plasmonic waveguides.[41–45] Microwave imaging by an array of thin metal wires that are perpendicular to the direction of the wave was considered;[46–48] in a recent paper,[49] the authors repeated the results of our work[45] by considering the image magnification in a system of trapped metal wires.

The plasmonic nature of the electromagnetic response in metals for optical and mid-IR frequencies is the main reason the original methodology of GLC resonances in the microwave and mid-IR spectral range is not extendable to higher frequencies. For the optical range, metamaterials with a negative refractive index were first demonstrated in the experimental works presented in Refs. 50–52. Authors of Refs. 50 and 51 experimentally verified their earlier theoretical prediction of negative refraction in an array of parallel metal nanorods.[53–55] In Ref. 52 the authors observed the negative real part of the refractive index at a wavelength of 2.0 μm in a system comprising two thin metal films, each film representing an array of subwavelength holes. The metal islands between the holes work as magnetic nanoantennas similar to the horseshoe nanoantennas.[56] The first experimental observations of negative n in the optical range were followed by other successful experiments.[57–62] For exam ple, Valentine, Zhang, Zentgraf, Ulin-Avil, Genov, Bartal, and X. Zhang[62] made the first optical prism composed of a metamaterial and observed negative refraction at ($\lambda \approx 1.5\,\mu$m). Negative permeability in optics ($\lambda \approx 0.4$–$0.5\,\mu$m) was also reported by Grigorenko et al.[63] However, the geometry of this experiment (vertical nanopillars that are perpendicular to the film plane) prevents excitation of a magnetic resonance. Indeed the electric field of the incident light is perpendicular to the axes of the nanopillars and cannot excite electric currents flowing back and forth in the pillars. In the absence of circular currents the magnetic response vanishes and the permeability equals one. Other problems with interpretation of the experiment[63] were discussed in Ref. 64

Note that loss becomes increasingly important as the wavelength approaches the optical range.[27,65] Moreover, nonzero loss inside the LHM superlens dramatically reduces the resolution of such a lens[66] and makes the goal of a flat superlens unattainable.

Plasmonic effects must be correctly accounted for in the design of metamaterials with optical magnetism. Below we discuss how the specifically arranged and shaped metal nanoparticles can support, along with the electrical SPR, a magnetic plasmon resonance (MPR). The MPR frequency can be made independent of the absolute characteristic structure size a. In fact, the only defining parameters of the MPR are the metal permittivity ε_m and the structure's geometry. Such structures act as optical nanoantennas by concentrating large electric and magnetic energies on the nanoscale, even at optical frequencies. The magnetic response is characterized by the magnetic polarizability α_M with the resonant behavior similar to the electric SPR polarizability α_E; the real part of α_M changes sign near the resonance and becomes negative, as required for NIMs. We show that an MPR must replace or strongly modify GLC resonances in the optical/mid-IR range if a strong magnetic response is desired. The idea of the MPR inducing optical magnetism is relatively new.[67–69] We propose using the horseshoe-shaped structures first suggested for optics by Sarychev and Shalaev.[56] These structures support strong magnetic moments at frequencies higher than the microwave and mid-IR range for which traditional split-ring resonators were proposed and demonstrated.[3,5,7,25–27] Conceptually, the horseshoe-shaped structures described here are distinct from the previously studied low-frequency structures that relied on the GLC resonance to produce a strong magnetic response. Horseshoe nanoantennas have a distinctively different magnetic response from split-ring antennas due to their "elongated" shape and concentration of the electromagnetic (EM) field inside the gap between the "arms" of the horseshoe. The plasmonic properties of the metal are very important when the sizes are small and the operational frequencies are high.

In the next section we consider the MPR in a simple horseshoe resonator filled with a gain medium. Excitation of nanowire plasmonic antennas is discussed in Secs. 13.3–13.4. There we will outline the derivation of the magnetic permeability of the metamaterial comprising such metallic nanoparticles. Horseshoe resonators filled with an active medium, plasmonic nanolasers, and spontaneous optical magnetization are considered in Secs. 13.5–13.7. In Sec. 13.5 we use the classical approximation and show its limitations. The new approach, which we call quantum plasmonics, is presented in Sec. 13.6. The plasmonic nanolaser is suggested in Sec. 13.7. Section 13.8 is devoted to conclusions. In all equations of the chapter we use the CGS unit system, where inductance and capacitance are measured in centimeters.

13.2 Magnetic Plasmon Resonance

We consider the interaction of a metallic horseshoe-shaped nanoantenna with a high gain medium. The gain medium is modelled as a collection of two-level amplifying systems (TLSs), which can be represented, for example, by quantum dots or dye molecules. In this chapter, we will attempt to derive the conditions under which such a metamaterial can demonstrate very low loss or even a gain leading to lasing in the LHM.[68,69] Lasing in a random medium without a cavity was predicted by Letokhov[71] and was first demonstrated experimentally by Lawandy et al.[72–75] (See also the excellent review by Cao.[76]) Recently, strong lasing was demonstrated in a dye solution containing 55-nm silver nanoparticles.[77] The possibility of a nonradiative transfer of energy from the active medium to the static plasmon modes of a metal nanoparticle was discussed by Bergman and Stockman.[78] Increased gain up to $2 \cdot 10^5\,\mathrm{cm}^{-1}$ was demonstrated in a medium comprising quantum dots.[79–81]

Propagation of a surface plasmon-polariton in the interface between metal and a gain medium was considered in many papers.[82–86] Recently the effect of a small increase in plasmon amplitude due to interaction with an active medium was observed in two experiments.[87,88] In the work presented in Ref. 89 the enhancement of fluorescence in a dye film deposited on a corrugated silver surface was investigated. Most recently, a new advance in improved gain of surface plasmon (SP) in an active medium was achieved. A sixfold gain was demonstrated by Noginov et al.[90] Increasing the resolution by elimination of absorption in a near-field lens via optical gain was discussed by Ramakrishna and Pendry.[91] A redshift of the excitonic peak and an enhanced local field in silver-coated quantum dots were observed by Je et al.[92] In the review provided in Ref. 60 the research on LHMs in a gain medium accomplished before 2006 is discussed. Authors of the study in Ref. 93 considered a layer of plasmonic nanoantennas deposited on a gain substrate. A selfconsistent approach to the LHM in a gain medium was very recently suggested.[94] Below we will consider optical properties of magnetic nanoantennas in an active medium with high gain. For this purpose we use quantum plasmonics.[68–70]

For the sake of simplicity, we assume that TLSs are placed inside the horseshoe resonator of a type shown in Fig. 13.1. An external pump provides the population inversion in the TLS. The pump mechanism may be optical or electrical. In the latter case, carriers are injected in a quantum dot from the bands of semiconductor material surrounding the embedded dot. In the equations, we will characterize only the pump rate and initial population inversion of the TLS interacting with the electromagnetic field inside the horseshoe. We will present calculations of the

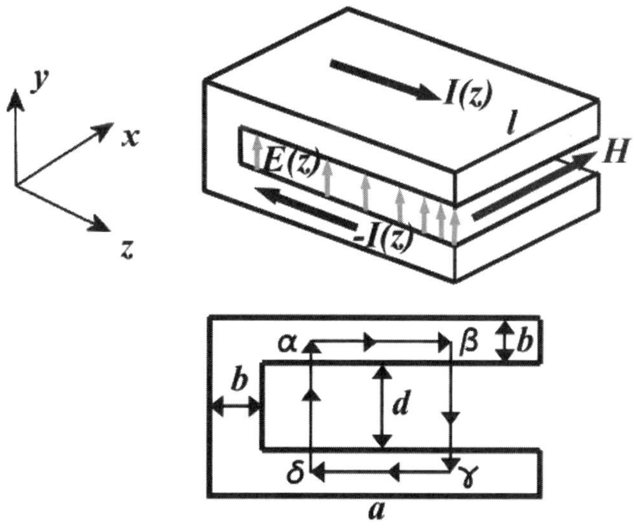

Figure 13.1 Horseshoe nanoantenna geometry; parameters used in modelling: $a = 300$ nm, $d = 70$ nm, $b = 34$ nm. (Reprinted from Ref. 69.)

electromagnetic field in a system comprising metallic nanoantennas with a resonant frequency lying in the visible or near-infrared band.

We consider a metal nanoantenna that has the shape of a horseshoe as shown in Fig. 13.1. For the sake of simplicity, we first consider the two-dimensional geometry in which the lateral (in the x direction) size of the horseshoe is much larger than other dimensions. The external magnetic field excites the electric current in the arms of the horseshoe as shown in Fig. 13.1. The magnetic moment associated with the circular current that flows in the antenna is responsible for the magnetic response of the nanoantenna. The external magnetic field $H = \{H_0 \exp(-i\omega t),\ 0,\ 0\}$ is applied in the plane of the horseshoe. The circular current $I(z)$, excited by the time-varying magnetic field, flows in opposite directions in each arm of the horseshoe as shown in the figure. The displacement current flowing between the horseshoe's arms completes the circuit. We neglect edge effects and assume that the currents and fields are independent of the coordinate x. To find the currents and the fields, we use the approach by Sarychev, Lagarkov, Shalaev, and Shvets.[3,67] Below we consider nano-horseshoes of a size much smaller than the wavelength of the external electromagnetic field. Hence, we can introduce a potential difference

$$U(z) = E_y d = -4\pi\left[Q(z) + P(z)\right] d \tag{13.1}$$

between the arms, where $Q(z)$ is the electric charge per unit area at the upper arm (i.e., arm $\alpha\beta$ in Fig. 13.1) and $P(z)$ is the polarization of the medium inside the horseshoe. Throughout the chapter we use the CGS unit system, where the electric displacement $\mathbf{D} = \mathbf{E} + 4\pi\mathbf{P}$. We also distinguish between the "free" electric charges in the metal and "bound" charges in the dielectric. The density q of the "bound" charges equals $\mathrm{div}\mathbf{P}$.

The electric current $I(z)$ generates the magnetic field $H(z) = 4\pi I(z)/c$ inside the horseshoe, where $I(z)$ is the surface density of the electric current at the upper arm (i.e., arm $\alpha\beta$ in Fig. 13.1) and c is the speed of light. Therefore, both electric and magnetic fields are present in the nanoantenna. This makes the quasistatic approximation, often used in treatment of plasmonic problems, inadequate. Below we show the proper approach to the treatment of nanoantennas with magnetic properties.

To find the current $I(z)$, we integrate the equation representing Faraday's law

$$\mathrm{curl}\mathbf{E} = -\frac{\partial}{c\,\partial t}\left(\mathbf{H} + \mathbf{H}_0\right) \tag{13.2}$$

over the contour $\{\alpha, \beta, \gamma, \delta\}$ in Fig. 13.1, where $\mathbf{H_0}$ is the amplitude of the external magnetic field of the electromagnetic wave with wave number $k = \omega/c = 2\pi/\lambda$. It is assumed that the horseshoe length a is much larger than the distance d between the arms. We also assume that $kd \ll 1$. Integration of Eq. (13.2) over the loop

$\{\alpha, \beta, \gamma, \delta\}$ results in the following differential equation:

$$[2I(z)Z - \partial U/\partial z]\,\Delta z = -\frac{d}{c}\left[4\pi \dot{I}(z)/c + \dot{H}_0\right]\Delta z, \tag{13.3}$$

where $I(z)$ is the surface current density, Δz is the distance between points α and β on the integration path in Fig. 13.1, $\dot{I}$ and $\dot{H}_0$ are time derivatives, $Z = 1/(\sigma b) = 4i\pi/(\varepsilon_m \omega b)$ is the surface impedance, and $\varepsilon_m = i4\pi\sigma_m/\omega$ is the metal's complex permittivity. We substitute Eq. (13.1) for the potential difference $U(z)$ in Eq. (13.3). Then we take the time derivative of both sides of Eq. (13.3). Taking into the account the charge conservation law $-\partial I/\partial z = \partial I_1/\partial z = \dot{Q}$, where I_1 is the surface current density in the down arm, we derive the differential equation for the current

$$\frac{\partial^2 I(z,t)}{\partial z^2} - \frac{\partial \dot{P}(z,t)}{\partial z} - \frac{Z}{2\pi d}\dot{I}(z,t) = \frac{1}{4\pi c}\left[\frac{4\pi}{c}\ddot{I}(z,t) + \ddot{H}_0(t)\right]. \tag{13.4}$$

To find the current and fields in the horseshoe we should add to this equation the constitutive law for the polarization P. We assume that the polarization P can be represented as the sum $P = P_1 + P_2$, where $P_1 = \chi_1 E_y$ is a regular, frequency-independent polarization, and P_2 is the "anomalous" polarization due to the resonant response of the TLS system; χ_1 is a regular susceptibility of the medium inside the horseshoe. We substitute $P = \chi_1 E_y + P_2$ in Eq. (13.1) and repeat the derivation of Eq. (13.4) obtaining it now in the following form:

$$\frac{\partial^2 I(z,t)}{\partial z^2} - \frac{\partial \dot{P}_2(z,t)}{\partial z} - \frac{Z\varepsilon_d}{2\pi d}\dot{I}(z,t) = \frac{\varepsilon_d}{4\pi c}\left[\frac{4\pi}{c}\ddot{I}(z,t) + \ddot{H}_0(t)\right], \tag{13.5}$$

where the regular polarization is renormalized into the "regular" part of the dielectric permittivity $\varepsilon_d = 1 + 4\pi\chi_1$.

We consider first the simplest case in which the anomalous (gain) TLS polarizability $P_2 = \chi_2 E_y$. Suppose that the external field H_0 is harmonic, i.e., $H = H_0 \exp(-i\omega t)$. Then Eq. (13.5) takes the following form:

$$\frac{d^2 I(z)}{dz^2} = -g^2 I(z) - \frac{\varepsilon_d \omega k}{4\pi} H_0, \tag{13.6}$$

where $0 < z < a$, and the coordinates $z = 0$ and $z = a$ correspond to the beginning and the end of the horseshoe so that $dI(0)/dz = I(a) = 0$; the parameter g is

$$g^2 = \varepsilon_d k^2 - \frac{2\varepsilon_d}{bd\varepsilon_m}, \tag{13.7}$$

where the dielectric permittivity now includes the regular and resonant parts $\varepsilon_d = 1 + 4\pi(\chi_1 + \chi_2)$. If we consider the limiting case with $k^2 bd|\varepsilon_m| \ll 1$, then $g = \sqrt{-2\varepsilon_d/(\varepsilon_m bd)}$ does not depend on the absolute length of the arms and does not depend explicitly on frequency (wavelength). The parameter g depends only on the metal permittivity ε_m and the thickness of the horseshoe arms. In order to have a sharp resonance in Eq. (13.6), the real part of g should be positive and the imaginary part should be small. Indeed, at IR/visible frequencies, ε_m is negative (with a small imaginary part) for typical low-loss metals such as Ag, Au, Al, etc. (see discussion at Eq. 13.19).

Solving Eq. (13.6) for the current yields

$$I(z) = \frac{\varepsilon_d \omega k}{4\pi g^2} H_0 \left[\frac{\cos(gz)}{\cos(ga)} - 1 \right]. \tag{13.8}$$

Magnetic and electric fields induced by the current and electric charge in the horseshoe nanoantenna can be represented as

$$H_x(z) = \frac{\varepsilon_d k^2}{g^2} H_0 \left[\frac{\cos(gz)}{\cos(ga)} - 1 \right], \tag{13.9}$$

$$E_x = 0, \qquad E_y(z) = -i\frac{k}{g} H_0 \frac{\sin(gz)}{\cos(ga)}, \quad \text{and} \tag{13.10}$$

$$E_z(y, z) = -iky H_0 \left[\frac{\cos(gz)}{\cos(ga)} - 1 \right]. \tag{13.11}$$

Equations (13.9)–(13.11) completely describe the electric and magnetic fields generated by a plasmon excited in the horseshoe by the external magnetic field H_0. We would like to emphasize that neither magnetic nor electric fields are irrotational in this type of nanoantenna. It is taken for granted that the magnetic field is solenoidal. More surprisingly, the electric field is also solenoidal in the horseshoe resonator, despite the fact that its size is much smaller than the wavelength. Indeed, it follows from Eqs. (13.10) and (13.11) that the electric field E_y depends on the coordinate z and the electric field E_z depends on the coordinate y. The subwavelength solenoidal electric field is the essence of MPR.

Since the external magnetic field excites circular electric currents in the horseshoe nanoantennas, it is natural to assume that the metamaterial comprising these nanoparticles has effective magnetic properties. We consider horseshoe nanoantennas with volume density p organized in a regular cubic lattice. We calculate the magnetic permeability μ of such a system using the approach originally developed by Vinogradov, Panina, and Sarychev[95] and discussed in Refs. 28 and 29. The expression for μ yields

$$\mu = 1 + \frac{p}{aH_0} \int_0^a H_{in}\, dz, \tag{13.12}$$

where a is the size (see Fig. 13.1) and H_{in} is the magnetic field generated by the horseshoe current. Substituting $H_{in} = H_x(z)$ from Eq. (13.9), we obtain the permeability

$$\mu = 1 + p\frac{k^2 ab\varepsilon_m}{2} \left[\frac{\tan(ga)}{ga} - 1 \right]. \tag{13.13}$$

This permeability exhibits Lorentz-shaped resonance with the central frequency defined by the following condition:

$$ga = \sqrt{-\frac{2\varepsilon_d a^2}{bd\varepsilon_m}} = \frac{\pi}{2}. \tag{13.14}$$

The magnetic field is shown in Fig. 13.2 for the horseshoe nanoantenna at a frequency slightly above the resonance. Note that the direction of the magnetic field inside the nanoantenna is opposite to the direction of the applied field. Therefore, the array of nanoantennas can operate as a metamaterial with a negative magnetic permeability.

We are particularly interested in the case in which the dielectric host medium filling the space inside the horseshoe is an active high gain medium. This situation can be modelled, to a first approximation, by the assumption that the imaginary part of ε_d is negative.[86,96] One can see that in different examples of metamaterials (Fig. 13.3), increased gain leads to narrowing of the absorption line of the horseshoe (dashed lines in Fig. 13.3). With large enough gain, the horseshoes behave as a set of nanolasers. At threshold, the permeability tends to infinity and becomes singular. In a real system, the permeability saturates. The process of saturation is discussed in Sec. 13.6 in the framework of the Maxwell–Bloch equations.

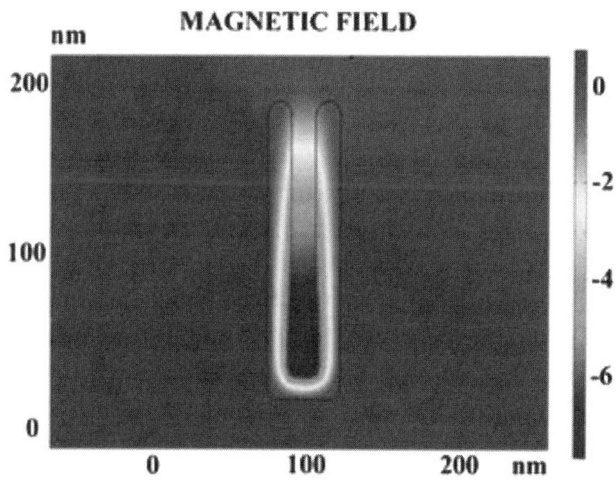

Figure 13.2 Magnetic plasmon resonance in a silver horseshoe nanoantenna placed in the maximal external magnetic field, directed perpendicular to the plane of paper. The incident wavelength is $\lambda = 1.5\,\mu m$; the silver permittivity is estimated from Eq. (13.19) and $\varepsilon_d = 2$. The magnetic field inside the horseshoe is opposite in direction to the external field. (Reprinted from Ref. 67.) (See color plate section.)

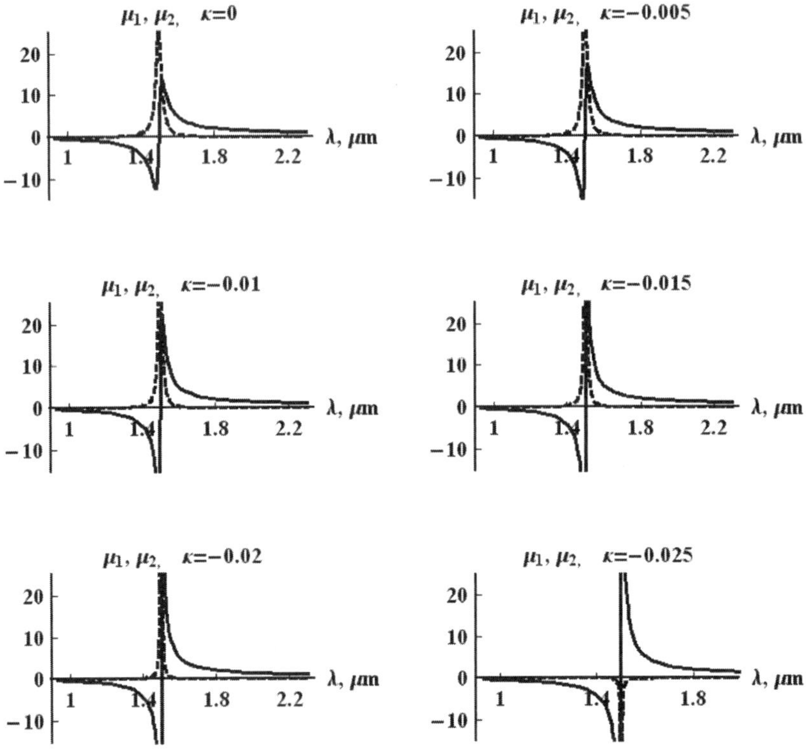

Figure 13.3 Effective permeability of a silver horseshoe metamaterial is shown. The effective permittivity actively pumped host medium has $\varepsilon_d = 4(1 + i\kappa)$, where the loss factor $\kappa < 0$. Parameters used in modelling are $a = 300$ nm, $d = 70$ nm, $b = 34$ nm, and the volume density of horseshoes $p = 0.3$. The real part of the permeability is shown by the solid lines and the imaginary part by dashed lines. (Reprinted from Ref. 69.)

13.3 Electrodynamics of a Nanowire Resonator

We will now discuss MPR in basic three-dimensional plasmonic structures. First, we consider a pair of parallel metallic nanorods. The external magnetic field excites the electric current in each pair of rods as shown in Fig. 13.4. The magnetic moment associated with the circular current flowing in the rods results in the magnetic response of the system. Suppose that an external magnetic field $\mathbf{H} = \{0,\ H_0 \exp(-i\omega t),\ 0\}$ is applied perpendicular to the plane of the pair. (We suppose that the magnetic field is along the y axis and the rods are in the $\{x, z\}$ plane.) The circular current $I\,(z)$, excited by the time-varying magnetic field, flows in opposite directions in the nanowires as shown in Fig. 13.4. Displacement current, flowing between the nanowires, closes the circuit. We then introduce the "potential drop" $U\,(z) = \int_a^\beta \mathbf{E} d\mathbf{l}$ between the pair where the integration is performed along the line $\{\alpha\,(z)\,, \beta\,(z)\}$.

We use the same approach as we used above deriving Eq. (13.4). To find the current $I\,(z)$, we integrate Faraday's Law $\mathrm{curl}\mathbf{E} = ik\,(\mathbf{H}_0{+}\mathbf{H})$ over the contour

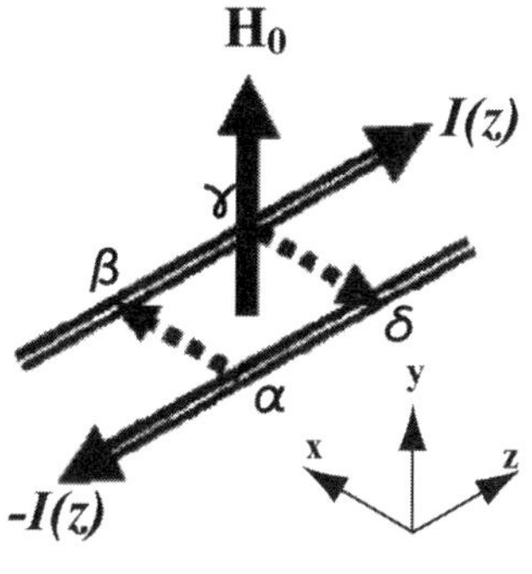

Figure 13.4 Currents in two parallel metal wires excited by an external magnetic field $\mathbf{H}_0$. Displacement currents "closing" the circuit are shown by dashed lines. Currents $I(z)$ and $-I(z)$ in the wires induce the magnetic field that is opposite to the applied field. (Reprinted from Ref. 67.)

$\{\alpha, \beta, \gamma, \delta\}$ in Fig. 13.4, where $\mathbf{H} = \mathrm{curl}\mathbf{A}$ is the magnetic field induced by the current. In our derivation, we assume that the nanowire length $2a$ is much larger than both the distance d between the nanowires and the radius b of each nanowire. We also assume that $kd \ll 1$. Under these assumptions, the vector potential $\mathbf{A}$ is primarily directed along the nanowires (z direction). Then the integral form of Faraday's Law yields

$$(2IZ - i2kA_z - dU/dz)\Delta z = -ikH_0 d\Delta z, \tag{13.15}$$

where $Z \simeq /(\sigma\pi b^2) \approx 8i/(\varepsilon_m b^2\omega)$ is the wire impedance, $\varepsilon_m = i4\pi\sigma/\omega \ll -1$ is the metal complex permittivity, and $\pm IZ/2$ are the electric fields on the surface of the nanowires. Equation (13.15) is the three-dimensional counterpart of Eq. (13.3) rewritten in terms of the vector potential. Note that the wire resistivity is explicitly taken into account. This distinguishes these calculations from the works on the resonances of conducting split-ring resonators[97] and conducting stick composites[3] because the plasmonic wire resonances are fully accounted for in our calculations.

The electric field $\mathbf{E}$ can be represented in terms of the vector potential $\mathbf{A}$ and electric potential ϕ, $\mathbf{E} = -\nabla\phi + ik\mathbf{A}$. For the standard Lorentz gauge, the electric potential ϕ is $\phi(\mathbf{r}_1) = \int \exp(ikr_{12})\, q(\mathbf{r}_2)/r_{12}\, d\mathbf{r}_2$ and the vector potential is $\mathbf{A}(\mathbf{r}_1) = c^{-1}\int \exp(ikr_{12})(\mathbf{j}(\mathbf{r}_2)/r_{12})\, d\mathbf{r}_2$, where $r_{12} = |\mathbf{r}_1 - \mathbf{r}_2|$, q and $\mathbf{j}$ are the charge and the current density, respectively. In the case of the two long wires, the currents flow inside the wires. Correspondingly, the vector potential $\mathbf{A}$ has a component only in the z direction: $\mathbf{A} = \{0, 0, A_z\}$. Since the vector potential $\mathbf{A}$ is perpendicular to the line $\{\alpha(z), \beta(z)\}$, the potential drop U in Eq. (13.15) is $U(z) = \int_\alpha^\beta \mathbf{E}d\mathbf{l} = \phi_\alpha - \phi_\beta$, where ϕ_α and ϕ_β are the electric potentials at the points $\alpha(z)$ and $\beta(z)$, respectively. Now, consider the excitation of an antisymmetric mode when the currents in the wires have the same magnitude but run in opposite directions (see Fig. 13.4), resulting in the electric charge per unit length being $Q(z) = Q_\beta = -Q_\alpha$. If we assume that the diameter b of the wire is much smaller than the interwire distance d, and the wire length is much bigger: $2a \gg d$, then the potential drop $U(z)$ between the pair can be written as $U(z) = -Q(z)/\tilde{C}$, where the interwire capacitance $\tilde{C}$ per unit length is independent of the coordinate z and is estimated in Sec. 13.4 ($\tilde{C} \simeq [4\ \ln(d/b)]^{-1}$). Below we denote the capacitance and inductance per unit length as $\tilde{C}$ and $\tilde{L}$, respectively. We show below that $\tilde{C}$ and $\tilde{L}$ are dimensionless quantities that are independent of the length of the wires. Full interwire capacitance C equals to $C = 2a\tilde{C}$, and full interwire inductance $L = 2a\tilde{L}$. Recall that $2a$ is the wire length.

The vector potential $A_z(z)$ is proportional to the electric current $A_z(z) = (\tilde{L}/c)\, I(z)/2$, where the wire pair inductance per unit length is estimated as $\tilde{L} \simeq 4 \ln(d/b)$ (as shown in the next section.) Note that both $\tilde{C}$ and $\tilde{L}$ are purely geometric factors that do not depend on the plasmonic nature of the metal rods. The entire $\tilde{L}\tilde{C}$ product can be estimated as $\tilde{L}\tilde{C} \simeq 1$. In the above consideration, we have not assumed that the wires were made of a perfect metal or of a metal with real conductivity (i.e., imaginary permittivity). Moreover, the nanowire pair (two nanoantennas) has an interesting behavior when the metal permittivity is real and negative. The reason is the plasmonic nature of the metal, which is elaborated below.

First, we substitute $U(z)$ and $A_z(z)$ into Eq. (13.15), taking into account the charge conservation law $dI/dz = i\omega Q(z)$, and obtain a differential equation for the current:

$$\frac{d^2 I(z)}{d z^2} = -g^2 I(z) - \frac{\tilde{C} d\omega^2}{c} H_0, \tag{13.16}$$

where $-a < z < a, \quad I(-a) = I(a) = 0$, and the parameter g is given by

$$g^2 = k^2 \left[\tilde{L}\tilde{C} - 8\tilde{C} \left((kb)^2 \varepsilon_m \right)^{-1} \right]. \tag{13.17}$$

It follows from Eq. (13.16) that the two-wire antenna is resonantly excited when $G = ga = \pi/2 + N\pi$, where N is an integer number. Note that the material properties of the metal enter the resonant parameter G through the metal permittivity ε_m. In the context of the wire pair, the earlier discussed GLC resonances,[3,5,25,53,97–99] occur when the wire radius b is much larger than the characteristic skin depth $(k^2 |\varepsilon_m|)^{-1/2}$. This approximation, typically valid for microwave and mid-IR frequencies, yields $g = k/\sqrt{\tilde{L}\tilde{C}} \cong k$ and the resonance condition $ka = \pi/2$. This resonance condition is equivalent to the condition $2a = \lambda/2$, which is also known as the antenna resonance.

Let us now consider the opposite ("electrostatic," as explained below) limit of $\left| 8\tilde{C} \left[(kb)^2 \varepsilon_m \right]^{-1} \right| \gg 1$. In the electrostatic regime, G depends only on the metal permittivity and the aspect ratio, namely,

$$G^2 \simeq -2(a/b)^2 \ln(d/b)/\varepsilon_m. \tag{13.18}$$

Note that it does not depend on the wavelength or the absolute wire length $2a$. Sharp resonance in Eq. (13.16) requires that G^2 be positive, possibly with a very small imaginary part. Indeed, ε_m is negative (with a smaller imaginary part) for IR/visible frequencies and typical low-loss metals (silver, gold, aluminum, etc). The metal permittivity ε_m can be approximated by the Drude formula

$$\varepsilon_m(\omega) \cong \varepsilon_b - (\omega_p/\omega)^2 / (1 - i\omega_\tau/\omega), \tag{13.19}$$

where ε_b is a "polarization" constant, ω_p is the plasma frequency, and $\omega_\tau = 1/\tau$ is the relaxation rate. For the silver nanoantennas considered here, $\varepsilon_b \approx 5$, $\hbar\omega_p \approx 9.1\,\text{eV}$, and $\hbar\omega_\tau \approx 0.02$ eV.[100,101] For example, the silver permittivity estimates as $\varepsilon'_m \approx -120$ with the loss factor $\kappa = \varepsilon''_m/|\varepsilon_m| \approx 0.025$ at $\lambda = 1.5\,\mu\text{m}$. We believe that it is convenient to measure optical frequency in electron volts since the characteristic energy in a metal is about a few electron volts (e.g., the Fermi energy for silver is approximately 5.5 eV). The frequency 1 eV corresponds to the wavelength $1.23984\ldots\,\mu\text{m}$.

We now consider the electric field in the system of two conducting rods, still assuming that the electrostatic limit holds when the propagation constant G is given by Eq. (13.18). The electric charge $Q(z)$ and the current $I(z)$, connected via charge conservation law $\left[Q(z) = (i\omega)^{-1} dI(z)/dz\right]$, are given by solution of Eq. (13.16):

$$Q(z) = Q_0 \frac{\sin(Gz/a)}{\cos G}, \tag{13.20}$$

$$I(z) = i\frac{Q_0 a\omega}{G}\left(1 - \frac{\cos(Gz/a)}{\cos G}\right), \tag{13.21}$$

where

$$Q_0 = ib\,d\,k\,H_0\sqrt{-\varepsilon_m}/\left[4\sqrt{2}\ln^{3/2}(d/b)\right]. \tag{13.22}$$

Using the Lorentz gauge, we can rewrite the equation for the electric potential as

$$\phi(r) = \int q(\mathbf{r}_1)\frac{\exp(ikR_1)}{R_1}d\mathbf{r}_1 - \int q(\mathbf{r}_2)\frac{\exp(ikR_2)}{R_2}d\mathbf{r}_2, \tag{13.23}$$

where $q(r_1)$ and $q(r_2)$ are electric charges distributed over the surface of the rods 1 and 2, $R_1 = |\mathbf{r} - \mathbf{r}_1|$, $R_2 = |\mathbf{r} - \mathbf{r}_2|$, and the integration is performed over both rods. We then consider the electric field in the space around the rods, assuming that $|x|, |y| \ll a$ and $|z| < a$, while the distances from the observation point $\mathbf{r} = \{x, y, z\}$ to the rods are:

$$d_1 = \sqrt{(x - d/2)^2 + y^2} \tag{13.24}$$

and

$$d_2 = \sqrt{(x + d/2)^2 + y^2}. \tag{13.25}$$

We exclude from consideration the very vicinity of the rods assuming that $d_1, d_2 \gg b$. The subsequent integration of Eq. (13.23) over the cross section of the rod provides a one-dimensional form

$$\phi(x,z) = \int_{-a}^{a} Q(z_1)\left[\frac{\exp(ikR_1)}{R_1} - \frac{\exp(ikR_2)}{R_2}\right] dz_1, \tag{13.26}$$

where the linear charge density $Q(z_1)$ is obtained from $q(r_1)$ by the integration over the rod circumference and is given by Eq. (13.20) $[Q(z_1) = -Q(z_2)]$, while the distances R_1 and R_2 take the forms $R_1 = \sqrt{d_1^2 + (z - z_1)^2}$ and $R_2 = \sqrt{d_2^2 + (z - z_1)^2}$. The two terms in the square brackets in Eq. (13.26) cancel when $|z - z_1| > d_1, d_2$. Since we assume that $kd \ll 1$ and $d \ll a$, we can approximate the exponential $\exp(ikR_1) \simeq \exp(ikR_1) \simeq 1$ and extend the integration in Eq. (13.26) from $z_1 = -\infty$ to $z_1 = \infty$. The resulting integral is solved explicitly, and we obtain an analytic equation for an electric potential in the system of two nanowires

$$\phi(\mathbf{r}) = 2Q(z)\left[K_0\left(\frac{Gd_1}{a}\right) - K_0\left(\frac{Gd_2}{a}\right)\right], \tag{13.27}$$

where K_0 is the modified Bessel function of the second kind and $Q(z)$ is given by Eq. (13.20). This equation simplifies when $d_1, d_2 \ll a$ and $G \sim 1$:

$$\phi(\mathbf{r}) = 2Q(z)\ln\left(\frac{d_2}{d_1}\right). \tag{13.28}$$

An extrapolation of this result to the surface of the wires gives the potential drop $U(z) = \phi(-d/2 + b, 0, z) - \phi(d/2 - b, 0, z) = -4\ \ln(d/b)\ Q(z)$. Thus, we ascertain that the interwire capacitance per unit length $\tilde{C} = -Q(z)/U(z) = [4 \ln (d/b)]^{-1}$ is indeed a constant and is independent of the coordinate z, which fully agrees with the estimate presented below in Sec. 13.4.

The vector potential $\mathbf{A} = \{0, 0, A\}$ is calculated in a similar way:

$$\begin{aligned} A(\mathbf{r}) &\simeq \frac{1}{c}\int_{-a}^{a} I(z_1)\left[\frac{\exp(ikR_1)}{R_1} - \frac{\exp(ikR_2)}{R_2}\right] dz_1 \\ &\simeq \frac{1}{c}\int_{-\infty}^{\infty} I(z_1)\left[\frac{1}{R_1} - \frac{1}{R_2}\right] dz_1 \\ &\simeq 2\frac{I(z)}{c}\frac{\left[K_0\left(\frac{Gd_2}{a}\right) - K_0\left(\frac{Gd_1}{a}\right)\right] G_z + 2\log\left(\frac{d_2}{d_1}\right)}{1 - G_z}, \end{aligned} \tag{13.29}$$

where d_1 and d_2 are given by Eqs.(13.24) and (13.25), and $G_z = \cos\left(\frac{Gz}{a}\right)/\cos(G)$. This equation also simplifies when $d_1, d_2 \ll a$ and $G \sim 1$:

$$A(\mathbf{r}) = 2\frac{I(z)}{c}\ln\left(\frac{d_2}{d_1}\right), \tag{13.30}$$

where the electric current is given by Eq. (13.21). Extrapolating the vector potential $\mathbf{A} = \{0, 0, A\}$ to the surface of the first wire ($x = d/2 - b$), we obtain $2cA = \tilde{L}I$, where the interwire inductance per unit length $\tilde{L} \simeq 4\ln(d/b)$. Note that the inductance $\tilde{L}$ is also independent of the coordinate z. Since the interwire specific capacitance $\tilde{C}$ and inductance $\tilde{L}$ both remain constant along the wire direction, Maxwell's equations reduce to an ordinary differential equation (13.16).

The electric field $\mathbf{E} = -\nabla\phi + ik\mathbf{A}$ is calculated from the potentials (13.28) and (13.30) as

$$E_x = Q_0 \frac{\sin\left(\frac{Gz}{a}\right)}{\cos(G)} \frac{\left(d_1^2 - d_2^2\right)^2 - d^2\left(d_1^2 + d_2^2\right)}{d d_1^2 d_2^2}, \tag{13.31}$$

$$E_y = Q_0 \frac{\sin\left(\frac{Gz}{a}\right)}{\cos(G)} \frac{\left(d_1^2 - d_2^2\right)\sqrt{\left(d^2 - (d_1 - d_2)^2\right)\left((d_1 + d_2)^2 - d^2\right)}}{d d_1^2 d_2^2}, \tag{13.32}$$

$$E_z = 2\log\left(\frac{d_1}{d_2}\right) \frac{G^2 G_z - (ak)^2 (G_z - 1)}{aG}, \tag{13.33}$$

where Q_0 is still given by Eq. (13.22), d_1 and d_2 are given by Eqs. (13.24) and (13.25), and correspondingly, $G_z = \cos\left(\frac{Gz}{a}\right)/\cos(G)$.

In the plane of the wires $\{y = 0\}$ electric field $E_y = 0$, while the fields E_x and E_z simplify to

$$E_x = -\frac{2Q_0\, d}{(d/2)^2 - x^2} \frac{\sin(G\,z/a)}{\cos(G)}, \tag{13.34}$$

$$E_z = -\frac{2Q_0}{a\,G} \ln\left|\frac{d/2 + x}{d/2 - x}\right| \left(G^2\, G_z + (a\,k)^2\,(1 - G_z)\right), \tag{13.35}$$

where we still assume that $|x| \ll a$, $|z| < a$, $|x - d/2| \gg b$, and $|x + d/2| \gg b$. The transverse electric field E_x changes its sign with the coordinate z, vanishing at $z = 0$. Yet, if the ratio $|E_x| / |E_z|$ is estimated near the resonance ($G \approx \pi/2$), the average $|E_x| / |E_z| \sim a/d \gg 1$, indicating that the transverse electric field is much larger than the longitudinal field at the MPR.

Recall that the problem of calculating the internal transverse field closely resembles the classical problem of finding the field induced in a dielectric cylinder by another charged cylinder placed parallel, with an elegant solution presented in Ref. 102, Sec.7. The ratio of the potential drop U_{in} across a wire to the potential drop $U(z)$ between the wires is then

$$\left|\frac{U_{in}(z)}{U(z)}\right| \simeq \frac{b}{d\,|\varepsilon_m|\ln(d/b)} \ll 1. \tag{13.36}$$

For all practical purposes we can neglect U_{in} in comparison with U, which allows us to reduce the problem of finding the charge and current distribution in the two-wire system to the ordinary differential Eq. (13.16) for the electric current $I(z)$.

Condition (13.36) is important for the analytical theory of MPR in the system of two thin rods. However, we can envision a system (for example, two closely packed metal nanowires or hemispheres) that exhibits MPR, but does not fulfill the condition in Eq. (13.36).

To clarify the nature of the MPR in the two-wire system, it is instructive to compute the ratio of the electric and magnetic resonance energies (recall that we use the CGS electromagnetic system):

$$\frac{\mathcal{E}_E}{\mathcal{E}_M} \sim c^2 \frac{\tilde{C}^{-1} \int |Q(z)|^2 dz}{\tilde{L} \int |I(z)|^2 dz} \approx \frac{g^2}{k^2} \approx 1 - \frac{2}{\ln(d/b) k^2 b^2 |\varepsilon_m|}, \tag{13.37}$$

where we have assumed that the spatial frequency g, given by Eq. (13.17), is close to the resonance ($Ga \approx \pi/2$) and we use the expressions for the specific capacitance $\tilde{C} \simeq [4\ln(d/b)]^{-1}$ and the specific inductance $\tilde{L} \simeq 4\ln(d/b)$ derived in the next section. In the electrostatic limit $\left|8\tilde{C}\left[(kb)^2\varepsilon_m\right]^{-1}\right| \gg 1$, we obtain $\mathcal{E}_E/\mathcal{E}_M \gg 1$, justifying the name given to the regime. Because of the symmetry of the electric potential problem, it is clear that the polarization cannot be induced by a uniform electric field. Therefore, the discussed resonance can be classified as a dark mode.[103]

We would like to emphasize that the electric field is not irrotational in this type of nanoantenna. It is taken for granted that the magnetic field is solenoidal. What is more surprising is that the electric field is also solenoidal in the two-stick resonator, despite the fact that its size is much smaller than the wavelength. Indeed, it follows from Eqs. (13.34) and (13.35), that the electric field E_x depends on the coordinate z, and electric field E_z depends on the coordinate x.

The electric current $I(z)$ is found from Eq. (13.16) and is used to calculate the magnetic moment of the wire pair: $\mathbf{m} = (2c)^{-1} \int [\mathbf{r} \times \mathbf{j}(\mathbf{r})]\, d\mathbf{r}$, where $\mathbf{j}(\mathbf{r})$ is the current density. After the integration is performed over both nanowires, we obtain

$$m = \frac{1}{2} H_0 a^3 (kd)^2 \ln(d/b) \frac{\tan G - G}{G^3}. \tag{13.38}$$

The metal permittivity ε_m has a large negative value in the optical/near-IR range, while its imaginary part is small; therefore, the magnetic moment, m has a resonance at $G \approx \pi/2$ (see Eq. 13.17) where m attains large values. For a typical metal, we can use the Drude formula (13.19) for ε_m, where the ratio $\omega_\tau/\omega \ll 1$. Then the normalized magnetic polarizability α_M near the MPR has the following form:

$$\alpha_M = \frac{4\pi m}{H_0 V} = \frac{16\, a\, d\, \omega_p}{\lambda^2\, \omega_r \sqrt{2\ln(d/b)}} \frac{1}{1 - \omega/\omega_r - i\omega_\tau/(2\omega_r)}, \tag{13.39}$$

where $V = 4abd$ is the volume and $\omega_r = b\pi\,\omega_p\sqrt{2\ln(d/b)}/(4a)$ is the MPR frequency. Note that the magnetic polarizability α_M contains a factor $a\,d/\lambda^2 \ll 1$ that is small for subwavelength nanostructures. However, near the resonance $G =$

$\pi/2$, the enhancement factor can be very large for optical and infrared frequencies because of the high quality of the plasmon resonances for $\omega_r \ll \omega_p$. Therefore, the total coefficient in Eq. (13.39) can be of the order of one, thereby enabling the excitation of a strong MPR.

Although the electric field near MPR is high, it is primarily concentrated in the direction perpendicular to the wires as was discussed following Eqs. (13.34) and (13.35). If the wavevector of the propagating field is in the wire plane but is perpendicular to the wire length, then the MPR described above does not strongly affect the electric field component, which is parallel to the wires. The integral of the electric field is an exact zero as follows from Eqs. (13.34) and (13.35). Envisioning a composite material that consists of similar wire pairs, one can expect that the MPR will not contribute to the dielectric permittivity in the direction parallel to the wires. Such a medium is therefore not bi-anisotropic[97] and can be described by two separate effective parameters: ε and μ.

Consider a metal nanoantenna having a horseshoe shape, obtained from a pair of nanowires by shorting their ends (see Fig. 13.5). When the quasistatic condition $\left|8\tilde{C}\left[(kb)^2\varepsilon_m\right]^{-1}\right| \gg 1$ holds in the horseshoe nanoantenna, the electric current $I(z)$ can be obtained from Eq. (13.16), where the boundary condition changes to $I_{z=a} = (dI/dz)_{z=0} = 0$ and, as above, $a \gg d \gg b$. It is easy to check in this case that the magnetic polarizability α_M is still given by Eq. (13.39), where a is now the total length of the horseshoe nanoantenna. The magnetic permeability for a metamaterial, having silver horseshoe nanoantennas oriented in one direction (z direction in Fig. 13.4) and organized in the periodic square lattice, is shown in Fig. 13.5. To simulate it, we have taken the optical parameters of silver from a

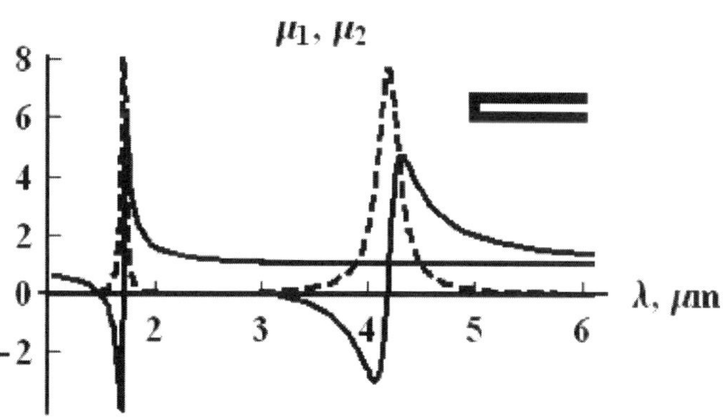

Figure 13.5 Optical magnetic permeability $\mu = \mu_1 + \mu_2$ (μ_1: continuous line, μ_2: dashed line) estimated from the Lorenz–Lorentz formula for the composite containing horseshoe-shaped silver nanoantennas; volume concentration $p = 0.3$; left curves: $a = 200$ nm, $d = 50$ nm, $b = 13$ nm; right curves: $a = 600$ nm, $d = 90$ nm, $b = 13$ nm, $d = 50$ nm, $b = 13$ nm. The dielectric susceptibility for silver is estimated from the Drude formula (13.19). (Reprinted from Ref. 67.)

classical work,[100] which almost coincides with a more recent study.[101] As seen from the figure, the negative magnetism can be observed in the near-infrared part of the spectrum, including the region of standard telecommunication wavelengths (1.5 μm).

13.4 Capacitance and Inductance of Two Parallel Wires

Below we derive equations for the capacitance per unit length $\tilde{C}$ and inductance per unit length $\tilde{L}$ between two parallel wires of the radius b and length a separated by the distance d. We assume that $b \ll d$ and $d \ll a$. We also assume that the permittivity of the wires is large in absolute value $|\varepsilon_m| \gg 1$, whereas the skin depth $\delta \sim 1/\left(k\sqrt{|\varepsilon_m|}\right) \gg b$. To find the capacitance $\tilde{C}$, we first calculate the electric potential Φ_α at the coordinate vector $\mathbf{r}_\alpha$, representing the wire surface (point α in Fig. 13.4):

$$\Phi_\alpha = \int q_1(\mathbf{r}_1)\frac{\exp(ikr_{\alpha 1})}{r_{\alpha 1}}d\mathbf{r}_1 + \int q_2(\mathbf{r}_2)\frac{\exp(ikr_{\alpha 2})}{r_{\alpha 2}}d\mathbf{r}_2, \tag{13.40}$$

where $r_{\alpha 1} = |\mathbf{r}_\alpha - \mathbf{r}_1|$, $r_{\alpha 2} = |\mathbf{r}_\alpha - \mathbf{r}_2|$, and q_1 and q_2 are the electric charges distributed over the wire surface; and we integrate over the surface of the first (α, δ) and second (β, γ) rods in Fig. 13.4. For further consideration, we choose the coordinate system $\{x, y, z\}$ with the z axis along the (α, δ) rod, the origin in the center of the system, and the x axis connecting the axes of the rods so that the y axis is perpendicular to the plane of the two rods. We introduce the vector $\mathbf{d} = \{d, 0, 0\}$ between the wires and the two-dimensional unit vector $\boldsymbol{\rho}(\phi) = \{\cos\phi,\ \sin\phi\}$ in the $\{x, y\}$ plane, where ϕ is the polar angle. Then the vectors in Eq. (13.40) can be written as $\mathbf{r}_\alpha(\phi_\alpha, z_\alpha) = \{b\,\boldsymbol{\rho}(\phi_\alpha) - \mathbf{d}/2,\ z_\alpha\}$, $\mathbf{r}_1(\phi_1, z_1) = \{b\,\boldsymbol{\rho}(\phi_1) - \mathbf{d}/2,\ z_1\}$, and $\mathbf{r}_2(\phi_2, z_2) = \{b\,\boldsymbol{\rho}(\phi_2) + \mathbf{d}/2,\ z_2\}$. As follows from the symmetry, $q_1(\phi,\ z) = q_1(-\phi,\ z) = -q_2(\pi - \phi,\ z) = -q_2(\phi + \pi,\ z)$. (Recall that we consider the antisymmetric mode, for which the electric currents are equal in the wires but flow in opposite directions.) We rewrite Eq. (13.40) by splitting Φ_α as follows:

$$\Phi_\alpha \equiv \Phi_\alpha^{(0)} + \Phi_\alpha^{(1)}, \tag{13.41}$$

$$\Phi_\alpha^{(0)}(z_\alpha, \phi_\alpha) = \int_{\phi=0}^{2\pi}\int_{z=-a}^{a} q(\phi, z)\left[-\frac{1}{\sqrt{\Delta z^2 + b^2\Delta\boldsymbol{\rho}_1^2}} + \frac{1}{\sqrt{\Delta z^2 + (b\Delta\boldsymbol{\rho}_2 - \mathbf{d})^2}}\right] dz\, bd\phi, \tag{13.42}$$

$$\Phi_\alpha^{(1)} = -\int q(\mathbf{r}_1)\frac{\exp(ikr_{\alpha1})-1}{r_{\alpha1}}d\mathbf{r}_1 + \int q(\mathbf{r}_2)\frac{\exp(ikr_{\alpha2})-1}{r_{\alpha2}}d\mathbf{r}_2, \qquad (13.43)$$

where $\Delta z = z_\alpha - z$, $\Delta\boldsymbol{\rho}_1 = \boldsymbol{\rho}_\alpha - \boldsymbol{\rho} = \{\cos\phi_\alpha - \cos\phi,\ \sin\phi_\alpha - \sin\phi\}$, $\Delta\boldsymbol{\rho}_2 = \boldsymbol{\rho}_\alpha + \boldsymbol{\rho} = \{\cos\phi_\alpha + \cos\phi,\ \sin\phi_\alpha + \sin\phi\}$. Note that dimensionless vectors $\Delta\boldsymbol{\rho}_1$ and $\Delta\boldsymbol{\rho}_2$ satisfy the condition $|\Delta\boldsymbol{\rho}_1|$, $|\Delta\boldsymbol{\rho}_2| < 2$, so that the second terms in the radicals in Eq. (13.42) are much less than a.

Electric currents in the wires (rods) and the corresponding electric charge q change with the z coordinate on the scale $\sim a$, which is much larger than the distance d between the rods. Therefore, we can neglect the variation of the electric charge along the coordinate z for $|\Delta z| < d$. On the other hand, the term in the square brackets in Eq. (13.42) vanishes as $\sim d^2/|\Delta z|^3$ for $|\Delta z| > d$. This allows us to replace the charge $q(z,\phi)$ by its value $q(z_\alpha,\phi_\alpha)$ at the observation point $\mathbf{r}_\alpha$ in Eq. (13.42), obtaining

$$\Phi_\alpha^{(0)}(z_\alpha,\phi_\alpha) = \int_{\phi=0}^{2\pi} q(z_\alpha,\phi)\int_{z=-a}^{a}\left[-\frac{1}{\sqrt{\Delta z^2 + b^2\Delta\boldsymbol{\rho}_1^2}} + \frac{1}{\sqrt{\Delta z^2 + (b\Delta\boldsymbol{\rho}_2 - \mathbf{d})^2}}\right]dz\, bd\phi. \qquad (13.44)$$

Error of this replacement is on the order of $(d/a)^2 \ll 1$. Since we consider the quasistatic limit where the distance between the rods $d \ll \lambda$ and the metal permittivity $|\varepsilon_m| \gg 1$, the potential lines in the (x,y) plane are close to the static case. Therefore, we can safely assume that the angular distribution of electric charge $q(z,\phi)$ is the same as it would be in the static case of two infinite metal cylinders, $q(z,\phi) = Q(z)\sqrt{(d/b)^2 - 4}/[2\pi(d+2b\cos\phi)]$, where $Q(z)$ is the electric charge per unit length of the rod $Q(z) = \int_{\phi=0}^{\phi=2\pi} q(z,\phi)\, b\, d\phi$. Then the integrals in Eq. (13.44) become

$$\Phi_\alpha^{(0)}(z) = -Q(z)\operatorname{arccosh}\left(\frac{d^2}{2b^2} - 1\right) + O(d/a)^2, \qquad (13.45)$$

where the second term includes all corrections to the integral (13.42) due to the finite size of the system. For thin wires considered here, the potential $\Phi_\alpha^{(0)}$ is approximated as

$$\Phi_\alpha^{(0)}(z) \simeq -Q(z)\, 2\left[\ln(d/b)\right]. \qquad (13.46)$$

The second term $\Phi_\alpha^{(1)}$ in Eq. (13.41) is small in the limit of $a \ll \lambda$, i.e., $(kr_{\alpha1}, kr_{\alpha2}) \ll 1$. The real part of $\Phi_\alpha^{(1)}$ gives a small correction on the order of $(d/a)^2$, which can be neglected. The imaginary part is important regardless of its absolute value since it gives radiative loss. To estimate the loss we assume that $b/d \ll 1$ and neglect the angular dependence of the charge distribution. We expand Eq. (13.43) in a power series of k's and linearly approximate $Q(z) \simeq q_1 z$ (recall that $Q(z)$ is an odd function of z) obtaining

$$\Phi_\alpha^{(1)}(z) \simeq iQ(z)(ak)^3(kd)^2/45, \tag{13.47}$$

where we neglect the higher-order terms in k as well as all of the terms of the form $(bk)^2$.

Due to the symmetry of the system, the potential drop $U = \Phi_\alpha - \Phi_\beta$ between points α and β (see Fig. 13.4; $z_\alpha = z_\beta$) is $U = 2\Phi_\alpha$. The capacitance per unit length $\tilde{C}$ defined as $\tilde{C} = -Q(z)/U(z)$ is then given by

$$\frac{1}{\tilde{C}} \simeq 2\operatorname{arccosh}\left[(d/b)^2/2 - 1\right] - i\tfrac{2}{45}(ak)^3(kd)^2 \simeq \tag{13.48}$$

$$4\ln\left(\tfrac{d}{b}\right) - i\tfrac{2}{45}(ak)^3(kd)^2, \tag{13.49}$$

where the first term represents the specific capacitance between two parallel infinite cylinders (see Ref. 102, Ch. 3), while the second term gives the radiative loss due to retardation effects.

To find the inductance $\tilde{L}$ between the wires, we first calculate the vector potential $\mathbf{A} = \{0, 0, A_\alpha\}$ in a wire. That is we calculate A_α at a space-vector $\mathbf{r}_\alpha$, representing the wire point. In this derivation, we neglect edge effects and assume that the vector potential as well as the currents are parallel to the wire axis, obtaining

$$A_\alpha = \frac{1}{c}\int j(\mathbf{r})\left(\frac{\exp(ikr_{\alpha1})}{r_{\alpha1}} - \frac{\exp(ikr_{\alpha2})}{r_{\alpha2}}\right)d\mathbf{r}, \tag{13.50}$$

where $r_{\alpha1} = |\mathbf{r}_\alpha - \mathbf{r}|$ and $r_{\alpha2} = |\mathbf{r}_\alpha - \mathbf{r} + \mathbf{d}|$, $j(\mathbf{r}_1)$ is the current density, and the integration is completed over the volume of the first wire. Since we are considering the quasistatic case in which the skin effect is small ($kb\sqrt{|\varepsilon_m|} \ll 1$), the electric current is distributed uniformly over the cross section of the wire, and $j(\mathbf{r}) = I(z)/(\pi b^2)$. Following the procedure above for calculating the electric potential, the vector potential is expressed as $A_\alpha = A_\alpha^{(0)} + A_\alpha^{(1)}$, where

$$A_\alpha^{(0)} = \frac{1}{c}\int \frac{I(z)}{\pi b^2}\left(\frac{1}{r_{\alpha1}} - \frac{1}{r_{\alpha2}}\right)d\mathbf{r}, \tag{13.51}$$

$$A_\alpha^{(1)} = \int \frac{I(z)}{\pi b^2}\left(\frac{\exp(ikr_{\alpha1}) - 1}{r_{\alpha1}} - \frac{\exp(ikr_{\alpha2}) - 1}{r_{\alpha2}}\right)d\mathbf{r}. \tag{13.52}$$

The term $A_\alpha^{(0)}$ can be estimated similarly to the way $\Phi_\alpha^{(0)}$ is estimated. As a result, we obtain the vector potential $A_\alpha^{(0)}$ averaged over the wire cross section:

$$A_\alpha^{(0)}(z) \simeq \frac{I(z)}{2c}\left[4\ln\left(\frac{d}{b}\right) + 1\right], \tag{13.53}$$

where $I(z)$ is the electric current and we have again neglected terms on the order of $(b/a)^2$ and $(d/a)^2$. The equation (13.52) can be expanded in a power series of k, obtaining a linear term that is zero; the k^2 term ($\sim (kd)^2$) gives a small correction to $A_\alpha^{(0)}$ and a third-order term gives the radiative loss, namely

$$A_\alpha^{(1)} \simeq i(kd)^2 k \frac{1}{c} \int I(z)\, dz \sim 2i \frac{I(z)}{c} (kd)^2 ka. \tag{13.54}$$

While deriving Eq. (13.54), we neglected the current variation over the wire length. The inductance per unit length $\tilde{L}$ is then obtained from the equation $A_\alpha - A_\beta = 2A_\alpha = (\tilde{L}/c) I(z)$:

$$\tilde{L} = 4 \ln\left(\frac{d}{b}\right) + 1 + 4i(kd)^2 ka. \tag{13.55}$$

Again, similar to the two-wire capacitance case, the first two terms represent a system of two parallel infinite wires; the self-inductance per unit length in this case can be found in Ref. 102, Ch. 34. This estimate, as well as Eq. (13.49), are certainly invalid near the rod ends; however, this does not affect the current distribution $I(z)$ and magnetic moment calculations.

We can now compare the radiation loss (given by the imaginary parts of the specific capacitance $\tilde{C}$ and specific inductance $\tilde{L}$) and the ohmic loss in the metal wires. In the red and near-infrared spectral region, the permittivity ε_m for a "good" optical metal (Ag, Au, etc.) can be estimated from the Drude formula (13.19) as $\varepsilon_m(\omega) \sim -(\omega_p/\omega)^2 (1 - i\omega_\tau/\omega)$, where ω_p is the plasma frequency and $\omega_\tau \ll \omega \ll \omega_p$ is the relaxation rate. The real part of the rod resistance $R_{ohm} \sim 8\left(\omega_\tau/\omega_p^2\right)\left(a/b^2\right)$ should then be compared with "radiation" resistance $R_{rad} \sim (kd)^2(ka)^2/c$. For the silver nanowires considered here, the ohmic loss is either larger ($R_{ohm} > R_{rad}$) or much larger ($R_{ohm} \gg R_{rad}$) than the radiation loss. Therefore, we can neglect the imaginary parts of the capacitance $\tilde{C}$ and inductance $\tilde{L}$:

$$\tilde{L} \simeq \frac{1}{\tilde{C}} \simeq 4 \ln\left(\frac{d}{b}\right). \tag{13.56}$$

This estimate has logarithmic accuracy, with an error on the order of $[4 \ln(d/b)]^{-1}$. Note that the radiation loss depends on the parameter ka in a crucial manner. The magnetic plasmon resonance becomes very broad when $ka > 1$, placing a rather severe constraint on the wire length $2a$.

Overall, we have described the new optical phenomenon of a magnetic plasmon resonance in metallic horseshoe-shaped nanoantennas. This resonance is distinctly different from the geometric LC resonance described earlier for split rings because it is determined by the plasmonic properties of the metal. MPR paves the way to designing metallic metamaterials that are magnetically active in the optical and near-infrared spectral ranges. The analytic calculations presented above and the two-dimensional numerical simulations reveal that resonantly enhanced magnetic moments can be induced in a very thin (thinner than the skin depth) horseshoe with typical dimensions much shorter than the wavelength (on the order of 100 nm). Periodic arrays of such horseshoe-shaped nanoantennas can be used to design left-handed metamaterials by exploiting the proximity of the electric resonances in the dielectric permittivity ε and magnetic permeability μ.

13.5 Lumped Model of a Resonator Filled with an Active Medium

The main features of horseshoe dynamics can be understood in terms of a simple equivalent model. We first consider the plain metal horseshoe nanoantenna whose lateral size is much larger than other dimensions. The nanoantenna is excited by the magnetic component H_0 of the incident electromagnetic field, as shown in Fig. 13.1. The electric current $J = lI$ flowing in the metal arms of the antenna is shorted by the displacement current (vertical arrows in Fig. 13.1). Note that l is the lateral size of the horseshoe, i.e., the size in the x direction in Fig. 13.1. The metal part of the nanoantenna can be represented as an inductance L. The gap between the two arms is modelled as a capacitance C. Then the horseshoe antenna can be described as an LCR circuit, shown in Fig. 13.6. The inductance L stands for the metal since the metal's permittivity is typically negative in the optical and IR range and is proportional to ω^{-2} (cf. Eq. 13.19). The resistance R presents the loss in the metal. The EMF "generator" $V = V_0 \cos(\omega t)$ in Fig. 13.6 represents the electromotive force induced by the external magnetic field H_0. For the equivalent circuit in Fig. 13.6, we obtain the following Kirchhoff's equation, which we write in terms of the electric charge $q(t)$:

$$\frac{L}{c^2}\ddot{q} + U + R\dot{q} = V, \tag{13.57}$$

where U is the potential drop in the capacitor C, and c is the speed of light. Recall that we use CGS units throughout the chapter. In the considered equivalent circuit approximation, the potential U, which is given by Eq. (13.1), is independent of the coordinate z. It can be written as $U = 4\pi d(Q - P_1 - P_2)$, where $Q = q/S = q/la$ is the charge density in the capacitor, $S = la$ is the area of a horseshoe arm, and $P_1 = \chi_1 E_y$ and P_2 are the regular and resonance (gain) polarizations of

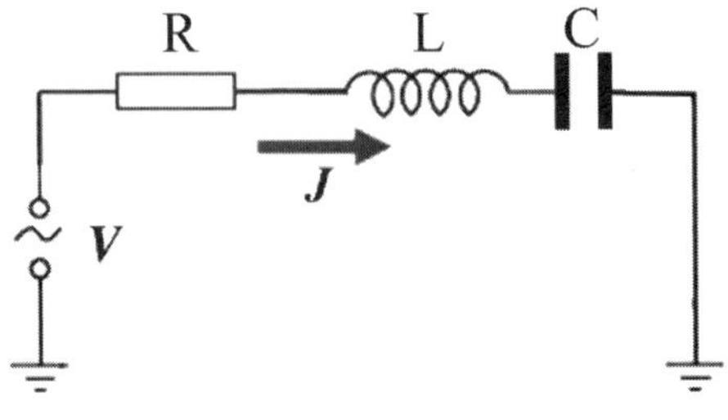

Figure 13.6 Equivalent LCR circuit of the horseshoe nanoantenna. (Reprinted from Ref. 69.)

the medium in the capacitor, respectively. We introduce the regular capacitance $C = \varepsilon_d S/(4\pi d)$, where $\varepsilon_d = 1 + 4\pi\chi_1$; then the potential equals

$$U = \frac{q}{C} - \frac{SN\hat{p}}{C}, \tag{13.58}$$

where N and $\hat{p}$ are the density and dipole moment of the TLS (e.g., quantum dots), respectively and $S = la$. Substituting Eq. (13.58) in Eq. (13.57), we obtain the equation

$$\frac{L}{c^2}\ddot{q} + \frac{q}{C} - \frac{SN\hat{p}}{C} + R\dot{q} = V. \tag{13.59}$$

This equation describes the charge and current oscillation in the horseshoe resonator in the presence of a gain medium. Note that the TLS dipole moment $\hat{p}$ in Eq. (13.59) is a quantum operator.

We first consider the classical approximation in which the polarization $P_2 = Np = \chi_2 E_y$ and susceptibility χ_2 do not depend on the electric field. That is, we neglect the depletion of the gain medium. Then Eq. (13.59) takes the usual form of Kirchhoff's equation

$$\left(L/c^2\right)\ddot{q} + q/C + R\dot{q} = V, \tag{13.60}$$

where $C = \varepsilon S/\left(4\pi d\right)$ and $\varepsilon = 1 + 4\pi\chi_1 + 4\pi\chi_2$ is the permittivity of the metamaterial. Thus, in the classic approximation, the permittivity includes both regular and "gain" susceptibilities. We suppose that the external electromagnetic field is switched on at the time $t = 0$ and rewrite Kirchhoff's equation as

$$\ddot{q} + q + \gamma\dot{q} = V_a \theta\left(t\right)\cos\left(\omega t\right), \tag{13.61}$$

where the time and frequency are measured now in terms of the resonance frequency $\omega_r = c/\sqrt{LC}$, $\theta\left(t\right)$ is the step function ($t < 0,\ \theta = 0;\ t > 0,\ \theta = 1$), the dimensionless relaxation parameter $\gamma = RC\omega_r$, and $V_a = CV_0$. It is easy to calculate the current $J \equiv \dot{q}$ in the circuit using Eq. (13.61). We assume that the loss factor $\gamma \ll 1$ and obtain

$$\begin{aligned} J\left(t\right) = V_a \,\mathrm{Re}\, \Bigg[& \exp\left(-i\,t\,\omega\right)\frac{i\,\omega}{\omega^2 + i\,\gamma\,\omega - 1} - \\ & \exp\left(-\,\gamma t/2\right)\frac{2\,i\,\omega\,\cos t + \left(2 - i\,\gamma\,\omega\right)\,\sin t}{2\,\left(\omega^2 + i\,\gamma\,\omega - 1\right)}\Bigg], \end{aligned} \tag{13.62}$$

where Re denotes the real part of an equation.

To estimate L, C, and R in terms of the horseshoe parameters, it is convenient to write the metal permittivity as $\varepsilon_m = -\left|\varepsilon_m\right|\left(1 - i\kappa_m\right)$ and the permittivity of the medium inside the horseshoe as $\varepsilon_d = \left|\varepsilon_d\right|\left(1 + i\kappa_d\right)$, where we take into account that $\mathrm{Re}\,\varepsilon_m$ is negative in optical and infrared spectral regions and that the metal loss factor $\kappa_m \ll 1$. We also suppose that the loss or gain is small in the dielectric, i.e., $|\kappa_d| \ll 1$ and include dielectric loss/gain in the resistor R, obtaining

$$L = \frac{8\pi ac^2}{\omega^2 |\varepsilon_m| bl} \approx \frac{8\pi ac^2}{\omega_p^2 bl}, \quad C = \frac{\varepsilon_d S}{4\pi d}, \quad \text{and} \quad R = L\frac{\omega\kappa_m}{c^2} + \frac{\kappa_d}{\omega C}, \tag{13.63}$$

where a, b, d, l, and S are the length, the arm thickness, the gap width, the lateral size, and the area of the horseshoe shown in Fig. 13.1, correspondingly. We use the Drude formula (13.19) for the second estimate of the inductance in Eq. (13.63), assuming that the frequency ω of the applied field is much less than the plasma frequency ω_p for the metal. Note that the equation for the resonance frequency

$$\omega_r = \frac{c}{\sqrt{LC}} \simeq \omega_p \frac{\sqrt{bd}}{a},$$

obtained in the LCR model, coincides, up to the factor $\pi/4$, with the exact Eq. (13.14).

From Eq. (13.63) we obtain the following relaxation parameter in Eq. (13.64):

$$\gamma = RC\omega_r = \kappa_m (\omega/\omega_r)^2 + \kappa_d. \tag{13.64}$$

The second term in Eq. (13.62) describes the transient response of the system. We assume that at the initial time $t = 0$, the current $J(0) = 0$, and the electric charge $q(0) = 0$. This term becomes irrelevant when $\gamma > 0$, and properties of the nanoantenna can be described in terms of the complex impedance

$$Z_a = -i\frac{\omega L}{c^2} + \frac{i}{\omega C} + R. \tag{13.65}$$

The condition $\operatorname{Im} Z_a = 0$ gives the resonance frequency ω_r, whereas the resistance R determines the line width. If the frequency of the incident light is larger than the resonant frequency, the current inside the horseshoe induces the magnetic field H_{in} with a direction opposite to the direction of the field H_0 of the incident wave. Therefore, the metamaterial, composed of horseshoe nanoantennas, has a negative magnetic susceptibility for $\omega > \omega_r$, and the permeability could also be negative if the concentration of horseshoe nanoantennas is large enough as demonstrated in Figs. (13.3) and (13.5).

Consider now an antenna in which the gap in the horseshoe is filled with an active medium, i.e., the capacitor has negative loss $\kappa_d < 0$ [see Eq. (13.64)]. For sufficiently large gain, the losses in the metal can be overcompensated and the parameter $\gamma < 0$. Then it follows from Eq. (13.62) that the electric current $J(t)$ exponentially increases in the antenna. Note that the instability takes place as soon as $\gamma < 0$, while the first term in Eq. (13.62) is irrelevant in this case. In the current literature, most authors use an approach that is equivalent to consideration of only the first term in Eq. (13.62), i.e., they use a complex impedance to describe the medium with gain (see, for example the very recent paper by Zheludev et al.[93]) We see that results obtained in this case have very limited meaning, if any, since the corresponding solution for Maxwell's equations becomes unstable.

Exponential growth of the electric current for $\gamma < 0$ does not last forever. Due to saturation of the active medium, the electric current and the field in the antenna cannot increase to infinity. To take into account the depletion of the active medium, we write the following phenomenological equation for the electric charge $q(t)$ in the horseshoe:

$$\ddot{q} + \left[\gamma_1 + \frac{\gamma_2}{1 + \dot{q}^2}\right] \dot{q} + q = V_a \theta(t) \cos(\omega t), \tag{13.66}$$

where $\gamma_2 < 0$ and $|\gamma_2| > \gamma_1$. Equation (13.66) cannot be solved in the general case; however, some predictions can be made immediately: Let us assume that at $t = 0$ there is no electric current or charge in the antenna, corresponding to the initial conditions $J(0) = 0$ and $q(0) = 0$. When the external magnetic field is turned on, the current starts to increase exponentially according to Eq. (13.62). Later, the electric current as well as the electric charge saturate at some level due to depletion of the active medium. In general, the saturation level will be proportional to the amplitude of the magnetic field in the incident electromagnetic wave. In our model it is denoted as the amplitude V_a of the electromotive force (EMF) in the equivalent LCR circuit depicted in Fig. 13.6. The dependence of the charge oscillation amplitude in the horseshoe nanoantenna on the EMF V_a is shown in Fig. 13.7, where the parameters are $\gamma_1 = 0.1$ and $\gamma_2 = -0.5$.

We can see that the amplitude increases linearly with sufficiently large V_a. It is the range of forced oscillations in which the amplitude is so large that the second term in the square brackets in Eq. (13.66) is negligible. We also notice in Fig. 13.7 the region of spontaneous oscillations; the electric current exists even in

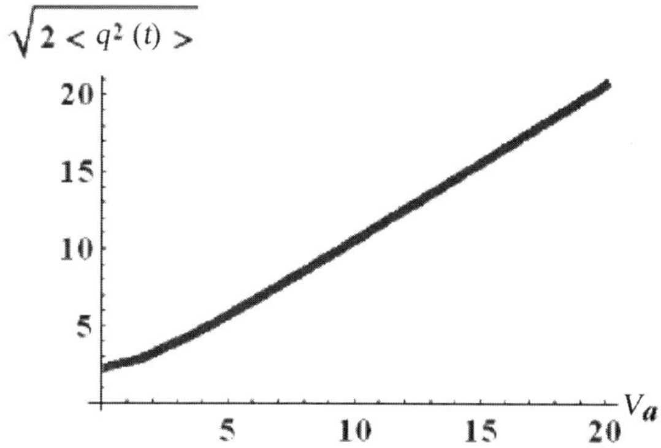

Figure 13.7 Time-averaged amplitude of the charge oscillation in the horseshoe that is filled with an active medium. The amplitude is obtained from solutions of Eq. (13.66). (Reprinted from Ref. 69.)

the absence of an external field. Therefore, the horseshoe antenna behaves as a *plasmonic nanolaser.*

13.6 Interaction of Nanoantennas with an Active Host Medium

In order to understand the origin of lasing in an active plasmonic medium, we will consider a microscopic model.[68–70] (See also consideration of a "dipole laser" by Protsenko et al.[104]) We use the quantum-mechanical derivation of the equations of motion for the system shown in Figs. 13.1 and 13.8, but neglect quantum correlations and fluctuations in our analysis. The Hamiltonian of the nanoantenna interacting with a TLS is given by the expression

$$H = H_0 + H_d + V_{int} + \Gamma, \tag{13.67}$$

where H_0 and H_d describe, respectively, the horseshoe and TLS. We assume that the energy of the TLS is $\hbar\omega_2$. The operator

$$V_{int} = -\hat{p}E \tag{13.68}$$

gives the interaction between the TLS and the nanoantenna, where E and $\hat{p}$ are the electric field of the magnetic plasmon excited in the horseshoe and the dipole moment of the TLS, respectively. The Γ term includes dissipation and pump effects. Electrons in the horseshoe nanoantenna couple to the local electric field and oscillate with the frequency ω, which is close to the magnetic plasmon resonance frequency ω_r. We will treat the electric charge

$$q(t) = q_1(t)\exp(-i\omega t) + q_1^*(t)\exp(i\omega t) \tag{13.69}$$

and the current

$$J(t) = \dot{q}(t) = J_1(t)\exp(-i\omega t) + J_1^*(t)\exp(i\omega t) \tag{13.70}$$

as classical objects, defined by their slowly varying amplitudes $q_1(t)$ and $J_1(t)$.

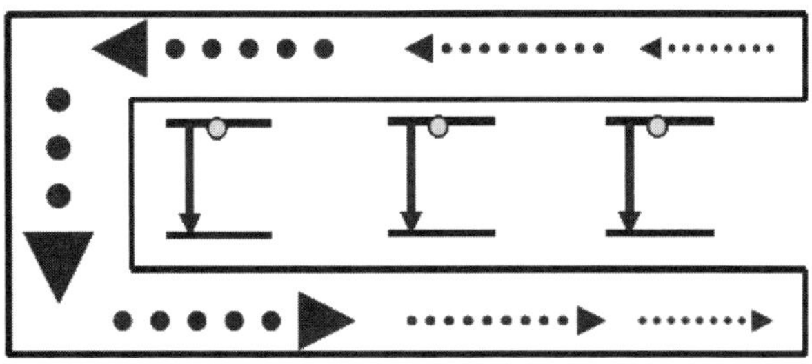

Figure 13.8 A plasmon propagates in a horseshoe nanoantenna (dashed line). Its amplitude increases due to the impact interaction with the TLS that supplies energy to the plasmon.

The TLS can be either in the ground state $|g\rangle$ or in the excited state $|e\rangle$ as is illustrated in Fig. 13.8. To find its time evolution we, as usual,[105] introduce the annihilation and creation operators η and η^+. These operators are defined by the equations

$$\eta\,|g\rangle = 0, \quad \eta\,|e\rangle = |g\rangle\,, \ \eta^+\,|g\rangle = |e\rangle\,, \text{ and } \eta^+\,|e\rangle = 0; \tag{13.71}$$

that is, these operators have the following matrix presentation:

$$\eta = \begin{bmatrix} 0 & 0 \\ 1 & 0 \end{bmatrix}, \ \eta^+ = \begin{bmatrix} 0 & 1 \\ 0 & 0 \end{bmatrix} \tag{13.72}$$

in the basis of $\{|e\rangle\,,\ |g\rangle\}$. Operators η and η^+ follow the usual anticommutation rule $\eta\eta^+ + \eta^+\eta = 1$, which is evident from Eqs. (13.72). In the Heisenberg representation, the time development of these operators is given by

$$\begin{aligned} \eta(t) &= \exp(iHt/\hbar)\,\eta\,\exp(-iHt/\hbar) \quad \text{and} \\ \eta^+(t) &= \exp(iHt/\hbar)\,\eta^+\,\exp(-iHt/\hbar), \end{aligned} \tag{13.73}$$

where the Hamiltonian H is given by Eq. (13.67).

To find the explicit equation for the time evolution of the system, it is constructive to introduce the "slow" operators

$$b(t) = \eta(t)\exp(i\omega t) \quad \text{and} \tag{13.74}$$

$$b^+(t) = \eta^+(t)\exp(-i\omega t). \tag{13.75}$$

The operators $b(t)$ and $b^+(t)$ also follow the anticommutation rule:

$$b(t)b^+(t) + b^+(t)b(t) = 1. \tag{13.76}$$

The physical meaning of the operators b and b^+ is clear from the following reasoning. Let us consider a TLS that is isolated from the surrounding world. The states $|e\rangle$ and $|g\rangle$ are eigenstates for the isolated system. Therefore, the operators $b(t)$ and $b^+(t)$ are $b(t) = \eta\exp(i(\omega - \omega_2)t)$ and $b^+(t) = \eta^+\exp(-i(\omega - \omega_2)t)$, where ω_2 is the resonance frequency of the TLS. We ascertain that operators $b(t)$ and $b^+(t)$ are slow operators when $\omega \approx \omega_2$. If we now turn on the interaction of the TLS with the horseshoe resonator, the operators $b(t)$ and $b^+(t)$ still slowly develop with time since we assume that the internal electric field in the TLS (atomic field) is much larger than the electric field in the horseshoe resonator and, therefore, $\hbar\omega_2 \gg \langle V_{int}\rangle$.

The operator of the dipole moment in Eq. (13.68) can be represented as

$$\hat{p}\,(t) = \Pi b\,(t)\exp\left(-i\omega t\right) + \Pi^* b^+\,(t)\exp\left(i\omega t\right), \tag{13.77}$$

where $\Pi \cong \langle g|\,re\,|e\rangle$ is the matrix element of the dipole operator between the excited and ground states of the TLS. We assume, for simplicity, that all losses are included in the last term in Eq. (13.67). We assume that the TLS oscillates between the upper and lower levels with the frequency close to the frequency ω_2, where ω_2 is the resonance frequency of the TLS. We express the electric field $E(t)$ in the horseshoe in terms of the electric charge $q(t)$ and the polarization $N\hat{p}(t)$ of the active medium, where N is the density of the TLS. Thus, we obtain

$$E(t) = \frac{1}{Cd}\left[q(t) - SN\hat{p}(t)\right], \tag{13.78}$$

where the capacitance C of the horseshoe is defined in Eq. (13.63) and d is the horseshoe gap (see Fig. 13.1).

We also introduce the population inversion operator

$$D(t) = n_g(t) - n_e(t), \tag{13.79}$$

where $n_e(t) = b^+b$ and $n_g(t) = bb^+$ are the operators describing the population of the excited and ground states of the TLS. Neglecting the fast oscillating terms $\propto \exp(\pm i2\omega t)$, we can express the Hamiltonian in terms of the following operators:

$$H_d = \hbar\omega_2 n_e \quad \text{and} \tag{13.80}$$

$$V_{int} = -\frac{1}{Cd}\left(\Pi^* q_1 b^+ + \Pi q_1^* b\right) + |\Pi|^2 SN, \tag{13.81}$$

where the second summand in Eq. (13.81) is a constant and, therefore, does not influence the dynamics of the system. We derive the Heisenberg equations of motion

$$i\hbar\dot{b} = [H, b] \text{ and } i\hbar\dot{D} = [H, D] \tag{13.82}$$

using the anticommutation rule (13.76), obtaining

$$\dot{b} = -(i\Delta + \Gamma)b + \frac{i}{\hbar Cd}\Pi^* q_1 D \quad \text{and} \tag{13.83}$$

$$\dot{D} = \frac{2i}{\hbar Cd}\left(\Pi q_1^* b - \Pi^* q_1 b^+\right) - \frac{D - D_0}{\tau}, \tag{13.84}$$

where $\Delta = \omega_2 - \omega$. We have included in Eqs. (13.83) and (13.84) terms involving $1/\tau$ and Γ to account for the relaxation and pump processes. D_0 would be the stationary value of the TLS population if it were not interacting with the horseshoe, i.e., when $q_1 = 0$. We assume that $D_0 < 0$ because the pumping process provides the initial population inversion in the TLS. By neglecting quantum fluctuations and correlation, D and b can be treated as complex variables with b and b^+ being replaced by b and b^*, respectively. Thus, we can use Eqs. (13.59), (13.83), and (13.84) as a full set of the differential equations that describe the dynamics of the horseshoe nanoantennas in an active host medium.

We have introduced the parameter D_0 in Eq. (13.84) to phenomenologically describe the excitation of a TLS (two-level) quantum system. Another approach is to use a four-level system as was done in the work of Fang, Koschny, Wegener, and Soukoulis.[94] The four-level system is more difficult for quantum mechanical consideration. Yet, it allows, in principle, one to investigate the evolution of the population of the second and third levels that are in resonance with the plasmonic nanoantenna. In the study by Fang et al.[94] it was suggested, as usual, that the relaxation between the fourth (most upper) level and third level is much faster than the annihilation/excitation process between the second and third levels. The rate of the transition from second to first (ground) level is also much faster than the annihilation/excitation process. It can be shown that under these conditions the system of four equations from Ref. 94 approximately reduces to Eqs. (13.83) and (13.84). That is, the kinetic parameters of the four-level system are renormalized to the single parameter D_0 in Eq. (13.83).

We now consider the lasing that is the natural oscillation of the electric charge in the horseshoe resonator in the absence of the external field $V = 0$ in Eq. (13.59). We suppose that the resonator oscillation does not change ($\dot{q}_1 = 0$, $\dot{b} = 0$, $\dot{D} = 0$), so the resonator moves over its limit cycle. Thus, the lasing is the auto-oscillation of the nanoantenna and gain medium system. Then Eqs. (13.59), (13.83), and (13.84) can be rewritten in the following form:

$$(i\delta + \gamma_1)\, q_2 - ib = 0, \tag{13.85}$$

$$(i\Delta_1 + \Gamma_1)\, b - iA_0 D q_2 = 0, \qquad \text{and} \tag{13.86}$$

$$-\frac{D - D_0}{\tau_1} + 2iA_0\, (q_2^* b - q_2 b^*) = 0, \tag{13.87}$$

where the dimensionless electric charge $q_2 = q_1/\,(SN\Pi)$, the dimensionless constant

$$A_0 = \frac{4\pi N\, |\Pi|^2}{\hbar\omega_r \varepsilon_d}, \tag{13.88}$$

$\Delta_1 = \Delta/\omega_r$, $\Gamma_1 = \Gamma/\omega_r$, and $\delta = 1 - (\omega/\omega_r)^2$; the plasmon resonance frequency $\omega_r = c/\sqrt{LC}$, and the relaxation parameter is equal to $\gamma_1 = (\varepsilon''_m/\,|\varepsilon_m|)\,(\omega\,/\omega_r)^2 = \kappa_m\,(\omega\,/\omega_r)^2$ (cf. Eq. 13.64). Equations (13.85)–(13.87) define the lasing in the horseshoe plasmonic resonator; they have a nonzero solution when the following conditions are fulfilled:

$$\frac{\Delta_1}{\Gamma_1} = -\frac{\delta}{\gamma_1} \quad \text{and} \tag{13.89}$$

$$\left(\frac{\delta}{\gamma_1}\right)^2 + 1 + \frac{A_0 D}{\Gamma_1\gamma_1} = 0. \tag{13.90}$$

The condition (13.89) gives the lasing frequency

$$\omega_L = \omega_r + \frac{\gamma_1\,(\omega_2 - \omega_r)}{\gamma_1 + 2\Gamma_1}, \tag{13.91}$$

which is always between the magnetic plasmon resonance frequency ω_r and the TLS resonance frequency ω_2. In the lasing condition (13.90), all terms are positive except the population D. Therefore, this condition holds only in the inverted medium $n_e > n_g$ where $D < 0$ [see Eq. (13.79)]. The population D cannot be smaller than -1, which corresponds to the case when the TLS is exactly in the excited state. We obtain the lasing condition for the horseshoe nanolaser:

$$\frac{A_0}{\Gamma_1 \gamma_1} = \frac{4\pi N |\Pi|^2}{\hbar \varepsilon_d \Gamma \gamma_1} > 1. \tag{13.92}$$

As soon as condition (13.92) is fulfilled, the interaction between the TLS and the plasmonic nanoantenna leads to coherent oscillations of the electric charge and the magnetic moment of the horseshoe, even in the absence of an external electromagnetic field.

13.7 Plasmonic Nanolasers and Optical Magnetism

In the previous section we found that metal nanoantennas filled by a gain medium can support spontaneous oscillation of the electric charge and electric current in the antenna. This effect corresponds to the spontaneous excitation of the magnetic plasmon resonance in the horseshoe resonator (see Figs. 13.1 and 13.8). To simplify the analysis of the spontaneous oscillations, we consider the horseshoe resonator where the magnetic plasmon frequency coincides with a TLS frequency, i.e., $\omega_r = \omega_2$. Then Eq. (13.89) gives the lasing frequency $\omega_L = \omega_r$. We also suppose that the active medium is pumped to such an extent that

$$D_0 < -\Gamma_1 \gamma_1 / A_0. \tag{13.93}$$

Recall that D_0 is the population of the TLS that does not interact with the horseshoe resonator. When the condition (13.93) is fulfilled, the amplitude of the charge oscillations q_2, the TLS polarization b, and the population D are

$$q_2 = \frac{ie^{i\varphi}}{2A_0}\sqrt{-\frac{A_0 D_0 + \gamma_1 \Gamma_1}{\gamma_1 \tau_1}}, \tag{13.94}$$

$$b = -i\gamma_1 q_2, \quad \text{and} \tag{13.95}$$

$$D = -\frac{\Gamma_1 \gamma_1}{A_0}, \tag{13.96}$$

where the amplitudes of the oscillations have an arbitrary phase φ. It is easy to check by direct substitution that Eqs. (13.94)–(13.96) indeed give the solution for Eqs. (13.85)–(13.87).

The amplitude of the spontaneous electric current in the horseshoe resonator can be estimated as $J_1 = -i\omega_L q_1 = -i\omega_L S N \Pi q_2$. We still assume that $\omega_r =$

$\omega_2 = \omega_L$. Making use of Eqs. (13.88) and (13.94), we obtain the electric current

$$J = \frac{\hbar\omega_r^2 e^{i\varphi} S\varepsilon_d}{8\pi\Pi^*}\sqrt{-\frac{A_0 D_0 + \gamma_1\Gamma_1}{\gamma_1\tau_1}} \tag{13.97}$$

and the magnetic moment of the horseshoe

$$m \simeq \frac{J_1 da}{c} = \frac{\hbar\omega_r^2 e^{i\varphi} aV\varepsilon_d}{8\pi c\Pi^*}\sqrt{-\frac{A_0 D_0 + \gamma_1\Gamma_1}{\gamma_1\tau_1}}, \tag{13.98}$$

where $V = Sd$ is the volume of the horseshoe. The magnetic moment exists even in the absence of an external electromagnetic field. Its "direction," i.e., the phase φ in Eq. (13.98), is undefined in this case. It is interesting to note that optical magnetism appears to be a relativistic quantum phenomenon. Indeed the magnetic moment m, given by Eq. (13.98), vanishes provided the speed of light tends to infinity and/or the Planck constant tends to zero. Note a similarity between optical magnetism and the usual ferromagnetism, which is a relativistic, quantum-mechanical effect.

Let us consider a metamaterial composed of a regular array of horseshoe nanoantennas filled by an active medium (see Figs. 13.1 and 13.9.) The volume concentration of the horseshoes is p. We suppose that the lasing condition (13.93) is fulfilled and the interaction between the nanoantennas results in synchronization, i.e., all of the phases φ are the same. Then the specific magnetic moment M of the metamaterial is $M = pm/V$, where the horseshoe moment m is given by Eq. (13.98). Therefore, the metamaterial has spontaneous high-frequency magnetization and operates similarly to an optical ferromagnetic. Dicke superradiance and other unusual optical phenomena can occur in this case. There is, however, an important difference between the usual ferromagnetism and the proposed optical magnetism. Static ferromagnetism is due to quantum mechanical, persistent currents and exists per se. Optical magnetism exists only if the active medium is pumped; even if made from a perfect metal and dielectric, the metamaterial radiates

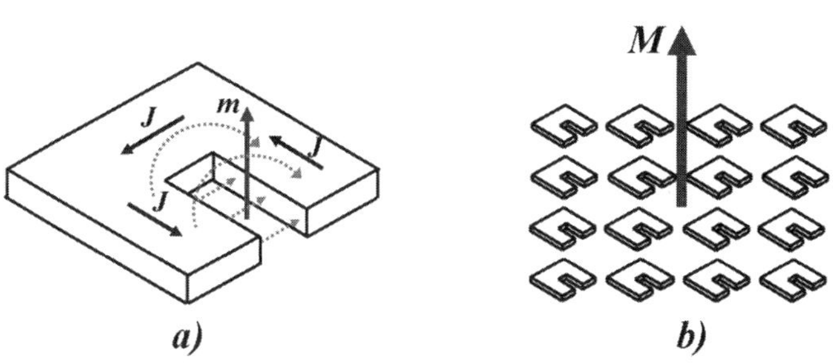

Figure 13.9 a) A flat horseshoe nanoantenna immersed in a gain medium. Spontaneous currents and magnetization are shown by arrows. b) An array of horseshoe nanoantennas; the arrow shows the magnetic moment.

when it has spontaneous magnetic moment. The emitted radiation does not need an optical cavity and, in the case of a single horseshoe, could be concentrated in a very small volume $\sim 10^{-3}$ μm^3.

We now consider the feasibility of a plasmonic laser and optical magnetism. It is instructive to express the lasing condition (13.92) in terms of the optical gain. Let us consider an electromagnetic field in the active medium, assuming, for simplicity, that the inversion does not change. That is, we assume that the population of the TLS remains constant with $D = D_0$ while the field oscillates. This assumption corresponds to the vanishing of the relaxation time τ in Eq. (13.84). Then the second term in Eq. (13.87) can be neglected. We substitute $D = D_0$ in Eq. (13.86) obtaining the amplitude b of the dipole moment of the TLS. Then we find the polarization P and the effective permittivity (the details of the procedure can be found in Ref. 106)

$$\varepsilon_e = \varepsilon_d + \frac{\varepsilon_d A_0 D_0}{\Delta_1 - i\Gamma_1} \tag{13.99}$$

of the gain medium. Next we consider the propagation of the electromagnetic wave in an infinite active medium (without any metal antennas) with the permittivity given by Eq. (13.99). The maximum gain is achieved when the frequency equals the resonance frequency ω_2 of the TLS ($\Delta_1 = 0$) and the active medium is completely inverted $D_0 = -1$. Then the intensity $|E|^2$ of the wave, propagating along, say the x direction, exponentially increases $|E(x)|^2 \propto \exp(Gx)$, where the optical gain

$$G = -\frac{4\pi}{\lambda}\operatorname{Im}\sqrt{\varepsilon_e} = 2\pi A_0 n_d/(\lambda\Gamma_1), \tag{13.100}$$

$n_d = \sqrt{\varepsilon_d}$, and we assume that the nonresonant (regular) dielectric permittivity ε_d is much larger than $|\operatorname{Im}\varepsilon_e|$. Therefore, we can express the lasing condition (13.92) in terms of the gain

$$\frac{G\lambda}{2\pi n_d \kappa_m} > 1, \tag{13.101}$$

recalling that $\kappa_m = \varepsilon''_m/|\varepsilon_m|$ is the metal loss factor. Note that the lasing condition depends only on the gain in the active medium and the loss in the metal. The lasing condition (13.101) is equivalent to the condition of SP excitation in a subwavelength metal particle immersed in a gain medium, obtained from quite different considerations.[107]

We assume that Eq. (13.101) is the general criterion that holds for any shape of the subwavelength-sized plasmonic laser. For instance, the silver nanolasers shown in Figs. 13.1, 13.4, and 13.9 would lase at a wavelength of 1.5 μm if the active medium can maintain an optical gain larger than the critical gain $G_c \approx 3\cdot 10^3$ cm^{-1}. Such a gain is typical of a "good" quantum well (QW) medium. The comparison of various QW lasers can be found in the Refs. 108 and 109. In a joint study by two groups from the Physical-Technical Institute of St. Petersburg and the Technical University of Berlin, very high material gain, up to 10^5 cm^{-1} was obtained in

GaAs-based quantum dot (QD) materials.[80,81] Recently, these results were repeated in the InAs QD laser (size $\sim 1 \times 0.3 \times 0.1\,\mu\text{m}$) created inside a photonic crystal.[79] Less than five QDs were enough to get lasing. The material gain $G = 2 \cdot 10^5\,\text{cm}^{-1}$ was reported, which is almost two orders of magnitude larger than the estimated critical gain for a nanolaser comprising a silver horseshoe nanoantenna. However, the typical surface density of QDs in currently synthesized materials is rather small ($p_{QD} \simeq$ 1–2%) and is limited mostly by the state of the art of QD array manufacturing.[80,110–113] QDs should fill about 5–10% of the internal volume of the horseshoe nanoantenna to meet the lasing condition, which can be formulated as the requirement that the modal gain is larger than the critical gain. Further development of nanotechnology will allow us to meet this goal. (See also the discussion in Ref. 86.) Note that we use the strong confinement of the electromagnetic field inside the horseshoe in our estimation of the modal gain.

Effects at the metal-semiconductor (horseshoe-QD) interface could be important (interface capacitance, electron injection, quenching, etc.) when the volume concentration of the QD inside the horseshoe increases. Experimental results[114] show that quenching of a dye molecule near the metal surface becomes important at distances less than 2 nm. The first theoretical and experimental investigations of QD covered by a silver shell show that full metal coverage does not substantially influence the profile of a QD emission line.[92] We speculate that interface effects will lead only to second-order corrections of the calculated optical properties. In summary, we believe that horseshoe nanoantennas filled with QW or QD gain elements can be developed into a plasmonic nanolaser. The horseshoe nanoantennas confine the electromagnetic field inside themselves and act as a cavity for the gain elements.

13.8 Conclusions

We have shown that horseshoe metallic nanoantennas can be used for building plasmonic metamaterials with both negative permittivity and permeability, which can lead to left-handed metamaterials in the near-infrared and optical band of the spectrum. Contrary to the previously well-established belief that most materials do not demonstrate any anomalous magnetic permeability at high optical frequencies, we have shown that specifically shaped nanoparticles can exhibit negative permeability and refraction in optics by exciting a magnetic plasmon resonance. It is well known that plasmonic materials will have very high absorption in the optical range, which makes them an unlikely candidate for superlenses and other practical optical devices. In this chapter we have derived the condition under which nanoantennas filled with a highly efficient gain medium can demonstrate low absorption or even gain sufficient for lasing. The proposed host medium should have initial gain greater than $10^3\,\text{cm}^{-1}$. There is still an open question regarding the quenching and spontaneous emission of such a high-gain medium in the near field of plasmonic particles. The set of self-consistent equations derived in this chapter allows for the calculation of those effects as well.

Consider now the pumping of a nanolaser that is excited by the magnetic component H of the impinging electromagnetic wave with frequency ω. The high-frequency magnetic field excites currents in the horseshoe and operates as a driving force. Without the driving force, the plasmonic nanolaser, which is a nonlinear oscillator, makes auto-oscillations and moves over its limit cycle given by Eqs. (13.85)–(13.87) with lasing frequency ω_L. (See Eq. 13.91.) When the driving force is applied, the plasmonic nanolaser still moves over the limit cycle but with the frequency ω of the driving force. The well-known effect of synchronization for nonlinear oscillations takes place: the nanolaser generates the frequency ω of the driving force. The incident electromagnetic wave retunes the nanolaser. Therefore, the nanolaser metamaterial can change the frequency and direction of the emitted coherent light under the action of an external light beam.

The author acknowledges stimulating discussions with A. N. Lagarkov, A. A. Pukhov, G. Tartakovsky, and V. G. Veselago. This work is supported by RFFI Grant 09-02-01519.

References

1. S. A. Schelkunoff and H. T. Friis, *Antennas: Theory and Practice,* Wiley, New York (1952).

2. A. N. Lagarkov, A. K. Sarychev, Y. R. Smychkovich, and A. P. Vinogradov, "Effective medium theory for microwave dielectric constant and magnetic permeability of conducting stick composites," *J. Electromagn. Waves Appl.* **6**, 1159 (1992).

3. A. N. Lagarkov and A. N. Sarychev, "Electromagnetic properties of composites containing elongated conducting inclusions," *Phys. Rev. B* **53**, 6318 (1996).

4. J. B. Pendry, A. J. Holden, W. J. Stewart, and I. Youngs, "Extremely low frequency plasmons in metallic mesostructures," *Phys. Rev. Lett.* **76**, 4773 (1996).

5. J. B. Pendry, A. J. Holden, D. J. Robbins, and W. J. Stewart, "Magnetism from conductors and enhanced nonlinear phenomena," *IEEE Trans. Microw. Theory Techn.* **47**, 2075 (1999).

6. A. N. Lagarkov, V. N. Semenenko, V. A. Chistyaev, D. E. Ryabov, S. A. Tretyakov, and C. R. Simovski, "Resonance properties of bi-helix media at microwaves," *Electromagnetics* **17**, 213 (1997).

7. M. C. K. Wiltshire, J. V. Hajnal, J. B. Pendry, D. J. Edwards, and C. J. Stevens, "Metamaterial endoscope for magnetic wires," *Optics Express* **11**, 709 (2003).

8. V. G. Veselago, "Electrodynamics of substances with simultaneously negative values of ε and μ," *Sov. Phys.: Uspekhi* **10**, 509 (1968).

9. D. R. Smith and N. Kroll, "Negative refractive index in left-handed materials," *Phys. Rev. Lett.* **85**, 2933 (2000).
10. J. B. Pendry, "Negative refraction makes a perfect lens," *Phys. Rev. Lett.* **85**, 3966 (2000).
11. A. A. Houck, J. B. Brock, and I. L. Chuang, "Experimental observations of a left-handed material that obeys Snell's Law," *Phys. Rev. Lett.* **90**, 137401 (2003).
12. C. G. Parazzoli, R. B. Greegor, K. Li, B. E. C. Koltenbah, and M. Tanielian, "Experimental verification and simulation of negative index of refraction using Snell's law," *Phys. Rev. Lett.* **90**, 107401 (2003).
13. A. Alu and N. Engheta, "Guided modes in a waveguide filled with a pair of single-negative (SNG), double-negative (DNG), and/or double-positive (DPS) layers," *IEEE Trans. Microw. Theory Techn.* **52**, 199 (2004).
14. H. O. Moser, B. D. F. Casse, O. Wilhelmi, and B. T. Saw, "Terahertz response of a microfabricated rod-split-ring-resonator electromagnetic metamaterial," *Phys. Rev. Lett.* **94**, 063901 (2005).
15. J. J. Chen, T. M. Grzegorczyk, B. I. Wu, and J. A. Kong, "Imaging properties of finite-size left-handed material slabs," *Phys. Rev. E* **74**, 046615 (2006).
16. M. Kafesaki, I. Tsiapa, N. Katsarakis, T. Koschny, C. M. Soukoulis, and E. N. Economou, "Left-handed metamaterials: the fishnet structure and its variations," *Phys. Rev. B* **75**, 235114 (2007).
17. M. A. Y. Abdalla, K. Phang, and G. V. Eleftheriades, "Printed and integrated CMOS positive/negative refractive-index phase shifters using tunable active inductors," *IEEE Trans. Microw. Theory Techn.* **55**, 1611 (2007).
18. D. Wang, L. Ran, H. Chen, M. Mu, J. A. Kong, and B. I. Wu, "Experimental validation of negative refraction of metamaterial composed of single side paired S-ring resonators," *Appl. Phys.* **90**, 254103 (2007).
19. A. N. Lagarkov and V. N. Kissel, "Near-perfect imaging in a focusing system based on a left-handed-material plate," *Phys. Rev. Lett.* **92**, 077401 (2004).
20. A. Grbic and G. V. Eleftheriades, "Overcoming the diffraction limit with a planar left-handed transmission-line lens," *Phys. Rev. Lett.* **92**, 117403 (2004).
21. V. N. Kissel and A. N. Lagarkov, "Superresolution in left-handed composite structures: from homogenization to a detailed electrodynamics description," *Phys. Rev. B* **72**, 085111 (2005).
22. K. Aydin and I. Bulu, "Focusing of electromagnetic waves by a left-handed metamaterial flat lens," *Optics Express* **13**, 8753 (2005).

23. M. C. K. Wiltshire, J. B. Pendry, and J. V. Hajnal, "Sub-wavelength imaging at radio frequency," *J. Phys. : Condens. Matter* **18**, L315 (2006).

24. K. Aydin, I. Bulu, and E. Ozbay, "Subwavelength resolution with a negative-index metamaterial superlens," *Appl. Phys. Lett.* **90**, 254102 (2007).

25. S. Linden, C. Enkrich, M. Wegener, J. Zhou, T. Koschny, and C. M. Soukoulis, "Magnetic response of metamaterials at 100 terahertz," *Science* **306**, 1351 (2004).

26. N. Katsarakis, T. Koschny, M. Kafesaki, E. N. Economy, and C. M. Soukoulis, "Electric coupling to the magnetic resonance of split ring resonators," *Appl. Phys. Lett.* **84**, 2943 (2004).

27. J. Zhou, T. Koschny, M. Kafesaki, E. N. Economou, J. B. Pendry, and C. M. Soukoulis, "Saturation of the magnetic response of split-ring resonators at optical frequencies," *Phys. Rev. Lett.* **95**, 223902 (2005).

28. A. K. Sarychev and V. M. Shalaev, "Electromagnetic field fluctuations and optical nonlinearities in metal-dielectric composites," *Phys. Rep.* **335**, 275 (2000).

29. A. K. Sarychev and V. M. Shalaev, *Electrodynamics of Metamaterials,* World Scientific (2007).

30. V. V. Batygin and I. N. Toptygin, *Problems in Electrodynamics,* Academic Press Inc., US (1978).

31. R. J. Blaikie and D. O. S. Melville, "Imaging through planar silver lenses in the optical near field," *J. Opt. A* **7**, S176 (2005).

32. N. Fang, H. Lee, C. Sun, and X. Zhang, "Theory of the transmission properties of an optical far-field superlens for imaging beyond the diffraction limit," *Science* **308**, 534 (2005).

33. T. Taubner, D. Korobkin, Y. Urzhumov, G. Shvets, and R. Hillenbrand, "Near-field microscopy through a SiC superlens," *Science* **313**, 15 (2006).

34. Z. Jacob, L. V. Alekseyev, and E. Narimanov, "Optical hyperlens: far-field imaging beyond the diffraction limit," *Optics Express* **14**, 8247 (2006).

35. A. Salandrino and N. Engheta, "Far-field subdiffraction optical microscopy using metamaterial crystals: theory and simulations," *Phys. Rev. B* **74**, 075103 (2006).

36. A. A. Govyadinov and V. A. Podolskiy, "Metamaterial photonic funnels for subdiffraction light compression and propagation," *Phys. Rev. B* **73**, 115108 (2006).

37. Z. Liu, H. Lee, Y. Xiong, C. Sun, and X. Zhang, "Far-field optical hyperlens magnifying sub-diffraction-limited objects," *Science* **315**, 1686 (2007).

38. H. Lee, Z. Liu, Y. Xiong, C. Sun, and X. Zhang, "Development of optical hyperlens for imaging below the diffraction limit," *Optics Express* **15**, 15886 (2007).

39. I. I. Smolyaninov, Y.-J. Hung, and C. C. Davis, "Magnifying superlens in the visible frequency range," *Science* **315**, 1699 (2007).

40. I. I. Smolyaninov, Y.-J. Hung, and C. C. Davis, "Imaging and focusing properties of plasmonic metamaterial devices," *Phys. Rev. B* **76**, 205424 (2007).

41. X. Fan, G. P. Wang, J. C. W. Lee, and C. T. Chan, "Imaging and focusing properties of plasmonic metamaterial devices," *Phys. Rev. Lett.* **97**, 073901 (2006).

42. H. J. Lezec, J. A. Dionne, and H. A. Atwater, "Negative refraction at visible frequencies," *Science* **316**, 430 (2007).

43. Z. Liu, S. Durant, H. Lee, Y. Xiong, Y. Pikus, C. Sun, and X. Zhang, "Far-field optical superlens," *Nano Lett.* **7**, 404 (2007).

44. Z. Liu, S. Durant, H. Lee, Y. Xiong, Y. Pikus, C. Sun, and X. Zhang, "Near-field Moire effect mediated by surface plasmon polariton excitation," *Opt. Lett.* **32**, 629 (2007).

45. G. Shvets, S. Trendafilov, J. B. Pendry, and A. K. Sarychev, "Guiding, focusing, and sensing on the subwavelength scale using metallic wire arrays," *Phys. Rev. Lett.* **98**, 107589 (2007).

46. G. Fedorov, S. I. Maslovski, A. V. Dorofeenko, A. P. Vinogradov, I. A. Ryzhikov, and S. A. Tretyakov, "Subwavelength imaging: resolution enhancement using metal wire gratings," *Phys. Rev. B* **73**, 035409 (2006).

47. P. A. Belov, Y. Zhao, S. Sudhakaran, A. Alomainy, and Y. Hao, "Experimental study of the subwavelength imaging by a wire medium slab," *Appl. Phys. Lett.* **89**, 262109 (2006).

48. P. Ikonen, P. Belov, C. Simovski, and S. Maslovski, "Experimental demonstration of subwavelength field channeling at microwave frequencies using a capacitively loaded wire medium," *Phys. Rev. B* **73**, 073102 (2006).

49. P. Ikonen, C. Simovski, S. Tretyakov, P. A. Belov, and Y. Hao, "Magnification of subwavelength field distributions at microwave frequencies using a wire medium slab operating in the canalization regime," *Appl. Phys. Lett.* **91**, 104102 (2007).

50. V. M. Shalaev, W. Cai, U. Chettiar, H.-K. Yuan, A. K. Sarychev, V. P. Drachev, and A. V. Kildishev, "Negative index of refraction in optical metamaterials," *Opt. Lett.* **30**, 3356 (2005).

51. V. M. Shalaev, W. Cai, U. Chettiar, H.-K. Yuan, A. K. Sarychev, V. P. Drachev, and A. V. Kildishev, "Experimental verification of an optical negative-index material," *Laser Phys. Lett.* **3**, 49 (2006).

52. S. Zhang, W. Fan, N. C. Panoiu, K. J. Malloy, R. M. Osgood, and S. R. J. Brueck, "Experimental demonstration of near-infrared negative-index metamaterials," *Phys. Rev. Lett.* **95**, 137404 (2005).

53. V. A. Podolskiy, A. K. Sarychev, and V. M. Shalaev, "Plasmon modes in metal nanowires and left-handed materials," *J. Nonlinear Opt. Phys. and Mat.* **11**, 65 (2002).

54. A. Sarychev, V. Drachev, H. Yuan, V. Podolskiy, and V. Shalaev, "Optical properties of metal nanowires" in *Nanotubes and Nanowires,* A. Lakhtakia, S. Maksimenko, Eds., *Proc. SPIE* **5219**, 92–98 (2003).

55. V. A. Podolskiy, A. K. Sarychev, and V. M. Shalaev, "Plasmon modes and negative refraction in metal nanowire composites," *Optics Express* **11**, 735 (2003).

56. A. Sarychev and V. Shalaev, "Magnetic resonance in metal nanoantennas," in *Complex Mediums V: Light and Complexity*, M. McCall, G. Dewar, Eds., *Proc. SPIE* **5508**, 128–137 (2004).

57. G. Dolling, C. Enkrich, M. Wegener, C. M. Soukoulis, and S. Linden, "Low-loss negative-index metamaterial at telecommunication wavelengths," *Opt. Lett.* **31**, 1800 (2006).

58. G. Dolling, C. Enkrich, M. Wegener, C. M. Soukoulis, and S. Linden, "Simultaneous negative phase and group velocity of light in a metamaterial," *Science* **312**, 892 (2006).

59. G. Dolling, M. Wegener, C. M. Soukoulis, and S. Linden, "Negative-index metamaterial at 780 nm wavelength," *Opt. Lett.* **32**, 53 (2007).

60. T. A. Klar, A. V. Kildishev, V. P. Drachev, and V. M. Shalaev, "Negative-index metamaterials: going optical," *IEEE J. Sel. Top. Quantum Electron.* **12**, 1106 (2006).

61. V. M. Shalaev, "Optical negative-index metamaterials," *Nature Photonics* **1**, 41 (2007).

62. J. Valentine, S. Zhang, T. Zentgraf, E. Ulin-Avil, D. A. Genov, G. Bartal, and X. Zhang, "Three-dimensional optical metamaterial with a negative refractive index," *Nature* **455**, 376 (2008).

63. A. N. Grigorenko, A. K. Geim, H. F. Gleeson, Y. Zhang, A. A. Firsov, I. Y. Khrushchev, and J. Petrovic, "Nanofabricated media with negative permeability at visible frequencies," *Nature* **438**, 335 (2005).

64. A. V. Kildishev, V. P. Drachev, U. K. Chettiar, V. M. Shalaev, D. H. Werner, and D. Kwon, "Comment on "Negative refractive index in artificial metamaterials," *Opt. Lett.* **32**, 1510 (2007).

65. J. O. Dimmock, "Losses in left-handed materials," *Optics Express* **11**, 2397 (2003).

66. V. A. Podolskiy and E. E. Narimanov, "Near-sighted superlens," *Opt. Lett.* **30**, 75 (2005).

67. A. K. Sarychev, G. Shvets, and V. M. Shalaev, "Magnetic plasmon resonance," *Phys. Rev. E* **73**, 036609 (2006).

68. A. K. Sarychev and G. Tartakovsky, "Magnetic plasmonic metamaterials in actively pumped host medium and plasmonic nanolaser," in *Complex Photonic Media,* G. Dewar, M. McCall, M. Noginov, and N. Zheludev, Eds., *Proc. SPIE* **6320**, 63200A (2006).

69. A. K. Sarychev and G. Tartakovsky, "Magnetic plasmonic metamaterials in actively pumped host medium and plasmonic nanolaser," *Phys. Rev. B* **75**, 085436 (2007).

70. A. K. Sarychev, A. A. Pukhov, and G. Tartakovsky, "Metamaterial comprising plasmonic nanolasers," in *PIERS 2007 Prague: Progress in Electromagnetics Research Symposium*, The Electromagnetics Academy, Cambridge, MA (2007).

71. V. S. Letokhov, "Generation of light by a scattering medium with negative resonance absorption," *Sov. Phys.: JETP* **26**, 835 (1968).

72. N. M. Lawandy, R. M. Balachandran, A. S. L. Gomes, and E. Sauvain, "Laser action in strongly scattering media," *Nature* **368**, 436 (1994).

73. N. M. Lawandy, "Localized surface plasmon singularities in amplifying media," *Appl. Phys. Lett.* **85**, 5040 (2004).

74. N. M. Lawandy, "Nano-particle plasmonics in active media," in *Complex Mediums VI: Light and Complexity,* M. McCall, G. Dewar, M. Noginov, Eds., *Proc. SPIE* **5924**, 59240G (2005).

75. A. Y. Smuk and N. M. Lawandy, "Spheroidal particle plasmons in amplifying media," *Appl. Phys. B* **84**, 125 (2006).

76. H. Cao, "Random lasers: development, features, and applications," *Opt. Photon. News* **16**, 24 (2005).

77. G. D. Dice, S. Mujumdar, and A. Y. Elezzabi, "Plasmonically enhanced diffusive and subdiffusive metal nanoparticle-dye random laser," *Appl. Phys. Lett.* **86**, 131105 (2005).

78. D. J. Bergman and M. I. Stockman, "Surface plasmon amplification by stimulated emission of radiation: quantum generation of coherent surface plasmons in nanosystems," *Phys. Rev. Lett.* **90**, 027402 (2003).

79. S. Strauf, K. Hennessy, M. T. Rakher, Y.-S. Choi, A. Badolato, L. C. Andreani, E. L. Hu, P. M. Petroff, and D. Bouwmeester, "Self-tuned quantum dot gain in photonic crystal lasers," *Phys. Rev. Lett.* **96**, 127404 (2006).

80. D. Bimberg, M. Grundmann, N. N. Ledentsov, M. H. Mao, C. Ribbat, R. Sellin, V. M. Ustinov, A. E. Zhukov, Z. I. Alferov, and J. A. Lott, "Novel infrared quantum dot lasers: Theory and reality," *Phys. Status Solidi B* **224**, 787 (2001).

81. D. Bimberg, N. Kirstaedter, N. N. Ledentsov, Z. I. Alferov, P. S. Kop'ev, and V. M. Ustinov "InGaAs-GaAs quantum-dot lasers," *IEEE J. Sel. Top. Quantum Electron.* **3**, 196 (1997).

82. B. Y. Kogan, V. M. Volkov, and S. A. Lebedev, "Superluminescence and generation of stimulated radiation under internal-reflection conditions," *JETP Lett.* **16**, 100 (1972).

83. G. A. Plotz, H. J. Simon, and J. M. Tucciarone, "Enhanced total reflection with surface plasmons," *J. Opt. Soc. Am.* **69**, 419 (1979).

84. A. N. Sudarkin and P. A. Demkovich, "Excitation of surface electromagnetic waves on the boundary of a metal with an amplifying medium," *Sov. Phys.: Techn. Phys.* **34**, 764 (1989).

85. A. Tredicucci, C. Gmachl, F. Capasso, A. L. Hutchinson, D. L. Sivco, and A. Y. Cho, "Single-mode surface-plasmon laser," *Appl. Phys. Lett.* **76**, 2164 (2000).

86. I. Avrutsky, "Surface plasmons at nanoscale relief gratings between a metal and a dielectric medium with optical gain," *Phys. Rev. B* **70**, 155416 (2004).

87. J. Seidel, S. Grafstrom, and L. Eng, "Stimulated emission of surface plasmons at the interface between a silver film and an optically pumped dye solution," *Phys. Rev. Lett.* **94**, 177401 (2005).

88. M. A. Noginov, V. A. Podolskiy, G. Zhu, M. Mayy, M. Bahoura, J. A. Adegoke, B. A. Ritzo, and K. Reynolds, "Compensation of loss in propagating surface plasmon polariton by gain in adjacent dielectric medium," *Optics Express* **16**, 1385 (2008).

89. T. Okamoto, F. H'Dhili, and S. Kawatab, "Towards plasmonic band gap laser," *Appl. Phys. Lett.* **85**, 3968 (2004).

90. M. A. Noginov, G. Zhu, M. Bahoura, J. Adegoke, C. Small, B. A. Ritzo, V. P. Drachev, and V. M. Shalaev, "The effect of gain and absorption on surface plasmons in metal nanoparticles," *Appl. Phys. B* **86**, 455 (2007).

91. S. A. Ramakrishna and J. B. Pendry, "Removal of absorption and increase in resolution in a near-field lens via optical gain," *Phys. Rev. B* **67**, 201101(R) (2003).

92. K. C. Je, H. Ju, M. Treguer, T. Cardinal, and S. H. Park, "Local field-induced optical properties of Ag-coated CdS quantum dots," *Optics Express* **14**, 8000 (2006).

93. N. I. Zheludev, S. L. Prosvirnin, N. Papasimakis, and V. A. Fedotov "Lasing spaser," *Nature Photonics* **2**, 351 (2008).

94. A. Fang and Th. Koschny and M. Wegener and C. M. Soukoulis "Self-consistent calculation of metamaterials with gain," *Phys. Rev. B* **79**, 241104(R) (2009).

95. A. P. Vinogradov, L. V. Panina, and A. K. Sarychev, "Method for the calculation of dielectric and magnetic-permeability of percolation systems," Doclady Akademii Nauk SSSR **306**, 847 (1989). [*Sov. Phys.: Doklady* **34**, 530 (1989).]

96. M. P. Nezhad, K. Tetz, and Y. Fainman, "Gain assisted propagation of surface plasmon polaritons on planar metallic waveguides," *Optics Express* **12**, 4072 (2004).

97. R. Marques, F. Medina, and R. Rafii-El-Idrissi, "Role of bi-anisotropy in negative permeability and left-handed metamaterials," *Phys. Rev. B* **65**, 144440 (2002).

98. L. V. Panina, A. N. Grigorenko, and D. P. Makhnovskiy, "Optomagnetic composite medium with conducting nanoelements," *Phys. Rev. B* **66**, 155411 (2002).

99. A. K. Sarychev and V. M. Shalaev, "Optical properties of metal–dielectric film," in *Introduction to Complex Mediums for Optics and Electromagnetics*, W. S. Weiglhofer and A. Lakhtakia, Eds., pp. 397–415, SPIE Press, Bellingham, WA (2003).

100. B. P. Johnson and R. W. Christy, "Optical constants of the noble metals," *Phys. Rev. B* **6**, 4370 (1972).

101. U. Kreibig and M. Volmer, *Optical Properties of Metal Clusters,* Springer-Verlag, Berlin (1995).

102. L. D. Landau, E. M. Lifshitz, and L. P. Pitaevskii, *Electromagnetics of Continuous Media*, 2nd ed., Pergamon, Oxford (1984).

103. M. I. Stockman, S. V. Faleev, and D. J. Bergman, "Localization versus delocalization of surface plasmons in nanosystems: can one state have both characteristics?" *Phys. Rev. Lett.* **87**, 167401 (2001).

104. I. E. Protsenko, A. V. Uskov, O. A. Zaimidoroga, V. N. Samoilov, and E. P. O'Reilly, "Dipole nanolaser," *Phys. Rev. A* **71**, 063812 (2005).

105. G. D. Mahan, *Many-Particle Physics,* Plenum Press, New York (1990).

106. P. Meystre and M. Sergeant, *Elements of Quantum Optics,* 3rd ed., Springer-Verlag, New York (1998).

107. F. Wang and Y. R. Shen, "General properties of local plasmons in metal nanostructures," *Phys. Rev. Lett.* **97**, 206806 (2006).

108. H. Carrere, X. Marie, J. Barrau, and T. Amand, "Comparison of the optical gain of InGaAsN quantum-well lasers with GaAs or GaAsP barriers," *Appl. Phys. Lett.* **86**, 071116 (2005).

109. H. Carrere, X. Marie, L. Lombez, and T. Amand, "Optical gain of InGaAsN/InP quantum wells for laser applications," *Appl. Phys. Lett.* **89**, 181115 (2006).

110. D. Bimberg, M. Grundmann, and N. N. Ledentsov, *Quantum Dot Heterostructure,* John Wiley, Chichester (1999).

111. J. M. Garcia, P. Mankad, P. O. Holtz, P. J. Wellman, and P. M. Petroff, "Electronic states tuning of InAs self-assembled quantum dots," *Appl. Phys. Lett.* **72**, 3172 (1998).

112. V. I. Klimov, A. A. Mikhailovsky, S. Xu, A. Malko, J. A. Hollingsworth, C. A. Leatherdale, H. J. Eisler, and M. G. Bawendi, "Optical gain and stimulated emission in nanocrystal quantum dots," *Science* **290**, 314 (2000).

113. T. Amano, T. Sugaya, and K. Komori, "Characteristics of 1.3 μm quantum-dot lasers with high-density and high-uniformity quantum dots," *Appl. Phys. Lett.* **89**, 171122 (2006).

114. P. Anger, P. Bharadwaj, and L. Novotny, "Enhancement and quenching of single-molecule fluorescence," *Phys. Rev. Lett.* **96**, 113002 (2006).

Biography

Andrey K. Sarychev received his B.S., M.S., and Ph.D. degrees in Physics from the Moscow Institute of Physics and Technology. He received the degree Doctor of Science in Applied Electrodynamics from the Russian Academy of Sciences in 1993. From 2001 to 2004 he worked as a Senior Research Scientist at Purdue University; from 2004 to 2007 he worked as a Senior R&D Engineer at Ethertronics Inc., San Diego, CA. In 2007 Dr. Sarychev joined the Institute of Theoretical and Applied Electrodynamics, Russian Academy of Sciences as a principal scientist. He is the author (together with V. M. Shalaev) of the book *Electrodynamics of Metamaterials* (World Scientific, 2007). His current research interests include: metamaterials and plasmonics, nano-optics and nanophotonics, nanocircuit and nanostructure modelling, optical chemical and biosensors, large-scale computer simulations in electrodynamics, antenna design, and materials science. Dr. Sarychev was nominated for the Russian Academy of Sciences in 2008. His is the author of 11 books and book chapters, 10 patents, and more than 120 international papers.

Chapter 14
Resonance Energy Transfer: Theoretical Foundations and Developing Applications

David L. Andrews
University of East Anglia Norwich, UK

14.1 Introduction

In the wide range of materials that are characterised by broad, relatively featureless optical spectra, the absorption of light in the ultraviolet–visible wavelength region is typically followed by rapid internal processes of dissipation and degradation of the acquired energy, the latter ultimately to be manifest in the form of heat. In more complex materials—those comprising a variety of light-absorbing atomic or molecular components (chromophores) with optically well characterised absorption and fluorescence bands—the absorption of light is commonly followed by a spatial translation of the absorbed electromagnetic radiation between different, though usually closely separated, chromophores. The process takes place well before the completion of any thermal degradation in such materials. This primary relocation of the acquired electronic energy, immediately following photo-excitation, is accomplished by a mechanism that has become known as *resonance energy transfer* (RET).[1–3] (At an earlier stage in the development of these ideas,[4] the term "resonance" was used to signify that no molecular vibrations were excited; however, such usage is now known to be relevant to few systems and has largely fallen into abeyance.) An alternative designation for the process is electronic energy transfer (EET); both terms are widely used, and in each case, the first letter of the acronym serves as a distinction from electron transfer.

In complex multichromophore materials, the singular properties of RET allow the flow of energy to exhibit a directed character. Because the process operates most efficiently between near-neighbor chromophores, the resonance propagation of energy through such a system generally takes the form of a series of short steps; an alternative process involving fewer long steps proves considerably less favourable. In suitably designed materials, the pattern of energy flow following optical absorption is thus determined by a sequence of transfer steps, beginning and ending at chromophores that differ chemically, or, if the chromophores are structurally equivalent, through local modifications in energy level structure reflecting the influence of their electronic environment. Hence, individual chromophores that act in the capacity of excitation acceptors can subsequently adopt the role of donors. This effect contributes to a crucial, property-determining characteristic for the channeling of electronic excitation in photosynthetic systems;[5] the same principles are emulated in synthetic energy harvesting systems such as the fractal polymers known as *dendrimers*.[6]

The observation and applications of RET extend well beyond the technology of light harvesting, as will be demonstrated in later sections of this chapter. The phenomenon has an important function in the operation of organic light-emitting diodes (OLEDs) and luminescence detectors; in crystalline solids and glasses doped with transition metal ions, mechanisms based on RET are also engaged for laser frequency conversion. In the fields of optical communications and computation, several optical switching and logic gate devices are founded on the same principle. As we shall see, those possibilities have been considerably extended by a recent discovery that electron spin can be transferred along with the energy. In the realm of molecular biology, the determination of protein structures and the characterisation of dynamical processes are furthered by studies of the transfer of energy between intrinsic or "tag" chromophores; other ultrasensitive molecular imaging applications are again based on the same underlying principle. Further applications include energy transfer systems designed to act as analyte-specific sensors and as sensitisers for photodynamic therapy. Last but not least, RET provides a rich ground for exploring the fundamental issues that arise from the nanoscale interplay of electromagnetism and quantum mechanics.

14.1.1 The nature of condensed phase energy transfer

In any nonhomogeneous dielectric material, the primary result of ultraviolet/visible absorption is the population of short-lived electronic excited states located in individual atomic/molecular/nanoscale centres. In general, this is immediately followed by one or more transfers of the acquired electronic excitation energy, commonly on an ultrafast timescale and with some associated losses such as vibrational dissipation. Given the broad compass of the term "condensed phase," it is really quite astonishing that the RET concept is so pervasive, and the fundamental theory so extensively applicable. In crystalline, semicrystalline, or glassy media, the centres of absorption (and subsequent re-emission) commonly take the form of ions, atoms, or colour centres; in other types of mediums they may be small molecules, electronically distinct parts of large molecules, or nanoparticles such as quantum dots. Where generality is intended in the following, the term "chromophore" is used, subsuming the term "fluorophore," which some employ to convey the frequently associated capacity to exhibit fluorescence. Although the energy flow that follows optical absorption is generally a multistep process, at the fundamental level each elementary transfer step is a radiationless pairwise interaction generally taking place between an electronically excited species termed the *donor* and an electronically distinct *acceptor* that is initially in its ground state. The theory is therefore based on pair couplings.

The primary equations for pair RET are based on the interactions of nonoverlapping transition dipoles. Before pursuing the detail, however, it is worth observing that other forms of coupling are also possible, though less relevant to most systems of interest in the following account. For example, the transfer of energy between particles or units with significantly overlapped wavefunctions is usually described in terms of Dexter theory,[7] where the coupling carries an exponential decay with distance, directly reflecting the radial form of the wavefunctions and electron distributions. Compared to materials in which the donor and acceptor orbitals do not spatially overlap, such systems are of less use for either device or analytical applications, largely because the coupled chromophores lose their electronic and optical integrity. This is the main reason that complex light-harvesting systems are commonly designed with nonconjugated linkages or spacer units between the chromophores, or else with the latter held on a host superstructure that prevents direct chromophore contact.

Parallels can be drawn with the way a dielectric layer operates in a simple electrical capacitor. In the alternative scenario, where spacer units act as a "conductive" bridge through delocalisation and mixing of their orbitals with the donor or acceptor orbitals, energy transfer is specifically expedited by the operation of a *superexchange* mechanism,[8,9] which, despite the efficiency gains, compromises diagnostic applications.

14.1.2 The Förster equation

The first theoretical formulation of pair transfer that successfully identified the inverse sixth power distance dependence (with which the process itself is now almost universally associated) was made by Förster[10] and experimentally verified by Latt et al.[11] Subsequently recast in quantum mechanical terms, this theory of "radiationless" energy transfer has been very successfully applied for well over half a century and remains widely valid, although subject to certain conditions that were not originally understood. Before proceeding with the detail of the Förster equation, a caveat is therefore due. This concerns a misunderstanding of the relationship between "radiationless" and "radiative" energy transfer (the latter signifying successive but distinct processes of fluorescence emission by the donor and capture of the ensuing photon by the acceptor). A full quantum electrodynamical treatment of the interaction was needed to clarify and resolve this issue (details are given in Sec. 14.2.2). However, the reader should be aware that some textbooks still obscure the subject, wrongly treating radiationless and radiative energy transfer between a given chromophore pair as separate, potentially competing processes.

To proceed, consider the pairwise transfer of excitation between two chromophores A and B. In the context of this elementary mechanism (which might be one RET component of a complex, multistep migration process), A is designated the donor and B the acceptor. Specifically, let it be assumed that prior excitation of the donor generates an electronically excited species A^*. Forward progress of the energy is then accompanied by donor decay to the ground electronic state. Acquiring the energy, the acceptor B undergoes a transition from its ground to its excited state, as illustrated in Fig. 14.1. The excited acceptor B^* subsequently decays either in a further transfer event, or by another means such as fluorescence.

Because the A^* and B^* states are real, having measurable lifetimes, the process of energy transfer itself is fundamentally separable from the initial electronic excitation of A and the eventual decay of B; the latter processes do not, therefore, enter into the theory of the pair transfer. However it will need to be registered that other dissipative processes may be engaged (such features will be discussed in detail in later sections). In a solid, the linewidth of optical transitions manifests a degree of coupling of individual optical centres with their electronic environment (which, in the case of strong coupling, may lead to the production of phonon side-bands). Similar effects in solutions or disordered solids represent

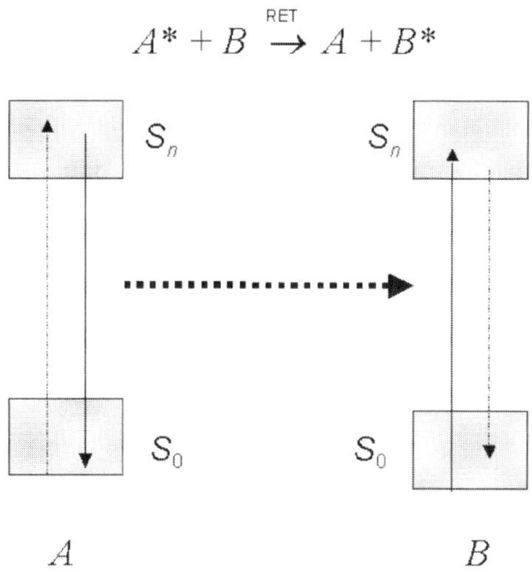

Figure 14.1 Energetics of single-step resonance energy transfer (RET), also showing the preceding excitation of the donor *A* and the concluding decay of the acceptor *B*. The boxes indicate lower (ground) and excited electronic states (here, designated as singlet states, though they need not be), and the vertical width represents a finite breadth on the energy scale. The transferred energy may be less than that initially acquired by the donor, due to dissipative processes; a similar remark applies to the acceptor.

inhomogeneous interactions with a solvent or host, while the broad bands exhibited by chromophores in complex molecular systems signify extensively overlapped vibrational levels, including those associated with skeletal modes of the superstructure. In each case, the net effect is to allow pair transfer to occur at any energy level within the region of overlap between the donor emission and acceptor absorption bands.

Restricting consideration to donor-acceptor separations R substantially smaller than the wavelengths of visible radiation, the Förster theory gives the following expression for the rate of pairwise energy transfer w_{F} for systems where the host material for the donor and acceptor has refractive index n (at the optical frequency corresponding to the mean transferred energy)

$$w_{\mathrm{F}} = \frac{9\kappa^2 c^4}{8\pi\tau_{A*} n^4 R^6} \int F_A(\omega)\sigma_B(\omega)\frac{d\omega}{\omega^4}\,. \tag{14.1}$$

In this expression, $F_A(\omega)$ is the fluorescence spectrum of the donor (normalised to unity); τ_{A*} is the associated radiative decay lifetime (related to the measured fluorescence lifetime τ_{fl} through the fluorescence quantum yield $\eta = \tau_{\mathrm{fl}} / \tau_{A*}$); $\sigma_B(\omega)$ is the linear absorption cross section of the acceptor; ω is an optical frequency in radians per second; and c is the speed of light. The spectral functions F_A and σ_B are mathematically defined and discussed in detail in Sec. 14.2.1. The κ factor in Eq. (14.1) depends on the orientations of the donor and acceptor, both with respect to each other and with respect to their mutual displacement unit vector $\hat{\mathbf{R}}$, as follows:

$$\kappa = (\hat{\boldsymbol{\mu}}_A \cdot \hat{\boldsymbol{\mu}}_B) - 3(\hat{\mathbf{R}} \cdot \hat{\boldsymbol{\mu}}_A)(\hat{\mathbf{R}} \cdot \hat{\boldsymbol{\mu}}_B). \tag{14.2}$$

For each chromophore, $\hat{\boldsymbol{\mu}}$ designates a unit vector in the direction of the appropriate transition dipole moment (see Sec. 14.2.1). Unfavourable orientations can reduce the rate of energy transfer to zero; others, including many of those found in nature, optimise the transfer rate. The angular disposition of chromophores is therefore a very important facet of energy transfer and one that invites careful consideration in the design of light-harvesting materials. To correct a common misconception, note that transfer is not necessarily precluded when the transition moments lie in perpendicular directions, provided that neither is orthogonal to $\mathbf{R}$ (= $R\hat{\mathbf{R}}$). The most striking features exhibited by Eq. (14.1) are the dependences on distance and on spectral overlap, both of which will be examined in detail in following sections.

In experimental studies of RET, it is usually significant that the electronically excited donor can in principle release its energy by spontaneous decay, and that the ensuing fluorescence radiation is detectable by any suitably placed photodetector. Because the alternative possibility (that of energy being transferred to another chromophore within the system) has such a sharp decline in efficiency as the distance to the acceptor increases, it is commonplace to introduce the concept of a critical distance R_0, a separation at which the theoretical rates of RET and spontaneous emission by the donor are equal (now known as the Förster distance). The Förster rate equation is often cast in an alternative form, exactly equivalent to Eq. (14.1), explicitly exhibiting this critical distance[12]

$$w_{\mathrm{F}} = \frac{3\kappa^2}{2}\frac{1}{\tau_{A^*}}\left(\frac{\bar{R}_0}{R}\right)^6 . \tag{14.3}$$

Here, $\bar{R}_0$ is defined as the Förster distance for which the orientation factor κ^2 assumes its isotropic average value, $\frac{2}{3}$.[13] For complex systems, the angular dependence is quite commonly disregarded, and the following simpler expression employed:

$$w_{\mathrm{F}} = \frac{1}{\tau_{A^*}}\left(\frac{R_0}{R}\right)^6 , \tag{14.4}$$

leading to a transfer efficiency Φ_{T} expressible as

$$\Phi_{\mathrm{T}} = \frac{1}{1+\left(R/R_0\right)^6} \ . \tag{14.5}$$

Typical values of the Förster radius range over a few nanometers. Thus, when a given electronically excited chromophore is within a distance R_0 of a suitable acceptor, RET will generally be the dominant decay mechanism; conversely, for distances beyond R_0, spontaneous decay will be the primary means of donor deactivation.

14.1.3 Established areas of application

In a host of multichromophore materials, RET represents a mechanism whose operation exerts a major influence on optical properties. Wide-ranging as these systems and applications are (several solid-state-device-oriented applications are to be discussed in later sections), the most widely studied and characterised examples are found in connection with light harvesting—both biological and synthetic. Following the capture of a photon by any such system, the RET mechanism dictates that the migration of the acquired energy from the site of the initial photoabsorption through to the site of its utilisation is at every stage subject to an inverse sixth-power dependence on distance. As a result, energy migration over distances beyond the Förster radius primarily operates through a series of short hops rather than one long hop. Commonly, these hops exhibit a "spectroscopic gradient,"[14] a term for the progressively longer wavelengths for absorption and fluorescence in successively visited chromophores (a feature to be studied in detail in Sec. 14.5.1). This is a property that makes a significant contribution to the high efficiency of photosynthetic and allied systems.

In experimental science, one obvious application of the strong distance dependence is the identification of motions in molecules, or parts of molecules, that can bring one chromophore into the proximity of another; prominent examples in biology are the traffic across a cell membrane and protein folding.[15] These and other such processes can be registered by selectively exciting one chromophore using laser light and monitoring either the decrease in fluorescence from that species or the rise in (again, generally longer-wavelength) fluorescence from the other chromophore as it enters into the role of acceptor. The judicious use of optical dichroic filters can make this fluorescence RET (FRET) technique perfectly straightforward (see Fig. 14.2). In cases where the two material components of interest do not display both absorption and fluorescence features in an appropriate wavelength range, molecular tagging with site-specific "extrinsic" (i.e., artificially attached) chromophores can solve the problem.

In some applications, the actual distance between the chromophore groups is of specific interest. When the same two chromophores feature (in spatially different configurations) in the chemical composition of two different systems (again, a common occurrence in biology), the relative displacements of the chromophores can be quantitatively assessed on the basis of comparisons

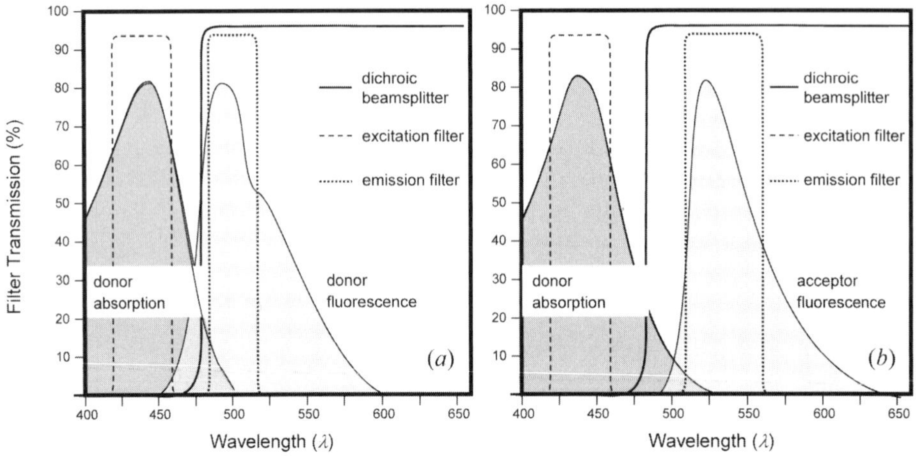

Figure 14.2 Schematic depiction of the spectral resolution of fluorescence from donor and acceptor species (based on cyan fluorescent protein donor and yellow fluorescent protein acceptor). (a) The transmission curve of a short-wavelength filter ensures initial excitation of the donor; a dichroic beam splitter and narrow emission filter ensure that only the (Stokes-shifted) fluorescence from the donor reaches a detector. (b) In the same system, a longer-wavelength emission filter ensures capture of only the acceptor fluorescence, following RET.

between the corresponding RET efficiencies. Such a technique is popularly known as a "spectroscopic ruler."[16] Such elucidations of molecular structure usually lack information on the relative *orientations* of the groups involved, and as such, the calculations usually ignore the kappa parameter [Eq. (14.2)]. The apparent crudeness of this approach becomes more defensible on realising that even if this approach were to introduce a factor of two inaccuracy, the deduced group spacing would still be in error by only 12% (since $2^{1/6} = 1.12$).

14.2 Electromagnetic Origins

In this section, the detailed electromagnetic origins of RET coupling are considered. First it is shown how the Förster formula signifies transition dipole coupling, and the possible influence of other multipoles is considered. An outline is then given of the modern and fully rigorous quantum electrodynamical derivation. Some distinctive features of the latter—in particular, the short- and long-range behaviour—are then examined in detail.

14.2.1 Coupling of transition dipoles

The spectral functions that were featured in Eq. (14.1) can be written in terms of the fundamental quantum properties of the chromophores. Assuming the usual Born-Oppenheimer separation of electronic and nuclear motions, the mathematical definitions are expressible as follows, using Dirac notation:

$$F_A(\omega) = \frac{\omega^3 \tau_{A*} \mu_A^2}{3\varepsilon_0 \pi c^3} \sum_{n,r} \rho_{A*}^{(n)} \left| \left\langle \varphi_A^{(r)} \middle| \varphi_{A*}^{(n)} \right\rangle \right|^2 \delta\left(E_{A_n^*} - E_{A_r} - \hbar\omega \right), \quad (14.6)$$

$$\sigma_B(\omega) = \frac{\pi \omega \mu_B^2}{3\varepsilon_0 c} \sum_{m,p} \rho_B^{(m)} \left| \left\langle \varphi_{B*}^{(p)} \middle| \varphi_B^{(m)} \right\rangle \right|^2 \delta\left(E_{B_p^*} - E_{B_m} - \hbar\omega \right). \quad (14.7)$$

Here, μ_A and μ_B are the magnitudes of the transition electric dipole moments for the donor decay and acceptor excitation, specifically given by

$$\boldsymbol{\mu}_A = \left\langle \psi_A \middle| \boldsymbol{\mu} \middle| \psi_{A*} \right\rangle; \; \boldsymbol{\mu}_B = \left\langle \psi_{B*} \middle| \boldsymbol{\mu} \middle| \psi_B \right\rangle, \quad (14.8)$$

where $\boldsymbol{\mu}$ is the dipole operator and each ψ is an electronic state wavefunction. Furthermore, the indices m, n, p, and r in Eq. (14.7) are generic labels denoting vibrational sublevels, with each φ representing an associated wavefunction (and E the corresponding energy), and ρ denoting a population distribution function for the initial state of each species. Comparing the above results with Eqs. (14.1) and (14.2) reveals the intrinsic quadratic dependence of the energy transfer rate on a coupling of the form

$$C = \frac{\mu_A \mu_B}{4\pi\varepsilon_0 R^3} \left[(\hat{\boldsymbol{\mu}}_A \cdot \hat{\boldsymbol{\mu}}_B) - 3(\hat{\mathbf{R}} \cdot \hat{\boldsymbol{\mu}}_A)(\hat{\mathbf{R}} \cdot \hat{\boldsymbol{\mu}}_B) \right], \quad (14.9)$$

which is, of course, the usual formula for the interaction of two static dipoles. However, the result given by Eq. (14.9) is not a conventional energy of interaction, but a quantum amplitude (strictly speaking, it is an off-diagonal matrix element connecting different initial and final states; only diagonal terms can directly signify energy). This is one of several important distinctions, the significance of which will become more apparent when the quantum electrodynamical theory is developed in the next section.

It is evident from Eq. (14.9) that the familiar inverse sixth-power distance dependence of RET owes its origin to the quadratic dependence of its rate on a coupling of transition electric dipoles. The result is, of course, applicable only when both the donor decay and the acceptor excitation transitions are electric dipole (E1) allowed. In general, the coupling is effected by a coupling between the lowest orders of multipoles (electric or magnetic) that can support the necessary transition. In the Förster range, the distance dependence exhibits the form $R^{-(P+Q+1)}$ for the coupling of two transition electric multipoles EP-EQ, or two magnetic multipoles MP-MQ; while for the coupling of an electric multipole with a magnetic multipole, EP-MQ, the distance dependence is $R^{-(P+Q)}$.[17,18] For example, the coupling of an electric dipole decay with an electric quadrupole excitation E1-E2 has an R^{-4} distance dependence within the Förster range. However, it should be kept in mind that each unit increase in multipolar order and each substitution of an electric transition by a magnetic counterpart lower the strength of the coupling by a factor of the order of 10^{-2} to 10^{-3}. The decreasing efficiency of successive multipole orders increasingly disfavours the role of RET in the decay of the donor, compared to other decay mechanisms.

14.2.2 Quantum electrodynamics

To develop a theory of RET from a fully rigorous basis, it makes sense to use a foundation in which quantum mechanics is used to describe the behaviour of both the matter and any electromagnetic fields that are involved. This is the framework of *quantum electrodynamics* (QED).[19,20] This theory has the advantage that it naturally accommodates not only quantum mechanical but also relativistic principles, so that it properly delivers retarded solutions (which will shortly prove to be a matter of key significance for RET). The wider successes of QED, such as its prediction of the Casimir effect, are well known; less well known is the fact that it alone can rigorously explain the much more familiar process of spontaneous emission. Moreover, a good case can be made that even the use of electric and magnetic multipoles is only defensible in the context of a fully quantum electrodynamical theory.[21] The development of a QED theory of RET, which began over forty years ago with pioneering work by Avery, Gomberoff, and Power,[22,23] culminated twenty years later in a unified theory[24] that has ramifications that continue to be explored today.[25] A concise exposition is presented below.

A suitable starting point for the analysis is the following (exact) multipolar Hamiltonian for the simple RET system comprising chromophores labeled A and B:

$$H = H_A + H_B + H_{\text{int}}(A) + H_{\text{int}}(B) + H_{\text{rad}}, \tag{14.10}$$

where the first two terms are the unperturbed Hamiltonian operators for the chromophores, and the two H_{int} operators represent interactions of the radiation field with A and B. The final term, H_{rad}, is the radiation Hamiltonian which, because it, too, is an operator, is always part of the sum, even when no photons are present. It becomes immediately evident that no single term in Eq. (14.10) links A with B; in other words, no static or *longitudinal* coupling occur. Hence, any form of coupling between the two chromophores has to be mediated by their individual interactions with the radiation field. This is an issue that it will be useful to revisit at the end of this section.

To continue, in the electric-dipole approximation, each $H_{\text{int}}(\xi)$ is given by the usual dipole coupling formula

$$H_{\text{int}}(\xi) = -\sum_{\xi} \boldsymbol{\mu}(\xi)\cdot\mathbf{e}^{\perp}(\mathbf{R}_{\xi}). \tag{14.11}$$

In Eq. (14.11), the electric-dipole moment operator $\boldsymbol{\mu}(\xi)$ operates on matter states, and the transverse electric-field operator $\mathbf{e}^{\perp}(\mathbf{R}_{\xi})$ on electromagnetic radiation states; $\mathbf{R}_\xi$ is the position vector of the chromophore labelled ξ. The electric field operator can be cast in the form of a mode summation, taken over optical wavevectors **p**, and polarisations λ, which is usually written as follows:[26]

$$\mathbf{e}^{\perp}\left(\mathbf{R}_{\xi}\right)=\mathrm{i}\sum_{\mathbf{p},\lambda}\left(\frac{\hbar cp}{2\varepsilon_{0}V}\right)^{1/2}\left[\mathbf{e}^{(\lambda)}\left(\mathbf{p}\right)a^{(\lambda)}\left(\mathbf{p}\right)\mathrm{e}^{\mathrm{i}\left(\mathbf{p}.\mathbf{R}_{\xi}\right)}-\overline{\mathbf{e}}^{(\lambda)}\left(\mathbf{p}\right)a^{\dagger(\lambda)}\left(\mathbf{p}\right)\mathrm{e}^{-\mathrm{i}\left(\mathbf{p}.\mathbf{R}_{\xi}\right)}\right]. \quad (14.12)$$

In this mode expansion, $\mathbf{e}^{(\lambda)}(\mathbf{p})$ is the polarisation unit vector (an overbar denoting complex conjugate), V is an arbitrary quantisation volume, and $a^{(\lambda)}(\mathbf{p})$ and $a^{\dagger(\lambda)}(\mathbf{p})$, respectively, are the photon annihilation and creation operators for the mode ($\mathbf{p}$, λ). These ladder operators act on the radiation states through the relations

$$\left.\begin{aligned} a^{(\lambda)}(\mathbf{p})\left|m(\mathbf{p},\lambda)\right\rangle &= \sqrt{m}\left|(m-1)(\mathbf{p},\lambda)\right\rangle \\ a^{\dagger(\lambda)}(\mathbf{p})\left|m(\mathbf{p},\lambda)\right\rangle &= \sqrt{m+1}\left|(m+1)(\mathbf{p},\lambda)\right\rangle \end{aligned}\right\}, \quad (14.13)$$

and as such, engage H_{int} in the fundamental processes of photon creation and annihilation.

The initial and final states for the RET process can be written as $\left|\varphi_{A*};\varphi_{B};0\right\rangle$ and $\left|\varphi_{A};\varphi_{B*};0\right\rangle$, respectively, each ket indicating the electronic states of the two chromophores and the number of photons present. For simplicity, vibrational levels are left out of the derivation, as they are readily incorporated at the end of the calculation. To effect the transition between the given system states, time-dependent perturbation theory is applied with Eq. (14.11) as the perturbation operator. The overall process can be achieved only by applying this operator twice, necessarily signifying the creation and annihilation of a photon; the leading contribution to the quantum amplitude M is therefore second order

$$M = \sum_{r}\frac{\left\langle f\left|H_{\mathrm{int}}\right|r\right\rangle\left\langle r\left|H_{\mathrm{int}}\right|i\right\rangle}{\left(E_{i}-E_{r}\right)}, \quad (14.14)$$

where i, f, and r denote initial, final, and intermediate states of the system and E signifies an energy. Since the photon will be unobservable and acts only in the capacity of effecting the coupling, it is termed a *virtual photon*, and its creation can take place at either A or B. Two time orderings therefore arise: (a) the virtual photon can be created at A (effecting the decay of the donor-excited state) and subsequently annihilated at B (effecting the acceptor excitation); (b) the virtual photon may be created at B (with the acceptor *excitation*) and annihilated at A (with the donor *decay*). These two possibilities are illustrated by Feynman diagrams in Fig. 14.3 and by a state-sequence diagram in Fig. 14.4. The counterintuitive nature of case (*b*) does not preclude its inclusion in the calculation; both here and in case (*a*), the mode expansion in the electric field operator Eq. (14.12) moreover requires summation over all optical frequencies. Physically, it can be understood that exact energy conservation is not imposed during the interval between creation and annihilation of the virtual photon, i.e., the ultrashort photon flight-time. This, a key feature of virtual photon behaviour, is entirely consistent with the time-energy uncertainty principle. When the whole system enters its final state, the balance of energy conservation is once again restored.

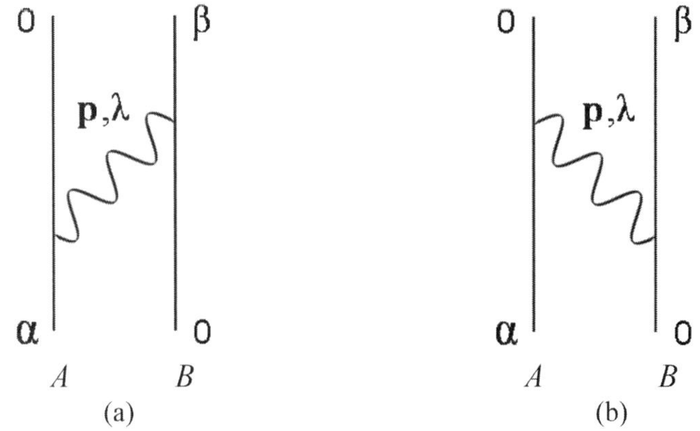

Figure 14.3 Feynman time-ordered diagrams corresponding to the two quantum amplitude contributions to RET, with time progressing upward. In (a) the virtual photon propagates from *A* to *B*, and in (b) it moves from *B* to *A*.

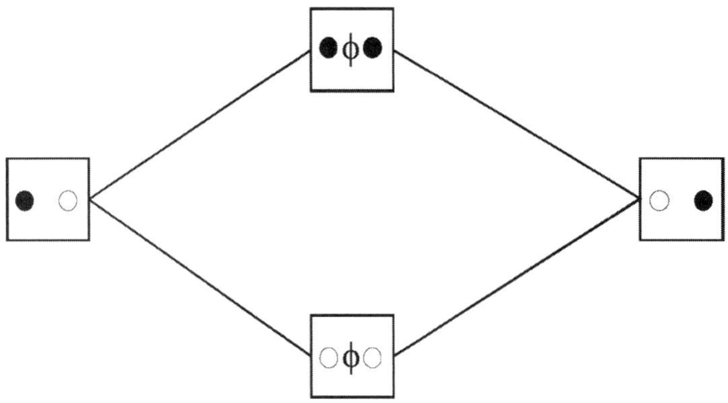

Figure 14.4 State-sequence diagram for RET, progressing from the initial system state of the left, through intermediate states, to the final state on the right. In each box, the two circles designate the states of *A* and *B*, a filled circle denotes an electronic excited state, and an open circle the ground state; ϕ denotes the presence of a virtual photon. The lower pathway corresponds to the Feynman graph (a), and the upper pathway to (b) in Fig. 14.3.

Figures 14.3 and 14.4 readily facilitate determination of the two contributions to the quantum amplitude Eq. (14.14), which emerge as

$$M_i = \sum_{\mathbf{p},\lambda}\left[\frac{\langle \varphi_A;\varphi_{B^*};0|H_{\text{int}}|\varphi_A;\varphi_B;1(\mathbf{p},\lambda)\rangle\langle \varphi_A;\varphi_B;1(\mathbf{p},\lambda)|H_{\text{int}}|\varphi_{A^*};\varphi_B;0\rangle}{\hbar ck-\hbar cp} + \frac{\langle \varphi_A;\varphi_{B^*};0|H_{\text{int}}|\varphi_{A^*};\varphi_{B^*};1(\mathbf{p},\lambda)\rangle\langle \varphi_{A^*};\varphi_{B^*};1(\mathbf{p},\lambda)|H_{\text{int}}|\varphi_{A^*};\varphi_B;0\rangle}{-\hbar ck-\hbar cp}\right], \quad (14.15)$$

where $\hbar cp$ is the virtual photon energy and k is defined through the equation for overall energy conservation:

$$E_{A^*}-E_A = E_{B^*}-E_B = \hbar ck\,. \quad (14.16)$$

The wavevector and polarisation summations in Eq. (14.15) require substantial manipulation for evaluation by any of several standard techniques that are detailed in the original papers and subsequent reviews.[27] Implementing the necessary tensor calculus, the result emerges in a form concisely expressible as follows:

$$M = \mu_{Ai}V_{ij}(k,\mathbf{R})\mu_{Bj}\,, \quad (14.17)$$

using the convention of implied summation over repeated vector and tensor indices i and j. In Eq. (14.17), the two transition dipole moments for the donor decay and acceptor excitation transitions are coupled by an E1-E1 coupling tensor defined by

$$V_{ij}(k,\mathbf{R}) = \frac{e^{ikR}}{4\pi\varepsilon_0 R^3}\left\{\left(\delta_{ij}-3\hat{R}_i\hat{R}_j\right)-\left(ikR\right)\left(\delta_{ij}-3\hat{R}_i\hat{R}_j\right)-\left(kR\right)^2\left(\delta_{ij}-\hat{R}_i\hat{R}_j\right)\right\}. \quad (14.18)$$

It is of immediate interest to note that Eqs. (14.17) and (14.18) can be interpreted as the interaction of a transition dipole at B with a *retarded* electric field E produced by a transition dipole source at A, as given by $E_j = \mu_{Ai}V_{ij}(k, \mathbf{R})$. This exactly correlates with the SI result delivered by classical electrodynamics:[28]

$$\mathbf{E} = k^2\left(\hat{\mathbf{R}}\times\boldsymbol{\mu}\right)\times\hat{\mathbf{R}}\frac{e^{ikR}}{4\pi\varepsilon_0 R}+\left[3\hat{\mathbf{R}}\left(\hat{\mathbf{R}}\cdot\boldsymbol{\mu}\right)-\boldsymbol{\mu}\right]\left(\frac{1}{4\pi\varepsilon_0 R^3}-\frac{ik}{4\pi\varepsilon_0 R^2}\right)e^{ikR}\,. \quad (14.19)$$

These are key results, the implications of which will now be examined in detail.

14.2.3 Near- and far-field behaviour

Before considering the rate result that ensues from the above QED treatment, some features of physical significance can be identified from the quantum amplitude. First, it is important to note that for short-range distances where $kR \ll 1$ (signifying a length significantly smaller than the wavelength of the donor decay transition), Eq. (14.18) couples with the transition dipoles, as in Eq. (14.17), to deliver a result that equates exactly with the classical expression for the coupling, Eq. (14.19). The additional terms identified by QED, which come into play at larger distances, signify the intrinsic accommodation of retardation—these are features that reflect the finite time taken for A and B to interact, in accordance with relativistic principles of causality.

A second feature is manifest on effecting the tensor contraction $V_{ij}R_iR_j$ (again, with implied summation over i and j), which identifies the longitudinal character of the coupling. It is evident that although the other terms in Eq. (14.18) deliver a finite result, the contribution from the last term is zero. Since the last term in Eq. (14.18) is in fact the long-range ($kR \gg 1$) asymptote, this result indicates that the far-field coupling is in fact fully *transverse* with respect to R, whereas near-field coupling is not (it has both transverse and longitudinal components).[29] This behaviour is to be distinguished from the completely transverse character of the coupling field with respect to the virtual photon propagation vector, over all distances.

Recalling the sum over wavevectors in the electric field expansion Eq. (14.12), it can therefore be concluded that only virtual photons whose wavevector **p** is essentially parallel to the separation vector **R** remain significant for energy transfer as the donor-acceptor separation increases toward infinity. In contrast, the short-range or near-field behaviour is consistent with an involvement of virtual photons propagating in various directions. This can be understood as a manifestation of position-momentum quantum uncertainty, as illustrated in Fig. 14.5. In other words, as the distance R increases, the virtual photon acquires an increasingly real character.[29] To quote from a well-known

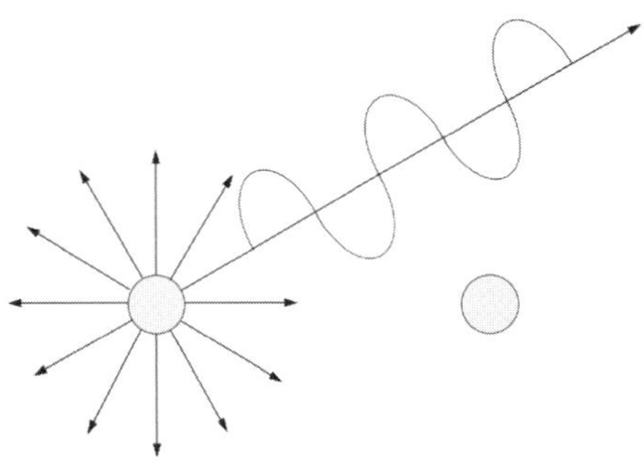

Figure 14.5 In the near field of an emitter-donor particle, virtual photons propagate in every direction, and quantum uncertainty allows their interaction with nearby absorber-acceptor particles. As the distance between emitter and absorber increases, the signal experienced by the latter is increasingly dominated by photons propagating directly toward it.

textbook of elementary particle physics: "In a sense every photon is virtual, being emitted and then sooner or later being absorbed."[30] It is only the comparatively long lifetime, on a quantum timescale, of the photons we commonly observe that results in a disappearance of their virtual traits.

14.2.4 Refractive and dissipative effects

The above analysis, developed from Eq. (14.10), rigorously applies to a system of a single donor and acceptor with no other matter present. However, as most cases of interest concern the condensed phase, the electronic influence of surrounding or host material cannot in general be ignored, and we should include *all* atoms and molecules (ξ) in a sum, writing

$$H = H_A + H_B + H_{\text{int}}(A) + H_{\text{int}}(B) + H_{\text{rad}} + \sum_{\xi \neq A,B} H_\xi + \sum_{\xi \neq A,B} H_{\text{int}}(\xi)\,. \tag{14.20}$$

The last three terms in Eq. (14.20) collectively represent an effective radiation field operator whose eigenstates signify modes in which the photons are "dressed" by the electromagnetic influence of the host. Strictly speaking, these are *polaritons*, though the distinction is not important if one is dealing with frequencies at which the host is relatively transparent. By a lengthy development of theory, the effect of making this correction is that the coupling tensor Eq. (14.18) emerges in the following modified form,[31] assuming Lorentz local field factors are assimilated into the expressions for the spectral functions $F_A(\omega)$ and $\sigma_B(\omega)$:

$$V_{ij}(k,\mathbf{R}) = \frac{1}{n(\omega)^2}\frac{e^{in(\omega)kR}}{4\pi\varepsilon_0 R^3}\left\{\begin{matrix}\left(\delta_{ij} - 3\hat{R}_i\hat{R}_j\right) - \left(in(\omega)kR\right)\left(\delta_{ij} - 3\hat{R}_i\hat{R}_j\right) \\ -\left(n(\omega)kR\right)^2\left(\delta_{ij} - \hat{R}_i\hat{R}_j\right)\end{matrix}\right\}, \tag{14.21}$$

where $n(\omega)$ is the complex refractive index for electromagnetic radiation with an optical frequency $\omega = ck$.

14.3 Features of the Pair Transfer Rate

By use of the Fermi rule,[32] the rate of energy transfer between a donor and acceptor pair has a quadratic dependence on the quantum amplitude. The quantum electrodynamical treatment of RET (discussed in the previous section) delivers a result that reveals important additional terms compared to the Förster Eq. (14.1), the extra terms arising from the influence of the bracketed terms, linear and quadratic in $n(\omega)kR$, in the quantum amplitude.[21] The full result is conveniently represented as follows:[33]

$$w = w_{\text{F}} + w_{\text{I}} + w_{\text{rad}}\,, \tag{14.22}$$

$$\left.\begin{aligned} w_{\mathrm{F}} &= \frac{9\kappa_3^2 c^4}{8\pi\tau_{A*}R^6}\int F_A(\omega)\sigma_B(\omega)\frac{d\omega}{\omega^4 n(\omega)^4} \\ w_{\mathrm{I}} &= \frac{9c^2}{8\pi\tau_{A*}R^4}(\kappa_3^2 - 2\kappa_1\kappa_3)\int F_A(\omega)\sigma_B(\omega)\frac{d\omega}{\omega^2 n(\omega)^2} \\ w_{\mathrm{rad}} &= \frac{9\kappa_1^2}{8\pi\tau_A R^2}\int F_A(\omega)\sigma_B(\omega)d\omega \end{aligned}\right\}. \tag{14.23}$$

In these expressions, the kappa orientational function of Eq. (14.2) is now generalised as

$$\kappa_j = (\hat{\boldsymbol{\mu}}_A \cdot \hat{\boldsymbol{\mu}}_B) - j(\hat{\mathbf{R}} \cdot \hat{\boldsymbol{\mu}}_A)(\hat{\mathbf{R}} \cdot \hat{\boldsymbol{\mu}}_B). \tag{14.24}$$

The first term of Eq. (14.22), w_{F}, is the usual Förster rate, identical to Eq. (14.1) if the refractive index is taken as a constant. The second contribution, w_{I}, is a correction that comes into play at distances beyond the near field, where the assumption $kR \ll 1$ no longer holds. The third term, w_{rad}, which dominates over both other contributions when $kR \gg 1$, equates exactly with the rate of acceptor excitation that results from the capture of a photon spontaneously emitted by the donor; i.e., a process of radiative transfer, its R^{-2} dependence fittingly designating the familiar inverse-square law. This radiative term manifests the long-range emergence of the coupling photon into a *real* character. Revisiting the underlying QED theory in order to disentangle the quantum pathways, recent work[34] has also shown that this long-range behaviour is completely identifiable with the physically more intuitive sense of propagation for the virtual photon—case (*a*) in Fig. 14.3 or the lower pathway in Fig. 14.4. In the same regime, the contribution from case (*b*) "backward" propagation drops off as R^{-8}. Nonetheless, it should be emphasised that (*a*) and (*b*) are *equally* important in the short range, where both run with R^{-6}.

The discovery that both Förster "radiationless" and radiative coupling are components of a single mechanism that operates over *all* distances (beyond wavefunction overlap) has resulted in a paradigm shift in the understanding of RET and is the reason that QED theory has been termed a *unified* theory.[24] This theory is valid over a span ranging from the nanoscale up to indefinitely large distances; Förster energy transfer is the short-range asymptote, and radiative transfer the long-range asymptote. Moreover, there is no competition between these processes, previously considered distinct, as they prove to be but aspects of a single coupling mechanism. However, the theory establishes more than this; it also addresses an intermediate range where neither the radiative nor the Förster mechanism is fully valid, because if $kR \sim 1$ to the nearest order of magnitude (suggesting distances in the hundreds of nanometers range), then all three terms in Eq. (14.23) contribute significantly. Although different kappa factors (κ_3 and

κ_I) characterise w_F and w_{rad}, both kappas are featured in the expression for w_I. This is the main reason that the intermediate range behaviour has eluded experimental identification—because it is difficult to envisage any technical circumstances where its contribution could be directly isolated. However, an interesting interplay of distance and orientational factors does arise in connection with fluorescence polarisation measurements, as will be shown below.

14.3.1 Distance dependence

The three principal determinants of the transfer efficiency, as given by Eqs. (14.22)–(14.24), are the separation, mutual orientation, and spectral overlap of the donor-acceptor pair, the effect of each which will now be considered. First, for simplicity in beginning to consider the dependence on distance, let it be assumed that the donor and acceptor have isotropically averaged orientations and that the refractive index is taken as unity. It then follows that the distance dependence of the full rate expression Eq. (14.22) factorises out in the following form,[24,35] which also identifies it with the tensor inner product of the QED coupling formula Eq. (14.18):

$$A(k,R)=\frac{1}{8\pi^2\varepsilon_o^{\ 2}R^6}\left\{3+(kR)^2+(kR)^4\right\}=V_{ij}(k,\mathbf{R})\bar{V}_{ij}(k,\mathbf{R}). \quad (14.25)$$

This excitation transfer function $A(k, R)$ is a scalar characterising the distance dependence of E1-E1 coupling in RET. The graph shown in Fig. 14.6 exhibits a log-log plot of the function over the range of 1 nm to 1 μm, for a value of $k = 9 \times 10^6$ m^{-1} (corresponding to a wavelength at the red end of the visible region). This figure gives a readily comprehensible representation of the Förster behaviour and "radiative" transfer as short- and long-range asymptotes, respectively, as shown by the change in gradient between the short- and long-range regions.

Most applications of RET relate to the Förster regime, i.e., systems in which energy-transfer steps occur between chromophores separated by less than the Förster distance, and therefore almost certainly within the short-range regime, $kR \ll 1$. Systems in which the mean transfer distance falls in the long-range regime, $kR \gg 1$, necessarily require the optically relevant species to be present in low concentrations; moreover, any diffusion processes that could produce transient short-range donor-acceptor juxtapositions should have a timescale significantly longer than the donor decay time, or else diffusion-limited Förster transfer would result. The *radiative* transport that ensues in the latter case (in a variety of systems, such as dilute dye solutions) leads into a distinct branch of the theory in which multiple scattering must also be considered, a detailed account of which is given by Berberan-Santos et al.[36] The possibility of experimentally verifying the *general* form of distance dependence given by Eq. (14.25), and, in particular, identifying the intermediate R^{-4} term, remains a currently unachieved but tantalising prospect.

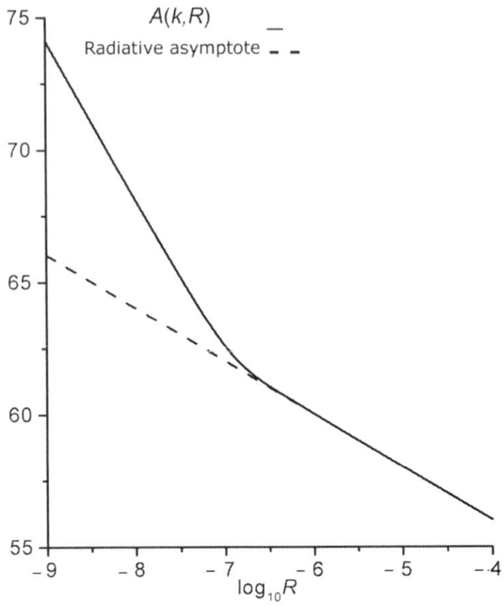

Figure 14.6 Log-log plot of the RET excitation transfer function $A(k, R)$, as defined by Eq. (14.25), against the donor-acceptor distance R (in nanometers) for $k = 9 \times 10^6\ \mathrm{m}^{-1}$.

14.3.2 Orientation of the transition dipoles

The kappa factors in Eq. (14.23), which depend on the mutual orientations of the donor and acceptor (see Fig. 14.7), represent another important facet of the energy transfer.[37–39] Certain orientations reduce the rate of transfer to zero; for others, they effect an "enhancement" of the energy transfer to its maximum possible rate. It is worth noting (since it is a fact that is not infrequently misreported), that energy transfer may be permissible even when the relevant donor and acceptor transition moments are at right angles to each other, if the second term in Eq. (14.24) is nonzero. In fact, the only case in which the transfer is necessarily forbidden, given the condition, $\boldsymbol{\mu}_A \perp \boldsymbol{\mu}_B$ is when one of these transition dipoles is also orthogonal to **R**. In this situation, all three terms of Eq. (14.23) vanish.

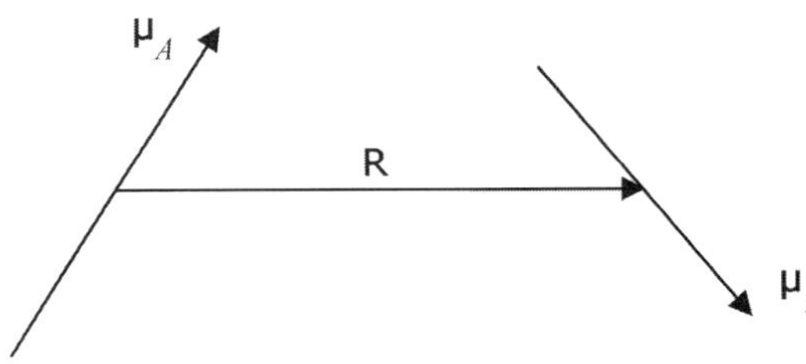

Figure 14.7 Typical noncoplanar orientations of the donor and acceptor transition dipoles, relative to the displacement vector.

Other factors can complicate the orientation dependence of RET. One is the fact that in any system that is to an extent fluid or disordered, the relative orientation of all donor-acceptor pairs may not be identical. It has already been noted that in the isotropic case (having completely uncorrelated orientations), the κ_3^2 factor in w_F averages to $2/3$; however, when a degree of orientational correlation is present, other results are possible (in the overall range of 0–4), the exact value characterising the detailed form of the angular distribution. Secondly, it can happen that either the donor or the acceptor transition moment is not unambiguously identifiable with a particular direction within the corresponding chromophore frame of reference. Specifically, the electronic transition may then relate to a degenerate state, as can occur with square planar complexes, for example. Alternatively, very rapid but orientationally confined motions might occur. The considerable complication that each of these effects brings into the trigonometric analysis of RET has been extensively researched and reported by van der Meer.[2]

It can at first sight be surprising to discover that the relative orientation of transition dipoles is manifest in readily discernible polarisation effects, even in media where the donor and acceptor orientations are uncorrelated. For example, when a single molecule in solution absorbs and then fluoresces, the angle α between the absorption and emission transition moments can be deduced from a determination of the fluorescence anisotropy, defined as[12,13]

$$r = \frac{I_{\parallel} - I_{\perp}}{I_{\parallel} + 2I_{\perp}}, \tag{14.26}$$

where $I_{\parallel}$ and $I_{\perp}$ respectively designate the intensities of fluorescence parallel and perpendicular to the polarisation of the excitation beam. It is relatively straightforward to show that

$$r_0 = \tfrac{1}{5}\left(3\cos^2\alpha - 1\right), \tag{14.27}$$

the subscript on the left indicating that no transfers of energy are involved. The value $r_0 = 0.4$ ensues when the absorption and emission moments are parallel. In a two-chromophore donor-acceptor system, if the pair were to be held in a fixed mutual orientation but could freely tumble as a pair, the result [Eq. (14.27)] would still be applicable, provided that within each chromophore there were parallel excitation and decay transition moments; the calculated value of α would then signify the angle between those differently oriented directions in the donor and acceptor.

If, however, the donor and acceptor are free to rotate independently, then the act of energy transfer significantly reduces the extent of the fluorescence anisotropy. It has long been known that one transfer step reduces the anisotropy to a value of $\frac{1}{25}r_0$, at least for conventional Förster transfer. However, the complete result, which first emerged from a detailed analysis based on the retarded coupling tensor Eq. (14.21), is as follows:[40]

$$r_1 = \frac{r_0}{25}\left(\frac{7\overline{k^4R^4} + \overline{k^2R^2} + 3}{\overline{k^4R^4} + \overline{k^2R^2} + 3}\right), \tag{14.28}$$

in which the overbars signify distributional averages over the k and R values for the transfer. A graph of the function r_1/r_0 is shown in Fig. 14.8. Clearly in the long-range asymptote corresponding to radiative transfer between the donor and acceptor, the anisotropy r_1 takes the value $\frac{7}{25}r_0$. The result is appealing because it manifests so dramatically the onset of retardation effects.

14.3.3 Spectral overlap

Finally, each component in the overall RET rate Eq. (14.23) involves an integral over a frequency-weighted product of the donor emission and acceptor absorption profiles. As such, it is not only the peak frequencies or wavelengths for emission and absorption that dictate the energy transfer conditions; so do the spectral shapes and bandwidths. Such features assume particular significance in connection with the multistep processes that take place in energy-harvesting materials. This subject will be further developed in that connection in Section 14.5.1.

Most theoretical work on the spectral overlap addressing the Förster regime ignores the dispersion of the refractive index that is featured in w_F. General analytical expressions for the frequency-weighted overlap have been determined for several cases of spectral line shape. Principal among these is the Gaussian case, which fits reasonably well the line profile of many molecular absorption and emission processes in the condensed phase, where the high densities of vibrational levels in the ground and excited electronic states produce a broad-

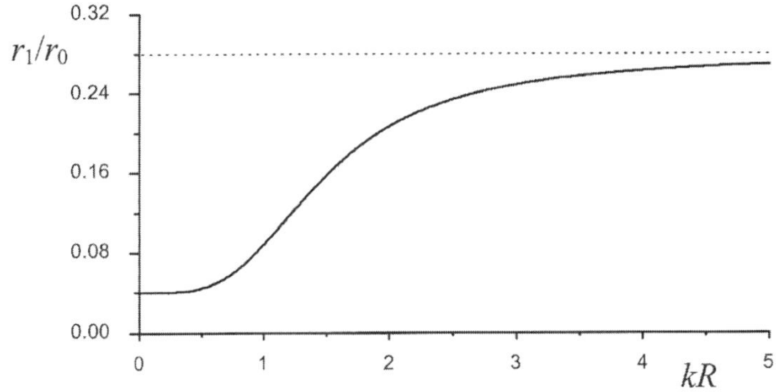

Figure 14.8 Rise in fluorescence anisotropy as a function of the distance between orientationally uncorrelated donor and acceptor particles.

ening that is similar in effect to a stochastic perturbation. For spectra that depart markedly from a completely symmetric Gaussian shape, a better fit is often afforded by a log-normal distribution. Typically, systems with the latter form of line-shape exhibit a sharper rise to the maximum on the high-frequency side of the absorption spectrum, and on the low-frequency side of the emission spectrum, together retaining the familiar ideal of a "reflection" relationship between absorption and emission spectra.

14.4 Energy Transfer in Heterogeneous Solids

A wide variety of materials can be engineered to exhibit photoactive properties based on RET. In this section, three types of systems are chosen and discussed to illustrate this range: doped solids, quantum dots, and multichromophore complexes. Not surprisingly, quite different attributes and applications are associated with each system type.

14.4.1 Doped solids

In optical materials dilutely doped with transition metals, RET represents a diffusive mechanism that moderates other processes such as stimulated emission (as another example, it influences laser efficiency). RET also plays an important role in the photophysics of rare-earth (lanthanide)-doped crystals, glasses, and fibers. These optically dilute materials display spectral features in the visible and near-visible regions that owe their origin to *f*-shell electronic transitions in the dopant.[41] The narrow linewidth of the energy levels, together with the coincidences of spacing between levels in different lanthanides, afford excellent opportunities for the design of materials that invoke not only conventional RET, but also higher-order effects.[42] A variety of proposals by Dexter[43] and Bloembergen[44] in the 1950s built on the premise of deploying RET to relay excitation between lanthanide ions. Following the development of lasers, the predicted higher-order effects were quickly brought to experimental fruition. Principal among the latter processes are stepwise or cooperative up-conversion, sensitisation, and down-conversion.[45,46]

The term *up-conversion* signifies processes in which low-frequency radiation is converted to a higher frequency, the output itself often being used as a basis for laser emission. In common with parametric processes involving the direct conversion of electromagnetic radiation in optically nonlinear crystals, the RET-based mechanism furnishes an output in which each photon is created at the expense of two input photons. In contrast to parametric processes, however, the input photons are not required to arrive simultaneously. The consequence is that the effect is achievable at lower intensities. The simplest scheme is one in which transfers of energy from two initially excited but spatially isolated donors, A^* and B^*, promote an acceptor C to a state with approximately twice the energy of the first excitation. One possibility is a stepwise process in which the twin donors sequentially deliver excitation energy to the acceptor, as shown in Fig. 14.9. In some of the literature, this is referred to as APTE (*addition de photons par* *transferts d'énergie*).[47] The mechanism can operate even when there is no suitable energy level to accommodate the initial transfer step. To satisfy the principle of energy conservation, the mechanism requires concurrent operation of the RET processes accommodated by the involvement of a virtual level in the acceptor; this latter process is often known as *cooperative* up-conversion. Chua and Tanner have shown how the theory can be extended to distance regimes where long-range radiative transfer is dominant.[48]

The QED theory of APTE up-conversion invokes fourth-order time-dependent perturbation theory, as befits the need to create and annihilate two virtual photons to effect the necessary couplings. Calculations lead to a quantum amplitude comprising three types of terms, each itself comprising 24 different contributions based on the allowed pathways through a state-sequence diagram,[49] as illustrated in Fig. 14.10. Of the three quantum amplitude components, one is designated *cooperative* to signify that each donor loses its energy directly to the acceptor; the other two are termed *accretive*, signifying that one donor passes its energy to the other, from which the sum energy is conveyed to the acceptor. Two cases arise due to the choice of the alternative roles played by each donor. The results are as follows:[50–52]

$$\left.\begin{aligned} M_i^{\text{coop}} &= \mu_{Ai} V_{ij}(k,\mathbf{R})\,\alpha_{Cjk}(-\omega,-\omega)\,V_{kl}(k,\mathbf{R}')\,\mu_{Bl} \\ M_i^{\text{acc1}} &= \mu_{Ai} V_{ij}(k,\mathbf{R}'')\,\alpha_{Bjk}(2\omega,-\omega)\,V_{kl}(2k,\mathbf{R}')\,\mu_{Cl} \\ M^{\text{acc2}} &= \mu_{Bi} V_{ij}(k,\mathbf{R}'')\,\alpha_{Ajk}(2\omega,-\omega)\,V_{kl}(2k,\mathbf{R})\,\mu_{Cl} \end{aligned}\right\}, \tag{14.29}$$

where $\omega = ck$, once again, k is as defined by Eq. (14.16), and each μ is a transition moment; $\alpha(-\omega, -\omega)$ is a two-photon absorption tensor, $\alpha(2\omega, -\omega)$ is formally an electronic anti-Stokes Raman tensor, and the three mutual displacement vectors are defined by $\mathbf{R} = \mathbf{R}_C - \mathbf{R}_A$, $\mathbf{R}' = \mathbf{R}_C - \mathbf{R}_B$, and $\mathbf{R}'' = \mathbf{R}_B - \mathbf{R}_A$.

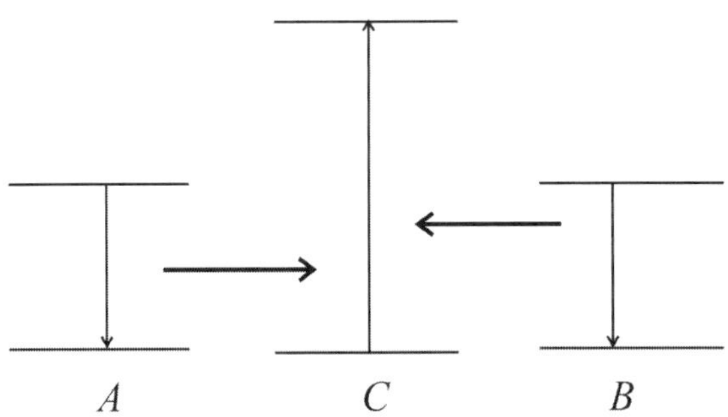

Figure 14.9 Diagram representing cooperative up-conversion.

For simplicity (conforming to the usual experimental case) it is assumed that *A* and *B* are chemically identical. From Eq. (14.29), the quadratic dependence of the up-conversion rate produces a result with three "diagonal" and six "off-diagonal" terms, as well as a complicated dependence on the dopant concentrations *C* of the donor and acceptor ions. For example, the third term of Eq. (14.29) contributes a diagonal rate term that in the short range runs as $R'^{-6}R''^{-6}$, producing a sum expressible as $\eta C_A C_B^2 C_C$ (with, for example, η = 64.39 for a regular lattice of cubic symmetry).[52]

Similar principles operate in *sensitisation* processes.[53] Here, the transfer of excitation from a donor ion A^* to an acceptor *C* engages a bridging species *B*, without which the transfer is ineffective. Once again, the term "up-conversion" is common for such observations, but "sensitisation" distinguishes it from the pooling processes described above. The other principal case of interest, *degenerate down-conversion* or *quantum cutting*, is a phenomenon in which excitation is simultaneously conveyed to two acceptors *A* and *B* from a single excited donor C^*, each transfer carrying approximately half of the energy of the donor excitation.[54] Thus, for example, an initial excitation of VUV radiation can lead to the emission of visible light. Systems exhibiting such effects represent the fulfillment of very early suggestions by Dexter of systems that can exhibit quantum yields greater than unity.[43]

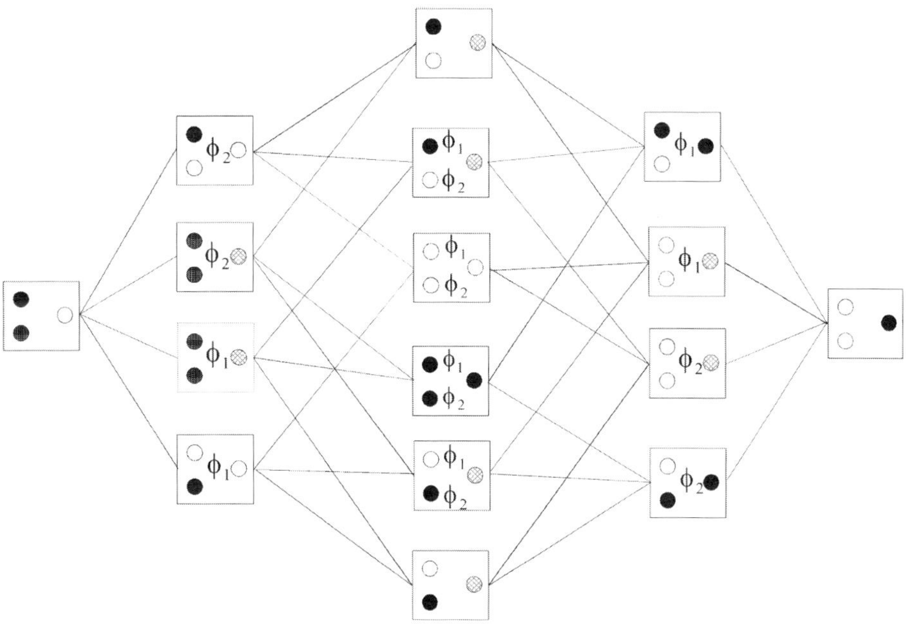

Figure 14.10 State-sequence diagram for cooperative up-conversion, accommodating 24 quantum amplitude contributions. In each box, the two circles on the left denote donors, and that on the right, an acceptor. Grey circles designate virtual states; black and white circles denote excited and ground states.

14.4.2 Quantum dots

There is growing interest in a variety of schemes that are based on the transfer of excitation between adjacent quantum dots[55–64] and are now being envisaged for performing quantum computation. Such proposals generally aim to exploit the discrete, size-tunable, and intense character of quantum dot exciton transitions, as well as the fact that these processes can be switched very rapidly using optical excitation. Moreover, it proves possible to code additional information by the use of circularly polarised excitation, since this enables specific exciton spin states to be populated.[65] This important question thus arises: Can the spin state of an exciton be transmitted or flipped through RET between suitably organised quantum dots? The answer is provided by a QED analysis[66] that separately identifies those contributions to the RET quantum amplitude as corresponding to left- and right-circular polarisations of the coupling virtual photons. The resulting plots in Fig. 14.11 show the effect of rotating one quantum dot relative to another (assuming the electronic coupling to be real). When the transition moments are parallel, the exciton orientation is transmitted unchanged from one quantum dot to another; when they are antiparallel, the geometry causes the exciton spin to flip. Energy migration down a column of quantum dots oriented in a common direction therefore proceeds with a full retention of spin orientation.

One advantage of the QED treatment of the problem is that not only the short-range but also the long-range behaviour is identified, even though the latter is of less interest from an application viewpoint. In the long-range asymptote, it is clear that angular momentum must be conserved about the propagation direction of the photon, which coincides with the mutual displacement vector of adjacent quantum dots and hence, the local columnar morphology. Thus, it transpires that the same feature operates in the technically important near-zone region, even though the coupling cannot in this case be ascribed to real photon propagation. Energy migration down a column of quantum dots with a common orientation absolutely preserves spin information; the observation of spin flipping between alternately inverted quantum dots is another manifestation of the same principle.

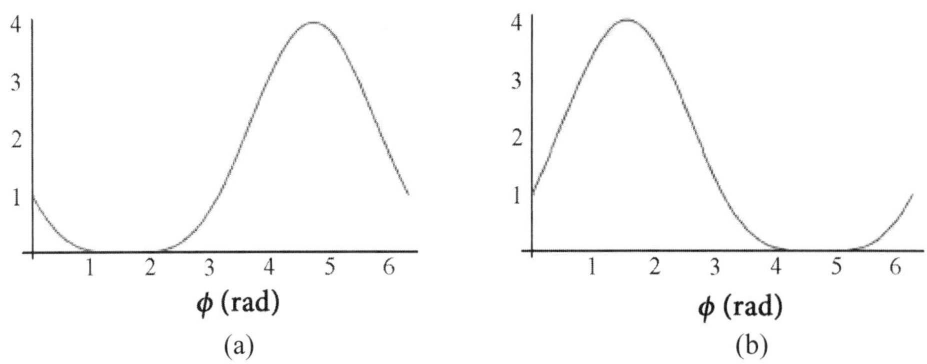

Figure 14.11 Quantum dot energy transfer. Variation of (a) spin-antiparallel and (b) spin-parallel transfer functions as a function of relative orientations.

14.4.3 Multichromophore complexes

RET is an extremely significant process in the operation of photosynthetic and biomimetic light-harvesting materials.[67–72] The photosynthetic systems of purple bacteria, in particular, have been extensively studied and characterised, and the emulation of their high efficiency is a key goal in the development of new polymer-based materials.

In order to most effectively utilise the sunlight that falls on them, photosynthetic organisms have a system of antenna complexes surrounding the reaction centres where photosynthesis takes place.[73,74] The complexes absorb sunlight, and the acquired energy migrates toward the reaction centre by a series of short-range, radiationless energy-transfer steps (see Fig. 14.12). In the overall migration of energy from the site of its initial deposition to the site of its chemical action, a spectroscopic gradient (see Sec. 14.5.1) is one of the key directional principles obviating random diffusion. Energy is quickly and efficiently directed toward a reaction centre. Not only does this allow an organism to harvest light incident on a large surface area, but by pooling energy from a large number of antenna chromophores, energy of a higher equivalent frequency can be produced. This is essential, since the majority of the incident light from the sun has too low a frequency for its individual photons to drive photosynthesis.

It is not only the spectroscopic properties of the chromophores that determine the character and direction of energy flow; the chromophore positioning and orientation are also important. Two-dimensional optical spectroscopy can unveil the intricate interplay between spectral and spatial overlap features in light-harvesting complexes, as beautifully exhibited in a recent study on bacterio-

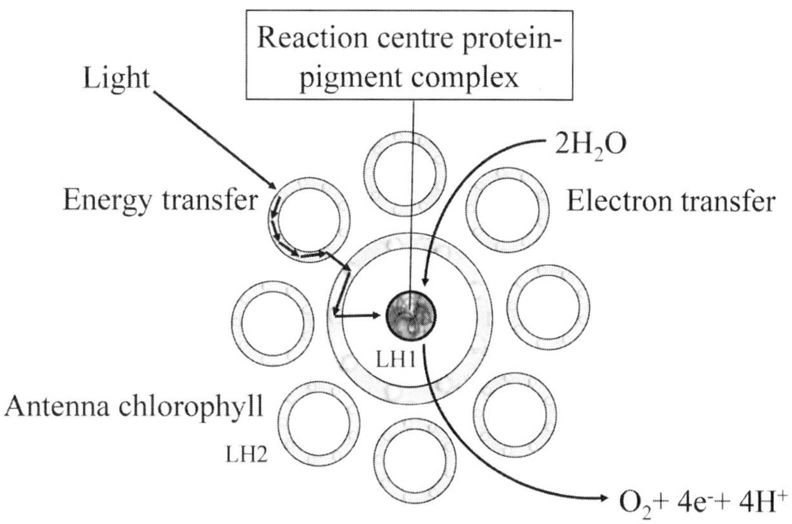

Figure 14.12 Energy flow in a bacterial photosystem for the oxidation of water. The outer rings of light-harvesting complex LH2 surround one inner ring of LH1 complex, near the middle of which is the reaction centre.

chlorophyll.[75] Interrogating the system with a sequence of ultrashort laser pulses, the optical response of the sample reveals linear absorption processes and couplings between chromophores, as well as dynamical aspects of the energy transfer. The results show that excitation relocation does not proceed simply by stepwise transfer from one energy state to another of nearest energy; it depends on strong coupling between chromophores, determined by the extent of their spatial overlap. Thus, excitation relocation may involve fewer intermediary chromophores than might otherwise be expected.

The efficiency of photosynthetic units has encouraged the design of a variety of synthetic light-harvesting systems.[76–79] The materials that have received the most attention are dendrimers,[80,81] macromolecules consisting of molecular units repeatedly branching out from a central core. Designed to act as an excitation trap, dendrimers are exemplified by the structure shown in Fig. 14.13. The outward branching leads to successive generations of structures, each with an increased number of peripheral antenna chromophores. In ideal cases, the requisite spectroscopic gradient is established through chemically similar chromophores in generationally different locations.[82,83] The most recent work on dendrimers has utilised branching motifs of threefold and fourfold local symmetry, based on trisubstituted benzene[84] and porphyrin rings,[85] respectively.

14.5 Directed Energy Transfer

There are a number of different ways in which a vectorial character can be produced or enhanced in RET.[86] In the operation of multichromophore complexes in particular, it is highly important to expedite the delivery of energy (acquired through photoabsorption) to an appropriate location, rather than allowing it to be subject to a long-path random walk with the attendant likelihood of dissipation as heat. Although the mechanisms for controlling energy flow are only just beginning to receive the full attention they deserve, a number of important principles have already been identified.

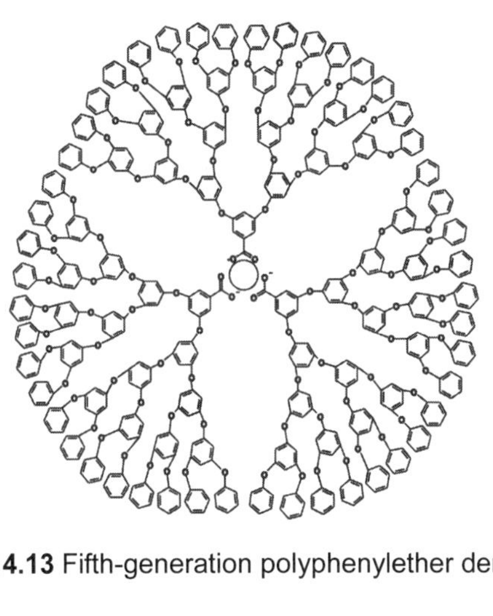

Figure 14.13 Fifth-generation polyphenylether dendrimer.

14.5.1 Spectroscopic gradient

In the context of energy-harvesting materials, it has been shown that any pair of chromophores that plays the roles of donor and acceptor will often do so as part of a sequence in which its donor unit acquires energy and the acceptor passes that energy further onward, through preceding and subsequent RET couplings. Commonly, the energy passes onward through a sequence of chromophores exhibiting progressively longer wavelengths of absorption because they differ either in their molecular structure or in their local electronic environment (the latter being a particularly prominent feature of photobiological systems). This progression through increasingly red-shifted chromophores, known as the spectroscopic gradient or "energy funnel," is the reason for the directed character of energy capture by the system.

In general, the excitation of each successive donor populates a vibrationally excited level within its electronic excited state, from which a degree of vibrational relaxation immediately occurs. The result is a slight degradation of the energy with each transfer event. Small enough not to be of major concern in energy efficiency terms, this feature of multistep energy migration is indeed the factor that largely determines its directed character. This is because, at each step, "forward" transfer is favored over "backward" transfer. The single parameter that effectively determines the extent of this favor is the ratio of spectral overlaps for forward and backward transfer, a parameter that exerts a significant influence on the overall trapping efficiency of any energy-harvesting complex. For example, in the case of dendrimers, the relative propensities for forward and backward transfer between different generation shells is a key factor in determining the directedness of energy flow toward the acceptor core.

The overall spectroscopic gradient experienced on the passage of excitation through a multichromophore system crucially depends on the relative propensities for the forward and backward transfer of energy at each individual step, as determined by the optical and photophysical properties of the relevant units acting as donor and acceptor.[87] To quantify such relative propensities, it is therefore convenient to introduce a dimensionless efficiency parameter ε

$$\varepsilon = \frac{\tau_{B^*}}{\tau_{A^*}} \times \frac{\int F_A(\omega)\sigma_B(\omega)\omega^{-4}d\omega}{\int F_B(\omega)\sigma_A(\omega)\omega^{-4}d\omega}. \tag{14.30}$$

The integrals in the numerator and denominator of Eq. (14.30) are respectively defined as the "forward" and "backward" spectral overlaps. Their role in determining the directionality of energy transfer at the pair-chromophore level is illustrated by Fig. 14.14. The peak frequencies of σ_B, F_A, F_B, and σ_A are ω_1, ω_2, ω_3, and ω_4, respectively, and each fluorescence peak is duly red shifted with respect to its absorption counterpart. In the forward transfer process, the excitation of A is followed by intramolecular vibrational redistribution (IVR) that dissipates part of the acquired energy. Following energy transfer, the same feature is observed in B. On comparing this process with the inverse transfer B to

A, it is clear that the forward case is favored because $\omega_4 - \omega_3$ exceeds $\omega_2 - \omega_1$ ($\delta_2 > \delta_1$ in Fig. 14.14); in general, it follows that the larger the value of δ_2, and the smaller the value of δ_1, the higher the relative efficiency. Two parameters expediently quantify the extent of the spectroscopic gradient: the absorption shift (the frequency displacement of the acceptor absorption maximum relative to that of the donor band), $\omega_G = \omega_4 - \omega_1$, and the corresponding fluorescence shift, $\omega'_G = \omega_2 - \omega_3$.

The general expression in Eq.(14.30) has been analysed in detail for a variety of complex spectral line-shapes, and the results are reported elsewhere.[87] Even the simplest case has interpretive value: when the relevant spectral functions are considered to be Gaussians of similar width (FWHM $\Delta\omega$) and height, the result for ε is expressible as follows:

$$\varepsilon \cong e^{2k(\delta_2^2 - \delta_1^2)} \cong e^{2k\omega_G\omega_S}, \tag{14.31}$$

where $k = 4 \ln 2 \cdot (\Delta\omega)^{-1}$ and in the expression on the right, written in terms of the Stokes shift, $\omega_s = \omega_4 - \omega_2 = \omega_1 - \omega_3 = \delta_2 - \delta_1$, the approximation is based on making ω_G and ω'_G equal. Notably, the ensuing result shows precisely the same functional dependence on the spectroscopic gradient and the state shift; the conclusion is that both properties are equally important in determining the directedness of the energy transfer.

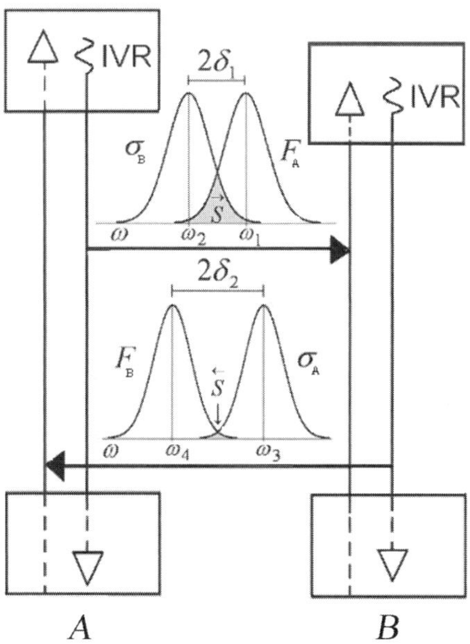

Figure 14.14 Energetics and spectral overlap features for forward and backward energy transfer between donor *A* and acceptor *B*.

14.5.2 Influence of a static electric field

The process of energy transfer can be significantly modified by interaction with a static electric field.[88] At a simple level, the influence can be understood as a consequence of effecting shifts in the electron distributions of the interacting chromophores, producing modified transition moments. In more detail, new contributions to the RET quantum amplitude [Eq. (14.15)] arise, featuring additional interactions with the static field. The most significant corrections entail linear coupling of the field with A or B, while higher-order correction involves field coupling with both chromophores. Significantly, when the static field engages with either the donor or the acceptor transition, different selection rules are invoked—those formally associated with a two-quantum transition. For example, if both the donor and acceptor transitions are electric-dipole-forbidden, only the higher-order interaction can mediate the energy transfer (except, conceivably, by involving a much weaker, higher-order multipole). The physical significance is that, in such a system, energy may not transfer without the presence of the static field; in a suitably designed system, the field-induced mechanism therefore allows electrically switchable control over the delivery of energy to the acceptor.[86]

14.5.3 Optically controlled energy transfer

The current pace of development in nanofabrication techniques has promoted an increasing interest in the specific effects of donor and acceptor placement in nanoscale geometries and periodic structures. However, the possibility of influencing the operation of RET by an optical field, through the input of an off-resonant auxiliary beam of laser radiation, has only very recently begun to receive consideration.

Attention was first focused on amplification effects that might be observed in systems where energy transfer can occur without an auxiliary beam. It was shown that, at the levels of intensity currently available from mode-locked solid state lasers, significant enhancements of the transfer rate could be expected.[89] Although the initial work anticipated effects that could be manifest in any donor-acceptor system, interest has subsequently refocused on structures tailored to exploit the laser-assisted phenomenon. Specifically, consideration has turned to systems in which each donor-acceptor pair has optical properties that satisfy the spectral overlap condition, but for which RET is designedly precluded by a customised geometric configuration.[90] For example, as was observed in Sec. 14.3.2, both short- and long-range RET is forbidden when the donor and acceptor undergo electric dipole transitions whose transition moments are perpendicular both to each other and to the donor-acceptor displacement vector. Through an optically nonlinear mechanism of optically controlled resonance energy transfer (OCRET), it transpires that the throughput of nonresonant laser pulses can facilitate energy transfer under such conditions where it would otherwise be rigorously forbidden; the system thus functions as an optical transistor, with excitation throughput switched on by the auxiliary beam. The laser systems most capable of delivering the necessary levels of irradiance prove to be precisely those that will also offer directly controllable ultrafast speeds of switching.

The mechanism for OCRET invokes fourth-order time-dependent perturbation theory. Each interaction is linear in the electric dipole interaction operator and thus, involves the absorption or emission of a photon. Specifically, in addition to the two virtual photon events (creation and annihilation) of normal RET, this process involves the absorption and the stimulated re-emission of a photon from the throughput off-resonant laser light. In common with the virtual photon interactions, each of the real photon events may occur at either the donor or the acceptor. In general, all four of the resulting possible combinations contribute to the overall quantum amplitude; moreover, each has 24 different associated time orderings. Figure 14.15 illustrates two of the 96 Feynman diagrams that arise. Again, the state-sequence method[49,91] represents a considerably more tractable basis for the QED calculations that are necessary to deliver an equation for the energy transfer rate.

Recently, detailed calculations have been performed on a prototype implementation of OCRET in planar nano-arrays (Fig. 14.16). The results, illustrated in Fig. 14.17, give encouragement that the mechanism affords a realistic basis for fabricating a configuration of optical switches with parallel processing capability, operating without significant crosstalk. The analysis also supports a view that a nano-array OCRET system may, in the longer term, come to represent a new and significant channel of progress toward reliable systems for use in optical computing and communications routing.[92]

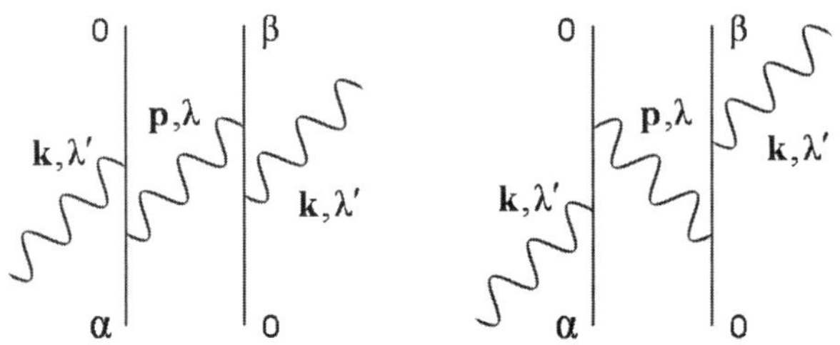

Figure 14.15 Two of the 96 Feynman time-ordered diagrams for OCRET.

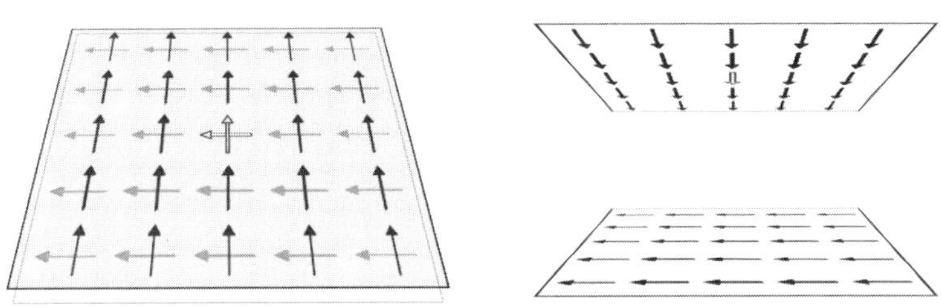

(a) (b)

Figure 14.16 Two views of the nano-array structure: (a) from above and (b) from between the layers with their separation exaggerated for clarity. Black arrows represent donor transition dipoles of the upper array, and grey the acceptor transition dipoles of the lower array; open arrows represent the excited donor and its corresponding ground-state acceptor.

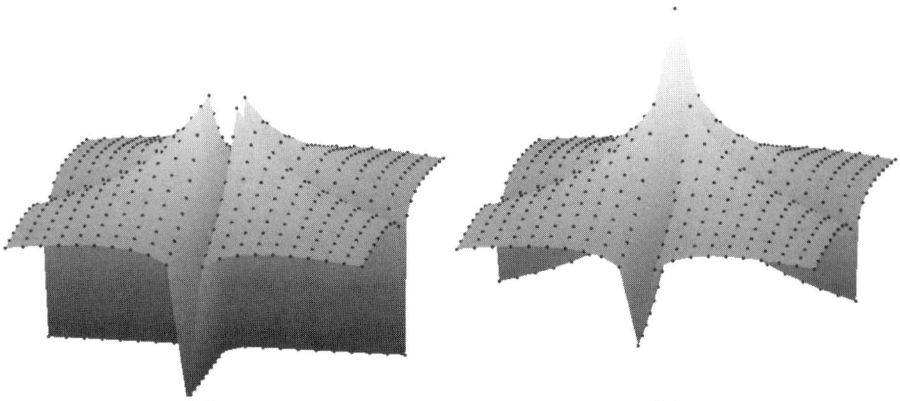

(a) (b)

Figure 14.17 Contour graphs depicting, on a logarithmic scale, the probability of energy transfer to the array of acceptors (a) in the absence and (b) in the presence of laser light. The efficiency of energy transfer is represented by the vertical scale, and the dots indicate the spatial locations of each acceptor.

14.6 Developing Applications

The mechanism of RET is involved in an ever-broadening range of applications, only some of which have been covered in this chapter. As mentioned earlier, the practice of measuring the fluorescence from chromophores excited through energy transfer (FRET) is one of the major techniques, and in this relatively mature field, many recent advances such as FRET imaging microscopy[93,94] reflect an exploitation of technical developments. Burgeoning applications are to be found in the biophotonics area, in particular.[95–97] Staying with molecular systems, the ongoing development of OLED emitters is also worth flagging as a prominent area of current activity. However, in connection with synthetic solid state systems where some of the more recently envisaged applications are being developed, one can see tangible signs of the emergence of another new research forefront.

One issue that would appear to merit much more consideration is the extent to which RET can be tailored, or even suppressed, in optical microcavity configurations. It is already well known that spontaneous emission is significantly influenced by inclusion of the donor in a cavity, through the restrictions that are thereby imposed on the sustainable optical modes. Barnes and others[98–100] have drawn attention to the operation of similar principles in connection with donor-acceptor systems. In an impressive series of works by Lovett et al.[62,63] (see also the references therein), several applications relating to quantum dot energy transfer processes have been considered, focusing in particular on the interplay of quantum state entanglement and Förster coupling. It has been suggested that these and other principles concerning the effect of static electric fields on coupled quantum dots may have an important role to play in quantum logic applications. Andrews has recently proposed a system in which the simultaneous delivery of laser pulses with two differing off-resonant frequencies might achieve a similar purpose,[101,102] while other studies currently underway are beginning to explore novel nanomechanical effects produced by the action of RET.[103]

14.7 Conclusion

Resonance energy transfer is a subject that has come of age. Having been studied already for more than half a century, it has many established applications, despite the fact that some of the fundamental principles have only quite recently become fully understood. Quantum electrodynamics has played an important part in consolidating the theory and has paved the way for a number of newer developments in which photonic and quantum optical aspects of light–matter coupling come into play. With such an extensive involvement in widely varying media, this is a subject that deserves to be more widely taught and understood; this chapter is offered as a contribution toward that objective.

Acknowledgments

It is a pleasure to acknowledge the involvement of several past and present members of my group in the work reported here, and I gladly acknowledge their invaluable contributions. In particular, I must thank Philip Allcock, David Bradshaw, Richard Crisp, Gareth Daniels, Robert Jenkins, Gediminas Juzeliūnas, and Justo Rodríguez. I also thank a number of these people, as well as Luciana Dávila Romero, and Jamie Leeder for producing most of the figures. I gratefully acknowledge funding support from the Engineering and Physical Sciences Research Council, the Leverhulme Trust, and the Royal Society.

References

1. V. M. Agranovich and M. D. Galanin, *Electronic Excitation Energy Transfer in Condensed Matter*, Elsevier/North-Holland, Amsterdam, (1982).

2. B. W. van der Meer, J. Coker III, and S.-Y. S. Chen, *Resonance Energy Transfer: Theory and Data*, VCH, New York (1982).

3. D. L. Andrews and A. A. Demidov, Eds., *Resonance Energy Transfer*, Wiley, New York (1999).

4. G. W. Robinson and R. P. Frosch, "Electronic excitation transfer and relaxation," *J. Chem. Phys.* **38,** 1187–1203 (1963).

5. H. van Amerongen, L. Valkunas, and R. van Grondelle, *Photosynthetic Excitons*, World Scientific, Singapore (2000).

6. D. M. Guldi, "Fullerene-porphyrin architectures; photosynthetic antenna and reaction center models," *Chem. Soc. Rev.* **31,** 22–36 (2002).

7. D. L. Dexter, "A theory of sensitized luminescence in solids," *J. Chem. Phys.* **21,** 836–850 (1953).

8. G. D. Scholes and K. P. Ghiggino, "Electronic interactions and interchromophore excitation transfer," *J. Phys. Chem.* **98,** 4580–4590 (1994).

9. E. K. L. Yeow and K. P. Ghiggino, "Electronic energy transfer in multichromophoric arrays. The effects of disorder on superexchange coupling and energy transfer rate," *J. Phys. Chem. A* **104,** 5825–5836 (2000).

10. T. Förster, "Zwischenmolekulare Energiewanderung und Fluoreszenz," *Annalen. Phys.* **2,** 55–75 (1948).

11. S. A. Latt, H. T. Cheung, and E. R. Blout, "Energy transfer: a system with relatively fixed donor-acceptor separation," *J. Am. Chem. Soc.* **87,** 995–1003 (1965).

12. J. R. Lakowicz, *Principles of Fluorescence Spectroscopy*, 2nd ed., Chapter 10, Kluwer Academic, New York (1999).

13. B. Valeur, *Molecular Fluorescence. Principles and Applications*, Chapter 9, Wiley-VCH, Weinheim, Germany (2002).

14. C. Devadoss, P. Bharathi, and J. S. Moore, "Energy transfer in dendritic macromolecules: molecular size effects and the role of an energy gradient," *J. Am. Chem. Soc.* **118,** 9635–9644 (1996).

15. C. G. dos Remedios and P. D. J Moens, "Fluorescence resonance energy transfer spectroscopy is a reliable "ruler" for measuring structural changes in proteins: dispelling the problem of the unknown orientation factor," *J. Struct. Biol.* **115,** 175–185 (1995).

16. L. Stryer and R. P. Haugland, Energy transfer: a spectroscopic ruler," *Proc. Natl. Acad. Sci. USA* **58,** 719–726 (1967).

17. G. D. Scholes and D. L. Andrews, "Damping and higher multipole effects in the quantum electrodynamical model for electronic energy transfer in the condensed phase," *J. Chem. Phys.* **107,** 5374–5384 (1997).

18. A. Salam, "A general formula for the rate of resonant transfer of energy between two electric multipole moments of arbitrary order using molecular quantum electrodynamics," *J. Chem. Phys.* **122,** 044112 (2005).

19. D. P. Craig and T. Thirunamachandran, *Molecular Quantum Electro-Dynamics*, Dover, New York (1998).

20. R. G. Woolley, "Charged particles, gauge invariance, and molecular electrodynamics," *Int. J. Quantum. Chem.* **74,** 531–545 (1999).

21. R. G. Woolley, "Gauge invariance in non-relativistic electrodynamics," *Proc. R. Soc. London, Ser. A* **456,** 1803–1819 (2000).

22. J. S. Avery, "Resonance energy transfer and spontaneous photon emission," *Proc. Phys. Soc.* **88,** 1–8 (1966).

23. L. Gomberoff and E. A. Power, "The resonance transfer of excitation," *Proc. Phys. Soc.* **88,** 281–284 (1966).

24. D. L. Andrews, "A unified theory of radiative and radiationless molecular energy transfer," *J. Chem. Phys.* **135,** 195–201 (1989).

25. G. J. Daniels, R. D. Jenkins, D. S. Bradshaw, and D. L. Andrews, "Resonance energy transfer: The unified theory revisited," *J. Chem. Phys.* **119,** 2264–2274 (2003).

26. D. L. Andrews and B. S. Sherborne, "Resonant excitation transfer: A quantum electrodynamical study," *J. Chem. Phys.* **86,** 4011–4017 (1987).

27. G. Juzeliūnas and D. L. Andrews, "Quantum electrodynamics of resonance energy transfer, *Adv. Chem. Phys.* **112,** 357–410 (2000).

28. J. D. Jackson, *Classical Electrodynamics*, 3rd Ed., Wiley, New York, p. 411 (1999).

29. D. L. Andrews and D. S. Bradshaw, "Virtual photons, dipole fields and energy transfer: A quantum electrodynamical approach," *Eur. J. Phys.* **25,** 845–858 (2004).

30. F. Halzen and A. D. Martin, *Quarks and Leptons*, Wiley, New York, p. 140 (1984).

31. G. Juzeliūnas and D. L. Andrews, "Quantum electrodynamics of resonant energy transfer in condensed matter," *Phys. Rev. B* **49,** 8751–8763 (1994).

32. E. Fermi, *Nuclear Physics*, University of Chicago Press, Chicago (1950).

33. D. L. Andrews and G. Juzeliūnas, "Intermolecular energy transfer: retardation effects," *J. Chem. Phys.* **96,** 6606–6612 (1992).

34. R. D. Jenkins, G. J. Daniels, and D. L. Andrews, "Quantum pathways for resonance energy transfer," *J. Chem. Phys.* **120,** 11442–11448 (2004).

35. L. Novotny and B. Hecht, *Principles of Nano-optics*, Cambridge University Press, Cambridge, 285–293 (2006).

36. M. N. Berberan-Santos, E. J. N. Pereira, and J. M. G. Martinho, "Stochastic theory of combined radiative and nonradiative transport," *J. Chem. Phys.* **107,** 10480–10484 (1997).

37. J. Baumann and M. D. Fayer, "Excitation transfer in disordered two-dimensional and anisotropic three-dimensional systems: effects of spatial geometry on time-resolved observables," *J. Chem. Phys.* **85**, 4087–4107 (1986).

38. B. Nickel, "Orientation factor in Förster energy-transfer with photoselection of donor and acceptor," *J. Chem. Phys.* **198,** 353–371 (1995).

39. B. W. van der Meer, in *Resonance Energy Transfer*, D. L. Andrews and A. A. Demidov, Eds., John Wiley and Sons, New York, 151–172 (1999).

40. D. L. Andrews and G. Juzeliūnas, "The range dependence of fluorescence anisotropy in molecular energy transfer," *J. Chem. Phys.* **95**, 5513–5518 (1991).

41. V. V. Ovsyankin, in *Spectroscopy of Solids Containing Rare Earth Ions,* A. A. Kaplyanskii and R. M. Macfarlane, Eds., Elsevier, Amsterdam, 343–480 (1987).

42. C. Candu, V. Ciobu, and V. Enaki, "Four-center resonance energy transfer in rare earth doped crystals," *J. Optoelectron. Adv. Mater.* **7,** 781–786 (2005).

43. D. L. Dexter, "Possibility of luminescent quantum yields greater than unity," *Phys. Rev.* **108**, 630–633 (1957).

44. N. Bloembergen, "Solid state infrared quantum counters," *Phys. Rev. Lett.* **2,** 84–85 (1959).

45. T. Miyakawa and D. L. Dexter, "Cooperative and stepwise excitation of luminescence: trivalent rare-earth ions in Yb^{3+}-sensitized crystals," *Phys. Rev. B* **1**, 70–80 (1970).

46. F. Auzel, "Upconversion processes in coupled ion systems," *J. Lumin.* **45**, 341–345 (1990).

47. F. Auzel, "Compteur quantique par transfert d'energie entre deux ions de terre rares dans un tungstate mixte et dans un verre," *Compt. Rend.* **262B**, 1016–1021 (1966).

48. M. Chua and P. A. Tanner, "Three-body energy transfer processes of lanthanide ions in crystals," *J. Lumin.* **66-7**, 203–207 (1995).

49. R. D. Jenkins, D. L. Andrews, and L. C. Dávila Romero, "A new diagrammatic methodology for non-relativistic quantum electrodynamics," *J. Phys. B: Atom. Molec. Opt. Phys.* **35**, 445–468 (2002).

50. R. D. Jenkins and D. L. Andrews, "Three-center systems for energy pooling: quantum electrodynamical theory," *J. Phys. Chem. A* **102**, 10834–10842 (1998).

51. D. L. Andrews, "An accretive mechanism for blue-shifted fluorescence in strongly-pumped systems: resonance energy transfer with Raman emission," *J. Raman Spectrosc.* **31**, 791–796 (2000).

52. D. L. Andrews and R. D. Jenkins, "A quantum electrodynamical theory of three-center energy transfer for upconversion and downconversion in rare earth doped materials," *J. Chem. Phys.* **114**, 1089–1100 (2001).

53. X. Zhang, J.-P. Jouart, and G. Mary, "Energy transfer upconversion in Ho^{3+} and Ho^{3+}, Yb^{3+} doped CdF_2 crystals," *J. Phys. Condens. Matter* **10**, 493–500 (1998).

54. R. T. Wegh, H. Donker, A. Meijerink, R. J. Lamminmăki, and J. Hölsä, "Vacuum-ultraviolet spectroscopy and quantum cutting for Gd^{3+} in $LiYF_4$," *Phys. Rev. B* **56**, 13841–13488 (1997).

55. C. R. Kagan, C. B. Murray, M. Nirmal, and M. G. Bawendi, "Electronic energy transfer in CdSe quantum dot solids," *Phys. Rev. Lett.* **76**, 1517–1520 (1997).

56. S. A. Crooker, J. A. Hollingsworth, S. Tretiak, and V. I. Klimov, "Spectrally resolved dynamics of energy transfer in quantum-dot assemblies: Towards engineered energy flows in artificial materials," *Phys. Rev. Lett.* **89**, 186802 (2002).

57. B. W. Lovett, J. H. Reina, A. Nazir, B. Kothari, and G. A. Briggs, "Resonant transfer of excitons and quantum computation," *Phys. Lett. A* **315**, 136–142 (2003).

58. A. O. Govorov, "Spin and energy transfer in nanocrystals without tunneling," *Phys. Rev. B* **68**, 075315 (2003).

59. H. Kamada and H. Gotoh, "Quantum computation with quantum dot excitons," *Semiconductor Sci. Tech.* **19**, S392–S396 (2004).

60. S. Sangu, K. Kobayashi, A. Shojiguchi, and M. Ohtsu, "Logic and functional operations using a near-field optically coupled quantum-dot system," *Phys. Rev. B* **69**, 115334 (2004).

61. A. R. Clapp, I. L. Medintz, H. T. Uyeda, B. R. Fisher, E. R. Goldman, M. G. Bawendi, and H. Mattoussi, "Quantum dot-based multiplexed fluorescence resonance energy transfer," *J. Am. Chem. Soc.* **127**, 18212–18221 (2005).

62. B. W. Lovett, J. H. Reina A. Nazir, and G. A. D. Briggs, "Optical schemes for quantum computation in quantum dot molecules," *Phys. Rev. B* **68**, 205319 (2003).

63. A. Nazir, B. W. Lovett, S. D. Barrett, J. H. Reina, and G. A. D. Briggs, "Anticrossings in Förster coupled quantum dots," *Phys. Rev. B* **71**, 045334 (2005).

64. L. Bányai and S.W. Koch, *Semiconductor Quantum Dots*, World Scientific, Singapore (1993).

65. J. M. Villas-Boas, S. E. Ulloa, and A. O. Govorov, "Photocurrent and spin manipulation in quantum dots," *Physica E: Low-dimensional Systems and Nanostructures* **34**, 333–335 (2006).

66. G. D. Scholes and D. L. Andrews, "Resonance energy transfer and quantum dots," *Phys. Rev. B* **72**, 125331 (2005).

67. G. D. Scholes, R. D. Harcourt, and G. R. Fleming, "Electronic interactions in photosynthetic light-harvesting complexes: The role of carotenoids," *J. Phys. Chem. B* **101**, 7302–7312 (1997).

68. B. P. Krueger, G. D. Scholes, R. Jimenez, and G. R. Fleming, "Electronic excitation transfer from carotenoid to bacteriochlorophyll in the purple bacterium *Rhodopseudomonas acidophila*," *J. Phys. Chem. B* **102**, 2284–2292 (1998).

69. S. Gnanakaran, R. Haran, R. Kumble, and R. M. Hochstrasser, "Energy transfer and localization: applications to photosynthetic systems," in *Resonance Energy Transfer*, D. L. Andrews and A. A. Demidov, Eds., Wiley, New York, 308–365 (1999).

70. R. van Grondelle and O. J. G. Samsen, in *Resonance Energy Transfer,* D. L. Andrews and A. A. Demidov, Eds., Wiley, New York, 366–398 (1999).

71. S. Savikhin, D. R. Buck, and W. S. Struve, in *Resonance Energy Transfer,* D. L. Andrews and A. A. Demidov, Eds., Wiley, New York, 399–434 (1999).

72. G. S. Engel, T. R. Calhoun, E. L. Read, T.-K. Ahn, T. Mancal, Y.-C. Cheng, R. E. Blankenship, and G. R. Fleming, "Evidence for wavelike energy transfer through quantum coherence in photosynthetic systems," *Nature* **446**, 782–786 (2007).

73. X. Hu and K. Schulten, "How nature harvests sunlight," *Phys. Today* **50** (8), 28–34 (1997).

74. T. Renger, V. May, and O. Kühn, "Ultrafast excitation energy transfer dynamics in photosynthetic pigment-protein complexes," *Phys. Rep.* **343**, 138–254 (2001)

75. T. Brixner, J. Stenger, H. M. Vaswani, M. Cho, R. E. Blankenship, and G. R. Fleming, "Two-dimensional spectroscopy of electronic couplings in photosynthesis," *Nature* **434**, 625–628 (2005).

76. A. Adronov and J. M. J. Fréchet, "Light-harvesting dendrimers," *Chem. Commun.*1701–1710 (2000).

77. P. Ball, "Natural strategies for the molecular engineer," *Nanotechnology* **13**, R15–R28 (2002).

78. D. L. Andrews, "Optical energy harvesting materials," in *Introduction to Complex Mediums for Optics and Electromagnetics*, W. S. Weiglhofer and A. Lakhtakia, Eds., SPIE Press, Bellingham, WA, 141–163 (2003).

79. D. L. Andrews, Ed., *Energy Harvesting Materials*, World Scientific, Singapore (2005).

80. A. Bar-Haim, J. Klafter, and R. Kopelmann, "Dendrimers as controlled artificial energy antennae," *J. Am. Chem. Soc.* **119**, 6197–6198 (1997).

81. A. Nantalaksakul, D. R. Reddy, C. J. Bardeen, and S. Thayumanavan, "Light harvesting dendrimers," *Photosynthesis Res.* **87**, 133–150 (2006).

82. C. Devadoss, P. Bharathi, and J. S. Moore, "Energy transfer in dendritic macromolecules: Molecular size effects and the role of an energy gradient," *J. Am. Chem. Soc.* **118**, 9635–9644 (1996).

83. S. F. Swallen, Z.-Y. Shi, W. Tan, Z. Xu, J. S. Moore, and R. Kopelman, "Exciton localisation hierarchy and directed energy transfer in conjugated linear aromatic chains and dendrimeric supermolecules," *J. Lumin.* **76-7**, 193–196 (1998).

84. A. Archut and F. Vögtle, "Functional cascade molecules," *Chem. Soc. Rev.* **27**, 233–240 (1998).

85. P. G. van Patten, A. P. Shreve, J. S. Lindsey, and R. J. Donohoe, "Energy-transfer modeling for the rational design of multiporphyrin light-harvesting arrays," *J. Phys. Chem. B* **102**, 4209–4216 (1998).

86. D. L. Andrews and R. G. Crisp, "Theory of directed electronic energy transfer," *J. Fluorescence* **16**, 191–199 (2006).

87. D. L. Andrews and J. Rodríguez, "Resonance energy transfer: Spectral overlap, efficiency and direction," *J. Chem. Phys.* **127,** 084509 (2007).

88. D. L. Andrews and A. M. Bittner, "Energy transfer in a static electric field," *J. Lumin.* **55**, 231–242 (1993).

89. P. Allcock, R. D. Jenkins, and D. L. Andrews, "Laser-assisted resonance-energy transfer, *Phys. Rev. A* **61**, 023812 (2000).

90. D. L. Andrews and R. G. Crisp, "Optically activated energy transfer: array implementation," *J. Opt. A: Pure Appl. Opt.* **8**, S106–S112 (2006).

91. B. W. Alligood and A. Salam, "On the application of state sequence diagrams to the calculation of the Casimir-Polder potential," *Mol. Phys.* **105**, 395–404 (2007).

92. D. S. Bradshaw and D. L. Andrews, "Optically controlled resonance energy transfer: Mechanism and configuration for all-optical switching," *J. Chem. Phys.* **128**, 144506 (2008).

93. M. Elangovan, H. Wallrabe, Y. Chen, R. N. Day, M. Barroso, and A. Periasamy, "Characterization of one- and two-photon excitation fluorescence resonance energy transfer microscopy," *Methods* **29**, 58–73 (2003).

94. S. K. Sekatskii, "Fluorescence resonance energy transfer scanning near-field optical microscopy, *Phil. Trans. R. Soc. A: Math. Phys. Eng. Sci.* **362**, 901–919 (2004).

95. R. MacColl, Allophycocyanin and energy transfer, *Biochim. Biophys. Acta Bioenergetics* **1657**, 73–81 (2004).

96. G. Ramanoudjame, M. Du, K. A. Mankiewicz, and V. Jayaraman, "Allosteric mechanism in AMPA receptors: A FRET-based investigation of conformational changes," *Proc. Natl. Acad. Sci. USA* **103**, 10473–10478 (2006).

97. X. You, A. W. Nguyen, A. Jabaiah, M. A. Sheff, K. S. Thorn, and P. S. Daugherty, "Intracellular protein interaction mapping with FRET hybrids," *Proc. Natl. Acad. Sci. USA* **103**, 18458–18463 (2006).

98. M. Cho, "Confinement-induced enhancement or suppression of the resonant dipole-dipole interaction," *J. Chem. Phys.* **110**, 4998–5010 (1999).

99. P. Andrew and W. L. Barnes, "Förster energy transfer in an optical microcavity," *Science* **290**, 785–788 (2000).

100. W. L. Barnes, "Spontaneous emission and energy transfer in the optical microcavity," *Contemp. Phys.* **41**, 287–300 (2000).

101. D. L. Andrews, "Optically switched energy transfer: Twin-beam off-resonance control," *Phys. Rev. Lett.* **99**, 023601 (2007).

102. D. L. Andrews, R. G. Crisp, and S. Li, "Single and dual beam optical switching of resonance energy transfer," *J. Chem. Phys.* **127**, 174702 (2007).

103. D. L. Andrews, D. S. Bradshaw, J. Leeder, and J. Rodríguez, "Dynamics of the dispersion interaction in an energy transfer system," *Phys. Chem. Chem. Phys.* **10**, 5250–5255 (2008).

Biography

David L. Andrews followed a degree in Chemistry at University College London with doctoral studies, then a postdoctoral appointment in Mathematics, before moving to a lectureship at University of East Anglia (UEA) in Norwich, UK, where he gained a Chair in Chemical Physics in 1996. Andrews was elected a Fellow of the Royal Society of Chemistry in 1988, a Fellow of the Institute of Physics in 1999, and a Fellow of SPIE in 2006. His research centres on molecular photophysics, utilising quantum electrodynamics to develop theory for energy transport and a range of multiphoton and other nonlinear optical phenomena. Andrews's early work led toward the unified theory of energy transfer accommodating both radiationless and radiative mechanisms. His group was the first to identify and predict the characteristics of two-photon resonance energy transfer, anticipating later experiments on biological systems. Currently his group's studies focus on nonlinear optical processes, optical binding, and energy harvesting in nanosystems. Andrews has over 200 papers and ten books to his name, including the textbook *Lasers in Chemistry,* Springer Publishing Co., 1997.

Chapter 15
Optics of Nanostructured Materials from First Principles

Vladimir I. Gavrilenko
Center for Materials Research, Norfolk State University, Norfolk, VA, USA

15.1 Introduction

The main distinctive feature of nanomaterials is that they have structures on a nanometer scale. The class of nanostructured materials (nanomaterials) includes a variety of novel inorganic (ceramic nanoparticles, semiconductor quantum dots), organic (carbon nanotubes, fullerenes, polymers, molecular aggregates, molecule-metallic nanoparticle assemblies), and biological nanomaterials. To the nanomaterials belong artificially fabricated structures,[1,2] photonic crystals,[3] ordered solid surface patterns at nanometer scale, and organic-inorganic combinations of nanostructures (wires, rods, dots, etc.).[4,5] Their optical properties are extremely fascinating and useful for a variety of applications.

Optics of nanostructured materials is a broad area covering different fields of both fundamental and applied science: optics of quantum confined systems, photonic crystals, nanoplasmonics, interaction of quantum particles with the optical incoherent and quantized field, etc. Nanostructured systems containing organic-inorganic interfaces open exciting possibilities for optical engineering.[5,6] Artificially fabricated nanostructured materials (the so-called *metamaterials*), as well as those combined from inorganic and organic substances, present unique properties and open new areas for applications.[7] Many features of nanomaterials optics can be understood within classic electrodynamics. For example, interactions with the optical local field generated due to plasma resonances in quantum confined systems can be well understood within classic Maxwell theory.[1,2,8,9]

Optical properties of nanomaterials are size dependent; they do not naturally occur in larger bulk materials. The first principle theories of chemical, physical, and optical properties of simple atoms and molecules are fairly well understood, predictable, and no longer considered overly complex. This chapter demonstrates that this contrasts markedly with the current state of knowledge of the first principle optics of nanostructures. Through comparative analysis of the size-dependent optical response from nanomaterials, it is shown that although strides have been made in computational chemistry and physics, bridging across length scales from nano to macro remains a major challenge.

The first principles theory of the optics of nanostructured materials is considered here separately for at least two significant areas, depending on the nature of the optical field interacting with the nanostructures. One area governs the interaction of nanostructured materials with random (noncorrelated) optical fields. Quantum confinement and quantization of the charged quasi-particle motions are the main features that modify the optical response of nanostructures.[4,5,9–11] The first principles theory of the optics of nanostructures has been actively studied within the last decade since the development of theoretical methods beyond the standard density functional theory (DFT).[5,11,12] This part of optics is addressed in Secs. 15.2, 15.4, and in Appendix I. Nonlocalities of the electron potential energy modify quasi-particle excitations, resulting in specific features of the optical response that are unique for nanostructures. One of the most remarkable and extensively studied features is the effect of the local field (LF) related to the plasmon resonance in nanoparticles.[13] This area is now extensively studied, and there are several excellent papers, reviews, and books addressing both fundamental and applied aspects of the plasmon resonance effects in nanostructures.[1,2,7,8,14] In Sec. 15.3 we describe how the LF effect can be modeled from first principles, and in Sec. 15.6.2 it is linked with the classic electrodynamics approach.

Another area of optics that is actively studied from first principles governs a material's interaction with a coherent optical field. This area corresponds to quantum electrodynamics (QED).[15,16] Since the discovery of the laser, QED has been very actively developed. QED offers exciting possibilities for fundamental and applied research when the light field is in resonance (quantized) within a microresonator (cavity) interacting with quantum particles. This specific area of QED is called cavity QED (CQED). The most fundamental work on CQED addresses the interactions of atoms (frequently described as the two-level system) with a quantized light field in a cavity. CQED studies allow direct proof of fundamental concepts of the quantum theory (e.g., entanglement[17,18]) and open new opportunities for applied research.[19,20] Relevant to the subject of the present chapter is a part of CQED that governs interaction of the quantized optical field with the nanostructured object (atomic clusters, quantum dots, etc.). This area has been rapidly developing within the past decade.[21–25] However, the first principles description of particle field and particle-particle interactions within a cavity of atomic clusters, molecules, and quantum dots is still challenging in view of the extreme complexity of the problem. Section 15.5 provides a brief introduction to the subject highlighting some active directions of future research.

As a part of the present volume, this chapter is written primarily for students and young researchers interested in first principle modeling and simulations in optics. It should not be considered as a review of the field. Citations of scientific results with illustrations are chosen only to explain basic ideas and approaches. The material starts with a theoretical description of the ground state and electron energy structure as a first step in optical modeling. The first paragraphs of the chapter (Secs. 15.2 and 15.3 with Appendices) are designed to provide an introductory overview, which seems reasonable considering the interdisciplinary nature of the field. Recent developments in the first principles description of electron energy structure and optical functions of nanostructured materials are presented in comparison between different methods of computational physics and chemistry. The improvements beyond the standard DFT, including the effects of microscopic local fields, are considered in conjunction with realistic predictions of optical functions and in comparison with a classic electrodynamics (ED) approach (Sec. 15.3). Other paragraphs present an overview of the first principles incoherent (Sec. 15.4) and coherent (Sec. 15.5) nano-optics: CQED of nanostructures, and fundamental (entanglement) and applied studies in quantum theory. First principles modeling aspects of Raman spectroscopy and surface-enhanced Raman scattering (SERS) of nanostructures are presented in Sec. 15.6, which also includes basics of the methods and progress in the area.

15.2 Optical Response from First Principles

The optics of materials can be understood as the interaction of atoms, molecules, or solids with external electromagnetic radiation. This interaction is described in terms of different kinds of excitations: appearance of additional induced charge due to electron transitions between available energy states, generation of lattice vibrations, electron-hole pairs, etc. The starting point for optical modeling from the first principles is to build a parameter-free theoretical model of the material that realistically reproduces its basic properties. Using the full potential of quantum theory and particle physics, one builds the atomic structure corresponding to the minimum of the free energy of the system, as follows from the laws of thermodynamics. Basics of an optical response can be understood fairly well by neglecting the effects of temperature, i.e., by modeling at zero temperature. In this case the free energy is equal to the total energy of the system. In order to study optics at a finite temperature one needs to calculate entropy, which substantially complicates the problem. Therefore the first step in first principles modeling of materials is the creation of an atomic structure that, at the minimum of the total energy (the *ground state*), realistically reproduces well-known measured properties (e.g., known atomic geometric parameters, bulk modulus for solids, etc.). After that, complete information about electronic states and their wave functions is required as a next step for quantitative analysis of the optical response of materials. In other words, before studying excitations (optics) by microscopic modeling, one needs to determine first the electron energy structure of the system in equilibrium (ground state). In modern materials science the ground state is realistically predicted by the DFT. Basic ideas of the DFT are explained in Appendix I.

In this section modeling of the optical functions of materials is presented assuming that eigenenergies and eigenfunctions of the system in equilibrium have already been obtained (e.g., from DFT calculations). In equilibrium all electronic materials consisting of the charged particles (electrons and nucleus) are neutral. Light illumination (or more generally, application of an external electromagnetic field) deforms electron density. This perturbation could be described in terms of induced charge density (*longitudinal* response) or induced current (*transversal* response). In a wide spectral range up to vacuum ultraviolet, both approaches are equivalent. Below we consider the longitudinal response case. Details of the calculations are given in Appendix II.

Equilibrium electron-charge density is defined through the density operator (using the definition of trace Tr as the sum of the diagonal elements)

$$n_{eq}(\mathbf{r}) = Tr[\rho_0, \delta(\mathbf{r} - \mathbf{r}_0)], \tag{15.1}$$

where index eq stands for equilibrium. In an external optical field when the light quanta strike electrically neutral atoms, the equilibrium is broken through the deformation of the electron clouds. Time-dependent changes of the electron charge density can be represented as a Taylor expansion. The number of terms to be included in the Taylor sum for the induced part of the charge is determined by the excitation intensity

$$\begin{aligned} n(\mathbf{r},t) &= n_{eq}(\mathbf{r}) + n_{ind}(\mathbf{r},t) = Tr[\rho,\delta] \\ &= Tr[\rho_0,\delta] + eTr[\rho^{(1)},\delta] + eTr[\rho^{(2)},\delta] + \dots \, . \end{aligned} \tag{15.2}$$

The first, second, and higher-order corrections ($\rho^{(1)}$, $\rho^{(2)}$, etc.) to the density operator (ρ_0) are determined from standard perturbation theory. Dynamic optical response is described through the time-dependent density operator, assuming that in an external electromagnetic field the perturbation is harmonic. The unperturbed density operator defined as Eq. (15.62) (see Appendix II) in matrix representation has the form

$$\rho_{ss'}^{(0)} = f(E_s)\delta_{ss'}. \tag{15.3}$$

At zero temperature, optical excitations occur between completely filled and empty states with Fermi functions equal to either 1 or 0, respectively. Consequently, first-order perturbation of the density operator describing linear optical response has the form

$$\rho_{ss'}^{(1)}(\omega) = \frac{f(E_{s'}) - f(E_s)}{E_s - E_{s'} - \hbar\omega} V_{ss'} = (E_s - E_{s'} - \hbar\omega)^{-1} V_{ss'}|_{T=0}. \tag{15.4}$$

The second-order perturbation of the density operator at $T = 0$ is given by

$$\begin{aligned} \rho_{ss'}^{(2)}(\omega) &= \frac{1}{E_s - E_{s'} - \hbar\omega} \sum_{s''} \left[V_{ss''}^{(1)} V_{s''s'}^{(1)} \left(\frac{1}{E_{s''} - E_{s'} - \hbar\omega} \right. \right. \\ &- \left. \left. \frac{1}{E_s - E_{s''} - \hbar\omega} \right) \right]. \end{aligned} \tag{15.5}$$

Equations (15.4) and (15.5) can now be used to obtain induced charge from Eq. (15.2) within first- and second-order perturbation, respectively. We use here the plane-wave representation. This approach is very convenient for evaluation of optical functions and is widely used in the literature. The wave function is given by

$$\psi_{n,\mathbf{k}}(\mathbf{r}) = \frac{1}{\sqrt{\Omega}} \sum_{\mathbf{G}} d_{n,\mathbf{k}}(\mathbf{G}) e^{i(\mathbf{q}+\mathbf{G})\mathbf{r}}, \tag{15.6}$$

where $d_{n,\mathbf{k}}$ are the expansion coefficients of the wave function characterized by the wave vector $\mathbf{k}$ and related to the n'th electron energy state, $\mathbf{G}$ is the reciprocal lattice vector, and $\mathbf{q}$ is the wave vector of light.

Evaluation of the induced charge within the representation given by Eq. (15.6) is described in detail in Appendices II and III. Here we present simplified results obtained in the limit of $\mathbf{q} \to 0$ (neglecting spatial dispersion) and neglecting the effects of the nonlocality (i.e., $\mathbf{G} = \mathbf{G}' = 0$). The result is given by

$$n_{ind}(\omega) = \hat{P}V(\omega), \tag{15.7}$$

where $V(\omega)$ is the external perturbation potential [see Eq. (15.68) in Appendix II]. Using Eq. (15.4) the tensor components ($\alpha, \beta = x, y, z$) of the polarization function $\hat{P}$ are given by

$$P_{\alpha,\beta}(\omega) = \frac{2}{\Omega N}\sum_{\mathbf{k}}\sum_{c,v}\frac{\mathbf{p}_{cv}(\mathbf{k})\mathbf{p}^*_{cv}(\mathbf{k})}{[E_c(\mathbf{k}) - E_v(\mathbf{k})]^2}F_{cv}(\omega, \mathbf{k}), \tag{15.8}$$

where the indices c and v run over conduction (empty) and valence (filled) electron states, respectively. The spectral function is given by

$$F_{cv}(\omega, \mathbf{k}) = \left\{\frac{1}{\omega + i\eta - E_c(\mathbf{k}) + E_v(\mathbf{k})} + \frac{1}{-\omega - i\eta - E_c(\mathbf{k}) + E_v(\mathbf{k})}\right\}. \tag{15.9}$$

The constant coefficient in Eq. (15.8) appears after summation over volume Ω of the homogeneous ambient of noninteracting N dipoles [the random phase approximation (RPA)]. A more general case describing the effect of the ambient in terms of dynamic interaction with surrounding light-induced dipoles (the LF effect) is addressed in Sec. 15.3. The full potential energy in materials can be separated into two parts, the external and induced potentials

$$V(\omega) = V_{ext}(\omega) + V_{ind}(\omega). \tag{15.10}$$

Equation (15.10) can be understood as a reduction (screening) of external potential through the induced charge in materials. This can be presented in terms of the dielectric function

$$V_{ext}(\omega) = \hat{\varepsilon}(\omega)V(\omega). \tag{15.11}$$

Tensor components of the dielectric function $\varepsilon(\omega)$ can be expressed now in terms of the polarization function [see also Eq. (15.96) in Appendix II]

$$\varepsilon_{\alpha,\beta}(\omega) = \delta_{\alpha,\beta} - 4\pi P_{\alpha,\beta}. \tag{15.12}$$

Equations (15.8) and (15.12) describe the optical polarization and dielectric functions. The imaginary part of Eq. (15.12) corresponds to the widely used *golden rule* formula.[26] In the plane-wave basis [see Eq. (15.6)] and in the limit of $\mathbf{q} \to 0$, Cartesian components of momentum matrix elements are given by

$$p^{cv}_{\alpha}(\mathbf{k}) = \sum_{\mathbf{G}'} d^*_{c\mathbf{k}}(\mathbf{G}')d_{v\mathbf{k}}(\mathbf{G}')(k_\alpha + G_\alpha). \tag{15.13}$$

Equation (15.13) represents a plane-wave evaluation of the momentum operator. Equations (15.8) and (15.12) are derived by evaluation of the induced charge dipole (and induced polarization) as the response to the external electric field. Equation (15.12) defines the *longitudinal* dielectric function. This is in contrast to the microscopic definition of *transversal* dielectric function which can be derived from the analysis of the induced current. The induced current is parallel to the electric field of light and therefore perpendicular (transversal) to the penetration of light.[27]

For historical reasons the definitions of longitudinal and transversal relate to the formulation of perturbation in quantum theory: if perturbation is caused by charge displacements due to momentum transformations (which are parallel to the wave propagation) the resulting function is called longitudinal. If the perturbation is taken as a current induced by the electric field component of the light (which is the perpendicular to the wave propagation) the calculated function is defined as transversal.[27] The longitudinal dielectric function determines the optical response of electron plasma. The transversal function determines the optical response to a transverse electric (TE) electromagnetic wave, but for a transverse magnetic (TM) wave, the contributions of both longitudinal and transversal dielectric functions are important. In the wavelength regions where the wavelength of the light is comparable with the characteristic dimensions of the elementary excitations (x ray for an electronic part or visible light for excitons), the difference between the two functions is significant. However, for electronic excitations in the visible and near-ultraviolet (UV) and/or infrared (IR) spectral regions, both definitions for the dielectric function are equivalent.[28]

15.3 Effect of the Local Field in Optics

The LF concept plays a very important role in the optics of nanomaterials. It can be understood as an interaction of the local light-induced dipoles with other dipoles from the ambient. In most cases the effect of the ambient results in additional screening and lowering of the electric field inside the materials.[29] For example, in solids in the static regime the dielectric constant decreases by 10 to 15% due to the LF effect.[28] However, the situation changes dramatically in the dynamic regime, in particular when the interaction with the ambient is resonant.[28] In that case the LF effect is substantially enhanced. Typical examples are systems where optically induced excitations are in resonance with plasmons.[13] In metallic nanoparticles the LF generated by the optical plasmons results in a dramatic enhancement of the radiated optical field that was observed in luminescence.[2,30] Another example is SERS (see Sec. 15.6). In this case the enhancement of the optical response by several orders of magnitude was first measured experimentally and then explained theoretically as an LF effect due to surface plasmon resonance.[31] The SERS phenomena is addressed in Secs. 15.6.2 and 15.6.3.

15.3.1 Local field effect in classical optics

It is instructive to consider first the LF effect within classical electrodynamics. The optical response described in terms of induced charge [Eq. (15.7)] and dielectric function [Eq. (15.12)] is determined through the total electric field, which includes the electric field of the light and the secondary electric field from the induced dipoles of surrounding atoms. Within the macroscopic theory this local electric field is given by[29]

$$\mathbf{E}_{loc} = \mathbf{E} + \left(\frac{4\pi}{3} + s \right) \mathbf{P}. \tag{15.14}$$

The factor s in Eq. (15.14) stands to account for the symmetry and differs from zero for symmetries other than cubic.[32] By calculation of the optical dielectric function one can take the field [Eq. (15.14)] instead of using only the electric field of incoming light. This will result in the well-known Clausius-Mossotti (Lorentz-Lorentz) formula for the dielectric function.[33] For a noncubic environment the coefficient of the polarization vector in Eq. (15.14) will be different. A detailed discussion of the derivation of the Clausius-Mossotti formula for different structures is given in several books.[26,33] In most cases this kind of correction to the LF leads to a reduction of the predicted static dielectric function.

Resonant excitation of plasmons can induce giant electromagnetic field enhancements in metallic nanostructures.[2] The incident light excites the plasma resonance in nanoparticles (or metallic clusters). The induced electric fields in every particle interact with each other, resulting in a strong LF enhancement. In addition, the field fluctuations in the randomly distributed array provide further contribution to the LF. The dipole-dipole interactions in inhomogeneously distributed arrays of nanoparticles (fractals) are not long range, and the optical excitations are strongly localized within the small areas (called *hot spots*) due to the fluctuations. These processes result in overall strong field enhancement in nanostructures that agrees very well with experiment.[2] The LF drastically enhances the cross section for optical spectroscopies such as SERS (see Sec. 15.6) and surface-enhanced infrared absorption (SEIRA).[34,35]

15.3.2 Optical local field effects in solids from first principles

In this section we consider the generalized evaluation of the dielectric susceptibility and the polarization functions by taking into account the LF effect. It is important to note that the LF are considered here on the atomic scale (in contrast to the previous section). Description of the LF effects is limited by optical excitations of bounded electrons (the LF effects caused by the free electron excitations or plasmons are well documented in literature[2,34,35]). In the independent particle picture (the RPA approximation) the microscopic formulation of the dielectric function neglects the effect of the neighboring induced dipoles, when considering for simplicity the induced charge (or current) only locally. The induced dynamic dipoles and their phases are taken independently (randomly). This is a key assumption of the independent particle picture or RPA in optics.[27] In the microscopic theory of optics based on standard DFT, the charge density is calculated within the RPA. Microscopic modeling of optical functions including the LF effect is a step beyond the standard DFT, which requires more extensive computational effort than the RPA model even for bulk cubic materials. Optics including the LF effects using first principle theory has been extensively studied in the last decade.[28,36,37] It has been shown[28,37] that the LF effect for bulk cubic semiconductors results in a relatively small (10 to 15%) reduction in static (macroscopic) dielectric function. However, in the region of bounded electron resonances the LF contribution may increase by up to 50%.[28] One might expect further enhancement of the LF effect in the bounded electron region with a decrease in the dimensions of the materials due to electron resonances. Here we present the evaluation of optical functions from first principles including the LF effect.

In the previous section an expression for the dielectric function [see Eq. (15.12)] was derived assuming a homogeneous ambient and by neglecting the interaction between surrounding light-induced charges. In this section the effect of this interaction (the LF effect) on the optical response is considered explicitly. Following the Adler-Wiser approach[38,39] all of the formulae are given in the $\mathbf{q} \to 0$ limit, neglecting the effects of spatial dispersion. More general cases have been studied.[28] In order to determine the LF effect on the optical dielectric function, it is convenient to present the dielectric susceptibility function [Eq. (15.12)] in the form of a square matrix

$$\tilde{\varepsilon}_{\mathbf{G},\mathbf{G}'}(\omega) = \begin{vmatrix} \varepsilon_{0,0}(\omega) & \varepsilon_{0,\mathbf{G}'}(\omega) \\ \varepsilon_{\mathbf{G},0}(\omega) & \varepsilon_{\mathbf{G},\mathbf{G}'}(\omega) \end{vmatrix}. \tag{15.15}$$

Evaluation of the components of the full dielectric matrix in Eq. (15.15) is explained in Appendices II and III. Here we provide the main results and focus primarily on the physics of this phenomenon. The leading diagonal element of the dielectric matrix in Eq. (15.15) is given by Eqs. (15.8), (15.12), and (15.9). Here we rewrite it in the form of the element of matrix Eq. (15.15)

$$\varepsilon_{0,0}(\omega) = 1 - \frac{8\pi}{\Omega_0}\frac{1}{N}\sum_{\mathbf{k}}\sum_{c,v}\frac{\mathbf{p}_{cv}(\mathbf{k})\mathbf{p}^*_{cv}(\mathbf{k})}{[E_c(\mathbf{k}) - E_v(\mathbf{k})]^2}F_{cv}(\omega,\mathbf{k}). \tag{15.16}$$

The normalization quantity Ω_0 has the dimension of volume in direct space. For properly defined wave vectors and summation limits this value is equal to 1. $\varepsilon_{0,0}$ represents the *local* part of the dielectric function. The LF effect is an essentially nonlocal contribution to ε. Another nonlocal effect that contributes to optical momentum and to predicted optical functions is the nonlocality of potential energy, which is discussed in earlier studies.[40,41] The off-diagonal elements and another diagonal element of the full dielectric matrix in Eq. (15.15) are given by

$$\varepsilon_{0,\mathbf{G}'}(\omega) = \frac{8\pi}{\Omega_0}\frac{1}{N}\sum_{\mathbf{k}}\sum_{c,v}\frac{B_{cv}^{\mathbf{kk}'}(\mathbf{G}')\mathbf{p}_{cv}^{*}(\mathbf{k})}{E_c(\mathbf{k})-E_v(\mathbf{k})}F_{cv}(\omega,\mathbf{k}), \qquad (15.17)$$

$$\varepsilon_{\mathbf{G},0}(\omega) = \frac{8\pi}{\Omega_0}\frac{1}{N}\sum_{\mathbf{k}}\sum_{c,v}\frac{\mathbf{p}_{cv}(\mathbf{k})B_{cv}^{*\mathbf{kk}'}(\mathbf{G})}{E_c(\mathbf{k})-E_v(\mathbf{k})}F_{cv}(\omega,\mathbf{k}), \qquad (15.18)$$

$$\varepsilon_{\mathbf{G},\mathbf{G}'}(\omega) = \frac{8\pi}{\Omega_0}\frac{1}{N}\sum_{\mathbf{k}}\sum_{c,v}\frac{B_{cv}^{\mathbf{kk}'}(\mathbf{G}')B_{cv}^{*\mathbf{kk}'}(\mathbf{G})}{E_c(\mathbf{k})-E_v(\mathbf{k})}F_{cv}(\omega,\mathbf{k}). \qquad (15.19)$$

Evaluation of Bloch integrals $B_{cv}^{\mathbf{kk}'}$ is given by Eqs. (15.98) and (15.99) in Appendix III. Equation (15.16) represents the dielectric function in RPA neglecting the LF effect.[27,42,43] In order to incorporate the LF effect into optical functions one needs to compute the full dielectric matrix in Eq. (15.15) including the off-diagonal [Eqs. (15.18) and (15.19)] and two diagonal [Eqs. (15.16) and (15.19)] elements.[28,37] The final value of the macroscopic dielectric function, including the LF effect, is computed according to[28]

$$\varepsilon^{M}(\omega) = \varepsilon_{0,0}(\omega) - \sum_{\mathbf{G},\mathbf{G}'\neq 0}\varepsilon_{0,\mathbf{G}}(\omega)\varepsilon^{-1}_{\mathbf{G},\mathbf{G}'}(\omega)\varepsilon_{\mathbf{G}',0}(\omega). \qquad (15.20)$$

The quantity $\varepsilon^{-1}_{\mathbf{G},\mathbf{G}'}$ is the inverse matrix of the dielectric matrix block given by Eq. (15.19). For practical computations of the optical functions including the LF effect, one needs to calculate all of the elements of the square matrix in Eq. (15.15) including a sufficient number of the reciprocal lattice values. This number is determined by the convergence of the macroscopic dielectric function.[28] This is illustrated by Fig. 15.1 for silicon.

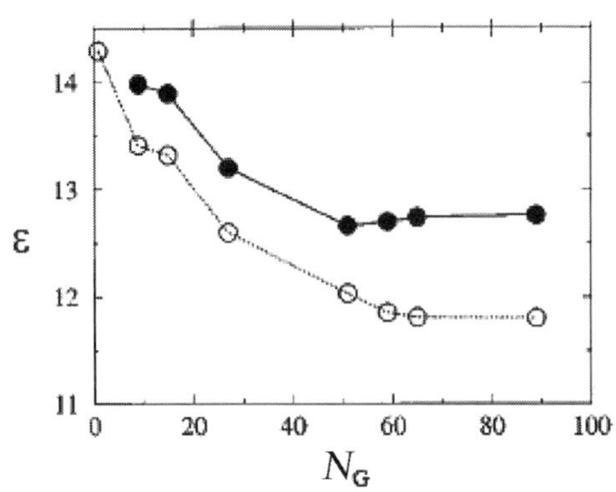

Figure 15.1 Macroscopic dielectric constant of silicon versus the number of reciprocal lattice vectors. Circles indicate the effect of LF correction, and dots show the effects of additional nonlocal (exchange-correlation) contributions. (Reprinted from Ref. 28.)

Results presented in Fig. 15.1 show rather slow convergence of the dielectric function with the number of reciprocal lattice vectors, which in low-dimensional systems requires even more lattice vectors. The important conclusion following from the above consideration is that LF contributions have a strong frequency dependence. The LF corrections in the static region are around 10 to 15% (see Fig. 15.1). The situation changes dramatically if the external optical field approaches induced dipole resonances. In the regions of strong electron resonances corresponding to the direct electronic transitions (e.g., the E_1 and E_2 transitions in Si), the LF contributions can reach 50% and higher.[28] Reduction of the symmetry in low-dimensional systems leads to a substantial increase of the number of nonzero elements in the full dielectric matrix Eq. (15.15). Consequently, the LF contributions in low-dimensional nanomaterials are expected to be further enhanced by the reduction of symmetry.

Strong enhancement of optical response due to the LF effect was observed on metallic nanoparticles and nanostructured materials in the spectral region, corresponding to surface plasmon resonances.[2,7,30] Plasma excitations in bulk solids are normally located in the ultraviolet spectral region and do not interact with visible light. However, due to confinement the effective mass of the plasmon increases, moving its excitation frequency to the visible range.[13] This very interesting phenomenon has both a fundamental aspect and numerous applications that are addressed in several recent reviews and books.[2,7,44] In Secs. 15.6.2 and 15.6.3 we address a part of this problem that is relevant to the subject of the present chapter: the enhancement of Raman scattering efficiency due to various induced dipole resonances by surface plasmons (electromagnetic mechanism) and by molecular adsorption due to electronic structure modifications and redistributions of oscillator strength (chemical mechanism).

15.4 Electrons in Quantum Confined Systems

One of the most important features affecting optical properties of nanostructured materials is the confinement of electron motion in real space.[9,13] Study of this phenomenon was started with the creation of quantum mechanics (the classic *particle-in-the-box* problem) and continued for more complex systems with the further development of quantum theory. For tutorial purposes we consider a few relevant examples in order to illustrate modifications of the optical functions of nanoparticles due to confinement.

15.4.1 Electron energy structure in quantum confined systems

It is instructive to start first with the well-known 1D system of the particle-in-the-box. Wave functions of an electron in infinitely high 1D potential well of the thickness L are given by[45]

$$\psi_n(x) = \sqrt{\frac{2}{L}} \sin \frac{\pi n x}{L}, \qquad n = 1, 2, 3 \ldots \ . \tag{15.21}$$

Solid slabs are the models most frequently used to study electron energy structure and optical properties of surfaces.[46] Electrons in such a thin slab are confined in one dimension thus showing quantum behavior. It is instructive to compare predictions made by analytic theory with direct computations. In Fig. 15.2 spatial distributions of electron density are shown for a 1D infinitely high potential well and for silicon slabs of different thicknesses.[47] Numerical calculations presented in Fig. 15.2 were performed for quantum Si-slabs having thickness of 8, 16, and 28 monolayers using the linear combination of atomic orbitals (LCAO) method. Calculated equilibrium thickness of one monolayer in bulk Si is 1.36 Å. Due to the atomic reconstruction on the monohydride Si(001)(2$\times$1) surface [shown in Fig. 15.2(b)] the interlayer distances change slightly near the surfaces. Numerical results clearly indicate that the surface-confined states localized within a few monolayers near the surfaces. Other *bulk-like* electron states are quantized along the z direction (perpendicular to the slab surfaces). These bulk-like states are reorganized into *cosine*- and *sine*-like states, in agreement with the analytic model in Fig. 15.2(a). In bulk Si, the top of the valence band is determined by the three-fold degenerate electrons. Due to the reduction of the symmetry in the slab, one state is split-off; however, two others are still doubly degenerate [see top-three highest occupied molecular orbital (HOMO) states in Fig. 15.2(b)]. The slight difference

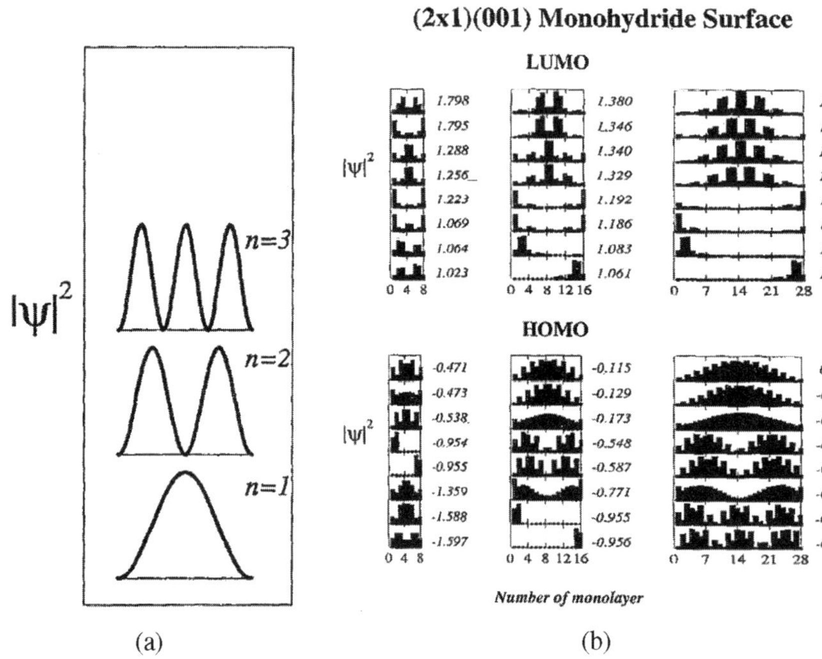

Figure 15.2 Spatial distribution of electron density in (a) an infinitely high 1D quantum well, and (b) within a silicon slab of different thickness. Right-scale numbers in panel (b) indicate absolute electron level energies (in eV) with respect to the top valence level of a 28-layer slab. (Reprinted from Ref. 47.)

in the energy positions of the nearly degenerate top HOMO states indicates the accuracy of eigenvalue convergence. (Note the misprint for the second HOMO top value of a 16-layer slab: the correct value is -0.119 eV).

Other important observations are the energy shifts of electron states due to quantum confinement [see Fig.15.2(b)]. One can see a substantial increase of the gap between HOMO and LUMO (lowest unoccupied molecular orbital) with reduction of the slab thickness due to confinement. On the other hand, the electrons localized on the surface are much less sensitive to quantum confinement [see surface states located near -0.9, 1.08, and 1.2 eV in Fig. 15.2(b)]. These states correspond to the Si-Si back bonds reconstructed on the surface. Other surface states corresponding to the Si-H bond are pushed away from the Fermi level because of the high bonding energy of the Si-H bond.[46]

15.4.2 Optical functions of nanocrystals

Within the last decade, the optical function of different nanostructures has been extensively studied from first principles.[48,49] Earlier studies were focused mainly on electron energy structure modifications due to quantum confinement.[9] More recent first principles calculations also predict redistributions of oscillator strengths. The direct computations from first principles of both eigenenergies and eigenfunctions allow detailcd understanding of the optical properties of nanostructures.[5,11,49,50] Such a modeling allows us to understand the evolution of an optical response from single atom to clusters and quantum dots.[11,49]

This section focuses on the optical properties of quantum dots (QDs) as a typical example of nanostructures. In contrast to bulk materials, optical spectra of QDs are characterized by quantum confinement, contribution of the surface, and redistribution of the oscillator strengths. Optical functions of Si and Ge QDs studied by the local density approximation (LDA) method within the DFT using *ab initio* pseudopotentials are strongly affected by quantum confinement.[5] The atomic structures of QDs are shell-like and are used for calculations of optical absorption spectra as shown in Fig. 15.3. Starting from a central Si (or Ge) atom the nanocrystals are built up following ideal tetrahedral configuration. The dangling bonds on the surfaces are passivated with hydrogen (the capping effect of the hydrogen was not considered). The free-standing nanocrystals with 17 or 41 Si (or Ge) atoms in the core (three or four core shells, respectively) with up to three capping shells were considered. The QDs studied contained 191 ($Si_{41}Ge_{42}H_{108}$), 295 ($Si_{41}Ge_{106}H_{148}$), and 459 ($Si_{41}Ge_{198}H_{220}$) atoms and had average diameters varying from 15 to 23 Å.[5]

These QD models allowed for study of the interplay between the effects of quantum confinement and changes in the chemical nature of atomic bonds in the optical absorption spectra. Optical spectra were calculated within the independent particle picture using basic DFT-LDA formalism close to that described above in this chapter.[5] Results are presented in Fig. 15.4. The lower panels of Fig. 15.4 demonstrate the effect of quantum confinement: increasing the average diameter of

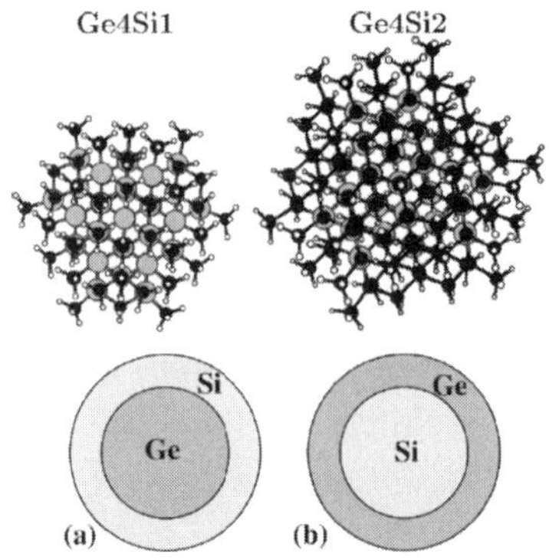

Figure 15.3 Structure of Si-capped Ge nanocrystals with one (Ge_4Si_1) and two (Ge_4Si_2) Si capping shells. The core of the structures contains 41 atoms (four shells). In the lower panels schematic representations of a Si-capped Ge- and Ge-capped Si nanocrystals are shown. [Reprinted with permission from Ref. 5. © (2005) by the American Physical Society.]

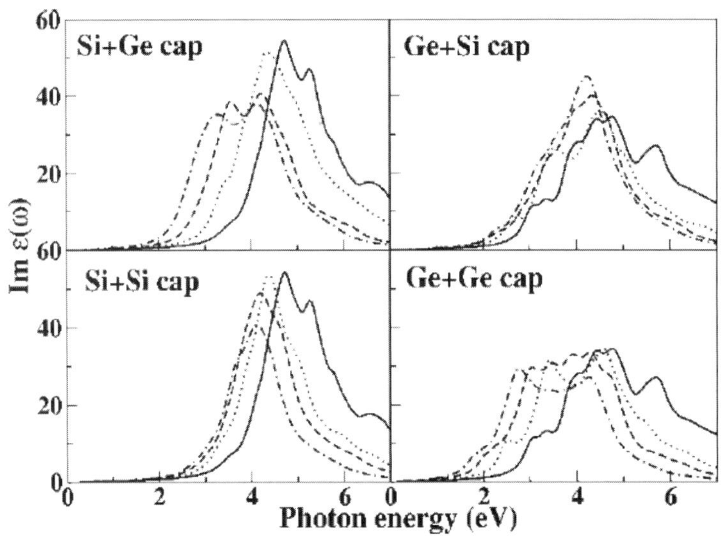

Figure 15.4 Optical absorption spectra of nanocrystals with Si (left panels) and Ge (right panels) core atoms. The upper panels correspond to capped structures, whereas the lower panels show spectra of uncapped nanocrystals. Different line spectra of 41 core atom nanocrystals without capping (solid), as well as with one (dotted), two (dashed), and three (dot-dashed) capping shells are shown for comparison. [Reprinted with permission from Ref. 5. © (2005) by the American Physical Society.]

QDs causes a red shift of the optical absorption spectra. However, if the nanocrystals are covered with atoms distinct from the core atoms, the red shift can be either enhanced or reduced. See the left or right upper panels of Fig. 15.4, respectively. These results clearly demonstrate the contributions of the surface chemical bond changes to the optical absorption of nanocrystals. They highlight the ways in which optical absorption engineering of nanoparticles can be used for future applications in nanophotonics.

The contribution of the interatomic interaction on nanoparticle surfaces due to molecular adsorption has been determined from light emission studies of these systems: metallic nanoparticles and organic dye molecules.[30,35] It has been demonstrated that adsorption of rhodamine 6G organic dye molecules on gold nanoparticles can lead to substantial (up to an order of magnitude) enhancement of the luminescence due to the electronic processes on the surface.[30]

The LF effect considered in Sec. 15.3 can substantially modify the optical properties of nanomaterials.[51,52] Accurate prediction of the LF effect in nanostructures is important for realistic modeling in nano-optics. It has been demonstrated that in addition to the plasmons, the resonances of bound electrons should be considered for detailed understanding of nanomaterials optics.[32,51] For the real system this problem is extremely complex. Within the first principles theory, the LF effects in optics were considered only for bulk materials.[12,28] On the other hand, the model two-level system incorporating optical excitations of bound electrons helps explain how the nature and basic features of the optical response from nanostructures (e.g., QD array) relates to the LF. In such systems the LF effect can be considered as an additional coupling (interaction) between induced dynamic dipoles of the nanoparticle and the ambient.[51] The effect of the LF is incorporated into the first-order perturbation term of the equation of motion [see Eq. (15.70)]. The dipole-dipole interaction within this model can be understood as a depolarization contribution in the Hamiltonian of the system. Solution of the equation of motion as described above shows changes in the resonance energy of the two-level system. Using a time-dependent technique it has been demonstrated that the ground state (which determines optical absorption) shows a blue shift of the resonant frequency, but the excited state (which characterizes optical emission) shows a red shift.[51] Note that in contrast to QDs, the LF effect in bulk materials does not show significant changes of electron resonance frequencies and normally results in redistribution of the oscillator strength.[12,28] The physical nature of the LF effect in QDs is the same as the effect of dipole-dipole intermolecular interaction on optical absorption of the large organic molecular aggregates (cf. dye with a built-in static dipole moment), where the most energetically favorable H-type (with antiparallel dipoles) configuration results in the predicted and measured blue shift of the optical absorption spectra.[53]

15.5 Cavity Quantum Electrodynamics

The previous paragraphs addressed nanostructured materials optics as a response to incoherent electromagnetic radiation (random optical field). In that case the optical properties of nanostructures are determined by the interaction of quantized electrons (or more generally of quasiparticles) with randomly incoming uncorrelated optical photons.

The material's interaction with a coherent electromagnetic (EM) field is the subject of QED, a science well developed since the discovery of the laser. Exciting possibilities for fundamental and applied research open up the possibility of studying optical phenomena when both electrons and interacting electromagnetic field are quantized, i.e., when the optical field is reconfigured due to internal resonances within cavities. Interactions of single atoms with electromagnetic fields in a microresonator (cavity) have been a central paradigm for the understanding, manipulation, and control of quantum coherence and entanglement.[17] Fundamentals of QED when a quantized light field interacts with atoms and molecules are well presented in the literature at both tutorial and more advanced levels.[15,16] QED is outside the scope of the present chapter. On the other hand, a part of QED, cavity QED (CQED), addresses interactions between materials and the quantized electromagnetic modes inside a cavity. Recent studies of optical phenomena related to the interactions of a quantized light field and nanostructures (e.g., QDs) show very interesting prospects for future fundamental and applied research in this field. In view of the relevance of this area to the subject of the present chapter, the related results are briefly reviewed in Sec. 15.5.2.

15.5.1 Interaction of a quantized optical field with a two-level atomic system

In the optical version of CQED one drives the cavity with a laser and monitors changes in the cavity transmission resulting from coupling to atoms falling through the cavity. One can also monitor the spontaneous emission of the atoms into transverse modes not confined by the cavity. It is not generally possible to directly determine the state of the atoms after they have passed through the cavity because the spontaneous emission lifetime is on the scale of nanoseconds. One can, however, infer information about the state of the atoms inside the cavity from real-time endform monitoring of the cavity's optical transmission.

Of particular importance is the regime of strong coupling, characterized by a reversible exchange of excitation between an atom and cavity fields. This coherent exchange can induce atom-photon, atom-atom, and photon-photon entanglement[15–17] and plays a key role in many quantum information processes. Strong coupling CQED has been realized with trapped atoms[19,20] and with solid state systems using QDs as artificial atoms.[21–24]

Interaction of a quantized light field with atoms can be described from first principles. Below we consider the basic ideas of CQED starting with the well-known model [the Jaynes-Cummings Model (JCM)] describing the interaction of a quantized light field with a two-level atomic system.[54,55]

The intensity of a coherent electromagnetic field in a microresonator (cavity) is spatially redistributed due to the resonances. It is convenient to represent the Hamiltonian of this quantized optical field in terms of second quantization operators, as described in Appendix IV. For the two-state atom (characterized by g and e

states) interacting with the quantized field of an optical cavity given by Eq. (15.116) in Appendix 15.8.4, the full Hamiltonian takes the form[17]

$$H = \hbar\omega_{eg}\sigma_z + \hbar\omega\left(\hat{a}\hat{a}^\dagger + \frac{1}{2}\right) - i\frac{1}{2}\hbar\Omega f(x)\left(\sigma_+\hat{a} - \sigma_-\hat{a}^\dagger\right), \tag{15.22}$$

where σ_z, σ_+, and σ_- are the Pauli matrices of the atomic pseudospin, the atomic coupling constant $\Omega/2$ is equal to $dE_0/\hbar$, E_0 is the rms vacuum field at the cavity center, and d is the dipole element for the electronic transition between $g-$ and $e-$states.[17] The real function $f(x) = exp\left(-x^2/w^2\right)$ describes the cavity mode Gaussian field, with $x \propto vt$ giving the atomic position along the beam and $x = 0$ on the cavity axis.

Assume that an atom in level e enters the cavity initially in its vacuum state $|0\rangle$. Let the cavity mode ω be equal to the $g \Leftrightarrow e$ transition frequency ω_{eg} [see Eq. (15.22)]. Through the electronic dipole transition, the lower atom-cavity state $|g, 1\rangle$ is coupled to the excited state $|e, 0\rangle$, which is described by the last term in Eq. (15.22).

Due to the coupling in the cavity, the probability P_e for detecting the atom in e state is now an oscillating function of the effective interaction time t_i. These oscillations are called *vacuum Rabi oscillations*.[15,16] Consider an atom at cavity center $x = 0$ with $v = 0$. If the system starts from $|e, 0\rangle$ state at time $t = 0$, its time-dependent state is given by[17]

$$|\Psi_e(t)\rangle = \cos\left(\frac{\Omega t}{2}\right)|e, 0\rangle + \sin\left(\frac{\Omega t}{2}\right)|g, 1\rangle. \tag{15.23}$$

If the system starts from $|g, 1\rangle$, its time dependence is given by

$$|\Psi_g(t)\rangle = \cos\left(\frac{\Omega t}{2}\right)|g, 1\rangle - \sin\left(\frac{\Omega t}{2}\right)|e, 0\rangle. \tag{15.24}$$

Equations (15.23) and (15.24) generally describe a time-varying entanglement between the atomic and cavity systems. This is a pure quantum effect that could be interpreted as a superposition principle applied to the atom-cavity system.

Rabi oscillations were measured experimentally[18] using *Rydberg atoms*. A Rydberg atom is an atom whose valence electrons are in states with a very large principal quantum number n. In such an atom, the many core electrons effectively shield the outer electron from the electric field of the nucleus. As a result, the outer electron generally interacts with a nucleus as with a proton. Consequently, it will behave much like the electron of a hydrogen atom. These atoms have a number of peculiar properties, including an exaggerated response to electric and magnetic fields, long decay periods, and electron wavefunctions that approximate under some conditions classical orbits of electrons about the nuclei.

The experimental setup is shown in Fig. 15.5. The velocity-selected Rb atomic beam is generated in oven O. Circular Rydberg atoms are produced in zone B.

Atoms cross the superconducting cavity C and are detected by the field ionization detector D. In order to minimize thermal noise the experimental setup is cooled down to 1 K.[17] Circular Rydberg states are the electronic orbitals with large principal n and maximum orbital l and magnetic m quantum numbers. The electronic transitions between neighboring circular states correspond to the millimeter-wave domain for $n = 50$. Experiments[18] were focused on states with $n = 49$, 50, and 51 denoted in the insert of Fig. 15.5 as i, g, and e, respectively.

The cavity was an open Fabry-Perot resonator, made of two high-quality polished spherical niobium mirrors. Each mirror had a diameter of 50 mm and a radius of curvature of 40 mm. The distance between mirrors was 27 mm. The cavity sustained a Gaussian TEM_{900} mode of frequency ω that was in resonance with $e \Leftrightarrow g$ transitions. The atoms could be prepared in a configuration characterized by the superposition of energy states before the interaction with the cavity C and mixed again after the interaction. This setup made it possible to prepare and analyze complex entangled states.[17]

The quantum Rabi oscillations in a vacuum were studied[18] by measuring the probability $P_e(t)$ that the atom remains in level e at time t_i. The results are shown in Fig. 15.6. The interaction time t_i is determined through the atomic velocity v and Gaussian mode waist w accordingly to $t_i = \sqrt{\pi}w/v$.

The probability $P_e(t)$ in an ideal case oscillates at frequency $\Omega/2\pi$ (vacuum Rabi frequency)[17]

$$P_e = \frac{1}{2}\left(1 + \cos \Omega t_i\right). \tag{15.25}$$

By properly chosing t_i one can obtain a variety of atom-cavity entangled states. If $\Omega t_i = \pi/2$ ($\pi/2$ Rabi rotation) the wave function of the atom-field state reads

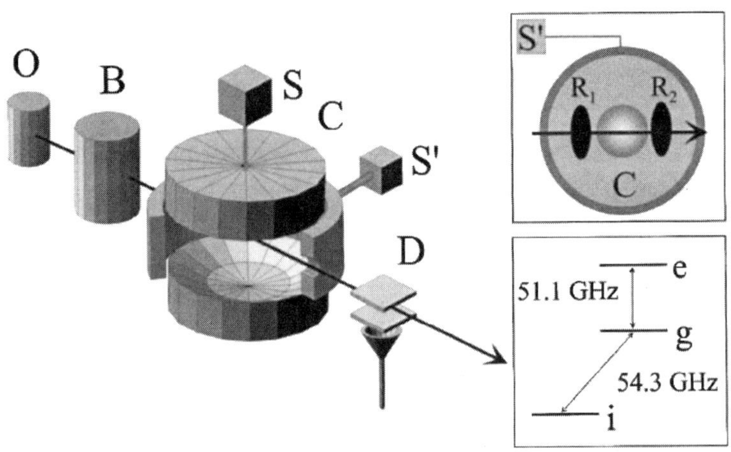

Figure 15.5 Experimental apparatus for a quantum entanglement study. [Reprinted with permission from Ref. 17. © (2001) by the American Physical Society.]

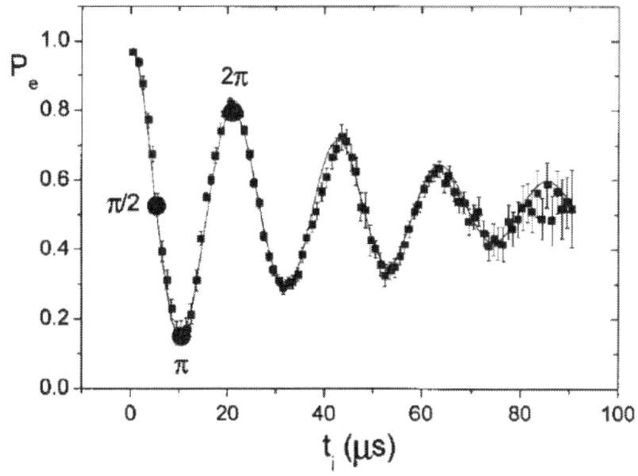

Figure 15.6 Measured vacuum Rabi oscillations. [Reprinted with permission from Ref. 17. © (2001) by the American Physical Society.]

$$| \Psi_{\pi/2}\rangle = \frac{1}{\sqrt{2}} \left(|e, 0\rangle + |g, 1\rangle \right). \tag{15.26}$$

In the entanglement experiment, atom A_1 underwent a $\pi/2$ Rabi rotation in an initially empty cavity.[17] The resulting atom-field state is described by Eq. (15.26). This entanglement persisted after A_1 left the cavity. However, detecting at some distance the atom in a given state instantaneously collapses the cavity mode in the correlated field state. In order to read out the field state one needs a second atom A_2, prepared in g state and undergoing a π-Rabi rotation in a single photon field. This atom will incorporate the cavity state including its entanglement with A_1. Thus, detecting A_1 field entanglement is equivalent to measuring the A_1–A_2 correlation.[17] These processes describe the underlying principles of quantum information and quantum computing.[17,19,23,24]

15.5.2 Interaction of a quantized optical field with quantum dots

The high degree of coherence that can be achieved in modern CQED experiments makes CQED a basis for quantum control and quantum computing. Quantum technology shows great potential for revolutionizing the methods for collecting and distributing information. In addition, a two-level system for interacting with a quantized light field in a cavity can be also considered as a simple emitter. Its basic properties can be dramatically modified if the photon lifetime (determined by the cavity quality) is long and the electric field per photon E_{vac} (responsible for coupling) is large. The radiation enhancement factor (the *Purcell* factor) is defined as the ratio of the spontaneous emission rate inside the cavity to that in free space.[24] Experiments based on a single atom injected into a high-quality cavity have achieved a strong-coupling regime of CQED as described in the previous paragraph.[18] One of the limitations in achieving a high Purcell factor is that typical cavity volumes are about four orders of magnitude larger than the fundamental limit for an optical field 3D quantization condition given by $(\lambda/2n)^3$, where n is the refractive index.

QDs embedded in photonic nanostructures overcome these limitations. As shown in Sec. 15.4 the confinement of electron motion in real space results in a discrete energy spectrum. Optical properties of QDs are determined by the electronic quantization in three dimensions (zero confinement). The optical response of QDs to an incoherent (nonquantized) photon field substantially differs from that of the bulk as demonstrated in Sec. 15.4.2. QDs are sometimes called artificial atoms. However, the confinement length in QDs extends over many lattice constants, in sharp contrast to their atomic counterparts.

When an atom is strongly coupled to a cavity mode it is possible to realize controlled coherent coupling and entanglement, which are key points for quantum information processing tasks (see Sec. 15.5.1). The QDs embedded in a cavity (the quantum dot CQED) open up new possibilities for fundamental and applied science.[24] The conventional practice is to incorporate many QDs into photonic nanostructures (such as *photonic crystals*).

Photonic crystals are periodically structured electromagnetic media, generally possessing photonic band gaps or ranges of frequency in which light cannot propagate through the structure.[3,56] Photonic crystals are composed of periodic dielectric or metallodielectric nanostructures that affect the propagation of electromagnetic waves in the same way that the periodic potential in a semiconductor crystal affects the electron motion by defining allowed and forbidden electronic energy bands. Essentially, photonic crystals contain regularly repeating internal regions of high and low dielectric constant. Ordered arrays of nanosized cavities within homogeneous dielectric solids that make up photonic crystals are extensively studied for modern applications in nanophotonics.[22] Photons (behaving as waves) propagate through this structure—or not—depending on their wavelength. Wavelengths of light (a stream of photons) that are allowed to travel are known as "modes" (for more details see Chapter 8 on photonic crystals by Hess et al.). Disallowed bands of wavelengths are called photonic band gaps. The physics of photonic crystals and its applications are fascinating.[3,56] Within this chapter we consider only one aspect of the applications of photonic crystals related to the CQED with QDs.

Even a few QDs as a gain medium could be sufficient to realize a photonic crystal laser containing high-quality nanocavities.[22] As demonstrated in Sec. 15.4.2, the electronic energy structures of QDs are characterized by both localized and quasi-continuous states. The quasi-continuous QD states become crucial since they provide an energy-transfer channel into the lasing mode, effectively leading to a self-tuned resonance for the gain medium. The lasing effect in that case occurs at an ultralow threshold.[22]

Important factors that limit QD cavity-QED practical applications are the required spatial and spectral matching of QD electronic resonances with high-quality cavity modes. The combination of a high-quality factor and extremely low mode volume that should be obtainable in point-defect microcavities makes the photonic crystal paradigm especially attractive for experiments in CQED. Optical microcavity design could be optimized for 2D photonic crystals for the purpose of strong coupling.[25] Progress in positioning a single QD in resonance with a photonic mode was recently demonstrated.[23] A buried single QD was positioned under the top half of a photonic crystal slab grown on the QD layer (see Fig. 15.7). The QDs were detected by atomic force microscopy (AFM) as 1- to 2-nm hills on the photonic crystal surface. The clear signature of a strong coupling regime in the system presented in Fig. 15.7 was confirmed by the time-resolved light emission measurements.[23] These results highlight some of the future developments of fundamental and applied research of mesoscopic physics and CQED.

15.6 Optical Raman Spectroscopy of Nanostructures

Raman spectroscopy is a powerful tool for characterization of nanostructures. Under Raman scattering one understands the inelastic scattering of light by molecular vibrations, which results in the generation of scattered light at new frequencies that are combinations of the primary photon and the vibration frequencies.[57] Today Raman spectroscopy is a standard spectroscopic tool for characterization of mate-

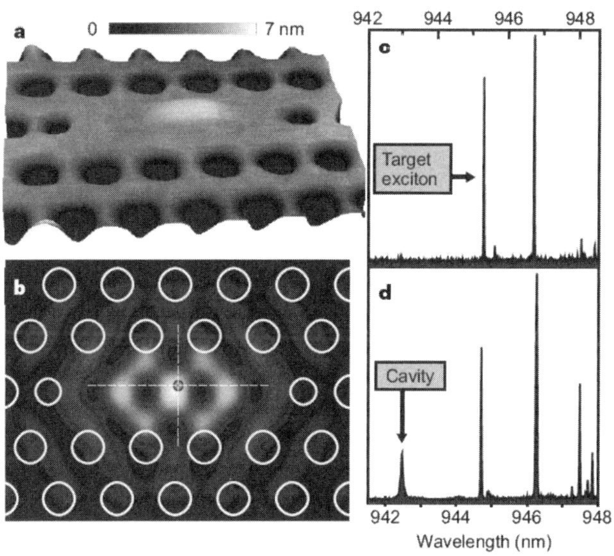

Figure 15.7 Positioning a photonic crystal cavity mode relative to a single buried QD. (a) AFM topography of photonic crystal nanocavity aligned to a hill from a single QD. Depth is depicted by the bar on the top. (b) Electric field intensity distribution of photonic cavity mode showing overlap field maximum with the QD. (c) Photoluminescence spectrum of a single QD (before cavity fabrication) showing bounded excitons excitations. (d) Photoluminescence spectrum after cavity fabrication showing emission from the cavity at 942.5 nm. [Reprinted with permission from Ref. 23. © (2007) by Nature Publishing Group.] (See color plate section.)

rials. Basic ideas of Raman spectroscopy can be understood from the following consideration.[57]

Polarization $\mathbf{P}(\mathbf{r}, t)$ of the material exposed to an external optical harmonic field $\mathbf{E}(\mathbf{r}, t)$ is determined by the susceptibility function $\chi(\omega)$ according to

$$\begin{aligned} \mathbf{P}(\mathbf{r}, t) &= \mathbf{P}(\mathbf{k}, \omega) \cos(\mathbf{kr} - \omega t), \\ \mathbf{E}(\mathbf{r}, t) &= \mathbf{E}_0(\mathbf{k}, \omega) \cos(\mathbf{kr} - \omega t), \\ \mathbf{P}(\mathbf{k}, \omega) &= \chi(\omega) \mathbf{E}_0(\mathbf{k}, \omega), \end{aligned} \tag{15.27}$$

where $|\mathbf{E}_0|^2$ is the excitation electric optical-field intensity. Normal modes of atomic vibrations are quantized to phonons. The atomic displacements associated with a phonon can be represented as a plane wave

$$\mathbf{Q}(\mathbf{r}, t) = \mathbf{Q}(\mathbf{q}, \Omega) \cos(\mathbf{qr} - \Omega t), \tag{15.28}$$

where $\mathbf{q}$ is a wave vector and Ω the phonon frequency. In the adiabatic approximation, when $\omega \gg \Omega$, the $\chi(\omega)$ function can be expanded in a Taylor series in $\mathbf{Q}(\mathbf{r}, t)$

$$\chi(\mathbf{k}, \omega, \Omega) = \chi_0(\mathbf{k}, \omega) + \left(\frac{\partial \chi}{\partial \mathbf{Q}}\right)_0 \mathbf{Q}(\mathbf{r}, t) + \ldots \quad . \tag{15.29}$$

Substituting Eq. (15.29) into Eq. (15.28), the polarization function reduces to[57]

$$\begin{aligned} \mathbf{P}(\mathbf{r}, t, \Omega) &= \mathbf{P}_0(\mathbf{r}, t) + \mathbf{P}_{ind}(\mathbf{r}, t, \Omega), \\ \mathbf{P}_0(\mathbf{r}, t) &= \chi_0(\mathbf{k}, \omega) \mathbf{E}_0(\mathbf{k}, \omega) \cos(\mathbf{kr} - \omega t), \\ \mathbf{P}_{ind}(\mathbf{r}, t, \Omega) &= \frac{1}{2} \left(\frac{\partial \chi}{\partial \mathbf{Q}}\right)_0 \mathbf{Q}(\mathbf{q}, \omega) \mathbf{E}_0(\mathbf{k}, \omega) \\ &\times \cos[(\mathbf{k} + \mathbf{q})\mathbf{r} - (\omega + \Omega)t)] \\ &+ \cos[(\mathbf{k} - \mathbf{q})\mathbf{r} - (\omega - \Omega)t)]. \end{aligned} \tag{15.30}$$

The induced part of the polarization function $\mathbf{P}_{ind}$ consists of two additional waves: a *Stokes* shifted wave with wavevector $\mathbf{k}_S = \mathbf{k} - \mathbf{q}$ and frequency $\omega_S = \omega - \Omega$, and an *anti-Stokes* shifted wave with wavevector $\mathbf{k}_{AS} = \mathbf{k} + \mathbf{q}$ and frequency $\omega_{AS} = \omega + \Omega$. The Raman frequencies ω_S and ω_{AS} are called *Stokes* and *anti-Stokes* shifts, respectively. Note that Eq. (15.30) represents one-phonon Raman frequencies. In the same way one can obtain more complex combinations of two-phonon (second-order Raman scattering) and higher-order Raman scattering.

15.6.1 Effect of quantum confinement

In Secs. 15.4 and 15.4.2 the effects of electron confinements on the optical response of nanostructures were discussed. Confinement of vibrational modes is a key point that strongly affects Raman spectra and results in their dependence on the average dimensions of the nanocrystals. Typical examples clearly demonstrating this phenomenon are the results of the Raman scattering study of PbSe QDs.[58] The size dependence of Raman spectra of PbSe QDs is shown in Fig. 15.8. The excitation wavelength was 750 nm. Size-dependent low-frequency peaks are observed in the range of the Raman shift from 8 cm^{-1} to 16 cm^{-1}. With decreasing dot size, the peak frequency increases, as does its width, reflecting the fact that the smaller dots have a larger spread of size dimensions.[58]

A theory of the vibration modes in spherical nanocrystals and related selection rules was developed previously.[59–61] According to these results, the vibrational mode within a nanosphere is classified into two categories: spheroidal modes and torsional modes. The selection rules derived from group theory dictate that the mode observed in the first-order Raman scattering was the only spheroidal mode with $l = 0$ or $l = 2$, where l is the spherical harmonic index.[58] The lowest mode spheroidal mode ($l = 0$) is known as the breathing mode, where expansion and shrinkage of the whole sphere occur. This mode is purely longitudinal, but $l = 2$ is a mixed mode and has both longitudinal and transverse components. Displacements in these modes are illustrated in the inset of Fig. 15.9.

Effects of the surface states on the nanoparticle surface were incorporated through an additional transition layer on the particle surface, thus modifying boundary conditions for vibrational modes.[59,60] The effect of the surface stress is demonstrated in Fig. 15.9. The frequencies of the spheroidal modes are expressed by $\nu_{ln} = \xi_{ln} V_l / R$, where ξ_{ln} is a coefficient depending on the ratio of longitudinal V_l and transverse V_t sound velocities, and R is the particle radius.[58] The data calculated under the assumption of zero stress due to the atomic relaxation on the surface (solid lines) agrees well with the measured frequency locations of the two spherical

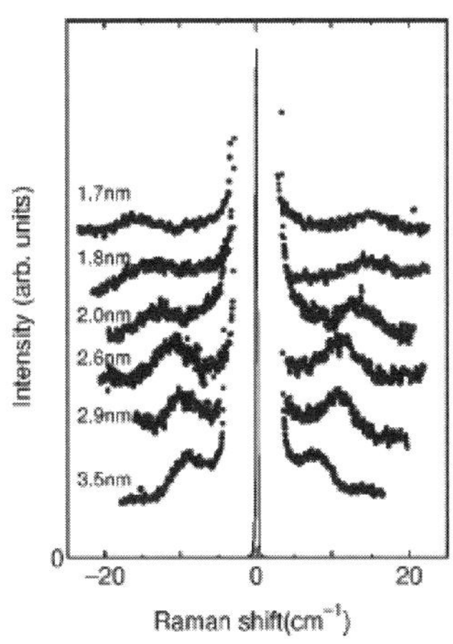

Figure 15.8 Raman spectra for various-sized PbSe QDs at room temperature. [Reprinted with permission from Ref. 58. © (2001) by American Physical Society.]

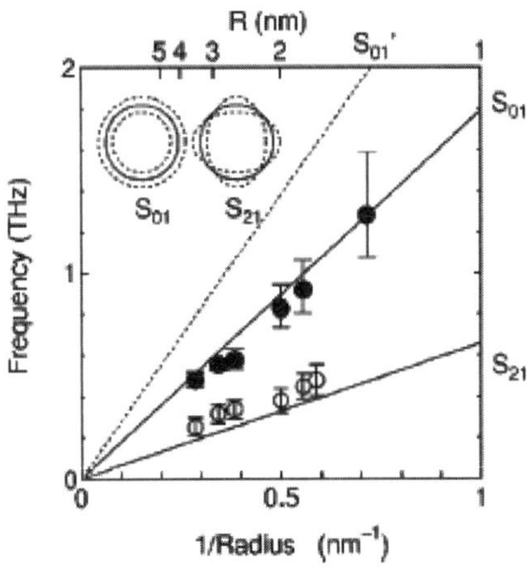

Figure 15.9 Frequencies of the Raman peaks as functions of dot size. Solid lines represent the calculation of the Raman shift including stress relaxation effect on the surface due to geometry optimization, while a dashed line represents the calculation with rigid boundary conditions. The inset represents the displacement in S_{01} and S_{21} spheroidal modes. [Reprinted with permission from Ref. 58. © (2001) by American Physical Society.]

vibrational modes S_{01} and S_{21}, which is a clear indication of the surface effects in Raman spectra of the nanoparticles.

15.6.2 Surface-enhanced Raman scattering: electromagnetic mechanism

Discovery of SERS almost three decades ago made a strong impact on modern optics.[31] The SERS effect represents a strong (several orders of magnitude) enhancement of the Raman radiation intensity of the molecules bounded to a nanostructured metallic surface (such as nanoparticles of noble metals). Several mechanisms were proposed to explain the SERS effect: electromagnetic (EM), charge transfer (CT), and chemical. The essential features of SERS can be well understood within classical electrodynamics in terms of surface plasmon resonance: the EM mechanism. In this section we focus on the SERS model based on surface plasmon resonance of the molecules and metals. This model is described in a review[34] that is recommended for advanced reading in the field. The chemical and charge transfer mechanisms are considered in the next section (15.6.3).

Consider metallic nanoparticles [characterized by the dielectric function $\varepsilon_m(\omega)$ embedded into a dielectric with a permittivity ε_d]. The SERS intensity enhancement through the LF $\mathbf{E}(\mathbf{r})$ for the system containing molecules and particles can be described by the factor[34]

$$g^R = \left(\frac{|\mathbf{E}(\mathbf{r})|^2}{|\mathbf{E}_0|^2}\right)^2, \tag{15.31}$$

where $|\mathbf{E}_0|^2$ is the excitation optical-field amplitude. Within the EM mechanism two main processes are responsible for the SERS of the molecule adsorbed on a metallic nanoparticle (or on a nanostructured rough metallic surface): the enhancement of the excitation rate producing the Raman radiation of the molecule through the increase of the LF due to plasmon resonance in the nanoparticle (considered as a dipole monomer), and additional enhancement of the Raman radiation through the interaction of the plasmon monomer fields generated by the neighboring metallic nanoparticles (in aggregates or fractals[2,34]).

Vector components ($\alpha, \beta = x, y, z$) of the local optical field responsible for generation of the Raman radiation are given by

$$E_\alpha(\mathbf{r}) = E_\alpha^0(\mathbf{r}) + \int_V G_{\alpha\beta}^r\left(\mathbf{r}, \mathbf{r}', \omega\right) \Theta(\mathbf{r}') d^3\mathbf{r}', \tag{15.32}$$

where $\mathbf{E}^0(\mathbf{r})$ is the external electric field, and the function $\Theta(\mathbf{r}) = 1$ for $\mathbf{r}$ within the particle and 0 outside the metal. Spectral representation of the retarded Green's function[43] through the LF potential eigenmodes [$\varphi_n(\mathbf{r})$ labeled by n] is given by[34]

$$G^r\left(\mathbf{r}, \mathbf{r}', \omega_R\right) = \frac{1}{s'(\omega)} \sum_n \frac{\varphi_n(\mathbf{r})\varphi_n(\mathbf{r}')^*}{\omega - \omega_n + i\Gamma_n}. \tag{15.33}$$

The field given by Eq. (15.32) is responsible for the locally induced dipole moment of the molecule $\mathbf{d}^R(\mathbf{r}_0)$ generating the primary Raman field.

The second process contributing to the SERS effect is the enhancement of the primary Raman field through the nonlocal interaction with the optical fields of the neighboring metallic nanoparticles. Vector components of the overall Raman optical field are expressed through the same Green's function of a metallic nanoparticle[34]

$$E_\alpha^R(\mathbf{r}) = \frac{4\pi}{\varepsilon_d} \int_V G_{\alpha\beta}^r\left(\mathbf{r}, \mathbf{r}', \omega_R\right) P_\beta^R\left(\mathbf{r}'\right) d^3\mathbf{r}'. \tag{15.34}$$

Taking into account the fact that Raman scattering is an incoherent (random-phase) optical process, the overall polarization of the molecular system will be a sum of single molecular-induced dipoles. Consequently, the molecular Raman polarization function $\mathbf{P}^R(\mathbf{r})$ can be expressed through the molecular dipole moment $\mathbf{d}^R(\mathbf{r}_0)$ for the molecule located at $\mathbf{r}_0$

$$\mathbf{P}^R(\mathbf{r}) = \delta\left(\mathbf{r} - \mathbf{r}_0\right) \mathbf{d}^R\left(\mathbf{r}_0\right). \tag{15.35}$$

This corresponds to the random phase approximation (RPA) in optics, which is addressed in Sec. 15.2. With the model function given by Eq. (15.35) the total Raman optical field of the single molecule adsorbed on a metallic nanoparticle including the nonlocal contribution is given by

$$E_\alpha^R(\mathbf{r}) = \frac{4\pi}{\varepsilon_d} G_{\alpha\beta}^r\left(\mathbf{r}, \mathbf{r}_0, \omega_R\right) d_\beta^R\left(\mathbf{r}_0\right). \tag{15.36}$$

In the case where many molecules are involved, Eq. (15.36) will contain a sum over the molecular array. Screening of the optical field in Eq. (15.36) is accounted for by $\varepsilon_m(\omega_R)$. The dielectric permittivity function can be taken to be of the form[34]

$$\chi_m(\omega_R) = \frac{1}{4\pi}\left[\varepsilon_m(\omega_R) - \varepsilon_d\right] = -\frac{\varepsilon_d}{4\pi s(\omega_R)}, \tag{15.37}$$

where $s(\omega_R)$ is the spectral parameter defined by Eq. (15.37). The total induced Raman dipole moment is now given by

$$\mathbf{D}^R = \mathbf{d}^R(\mathbf{r}_0) - \chi_m(\omega_R)\int_V \Theta(\mathbf{r})\,\mathbf{E}^R(\mathbf{r})\,d^3\mathbf{r}, \tag{15.38}$$

where the factor $\Theta(\mathbf{r})$ indicates that integration is performed within the metal only.[34]

The EM mechanism[2,34] explains the huge enhancement of the Raman radiation observed in metallic nanoparticles (up to 10^8) and metallic aggregates (up to 10^{12}). At the same time, the contribution of the chemical mechanism (due to the surface-molecular bonding) can be significant, which was observed in SERS experiments.[62,63]

15.6.3 Surface-enhanced Raman scattering: chemical mechanism

In the EM mechanism, the induced surface plasmon excitations (for which the surface roughness seems important) at or near the resonance gives rise to an electric field enhancement. Since the typical EM models treat the adsorbed molecule as a simple dipole or ignore it completely in accounting for the response of the metal surface, they cannot distinguish between different adsorbed molecules. Furthermore, the dipole approximation fails when the molecule has a short-range interaction with the surface. In fact, many theoretical studies show that the short-range chemical bond is essential for chemisorption. On the other hand, CT theory uses the specific ionized or affinity level of the free molecule. The resonance scattering caused by the CT transition between the surface and the adsorbed molecule is the basis of this mechanism. However, this treatment cannot explain the differences due to the orientation of the adsorbed molecule and the distance from the surface, since the interaction between the surface and the molecule is not taken into account.

The chemical (or electronic) mechanism of the SERS is due to the resonant electronic transition, in which the surface polarization and the surface-molecule interaction are important. This mechanism explains the orientation and distance dependencies of the adsorbed molecule. The Raman scattering is a linear optical two-photon process, which is described by the second-order perturbation theory for the Hamiltonian representing interaction between the electron and the radiation field. The intensity is given by[64]

$$I_{nm} = \frac{8\pi}{9c^4} I_0 (\omega_0 + \omega_m - \omega_n)^4 \sum_{\alpha,\beta} \left|\mathcal{R}_{\alpha,\beta}\right|^2, \tag{15.39}$$

where I_0 is the intensity of the incident light, ω_0 is its frequency, I_{mn} is the intensity of the scattered light with the transition from vibronic state $|m\rangle$ to $|n\rangle$, $\omega_{mn} = (\omega_0 - \omega_m - \omega_n)$ is its frequency, and c is the speed of light. The sum goes over $\alpha, \beta = x, y, z$ independently. The matrix elements of the Raman polarizability tensor $\hat{\mathcal{R}}$ (see Sec. 15.6) for the transition $|m\rangle$ to $|n\rangle$, are given by[64]

$$\mathcal{R}_{\alpha,\beta} = \sum_{l \neq m,n} \left\{ \frac{\langle m|r_\alpha|l\rangle\langle l|r_\beta|n\rangle}{\hbar\left(\omega_l - \omega_m - \omega_0\right) - i\Gamma_l} + \frac{\langle m|r_\alpha|l\rangle\langle l|r_\beta|n\rangle}{\hbar\left(\omega_l - \omega_n + \omega_0\right) - i\Gamma_l} \right\}, \quad (15.40)$$

where r_α is the α'th component of the dipole moment operator, and Γ_l is the damping constant, which is related to the lifetime τ_l of the intermediate state $|l\rangle$ by $\tau_l = \hbar/\Gamma_l$.

For practical measurements of Raman polarizabilities and scattering efficiencies one uses reference samples. In the case of Raman scattering by one longitudinal optical (LO) phonon, the counting rates R'_s outside the crystal should be corrected for absorption, reflectivity, and refractive index according to[65]

$$R'_s = \frac{T_s T_l \omega_s^3 [n(\omega_0) + 1]}{(\alpha_l + \alpha_s) n_l n_s M^* \omega_0 V_c} \frac{P'_l \Delta\Omega'}{2c^4} |\mathbf{E}_s \cdot \hat{\mathcal{R}} \cdot \mathbf{E}_i|^2, \quad (15.41)$$

where P'_l is the incident power; $\Delta\Omega'$ is the solid angle of collection outside the crystal; $\mathbf{E}_i(\mathbf{E}_s)$, $\alpha_l(\alpha_s)$, $n_l(n_s)$, and $T_l(T_s)$ denote the polarization vector, absorption coefficient, refractive index, and transmission coefficient at the frequency $\omega_l(\omega_s)$ of the incident (scattered) light, respectively; c is the speed of light in vacuum; V is the unit-cell volume; M^* is the reduced mass; and $n(\omega_0)$ is the LO-phonon occupation number.[65] Details of the polarizability tensor calculations [see Eq. (15.40)] for solids and solid nanoparticles are described in Sec. 15.2 and Appendix I; for molecular systems the procedure is given elsewhere.[40]

The SERS chemical mechanism has been studied in the system containing CO molecules adsorbed on a cluster containing 10 Ag atoms.[64] Since the d orbitals are found in Ref. 64 to be inactive for the SERS effect, the one-electron effective core potential (ECP) and the $[3s2p]$ outer electron set for the silver atoms were used except for the central Ag atom, for which the 11-electron ECP and $[3s2p2d]$ set were used. The square of the polarizability derivative (SPD) curves with $\Gamma = 0$ and $\Gamma = 0.01$ eV are shown in Fig. 15.10 by the solid and dotted-broken lines, respectively. For comparison, the SPD curves of the free CO molecule are shown by the broken lines.

In Fig. 15.10, there is a peak predicted at 2.23 eV for the SPD curve of $\Gamma = 0$ eV. This peak is broad, and the intensity is still very large even when $\Gamma = 0.01$ eV. The SPD value at the peak is about seven orders of magnitude larger than that of the free CO molecule. This peak of the SPD curve is related to the 2^1A_1 excited states of the $Ag_{10}CO$ system, whose main configuration is the inner excitation of the Ag_{10} cluster. The population analysis and the net charge of the C and O atoms

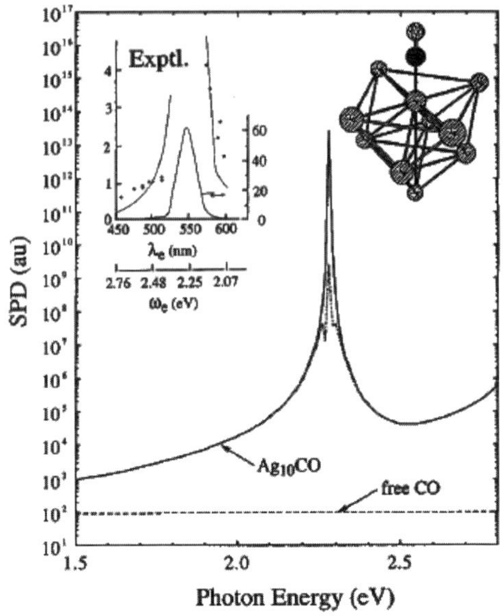

Figure 15.10 Frequency dependence of the SPD in the $Ag_{10}CO$ system. The solid and dotted lines are the results for $\Gamma = 0$ and $\Gamma = 0.01$ eV, respectively. The dotted line shows the SPD for free CO. The inserted figures show the geometry of the $Ag_{10}CO$ system and the experimental spectrum for the wavelength dependence of the Raman intensity.[62] [Reprinted with permission from Ref. 64. © (1995) American Institute of Physics.]

of $Ag_{10}CO$ in the 1^1A_1 ground and the 2^1A_1 excited states have been studied.[64] Since the net charges on C and O did not change much by the 1^1A_1 to 2^1A_1 transition, the resonance states did not correspond to the CT between CO and Ag_{10}.[64] These results showed that the SERS was mainly generated in this system due to the electron polarization of the entire metal-molecule system and that the CT provided a minor contribution to SERS.

15.7 Concluding Remarks

Recent developments in the first principles description of the optical functions of nanostructured materials have been presented to compare different methods of computational physics and chemistry. Improvements beyond the standard DFT included consideration of the effects of microscopic local fields in conjunction with realistic predictions of optical functions. Also presented were introductory overviews of new rapidly developing areas of the first principles nano-optics: CQED of nanostructures, nonlinear optics of nanostructures (surfaces and interfaces), Raman spectroscopy, and surface-enhanced Raman spectroscopy (SERS) of nanostructures.

15.8 Appendices

15.8.1 Appendix I: Electron energy structure and standard density functional theory

As stated at the beginning of Sec. 15.2, the total energy is a key function describing basic physical and chemical properties of materials in the ground state. It consists of both kinetic (describing the motion) and potential energy parts. To make the theoretical model realistic it is very important to incorporate all of the most significant contributions to both parts of the total energy. In view of the large number of particles involved in the model, a first principles theory is very challenging. Different approximations are applied in order to achieve a trade-off between complexity and accuracy. The density functional theory (DFT) is very succesful in realistic modeling of the ground state. Within this chapter we present basic ideas of the DFT and demonstrate both advantages and problems for optics using this method.

Initially, in the 1920s Thomas and Fermi (TF) suggested describing atoms as uniformly distributed electrons (negatively charged clouds) around nuclei in a 6D phase space (momentum and coordinates). This is an enormous simplification of the actual many-body problem. It is instructive to consider the basic ideas of the TF approximation before starting with a more accurate theory, the DFT. The basic ideas and results of the TF model in application for atoms are provided here.

Following the TF approach the total energy of the system can be presented as a function of electron density.[66,67] Each h^3 of the momentum space volume (h is the Planck constant) is occupied by two electrons, and the electrons are moving in an effective potential field that is determined by the nuclear charge and by assumed uniform distribution of electrons. The density of ΔN electrons in real space within a cube (nanoparticle) with a side l is given by

$$\rho(\boldsymbol{r}) = \frac{\Delta N}{v} = \frac{\Delta N}{l^3}. \tag{15.42}$$

The electron energy levels in this 3D infinite well are given by

$$E = \frac{h^2}{8ml^2}(n_x^2 + n_y^2 + n_z^2) = \frac{h^2}{8ml^2}\tilde{R}^2, \qquad n_x, n_y, n_z = 1, 2, 3, \ldots\ . \tag{15.43}$$

The radius $R = \tilde{R}_{max}$ of the sphere in the space (n_x, n_y, n_z) covering all occupied states determines the maximum energy of electrons: the Fermi energy F. The number of energy levels within this maximum value at zero temperature is given by

$$N_F = \frac{1}{2^3}\frac{4\pi R^3}{3} = \frac{\pi}{6}\left(\frac{8ml^2 F}{h^2}\right)^{3/2}. \tag{15.44}$$

The density of states is defined as

$$g(E)dE = N_F(E + dE) - N_F(E) = \frac{\pi}{4}\left(\frac{8ml^2}{h^2}\right)^{3/2} E^{1/2}dE. \tag{15.45}$$

At zero temperature all energy levels below the Fermi energy are occupied:

$$f(E) = \left\{ \begin{array}{cc} 1 & E \leq F \\ 0 & E > F \end{array} \right\}. \tag{15.46}$$

Consequently, the total energy of the electrons in one cell will be given by

$$\begin{aligned} E &= \int_0^F Ef(E)g(E)dE = \frac{4\pi}{h^3}(2m)^{3/2}l^3 \int_0^F E^{3/2}dE \\ &= \frac{\pi}{5}\left(\frac{2l}{h}\right)^3 (2m)^{3/2}F^{5/2}. \end{aligned} \tag{15.47}$$

The Fermi energy F can be obtained from the total number of electrons ΔN in a cell:

$$\Delta N = 2\int_0^F f(E)g(E)dE = \frac{\pi}{h^3}\left(\frac{2l}{h}\right)^3 (2m)^{3/2}F^{3/2}. \tag{15.48}$$

Combining Eqs. (15.47) and (15.48), the energy of the electrons in one cell is given by

$$\begin{aligned} E &= \frac{3}{10}(3\pi^2)^{2/3}\frac{l^3}{(2\pi)^2}\rho^{5/3} = C_F\frac{l^3}{(2\pi)^{5/3}} \\ C_F &= \frac{3}{10}\left(3\pi^2\right)^{2/3} = 2.871. \end{aligned} \tag{15.49}$$

In Eq. (15.49) we reverted to atomic units $e = h = m_0 = 1$. The electron density is a smooth function in real space. For systems without translational symmetry it is different for different cells. However, for spatially periodic systems it is only necessary to consider one unit cell since all unit cells are equivalent. Now adding the contributions from all cells with energies within F we obtain

$$T_{TF}[\rho] = C_F \int \rho^{5/3}(\mathbf{r})d^3\mathbf{r}. \tag{15.50}$$

Equation (15.50) represents the well-known TF kinetic energy functional, which is a function of the local electron density. The functional Eq. (15.50) can be applied to electrons in atoms encountering the most important idea of modern DFT, the local density approximation (LDA).[27] Adding to Eq. (15.50), the classical electrostatic energies of electron-nucleus attraction and electron-electron repulsion arrives at the energy functional of the TF theory of atoms:

$$\begin{aligned} E_{TF}[\rho(\mathbf{r})] &= C_F \int \rho^{5/3}(\mathbf{r})d^3\mathbf{r} - Z\int \frac{\rho(\mathbf{r})}{\mathbf{r}}d^3\mathbf{r} \\ &+ \frac{1}{2}\int \frac{\rho(\mathbf{r}_1)\rho(\mathbf{r}_2)}{|\mathbf{r}_1 - \mathbf{r}_2|}d^3\mathbf{r}_1 d^3\mathbf{r}_2. \end{aligned} \tag{15.51}$$

Note that the nucleus charge Z is measured in atomic units. The energy of the ground state and electron density can be found by minimizing the functional Eq. (15.51) with the constraint condition

$$N = \int \rho(\mathbf{r}) d^3\mathbf{r}. \tag{15.52}$$

The electron density in Eq. (15.51) has to be calculated in conjunction with Eq. (15.52) from the following equation for chemical potential, defined as the variational derivative according to

$$\mu_{TF} = \frac{\delta E_{TF}[\rho]}{\delta \rho(\mathbf{r})} = \frac{5}{3} C_{TF} \rho^{5/3}(\mathbf{r}) - \frac{Z}{\mathbf{r}} + \int \frac{\rho(\mathbf{r}_2)}{|\mathbf{r}_1 - \mathbf{r}_2|} d^3\mathbf{r}_2. \tag{15.53}$$

The TF model provides reasonably good predictions for atoms. It has been used to study potential fields and charge density in metals and the equation of states of elements.[68] However, this method is considered rather crude for more complex systems because it does not incorporate the actual orbital structure of electrons. In view of the modern DFT theory, the TF method could be considered as an approximation to the more accurate theory.

For systems such as molecules and solids much better predictions are provided by the DFT. Searching for the ground state within the DFT follows the rules that the electron density is a basic variable in the electronic problem (the first theorem of Hohenberg and Kohn[69]) and that the ground state can be found from the energy variational principle for the density (the second theorem of Hohenberg and Kohn[70]). According to the DFT the total energy could be written as

$$E[\rho] = T[\rho] + U[\rho] + E_{XC}[\rho], \tag{15.54}$$

where T is the kinetic energy of the system of noninteracting particles and U is the electrostatic energy due to Coulomb interactions. The most important part in the DFT is E_{XC}, the exchange and correlation (XC) energy that includes all many-body contributions to the total energy. The charge density is determined by the wave functions, which for practical computations can be constructed from single orbitals ϕ_j (e.g., antisymmetrized products—the Slater determinants, atomic or Gaussian orbitals, linear combinations of plane waves, etc.). The charge density is given by

$$\rho(\mathbf{r}) = \sum_j |\phi_j(\mathbf{r})|^2, \tag{15.55}$$

where the sum is taken over all occupied j orbitals. In the spin-resolved case there will be orbitals occupied with spin-up and spin-down electrons. Their sum gives the total charge density, and their difference gives the spin density. In terms of the electron orbitals, the energy components are given in atomic units as

$$T = -\frac{1}{2}\int\sum_{j}\phi_j^*(\mathbf{r})|\boldsymbol{\nabla}^2|\phi_j(\mathbf{r})d^3\mathbf{r}, \tag{15.56}$$

$$\begin{aligned} U &= -\sum_{j}^{n}\sum_{\alpha}^{N}\int\phi_j^*(\mathbf{r})\left|\frac{Z_\alpha}{(\mathbf{R}_\alpha-\mathbf{r})}\right|\phi_j(\mathbf{r})d^3\mathbf{r} \\ &+ \frac{1}{2}\sum_{i,j}\int\phi_i^*(\mathbf{r}_1)\phi_j^*(\mathbf{r}_2)\frac{1}{(\mathbf{r}_1-\mathbf{r}_2)}\phi_i(\mathbf{r}_1)\phi_j(\mathbf{r}_2)d^3\mathbf{r}_1d^3\mathbf{r}_2 \\ &+ \sum_{\alpha}^{N}\sum_{\beta<\alpha}^{N}\frac{Z_\alpha-Z_\beta}{|\mathbf{R}_\alpha-\mathbf{R}_\beta|}. \end{aligned} \tag{15.57}$$

The first term in potential energy [Eq. (15.57)] stands for the electron-nucleus attraction, the second term describes electron-electron repulsion, and the third term represents nucleus-nucleus repulsion. In Eq. (15.57), Z_α refers to the charge on nucleus α of the N atom system.

The third term in Eq. (15.54) describes the exchange and correlation energy. The LDA is rather simple for computations but is nevertheless a surprisingly good approximation, which assumes that the charge density varies slowly on the atomic scale; i.e., the effect of other electrons on a given (local) electron density is described as a uniform electron gas. The XC energy can be obtained by integrating with the uniform gas model:[71]

$$E_{XC} \cong \int\rho(\mathbf{r})\tilde{E}_{XC}[\rho(\mathbf{r})]d^3(\mathbf{r}), \tag{15.58}$$

where $\tilde{E}_{XC}[\rho(\mathbf{r})]$ is the XC energy per particle in a uniform electron gas. For many systems a good approximation provides an analytic expression for $\tilde{E}_{XC}[\rho(\mathbf{r})]$.[72] In practical calculations through minimization of the total energy [Eq. (15.54)] one selfconsistently determines the electron density and the actual XC part. A variational minimization procedure leads to a set of coupled equations proposed by Kohn and Sham:[73]

$$\left[-\frac{1}{2}\boldsymbol{\nabla}^2 - V_N + V_e + \mu_{XC}(\rho)\right]\phi_j = E_j\phi_j, \tag{15.59}$$

with

$$\mu_{XC} = \frac{\partial}{\partial\rho}\left(\rho E_{XC}\right). \tag{15.60}$$

Solution of the Kohn-Sham equation provides equilibrium geometry and the ground-state energy of the system. However, eigenfunctions and eigenenergies of the Kohn-Sham equation can not be interpreted as the quasi-particle quantities needed for optics. The definition of quasi-particle refers to a particle-like entity arising in certain systems of interacting particles. If a single particle moves through the system surrounded by a cloud of other interacting particles, the entire entity moves along somewhat like a free particle (but slightly different). The quasi-particle concept is one of the most important in materials science because it is one of the few known ways of simplifying the quantum mechanical many-body problem describing excitation state and is applicable to an extremely wide range of many-body systems.

Calculation of the ground state from the Kohn-Sham equation does not result automatically in correct prediction of excitation energies required for optics. For example, in nonmetallic systems the predicted value of the energy difference (energy gap) between highest occupied molecular orbital (HOMO) and lowest unoccupied molecular orbital (LUMO) in most cases is underestimated (gap problem). Special corrections [quasi-particle (QP) corrections] are required to obtain more accurate excitation energies.[42] Without corrections in semiconductors and insulators, the LDA substantially underestimates forbidden gap values. In this chapter we present LDA results for optics with different QP corrections, avoiding detailed analysis of theoretical methods. For more advanced reading of the DFT one can read the original papers[69–71] and monographs.[27,66,67]

15.8.2 Appendix II: Optical functions within perturbation theory

In this section we describe evaluation of the light-field-induced charge within perturbation theory using the plane-wave representation that is used in this chapter for calculations of optical functions (see Sec. 15.2). Equilibrium electron charge density is defined through the density operator (using the definition of trace Tr as a sum of the diagonal elements):

$$n_{eq}(\mathbf{r}) = eTr[\rho_0, \delta(\mathbf{r} - \mathbf{r}_0)]. \tag{15.61}$$

Without illumination, if the system is periodic (at least in one dimension) the density operator can be defined in the energy representation on a set of Bloch functions according to[43]

$$\rho|s\rangle = \rho_0|\mathbf{k}, l\rangle = f(E_{\mathbf{k},l})|\mathbf{k}, l\rangle, \tag{15.62}$$

where the equilibrium Fermi distribution function is given by

$$f(E_s) = \left(e^{\frac{F-E_s}{kT}} - 1\right)^{-1}. \tag{15.63}$$

The Bloch functions

$$|s\rangle = |\mathbf{k}, l\rangle = u_{\mathbf{k},l}(\mathbf{k})e^{i\mathbf{kr}} \tag{15.64}$$

are solutions of the unperturbed Schrödinger equation with periodic potential

$$H_0|\mathbf{k}, l\rangle = \left[-\frac{1}{2}\boldsymbol{\nabla}^2 + V_0(\mathbf{r})\right]|\mathbf{k}, l\rangle = E_{\mathbf{k},l}|\mathbf{k}, l\rangle. \tag{15.65}$$

In an external optical field, when the light quanta strike electrically neutral atoms, the equilibrium is broken through the deformation of electron clouds. Time-dependent changes of the electron charge density can be represented as a Taylor expansion. The number of the terms to be included in the Taylor sum for the induced part of the charge depends on the excitation intensity

$$\begin{aligned} n(\mathbf{r},t) &= n_{eq}(\mathbf{r}) + n_{ind}(\mathbf{r},t) = eTr[\rho,\delta] \\ &= eTr[\rho_0,\delta] + eTr[\rho^{(1)},\delta] + eTr[\rho^{(2)},\delta] + \dots \; . \end{aligned} \tag{15.66}$$

The first- and higher-order corrections to the density operator are determined from the standard perturbation theory

$$i\hbar\frac{d\rho}{dt} = [H,\rho] = (H\rho - \rho H), \tag{15.67}$$

with

$$\begin{aligned} H &= H_0 + V^{(1)} + \dots, \\ \rho &= \rho_0 + \rho^{(1)} + \rho^{(2)} + \dots \; . \end{aligned} \tag{15.68}$$

In Eq. (15.67), for simplicity we neglected the effect of the energy dissipation, which could be included through the relaxation time. Substituting Eq. (15.68) into Eq. (15.67) and equating terms of the same order on both left and right sides of the equation of motion for the density operator, Eq. (15.67) splits into a series of equations for zero, first, second, etc. orders of perturbations, respectively:

$$\begin{aligned} i\hbar\dot{\rho}_0 &= [H_0,\rho_0], \\ i\hbar\dot{\rho}^{(1)} &= \left[H_0,\rho^{(1)}\right] + \left[V^{(1)},\rho_0\right], \\ i\hbar\dot{\rho}^{(2)} &= \left[H_0,\rho^{(2)}\right] + \left[V^{(1)},\rho^{(1)}\right]. \end{aligned} \tag{15.69}$$

Dynamic optical response is described through the time-dependent density operator. In an external electromagnetic field the perturbation is harmonic, i.e.,

$$\begin{aligned} \rho(t) &= \rho(0)e^{i\omega t}, \\ i\hbar\dot{\rho}(\omega) &= -\hbar\omega\rho(\omega). \end{aligned} \tag{15.70}$$

It is convenient now to switch to the matrix representation in Eq. (15.70) by projecting the relevant quantities onto a set of Bloch functions, Eq. (15.6). To this end one should multiply every term in Eq. (15.70) by the function Eq. (15.6); the complex conjugate of Eq. (15.6) is then multiplied on the left and right sides of the relevant equation. By integrating over the entire space and by taking into account the orthonormality conditions for Bloch functions, one obtains the following expression for the first-order terms:

$$-\hbar\omega\rho^{(1)}_{ss'} = (E_s - E_{s'})\rho^{(1)}_{ss'}, + \sum_t V^{(1)}_{st}\rho^0_{ts'} - \sum_p \rho^0_{sp}V^{(1)}_{ps'}. \tag{15.71}$$

The density operator defined as Eq. (15.62) in matrix representation has the form

$$\rho_{ss'}^{(0)} = f(E_s)\delta_{ss'}. \tag{15.72}$$

Equation (15.71) is now transformed into

$$-\hbar\omega\rho_{ss'}^{(1)} = (E_s - E_{s'})\rho_{ss'}^{(1)} + [f(E_{s'}) - f(E_s)]\,V_{ss'}. \tag{15.73}$$

At zero temperature optical excitations occur between completely filled and empty states with Fermi functions equal to either 1 or 0, respectively. Consequently,

$$\rho_{ss'}^{(1)}(\omega) = \frac{f(E_{s'}) - f(E_s)}{E_s - E_{s'} - \hbar\omega}V_{ss'} = (E_s - E_{s'} - \hbar\omega)^{-1}V_{ss'}|_{T=0}. \tag{15.74}$$

For the second-order perturbation one needs to use the first-order solution [Eq. (15.74)]. Substituting it into Eq. (15.70) leads, after some algebra, to the following expressions at $T = 0$:

$$-\hbar\omega\rho_{ss'}^{(2)} = (E_s - E_{s'})\rho_{ss'}^{(2)} + \sum_{s''} V_{ss''}^{(1)}\rho_{s''s'}^{(1)} - \sum_{s''}\rho_{ss''}^{(1)}V_{s''s'}^{(1)}, \tag{15.75}$$

or

$$\begin{aligned}\rho_{ss'}^{(2)}(\omega) &= \frac{1}{E_s - E_{s'} - \hbar\omega}\sum_{s''}\left[V_{ss''}^{(1)}V_{s''s'}^{(1)}\left(\frac{1}{E_{s''} - E_{s'} - \hbar\omega}\right.\right.\\ &- \left.\left.\frac{1}{E_s - E_{s''} - \hbar\omega}\right)\right].\end{aligned} \tag{15.76}$$

Equations (15.74) and (15.76) can be now used to obtain the induced-charge density from Eq. (15.66) to the first and second orders of external perturbation, respectively. The first- and second-order contributions to the induced-charge density in Eq. (15.66) follow from Eqs. (15.74) and (15.76), respectively. In a spatially periodic system the perturbation potential is given by

$$V(\mathbf{r},t) = \sum_{\mathbf{qG}}\int_{-\infty}^{\infty} V(\mathbf{q}+\mathbf{G},\omega)e^{i(\mathbf{q}+\mathbf{G})\mathbf{r}}d\omega, \tag{15.77}$$

where $\mathbf{G}$ is a reciprocal lattice vector. The Fourier transform of the potential is

$$V(\mathbf{r},t) = \int_{-\infty}^{\infty} V(\mathbf{r},\omega)e^{i\omega t}. \tag{15.78}$$

The expansion of the potential is given by

$$V(\mathbf{r},\omega) = \sum_{\mathbf{qG}} V(\mathbf{q}+\mathbf{G},\omega)e^{i(\mathbf{q}+\mathbf{G})\mathbf{r}}. \tag{15.79}$$

In a periodic system with all equivalent atoms separated by $\mathbf{R}_i$, one has for the charge

$$n_{eq}(\mathbf{r}_0 + \mathbf{R}_i) = n_{eq}(\mathbf{r}_0), \tag{15.80}$$

with

$$n(\mathbf{r}) = \sum_{\mathbf{q}} e^{i\mathbf{qr}} n(\mathbf{q}) = \sum_{\mathbf{q}} e^{i\mathbf{qr}} \sum_{\mathbf{G}} \delta_{\mathbf{qG}} n(\mathbf{G}). \tag{15.81}$$

For the induced-charge density in Eq. (15.66) we have

$$n_{ind}(\mathbf{r}, \omega) = \sum_{\mathbf{qG}} n_{ind}(\mathbf{q} + \mathbf{G}, \omega) e^{i(\mathbf{q}+\mathbf{G})\mathbf{r}}. \tag{15.82}$$

Where the Fourier transform of the induced charge is given by the Fourier integral,

$$n_{ind}(\mathbf{q} + \mathbf{G}, \omega) = \int n_{ind}(\mathbf{r}, \omega) e^{-i(\mathbf{q}+\mathbf{G})\mathbf{r}} d^3\mathbf{r}. \tag{15.83}$$

The linear part of the induced charge in Eq. (15.83) follows from Eq. (15.66):

$$\begin{aligned} n_{ind}(\mathbf{q} + \mathbf{G}, \omega) &= e \int Tr \left[\rho^{(1)}(\omega), \delta(\mathbf{r}' - \mathbf{r})\right] e^{-i(\mathbf{q}+\mathbf{G})\mathbf{r}'} d^3\mathbf{r}' \\ &= eTr \left[\rho^{(1)}(\omega), e^{-i(\mathbf{q}+\mathbf{G})\mathbf{r}}\right]. \end{aligned} \tag{15.84}$$

The trace of the operator product is calculated according to

$$Tr\left(\hat{A}\hat{B}\right) = \sum_m \langle m|\hat{A}\hat{B}|m\rangle = \sum_m \sum_n A_{mn} B_{nm}. \tag{15.85}$$

Equation (15.84) projected onto the plane-wave basis [Eq. (15.6)] can be represented as

$$\begin{aligned} n_{ind} &= n_{ind}(\mathbf{q} + \mathbf{G}, \omega) \\ &= e \sum_{\mathbf{k}+\mathbf{q}} \sum_{\mathbf{k},l} \langle l', \mathbf{k} + \mathbf{q}|\rho_\omega^{(1)}|\mathbf{k}, l\rangle \langle l, \mathbf{k}|\delta(\mathbf{r}' - \mathbf{r})|\mathbf{k} + \mathbf{q}, l'\rangle e^{-i(\mathbf{q}+\mathbf{G})\mathbf{r}'} \\ &= e \sum_{\mathbf{k}+\mathbf{q}} \sum_{\mathbf{k},l} \langle l', \mathbf{k} + \mathbf{q}|\rho_\omega^{(1)}|\mathbf{k}, l\rangle \langle l, \mathbf{k}| \sum_{\mathbf{G}'} e^{i(\mathbf{r}-\mathbf{r}')\mathbf{G}'} e^{-i(\mathbf{q}+\mathbf{G})\mathbf{r}'} |\mathbf{k} + \mathbf{q}, l'\rangle \\ &= e \sum_{\mathbf{k}+\mathbf{q}} \sum_{\mathbf{k},l} \langle l', \mathbf{k} + \mathbf{q}|\rho_\omega^{(1)}|\mathbf{k}, l\rangle \langle l, \mathbf{k}|e^{-i(\mathbf{q}+\mathbf{G})\mathbf{r}}|\mathbf{k} + \mathbf{q}, l'\rangle. \end{aligned} \tag{15.86}$$

In Eq. (15.86) the bra–ket notation is used for the wave functions. We also use the definition of the δ function:

$$\delta\left(\mathbf{r} - \mathbf{r}'\right) = \sum_{\mathbf{G}} e^{i(\mathbf{r}-\mathbf{r}')\mathbf{G}}. \tag{15.87}$$

Now Eq. (15.74) can be written as

$$\begin{aligned}\langle l', \mathbf{k}+\mathbf{q}|\rho_\omega^{(1)}|\mathbf{k}, l\rangle &= \frac{f\left[E_{l'}(\mathbf{k}+\mathbf{q})\right] - f\left[E_l(\mathbf{k})\right]}{E_{l'}(\mathbf{k}+\mathbf{q}) - E_l(\mathbf{k}) - \omega - i\eta}\langle l', \mathbf{k}+\mathbf{q}|V(\mathbf{r},\omega)|\mathbf{k}, l\rangle \\ &= \frac{f\left[E_{l'}(\mathbf{k}+\mathbf{q})\right] - f\left[E_l(\mathbf{k})\right]}{E_{l'}(\mathbf{k}+\mathbf{q}) - E_l(\mathbf{k}) - \omega - i\eta} \\ &\times \sum_{\mathbf{q},\mathbf{G}} V(\mathbf{q}+\mathbf{G},\omega)\langle l', \mathbf{k}+\mathbf{q}|e^{i(\mathbf{q}+\mathbf{G})\mathbf{r}}|\mathbf{k}, l\rangle. \qquad (15.88)\end{aligned}$$

A complex part of the energy in the denominator of Eq. (15.88) is introduced to prevent unphysical divergences at resonance frequencies. Plugging Eq. (15.88) into Eq. (15.86), we arrive at the following expression for the induced charge:

$$n_{ind}(\mathbf{q}+\mathbf{G},\omega) = e\sum_{\mathbf{G}'} P_{\mathbf{G},\mathbf{G}'} V(\mathbf{q}+\mathbf{G},\omega). \qquad (15.89)$$

Using the notation $\mathbf{k}' = \mathbf{k}+\mathbf{q}$, the polarization function is defined as

$$\begin{aligned}P_{\mathbf{G},\mathbf{G}'}(\omega) &= \sum_{\mathbf{k}',\mathbf{k}}\sum_{l',l}\langle l', \mathbf{k}'|e^{i(\mathbf{q}+\mathbf{G})\mathbf{r}}|\mathbf{k}, l\rangle\langle l, \mathbf{k}|e^{i(\mathbf{q}+\mathbf{G}')\mathbf{r}}|\mathbf{k}', l'\rangle \\ &\times \frac{f\left[E_{l'}(\mathbf{k}+\mathbf{q})\right] - f\left[E_l(\mathbf{k})\right]}{E_{l'}(\mathbf{k}+\mathbf{q}) - E_l(\mathbf{k}) - (\omega + i\eta)}. \qquad (15.90)\end{aligned}$$

Evaluation of the full polarization function is given in Appendix III. The full potential in materials can be separated into two parts, the external and induced potentials:

$$V(\mathbf{q}+\mathbf{G},\omega) = V_{ext}(\mathbf{q}+\mathbf{G},\omega) + V_{ind}(\mathbf{q}+\mathbf{G},\omega). \qquad (15.91)$$

Equation (15.91) can be understood as a reduction (screening) of the external potential through the induced charge in materials. This can be presented in terms of the dielectric function

$$V(\mathbf{q}+\mathbf{G},\omega) = \sum_{\mathbf{G}'} \varepsilon^{-1}_{\mathbf{G},\mathbf{G}'} V_{ext}(\mathbf{q}+\mathbf{G}',\omega), \qquad (15.92)$$

or

$$V_{ext}(\mathbf{q}+\mathbf{G}',\omega) = \sum_{\mathbf{G}} \varepsilon_{\mathbf{G},\mathbf{G}'} V(\mathbf{q}+\mathbf{G},\omega). \qquad (15.93)$$

Equations (15.92) and (15.93) can be considered as the microscopic definition of the dielectric function. The described computation of ε presents a transformation from the microscopic (atom-related) quantities to the macroscopic values used in classic electrodynamics theory. For advanced reading related to the definition of the optical functions within the first priciples theory, one can read monographs[26,27] or original papers.[28,42]

The induced potential satisfies the Poisson equation

$$V_{ind}(\mathbf{q}+\mathbf{G},\omega) = \frac{4\pi}{|\mathbf{q}+\mathbf{G}|^2} n_{ind}(\mathbf{q}+\mathbf{G},\omega). \tag{15.94}$$

From Eqs. (15.89), (15.91), (15.93), and (15.94) we obtain

$$\begin{aligned} V(\mathbf{q}+\mathbf{G},\omega) &= \sum_{\mathbf{G}'} \left[\varepsilon_{\mathbf{G},\mathbf{G}'}(\omega) + \frac{4\pi}{|\mathbf{q}+\mathbf{G}|^2} P_{\mathbf{G},\mathbf{G}'} \right] V(\mathbf{q}+\mathbf{G},\omega) \\ &= \sum_{\mathbf{G}'} \delta_{\mathbf{G},\mathbf{G}'} V(\mathbf{q}+\mathbf{G},\omega). \end{aligned} \tag{15.95}$$

The dielectric function can be expressed now in terms of the polarization function

$$\varepsilon_{\mathbf{G},\mathbf{G}'}(\omega) = \delta_{\mathbf{G},\mathbf{G}'} - \frac{4\pi}{|\mathbf{q}+\mathbf{G}|^2} P_{\mathbf{G},\mathbf{G}'}. \tag{15.96}$$

15.8.3 Appendix III: Evaluation of the polarizaton function including the local field effect

The formula for the polarization function [see Eq. (15.90)] can be represented as

$$\begin{aligned} P_{\mathbf{G},\mathbf{G}'}(\omega) &= \frac{2}{\Omega} \sum_{\mathbf{k}',\mathbf{k}} \sum_{n',n} B^{\mathbf{k}',\mathbf{k}}_{n',n}(\mathbf{q}+\mathbf{G}) B^{*\mathbf{k},\mathbf{k}'}_{n,n'}(\mathbf{q}+\mathbf{G}') \\ &\times \frac{f\left[E_{l'}(\mathbf{k}+\mathbf{q})\right] - f\left[E_l(\mathbf{k})\right]}{E_{l'}(\mathbf{k}+\mathbf{q}) - E_l(\mathbf{k}) - (\omega + i\eta)}. \end{aligned} \tag{15.97}$$

The Bloch integrals in Eq. (15.97) are defined as

$$\begin{aligned} B^{\mathbf{k}',\mathbf{k}}_{n',n}(\mathbf{q}+\mathbf{G}) &= \langle n',\mathbf{k}'|e^{i(\mathbf{q}+\mathbf{G})\mathbf{r}}|\mathbf{k},n\rangle \\ &= \frac{1}{\Omega} \int \psi^*_{n',\mathbf{k}'}(\mathbf{r}) e^{i(\mathbf{q}+\mathbf{G})\mathbf{r}} \psi_{n,\mathbf{k}} d^3\mathbf{r}. \end{aligned} \tag{15.98}$$

In the plane-wave representation [Eq. (15.6)] neglecting the umklapp processes (the nonconserving crystal momentum electron-electron scattering)[46] and in the limit of $\mathbf{q} \to 0$, the Bloch integrals have an extremely simple form given by

$$B^{\mathbf{k}',\mathbf{k}}_{c,v}(\mathbf{G}) = \sum_{\mathbf{G}_1} d^*_{c,\mathbf{k}'}(\mathbf{G}_1) d_{v,\mathbf{k}}(\mathbf{G}_1 - \mathbf{G}). \tag{15.99}$$

Indices c and v in Eq. (15.99) denote empty antibonding (*conducting*) and filled bonding (*valence*) electron states, respectively, at zero temperature. In deriving Eq. (15.99) we used the following properties of direct and reciprocal lattice vectors:

$$\begin{aligned} \mathbf{G}_i \mathbf{r}_j &= 2\pi\delta_{ij}, \\ e^{i\mathbf{G}_j\mathbf{r}_j} &= 1 \\ \sum_{\mathbf{r}_i} e^{i(\mathbf{k}'-\mathbf{k}+\mathbf{q})\mathbf{r}_i} &= \sum_{\mathbf{G}_i} \delta_{\mathbf{k}'-\mathbf{k}+\mathbf{q},\mathbf{G}_i}. \end{aligned} \tag{15.100}$$

Equation (15.97) at zero temperature takes the following form:

$$P_{\mathbf{G},\mathbf{G}'}(\omega) = \frac{2}{\Omega} \sum_{\mathbf{k}',\mathbf{k}} \sum_{n',n} \frac{B^{\mathbf{k}',\mathbf{k}}_{n',n}(\mathbf{q}+\mathbf{G}) B^{*\mathbf{k},\mathbf{k}'}_{n,n'}(\mathbf{q}+\mathbf{G}')}{E_{l'}(\mathbf{k}+\mathbf{q}) - E_l(\mathbf{k}) - (\omega + i\eta)}. \tag{15.101}$$

From the orthonormality of the wave functions follows the properties of Bloch integrals:

$$\begin{aligned} B^{\mathbf{k}',\mathbf{k}}_{c,v}(0) = \sum_{\mathbf{G}_1} d^*_{c,\mathbf{k}'}(\mathbf{G}_1) d_{v,\mathbf{k}}(\mathbf{G}_1) &= 0, \\ \sum_{n'=1}^{N} \sum_{\mathbf{k}'} \left| B^{\mathbf{k}',\mathbf{k}}_{n',n}(\mathbf{q}+\mathbf{G}) \right|^2 &= 1. \end{aligned} \tag{15.102}$$

For $\mathbf{G} = 0$ and in the limit $\mathbf{q} \to 0$, the Bloch integrals have the following properties:

$$\begin{aligned} \lim_{\mathbf{q}\to 0} B^{\mathbf{k}',\mathbf{k}}_{c,v}(\mathbf{q}) &= \delta_{\mathbf{k},\mathbf{k}'} \lim_{\mathbf{q}\to 0} i \sum_{\alpha=1}^{3} q_\alpha \langle c,\mathbf{k}|r_\alpha|\mathbf{k},v\rangle \\ &= \frac{\delta_{\mathbf{k},\mathbf{k}'}}{E_c(\mathbf{k}) - E_v(\mathbf{k})} \lim_{\mathbf{q}\to 0} \sum_{\alpha=1}^{3} q_\alpha \langle c,\mathbf{k}|v_\alpha|\mathbf{k},v\rangle, \\ \lim_{\mathbf{q}\to 0} \frac{B^{\mathbf{k}',\mathbf{k}}_{c,v}(q_\alpha)}{|q_\alpha|} &= \frac{1}{E_c(\mathbf{k}) - E_v(\mathbf{k})} \langle c,\mathbf{k}|v_\alpha|\mathbf{k},v\rangle. \end{aligned} \tag{15.103}$$

Here we used the general definition of the velocity (or momentum)[41] given by

$$\mathbf{v} = \lim_{\mathbf{q}\to 0} \frac{1}{\mathbf{q}} \left[H, e^{i\mathbf{q}\mathbf{r}} \right]. \tag{15.104}$$

At the limit, the velocity is given by

$$\mathbf{v} = i\,[H,\mathbf{r}]. \tag{15.105}$$

After projecting Eq. (15.105) on the full set of eigenfunctions of the Hamiltonian, it follows that

$$\begin{aligned} \langle n\mathbf{k}|\,[H,r_\alpha]\,|n'\mathbf{k}'\rangle &= \sum_{m,l} \langle n\mathbf{k}|H|m,l\rangle\langle m,l|r_\alpha|n'\mathbf{k}'\rangle \\ &- \sum_{m',l'} \langle n\mathbf{k}|r_\alpha|m',l'\rangle\langle m',l'|H|n'\mathbf{k}'\rangle \\ &= (E_{n\mathbf{k}} - E_{n'\mathbf{k}'})\langle n\mathbf{k}|r_\alpha|n'\mathbf{k}'\rangle. \end{aligned} \tag{15.106}$$

Equation (15.106) represents the relationship between the matrix elements of the velocity and the matrix element of the induced dipole momentum, which can be used to obtain a relationship between the optical functions calculated in velocity and length gauges.[28]

15.8.4 Appendix IV: Optical field Hamiltonian in second quantization representation

If there are resonances of the electromagnetic field within a cavity, the entire field can be represented as a superposition of single modes in the following form:[54]

$$\mathbf{E}(\mathbf{r},t) = -\sqrt{4\pi}\sum_j p_j(t)\mathbf{E}_j(\mathbf{r}), \tag{15.107}$$

$$\mathbf{H}(\mathbf{r},t) = \sqrt{4\pi}\sum_j \omega_j q_j(t)\mathbf{H}_j(\mathbf{r}). \tag{15.108}$$

The total energy of the field is given by:

$$H = \frac{1}{8\pi}\int (|\mathbf{E}|^2 + |\mathbf{H}|^2)d^3\mathbf{r} = \frac{1}{2}\sum_j (p_j^2 + \omega_j^2 q_j^2). \tag{15.109}$$

The Hamiltonian equations of motion are given by

$$\begin{aligned} \dot{q_j} &= \frac{\partial H}{\partial p_j} = p_j, \\ \dot{p_j} &= -\frac{\partial H}{\partial q_j} = -\omega_j^2 q_j. \end{aligned} \tag{15.110}$$

Mathematically, the quantization of the field is represented by the commutation rules for the canonically conjugated coordinates and momenta:

$$[q_i, q_j] = 0, \quad [p_i, p_j] = 0, \quad [q_i, p_j] = i\hbar\delta_{ij}. \tag{15.111}$$

The Hamiltonian of the optical field is conveniently represented in terms of the second quantization operators, the Bosonic operators of creation $\hat{a}_j^\dagger$, and annihilation $\hat{a}_j$ of photons defined as[74]

$$p_j = \sqrt{\frac{\hbar\omega_j}{2}}\left(\hat{a}_j + \hat{a}_j^\dagger\right), \tag{15.112}$$

$$\omega_j q_j = i\sqrt{\frac{\hbar\omega_j}{2}}\left(\hat{a}_j - \hat{a}_j^\dagger\right), \tag{15.113}$$

with the commutation rule

$$[\hat{a}_i, \hat{a}_j^\dagger] = \delta_{ij}. \tag{15.114}$$

The properties of $\hat{a}_i$ operators are described through their action on the state vector $\phi(n_1, n_2, \ldots n_i \ldots)$ according to

$$\begin{aligned} \hat{a}_i\phi(n_1, \ldots n_{i-1}, n_i, n_{i+1} \ldots) &= \sqrt{n_i}\phi(n_1, \ldots n_{i-1}, n_{i-1}, n_{i+1} \ldots), \\ \hat{a}_i^\dagger\phi(n_1, \ldots n_{i-1}, n_i, n_{i+1} \ldots) &= \sqrt{n_i + 1}\phi(n_1, \ldots n_{i-1}, n_{i+1}, n_{i+1} \ldots). \end{aligned} \tag{15.115}$$

The Hamiltonian of the quantized optical field is now given by

$$H = \sum_i \hbar\omega_i \left(\hat{a}_i \hat{a}_i^{\dagger} + \frac{1}{2} \right). \tag{15.116}$$

References

1. J. B. Pendry, "Negative refraction makes a perfect lens," *Phys. Rev. Lett.* **85**, 3966 (2000).
2. V. M. Shalaev, *Nonlinear Optics of Random Media: Fractal Composites and Metal-Dielectric Films*, vol. 158 of *Springer Tracts in Modern Physics*, Springer, Berlin, New York (2000).
3. E. Yablonovitch, "Photonic crystals: semiconductors of light," *Scientific American* **N12**, 47 (2001).
4. U. Banin and O. Millo, "Tunneling and optical spectroscopy of semiconductor nanocrystals," *Annu. Rev. Phys. Chem.* **54**, 465 (2003).
5. L. E. Ramos, J. Furthmüller, and F. Bechstedt, "Quantum confinement in Si- and Ge-capped nanocrystallites," *Phys. Rev. B* **72**, 45351 (2005). http://link.aps.org/doi/10.1103/PhysRevB.72.045351
6. Y. Yin and P. Alivisatos, "Colloidal nanocrystal synthesis and the organic-inorganic interface," *Nature* **437**, 665 (2005).
7. W. Cai, U. K. Chettiar, V. Kildishev, and V. M. Shalaev, "Optical cloaking with metamaterials," *Nature Photonics* **1**, 224 (2007).
8. A. B. Evlyukhin, S. I. Bozhevolnyi, A. L. Stepanov, and J. R. Krenn, "Splitting of plasmon polariton beam by chains of nanoparticles," *Applied Physics B* **84**, 29 (2006).
9. A. L. Efros and M. Rosen, "The electronic structure of semiconductor nanocrystals," *Annu. Rev. Mater. Sci.* **30**, 475 (2000).
10. L. Brus, "Chemical approaches to semiconductor nanocrystals," *J. Phys. Chem. Solids* **59**, 459 (1998).
11. I. Vasiliev, S. Öğüt, and J. R. Chelikowsky, "First-principles density-functional calculations for optical spectra of clusters and nanocrystals," *Phys. Rev. B* **65**, 115416 (2002).
12. R. Leitsmann, W. G. Schmidt, P. H. Hahn, and F. Bechstedt, "Second-harmonic polarizability including electron-hole attraction from band-structure theory," *Phys. Rev. B* **71**, 195209 (2005).
13. J. B. Pendry, A. J. Holden, and W. J. S. I. Youngs, "Extremely low frequency plasmons in metallic mesostructures," *Phys. Rev. Lett.* **76**, 4773 (1996).

14. V. M. Shalaev, "Optical negative-index metamaterials," *Nature Photonics* **1**, 41 (2007).

15. M. O. Scully and M. S. Zubairy, *Quantum Optics*, Cambridge University Press, Cambridge (1997).

16. C. Gerry and P. Knight, *Introductory Quantum Optics*, Cambridge University Press, Cambridge (2005).

17. J. M. Raimond, M. Brune, and S. Haroche, "Colloquium: manipulating quantum entanglement with atoms and photons in a cavity," *Rev. Mod. Phys.* **73**, 565 (2001). http://link.aps.org/doi/10.1103/RevModPhys.73.565

18. M. Brune, E. Hagley, J. Dreyer, X. Maitre, A. Maali, C. Wunderlich, J. M. Raimond, and S. Haroche, "Observing the progressive decoherence of the meter in a quantum measurement," *Phys. Rev. Lett.* **77**, 4887 (1996).

19. A. Boca, R. Miller, K. M. Birnbaum, A. D. Boozer, J. McKeever, and H. J. Kimble, "Observation of the vacuum Rabi spectrum for one trapped atom," *Phys. Rev. Lett.* **93**, 233603 (2004).

20. P. Maunz, T. Puppe, I. Schuster, N. Syassen, P. W. H. Pinkse, and G. Rempe, "Normal mode spectroscopy of a single-bound-atom-cavity system," *Phys. Rev. Lett.* **94**, 033002 (2005).

21. Y.-S. Park, A. K. Cook, and H. Wang, "Cavity QED with diamond nanocrystals and silica microspheres," *Nano Letters* **6**, 2075 (2006).

22. S. Strauf, K. Hennessy, M. T. Rakher, Y.-S. Choi, A. Badolato, L. C. Andreani, E. L. Hu, P. M. Petroff, and D. Bouwmeester, "Self-tuned quantum dot gain in photonic crystal lasers," *Phys. Rev. Lett.* **96**, 127404 (2006).

23. K. Hennessy, A. Badolato, M. Winger, D. Gerace, M. Atatüre, S. Gulde, S. Fält, E. L. Hu, and A. Imamoglu, "Quantum nature of a strongly coupled single quantum dot-cavity system," *Nature* **445**, 896–889 (2007). http://www.nature.com/nature/index.htm

24. A. Kiraz, C. Reese, B. Gayral, L. Zhang, W. V. Schoenfeld, B. D. Gerardot, P. M. Petroff, E. L. Hu, and A. Imamoglu, "Cavity-quantum electrodynamics with quantum dots," *J. Opt. B* **5**, 129 (2003).

25. J. Vučković, M. Lončar, H. Mabuchi, and A. Scherer, "Design of photonic crystal microcavities for cavity QED," *Phys. Rev. A* **65**, 016608 (2001).

26. P. Y. Yu and M. Cardona, *Fundamentals of Semiconductors*, Springer, Berlin (2001).

27. R. M. Martin, *Electronic Structure: Basic Theory and Practical Methods*, Cambridge University Press, New York (2004).

28. V. I. Gavrilenko and F. Bechstedt, "Optical functions of semiconductors beyond density functional theory," *Phys. Rev. B* **55**, 4343 (1997).

29. J. D. Jackson, *Classic Electrodynamics*, 2nd ed., Wiley, New York (1975).

30. M. A. Noginov, G. Zhu, M. Bahoura, J. Adegoke, C. E. Small, B. A. Ritzo, V. P. Drachev, and V. M. Shalaev, "Enhancement of surface plasmons in Ag aggregates by optical gain in a dielectric medium," *Optic. Letters* **31**, 3022 (2006).

31. M. Moskovits, "Surface-enhanced spectroscopy," *Rev. Mod. Phys.* **57**, 783 (1985).

32. C. M. Bowden and J. P. Dowling, "Near-dipole-dipole effects in dense media: generalized Maxwell-Bloch equations," *Phys. Rev. A* **47**, 1247 (1993).

33. G. Burns, *Solid State Physics*, Academic Press, New York (1985).

34. M. I. Stockman, "Electromagnetic theory of SERS," in *Surface Enhanced Raman Scattering: Physics and Applications*, K. Kneipp, H. Kneipp, and M. Moskovits, Eds., 47, Springer, Berlin, Heidelberg (2006).

35. M. A. Noginov, M. Vondrova, S. M. Williams, M. Bahoura, V. I. Gavrilenko, S. M. Black, V. P. Drachev, V. M. Shalaev, and A. Sykes, "Spectroscopic studies of liquid solutions of Rh6G laser dye and Ag nanoparticle aggregates," *J. Opt. A* **7**, 219–229 (2005).

36. W. R. L. Lambrecht and S. N. Rashkeev, "From band structures to linear and nonlinear optical spectra in semiconductors," *Physica Status Solidi (B)* **217**, 599 (2000).

37. B. Arnaud and M. Alouani, "Local-field and excitonic effects in the calculated optical properties of semiconductors from first principles," *Phys. Rev. B* **63**, 85208 (2001).

38. S. L. Adler, "Quantum theory of the dielectric constant in real solids," *Phys. Rev.* **126**, 413 (1962).

39. N. Wiser, "Dielectric constant with local field effects included," *Phys. Rev.* **129**, 62 (1963).

40. V. I. Gavrilenko, "Ab initio modeling of optical properties of organic molecules and molecular complexes," in *Lecture Notes in Computational Science*, V. N. Alexandrov, G. D. van Albada, P. M. A. Sloot, and J. Dongarra, Eds., *ICCS LNCS 3993* Part III, 86, Springer (2006).

41. B. Adolph, V. I. Gavrilenko, K. Tenelsen, F. Bechstedt, and R. D. Sole, "Nonlocality and many-body effects in the optical properties of semiconductors," *Phys. Rev. B* **53**, 9797 (1996).

42. G. Onida, L. Reining, and A. Rubio, "Electronic excitations: density functional versus many-body Green's-function approach," *Rev. Mod. Phys.* **74**, 601–656 (2002).

43. A. S. Davydov, *Solid State Theory*, Academic Press, New York (1980).

44. M. A. Noginov, *Solid-State Random Lasers*, Springer, New York (2005).

45. L. D. Landau and E. M. Lifshits, *Quantum Mechanics*, Academic Press, New York (1980).

46. F. Bechstedt, *Principles of Surface Physics*, Springer, Berlin (2003).

47. V. I. Gavrilenko and F. Koch, "Electronic structure of nanometer-thickness Si(001) film," *J. Appl. Phys.* **77**, 3288 (1995).

48. J. Goniakowski, F. Finocchi, and C. Noguera, "Polarity of oxide surfaces and nanostructures," *Rep. Prog. Phys.* **71**, 016501 (2008).

49. A. J. Williamson and A. Zunger, "InAs quantum dots: predicted electronic structure of free-standing versus GaAs-embedded structures," *Phys. Rev. B* **59**, 15819 (1999).

50. M. U. Kahaly, P. Ghosh, S. Narasimhan, and U. V. Waghmare, "Size dependence of structural, electronic, elastic, and optical properties of selenium nanowires: a first principles study," *J. Chem. Phys.* **128**, 044718 (2008).

51. G. Y. Slepyan, S. A. Maksimenko, A. Hoffmann, and D. Bimberg, "Quantum optics of a quantum dot: local-field effect," *Phys. Rev. A* **66**, 63804 (2002).

52. D. Barchiesi, B. Guizal, and T. Grosges, "Accuracy of local field enhancement models: toward predictive models?," *Applied Physics B* **84**, 55 (2006).

53. V. I. Gavrilenko and M. A. Noginov, "Ab initio study of optical properties of Rhodamine 6G molecular dimers," *J. Chem. Phys.* **124**, 44301 (2006).

54. E. T. Jaynes and F. W. Cummins, "Comparison of quantum and semiclassical radiation theories with application to the beam maser," *Proc. IEEE* **51**, 89 (1963).

55. F. W. Cummings, "Stimulated emission of radiation in a single mode," *Phys. Rev.* **140**, A1051 (1965).

56. J. D. Joannopoulos, R. D. Meade, and J. N. Winn, *Photonic Crystals: Moulding the Flow of Light*, Princeton University Press (1995).

57. M. Cardona, "Resonance phenomena," in *Light Scattering in Solids II*, M. Cardona and G. Güntherodt, Eds., 19, Springer, Berlin, Heidelberg (1982).

58. M. Ikezawa, T. Okuno, Y. Masumoto, and A. A. Lipovskii, "Complementary detection of confined acoustic phonons in quantum dots by coherent phonon measurement and Raman scattering," *Phys. Rev. B* **64**, 201315R (2001). http://link.aps.org/doi/10.1103/PhysRevB.64.201315

59. A. Tamura, K. Higeta, and T. Ichinokawa, "Lattice vibrations and specific heat of a small particle," *J. Phys. C* **15**, 4975 (1982).

60. A. Tamura, K. Higeta, and T. Ichinokawa, "The size dependence of vibrational eigenfrequencies and the mean square vibrational displacement of a small particle," *J. Phys. C* **16**, 1585 (1983).

61. E. Duval, "Far-infrared and Raman vibrational transitions of a solid sphere: selection rules," *Phys. Rev. B* **46**, 5795 (1992).

62. D. P. DiLella, A. Gohin, R. H. Lipson, P. McBreen, and M. Moskovitz, "Enhanced Raman spectroscopy of CO adsorbed on vapor-deposited silver," *J. Chem. Phys.* **73**, 4282 (1980).

63. K. Kneipp, Y. Wang, H. Kneipp, L. T. Perelman, I. Itzkan, R. R. Dasari, and M. S. Feld, "Single molecule detection using surface-enhanced Raman scattering (SERS)," *Phys. Rev. Lett.* **78**, 1667 (1997).

64. H. Nakai and H. Nakatsuji, "Electronic mechanism of the surface enhanced Raman scattering," *J. Chem. Phys.* **103**, 2286 (1995).

65. V. I. Gavrilenko, D. Martinez, A. Cantarero, M. Cardona, and C. Trallero-Giner, "Resonant first- and second-order Raman scattering of AlSb," *Phys. Rev. B* **42**, 11718 (1990).

66. D. A. McQuarrie, *Statistical Mechanics*, Harper and Row, New York (1976).

67. R. G. Parr and W. Yang, *Density Functional Theory of Atoms and Molecules*, Oxford University Press, New York (1989).

68. R. P. Feynman, N. Metropolis, and E. Teller, "Equations of state of elements based on the generalized Thomas-Fermi theory," *Phys. Rev.* **75**, 1561–1573 (1949).

69. P. Hohenberg and W. Kohn, "Inhomogeneous electron gas," *Phys. Rev.* **136**, B864–B871 (1964).

70. W. Kohn, "Electronic structure of matter - wave functions and density functionals," *Rev. Mod. Phys.* **71**, 1253–1266 (1999).

71. D. M. Ceperley and B. J. Adler, "Ground state of electron gas by a stochastic method," *Phys. Rev. Lett.* **45**, 566–569 (1980).

72. J. P. Perdew and Y. Wang, "Accurate and simple analytic representation of the electron-gas corelation energy," *Phys. Rev. B* **45**, 13244–13249 (1992).

73. W. Kohn and L. J. Sham, "Self-consistent equations including exchange and correlation effects," *Phys. Rev.* **140**, A1133 (1965).

74. A. S. Davydov, *Quantum Mechanics*, 2nd ed., Pergamon Press, New York (1976).

Biography

Vladimir I. Gavrilenko received his M.S. degree from T. Shevchenko National University of Kiev (Ukraine), and his Ph.D. and D.Sc. (habilitation) from the National Academy of Sciences of Ukraine. He performs research in the area of computational materials science, including first principles modeling and simulations of nanomaterials at the Center for Materials Research, Norfolk State University, Norfolk, Virginia. Prof. Gavrilenko has authored and coauthored more than 100 scientific publications in peer-reviewed journals, two books, and five book chapters. His current research interests lie in linear and nonlinear optics of solid surfaces, organic-inorganic interfaces, and nanostructures.

Chapter 16
Organic Photonic Materials

Larry R. Dalton, Philip A. Sullivan, Denise H. Bale, Scott R. Hammond, Benjamin C. Olbricht, Harrison Rommel, Bruce Eichinger, and Bruce H. Robinson
University of Washington, Seattle, WA, USA

16.1 Preface

Electro-optic (EO) devices are critical to telecommunications, computing, defense, sensing, and transportation technology sectors. Indeed, they will play a central role in the coming technology revolution of chipscale electronic/photonic integration, which is well recognized as important to future computer technology. Radio-frequency (RF) photonics is also well recognized as an important option for the long-distance delivery of RF/microwave/millimeter-wave signals. Electro-optic device technology is currently dominated by inorganic materials, such as lithium niobate [which exhibits a useable EO activity of approximately 30 pm/V (picometers/volt) and an intrinsic optical loss of 0.1–0.2 dB/cm]. Unfortunately, there is little possibility for further improvement of such ionic crystalline materials, where EO activity arises as a result of electric-field-induced displacement of ionic charges. Moreover, the mass of these charges limits the response time of the EO effect for such materials, and a substantial velocity mismatch exists between electrical and optical waves propagating in lithium niobate. This velocity mismatch limits both the bandwidth and the drive voltage that can be achieved. In contrast, EO activity in organic π-electron materials is determined by applied electric-field-induced electronic charge perturbation, and the lower mass of the coupled electron system leads to very fast (tens of femtoseconds) response times. Moreover, EO activity in organic materials can be systematically improved by modification of the structure and organization of the π-electron molecules (called chromophores because they are "colored" due to absorption of visible wavelength light) that make up organic EO materials. A question of major interest to technology sectors dependent on EO devices is: How much and how soon can organic EO materials be improved to enable new and dramatically improved applications? This chapter addresses that concern.

Theoretically inspired design of chromophores with exceptional molecular first hyperpolarizability, nanoscopically engineered to control intermolecular electrostatic interactions for enhanced acentric chromophore order, has produced dramatic improvements in EO activity to values more than 15 times that of lithium niobate. The large EO coefficients of solution-processed organic EO materials permit drive voltages in devices to be reduced to values of less than 1 V. Such voltages are critical to the realization of gain in RF/ microwave/millimeter-wave photonic applications (e.g., amplification of electrical signals in the electrical-optical-electrical signal transduction process). Organic EO materials afford a number of other advantages, including exceptional bandwidth (> 100 GHz), low temperature solution (spin casting) and melt processing (nano-imprint and soft lithography) of thin film devices, adaptability to production of conformal and flexible device structures, and compatibility with a diverse array of materials and material technologies including silicon photonics. This last advantage has permitted integration of organic EO materials with silicon photonic circuitry, resulting in low-power high-bandwidth EO modulation, optical rectification, and all-optical modulation. The last two phenomena relate to the concentration of optical power in the reduced dimensions of silicon photonic circuitry. Another recently demonstrated advantage of organic EO materials relates to terahertz applications, where the combination of large EO coefficients and the absence of attenuation by phonon modes permit development of highly efficient and broad-bandwidth terahertz sources and detectors.

This chapter focuses on introducing the reader to critical concepts in the design of organic EO materials and the integration of these materials into novel device platforms that promise a new generation of applications for EO technology. Advances in quantum and statistical mechanical numerical methods have facilitated the design of improved materials. The focus of this chapter will not be on the mathematical details of such calculations but rather on the definition of the critical structure/function relationships derived from these calculations. The objective is to demonstrate clear paradigms for further improvement of EO activity.

In the course of the discussion, a new class of organic electroactive materials, binary chromophore organic glasses (BCOGs), will be introduced. These materials exhibit enhanced EO activity, reduced optical loss, and enhanced stability (both thermal and photochemical) in agreement with the predictions of theory. BCOGs permit control of both electronic charge perturbation [relevant to nonlinear optical (NLO) behavior] and transport (relevant to electronic, photovoltaic, light-emitting, and photorefractive behavior) and as such may have broader relevance to electronic, photonic, and optoelectronic behavior. Electrical conductivity in EO materials is critical to several aspects of EO device performance, including bias voltage stability.

This chapter provides a tutorial on intermolecular electrostatic interactions in non-ionic materials and should be of interest and utility to advanced undergraduate, graduate, and postgraduate students in chemistry, physics, and materials science for understanding a variety of photonic, optoelectronic, and electronic phenomena. The device concepts and material specifications should be of interest to electrical and optical engineers, particularly those focused on "Beyond Moore's law" information technology and a variety of technologies ranging from phased-array radar to embedded network sensing.

16.2 Introduction

An often stated advantage of organic materials for nonlinear optics is the virtually limitless ability to improve their performance by synthetic modification of both chromophores (individual π-molecules that are the active component of organic NLO materials) and material structures (the supermolecular or nanoscopic organization of molecules). However, for synthetic improvement of NLO materials to be accomplished in a timely manner, material design must be driven by a quantitative knowledge of critical structure/function relationships. Quantum and statistical mechanical calculations are required for a "first principles" definition of structure/function relationships at both the molecular and macroscopic levels. Quantum mechanical calculations are most commonly carried out considering isolated molecules; properties such as molecular first hyperpolarizability (the tendency for the electron distribution of molecules to be perturbed or polarized by external electric fields) are calculated in the long-wavelength limit. In contrast, experimental measurement of molecular optical nonlinearity (e.g., molecular first hyperpolarizability—see the Appendix for an introduction to the basic concepts and nomenclature of nonlinear optics) is typically carried out on molecules in dielectric media and at finite measurement (optical) frequencies. Moreover, macroscopic NLO properties, such as EO activity, have commonly been theoretically estimated with the assumption that chromophores behave as independent particles, i.e., they experience no intermolecular electrostatic interactions. In practice, molecules with substantial hyperpolarizability usually experience strong intermolecular electrostatic interactions that strongly influence the macroscopic order and dielectric permittivity of materials. Thus, meaningful direct correlation of experiment and theory has not been possible in the past, inhibiting the systematic improvement of materials. Here we are interested in understanding the role of such strong intermolecular electrostatic interactions in defining the properties of complex photonic materials and using that knowledge to prepare new materials with dramatically improved properties. Stated in other terms, we are interested in a first principles understanding of optical nonlinearity from molecules to materials.

We are also interested in understanding the role of device structure in defining performance in NLO applications. In particular, we are interested in resonant device structures, silicon photonics, and sub-λ photonic phenomena. We will demonstrate that organic NLO materials can be combined with silicon photonic waveguides of nanoscopic cross-sectional dimensions to achieve dramatically improved performance related to EO modulation, optical rectification, and all-optical modulation.

To make our discussion as concrete and concise as possible, we largely limit ourselves to second-order NLO materials (see Appendix) and particularly to dipolar chromophores and EO materials prepared by the electrical poling of dipolar chromophores in various material lattices at temperatures near the glass transition temperatures of the materials. Such second-order materials are particularly attractive in that they have only two nonzero components for the second-order NLO tensors (e.g., r_{33} and r_{13} for the EO tensor relevant to the interaction of optical and RF fields in a material). For second-order optical nonlinearity to be nonzero for such materials, noncentrosymmetric (acentric) symmetry must apply at both the molecular and macroscopic levels. Indeed, the principal element of the EO tensor r_{33} can be related to an acentric order parameter $\langle \cos^3 \theta \rangle$ by

$$r_{33} = \beta(\omega, \varepsilon) N \langle \cos^3 \theta \rangle \text{constant}(n, \varepsilon), \tag{16.1}$$

where $\beta(\omega,\varepsilon)$ is the molecular first hyperpolarizability (which depends on optical frequency ω and material dielectric permittivity ε), N is the chromophore number density (molecules/cc), and n is the index of refraction. The three angles of the acentric order parameter relate the principal axes of the high-frequency optical field, the low-frequency RF field, and the material EO tensor. In Eq. (16.1), chromophore number density and acentric order parameter are not independent; rather, the dependence of $\langle \cos^3 \theta \rangle$ on N is defined by intermolecular electrostatic interactions. Thus, the problem of maximizing r_{33} can be thought of as a two-fold process of optimizing $\beta(\omega,\varepsilon)$ and optimizing $N\langle \cos^3 \theta \rangle$. Of course, ε depends on N, so the variables of Eq. (16.1) are quite interdependent. Indeed, $\langle \cos^3 \theta \rangle$ can also be influenced by material conductivity, which, in turn, can depend in a complex manner on N. Conductivity acts to influence the effective poling field felt by the chromophores. Device performance, e.g., bias voltage drift, can also be influenced by material conductivity. It is evident from these comments that a unified and quantitative understanding of the factors that influence EO activity is important.

Quantum mechanical calculations, including density functional theory (DFT) and time-dependent density functional theory (TD-DFT), have proven useful in predicting trends in linear and NLO properties.[1–10] In the following sections, we focus on understanding the dielectric and frequency effects necessary to understand the absolute values (as opposed to the trends) of EO coefficients for different materials.

State-of-the-art organic EO materials are most commonly composed of chromophores with ground-state dipole moments between 10 and 15 debye. Clearly, strong intermolecular electrostatic interactions will exist among these high-dipole-moment chromophores at moderate ($> 2–3 \times 10^{20}$ molecules/cc) chromophore number densities. A number of approximate treatments have been developed for considering intermolecular electrostatic interactions that are comparable to chromophore-poling field interactions and to thermal energies. One approach has been to develop analytical expressions for the potential function describing the effective electrostatic field felt by a reference chromophore from a surrounding ensemble of strongly interacting chromophores.[11–13] This leads to a multiplicative attenuation factor for the acentric order parameter given by $[1 - L(W/kT)^2]$, where L is the Langevin function, W is the intermolecular electrostatic interaction energy, and kT is the thermal energy.

In this approach, we have followed the work of Piekara;[14] however, we have used analytical potential functions with approximations ranging from the point-dipole approximation, to a hard object treatment of nuclear repulsive effects, to classic 6–12 potential function treatments. Even our most approximate (point-dipole) treatments correctly predict the qualitative variation of EO activity with chromophore number density, namely, that EO activity is predicted and observed to go through a maximum with increasing number density. The maximum in the plot of r_{33} versus N is predicted and observed to vary as $N_{\max} \propto kT/\mu^2$. Incorporation of the effect of nuclear repulsive interactions into calculations is required to quantitatively predict the behavior of acentric order parameter with number density. More recently, the effect of intermolecular electrostatic interactions on the order of electrically poled organic materials has been addressed by atomistic Monte Carlo/molecular dynamics[15–17] and pseudo-atomistic Monte Carlo[18–21] calculations. There is reasonable agreement between various approaches with respect to reproducing trends in the variation of EO activity with structure and with chromophore number density. The recent research of Robinson and Rommel[20,21] is particularly enlightening in predicting that spherical chromophores represent an optimum shape for maximizing the EO activity of chromophore/polymer composite materials. These researchers have considered both "on-lattice" (where a uniform spatial distribution of chromophores is an imposed condition of the calculation) and "off-lattice" (where chromophores are permitted to assume a nonuniform spatial distribution) calculations of the variation of EO activity with number density for different chromophore shapes. Both on-lattice and off-lattice calculations predict that maximum-achievable EO activity will increase as the major- to minor-axis ratio of prolate ellipsoid chromophores is decreased to unity (the case of a sphere). The predictions deriving from on-lattice and off-lattice calculations differ when progressing from spheres to oblate ellipsoids. On-lattice calculations predict oblate-ellipsoid-shaped chromophores to exhibit larger maximum-achievable EO activity than is possible with spheres, while the converse is true for off-lattice calculations. Unfortunately, to the present time, chemists specializing in the synthesis of chromophores have not produced chromophores with oblate ellipsoidal shapes that represent good approximations to the theoretically considered shapes. Thus, a good test of the predictability of on-lattice and off-lattice calculations for oblate chromophores does not exist at this time. The trends observed in the transition from prolate shapes to spheres appear reasonably well reproduced.

While studies aimed at understanding chromophore/polymer composite materials have provided good insight into optimizing EO activity for these materials, quantitative prediction of EO activity has been elusive due to the influence of medium dielectric permittivity on EO activity and the variation of EO activity with optical frequency (dispersion effects). In this chapter, we will demonstrate how such effects can be rigorously considered. Moreover, we will extend theoretical consideration to materials more complex than simple chromophore/polymer composites. In particular, we will consider multi-chromophore-containing dendrimer materials (for which the presence of covalent bonds influence poling-induced order), BCOGs, and complex organic EO materials prepared by optically assisted poling. Although we have not (at this time) rigorously investigated the latter two classes of materials, we can report preliminary experimental measurements (and some coarse-grained theoretical calculations) that are highly suggestive concerning routes to optimizing EO activity. Indeed, these preliminary studies have led to EO coefficients (measured at telecommunication wavelengths) as high as 450 pm/V (and even more recently as high as 550 pm/V); these measurements suggest a route to further substantial improvement of EO activity and auxiliary properties such as optical loss.

The chronological improvement of molecular and macroscopic second-order optical nonlinearity is shown in Fig. 16.1. In this figure, data are limited to materials that have been used to fabricate prototype devices. Marks and coworkers[7] have recently reported "twisted" chromophores exhibiting even larger values of molecular first hyperpolarizability than the chromophores in the materials of Fig. 16.1. However, the chromophores of Marks and coworkers have not been incorporated into device-relevant materials at the time of this writing. From Fig. 16.1, it is clear that the improvement rate of EO activity exceeds that expected from a Moore's law plot. Recently, the Defense Advanced Projects Research Agency (DARPA) launched the supermolecular photonics (MORPH) program. The phase II goal of this program is an EO activity of 600 pm/V (indicated by the star in Fig. 16.1). The structures of representative EO chromophores are shown in Fig. 16.2, and the structure of several chromophore-containing dendrimers and polymers are shown in Fig. 16.3. BCOGs are

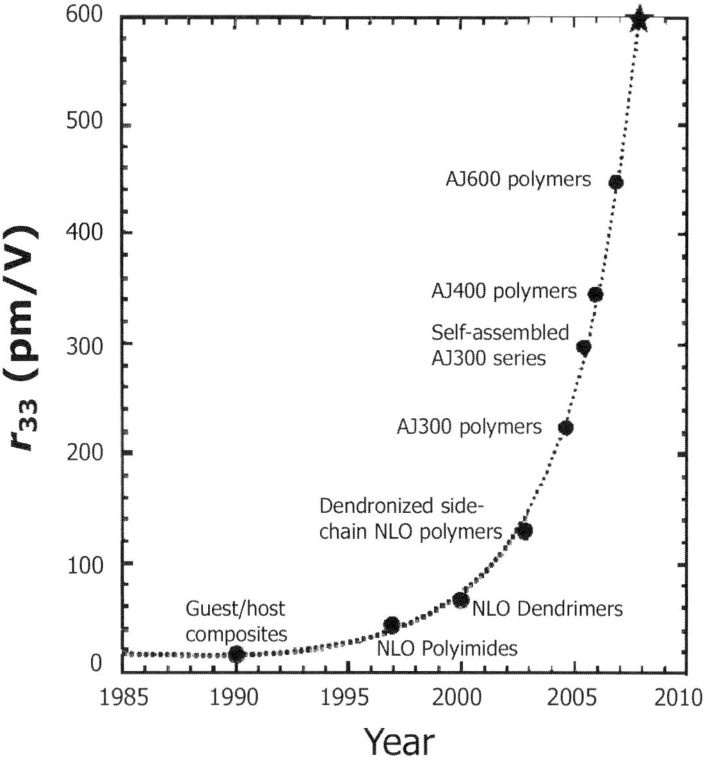

Figure 16.1 The chronological evolution of electro-optic (EO) activity for materials used to fabricate prototype devices. The star indicates the DARPA MORPH phase II goal. The rate of increase is greater than that predicted by a Moore's law plot.

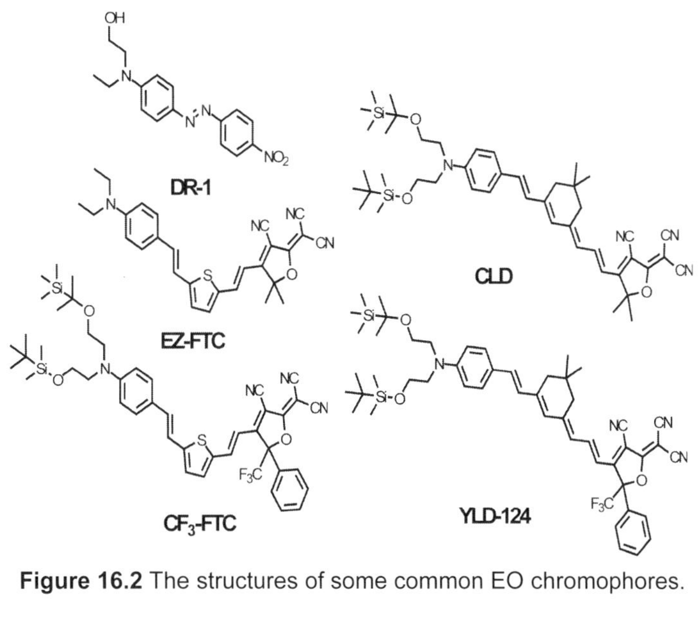

Figure 16.2 The structures of some common EO chromophores.

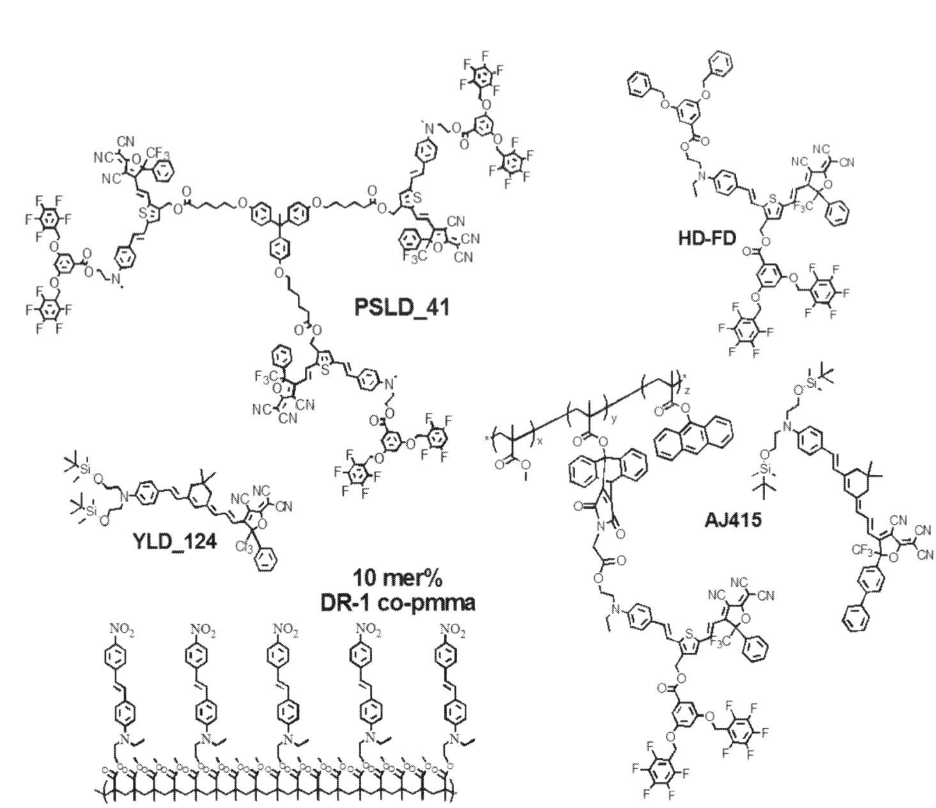

Figure 16.3 Representative guest chromophore and chromophore-containing-host materials used to fabricate binary-chromophore-containing organic glasses (BCOGs).

formulated by combining a chromophore guest with a chromophore-containing host; several examples are shown in Fig. 16.3 and will be discussed in Sec. 16.5.

16.3 Effects of Dielectric Permittivity and Dispersion

In this section, we demonstrate that DFT methods lead to a quantitative understanding of the dependence of β on ε and ω. As noted above, an obstacle to the quantitative correlation of theory and experiment for molecular first hyperpolarizability β and EO activity r_{33} is the dependence of the properties on dielectric permittivity ε and optical operating frequency ω. We have recently carried out experimental and theoretical investigations of these dependences. Femtosecond-time-resolution wavelength-agile hyper-Rayleigh-scattering (HRS) measurements[22] on chromophores in different media were used to investigate the dependence of β on ε and ω. These measurements were complemented by electric-field-induced second-harmonic (EFISH) measurements[23,24] of the product of chromophore dipole moment μ and molecular first hyperpolarizability β. Modified Teng-Man ellipsometry[25,26] and attenuated total reflection (ATR)[26,27] were used to measure EO coefficients (tensor elements) (see Fig. 16.4).

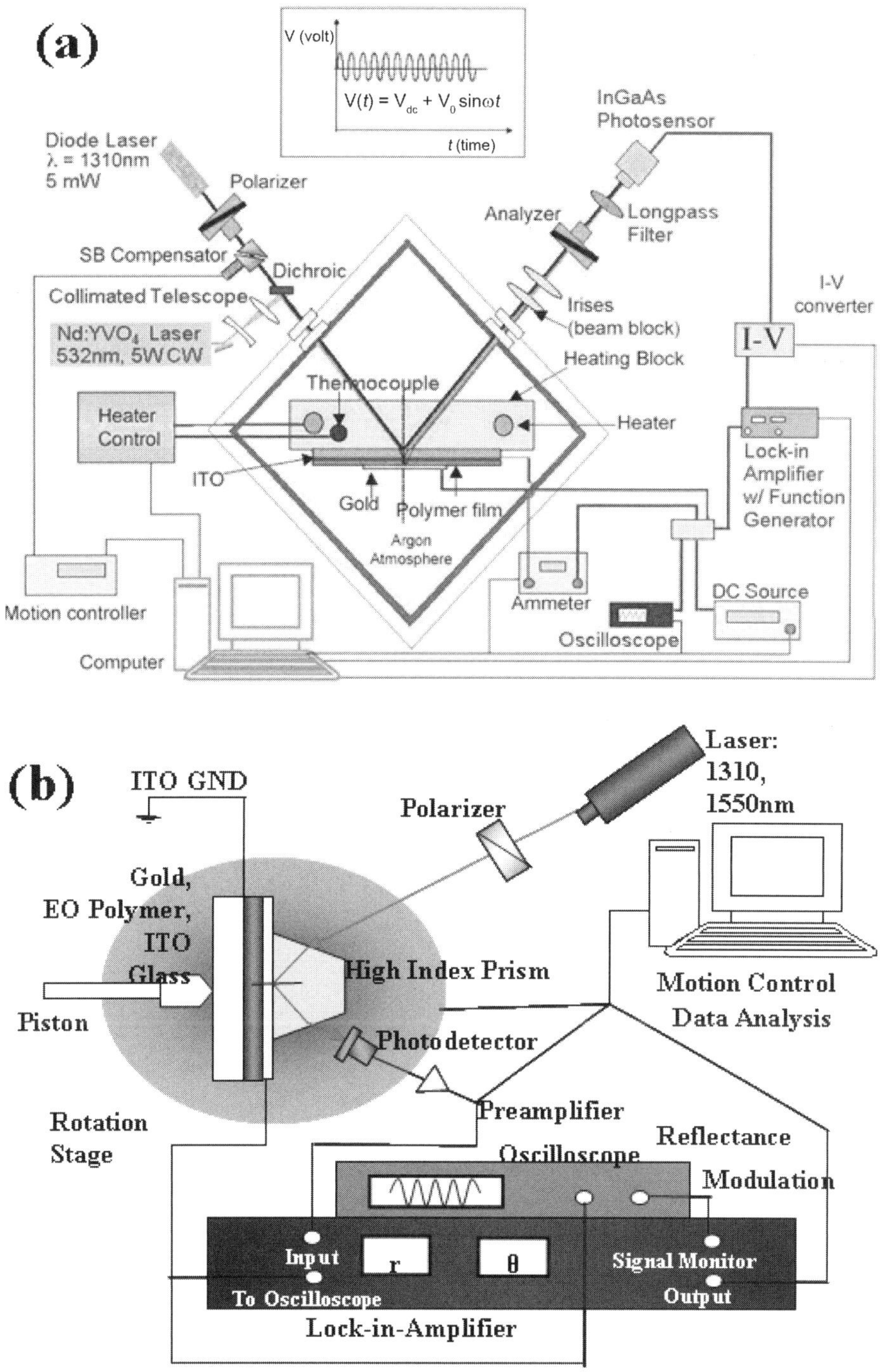

Figure 16.4 (a) A schematic diagram of a Teng-Man apparatus modified for *in situ* monitoring of poling and relaxation of poling-induced order and for laser-assisted poling experiments. (b) A schematic diagram of an attenuated total reflection (ATR) apparatus for the measurement of all elements of the EO tensor.

Measurements of EO coefficients were also carried out in device geometries including Mach Zehnder interferometers and ring microresonators.[28–31] Calculations of β were carried out employing DFT or TD-DFT methods. Calculations of acentric order were effected employing pseudo-atomistic Monte Carlo methods.

The chromophores and macromolecular materials that are the focus of this chapter are summarized in Figs. 16.2 and 16.3. We have used two DFT approaches (Gaussian B3LYP/3-21g*PCM and DMol using PBE/dnp/Cosmo) to investigate the dependence of β on ε. $\beta_{zzz}(\varepsilon)$ is predicted to vary linearly with $(\varepsilon - 1)/(\varepsilon + 2.5)$ close to the predicted Onsager dependence, $\beta_{zzz}(\varepsilon) \propto (\varepsilon - 1)/(\varepsilon + 2)$. The two DFT methods predict a linear dependence for the variation of dipole moment μ with $(\varepsilon - 1)/(\varepsilon + 1)$. A discussion of dielectric permittivity measurements is given in the thesis of Scott Hammond.[32] DFT calculations are also used to predict the dependence of molecular first hyperpolarizability on measurement frequency ω (see Fig. 16.5). The ability of density functional theory methods to simulate the variation of μβ with ε is illustrated in Table 16.1, which also illustrates how the understanding of the dependence of β on ε and ω translates to improved correlation of theory and experimental data.

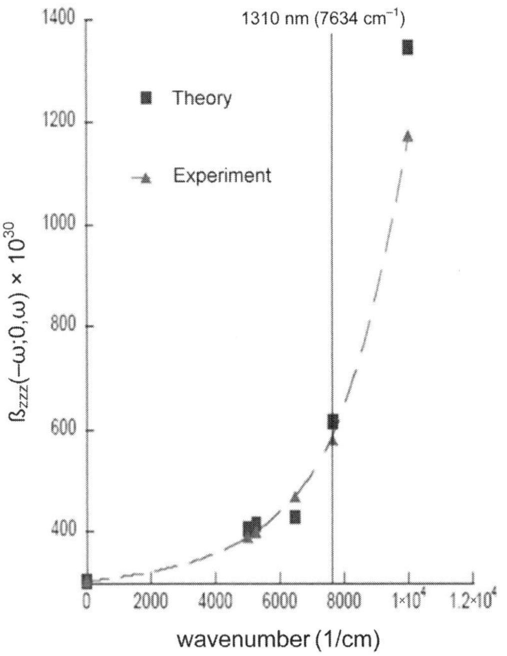

Figure 16.5 The theoretical and experimental frequency dependence of the low-frequency (EO) component molecular first hyperpolarizability $\beta_{zzz}(-\omega; 0, \omega)$ is shown for the CF_3-FTC chromophore (see Fig. 16.2). The theoretical data are indicated by solid squares and were calculated employing time-dependent density functional theory methods. The experimental data were recorded using wavelength-agile femtosecond hyper-Rayleigh-scattering measurements and are indicated by solid triangles and a dashed line (to guide the eye).

Table 16.1 Electric-field-induced second-harmonic data for EZ-FTC (see Fig. 16.2 for chromophore structure).

Solvent	ε	λ_{max}/nm	$\mu\beta_\omega/10^{-69}$ SI (exp.)	$\mu\beta_0/10^{-44}$ esu(exp.)	$\mu\beta_0/10^{-44}$ esu (theory)
Dioxane	2.21	622	9291	3.41	4.9
Chloroform	4.81	675	27280	8.52	8.9

16.4 Complex Dendrimer Materials: Effects of Covalent Bonds

Covalent bond potentials have a strong effect on the rotation and spatial organization of chromophores in complex structures such as dendrimers (see Fig. 16.3). Fully atomistic Monte Carlo/molecular dynamic methods[15,16] provide an attractive path to consideration of the effects of covalent bonds; however, when applied to systems of the complexity of those in Fig. 16.3, such calculations become difficult to execute in a time- and cost-effective manner. This dilemma can be circumvented by adapting the coarse-grained models of our previous work to a "pseudo-atomistic" Monte Carlo approach.[18–21] Since π-conjugation inhibits internal rotation, π-electron segments can be effectively treated within the united atom approximation (see Fig. 16.6). Quantum mechanics are employed to define the charge distributions over these "united atom ellipsoids" in a manner analogous to fully atomistic methods. Sigma-bonded regions of dendritic and polymeric structures are treated in a fully atomistic manner (Fig. 16.6); thus, we refer to this computational methodology as a "pseudo-atomistic" approach. The details of the computations are presented elsewhere.[20,21,33] The focus in this chapter is on the important conclusions derived from such calculations. An obvious requirement of theory is that it must correctly predict electric-field poling behavior; indeed, this is seen to be the case in Fig. 16.7, where a linear dependence on poling voltage is predicted. A linear dependence is experimentally observed for the PSLD-33 dendrimer material (Note that the structure of the PSLD-33 dendrimer is the same as the PSLD-41 structure, without the outer Frechet dendrons). Because EO activity (r_{33}) increases linearly with electric-field poling strength (E_p), it is useful to report the variation of r_{33}/E_p with number density N, where each r_{33}/E_p has been obtained by linear least-squares fitting of the variation of r_{33} with E_p.

The results of applying pseudo-atomistic Monte Carlo methods to the two dendrimers (PSLD-33 and PSLD-41) of Figs. 16.3 and 16.6 are shown in Fig. 16.8. Chromophores within these dendrimer materials act as independent particles; i.e., EO activity increases linearly with chromophore number density, implying that the acentric order parameter is independent of chromophore concentration. Intra-dendrimer chromophore organization is strongly influenced by covalent bond

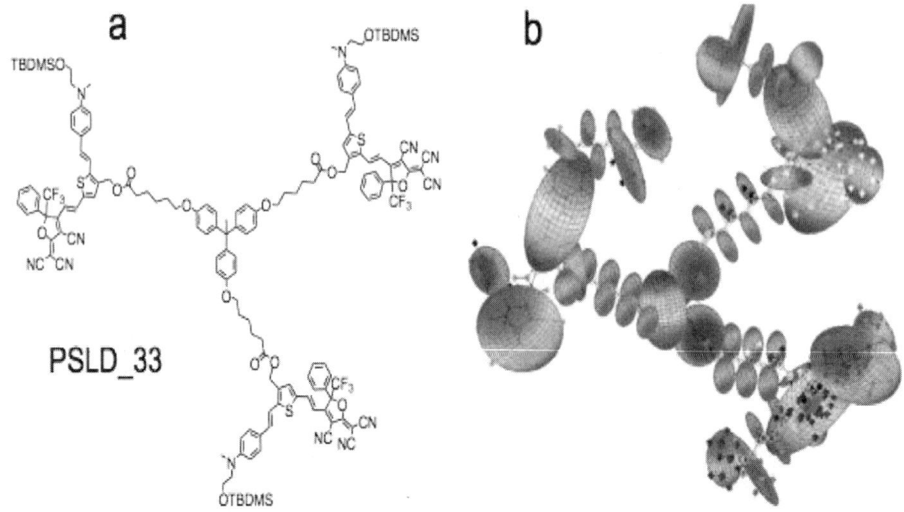

Figure 16.6 (a) The chemical structure of the PSLD-33 dendrimer. (b) The pseudo-atomistic Monte Carlo representation of the PSLD-33 dendrimer.

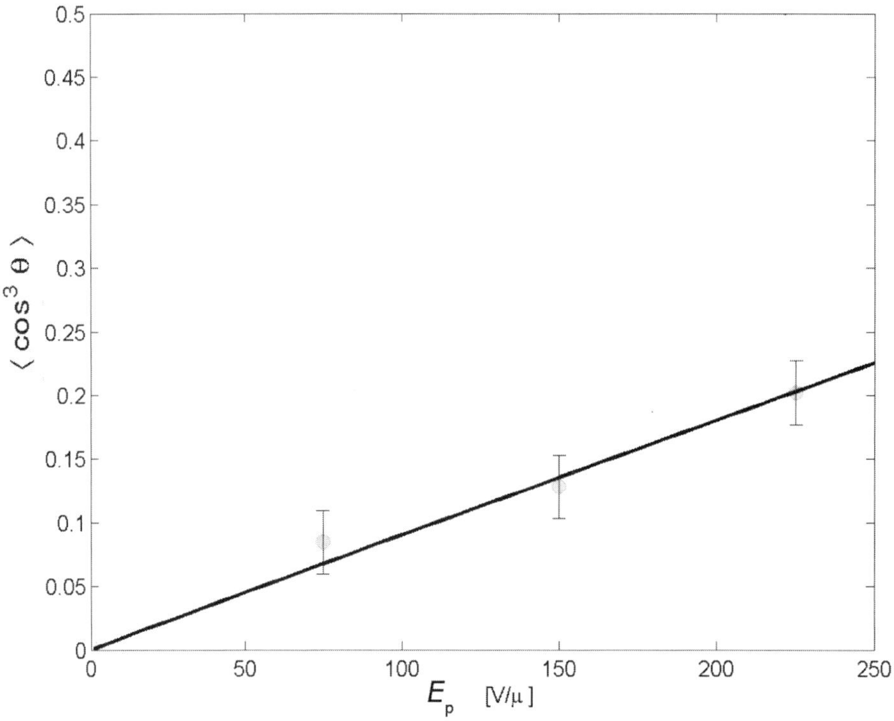

Figure 16.7 Both theory and experiment suggest a linear dependence for the variation of EO activity (r_{33}) with poling voltage (E_p). Data are shown for the PSLD-33 dendrimer.

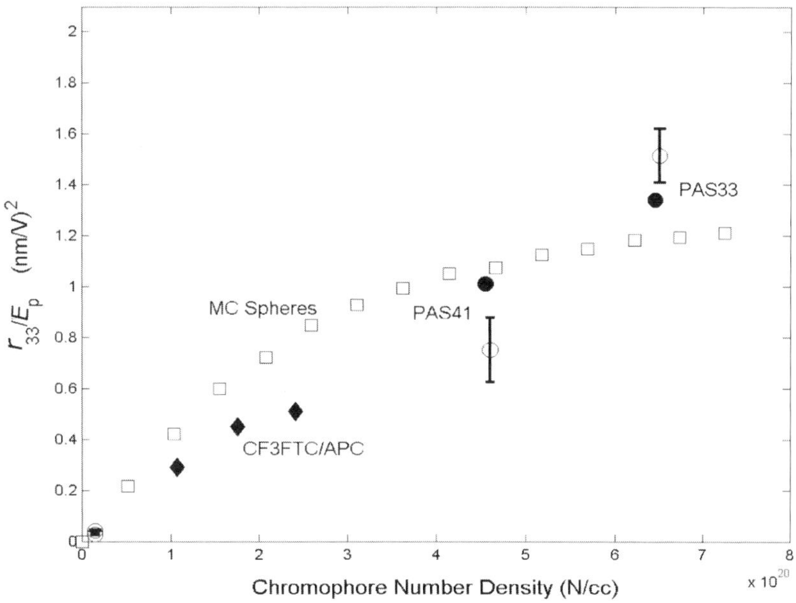

Figure 16.8 First principles theoretical data (open circles) and experimental data (solid circles) for PSLD-41 and PSLD-33 multichromophore-containing dendrimers. Also shown are the experimental data (solid diamonds) for the same chromophore in APC composite materials. Note that the maximum EO activity for the CF3FTC/APC composite materials occurs at a number density of 2.5 × 10^{20} chromophores/cc; the EO activity decreases at higher loading, and phase separation starts to occur. Substantially higher loading is achieved for the multichromophore-containing dendrimer materials; the chromophores act as independent particles (note the linear dependence on number density). Also shown are the Monte Carlo theoretical results for the CF3FTC chromophore treated as a spherical object, which can be considered the optimum EO activity for a single chromophore/polymer composite material.

potentials, and acentric order at low concentrations is less than that observed for the same chromophore (an FTC-type chromophore—see Fig. 16.2) in a polymer such as amorphous polycarbonate (APC). We have used polarized absorption spectroscopy[34] to independently measure order for the materials of Fig. 16.8. The results are consistent with the theoretical predictions shown in the same figure, i.e., with a $\langle \cos^3 \theta \rangle$ ~ 0.08–0.12, depending on poling voltage. Note also that $r_{33}/E_p = 1.33$ and $\langle P_2 \rangle /E_p = 1.33$ for PSLD-33, indicating the typical agreement between EO and polarized absorption spectroscopy for determining the field dependence of chromophore order. The measured ratio $r_{33}/r_{13} \approx 3$ (from ATR measurements) is also consistent with low acentric order.

Multichromophore-containing dendrimers do permit high chromophore loading to be achieved without phase separation or centrosymmetric crystallization. Indeed, the maximum EO activity that can be achieved with these materials exceeds the maximum that can be obtained for the same chromophore in chromophore-polymer composite materials (because the roll-off with increasing N is avoided). However, results to the present do not suggest that multichromophore-containing dendrimer materials can yield EO coefficients exceeding the Langevin limit (chromophores behaving as independent particles in a 3D lattice).

Multichromophore-containing dendrimers permit the dielectric environment of chromophores to be systematically tuned, a characterisitic that can be used to influence both EO and optical absorption (and index of refraction) properties. When functionalized with cross-linking moieties, dendrimers can lead to more homogeneous cross linking (and thus reduced light scattering) than is observed for chromophore/polymer materials (where a distribution of void volumes is an inherent problem).

The experimental results and conclusions for these multichromophore-containing dendrimers are not general for other dendrimer systems. For example, if the length of the flexible spacer between the dendrimer core and the chromophores is increased, then the EO behavior approaches that observed for chromophore/polymer composites. These results emphasize that EO activity in complex nanostructured materials such as dendrimers can be quantitatively predicted from first principles calculations. For the sake of brevity, we have not discussed the details of the synthesis and characterization of the dendrimers discussed above. The reader is referred elsewhere for this detail.[26,35,36]

The important conclusion of this section is that the large EO activities observed for multichromophore-containing dendrimers do not arise from high acentric order, but rather from high chromophore number density and enhancement of molecular first hyperpolarizability by the dielectric permittivity of the environment of the dendrimer materials. Indeed, the order parameter is quite small, suggesting that further substantial enhancement of EO activity can be achieved by increasing acentric order.

In the preceding sections, we have demonstrated that strong electronic dipolar interactions, strong nuclear repulsive (steric) interactions, and covalent bond potentials (all of which are spatially anisotropic interactions) can be used to influence the assembly and organization of charge-transfer chromophores. We have further demonstrated that the resulting order can be quantitatively understood from first principles quantum and statistical mechanical computations.

16.5 Binary Chromophore Organic Glasses (BCOGs)

16.5.1 Optimizing EO activity and optical transparency

Binary chromophore organic glasses (BCOGs) are a new class of organic EO materials that affords important control of dielectric permittivity and chromophore acentric order. BCOGs permit very high total chromophore concentrations (number densities) to be realized without the consequences of unwanted phase separation or problems associated with dielectric permittivity changing with chromophore concentration. The absence of chromophore-concentration-dependent solvato-chromic (bathochromic or "red") shifts for BCOGs minimizes absorption loss at telecommunication wavelengths. The absence of phase separation and index of refraction heterogeneity minimizes optical loss due to light scattering. Indeed, BCOGs may ultimately afford a route to "organic crystal engineering" based on solution-processed (e.g., spin-coated) materials.

A BCOG consists of a "guest" chromophore doped into a chromophore-containing "host" material. One might well expect such doping to lead to greater guest–host compatibility than occurs when doping chromophores into traditional polymer materials such as APC or poly(methylmethacrylate) (PMMA). This is indeed the case. The free energy of mixing of guest and host components of BCOGs should be more favorable from both enthalpic and entropic considerations. With BCOGs, a polar chromophore guest is dissolved in a polar host, in contrast to the traditional composite material that involves dissolving a polar chromophore guest into a nonpolar host. The shapes of both guest and host chromophores can be engineered to control steric interactions and thus control chromophore packing and order.

As shown in Fig. 16.9, for the case of doping the YLD-124 chromophore into PSLD-41, EO activity is observed to increase rapidly (and linearly) with added guest chromophore concentration. The rate of increase is a factor of 2–3 times greater than observed for doping the same guest chromophore into a traditional polymer such as APC. Logically, this increase in EO activity arises from an increase in $\beta(\varepsilon,\omega)$ or $\langle \cos^3\theta \rangle$, or a combination of these two effects. To discriminate among these possibilities, we have carried out detailed studies of changes of dielectric permittivity, solvatochromism, and spectral line broadening with guest chromophore doping (number density). The latter two studies involved the use of spectral deconvolution techniques for the guest and host chromophores when doping YLD-124 into PSLD-41. Because the conclusions of these studies are dramatically reinforced by the studies reported in the next section involving more first order analysis, we will not discuss spectral deconvolution results here other than to note that essentially no significant solvatochromic shifts or spectral line broadening were observed with increasing guest (YLD-124) chromophore concentration. This is in marked contrast to doping of chromophores into APC as discussed at length by researchers at Lockheed Martin Corp.[37,38] Moreover, we observed the changes in bulk dielectric permittivity to be too small to account for the observed changes in EO activity, although a detailed investigation of such effects continues. The details of these measurements and analysis will be discussed elsewhere, but the results are consistent with the lack of solvatochromic shifts discussed in the next section.

The EO behavior observed in Fig. 16.9 (and for many other BCOG systems[39–41]) is striking and implies that the acentric order parameter $\langle \cos^3\theta \rangle$ is either independent of concentration or increases linearly with concentration. Moreover, the rate of increase of EO activity is greater than that predicted from the independent particle approximation; thus, intermolecular electrostatic

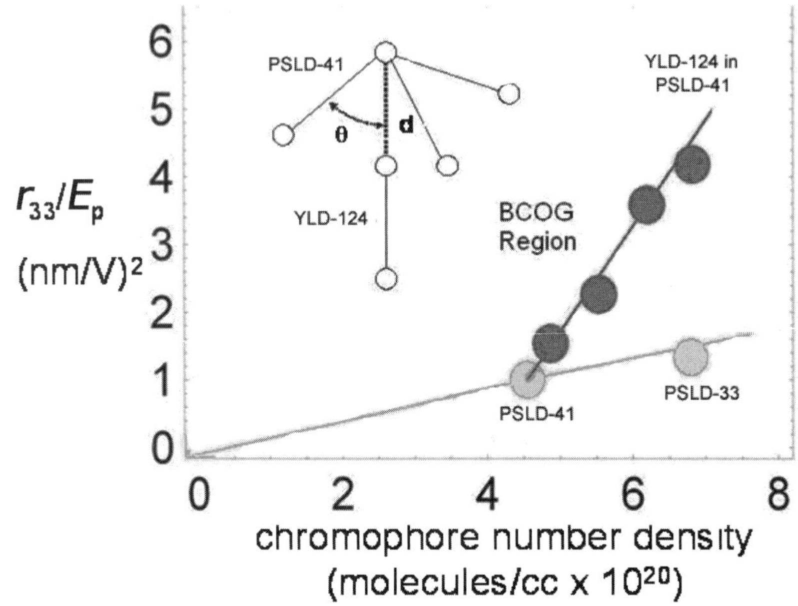

Figure 16.9 Experimental (dark solid circles) EO (r_{33}/E_p) data for binary chromophore organic glasses (BCOGs) formed by dissolving the YLD-124 chromophore into the PSLD-41 multichromophore-containing dendrimer (see Fig. 16.3). The course-grained Monte Carlo calculation results are indicated by the dark solid line. The inset is an illustrated representation of one particularly favorable interaction configuration between the two chromophore components of the BCOG. Also shown in this figure is the experimental data for the pure dendrimers (PSLD-33 and PSLD-41).

interactions must be acting to enhance poling-induced order if the increases in EO activity are to be assigned to increasing acentric order. We are currently investigating a variety of BCOGs employing the same pseudo-atomistic Monte Carlo approaches discussed in the preceding section. The completion of such studies is critical to a truly meaningful discussion of the various contributing factors to the dramatically improved EO activity observed for BCOGs. However, we have already utilized very coarse-grained Monte Carlo calculations based on the earlier mean field work of Prezhdo and coworkers[42] to analyze the results presented in Fig. 16.9. The initial coarse-grained Monte Carlo results are indicated by a dark solid line in Fig. 16.9 and show surprising agreement with the experimental data denoted by dark circles. These results have been mapped back to fully atomistic Monte Carlo calculations. The essential conclusions are summarized in the following paragraph.

Theory suggests that specific spatially anisotropic intermolecular electrostatic interactions among guest and host chromophores result in a significant increase in $\langle \cos^3 \theta \rangle$. We have examined snap shot pictures of equilibrium chromophore spatial distributions calculated by Monte Carlo methods. One very favorable interaction between guest and host chromophores involves the chromophore-containing host dendrimer (PSLD-41) forming an umbrella or pyramidal shape, and the guest chromophore (YLD-124) approaching from the base of the pyramid along the normal to the base. The donor region of the guest chromophore thus experiences a favorable interaction with the acceptor regions of the host chromophores. The strength of this interaction is many times that of the thermal energy *kT*. We must immediately note that this is just one of an enormous number of observed interactions, and the moderate order observed and calculated cautions against putting too much emphasis on such particularly favorable interactions. Indeed, it might be argued that the improvement in order is simply a result that is defined by nuclear repulsive interactions or lattice symmetry effects (see Fig. 16.10). As shown in Fig. 16.10, reducing the symmetry of the lattice that the chromophore experiences from 3D to 2D to 1D is predicted to result in an increase in $\langle \cos^3 \theta \rangle$.

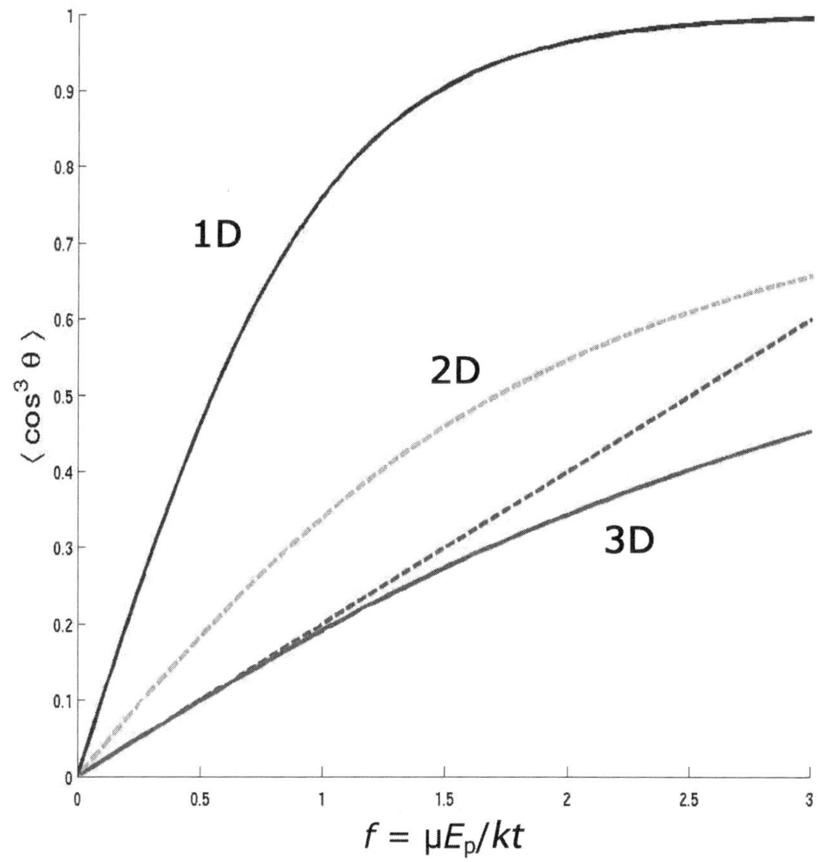

Figure 16.10 The theoretically predicted variation of acentric order parameter with poling energy is shown as a function of the symmetry of the lattice in which the chromophore is embedded. Acentric order, and thus EO activity, is predicted to increase in progressing from 3D (Langevin) to 2D (Bessel) to 1D (Ising) symmetry. The straight dotted line is the result predicted for independent particles (chromophores experiencing no intermolecular electrostatic interactions).

The one example where a specific spatial anisotropic interaction may clearly play a significant role is the case of quadrupolar aromatic-H•••aromatic-F dendron interactions reported by Jen and coworkers.[39,40] (The dots indicate the quadrupolar interaction that exists between the protonated and fluorinated aromatic dendrons.) In these dendrimer materials, long-range "chiral-like" interactions (orthogonal quadrupolar and dipolar interactions) may enhance noncentrosymmetric order. Such interactions also appear to influence material glass transition temperatures and the stability of poling-induced acentric order. This was the first BCOG to exhibit EO coefficients in excess of 300 pm/V, while also exhibiting optical loss of less than 2 dB/cm and glass transition temperatures greater than 200°C (when cross linked). To test the effect of host lattice order on guest chromophore orientation under a poling field and vice versa, we developed an experiment where the order of the host chromophore could be increased by laser-assisted poling (LAP).[43]

16.5.2 Laser-assisted poling (LAP)

The disperse red 1 chromophore-containing polymethylmethacrylate polymer host (DR-1)-co-PMMA used for laser-assisted poling (LAP) is shown in Fig. 16.3, and the YLD-124 guest chromophore used in these experiments is shown in both Figs. 16.2 and 16.3. The modification of our Teng-Man apparatus for LAP experiments using a linearly polarized optical field is shown in Fig. 16.4. Figure 16.11 shows a typical experiment demonstrating the dramatic increase in EO activity with LAP. LAP should have little impact on bulk dielectric permittivity; however, it is well known to increase the order of DR-1.[43,44] A comparison

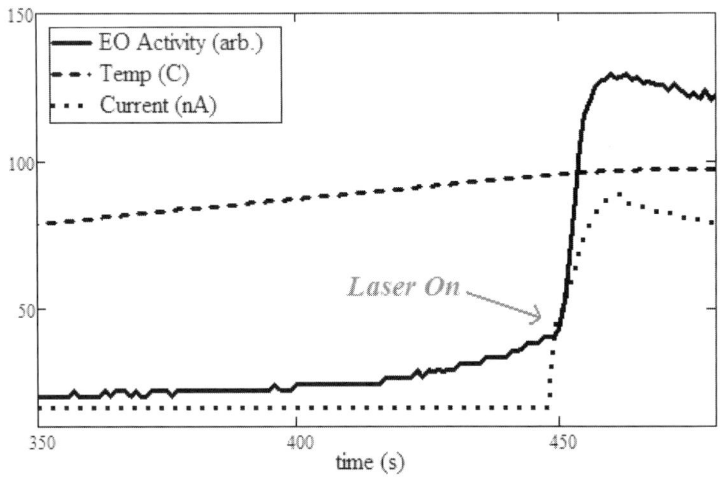

Figure 16.11 A representative real-time result from a laser-assisted poling experiment. The solid line represents the temporal behavior of EO activity (arbitrary units) measured using the Teng-Man apparatus of Fig. 16.4. The dashed line represents the variation of temperature (heating) with time. Note the dramatic increase in EO activity when the polarized laser field is turned on. Systematic investigation of the dependence of the effect on laser power and temperature has been carried out.

of the variation of EO activity with poling field is shown in Fig. 16.12 with and without LAP. In Fig. 16.13, we show the variation of EO activity with YLD-124 concentration with and without LAP. Although detailed simulations of this behavior have not yet been completed, there is little doubt that the observed phenomena reflect the influence of guest–host interactions and that LAP provides a systematic way of increasing host chromophore order. More detailed studies should shed greater light on the exact nature of these interactions and on the relationship between guest and host order.

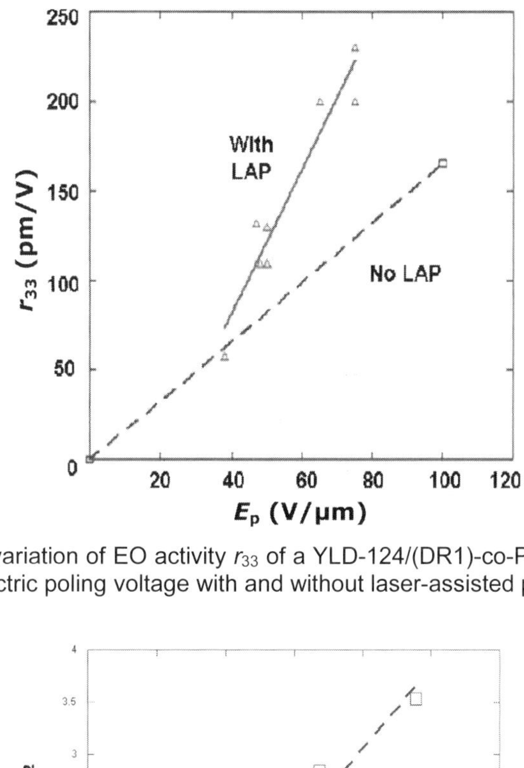

Figure 16.12 The variation of EO activity r_{33} of a YLD-124/(DR1)-co-PMMA BCOG shown as a function of electric poling voltage with and without laser-assisted poling.

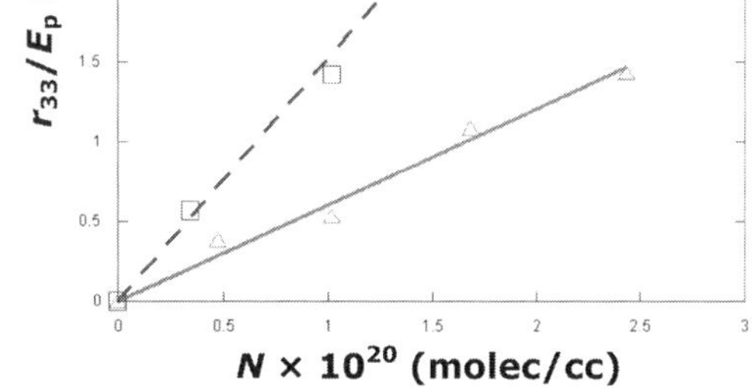

Figure 16.13 The variation of EO activity (r_{33}/E_p) of the YLD-124/(DR1)-co-PMMA BCOG shown as a function of dopant YLD-124 chromophore number density with laser-assisted poling (squares and dashed line) and without laser-assisted poling (triangles and solid line).

Figure 16.14 illustrates another reason for studying the YLD-124/(DR-1)-co-PMMA system. The charge transfer absorption bands of DR-1 and YLD-124 do not overlap, so the absence of solvatochromic shifts and spectra line broadening can be directly observed without resorting to spectral deconvolution methods. Enhanced EO activity is observed for BCOGs, while at the same time reduced optical loss is observed. This can be explained in terms of the better compatibility between guest and host materials, which leads to improved free energy of mixing and thus less phase separation at high chromophore loading. The weak dependence of spectral features on guest chromophore concentration results in reduced contributions to optical loss from both absorption and scattering. Spectral line positions and widths are essentially independent of chromophore concentration for BCOGs. Total material optical loss of 2 dB/cm or less is obtained even for materials with high chromophore loading. For example, the intrinsic thin film optical loss of the AJ415 BCOG (see Fig. 16.3) is approximately 1.4 dB/cm at 1.55-μm wavelength. Hydrogen vibrational overtone absorptions typically result in optical loss on the order of 1 dB/cm, so very little excess loss due to other mechanisms is observed for the AJ415 system.

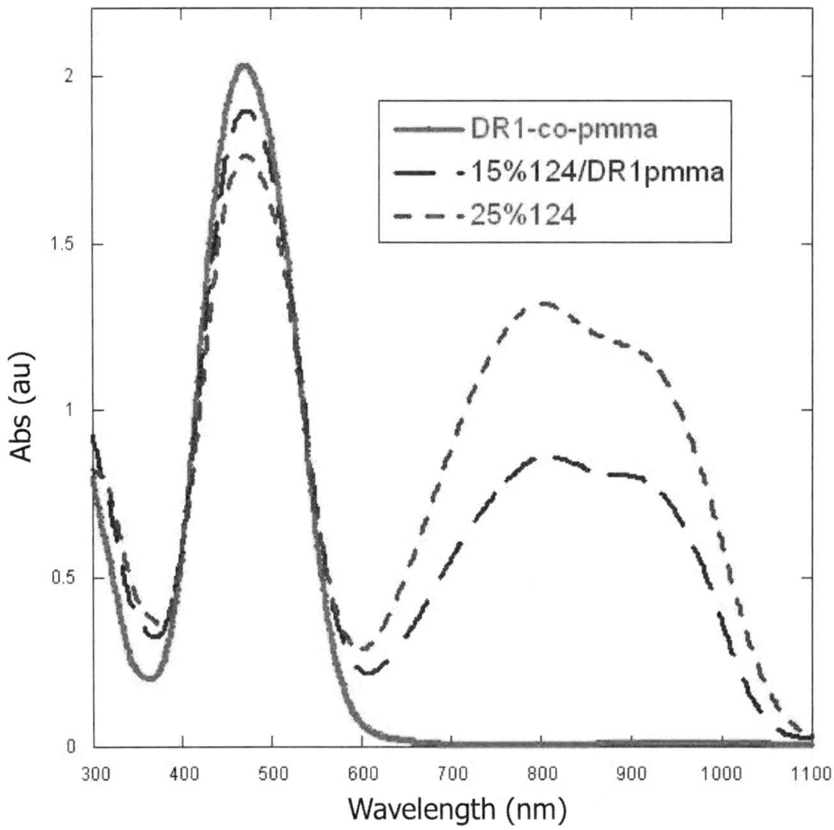

Figure 16.14 The linear absorption spectra of YLD-124/(DR1)-co-PMMA BCOGs shown as a function of YLD-124 concentration.

The LAP YLD-124/(DR1)-co-PMMA materials discussed in the preceding paragraphs are not appropriate for serious device applications because the material glass transition temperatures are too low. We would need to introduce cross linking to harden the final material lattice to an acceptable level for device applications. The general principles of lattice hardening will be discussed in Sec. 16.6 of this chapter. While the above system does lead to low optical loss (DR-1 does not make a significant contribution to absorption loss), the EO activity does not exceed that of other BCOGs. The primary reason for this is that the DR-1 chromophore does not make a significant contribution to EO activity even when its acentric order is increased by LAP and interaction with YLD-124 chromophores. Moreover, the aspect ratio of the DR-1 chromophore is not sufficiently large to have an optimum effect on increasing overall order and thus EO activity. Production of practical (for device application) LAP BCOGs will likely require development of improved LAP host materials that contribute significantly to total EO activity and produce further enhancement of total acentric order. Nevertheless, the current materials contribute appreciably to the understanding of the effect of guest–host intermolecular electrostatic interactions on EO activity.

LAP BCOG materials may be the most promising route to solution-processed EO materials with exceptional EO activity and desirable auxiliary properties that permit the production of devices with exceptional performance and prerequisite stability. Before we leave this section, let us consider, in the most general terms, the use of LAP to optimize EO activity. Certain types of chromophores dispersed in media that limit rotational diffusion can be oriented by optical poling. Setting aside a discussion of the detailed mechanism of the action of the laser field to produce trans-cis isomerization of the DR1 chromophore and the rapid relaxation of the cis isomer back to the trans-isomer conformation, the effect can be understood through consideration of Le Chatelier's principle: When a stress is applied to a system, the system adjusts itself so as to relieve the stress. When a laser field illuminates a sample containing a chromophore that is capable of absorbing the light and converting the energy into heat, the system will adjust itself so as to minimize the heating. Since the probability for absorption of energy is proportional to the scalar product $\mu_{ge}E$ of the transition moment μ_{ge} and the applied electric field E, the system can minimize heating if the molecules reorient themselves so that their transition dipoles are orthogonal to the applied field.

In order for this reorientation to lead to a useful ordering of the EO materials, the following five conditions are relevant: (1) The excited state must relax back to the ground state via librational-vibrational motions. In so doing, the inert matrix surrounding the dipole (chromophore) will undergo local heating, thus facilitating rotational (reorienting) diffusion of the chromophores. While trans-cis isomerization is implicated as the mechanism for conversion of electronic excitation energy into nuclear motions, other types of relaxation mechanisms, e.g., intersystem crossing, may be able to accomplish the same conversion. (2) The angle between the transition dipole and ground-state dipole vectors is a critical parameter. Assuming for the moment that the optical poling is perfect, the orientation of the molecules relative to the optical field could be anywhere. In practice, the orientation of the electric field component of the optical poling field should be independently optimized to maximize the desired EO enhancement. The transition dipole is calculable with quantum methods. (3) The temperature of the matrix should be as cold as possible while allowing the poling to occur. The local heating that results from absorption of a photon with energy of about 2 eV ≈ 46 kcal/mol is sufficient to heat about 500 atoms to a temperature on the order of 45 K. By maintaining the temperature of the sample this far below the glass transition temperature of the matrix, the poling efficiency should be maximized. This order of magnitude estimate seems to be reasonable relative to experiments—for systems investigated to date, the optimum temperature appears to be approximately 30 K below the glass transition temperature. Once the chromophore has rotated, the matrix must freeze around it so as to obstruct reversion. Optimally, all of the chromophores will have rotated "into the shade" so as to align their transition dipoles orthogonally to the applied optical field; they will not relax back when the optical field is removed because the matrix is sufficiently rigid to prevent rotational diffusion. Experimentally, the optimum poling temperature is defined by investigation of the temperature and optical power that leads to maximum EO activity using *in situ* monitoring of the experiment (see Fig. 16.4). (4) The optical poling must be done in the presence of an external static electric field so as to break the symmetry of the applied laser field. (5) Plane-polarized light will orient the transition dipoles in a plane orthogonal to the plane of the electric vector (2D Bessel order). If circularly polarized light is used, the transition dipole will be oriented orthogonal to the plane of the rotation (1D Ising order of the transition dipoles), although in a real system (due to higher-order symmetry considerations) the resultant order will lie somewhere between Bessel and Ising lattices, as confirmed by experiment.

16.5.3 Conductivity issues

James Grote and coworkers[45-50] have, for some time, pointed out the critical issue of the relative conductivities (or conversely, resistivities) of EO core and cladding materials for the fabrication of multistack (bottom electrode / bottom cladding / EO core / top cladding / top electrode) EO devices. The resistivity of the EO core must be higher than that of the cladding materials if voltage is to be dropped across the core material. If this condition is not met, poling efficiency will be compromised, and drive voltage requirements of devices will be increased.

Under the best of circumstances, this is not an easy requirement to satisfy. The index of refraction for cladding materials must be less than that of the EO waveguide materials so that light propagating in the EO core material is tightly confined in the core and prevented from interacting with metal electrodes. Such interaction would lead to unacceptable optical loss. The index of refraction condition is easily met due to the presence of π-electron chromophores in the core material. However, these π-electron chromophores can also lead to electrical conductivity under poling conditions; thus, it is difficult to simultaneously satisfy relative conductivity and index of refraction requirements for core and cladding materials. Efforts to increase the conductivity of the cladding layers have typically been accompanied by increased optical loss, which is unacceptable. Among the most successful efforts to improve the conductivity of cladding materials without unacceptable side effects is that of Peyghambarian and coworkers working with sol gel glass materials; these researchers have also addressed this issue by exploiting novel device structures.[51–54] Another route to dealing with core-cladding issues is to pursue electric-field poling and device operation using coplanar electrodes, techniques that have been investigated by Steier and coworkers.[55]

An effort has also been made to address the core-cladding issue by developing improved transparent metal oxide conducting electrode materials.[56,57] However, the conductivity of such electrodes is low compared to that of gold or copper electrodes, so this method can result in severe bandwidth limitations for devices, as bandwidth is currently defined by the resistivity of drive electrodes. Moreover, the optical loss of transparent metal oxide electrodes can be unacceptably high at telecommunication wavelengths. This, like the development of conducting cladding materials, remains a critical research challenge for realization of high-performance EO devices based on utilization of organic EO core materials.

The problem of conductivity is particularly problematic for BCOG materials and, indeed, currently limits the performance of these materials, which would be truly spectacular in the absence of this problem. This problem is illustrated in Figs. 16.15 and 16.16, where conductivity is shown to limit poling efficiency by defining the maximum effective electric poling field that can be realized across the EO thin film material. As seen in Fig. 16.16, the electric field at which "runaway" conductivity occurs decreases with increasing chromophore concentration (number density) for BCOG materials. Note that maximum-achievable EO activity is defined by the chromophore concentration and poling voltage at which runaway conductivity occurs. Poling voltage cannot be further increased beyond the value at which runaway current develops, as the conductivity simply drops the poling field felt by the chromophores. Conductivity in BCOG materials results from variable-range charge hopping between chromophores, which is obviously related to the spatial separation and relative orientation of chromophores. One potential method to decease this unwanted effect for materials with high chromophore number densities may be to sterically protect chromophores with "insulating" substituents.

To illustrate the severity of conductivity at the present, it can be noted that EO activity would be doubled in BCOGs if poling fields could be increased to normal values of 120 volts/μm. The problem is even more severe for poling of multistack device structures. If this problem could be solved, it is clear that high-bandwidth EO devices could be fabricated with operating voltages on the order of 100 mV, with no further improvements in chromophore molecular first hyperpolarizability. Recently, some improvement in poling efficiency has been achieved by adding thin (50–150 nm) layers of TiO_2 between indium tin oxide (ITO) electrodes and the organic EO material.

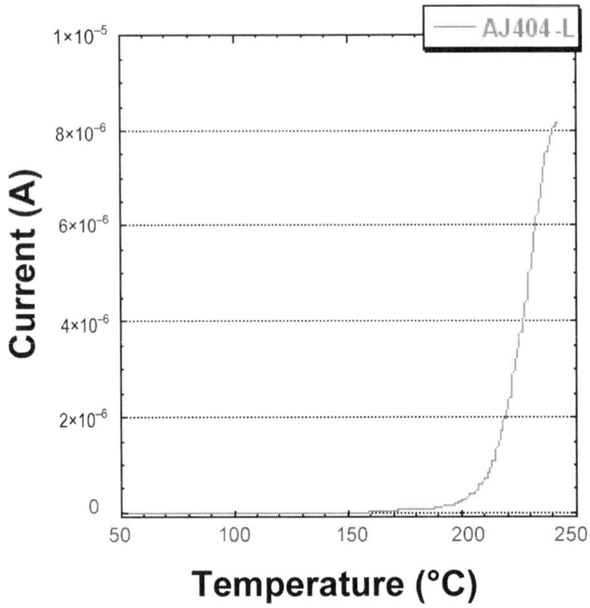

Figure 16.15 The variation of current with temperature is shown for a cross-linked BCOG. The onset of current flow indicates lattice softening and the movement of chromophores. This current flow could be due either to the diffusion of ionic impurities or to the onset of variable-range hopping of charge between the chromophores. Materials are typically purified to the point where no further change in current versus temperature curves is observed. Although ionic conductivity cannot be ruled out, it is our strong suspicion that the observed conductivity involves charge hopping among the chromophores.

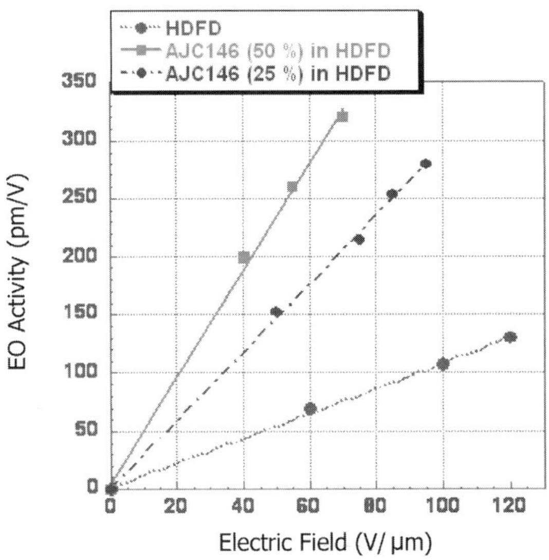

Figure 16.16 The variation of poling profiles with chromophore concentration is shown. The maximum-achievable poling voltage is defined by the onset of conductivity. Note that the rate of change of EO activity with electric poling voltage together with the maximum-achievable poling voltage define the maximum-achievable EO activity.

16.6 Thermal and Photochemical Stability: Lattice Hardening

Materials having glass transition temperatures between 100 and 200°C afford many advantages for processing. Not only are such temperatures attractive for avoiding chromophore decomposition during electric-field poling, but such temperatures permit use of soft and nano-imprint lithography techniques for the fast and low-cost fabrication of complex optical circuitry. However, such relatively "soft" materials are not suitable for producing devices that surpass Telcordia standards (http://www.telcordia.com/services/tsk.html), e.g., long-term operational stability at 85°C.

To satisfy Telcordia standards for thermal stability, a high glass transition temperature material is required. Typically, this condition has been satisfied by employing a high glass transition temperature host, such as a polyimide or polyquinoline polymer, to fabricate composite materials or by effecting cross linking of materials subsequent to induction of acentric order by electric-field poling. Cross-linking chemistries are also frequently used to harden cladding materials after spin-casting deposition; e.g., UV-curable epoxies are commonly used as cladding materials.

Use of a high glass transition temperature polymer as a host for the preparation of EO composites has a number of drawbacks. The high processing temperatures (200°C and greater) required for such materials can lead to sublimation of chromophores and to chromophore decomposition. Moreover, harsh solvents are often required for spin casting of such materials, leading to poor optical quality films and high optical loss.

Lattice hardening subsequent to poling has become an attractive option, permitting low-temperature processing through spin casting and poling stages and then permitting a high glass transition temperature material to be generated at the end of the poling process. Prior to 2000, most cross-linking protocols focused on condensation reactions (e.g., urethane chemistry) or radical addition reactions.[58,59] Light-induced cross-linking chemistries were pursued without much success because the EO chromophores compete with the cross-linking initiators for light.

A breakthrough in lattice hardening occurred with the introduction of two cycloaddition reactions: (1) The Diels-Alder/Retro-Diels-Alder reaction[36,40,60,61] involving dienes and dienophiles and (2) the thermally initiated soft free-radical reaction of the fluorovinyl moiety to yield cyclobutyl cross links.[62-65] The latter reaction was popularized by Dennis Smith and his colleagues at Clemson University.[65] Both cycloaddition chemistries are thermally activated. Both approaches have yielded hardened EO materials exhibiting glass transition temperatures on the order of 200°C. The resulting hardened materials meet and surpass Telcordia standards for thermal stability. Such lattice hardening also improves photostability.

An advantage of the Diels-Alder/Retro-Diels-Alder reaction is that the processing temperature and the reversibility of the cycloaddition reaction can be controlled by the choice of diene and dienophile reactants (see Fig. 16.17). The

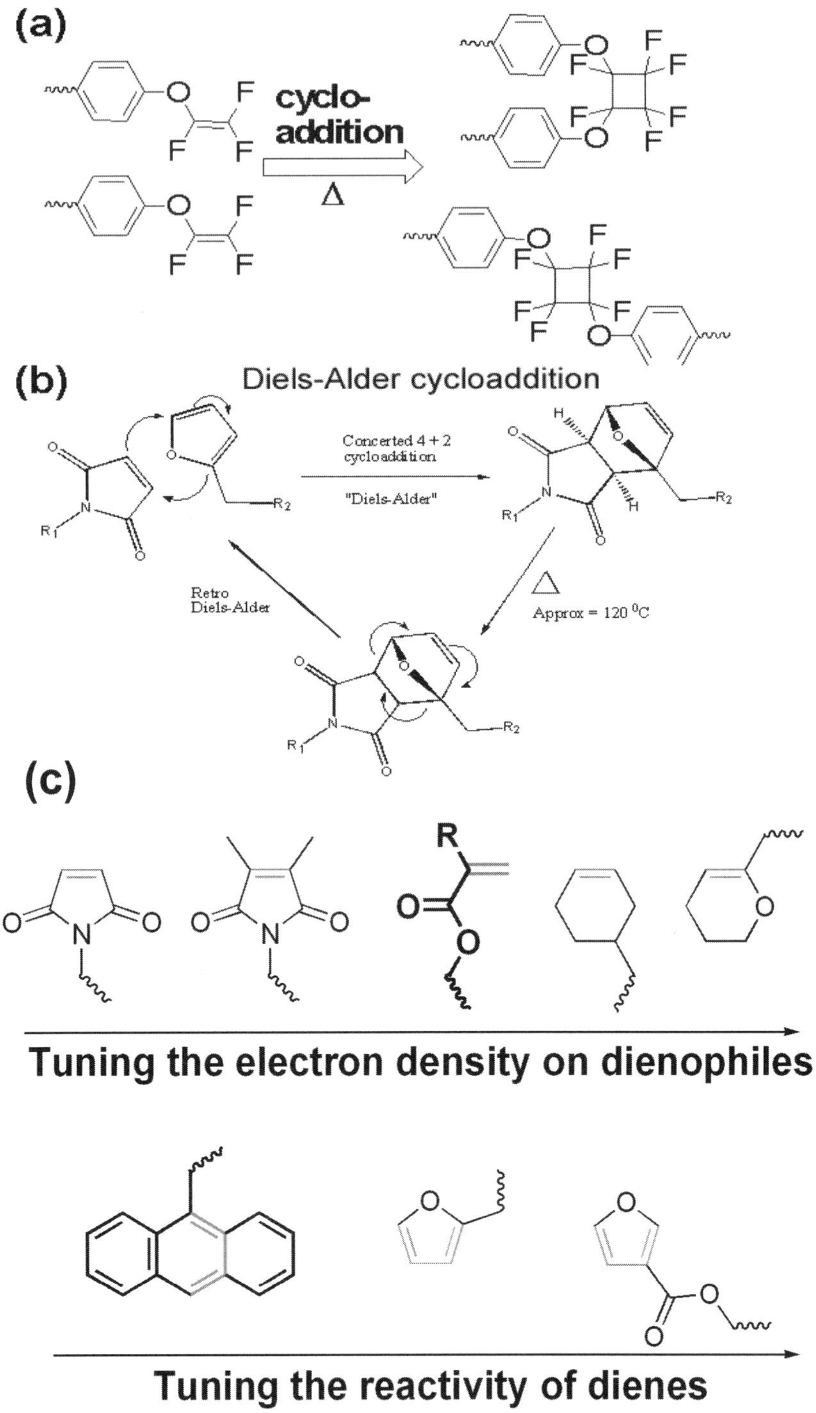

Figure 16.17 (a) Cycloaddition cross linking by reaction of a fluorovinyl ether moiety to form cyclobutyl cross links. (b) Diels-Alder/Retro Diels Alder cross linking. (c) Representative diene and dienophiles.

UV curing of cladding materials in triple-stack (lower cladding / EO core / upper cladding) devices can be problematic for EO materials containing cycloaddition precursors. For example, reaction of the radicals from the cladding materials with the anthracene diene can produce anthracenyl radicals that can attack chromophores and disrupt the stoichiometry of the Diels-Alder reaction.

Another problem associated with the use of cross-linkable EO materials arises when poling is carried out through cladding layers. Because the core EO material is not hardened when the upper cladding layer is deposited, the deposition process associated with the upper cladding layer can influence the surface smoothness of the EO core waveguide, leading to light scattering and unacceptable optical loss. Moreover, if the glass transition temperature of the cladding layer is not much higher than that of the core, the two may interdiffuse during poling. Several options exist for circumventing this problem, including corona poling and lattice hardening, followed by cladding layer deposition. Alternatively, electrode (parallel plate or coplanar) poling can be carried out and the EO material hardened, followed by removal of the electrodes and deposition of the upper cladding layer and the upper electrodes.

When discussing thermal stability, a distinction should be made between the stability of poling-induced order (which relates to the final glass transition temperature of the material) and the thermochemical stability of molecular components of the EO material. Normally, for use in the production of EO materials, chromophores are required to exhibit thermochemical stability greater than 250°C. Thus, the thermal stability of devices usually relates to the glass transition temperature of the EO material, which is on the order of 200°C or somewhat lower. Photochemical stability is equally important as thermal stability; organic EO materials must be capable of withstanding power levels used in telecommunications (currently, 10–20 mW) for many years.

16.7 Thermal and Photochemical Stability: Measurement

The modified Teng-Man apparatus of Fig. 16.4 is highly effective for measuring thermal stability, since the temporal stability of EO activity can be continuously measured at various elevated temperatures. Thermal stability is also conveniently measured by ramping temperature and observing the temperature at which EO activity is first observed to decrease (see Fig. 16.18 for an example).[35,66] Such measurements permit the activation energies for rotational relaxation to be quantitatively defined, as discussed elsewhere.[66] Insight into thermal stability can also be obtained by measurements of material glass transition temperature using thermal analysis [differential scanning calorimetry (DSC)] methods and by measuring the temperature dependence of conductivity (see Fig. 16.15), which reflects the temperature at which the lattice first begins to soften. The glass transition temperature measured by DSC is frequently observed to be higher than that measured by Teng-Man or by conductivity measurements. This is because the former reflects overall melting while the latter can reflect local melting

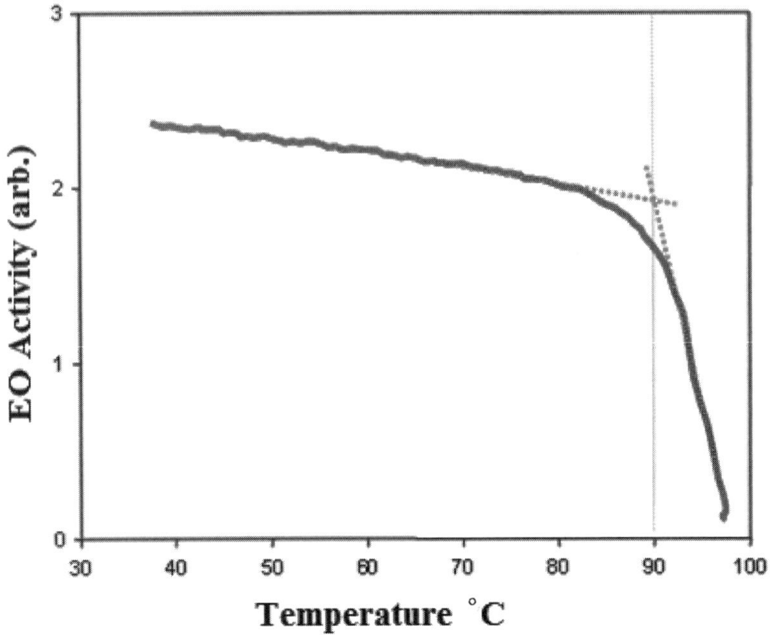

Figure 16.18 The variation of EO activity (measured using the *in situ* Teng Man apparatus of Fig. 16.4) with thermal ramping. The material used in this example is a non-cross-linked multichromophore-containing dendrimer (similar to the structures shown in Fig. 16.3 but with different spacer moieties).

(molecular motion, such as torsional motion). Such glass transitions and pre-transitions are common in complex organic materials such as polymers and lipid bilayer materials. The pre-transition is more relevant to definition of the thermal stability of EO activity, as the pre-transition can lead to significant loss of EO activity. With cross-linked materials, it is not uncommon to observe a stepped loss of EO activity in thermal ramping experiments; this observation typically reflects heterogeneity in cross-link density.

Stability studies can also be carried out on devices; recently, Ashley, Lindsay, and coworkers[67] conducted an interesting study of the FTC and CLD chromophores (see Fig. 16.2) incorporated into APC and polyimide polymer hosts. In their studies, the change of drive voltage (V_π) is recorded as a function of time. Their conclusion is that multiyear operational stability is possible with the materials used in their research.

Photochemical stability can also be assessed in device structures; however, a more useful method appropriate for mechanistic studies uses pump and probe lasers, as shown in Fig. 16.19. Some representative results are presented in Table 16.2. The pump-probe method permits accelerated photodegradation studies to be conducted using pump powers up to 1 W. Even at the highest pump power levels, rates of photochemical decay are slow, requiring single experiments to be conducted over periods of days. Because of long measurement times, care must be exercised to ensure that the optical probe power does not influence measurements.

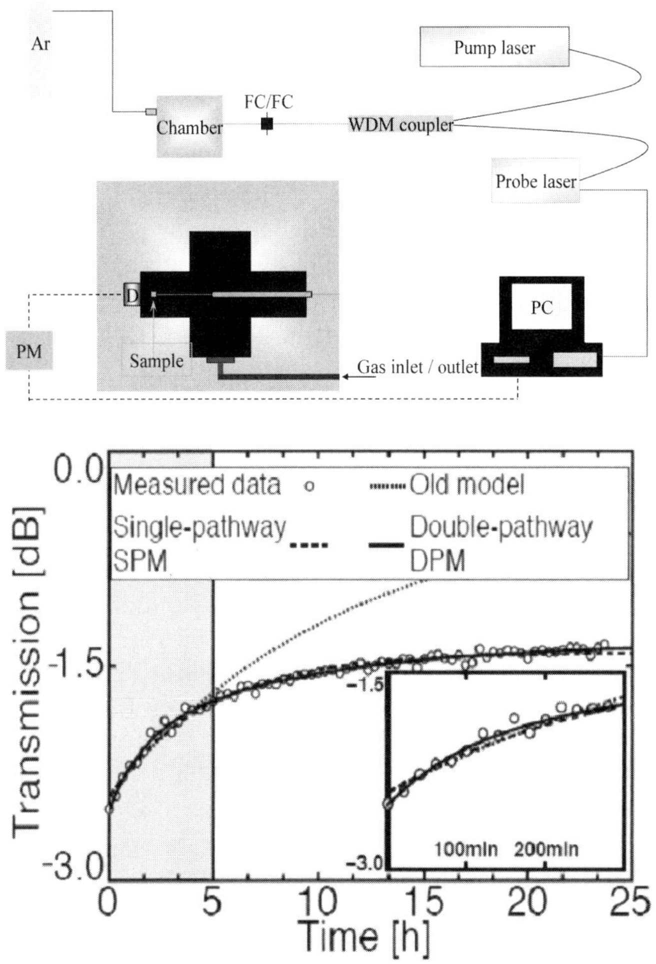

Figure 16.19 A schematic diagram of the apparatus used to define photostability is shown together with a typical kinetic trace. The experimental data is for a 25% FTC sample in APC.

Table 16.2 Representative photostability data for FTC and CLD chromophores in APC.

Sample	B/σ^a x10^{32} (m^{-2})	B/σ^b x10^{32} (m^{-2})	B/σ^c x10^{32} (m^{-2})
CLD-1 in APC	3.2 ± 0.6	0.18 ± 0.03	2.3 ± 0.6
CLD-1 + Bis(dithiobenzil)nickel in APC	8 ± 2	2.2 ± 0.7	19 ± 5
FTC in APC	7.2 ± 0.7	0.8 ± 0.3	8 ± 2
FTC + Bis(dithiobenzil)nickel in APC	66 ± 33	9 ± 10	165 ± 136

[a]$(B/\sigma)^a$ was determined by fitting of data to a single exponential following Stegeman and coworkers.
[b]$(B/\sigma)^b$ was determined by fitting of data to a two-exponential expression. The figure of merit corresponds to the fast decay described by the first exponential.
[c]$(B/\sigma)^c$ was determined by fitting of data to a two-exponential expression. The figure of merit corresponds to the slow decay described by the second exponential.

It is generally recognized that the dominant mechanism of photochemical decay in organic EO materials (and other organic electroactive materials) involves singlet oxygen chemistry.[68–75] No direct bond breaking or two-photon-activated processes are observed with currently employed power levels (or even in high-pulse-power femtosecond pulse experiments). Stegeman and coworkers[69–72] have demonstrated most of the critical features of the photodecomposition of organic EO materials, including the absence of contributions from multiphoton absorption. They have defined a single photostability figure of merit B/σ, where B^{-1} is the probability of photodecay from the lowest unoccupied molecular orbital (LUMO) charge transfer state and σ is the interband (charge transfer) absorption coefficient. This definition has been used by subsequent researchers, although the data analysis of Stegeman and coworkers may have been somewhat overly simplistic with the consequence of over-estimating photo-instability. For example, least-squares analysis of decay data with two exponentials yields improved fitting relative to the use of a single exponential model[75]; also, the observer power in the Stegeman pump-probe experiments may lead to artificially faster decay by contributing to photodegradation rather than simply observing it. Moreover, Stegeman and coworkers failed to carry out measurements at telecommunication wavelengths; this was addressed in subsequent work by researchers at Corning[73,74] and elsewhere.[75] The Corning group demonstrated that the photostability figure of merit for a given chromophore structure could vary over four orders of magnitude depending on conditions that influence singlet oxygen chemistry. Even larger variation has been observed by other groups, and photostability has been shown to improve with use of small quantities of singlet oxygen quenchers (see Table 16.2). In addition to the pump-probe experiments (carried out by Stegeman and coworkers, researchers at Corning, Gunter and coworkers, and Dalton and coworkers), photostability has also been investigated in operating Mach Zehnder devices by Steier and coworkers and by Ashley and coworkers. Again, in these studies, photo-instability could be attributed to singlet oxygen chemistry, with improved photostability being observed for materials and devices in which this chemistry was partially inhibited (by partial exclusion of oxygen, i.e., by rudimentary packaging).

In summary, it appears that reasonable photostability can be achieved with appropriate materials modification or with appropriate packaging of devices to minimize the presence of oxygen. In this latter regard, the problems faced with organic EO materials are analogous to those faced with organic light-emitting diode (OLED) materials. It should be noted that dense crystals such as DAST (4-dimethylamino-N-methyl-4-stibazolium tosylate) exhibit excellent photostability, again consistent with the role played by singlet oxygen chemistry in photodecomposition. The dense lattices of crystalline materials inhibit oxygen indiffusion.

When space applications of EO materials are considered, radiation hardness may be important (depending on how EO devices are packaged and where they are located in satellites). Few studies of the stability of organic EO materials in the presence of high-energy radiation have been published. However, the one published study suggests reasonable stability in the presence of high-energy gamma rays and protons.[76]

16.8 Devices and Applications

A distinct advantage of organic EO materials relative to their crystalline inorganic counterpart is their processability and their ability to be integrated with a diverse range of materials. Conformal and flexible devices have been fabricated by lift-off techniques. These devices exhibit excellent retention of performance properties (drive voltage, bias voltage, and insertion loss) with repeated and extreme flexing and bending.[77]

In addition to standard reaction ion etching (RIE)[78] and photolithographic techniques,[58] devices (e.g., Mach-Zehnder modulators and ring microresonators) can be fabricated employing nano-imprint and soft lithography techniques.[79,80] Organic EO materials have also been incorporated into silicon photonic devices, including ring microresonators[31] and Mach-Zehnder device structures.[81]

Three fundamental device structures have been commonly investigated. These include (1) stripline waveguide structures such as Mach-Zehnder and birefringent modulators, (2) resonant structures such as ring microresonators and etalons, and (3) prisms including superprism structures. With stripline devices, a critical device performance parameter is the voltage required to effect modulation or switching. For a Mach-Zehnder modulator, the parameter is the drive or V_π voltage (the voltage required to produce a π phase shift in light passing through the device). This is the voltage required to effect optimum transduction of electrical signal information onto an optical carrier as an amplitude modulation. The operational equation is

$$V_\pi = \frac{\lambda h}{n^3 r_{33} L \Gamma}, \tag{16.2}$$

where λ is the operating wavelength, h is the electrode spacing, L is the electrode length, and Γ is the modal overlap parameter. If a push-pull Mach-Zehnder interferometer structure is employed, a factor of 2 should be added to the denominator of the V_π equation. Because of the dependence of bandwidth on the resistivity of metal electrodes, choice of electrode length will impact both bandwidth and drive voltage. Device length will also influence insertion loss. A more detailed discussion of performance trade-offs with device designs is given elsewhere,[82] but if one assumes for typical material and device dimension values r_{33} = 300 pm/V, h = 8 μm, microwave electrode (gold) loss = 0.75 dB(GHz)$^{1/2}$/cm, fiber coupling loss = 0.8 dB/facet, material waveguide loss = 2 dB/cm, and L = 5 mm, then V_π = 0.75 V, 3-dBe bandwidth = 90 GHz, and total insertion loss = 2.6 dB. Clearly, the large EO activity coefficients of organic materials permit superior performance to be realized in each of the critical areas of device performance (drive voltage, bandwidth, and insertion loss). Significantly reduced device dimensions also afford advantage, particularly when integrating EO device technology with electronic and photonic circuitry. Drive voltage expressions for other stripline device configurations are discussed elsewhere.[83]

An analogous equation exists for prism and cascaded prism device structures. For such devices, the critical performance parameter is the beam steering angle, which is given by $\theta = (n)^3\, r_{33}\, (VL/dh)$, where L and h are the length and width of the prism or the array of prisms,[84,85] and V is the voltage applied across an overall thickness d (or the electrode spacing). In a cascaded prism device, L is the length of the base of the prism cascade. Again, it can be seen that smaller devices can be used if EO activity is sufficiently high.

For resonant devices such as ring microresonators[86,87] and etalons[53], drive voltage and bandwidth performance cannot be separated. The quality Q factor (number of times light transits the resonant structure before being lost) influences both; drive voltage and bandwidth decrease with increasing quality factor. The critical performance factor for resonant devices is thus the bandwidth/voltage sensitivity factor. The best value for this factor currently obtained with organic EO materials (incorporated into silicon photonic ring microresonators) is approximately 18 GHz/V, which means that the bandpass notch (see Fig. 16.20) associated with the ring microresonator is tuned by 18 GHz with application of a voltage of 1 V. To achieve optical amplitude modulation employing resonant device structures, the bandpass (notch filter) must be shifted on the order of a full width at half maximum. Enhanced performance is observed for organic EO/silicon photonic ring microresonators relative to all-organic devices, due to optical field concentration and reduced electrode spacings.[87] Indeed, optical field intensities are sufficiently large in organic-EO-material-filled 70-nm slots of slotted silicon waveguides that optical rectification is observed with milliwatt and even microwatt input powers.[87] Millivolt EO modulation[88] is possible with such devices. These devices can be used as the fundamental active elements of reconfigurable optical add/drop multiplexers/demultiplexer (ROADM) chipscale routing systems.[89]

Another example of an application exploiting optical rectification is terahertz electromagnetic generation and detection.[90,91] While lithium niobate (device-relevant EO coefficient of 30 pm/V) is the benchmark for EO modulation applications, zinc telluride (device-relevant EO coefficient of 4 pm/V) is the benchmark for terahertz technology, based on high-pulse-power femtosecond pulses. In addition to affording significant advantage associated with greater EO activity, organic materials afford dramatically improved bandwidth (currently, terahertz generation and detection is demonstrated to 12 THz, but 30 THz may be possible). The improved bandwidth performance is associated with the absence of loss in specific THz spectral regions associated with the phonon modes of crystalline ZnTe. Organic EO materials also afford more facile phase matching of optical and THz waves, leading to dramatically improved sensitivity. The most serious difficulty currently experienced in using organic EO materials for THz applications is the requirement of fabricating thick (millimeter rather than micron thick) films. If this processing challenge can be overcome, organic EO materials could facilitate the production of compact THz spectrometers.

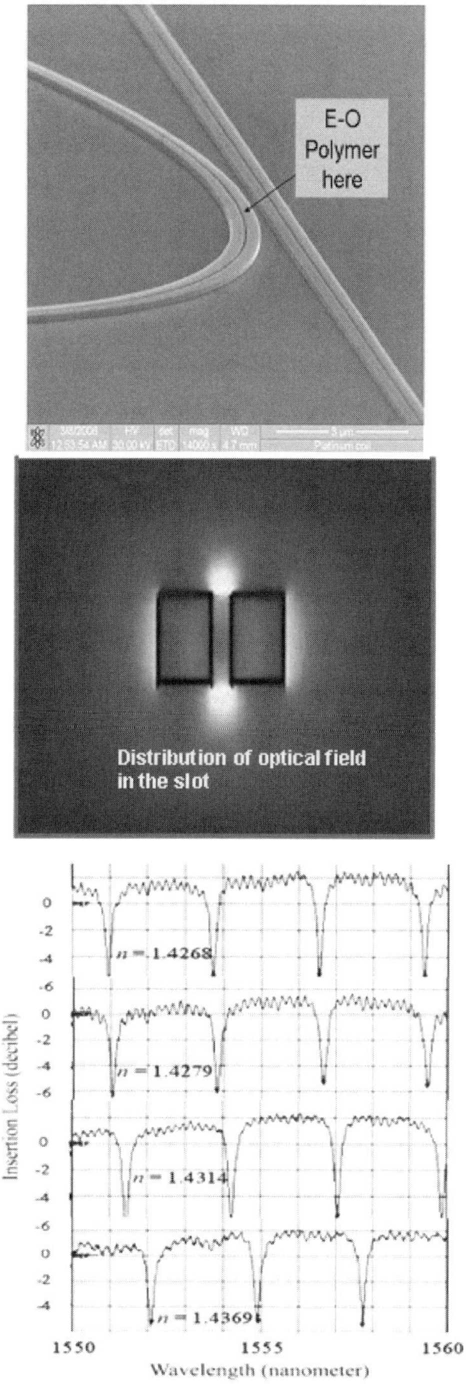

Figure 16.20 A micrograph of a slotted ring resonator (top) is shown together with EO tuning of the bandpass notches (bottom). The middle figure shows the concentration of the optical field in the silicon waveguide slot where the organic EO material is located.

Applications of organic EO materials are broad, ranging from optical gyroscopes, to phased-array radar, to GHz A/D conversion, to digital signal processors, to acoustic spectrum analyzers, to sensors.[92–98] Sensing applications are receiving increasing attention.[99]

16.9 Summary and Conclusions

Organic electro-optic (EO) materials currently exhibit a number of advantages relative to their inorganic counterparts (e.g., lithium niobate, zinc telluride, etc.). These include greater EO activity, faster response time, superior processability facilitating the mass production of sophisticated and highly integrated (including 3D) circuitry, improved compatibility with a diverse range of materials, and potentially lower cost. Crystalline inorganic EO materials retain advantage with respect to optical loss and stability. Note that these two advantages of inorganic EO materials do not hold for inorganic electro-absorptive materials.

A distinct advantage of organic EO materials is that properties can be further dramatically improved by molecular and supermolecular engineering. For example, it is likely that the hyperpolarizability of chromophores will be further improved, perhaps by more than an order of magnitude. In like manner, the acentric order of organic EO materials can obviously be further improved, and perhaps "crystalline" organic EO materials can be produced by theoretically inspired design. In this communication, we have presented several paradigms for the improvement of EO activity, including the special intermolecular electrostatic interactions of binary chromophore organic glasses (BOCGs), laser-assisted electrically poled BOCGs, and self-assembly/sequential synthesis of materials in nano-slot silicon waveguides. It is within the realm of possibility that EO coefficients of "purely electronic" organic EO materials can be increased to values competitive with liquid crystalline materials, while retaining the very fast response times and advantages of robust thin solid (hardened) films. Auxiliary properties of organic EO materials are also likely to be further improved. Optical loss values as low as 0.1–0.2 dB/cm have been observed for "proton-deficient" dendrimer materials exhibiting optical nonlinearity comparable to lithium niobate. The ability to fine tune diverse properties of organic EO materials should permit the tailoring of materials to specific device application requirements. Obviously, there are upper limits to the improvement of organic EO materials; even if crystalline materials can be produced by design, thermal stability will likely be limited to temperatures below 300°C.

In addition to commenting on the potential for future dramatic improvements in material performance, it is useful to speculate on the performance likely achievable within the next couple of years. Electro-optic coefficients in the range of 500–1000 pm/V (at telecommunication wavelengths) are highly probable for materials exhibiting optical loss of less than 2 dB/cm and exhibiting stability that satisfies Telcordia standards. Such materials should permit the fabrication of devices with operating voltages on the order of 0.1 V (and with bandwidths of > 20 GHz). Indeed, Lumera Corporation (Bothell, WA) has already demonstrated modulators with drive voltages of 0.3–0.4 volts and > 20 GHz bandwidths. A strong driver for pushing toward 0.1-V drive voltages is the potential for realizing gain in RF photonic applications. For resonant devices, bandwidth/sensitivity factors of greater than 20 GHz/V should be obtainable. The bandwidth and sensitivity gains to be realized for optical rectification (terahertz sources and detectors) should be dramatic and should facilitate the development of lightweight compact terahertz systems including spectrometers. The integration of organic EO materials with silicon photonics appears to be especially attractive, affording the concentration of both optical and electric field, due to reduced device dimensions. The smaller dimensions of such device structures may permit new processing techniques (layer-by-layer deposition and cross linking of self-assembling materials) to be effectively implemented.

The importance of organic EO materials does not lie in simply replacing inorganic EO materials in simple device structures such as Mach-Zehnder interferometers, but rather in the stimulation of new applications and device concepts. A significant opportunity for organic EO materials is to facilitate the production of highly integrated electronic/photonic platforms (chipscale integration and higher). The integration of organic EO materials with silicon photonics is particularly attractive and provides a route to utilization of the tremendous resource of existing CMOS foundries.

While significant molecular, supramolecular, and device engineering challengcs cxist, the field of organic EO materials appears to be entering an exciting new era that will likely see dramatic improvement in materials properties and device performance. The lessons learned in EOs are likely to be relevant to the engineering of improved properties for other organic electroactive materials and device applications (electronic, photovoltaic, photorefractive, and light emitting).

16.10 Appendix: Linear and Nonlinear Polarization

In this Appendix we attempt to provide a very brief introduction to the fundamental concepts related to organic nonlinear optical (NLO) molecules (chromophores) and materials. Nonlinear optical effects result from the nonlinear polarization of molecules and materials under the influence of either static electric fields or the oscillating electric fields associated with electromagnetic radiation. Before discussing nonlinear polarization, we briefly review some concepts regarding linear polarization. Application of an electric field E to a system of charges can result in a charge separation; for an organic molecule with extended π-conjugation, this charge separation will be dominated by the electrons in the more polarizable π-system. Considering only linear polarization, the polarization induced by an electric field E, i.e., the induced dipole moment μ, is proportional to the field

$$(\text{Polarization})_i = \mu_i = \alpha_{ij} E_j , \tag{16.3}$$

where i and j refer to components in the molecular frame. The constant α_{ij} is the relevant component of the linear polarizability α, which is a second-rank tensor that relates the applied electric field to the induced polarization in a molecule (the two subscripts relate to the molecular and applied field axis systems). For example, the component of the dipole moment induced in the x direction by an electric field oriented along the x direction is determined by α_{xx}, whereas the component induced along the y axis is determined by α_{yx}, and so on. The linear polarization per unit volume in a bulk material composed of these molecules is given by an analogous equation

$$P_i = \chi_{ij} E_j , \tag{16.4}$$

where χ_{ij} is the linear susceptibility tensor of the material. When no significant intermolecular interactions occur, the bulk susceptibility χ_{ij} is related to the sum of individual molecular polarizabilities α_{ij}. While Eqs. (16.3) and (16.4) are reasonable approximations for small applied electric fields, in general, polarization is not a linear function of field. In more intense fields, such as those associated with laser light, nonlinear polarization must also be considered to adequately describe the overall polarization. The nonlinear dependence of dipole moment on field can be expressed using the following Taylor expansion series:

$$\mu_{i(E)} = \mu_{i(0)} + \left.\frac{\partial \mu_i}{\partial E_j}\right|_0 E_j + \frac{1}{2!}\left.\frac{\partial^2 \mu_i}{\partial E_j \partial E_k}\right|_0 E_j E_k + \frac{1}{3!}\left.\frac{\partial^3 \mu_i}{\partial E_j \partial E_k \partial E_l}\right|_0 E_j E_k E_l + \ldots, \tag{16.5}$$

where the 0 subscripts indicate the values of the differentials at $E = 0$, and i, j, k ... refer to components in the molecular frame. The first term, $\mu_i(0)$, is the permanent dipole moment (in the i direction) of the molecule in the absence of an applied electric field. The second term is equivalent to the linearly induced dipole moment of Eq. (16.1). The remaining terms describe *nonlinear polarization*. The values of the first, second, and third differentials of dipole moment with respect to the electric field are the linear polarizability α, the first hyperpolarizability β (a third-rank tensor), and the second hyperpolarizability γ (a fourth-rank tensor), respectively; i.e., one can rewrite Eq. (16.5) as

$$\mu_{i(E)} = \mu_{i(0)} + \alpha_{ij} E_j + \frac{1}{2!}\beta_{ijk} E_j E_k + \frac{1}{3!}\gamma_{ijk} E_j E_k E_l + \ldots. \tag{16.6}$$

It is important to note that there are several alternative definitions of β and γ (and higher hyperpolarizabilities); often, it is not clearly stated in the literature which convention is being followed. The use of different conventions can lead to confusion when comparing different values of hyperpolarizabilites reported in different papers. A fairly common definition is to equate the $(n-1)^{\text{th}}$ order hyperpolarizability to $(1 / n!) \times (\partial^n \mu / \partial E^n)$, rather than simply to the differential.

For the bulk material, nonlinear polarization density is given by an expression analogous to Eq. (16.6):

$$P_{i(E)} = P_{i(0)} + \chi_{ij}^{(1)} E_j + \chi_{ijk}^{(2)} E_j E_k + \chi_{ijkl}^{(3)} E_j E_k E_l \ldots, \tag{16.7}$$

where $\chi^{(n)}$ are linear ($n = 1$) and nonlinear ($n > 1$) susceptibilities, and $P(0)$ is the intrinsic static dipole-moment density of the sample.

Before considering how second-order NLO effects are related to second-order nonlinear polarization, it is worth noting that there is an important symmetry restriction on second-order NLO properties. Equation (16.7) indicates that $P(E) = P(0) + \chi^{(1)}E + \chi^{(2)}E^2 + \chi^{(3)}E^3 + \ldots$, and that $P(-E) = P(0) - \chi^{(1)}E + \chi^{(2)}E^2 - \chi^{(3)}E^3 + \ldots$. Clearly, if $\chi^{(2)} \neq 0$, then $P(E) \neq P(-E)$. However in a centrosymmetric material $P(E)$ is, by definition, equal to $P(-E)$; therefore, $P(0)$, $\chi^{(2)}$, and other even-order terms must be zero. Hence, effects arising from second-order polarization can be observed only in molecules or materials that are non-centrosymmetric. In contrast, no such restrictions apply to the observation of odd-order phenomena, such as linear polarizability and third-order NLO effects.

When the polarizable electrons of the NLO material interact with two electric fields E_1 and E_2 that potentially have different polarizations and frequencies ω_1 and ω_2, respectively, second-order NLO effects can be observed. For example, consider the interaction of the material with two laser beams of different frequencies. The second-order term of Eq. (16.7) becomes

$$\chi^{(2)} E_1 \cos(\omega_1 t) E_2 \cos(\omega_2 t), \tag{16.8}$$

which, according to trigonometry, can be rewritten as

$$\left(\frac{1}{2}\right)\chi^{(2)} E_1 E_2 \cos[(\omega_1 + \omega_2)t] + \left(\frac{1}{2}\right)\chi^{(2)} E_1 E_2 \cos[(\omega_1 - \omega_2)t]. \tag{16.9}$$

Thus, nonlinear polarization will be induced at sum ($\omega_1 + \omega_2$) and difference ($\omega_1 - \omega_2$) frequencies when two electromagnetic beams of frequencies ω_1 and ω_2 interact in an NLO material. Since one can regard the oscillating polarization as a classical oscillating dipole that emits radiation at all of its oscillation frequencies, light will be emitted at the sum and difference frequencies (in addition to the two frequencies ω_1 and ω_2); accordingly, this effect is called sum (or difference) frequency generation (SFG). Where the two field oscillation frequencies are the same, i.e., $\omega_1 = \omega_2$, light is generated at the second harmonic of ω_1, $2\omega_1$, this effect being referred to as second-harmonic generation (SHG). The difference gives a DC electric field; this is known as optical rectification.

A different second-order NLO effect is obtained in the case where one of the fields in Eq. 16.9 is a DC electric field E_2 applied to the material, i.e., $\omega_2 = 0$. In this case the second-order polarization is

$$P_{\text{opt}}^{(2)} = \chi^{(2)} E_1 E_2 (\cos \omega_1 t), \tag{16.10}$$

and the total optical polarization, only including terms to second order, is given by

$$P_{\text{opt}} = \chi^{(1)} E_1 (\cos \omega_1 t) + \chi^{(2)} E_1 E_2 (\cos \omega_1 t) = \left(\chi^{(1)} + \chi^{(2)} E_2\right) E_1 (\cos \omega_1 t). \quad (16.11)$$

The quantity ($\chi^{(1)} + \chi^{(2)}E_2$) describes the polarization of the material by the light field E_1; i.e., it corresponds to an *effective* linear susceptibility that is dependent on the applied field E_2. Since the linear susceptibility is related to the dielectric constant and, therefore, to the refractive index, the material has an effective refractive index that is field dependent, and the polarization or phase of light passing through the material can be modulated by changing an applied voltage. This effect is known as the linear electro-optic (EO) or Pockels effect. An important equation representing the Pockels effect is

$$\Delta\left(\frac{1}{\varepsilon}\right)_{IJ} = \Delta\left(\frac{1}{n^2}\right)_{IJ} = r_{I,J,Z}(\omega) \cdot E_Z(\omega_m) + \text{higher-order terms}, \quad (16.12)$$

which can be rewritten (to express the voltage-induced change in index of refraction) as

$$\Delta n_{ZZ} \approx -\frac{1}{2} \cdot n_{ZZ}^3 \cdot r_{Z,Z,Z} \cdot E_z. \quad (16.13)$$

In like manner, the voltage-induced phase shift of light passing through the EO material can be expressed as

$$2\pi\left(\frac{L}{\lambda}\right)\Delta n_{ZZ}. \quad (16.14)$$

For dipolar chromophores that are ordered by electric-field poling, only two nonzero components of the EO tensor $r_{I,J,Z}$ exist. These are

$$\begin{aligned} r_{33} &= 2\beta_{zzz} F_L(\omega) N \left\langle \cos^3\theta \right\rangle \\ r_{13} &= \beta_{zzz} F_L(\omega) N \left\langle \sin^2\theta \cos\theta \right\rangle \\ \left\langle \cos^3\theta \right\rangle &= \left\langle \cos\theta \right\rangle - \left\langle \sin^2\theta\cos\theta \right\rangle \\ \frac{r_{13}}{r_{33}} &= \frac{1}{2}\left\{\frac{\left\langle \cos\theta \right\rangle}{\left\langle \cos^3\theta \right\rangle} - 1\right\}, \end{aligned} \quad (16.15)$$

where $F_L(\omega)$ is the Debye-Onsager factor accounting for the dielectric environment in which the chromophore exists. The $\left\langle \cos^n\theta \right\rangle$ represent different order parameters. The ratio of major (r_{33}) and minor (r_{13}) tensor elements depends on the spatial distribution of chromophores and hence, on lattice symmetry. For the limiting cases of chromophores confined in lattices of 3D, 2D, and 1D, the ratios take on the following limiting values:

$$\begin{aligned} &\text{Ising order (1D)} && \frac{r_{13}}{r_{33}} = 0 \\ &\text{Bessel order (2D)} && \frac{r_{13}}{r_{33}} = \frac{1}{6} \\ &\text{Langevin order (3D)} && \frac{r_{13}}{r_{33}} = \frac{1}{3}. \end{aligned} \tag{16.16}$$

For a discussion of measurement of r_{33} and r_{13} by the attenuated total reflection (ATR) method and the measurement of different order parameters $\langle \cos^n \theta \rangle$ by various methods, the reader is referred to Ref. 34 and to the following website: http://depts.washington.edu/eooptic/. As a final comment, it should be noted that considerable contraction of notation is common in discussing nonlinear optics. For example, the EO tensor element, r_{33} or r_{ZZ}, actually should be written as r_{333} or r_{ZZZ}, representing the tensor element corresponding to parallel alignment (pointing in the same direction) of the principal elements of the optical (high-frequency) field, electrical (low-frequency) field, and molecular vectors/tensors. These fields refer to operation of an EO device or to the EO phenomenon. Confusion sometimes arises because one also speaks of the electric poling field associated with processing of materials. The poling field is considered to be applied in the "*Z*" or "3" direction. The "low-frequency" electrical field used in operation of EO devices is normally applied in the "*Z*" or "3" direction as well. For TM (transverse magnetic) light, the electric-field vector of the light points in the "*Z*" or "3" direction; for TE (transverse electric) light, the electric-field vector points in the "*X*" or "1" direction. Finally, we note that second-order NLO susceptibility can be related to EO activity by the relationship

$$\chi^{(2)}(-\omega;0,\omega)_{ZZZ} = -n^4(\omega)r_{ZZZ}(\omega) = -n^4(\omega)r_{33}(\omega), \tag{16.17}$$

where the "*ZZZ*" or "33" tensor elements have been chosen as an example.

Acknowledgments

The authors gratefully acknowledge financial support provided by the National Science Foundation (DMR-0551020 and DMR-0120967), by the Air Force Office of Scientific Research, and by the DARPA MORPH program. Helpful discussions and technical assistance provided by members of the Prezhdo, Scherer, and Jen research groups are also gratefully acknowledged, as are many helpful discussions with Oleg Prezhdo (University of Washington), Michael Hayden (University of Maryland, Baltimore County), Axel Scherer (California Institute of Technology) and Alex Jen (University of Washington).

References

1. L. Liao, B. E. Eichinger, K. A. Firestone, M. Haller, J. Luo, W. Kaminsky, J. B. Benedict, P. J. Reid, A. K.-Y. Jen, L. R. Dalton, and B. H. Robinson, "Systematic study of the structure-property relationship of a series of ferrocenyl nonlinear optical chromophores", *J. Am. Chem. Soc.* **127**, 2758–2766 (2005).
2. C. M. Ishorn, A. Lerlercq, F. D. Vila, L. R. Dalton, J. L. Bredas, B. E. Eichinger, and B. H. Robinson, "Comparison of static first hyperpolarizabilities calculated with various quantum mechanical methods," *J. Phys. Chem. A* **111**, 1319–1327 (2007).
3. S. H. Jang, J. Luo, N. M. Tucker, A. Leclercq, E. Zojer, M. A. Haller, T. D. Kim, J. W. Kang, K. Firestone, D. Bale, D. Lao, J. B. Benedict, D. Cohen, W. Kaminsky, B. Kahr, J. L. Bredas, P. Reid, L. R. Dalton, and A. K. Y. Jen, "Pyrroline chromophores for electro-optics," *Chem. Mater.* **18**, 2983–2998 (2006).
4. T. Kinnibrugh, S. Bhattacharjee, P. Sullivan, C. Isborn, B. H. Robinson, and B. E. Eichinger, "Influence of isomerization on nonlinear optical properties of molecules," *J. Phys. Chem. B* **110**, 13512–13522 (2006).
5. E. R. Davidson, B. E. Eichinger, and B. H. Robinson, "Hyperpolarizability: calibration of theoretical methods for chloroform, water, acetonitrile, and p-nitroaniline," *Opt. Mater.* **29**, 360–364 (2006).
6. C. S. Isborn and B. H. Robinson, "*Ab initio* diradical/zwiterionic polarizabilities and hyperpolarizabilities in twisted diradicals," *J. Phys. Chem. A* **110**, 7189–7196 (2006).
7. H. Kang, A. Facchetti, H. Jiang, E. Cariati, S. Righetto, R. Ugo, C. Zuccacccia, A. Macchioni, C. L. Stern, Z. Liu, S. T. Ho, E. C. Brown, M. A. Ratner, and T. J. Marks, "Ultralarge hyperpolarizability twisted π-electron system electro-optic chromophores: Synthesis, solid-state and solution-phase structural characteristics, electronic structures, linear and nonlinear optical properties, and computational studies," *J. Am. Chem. Soc.* **129**, 3267–3286 (2007).
8. Y. Wang, "Theoretical Design of Molecular Photonic Materials," Ph.D. thesis, Royal Institute of Technology, Stockholm, Sweden, 2007.
9. C. K. Wang, Y. H. Wang, Y. Su, and Y. Luo, "Solvent dependence of solvatochromic shifts and the first hyperpolarizability of para-nitroaniline: A nonmonotonic behavior," *J. Chem. Phys.* **119**, 4409–4412 (2003).
10. S. Ando, T. Fujigaya, and M. Ueda, "Density functional theory calculation of photoabsorption spectra of organic molecules in the vacuum ultraviolet region," *Jpn. J. Appl. Phys.* **41**, L105–L108 (2002).

11. L. R. Dalton, A. W. Harper, and B. H. Robinson, "The role of London forces in defining noncentrosymmetric order of high dipole moment-high hyperpolarizability chromophores in electrically poled polymeric thin films," *Proc. Natl. Acad. Sci. USA* **94**, 4842–4847 (1997).

12. L. R. Dalton, B. H. Robinson, A. K. Y. Jen, W. H. Steier, and R. Nielsen "Systematic development of high bandwidth, low drive voltage organic electro-optic devices and their applications," *Opt. Mater.* **21**, 19–28 (2003).

13. R. D. Nielsen, H. L. Rommel, and B. H. Robinson, "Simulation of the loading parameter in organic nonlinear optical materials," *J. Phys. Chem. B* **108**, 8659–8667 (2004).

14. A. Piekara, "A theory of electric polarization, electro-optical Kerr effect and electric saturation in liquids and solutions," *Proc. Roy. Soc. Lon. A* **172**, 360–383 (1939).

15. W. K. Kim and L. M. Hayden, "Fully atomistic modeling of an electric field poled guest-host nonlinear optical polymer," *J. Chem. Phys.* **111**, 5212–5222 (1999).

16. M. R. Leahy-Hoppa, P. D. Cunningham, J. A. French, and L. M. Hayden, "Atomistic molecular modeling of the effect of chromophore concentration on the electro-optic coefficient in nonlinear optical polymers," *J. Phys. Chem. A* **110**, 5792–5797 (2006).

17. M. Makowska-Janusik, H. Reis, M. G. Papadopoulos, I. G. Economou, and N. Zacharopoulos, "Molecular dynamics simulations of electric field poled nonlinear optical chromophores incorporated in a polymer matrix," *J. Phys. Chem. B* **108**, 588–596 (2004).

18. B. H. Robinson and L. R. Dalton, "Monte Carlo statistical mechanical simulations of the competition of intermolecular electrostatic and poling field interactions in defining macroscopic electro-optic activity for organic chromophore/polymer materials," *J. Phys. Chem.* **104**, 4785–4795 (2000).

19. L. R. Dalton, A. K. Y. Jen, P. Sullivan, B. Eichinger, B. H. Robinson, and A. Chen, "Theoretically inspired rational design of electro-optic materials," *Nonlinear Optics and Quantum Optics*, **35**, 1–19 (2006).

20. H. L. Rommel and B. H. Robinson, "Orientation of electro-optic chromophores under poling conditions: A spheroidal model," *J. Phys. Chem. B* **111**, 18765–18777 (2004).

21. H. L. Rommel, "Determining the Order Parameters of Organic Nonlinear Optical Chromophores by Monte Carlo Methods," Ph.D. thesis, University of Washington, Seattle, USA, 2007.

22. K. A. Firestone, D. Bale, Y. Liao, D. M. Casmier, O. Clot, L. R. Dalton, and P. J. Reid, "Frequency-agile hyper-Rayleigh scattering studies of electro-optic chromophores," *Proc. SPIE* **5935**, 59350P (2005).

23. O. Ostinelli, "Solvent dependence of microscopic optical nonlinearities of the bithiophene molecule CC172 and investigation of poling processes for polyimide AM3 148.02," M.S. thesis, Swiss Federal Institute of Technology (ETH), Zurich, Switzerland, 2000.

24. EFISH measurements were carried out by Denise Bale in the laboratory of Professor P. Gunter, ETH, Zurich, Switzerland in 2006. A detailed account of these measurements is published in: Denise H. Bale, "Nonlinear Optical Materials' Characterization Studies Employing Photostability, Hyper-Rayleigh Scattering, and Electric Field Induced Second Harmonic Generation Techniques." Ph.D. thesis, University of Washington, Seattle, WA, USA, 2008.

25. C. C. Teng and H. T. Man, "Simple reflection technique for measuring the electro-optic coefficient of poled polymers," *Appl. Phys. Lett.* **56**, 1734–1736 (1990).

26. P. A. Sullivan, "Theory Guided Design and Molecular Engineering of Organic Materials for Enhanced Second-Order Nonlinear Optical Properties," Ph.D. thesis, University of Washington, Seattle, WA, USA, 2006.

27. A. Chen, V. Chuyanov, S. Garner, W. H. Steier, L. R. Dalton, "Modified attenuated total reflection for the fast and routine electrooptic measurements of nonlinear optical polymer thin films," in *Organic Thin Films for Photonic Applications*, Vol. 14, OSA Technical Digest Series (Optical Society of America, Washington DC, 1997) 158–9. (This instrument has been further modified by incorporation of a rutile prism permitting measurements to be carried out at both 1.3 and 1.55 μm and facilitating the measurement of both r_{33} and r_{13}.)

28. A. Chen, V. Chuyanov, H. Zhang, S. Garner, W. H. Steier, J. Chen, J. Zhu, M. He, S. S. H. Mao, and L. R. Dalton, "Demonstration of the full potential of second order nonlinear optic polymers for electrooptic modulation using a high $\mu\beta$ chromophore and a constant bias field," *Optics Lett.* **23**, 478–480 (1998).

29. N. P Bhatambrekar, L. R. Dalton, J. Luo, A. K. Y. Jen, and A. Chen, "Third order nonlinearity contribution to electro-optic activity in polymer materials in a constant DC bias field," *Appl. Phys. Lett.* **88**, 041115-1 (2005).

30. P. Rabiei, W. H. Steier, C. Zhang, and L. R. Dalton, "Polymer micro-ring filters and modulators," *J. Lightwave Technol.* **20**, 1968–1975 (2002).

31. T. Baehr-Jones, M. Hochberg, G. Wang, R. Lawson, Y. Liao, P. A. Sullivan, L. R. Dalton, A. K. Y. Jen, and A. Scherer, "Optical modulation and detection in slotted silicon waveguides," *Opt. Exp.* **13**, 5216–5226 (2005).

32. S. Hammond, "Molecular and Nanoscale Engineering for Enhanced Order in Organic Electro-optic Materials," Ph.D. thesis, University of Washington, Seattle, WA, USA, 2007.

33. P. A. Sullivan, H. Rommel, Y. Liao, B. C. Olbricht, A. J. P. Akelaitis, K. A. Firestone, J. W. Kang, J. Luo, D. H. Choi, B. E. Eichinger, P. J. Reid, A. Chen, A. K. Y. Jen, B. H. Robinson, and L. R. Dalton, "Theory guided design and synthesis of multi-chromophore dendrimers and computational investigation of the origins of their improved electro-optic activity," *J. Amer. Chem. Soc.* **129**, 7523–7530 (2007).

34. P. A. Sullivan, and L. R. Dalton, "Theory-inspired development of organic electro-optic materials," *Acc. Chem. Res.* (submitted).

35. P. A. Sullivan, A. J. P. Akelaitis, S. K. Lee, G. McGrew, S. K. Lee, D. H. Choi, and L. R. Dalton, "Novel dendritic chromophores for electro-optics: Influence of binding mode and attachment flexibility on EO behavior," *Chem. Mater.* **18**, 344–351 (2006).

36. P. A. Sullivan, B. C. Olbricht, A. J. P. Akelaitis, A. A. Mistry, Y. Liao, and L. R. Dalton, "Tri-component Diels-Alder polymerized dendrimer glass exhibiting large, thermally stable, electro-optic activity," *J. Mater. Chem.* **17**, 2899–2903 (2007).

37. R. Barto, Jr., P. V. Bedworth, C. W. Frank, S. Ermer, and R. E. Taylor, "Near-infrared optical-absorption behavior in high-beta nonlinear optical chromophore-polymer guest-host materials. II. Dye spacer length effects in an amorphous polycarbonate copolymer host," *J. Chem. Phys.* **122**, 234907, 1–14 (2005).

38. R. Barto, Jr., C. W. Frank, P. V. Bedworth, S. Ermer, and R. E. Taylor, "Near-infrared optical absorption behavior in high-β nonlinear optical chromophore-polymer guest-host materials. 1. Continuum dielectric effects in polycarbonate hosts," *J. Phys. Chem. B* **108**, 8702–8715 (2004).

39. T. D. Kim, J. W. Kang, J. Luo, S. H. Jang, J. W. Ka, N. Tucker, J. B. Benedict, L. R. Dalton, T. Gray, R. M. Overney, D. H. Park, W. N. Herman and A. K.-Y. Jen, "Ultralarge and thermally stable electro-optic activities from supramolecular self-assembled molecular glasses," *J. Am. Chem. Soc.* **129**, 488–489 (2007).

40. T. D. Kim, Z. Shi, J. Luo, S. H. Jang, Y. J. Cheng, X. Zhou, S. Huang, L. R. Dalton, W. Herman, and A. K. Y. Jen, "New paradigm for ultrahigh electro-optic activity: Through supramolecular self-assembly and novel lattice hardening," *Proc. SPIE* **6470**, 64700D (2007).

41. L. R. Dalton, B. H. Robinson, A. K. Y. Jen, P. Reid, B. Eichinger, P. Sullivan, A. Akelaitis, D. Bale, M. Haller, J. Luo, S. Liu, Y. Liao, K. Firestone, N. Bhatambrekar, S. Bhattacharjee, J. Sinness, S. Hammond, N. Buker, R. Snoeberger, M. Lingwood, H. Rommel, J. Amend, S. H. Jang, A. Chen, and W. Steier, "Acentric lattice electro-optic materials by rational design," *Proc. SPIE* **5912**, 43–54 (2005).

42. Y. V. Pereverzev, O. V. Prezhdo, and L. R. Dalton, "Macroscopic order and electro-optic response of dipolar chromophore-polymer materials," *Chem. Phys. Chem.* **5**, 1–11 (2004).

43. A. Natansohn and P. Rochon, "Photoinduced motions in azo-containing polymers," *Chem. Rev.* **102**, 41394175 (2002).

44. J. A. Delaire and K. Nakatani, "Linear and nonlinear optical properties of photochromic molecules and materials," *Chem. Rev.* **100**, 1817–1845 (2000).

45. J. G. Grote, J. S. Zetts, C. H. Zhang, R. L. Nelson, L. R. Dalton, F. K. Hopkins, and W. H. Steier, "Conductive Cladding Layers for Electrode Poled Nonlinear Optic Polymer Electro-Optics," *Proc. SPIE* **4114**, 101–109 (2000).

46. J. G. Grote, J. S. Zetts, R. L. Nelson, F. K. Hopkins, L. R. Dalton, C. Zhang, and W. H. Steier, "Effect of conductivity and dielectric constant on the modulation voltage for optoelectronic devices based on nonlinear optical polymers," *Opt. Eng.* **40**, 2464–2473 (2001).

47. J. G. Grote, J. S. Zetts, R. L. Nelson, F. K. Hopkins, J. B. Huddleston, P. P. Yaney, C. H. Zhang, W. H. Steier, M.-C. Oh, H. R. Fetterman, A. K. Jen, and L. R. Dalton, "Advancements in conductive cladding materials for nonlinear-optic-polymer-based optoelectronic devices," *Proc. SPIE* **4470**, 10–19 (2001).

48. J. G. Grote, J. S. Zetts, R. L. Nelson, F. K. Hopkins, P. P. Yaney, C. Zhang, W. H. Steier, M. C. Oh, H. R. Fetterman, A. K. Y. Jen, and L. R. Dalton, "Conductive cladding materials for nonlinear optic polymer based opto-electronic devices," *Proc. GOMAC 2002*, 161–165 (2002).

49. M. Leovich, P. P. Yaney, C. H. Zhang, W. H. Steier, M.-O. Oh, H. R. Fetterman, A. K. Y. Jen, L. R. Dalton, J. G. Grote, R. L. Nelson, J. S. Zetts, and F. K. Hopkins, "Optimized cladding materials for nonlinear-optic polymer-based devices," *Proc. SPIE* **4652**, 97–103 (2002).

50. J. A. Hagen, J. G. Grote, J. S. Zetts, D. E. Diggs, R. L. Nelson, F. K. Hopkins, P. P. Yaney, A. K. Jen, and L. R. Dalton, "Effects of the electric field poling procedure on electro-optic coefficient for guest-host nonlinear optic polymers," *Proc. SPIE* **5724**, 217–223 (2005).

51. Y. Enami, C. T. DeRose, C. Loychik, D. Mathine, R. A. Norwood, J. Luo, A. K. Y. Jen, and N. Peyghambarian, "Low half-wave voltage and high electro-optic effect in hybrid polymer/sol-gel waveguide modulators," *Appl. Phys. Lett.* **89**, 143506 1–3 (2006).

52. C. T. DeRose, Y. Enami, C. Loychik, R. A. Norwood, D. Mathine, M. Fallahi, N. Peyghambarian, J. D. Luo, A. K. Y. Jen, M. Kathaperumal, and M. Yamamoto, "Pockel's coefficient enhancement of poled electro-optic polymers with a hybrid organic-inorganic sol-gel cladding layer," *Appl. Phys. Lett.* **89**, 131102 (2006).

53. H. Gan, H. Zhang, C. T. DeRose, R. A. Norwood, N. Peyhambarian, M. Fallahi, J. Luo, B. Chen, and A. K. Y. Jen, "Low drive voltage Fabry-Perot etalon device tunable filters using poled hybrid sol-gel materials," *Appl. Phys. Lett.* **89**, 041127 (2006).

54. Y. Enami, C. T. DeRose, D. Mathine, C. Loychik, C. Greenlee, R. A. Norwood, T. D. Kim, J. Luo, Y. Tian, A. K. Y. Jen, and N. Peyghambarian, "Hybrid polymer/sol-gel waveguide modulators with exceptionally large electro-optic coefficients," *Nature Photonics* **1**, 180–185 (2007).

55. R. Song, A. Yick, and W. H. Steier, "Conductivity-dependency-free in-plane poling for Mach-Zehnder modulator with highly conductive electro-optic polymer," *Appl. Phys. Lett.* **90**, 191103 (2007).

56. G. Xu, J. Ma, Z. Liu, B. Liu, S. T. Ho, P. Zhu, L. Wang, Y. Yang, T. J. Marks, J. Luo, N. Tucker, and A. K. Y. Jen, "Low-voltage organic electro-optic modulators using transparent conducting oxides as electrodes," *Opt. Exp.* **13**, 7380–7385 (2005).

57. L. Wang, Y. Yang, and T. J. Marks, "Near-infrared transparent electrodes for precision Teng-Man electro-optic measurements. In_2O_3 thin film electrodes with tunable near-infrared transparency," *Appl. Phys. Lett.* **87**, 161107 (2005).

58. L. R. Dalton, A. W. Harper, R. Ghosn, W. H. Steier, M. Ziari, H. Fetterman, Y. Shi, R. Mustacich, A. K.-Y. Jen, and K. J. Shea, "Synthesis and processing of improved second order nonlinear optical materials for applications in photonics," *Chem. Mater.* **7**, 1060–1081 (1995).

59. S. S. H. Mao, Y. Ra, L. Guo, C. Zhang, L. R. Dalton, A. Chen, S. Garner, and W. H. Steier, "Progress towards device-quality second-order nonlinear optical materials: 1. Influence of composition and processing conditions on nonlinearity, temporal stability and optical loss," *Chem. Mater.* **10**, 146–155 (1998).

60. M. Haller, J. Luo, Hongxian Li, T. D. Kim, Y. Liao, B. Robinson, L. R. Dalton, and A. K. Y. Jen, "A novel lattice-hardening process to achieve highly efficient and thermally stable nonlinear optical polymers," *Macromolecules* **37**, 688–690 (2004).

61. N. Tucker, H. Li, H. Tang, L. R. Dalton, Y. Liao, B. H. Robinson, A. K. Jen, J. Luo, S. Liu, M. Haller, J. Kang, T. Kim, S. Jang and B. Chen, "Recent progress in developing highly efficient and thermally stable nonlinear optical polymers for electro-optics," *Proc. SPIE* **5351**, 36–43 (2004).

62. C. Zhang, H. Zhang, M. Oh, L. Dalton, and W. Steier, "What the ultimate polymeric electro-optic materials will be: Guest-host, crosslinked, or side-chain," *Proc. SPIE* **4991**, 537–551 (2003).

63. D. Jin, T. Londergan, D. Huang, N. Wolf, S. Condon, D. Tolstedt, H. Guan, S. Cong, E. Johnson, and R. Dinu, "Achieving large electro-optic response: DH-type chromophores in both crosslinked systems and linear high Tg systems," *Proc. SPIE* **5351**, 44–56 (2004).

64. J. Luo, M. Haller, H. Ma, S. Liu, T. D. Kim, Y. Tian, B. Chen, S. H. Jang, L. R. Dalton, and A. K. Y. Jen, "Nanoscale architectural control and macromolecular engineering of nonlinear optical dendrimers and polymers for electro-optics," *J. Phys. Chem. B* **108**, 8523–8530 (2004).

65. S. Suresh, S. Chen, C. M. Topping, J. M. Ballato, and D. W. Smith, Jr., "Novel perfluorocyclo-butyl (PFCB) polymers containing isophorone derived chromophore for electro-optic [EO] applications," *Proc. SPIE* **4991**, 530–536 (2003).

66. P. A. Sullivan, B. C. Olbricht, A. J. P. Akelaitis, A. A. Mistry, Y. Liao, and L. R. Dalton, "Tri-component Diels-Alder polymerized dendrimer glass exhibiting large, thermally stable, electro-optic activity," *J. Mater. Chem.* **17**, 2899–2903 (2007).

67. G. A. Lindsay, A. J. Guenthner, M. E. Wright, M. Sanghadasa, and P. R. Ashley, "Long-term alignment stability of CLD and FTC chromophores in polycarbonate and polyimide poled glassy films at elevated temperatures," private communication.

68. C. Zhang, L. R. Dalton, M. C. Oh, H. Zhang, and W. H. Steier, "Low V_π electrooptic modulators from CLD-1: Chromophore design and synthesis, materials processing, and characterization," *Chem. Mater.* **13**, 3043–3050 (2001).

69. Q. Zhang, M. Canva, and G. Stegeman, "Wavelength dependence of 4-dimethylamino-4'-nitrostilbene polymer thin film photodegradation," *Appl. Phys. Lett.* **73**, 912–914 (1998).

70. A. Galvan-Gonzalez, M. Canva, G. I. Stegeman, R. J. Twieg, T. C. Kowalczyk, and H. S. Lackritz, "Effect of temperature and atmospheric environment on the photodegradation of some Disperse Red 1 type polymers," *Opt. Lett.* **24**, 1741–1743 (1999).

71. A. Galvan-Gonzalez, M. Canva, G. I. Stegeman, R. J. Twieg, K. P. Chang, T. C. Kowalczyk, X. Q. Zhang, H. S. Lackritz, S. Marder, and S. Thayumanavan, "Systematic behavior of electro-optic chromophore photostability," *Opt. Lett.* **25**, 332–334 (2000).

72. A. Galvan-Gonzalez, M. Canva, G. I. Stegeman, L. Sukhomilinova, R. J. Twieg, K. P. Chang, T. C. Kowalczyk, and H. S. Lackritz, "Photodegradation of azobenzene nonlinear optical chromophores: The influence of structure and environment," *J. Opt. Soc. Am. B* **17**, 1992–2000 (2000).

73. M. E. DeRosa, M. He, J. S. Cites, S. M. Garner, and Y. R. Tang, "Photostability of high $\mu\beta$ electro-optic chromophores at 1550 nm," *J. Phys. Chem. B* **108**, 8725–8730 (2004).

74. M. He, T. Leslie, S. Garner, M. E. DeRosa, and J. Cites, "Synthesis of new electrooptic chromophores and their structure-property relationship," *J. Phys. Chem. B* **108**, 8731–8736 (2004).

75. D. Rezzonico, M. Jazbinsek, C. Bosshard, P. Gunter, D. H. Bale, Y. Liao, L. R. Dalton, and P. J. Reid, "Photostability studies of π-conjugated chromophores with resonant and nonresonant excitations," *J. Opt. Soc. B* **24**, 2199–2207 (2007).

76. E. W. Taylor, E, J. E. Nichter, F. D. Nash, F. Haas, A. A. Szep, R. J. Michalak, B. M. Flusche, P. R. Cook, T. A. McEwen, B. F. McKeon, P. M. Payson, G. A. Brost, A. R. Pirich, C. Castaneda, B. Tsap, and H. R. Fetterman, "Radiation resistance of electro-optic polymer-based modulators," *Appl. Phys. Lett.* **86**, 201122 (2005).

77. H. C. Song, M. C. Oh, S. W. Ahn, and W. H. Steier, "Flexible low-voltage electro-optic polymer modulators," *Appl. Phys. Lett.* **82**, 4432–4434 (2003).

78. L. R. Dalton, A. W. Harper, A. Ren, F. Wang, G. Todorova, J. Chen, C. Zhang, and M. Lee, "Polymeric electro-optic modulators: From chromophore design to integration with semiconductor VLSI electronics and silica fiber optics," *Ind. Eng. Chem. Res.* **38**, 8–33 (1999).

79. A. Yariv, C. Zhang, L. R. Dalton, Y. Huang, and G. T. Paloczi, "Fabrication and replication of polymer integrated optical devices using electron-beam lithography and soft lithography," *J. Phys. Chem. B* **108**, 8006–8013 (2004).

80. G. T. Paloczi, Y. Huang, A. Yariv, J. Luo, A. Jen, "Replica-molded electro-optic polymer Mach-Zehnder modulator," *Appl. Phys. Lett.* **85**, 1662–1664 (2004).

81. M. Hochberg, T. Baehr-Jones, G. Wang, J. Parker, K. Harvard, J. Liu, B. Chen, Z. Shi, R. Lawson, P. Sullivan, A. K. Y. Jen, L. R. Dalton, and A. Scherer, "All optical modulator in silicon with terahertz bandwidth," *Nature Materials* **5**, 703–709 (2006).

82. L. R. Dalton, "Organic electro-optic materials," in *Conjugated Polymers: Processing and Applications*, eds. T. A. Skotheim and J. R. Reynods, CRC Press, Boca Raton, FL, 6-1–6-39 (2007).

83. L. R. Dalton, A. W. Harper, B. Wu, R. Ghosn, J. Laquindanum, Z. Liang, A. Hubbel, and C. Xu, "Polymeric electro-optic modulators: Materials synthesis and processing," *Adv. Mater.* **7**, 519–540 (1995).

84. J. H. Kim, L. Sun, C.-H. Jang, D. An, J. M. Taboada, Q. Zhou, X. Lu, R. T. Chen, X. Han, S. Tang, H. Zhang, W. H. Steier, A. Ren, and L. R. Dalton, "Polymeric waveguide beam deflector for electro-optic switching," *Proc. SPIE* **4279**, 37–44 (2001).

85. L. Sun, J. Kim, C. Jang, D. An, X. Lu, Q. Zhou, J. M. Taboada, R. T. Chen, J. J. Maki, S. Tang, H. Zhang, W. H. Steier, C. Zhang, and L. R. Dalton, "Polymeric waveguide prism based electro-optic beam deflector," *Opt. Eng.* **40**, 1217–1222 (2001).

86. P. Rabiei, W. H. Steier, C. Zhang, and L. R. Dalton, "Polymer micro-ring filters and modulators," *J. Lightwave Technology* **20**, 1968–1975 (2002).

87. T. Baehr-Jones, M. Hochberg, G. Wang, R. Lawson, Y. Liao, P. A. Sullivan, L. R. Dalton, A. K. Y. Jen, and A. Scherer, "Optical modulation and detection in slotted silicon waveguides," *Opt. Exp.* **13**, 5216–5226 (2005).

88. T. Baehr-Jones, P. Boyan, J. Huang, P. A. Sullivan, J. Davies, J. Takayesu, J. Luo, T.-D. Kim, L. R. Dalton A. K.-Y. Jen, M. Hochberg, and A. Scherer, "Nonlinear polymer-clad silicon slot waveguide modulator with a half wave voltage of 0.25 V," *Appl. Phys. Lett.* **92**, 163303 (2008).

89. J. Takayesu, M. Hochberg, T. Baehr-Jones, E. Chan, D. Koshniz, G. Wang, P. Sullivan, Y. Liao, J. Davies, L. Dalton, A. Scherer, and W. Krug, "Hybrid electro-optic microring resonator-based 1x4x1 ROADM for wafer scale optical interconnects," *J. Lightwave Technology* **27**, 440–448 (2008).

90. A. M. Sinyukov and L. M. Hayden, Efficient electrooptic polymers for THz applications," *J. Phys. Chem. B* **108**, 8515–8522 (2004).

91. A. M. Sinyukov, M. R. Leahy, L. M. Hayden, M. Haller, J. Luo, A. K. Y. Jen, and L. R. Dalton, "Resonance enhanced THz generation in electro-optic polymers near the absorption maximum," *Appl. Phys. Lett.* **85**, 5827–5829 (2004).

92. L. R. Dalton, "Nonlinear optical polymeric materials: From chromophore design to commercial applications," in *Advances in Polymer Science* **158**, Springer-Verlag, Heidelberg, 1–86 (2001).

93. L. R. Dalton, "Novel polymer-based, high-speed electro-optic devices," in *Proceedings 23rd European Conference on Optical Communications/14th International Conference on Integrated Optics and Optical Fibre Communication*, vol. 2, 346–349 (2003).

94. J. H. Bechtel, Y. Shi, H. Zhang, W. H. Steier, C. H. Zhang, and L. R. Dalton, "Low-driving-voltage electro-optic polymer modulators for advanced photonic applications," *Proc. SPIE* **4114**, 58–64 (2000).

95. H. R. Fetterman, D. H. Chang, H. Erlig, M. Oh, C. H. Zhang, W. H. Steier, and L. R. Dalton, "Photonic time-stretching of 102 GHz millimeter waves using 1.55 μm polymer electro-optic modulator," *Proc. SPIE* **4114**, 44–57 (2000).

96. S. S. Lee, A. H. Udupa, H. Erlig, H. Zhang, Y. Chang, C. Zhang, D.H. Chang, D. Bhattacharya, B. Tsap, W. H. Steier, L. R. Dalton, and H. R. Fetterman, "Demonstration of a photonically controlled RF phase shifter," *IEEE Microwave and Guided Wave Letters* **9**, 357–359 (1999).

97. A. Yacoubian, A., V. Chuyanov, S. M. Garner, W. H. Steier, A. S. Ren, and L. R. Dalton, "EO polymer-based integrated-optical acoustic spectrum analyzer," *IEEE J. Sel. Topics in Quantum Electron.* **6**, 810–816 (2000).

98. H. Sun, A. Pyajt, J. Luo, Z. Shi, S. Hau, A. K. Y. Jen, L. R. Dalton, and A. Chen, "All-dielectric electrooptic sensor based on a polymer microresonator coupled side-polished optical fiber," *IEEE Sensors J.* **7**, 515–524 (2007).

99. B. Bhola, H. C. Song, H. Tazawa, and W. H. Steier, "Polymer microresonator strain sensors," *IEEE Phot. Tech. Lett.* **17**, 867–870 (2005).

Biographies

Larry R. Dalton is the George B. Kauffman Professor of Chemistry and Electrical Engineering, the B. Seymour Rabinovitch Chair of Chemistry, and the Director of the National Science Foundation (NSF) Science and Technology Center on Materials and Devices for Information Technology Research at the University of Washington. He is a member of the Nanotechnology Technical Advisory Group of the President's Council of Advisors for Science and Technology, the Defense Science Board Advisory Group on Electronic Devices, and the NSF Mathematical and Physical Sciences Directorate Advisory Committee. He is a recipient of an SPIE Lifetime Achievement Award (2008), the 2003 American Chemical Society Award in the Chemistry of Materials, the 2006 IEEE/LEOS William Streifer Scientific Achievement Award, and the 1996 Richard C. Tolman Medal of the American Chemical Society. He has co-authored more than 550 publications.

Philip A. Sullivan is currently a Research Assistant Professor in the Department of Chemistry at the University of Washington, Seattle. He received his B.S. degree in 2001 from Montana State University, Bozeman and his Ph.D. in 2006 from the University of Washington. He has coauthored more than 20 publications. His current research interests focus on organic materials for optoelectronic and sensor applications.

Denise H. Bale is currently a Postdoctoral Fellow in chemistry at the University of Washington. She received a B.S. (Honors) in chemistry from Western Washington University, Bellingham, WA (2002), followed by M.S. (2004) and Ph.D. (2007) degrees in chemistry from the University of Washington. Her research focuses on characterization of electro-optic materials including photostability and hyperpolarizability of chromophores through femtosecond wavelength-agile hyper-Rayleigh scattering measurements.

Scott R. Hammond is a Research Associate at the National Renewal Energy Laboratory (NREL). He received his B.S. degree in chemistry with high honors from the University of California, Berkeley in 2001, and his Ph.D. in organic/materials chemistry and nanotechnology from the University of Washington in 2007. He is coauthor on more than 10 journal articles. His research interests are in organic nanostructured materials with new or improved photonic and optoelectronic properties.

Benjamin C. Olbricht is a Ph.D. graduate student in the Department of Chemistry at the University of Washington. He received his M.S. degree in chemistry from the University of Washington in 2007 and a B.A. in chemistry from Albion College in 2005. He is currently studying electro-optic materials, including the processing of materials by laser-assisted poling.

Harrison Rommel received his B.S. in biochemistry from the University of New Mexico. He received his Ph.D. in physical chemistry from the University of Washington in 2007. His current research interests include investigation of the field-induced ordering of bulk organic NLO materials by Monte Carlo methods.

Bruce Eichinger is a Staff Scientist in the Department of Chemistry at the University of Washington. He received a B.S. degree in chemistry from the University of Minnesota and a Ph.D. from Stanford University. He came to his present position after working in the private sector for 13 years; prior to that time he served for 21 years on the chemistry faculty at the University of Washington. He has published over 120 papers in the areas of polymer science and electro-optic materials. His current research is aimed at theoretical understanding and prediction of the electro-optic properties of molecules.

Bruce H. Robinson is Professor of Chemistry at the University of Washington and Associate Director of the NSF-STC on Materials and Devices for Information Technology Research. He received his B.S. in chemistry from Princeton University and his Ph.D. in chemistry from Vanderbilt University. His current research interests include biological magnetic resonance and the theory of intermolecular electrostatic interactions relevant to optimizing the performance of organic electro-optic materials.

Chapter 17
Charge Transport and Optical Effects in Disordered Organic Semiconductors

Harry H. L. Kwok
University of Victoria, Victoria, BC, Canada

You-Lin Wu and Tai-Ping Sun
National Chi-Nan University, Nantou, Taiwan

17.1 Introduction

Charge transport in disordered organic semiconductors is receiving a great deal of attention because, as with inorganic semiconductors, conductivity in organic semiconductors can be changed through doping.[1] In addition, a few disordered organic semiconductors also have "high" carrier mobilities.[2] The discovery of these organic semiconductors marks a major change in the research outlook for these semiconductors, since they have the potential of replacing more expensive single-crystal semiconductors in devices. Organic materials possess other advantages as well.[3] A few organic semiconductors have already been used in industrial applications, such as the productio of light-emitting devices (LEDs), transistor circuits built on flexible substrates, and a variety of thin-film sensors and solar cells. In all of these applications, cleverly devised structures were adopted while, in many cases, advantages in manufacturing and cost effectiveness remained to be exploited. Indeed, some electronic display devices based on disordered organic semiconductors have already been successfully marketed in products such as cell phones and television screens. This chapter focuses on the study of charge transport and the optical properties of disordered organic semiconductors.

17.2 Charge Transport

We first raise the question of how disordered organic semiconductors conduct current. The answer requires that the organic semiconductors possess a substantial number of "free" carriers (free carriers include electrons and holes). In organic semiconductors, the fact that free carriers can be produced from the distributed π-bonds in a chain molecule is well known. In addition, the free carriers must be able to move effectively inside the organic semiconductors. This requires the presence of "transport sites," which are the locations within the molecules that act as intermediate stopping points as the free carriers move. A proper description of transport sites requires one to include site energies. At thermal equilibrium, the free carriers have a distribution of energies; how they interact with the transport sites depends on the affiliated energy correlation. Free carriers essentially move from one transport site to another with a weighted probability based on energy correlation. This probability may also be affected by atomic/molecular vibrations.

In addition to the transport sites, free carriers may also be captured by traps that are usually associated with defects and/or grain boundaries. The release of a carrier from a trap involves energy exchange and a time delay. A trapped carrier may at times recombine with a free carrier of the opposite polarity to release the excess energy in the form of heat or light. Assuming that the trap density is small (usually justifiable in the case of a high-quality organic semiconductor), one may consider the density of transport sites to be the density of states. Obviously, such an assumption ignores the effect of the trapped charges on the molecular potentials. In some disordered molecular semiconductors, deformation associated with trapped charges cannot be ignored, and traps are known to create a dipole field extending many atomic radii. When coupled to a self-induced structural deformation, a charge will form an entity known as a polaron.[4] A polaron can therefore be viewed as a charge localized in the potential minimum formed by a molecular deformation. It is interesting to note that polarons (sometimes considered to be "dressed" charges) are capable of migrating across molecules.

17.2.1 Energy bands

Strictly speaking, the concept of "energy bands" is applicable only to crystalline solids when charges move collectively within a periodic lattice. This is not the case for a charge migrating in a disordered organic semiconductor, as there will be a high degree of randomization when the charge moves from one transport site to another. Nevertheless, to mark the location of the energy states and their distribution, energy levels known as the lowest unoccupied molecular orbital (LUMO) and the highest occupied molecular orbital (HOMO) are chosen. LUMO and HOMO are not energy bands in the conventional sense that they define the energy-momentum relationship of propagating wave functions; rather, they exist to serve as energy references to demarcate ground states and activated states.

17.2.2 Dispersive charge transport

One important observation regarding disordered organic semiconductors has been dispersive transport. Basically, this refers to the fact that charges injected at one point in an organic semiconductor move (say, under a constant applied force) at different velocities and arrive separately in time at a receiving electrode. Dispersive charge transport has been studied extensively and may be described in the context of the Poole-Frenkel effect. The carrier mobility μ in a disordered semiconductor has the form

$$\mu \sim \exp\left(\beta\sqrt{E}\right), \tag{17.1}$$

where β is the disorder parameter, and E is the electric field.

Such a field-dependent mobility has often been interpreted as associated with charge transport via hopping among localized states characterized by a Gaussian distribution of site energies[5] in what is known as the Gaussian disorder model (GDM). Using the GDM, simulations have been used to fit field-related experimental data to extract relevant transport parameters. Extensions have also been made to include energetic interactions with the permanent dipoles present in a solid to determine the energy correlation between the different transport sites.[6] The latter is examined using the correlated disorder model (CDM), which has been shown to provide the mobility's observed electric-field dependence over a wide field range. In either case, both models have been successfully used to explain the electric-field dependence of mobility in disordered solids.

17.2.3 Hopping mobility

In addition to dispersion, "hopping" is also well defined in disordered solids, and the conductivity σ_C is described in terms of Mott's law, which has the form[7]

$$\sigma_c(T) = \sigma_0 \exp\left(\frac{-T_0}{T}\right)^{\gamma}, \tag{17.2}$$

where γ has the value of ¼ in three dimensions and changes to ½ for one-dimensional hopping. Equation 17.2 also applies to tunneling through nonconducting materials separated by mesoscopic metallic "islands."

Mathematical models have evolved to compute charge mobility due to hopping. The GDM, as the name implies, assumes a Gaussian energy distribution for the transport sites, giving the following expression for the resistivity:

$$\rho(\varepsilon) = \frac{1}{\sqrt{2\pi\sigma^2}} \exp\left(\frac{-\varepsilon^2}{2\sigma^2}\right), \tag{17.3}$$

where ε is the site energy away from the peak value, and σ is the rms spread in energy of those sites.

This model also assumes a Miller-Abrahams type of jump rate, resulting in a conductivity of the form

$$\sigma_c = \nu \exp\left[\frac{-2R}{\xi} - \frac{\varepsilon_0(R)}{kT}\right], \tag{17.4}$$

where ν is the jump frequency, R is the site spacing, $\varepsilon_0(R)$ is the energy difference between sites, k is the Boltzmann constant, and T is the absolute temperature.

The associated mobility has the form

$$\mu = \mu_0 \exp\left\{-\left(\frac{2\sigma}{3kT}\right)^2 + C'\left[\left(\frac{\sigma}{kT}\right)^2 - \Sigma\right]E^{0.5}\right\}, \tag{17.5}$$

where $cm^{½}/V^{½}$ ~ 2.9×10^{-4} $\sqrt{cm^{1/2}/V^{1/2}s}^{1/2}$ and Σ ~ 2.25 are evaluated numerically. Under specific conditions such as when σ ~ 0.1 eV at room temperature, all carriers will be located at an energy ~ σ^2/kT. When $\varepsilon_0 = 0$, the carrier mobility becomes

$$\mu(T) = \mu_0 \exp\left[-\left(\frac{\sigma}{kT}\right)^2\right]. \tag{17.6}$$

Such a temperature dependence is frequently observed experimentally in μ versus $1/T$ plots. In practice, the GDM is valid over a somewhat limited temperature range. Off-diagonal charge transport (away from the direction of the electric field due to adverse energetics) has been used to explain the decrease in charge mobility with E at low field, which is frequently encountered in experiments [see Eq. (17.1)].

A further refinement of the GDM was proposed to take into account energetic fluctuations in a medium possessing permanent dipole moments. This required the inclusion of an energy correlation function of the form

$$C(R) = \langle U_d(0) U_d(R) \rangle \sim \sqrt{\langle U_d^2 \rangle} \frac{a'}{R}, \tag{17.7}$$

where $\langle U_d \rangle$ is the energetic disorder, and a' is a minimal charge-dipole separation. Such a CDM, through simulations, has been shown to better retain the Poole-Frenkel dependence at low field as well as to accommodate symmetric hopping, a characteristic feature of the small polaron. The empirical relationship observed is

$$\mu = \mu_0 \exp\left\{ -\left(\frac{3\sigma}{5kT}\right)^2 + C_0 \left[\left(\frac{\sigma}{kT}\right)^{1.5} - \Gamma \right] \sqrt{\frac{qa'E}{\sigma}} \right\}, \tag{17.8}$$

where C_0 (= 0.78) and Γ (= 2) have been determined numerically. In addition to the GDM and the CDM, other models have also been used to derive the temperature dependence of mobility. One such expression has the form

$$\mu(T) = \mu_0 \exp\left(-\frac{\Delta}{kT} \right) \exp\left[B \left(\frac{1}{kT} - \frac{1}{kT_0} \right) \sqrt{E} \right], \tag{17.9}$$

where Δ, B, and T_0 are empirical constants. Reasonable agreement with experiments can be achieved. The $\sqrt{E}$ dependence can also be obtained from a space-charged barrier-height-lowering effect, assuming a barrier-height-limited charge transport in the disordered organic semiconductor.

17.2.4 Density of states

In principle, a fair amount of similarity exists between a disordered organic semiconductor and a disordered inorganic semiconductor, such as amorphous silicon, which can be structurally represented by a continuous random network. Electronically, both have low carrier mobility. One major difference, however, is the fact that disordered organic semiconductors are solids with weak molecule-to-molecule interaction and are sensitive to localized charges. Understanding the effect of the localized charge states therefore becomes very important. Indeed, reports have suggested that steep (exponential) donor states and shallower acceptor states between the LUMO and the HOMO exist in a field-effect device structure,[8] which is somewhat similar to the case of amorphous silicon. Previously mentioned models that are used to study charge transport assuming a Gaussian density of (localized) states may indeed be valid under conditions such as when energies are far away from the Gaussian peak. It is nevertheless important that the density of states in disordered organic semiconductors be properly characterized.

In Ref. 8, reported measurements on the density of states for organic amorphous thin films using Kelvin probe-force microscopy on doped and undoped samples suggested a broadening of the density of states distribution as well as the existence of sharp induced peaks found in the doped samples, as shown in Fig. 17.1. In addition, three distinct regions appeared in the undoped samples as the carrier density increased (with the population of lower energy states filled by holes). Nearest to the HOMO, there is a Gaussian distribution of states, interrupted by a single peak; further from the HOMO, extended tail states occur. The state density in the sample was estimated to be 1.2×10^{27} m^{-3}, with σ = 100 meV at 0.84 eV away from the LUMO. This behavior of the density of states function deviates from a Gaussian distribution away from the HOMO (at the right-hand side of Fig. 17.1) and is better fitted with an exponential distribution. The doped sample, on the other hand, shows an exponential distribution of the density of states with a number of separated peaks. There is evidence that many of the peaks and their fluctuations might be associated with the interface states linked to the field-effect structure.

17.3 Impedance Spectroscopy: Bias and Temperature Dependence

Impedance spectroscopy[9] is a popular means of studying ac conductivity. This measurement method has been used to extract material parameters such as carrier density and mobility in organic semiconductors at different biases and temperature, and offers an effective means to verify theoretical charge transport models. In many organic semiconductors, a good match between theory and experiment has been obtained, and most point to the presence of a trap-related charge transport. However, not all disordered organic semiconductors behave properly, and in some, a negative capacitance effect has been observed. Negative

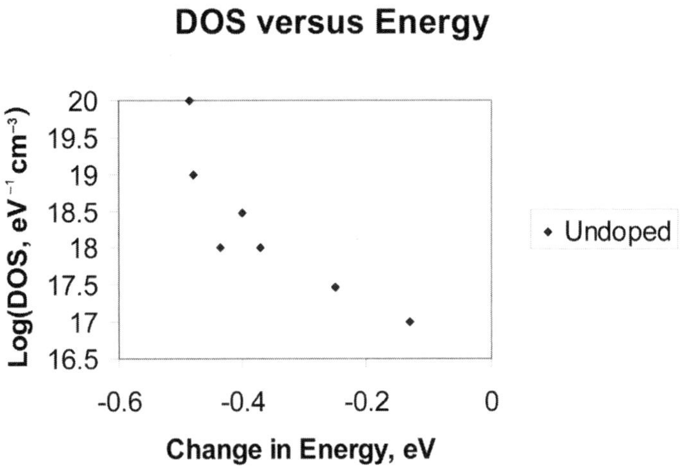

Figure 17.1 Plot of the density of states distribution versus energy. The HOMO energy is located at 0 (extreme right). [Reprinted with permission from Ref. 8. © (2005) by the American Physical Society.]

capacitance can be derived from the Drude model[10] if one assumes that the solid consists of a collection of charges and ions in a plasma that is under the influence of an oscillating electric field E. Energy relaxation is associated with γ_0, a parameter proportional to the rate of inelastic collisions per unit charge. In addition, charge will oscillate with a Hooke's law force constant K_0, which varies linearly with the applied force, as it does in the case of a valence charge closely coupled to the lattice ions. Considering holes only, the model equation[11] has the form

$$\frac{m_h d^2x}{dt^2} + \frac{qg_0 dx}{dt} + K_0 x = qE\,, \tag{17.10}$$

where m_h is the hole mass, x is the average position of the charges, and q is the electronic charge. Using an oscillating electric field of the form $E = E_0 e^{jwt}$, where E_0 is the amplitude and ω is the angular frequency, the response is $x = x_0 e^{jwt}$. Equation (17.10) can be solved, and the average hole position becomes

$$x_0 = \frac{\left(-\dfrac{qE_0}{m_h}\right)}{(\omega^2 - \omega_0{}^2) - \dfrac{j\omega q\gamma}{m_e}}, \tag{17.11}$$

where $\omega_0{}^2 = K_0/m_h$ (ω_0 is the resonant frequency). The polarization vector P is given by

$$P = pqx_0 = \frac{\left(\dfrac{pq^2E_0}{m_h}\right)}{\left(\omega_0{}^2 - \omega^2\right) + \dfrac{j\omega\gamma q}{m_h}}, \tag{17.12}$$

where p is the hole density. Assuming a parallel plate capacitor with a cross-sectional area A and a plate separation L, the capacitance C becomes

$$C = \mathrm{Re}\left\{\left(\frac{P}{E_0}\right)\left(\frac{A}{L}\right)\right\} = \mathrm{Re}\left\{\frac{\left(\dfrac{pq^2A}{Lm_h\omega_0{}^2}\right)}{1 - \left(\dfrac{\omega}{\omega_0}\right)^2 + \dfrac{j\omega q\gamma}{m_h\omega_0{}^2}}\right\}. \tag{17.13}$$

For a dispersive medium, γ_0 is complex and may be replaced by $\gamma_1 + j\gamma_2$. After substitution, Eq. (17.13) becomes

$$C = \left(\frac{pq^2 A}{Lm_h \omega_0^{\,2}}\right)\left[\frac{\dfrac{1 - \varpi^2 - \varpi q\gamma_2}{m_h\omega_0}}{\left(\dfrac{1 - \varpi^2 - \varpi q\gamma_2}{m_h\omega_0}\right)^2 + \left(\dfrac{\varpi q\gamma_1}{m_h\omega_0}\right)^2}\right], \tag{17.14}$$

where we have denoted (ω/ω_0) by ϖ. C is computed from Eq. (17.14), provided the parameters p, γ_0, ω_0, and the dimensions of the capacitor are known. γ_0 is associated with the carrier mobility μ. By setting $\gamma_0 = 1/\mu = 1/(\mu_1 + j\mu_2)$, where μ_1 and μ_2 are the real and the imaginary parts of μ respectively, we have the relationship: $\mu_1 = \gamma_1/(\gamma_1^{\,2} + \gamma_2^{\,2})$ and $\mu_2 = -\gamma_2/(\gamma_1^{\,2} + \gamma_2^{\,2})$.

For the case in which dispersion is absent, i.e., $\mu_2 = \gamma_2 = 0$, we have [from Eq. (17.14)]

$$C = \frac{\left(\dfrac{pq^2 A}{Lm_h\omega_0^{\,2}}\right)\left(1 - \varpi^2\right)}{\left(1 - \varpi^2\right)^2 + \left(\dfrac{\varpi q\gamma_1}{m_h\omega_0}\right)^2}. \tag{17.15}$$

In this limit, C becomes negative, assuming $\varpi > 1$. For electronic polarization, this usually does not happen, since $\varpi << 1$. The sign of C, according to Eq. (17.14), hinges on whether $(1 - \varpi^2 - \varpi q\gamma_2/m_h\omega_0)$ is larger than, or less than, zero. At a moderate frequency, $\varpi^2 << 1$ and the capacitance is

$$C \approx \left(\frac{pq^2 A}{Lm_h\omega_0^{\,2}}\right)\left[\frac{\dfrac{1 - \varpi q\gamma_2}{m_h\omega_0}}{\left(\dfrac{1 - \varpi q\gamma_2}{m_h\omega_0}\right)^2 + \left(\dfrac{\varpi q\gamma_1}{m_h\omega_0}\right)^2}\right]. \tag{17.16}$$

Equation (17.16) is negative when $\varpi q\gamma_2/m_h\omega_0 > 1$, and in this intermediate range,

$$C \approx \left(\frac{pqA}{L\omega}\right)\left(\frac{-\gamma_2}{\gamma_1^2 + \gamma_2^2}\right) = \frac{pqA\mu_2}{L\omega}. \tag{17.17}$$

Note that C is inversely proportional to ω. More interestingly, Eq. (17.17) allows one to evaluate μ_2, which is negative when C is negative. The determination of μ_1 and μ_2, or γ_1 and γ_2, requires detailed measurements of C, p, and ω_0 over a broad frequency range. In general, these parameters vary with bias, frequency of the oscillating field, and quite often with different samples depending on the preparation conditions.

Figure 17.2(a) shows a series of capacitance-versus-frequency curves obtained from Eq. (17.14) with different values of μ_1 ($\mu_2 = 0$) based on the parameters extracted from a poly[2-(3′7′-dimethyloctyloxy)-5-methoxy-1,4-phenlyene-vinylene], or OC_1C_{10}-PPV, sample. Figure 17.2(b) shows similar curves obtained from Eq. (17.16) when μ_1 is a constant and μ_2 is allowed to vary. The parameters used in the calculations are listed in Table 17.1 (m_h is assumed to be the electron rest mass).

From the figures, one can see that a plot of the capacitance change with log(frequency) results in an S-shaped curve. At low frequency, the saturation values change as the carrier mobility decreases. This is in basic agreement with the fact that damping (caused by collisions with trap states) has a direct negative impact on the polarization of the plasma, causing the capacitance to be lowered. The transition region between the two saturation levels (at high and low frequencies) is sensitive to the value of μ_1. It shifts toward the lower frequencies as μ_1 is decreased. The capacitance-versus-log(frequency) curves also change shape

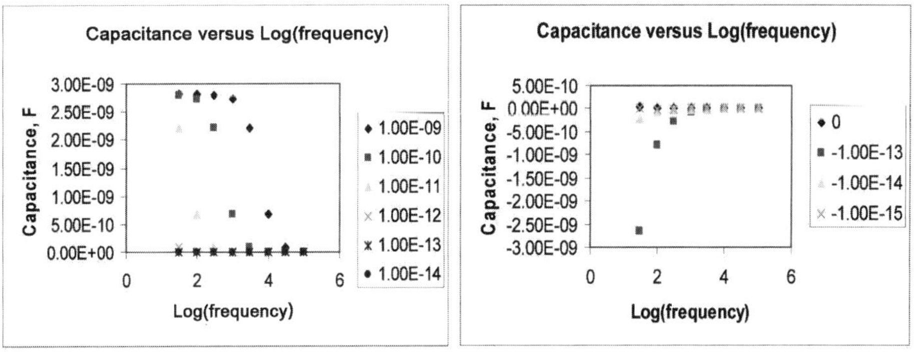

Figure 17.2 (a) Capacitance-versus-frequency curves when μ_1 varies between 1×10^{-9} and 1×10^{-13} $m^2/V{\cdot}s$, and $\mu_2 = 0$. (b) Capacitance-versus-frequency curves when $\mu_1 = 1 \times 10^{-13}$ $m^2/V{\cdot}s$ and μ_2 varies between -1×10^{-13} and -1×10^{-15} $m^2/V{\cdot}s$. (Reprinted from Ref. 11 with permission from Elsevier Science Ltd.)

Table 17.1 Model parameters used in the simulations. (Data used from Ref. 11 with permission from Elsevier Science Ltd.)

	Figure 17.2(a)	Figure 17.2(b)
ω (rad/s)	300 to 100000	300 to 100000
A/L (m)	100	100
p (m^{-3})	10^{21}	10^{21}
ω_0 (rad/s)	10^{12}	10^{12}
μ_1($m^2/V{\cdot}s$)	10^{-9} to 10^{-14}	10^{-13}
μ_2($m^2/V{\cdot}s$)	0	0, -10^{-13} to -10^{-15}

when the carrier mobility becomes complex [Fig. 17.2(b)]. For a small μ_2, the capacitance changes sign and becomes negative. In addition, negative capacitance also intensifies as μ_2 increases, confirming the fact that it arises from dispersion, a feature frequently observed in transit time measurements of organic polymers.

It would be interesting to research whether Eq. (17.14) is applicable to other types of organic semiconductors such as small molecules. Recent work[12] reports on impedance measurements on Tris(8-quinolionoato) aluminum, Alq_3, with Al and Ca contacts. The Alq_3 exhibited a negative capacitance effect under voltage bias and illumination. These results were compared to simulations[13] based on a field-dependent carrier mobility and Eq. (17.14). That report successfully normalized the simulated impedance-frequency curves under different biases with respect to the frequency of the minimum capacitance point, and reduced an entire family of C-V curves to a single curve. Such a shape invariance of the C-V curves was suggested to reflect a characteristic feature of dispersive charge transport known as the "universality of photocurrent transients." Similar observations were found for the normalization process applied to C-V curves taken at different temperatures. The report also suggests that negative capacitance can be associated with the dominance of positional off-diagonal disorder in Alq_3, and deduced that trapping and hopping processes are relatively independent of each other. A semilog plot of the hole mobility data versus the inverse temperature squared gave a straight line fit to the data with a characteristic temperature T_0 of ~550 K. Such a temperature dependence is compatible with a disordered solid having a Gaussian density of states [see Eq. (17.6)]. Table 17.2 lists the values of the parameters used in the calculations. The field-dependent complex carrier mobility was computed using Eq. (17.1).

Figure 17.3(a) shows the simulated capacitance-frequency curves for voltage bias between 0 and 10 V. These curves closely resemble those reported in Ref. 12. The capacitance and frequency values at the capacitance minimum are denoted by C_{min} and f_{min}, respectively. Figure 17.3(b) shows a plot of f_{min} versus the voltage bias. Also included in the figure are the frequencies f_{tr} at the inflection points. Next, the capacitance-frequency curves in Fig. 17.3(a) were normalized with respect to f_{min} at different biases. The results of this normalization are shown in Fig. 17.4.

Figure 17.5 shows a plot of the magnitude of the complex carrier mobility at different voltage biases (values used in the simulations) versus f_{min}. The results indicate that f_{min} is proportional to the magnitude of the complex carrier mobility. In essence, the frequency shifts in the capacitance-frequency curves were the result of a change in the complex carrier mobility.

Table 17.2 Additional model parameters (μ_1 and μ_2 refer to zero-field values). (Data used from Ref. 11 with permission from Elsevier Science Ltd.)

ω (rad/s)	A/L (m)	p (m^{-3})	ω_0 (rad/s)	μ_1(m^2/V·s)	μ_2(m^2/V·s)	β
0.001 to 40,960	2.45	10^{20}	0.8×0^{12}	3×10^{-15}	-3×10^{-15}	10^{-3}

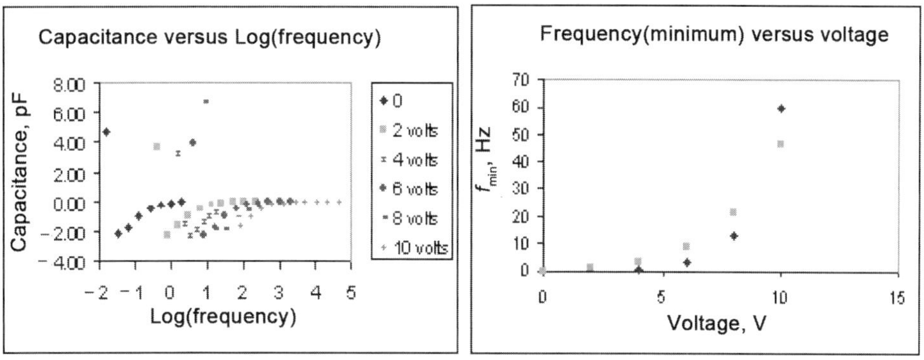

(a) (b)

Figure 17.3 (a) Simulated capacitance-frequency curves using parameter values listed in Table 17.2. (b) Plot of f_{min} (diamonds) and f_{tr} (squares) versus voltage bias. (Reprinted from Ref. 13.)

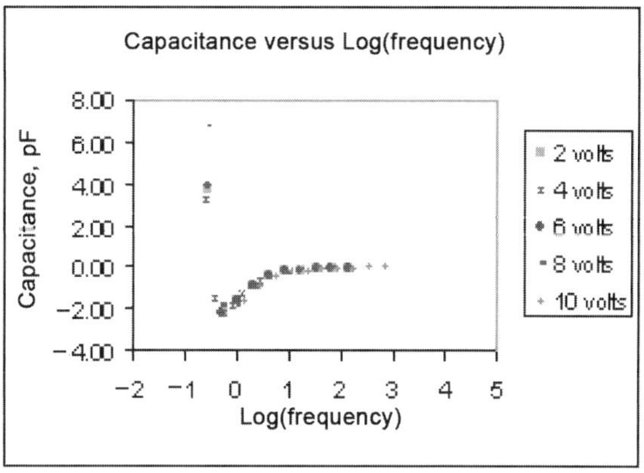

Figure 17.4 Normalized capacitance-frequency curve. (Reprinted from Ref. 13.)

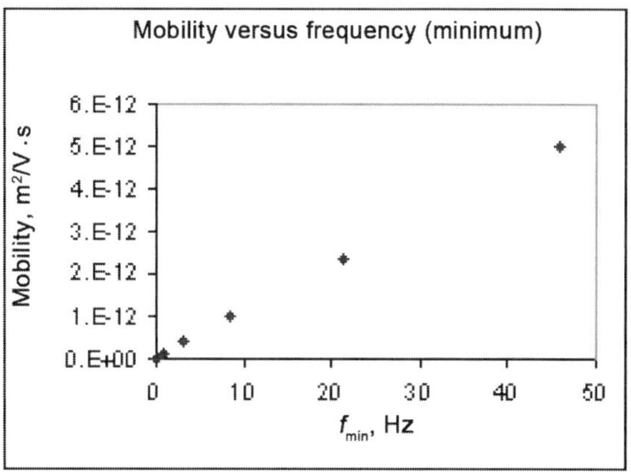

Figure 17.5 Plot of the magnitude of the complex carrier mobility versus $f_{min.}$ (Reprinted from Ref. 13.)

Normalization was performed on the reported capacitance-frequency curves measured at different temperatures in Ref. 12. From these curves, one could extract the values of f_{min} and the corresponding values of the carrier mobility that allow one to make a plot of the carrier mobility versus the inverse temperature squared, as shown in Fig. 17.6. The result suggests a temperature dependence of the form $\mu \sim \exp[-(T_0/T)^2]$ [similar to what was given in Eq. (17.6)], as is routinely observed in disordered solids. Figures 17.7(a) and (b) show the capacitance-frequency curves at different temperatures before and after normalizing, as was shown in Fig. 17.4.

These results show that computations based on the Drude model are in agreement with the concept of "universality of photocurrent transients," provided that bias and temperature effects have been properly accounted for. With the Drude model, detailed processes such as trapping and their energy distribution, the hopping mechanism, etc., are not explicitly stated but are collectively included in the phase and magnitude of the complex carrier mobility. Correlation

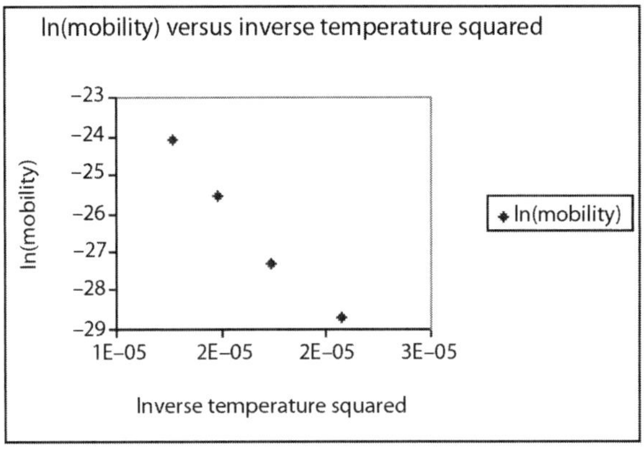

Figure 17.6 Semilog plot of the complex carrier mobility versus the inverse temperature squared. (Reprinted from Ref. 13.)

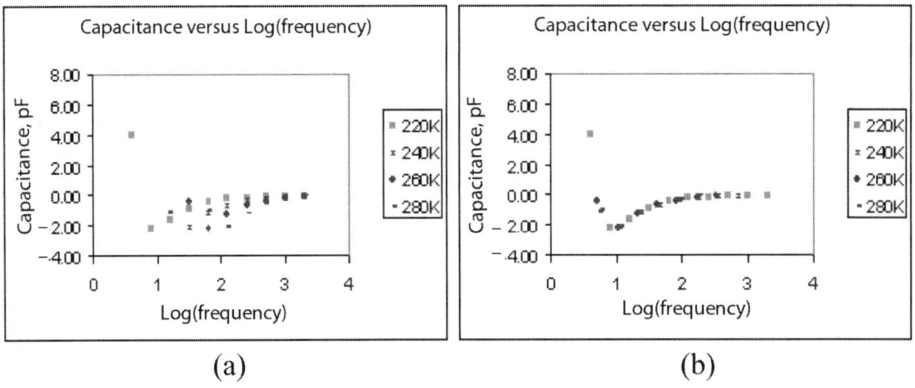

Figure 17.7 (a) Simulated capacitance-frequency curves using the parameter values listed in Table 17.2. (b) Normalized capacitance-frequency curve. (Reprinted from Ref. 13.)

between the experimental and theoretical frequencies at the capacitance minima is pivotal to the normalization process after taking into account the mobility dependence on bias and temperature. According to Ref. 12, disorder in organic semiconductors is expected to affect the phase of the complex carrier mobility in a manner similar to the parameter α in the expression for the hopping mobility $\mu'(\omega) = [1 + M(i\omega\tau_{tr})^{1-\alpha}]$. The lack of evidence requiring a phase shift upholds the argument needed to support the validity of the universality of photocurrent transients in Alq_3. Furthermore, the observed relationship $\ln(\mu) \sim -(T_0/T)^2$ is in basic agreement with what is expected for disordered materials. Shape invariance of the capacitance-frequency plots at different temperatures after normalization confirms the postulate that positional off-diagonal disorder predominates. This is equivalent to the fact that the phase of the complex carrier mobility remains unchanged with changing temperature. Along this line of reasoning, one can argue that the processes of trapping and hopping act independently. This argument is based on the premise that trapped carriers are thermally activated. The observed phase invariance with changing temperature in our computations does not preclude that processes such as the image force effect are entirely absent.

Our previous treatment suggests that the "negative" capacitance effect can be interpreted in terms of trapping. Other mechanisms may also give rise to "negative" capacitance. In the following, we use an equivalent circuit that mimics the trapping process and explains the results in terms of a time delay and/or phase shift. If one replaces $q\ dx/dt$ in Eq. (17.10) with i/A, where i is the current and A is the cross-sectional area, and assumes that the first term on the left-hand side (i.e., away from the inflection points) is small, one obtains

$$iR + \frac{1}{C_0}\int i\,dt \approx V, \tag{17.18}$$

where we have set $R = q\gamma_0$, $C_0 = 1/K_0$, and $q^2EA = V$. Equation (17.18) is a first-order differential equation that can be obtained from a series resistor-capacitor circuit.[14] This is shown in Fig. 17.8, assuming V is the driving voltage.

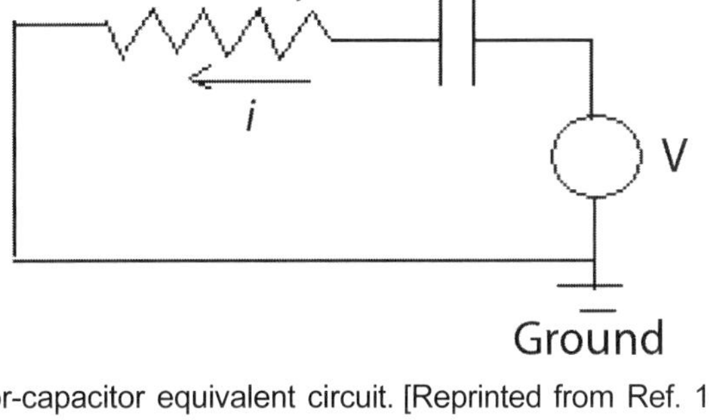

Figure 17.8 Resistor-capacitor equivalent circuit. [Reprinted from Ref. 14. © (2008) Wiley-VCH Verlag GmbH & Co. KGaA. Reproduced with permission.]

For a sinusoidal input with an angular frequency ω, Eq. (17.18) becomes

$$i/V = \frac{1}{\left(\frac{R+1}{j\omega C_0}\right)}. \tag{17.19}$$

In order for "negative" capacitance to appear, one can impose the condition that $R = 1/(G_1 - jG_2)$, where G_1 and G_2 are the real and imaginary components of the conductance. Note that the term $-jG_2$ is negative. Manipulation of Eq. (17.19) gives an equivalent capacitance C_{eq} (= $i/j\omega v$) for this circuit given by

$$C_{eq} = \frac{C_0\left[-G_1\omega C_0 + j\left(G_1^2 + G_2^2 - G_2\omega C_0\right)\right]}{G_1^2 + \left(G_2 - \omega C_0\right)^2}. \tag{17.20}$$

Figure 17.9(a) shows a plot of C_{eq}/C_0 as a function of G_2 (note that G_2 is positive). In the calculations, we have set $L/A = 100\ \text{m}^{-1}$, $p = 1 \times 10^{21}\ \text{m}^{-3}$, $\omega = 10$ rad/s, $C_0 = 1\ nF$, and $\mu = 1 \times 10^{-13}\ \text{m}^2/\text{V·s}$. Figure 17.9(b) shows a plot of the normalized capacitance versus frequency.

One can determine the frequency of the capacitance minimum ω_{peak} by computing $dC_{eq}/d\omega$ and setting it equal to zero. After some algebraic steps, we obtain the following quadratic equation in ω' (= $\omega_{peak}C_0$):

$$\begin{aligned}&\left[\left(G_1^2 + G_2^2\right) - 2G_2\omega' + \omega'^2\right]\left(G_2 + jG_1\right) - 2\left(G_2 - \omega'\right)\\&\times\left(G_1^2 + G_2^2 - \omega' G_2 - jG_1\omega'\right) = 0.\end{aligned} \tag{17.21}$$

The solution is

$$j\omega' = G_1 - jG_2. \tag{17.22}$$

Equation (17.22) suggests that the frequency at the capacitance peak is directly proportional to the complex conductance G ($=G_1 - jG_2$). This implies that frequency at the capacitance minimum is proportional to the complex carrier mobility. Such a relationship has been observed previously and suggests a linear relationship between $\log(\omega_{peak})$ and $\sqrt{E}$. The important question to address is: What is the physical origin of the observed complex carrier mobility? For a disordered semiconductor with a large number of traps, charge transport will be delayed due to retention at the trap states. This is qualitatively illustrated in Fig. 17.10. Under such conditions, the phase relationship between current and voltage will deviate from what is expected for an ideal capacitor. Instead of having an "instantaneous" flow of current, a delayed current will be experienced, and the

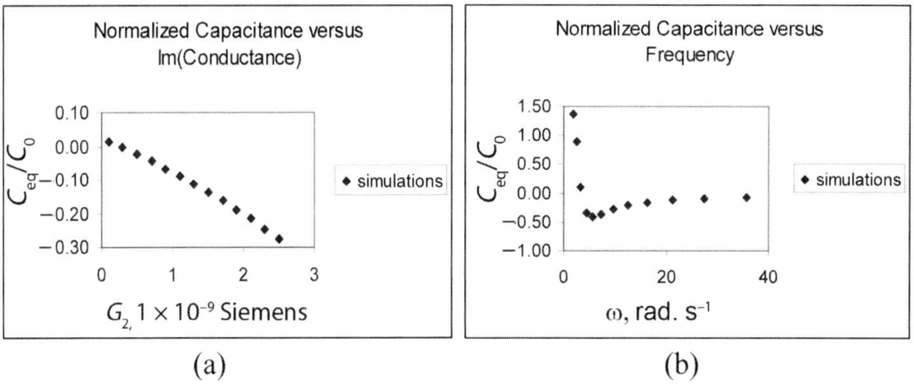

Figure 17.9 (a) Normalized capacitance versus the imaginary component of the conductance (in 1 × 10^{-9} Siemens). (b) Normalized capacitance versus frequency assuming $|G_1| = |G_2|$. [Reprinted from Ref. 14. © (2008) Wiley-VCH Verlag GmbH & Co. KGaA. Reproduced with permission.]

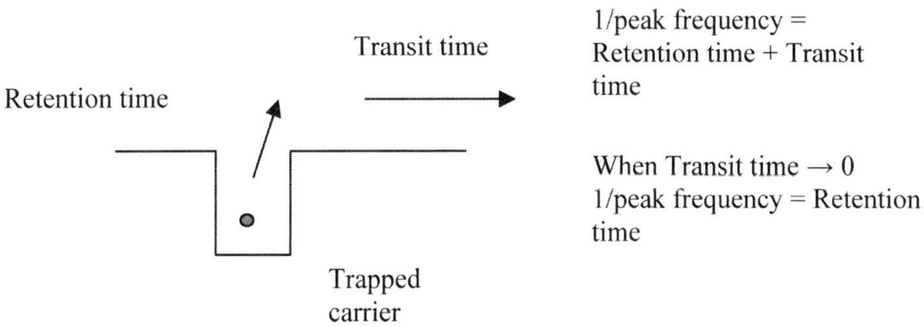

Figure 17.10 Schematic showing the physical processes linked to charge retention and transit. [Reprinted from Ref. 14. © (2008) Wiley-VCH Verlag GmbH & Co. KGaA. Reproduced with permission.]

time delay depends on the time interval that the carriers are trapped. Such time delay is responsible for the observed "negative" capacitance effect, which as far as the impedance-frequency characteristics are concerned, resembles the behavior of an inductor. For a series resistor-capacitor circuit, the expected phase delay is between 0 and 90 deg [our computations give a phase delay of 45 deg due the assumptions made earlier—this is also reflected in the phase factor of $(1 - j)$ appearing in Eq. (17.22)]. One can also estimate the value of ω_{peak}. Using typical values obtained from OC_1C_{10}-PPV samples, $G_1 \approx 5 \times 10^{-9}$ S. Since $C_0 \approx 1$ nF, we have: $\omega_{max} \approx 10$ rad/s, in essential agreement with the reported frequency range of between 0.1 to 100 Hz when a "negative" capacitance effect has been observed. Another interesting question to ask is: What happens to ω_{peak} if G is large, as is the case for a high mobility organic semiconductor such as pentacene? According to theory, ω_{peak} ought to increase. This would have been the case were it not for the finite retention time of the trapped carriers. At higher frequencies when the retention time exceeds the period of the ac signal, the trapped carriers no longer respond to the driving voltage and the imaginary component of the conductance vanishes. As a consequence, the conditions needed for "negative" capacitance to develop are no longer in place.

17.4 Transient Spectroscopy

Recent research also involves the study of dispersive organic solids using transient spectroscopic techniques. In contrast to inorganic semiconductors, charge transport in organic semiconductors tends to be local, meaning that charge flow results from correlated/uncorrelated hopping events occurring over short distances. Phenomenally, hopping involves two identifiable steps. The first is an escape process (usually thermally activated); while the second relates to charge transfer from one transport site to another. In most organic solids, these steps occur infrequently and over short distances. As an example, in the geometric model[15] developed to study charge transport, the random migration of charges was found to peak at a distance 1.4× the site spacing. For a solid with 1×10^{24} transport sites per cubic meter, this suggests a migration distance of ~ 14 nm, a distance many orders of magnitude smaller than the mean free path of carriers in a crystalline semiconductor. A short migration distance also implies lower carrier mobility. As mentioned earlier, the computation of the carrier mobility in dispersive semiconductors usually relies on the GDM or the CDM. The latter also takes into account energy correlation between the localized states in the presence of a dipole field. In addition, both models rely on the knowledge of material parameters at the microscopic scale.

One problem with these models is that they are valid over a limited temperature range. Not all dispersive solids are disordered even though many single-crystal organic semiconductor samples exhibit dispersive transport properties and stretched exponential relaxation. Such observations, we believe, arise from the presence of induced dipoles/residual charges during photoexcitation. It is therefore often advantageous to examine simultaneously the relationship between charge relaxation after photoexcitation and charge transport. One approach is to identify the material parameters that are common to the two processes and attempt to analyze them interactively. This is where the study of transient spectroscopy becomes important. This technique allows one to obtain and refine material parameters governing relaxation and charge transport. This may shed light upon the mechanisms linking the two processes.

The theories relating relaxation and charge transport are sometimes treated separately even though modern first-principle theories based on time-dependent density functional theory are capable of treating these processes consistently.[16] We follow the same route and will try to identify the material parameters common to the two processes. For charge relaxation, a dispersive semiconductor with transport sites randomly distributed is assumed. This applies to single-crystal organic semiconductors provided the site locations are not specific to any particular regions of the molecules. Other issues such as energy correlation also affect the charge transport mechanism. In addition, we also assume the semiconductor to be polar (either induced or due to geometric fluctuations[17]). In the case of induced dipoles, the transport sites may or may not be those locations where the excited carriers are generated. Similar to the treatment in Ref. 15, there is a high density of transport sites and the traps are at the locations where the excited carriers are formed. As a result, the density of the transport sites N_T is equal to or greater than the excited carrier density N_0. The trap density is assumed to be equal to N_0. In three dimensions the spacing of the transport sites is either smaller than or equal to $(N_0)^{-1/3}$. For the case in which hopping dominates, the geometric model suggested a decay function given by

$$f = \int_0^\infty F(R)\ \exp\left[-W(R)t\right]dR, \tag{17.23}$$

where $F(R)$ is a geometric probability function, R is the site spacing, $W(R)$ [= ν exp(–2R/L)] is the charge escape rate, ν is the escape frequency, L is the localization length, and t is time. $F(R)$ can be determined from experiment for different ratios of (transport) site density to trap density. Assuming R to be strictly a geometric parameter, $F(R)$ will be maximized in the evaluation of the decay function [Eq. (17.23)]. A typical plot of the decay function[18] is shown in Fig. 17.11 (with $N_T/N_0 = 1$; $N_0 = 5 \times 10^{23}$ m^{-3}; $\nu = 1 \times 10^{12}$ s^{-1}; and L between 3 and 10 nm). The temporal variation of the decay function behaves very similarly to the stretched exponential function observed during charge relaxation in disordered organic semiconductors. Theoretically, the stretching index β can be estimated from the slopes of log[ln(f)] versus log(t) plots.

To compute the carrier mobility, one may use a model similar to the one that has given rise to Eq. (17.9), i.e.,

$$\mu = \mu_\infty \exp\left(\frac{-\Delta}{kT}\right)\exp\sqrt{-\alpha E^{0.5}}, \tag{17.24}$$

where μ_∞ is the low-field carrier mobility; Δ is the activation energy, α is a "disorder" parameter, and E is the electric field. Furthermore, the low-field carrier mobility has the form [see Eq. (17.4)]

$$\mu_\infty = \frac{qR_0^{\ 2}\nu}{\sigma}\exp\left(\frac{-2R}{L}\right), \tag{17.25}$$

where q is the electron charge, ν is the escape frequency, σ is the rms width of the density of (transport) states, and L is the localization length. Using the Poole-Frenkel relation, it is possible to show that $\alpha = q\sqrt{(q/4\pi\varepsilon_s)}/kT_0$ where $T_0 = 385$ K for pentacene and ε_s is the semiconductor permittivity (assumed to be 2.5 times the permittivity of free space). Based on the above equations, the material parameters needed to compute β and μ are: ν, L, σ, N_0 and N_T.

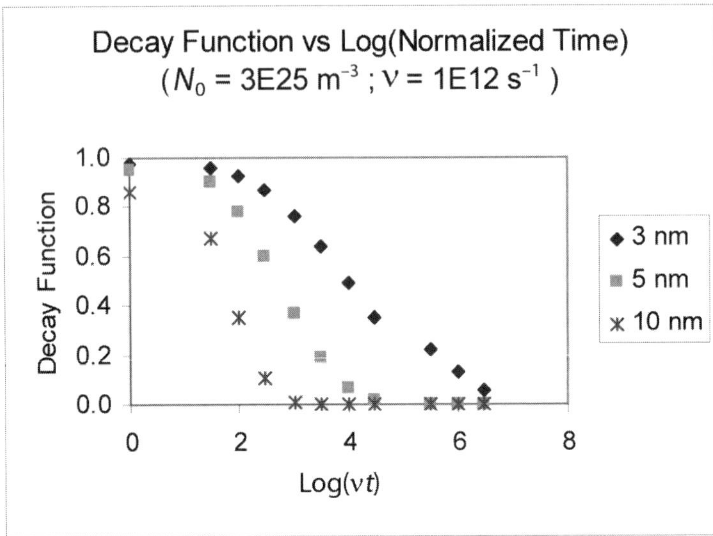

Figure 17.11 Decay functions for different values of the localization length *L* according to Eq. (17.23). (Reprinted from Ref. 18.)

While there have been occasional reports on the estimated values of ν, *L*, and σ, these parameters are material specific and have to be measured for the samples under consideration. According to Hegman et al.[19] their functionalized pentacene samples had a carrier mobility of ~ 0.2 cm^2/V·s at 300 K and an average carrier density of ~ 5×10^{23} m^{-3}. This implied R ≤ 12.6 nm. Using this value of N_0 and Eqs. (17.23) and (17.24), we computed μ as a function of σ, the rms width of the density of (trap) states for different values of N_0/N_T, ν, and *L*. The results are shown in Figs. 17.12(a)–(d). The dotted lines in the figures correspond to mobility μ = 0.2 cm^2/V·s. From the figures, we estimate that N_T ought to be at least equal to N_0. (This would not be true in the unlikely event that either ν significantly exceeds the value 1×10^{14} s^{-1} or *L* >> 9 nm.)

One may restrict consideration to the two cases in which (a) $N_T/N_0 = 1$; $\nu = 1 \times 10^{12}$ s^{-1}, and (b) $N_T/N_0 = 40$; $\nu = 1 \times 10^{14}$ s^{-1}. Using these parameter values, we match the reported transient photoconductivity data to the simulated decay function [Eq. (17.23)] and determine the values of *L*. (It should be pointed out the reported transient photoconductivity data in Ref. 19 were measured at 240 K while the mobility data were estimated at 300 K.) The results are plotted in Figs. 17.13(a) and 17.13(b). Based on the figures, *L* is ~ 9 nm for $\nu = 1 \times 10^{12}$ s^{-1} (with $N_T/N_0 = 1$) and ~ 5 nm for $\nu = 1 \times 10^{14}$ s^{-1} (with $N_T/N_0 = 40$). Since earlier results also suggested that *L* had to be ~ 9 nm to give the reported carrier mobility [see Fig. 17.12(d)], the most likely description of the pentacene samples is that they had a transport site spacing of ~ 12.6 nm and an escape frequency ~ 1×10^{12} s^{-1}. One can extend the simulations to consider changes in the carrier mobility due to N_T and ν. These are shown in Figs. 17.14(a) and 17.14(b). Based on these figures, increasing N_T results in a moderate increase in μ, but a much greater increase is observed when ν is increased.

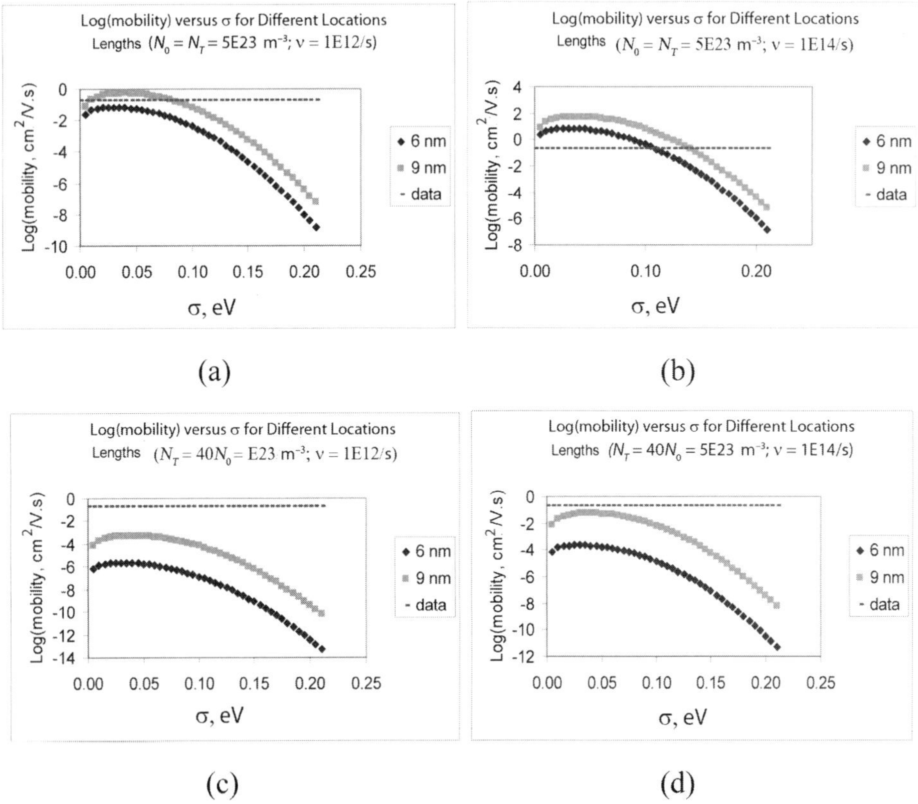

Figure 17.12 (a)–(d): Plots of log(μ) versus σ for pentacene samples at T = 300 K using different values of N_0/N_T, ν, and L. (Reprinted from Ref. 18.)

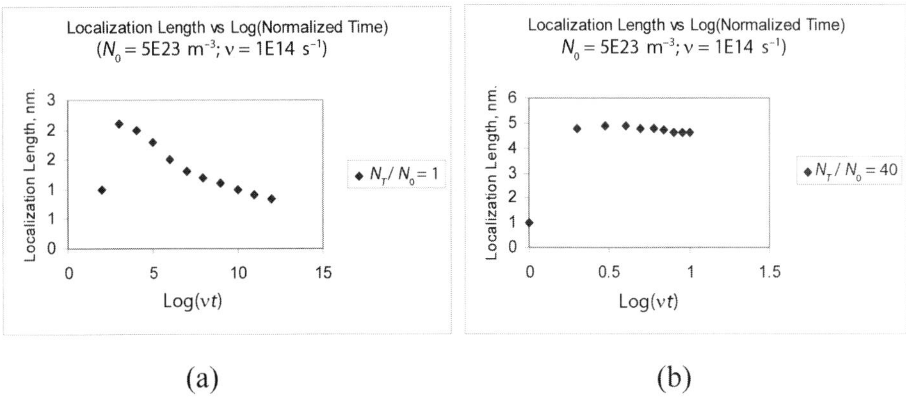

Figure 17.13 Plots of L versus log(ν t) for two different values of N_T/N_0. (Reprinted from Ref. 18.)

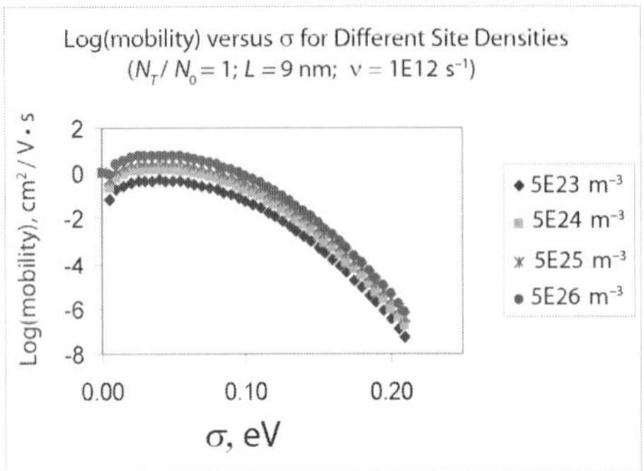

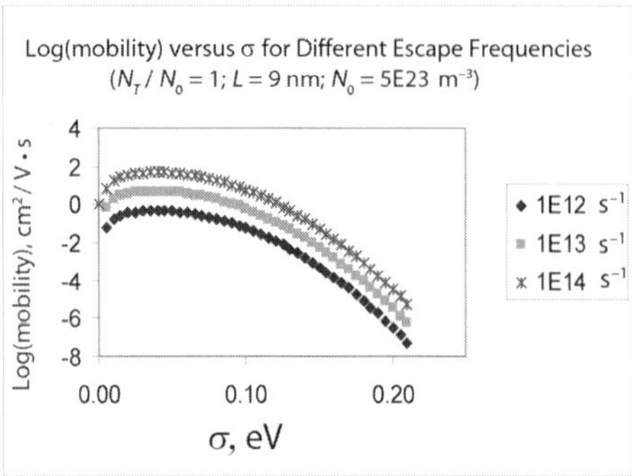

(a) (b)

Figure 17.14 Log(μ) versus σ for different values of N_0 and ν (N_T/N_0 = 1 and L = 9 nm). (Reprinted from Ref. 18.)

Using the decay function [Eq. (17.23)] it is possible to extract the stretching index using the material parameters obtained above. In Figs. 17.15(a)–(c), we plot the simulated values of the stretching index β as a function of log(t) when N_T, ν and L are varied. We see that β changes somewhat for small values of t but otherwise converged for $t > 10$ ps [according to the model, Eq. (17.23) is only valid within a finite time interval, even though recent studies indicated that the electron mobility in [6,6]-phenyl-C61-butyric acid methyl ester (PCBM) can remain stationary within the first 10 ps after excitation). In general, β increases with increasing N_T and L but not ν; increasing ν actually decreases β.

We are now in a position to extract some common features from our simulations. First of all, we note that there is a special relationship between the site spacing N_0 and the localization length L [see the expression for $W(R)$]. For the pentacene samples, the observed mobility can only exist under the conditions that $N_0 = N_T$, $L \approx 9$ nm and $\nu \approx 1 \times 10^{12}$ s^{-1}. As the simulations show, the most effective means to increase the carrier mobility is to change ν (less so by changing the site density N_T). When the value of μ is large, the effect of changing the localization length L diminishes [see Fig. 17.13(b)] suggesting the value for μ saturates. Localization length also affects the charge relaxation process. An examination of the plots of the stretching index β [see Figs. 15(a)–(c)] reveals that the initial values of β are in the range of 0.4 to 0.6, not appreciably different from values reported by other authors. β is more sensitive to changes in N_T and ν (less so with changes in L). According to these figures, β decreases significantly when N_T is lowered (say, at the value of 1×10^{22} m^{-3}). These simulation plots imply that changes in the values of β could be the direct consequence of changes in the transport mechanisms. For instance, in Fig. 17.15(a), the initial value of β decreases from a value of 0.5 to a value less than 0.2. This can be explained in terms of an increase in the site spacing (transport-limited) resulting in a changeover from short-range molecular interaction to the more distant coulomb

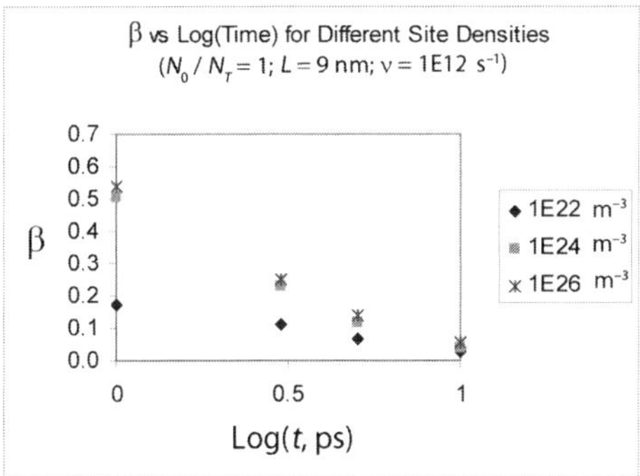

(a)

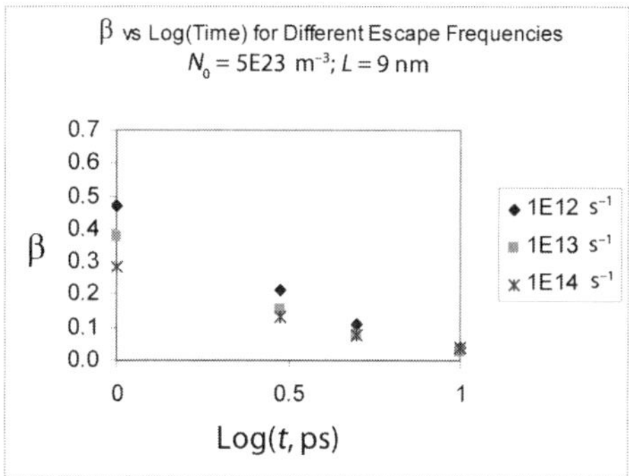

(b)

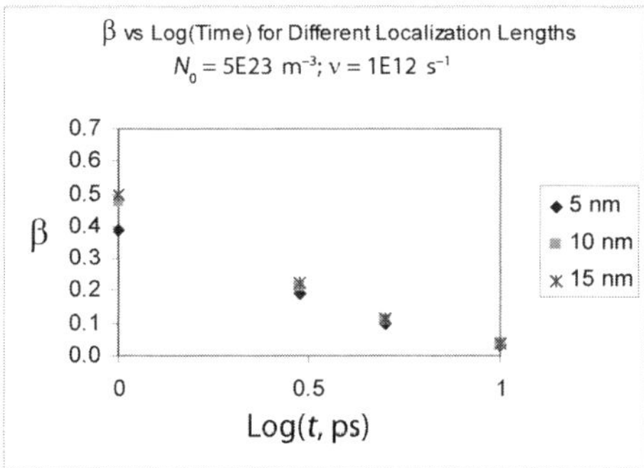

(c)

Figure 17.15 β versus log(t) for different values of (a) N_T, (b) ν, and (c) L. (Reprinted from Ref. 18.)

interaction. The same argument may be extended to the situation for which the escape frequency is large, suggesting a reduction in the molecular forces that confine the carriers (escape-limited). In any case, the sensitivity of β to the material parameters important to carrier transport is an indication that charge relaxation is affected by carrier release and migration. Further experiments are required to support or dispute this hypothesis.

17.5 Thermoelectric Effect

The thermoelectric effect known as the Seebeck effect is a useful technique to explore the intricate relationship between charge transport and heat flow. In the presence of a thermal temperature gradient ΔT, charges tend to migrate from the hot junction to the cold junction resulting in an imbalance of the charge densities. This results in a voltage differential known as the thermoelectric voltage. Mathematically, tilting of the Fermi level $E_C - E_F$ for an n-type semiconductor yields

$$E_C - E_F = kT\left[\ln\left(N_C\right) - \ln\left(n\right)\right], \tag{17.26}$$

where N_C is the effective density of states in the conduction band, and n is the electron density. Differentiating Eq. (17.26) gives:

$$\frac{d(E_C - E_F)}{dT} = \frac{(E_C - E_F)}{T} + \frac{3k}{2} - kT\frac{d[\ln(n)]}{dT}. \tag{17.27}$$

Under open-circuit conditions (through a balancing of the drift and diffusion currents) in one dimension, it can be shown that the induced electric field is

$$E = \frac{-\left[\frac{d(D_n n)}{dx}\right]}{\mu n} = \frac{-kT}{q}\left\{\frac{d[\ln(n)]}{dx} + \frac{d[\ln(D_n)]}{dx}\right\}, \tag{17.28}$$

where D_n is the diffusivity, x is the direction of heat (and charge) flow, q is the electron charge, k is the Boltzmann constant, and T is absolute temperature.

Assuming the carrier mobility $\mu = q\,\tau/m^*$, where m^* is the effective mass, and τ is the carrier lifetime with $\tau \sim kT^{q'}$, one arrives at (after using Einstein's relation)

$$E = \frac{-kT}{q}\left[\frac{d[\ln(n)]}{dx} + \frac{(q' + 1)}{T}\left(\frac{dT}{dx}\right)\right]. \tag{17.29}$$

Since $\Delta V = E\,\Delta x$, It can be shown that

$$\Delta V = \frac{-1}{qT}\left[(E_C - E_F) + kT\left(\frac{5}{2} + q'\right)\right]\Delta T. \tag{17.30}$$

The Seebeck coefficient (or thermoelectric power) S becomes

$$S = \frac{-1}{qT}\left[(E_C - E_F) + kT\left(\frac{5}{2} + q'\right)\right]. \tag{17.31}$$

Theoretically, q' changes from –3/2 for lattice scattering to 1/2 for impurity scattering. For a disordered semiconductor, the following expression has often been used:

$$S = \left(\frac{k^2}{2q}\right)\sqrt{\frac{T_0}{T}}\left\{\frac{d(\ln(N(E)))}{dE}\right\}\Bigg|_{E_F}, \tag{17.32}$$

where $N(E)$ is the density of states. Figure 17.16 shows a plot of S versus the log of conductivity for polyacetylene samples.[20] The variation of S is more in line with

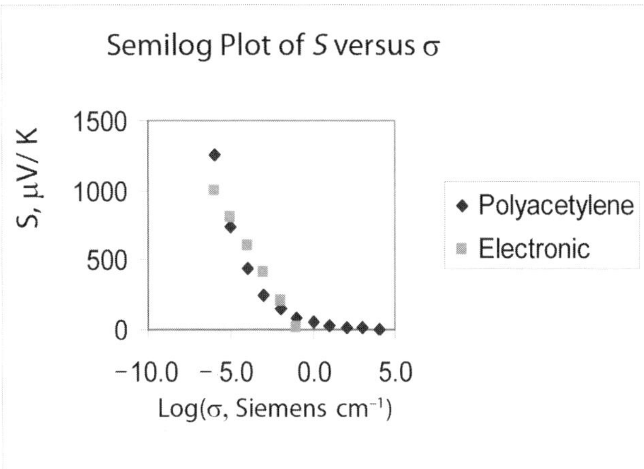

Figure 17.16 Thermoelectric power versus log(σ) for polyacetylene samples. The theoretical curve (electronic contribution) was computed using Eq. (17.29). (Reprinted from Ref. 20 with permission from IOP Publishing Ltd.)

that of a metal at high conductivity. However, at low conductivity the variation of S appears to follow the changes in the carrier density [see Eq. (17.29)].

17.6 Exciton Formation

Recent research in organic luminescent devices has focused on the role played by the charged carriers in exciton formation and recombination. In particular, it is useful to develop a model addressing exciton formation and recombination for different charge density profiles. One such model[21] stipulates a characteristic length that describes the localization of the charged carriers and exciton formation when the electrons and the holes approach one another. The rate of exciton formation then peaks when the charged carriers are at an "optimal" separation and declines exponentially with distance for a departure from this optimal separation. The rate of exciton formation also depends on other transport parameters. Such dependence is not explicitly included but will be taken care of through the dependence of characteristic length on the field-dependent carrier mobility. Because of the distributed nature of the model, one expects it to address different forms of position-dependent luminescence issues, such as, for example, when the device has been selectively doped.

Organic electroluminescent devices differ from inorganic light-emitting devices in that minority carriers are absent at equilibrium. This implies that recombination can occur via injected carriers. The entities in organic semiconductors responsible for radiative recombination have been the excitons, of which roughly one-quarter are linked to the light-emission process. Notwithstanding the argument that minority carriers are absent, donors and acceptors exist and they form the background space charge. None of the other types of space charge have been included in the model, although it has been suggested that dipoles (and their interaction with the carriers) might be present.

Usually, organic electroluminescent devices operate at a higher bias voltage than inorganic homojunction devices; this is attributed to field emission across the interface which is commonly observed in disordered organic semiconductors. To be specific, our model focuses on the ITO/CuPc/NPB/Alq_3/Mg system (ITO: indium tin oxide, CuPc: copper phthalocyanine, NPB: n-propyl bromide). This system exhibits most, if not all, of the features mentioned earlier. In addition, one can find degradation data on selective doping using In atoms which points to localized degradation in the active layers. Similar behavior is found in the I-V characteristics. Using a planar ITO/CuPc/NPB/Alq_3/Mg structure, it is possible to set up the equations for the injected carrier density profiles. In the presence of diffusion and recombination only, one would expect the positional dependence of these profiles to be exponential. Charged carriers in the organic semiconductors do not necessarily recombine in the active layer but are "captured" repeatedly at hopping sites. Assuming the rate of charge capture is high compared with the rate of exciton formation, the decline in the carrier density ought to follow a pattern similar to that of the diffusing minority carriers. Instead of using diffusion length, we consider characteristic length. Not all hopping sites result in exciton formation and we assume that excitons are most readily formed when the separation between the charged carriers reaches a critical value—this may be one or more times the separation between the hopping sites. To quantify this process, we define a field-emission current per unit area given by[22]

$$\frac{I_e}{E} = n_{\mathrm{eff}} q \mu \left[\exp\left(\frac{qV}{\Phi_0} \right) - 1 \right], \tag{17.33}$$

where n_{eff} is the effective charge carrier density, q is the electron charge, μ is the carrier mobility, V is the applied voltage, Φ_0 is an energy parameter, and E is the electric field. Equation (17.33) can be transformed to describe a space-charge-limited current if $n_{\mathrm{eff}} q^2/\Phi_0$ is replaced by $9\varepsilon/8L^2$ at low bias (ε being the semiconductor permittivity, and $L = V/E$). The term n_{eff} $[\exp(qV/\Phi_0) - 1]$ represents the excess carrier density. These excess carriers can either recombine nonradiatively or form excitons. As mentioned earlier, in the active layer the injected carrier densities (for electrons and holes) have the following profiles:

$$n = n_0 \exp\left(\frac{x - d_0}{L_n} \right)$$

and

$$p = p_0 \exp\left(\frac{-x}{L_p} \right), \tag{17.34}$$

where n and p are the respective electron and hole densities; n_0 and p_0 are their values at the edges of the active layer, i.e., at $x = 0$ (at the NPB/Alq_3 interface, where NPB is n-propyl bromide), at $x = d_0$ (at the Alq_3/Mg interface); and L_n and L_p are the respective electron and hole characteristic lengths. The characteristic lengths are analogous to the minority carrier diffusion lengths but instead of being recombination limited, they depend on the localization of the charge carriers at the hopping sites. It has been reported the characteristic lengths in PPV can be as short as 0.3 to 0.5 nm.[23]

As is well known, excitons are responsible for light emission in organic semiconductors. Usually, the formation of excitons depends on the charge separation and one may postulate that there exists an optimal separation a_0. If the electrons and holes are to be spaced out evenly inside the semiconductor, then the optimal separation between the electrons or the holes becomes $2a_0$. As expected, the smallest values of a_0 ought to be no less than the exciton Bohr radius ($= \varepsilon_0 h^2/4\pi^2 m_r q^2$) even though a_0 should really be a multiple of the site spacing which has been reported to be 1 to 2 nm in phenylene vinylene (PPV). We now assume the peak exciton formation rate to be given by

$$R_{\text{peak}} = \frac{1}{8a_0^3\tau}, \tag{17.35}$$

where τ is a generation lifetime parameter, and $1/(8a_0^3)$ is the optimal carrier density for excitons to form. Deviation from the optimal carrier separation $2a_0$ results in a reduced formation rate. Based on Eq. (17.35), the exciton formation rate R is

$$R = P_{\text{peak}}\left[\exp\frac{-\left|\dfrac{1}{n^{1/3}} - 2a_0\right|}{2a_0}\right]\left[\exp\frac{-\left|\dfrac{1}{p^{1/3}} - 2a_0\right|}{2a_0}\right]. \tag{17.36}$$

Equation (17.36) imposes the condition that $R = R_{\text{peak}}$ when the electrons and the holes are separated by $2a_0$. Otherwise, $R < R_{\text{peak}}$. Combining Eqs. (17.34) and (17.36) gives:

$$R = P_{\text{peak}}\exp\left[-\left|\frac{\exp\left(\dfrac{x}{3L_p}\right)}{2a_0 p_0^{1/3}} - 1\right|\right]\exp\left\{-\left|\frac{\exp\left[\dfrac{-(x-d_0)}{3L_n}\right]}{2a_0 n_0^{1/3}} - 1\right|\right\}. \tag{17.37}$$

Equation (17.37) describes how R varies with the distance x away from the NPB/Alq_3 interface (where $x = 0$). To a first order of approximation, one may assume the exciton formation rate R to be the same as the emission rate (except for a factor of ¼, which accounts for the fact that radiative recombination occurs only with the singlet states). It is then possible to put in some realistic values for the parameters. For simplicity, one may assume $n_0 = p_0$. In general, these two values differ since the barrier heights at the semiconductor contacts are different. The magnitude of n_0, as reported, can be as high as 10^{23} m^{-3}. Furthermore, one may also allow n_0 and p_0 to vary from 10^{16} m^{-3} to 10^{21} m^{-3}. As far as the characteristic lengths are concerned, their values are related to the measured carrier mobilities and the Einstein relation. Recent work reported a mobility ratio of electrons to holes to be roughly five (even at high field) and we adopted the same ratio. In our simulations, L_n was allowed to vary from 0.5 nm to 20 nm. This is important because μ_n and μ_p do vary with preparation conditions. The lower limits on L_n and L_p were chosen to be 0.5 nm because R becomes negligible for smaller values of L_n. For easy comparison, the thickness of the $\mathrm{Alq_3}$ layer d_0 was assumed to be 75 nm. The remaining variables are x and a_0; a_0 was allowed to vary from 1 nm to 100 nm. Note that the exciton diffusion length is sometimes quoted to be 1 nm. Using the above values, we computed $\log(R/R_{\mathrm{peak}})$ as a function of x/d_0, the distance from the NPB/$\mathrm{Alq_3}$ interface. The results for $n_0 = 1 \times 10^{21}$ m^{-3}, 1×10^{19} m^{-3}, and 1×10^{18} m^{-3} and assuming $L_n = 2$ nm are shown in Figs. 17.17(a) and (b). The boxed numbers in the figures are the values of a_0. Here one can clearly see the decrease in $\log(R/R_{\mathrm{peak}})$ away from the NPB/$\mathrm{Alq_3}$ interface and the fact that the decrease intensifies for smaller values of n_0 and a_0.

For a_0 less than 10 nm, most excitons are formed within a distance $0.2d_0$ of the injection interface, even at the highest charge density $n_0 = 10^{21}$ m^{-3}. From these results, we can also observe the localization effect when a_0 is only a few nanometers and n_0 is small. Figures 17.18(a)–17.18(c) show the averaged values of R/R_{peak} as a function of $\log(a_0)$ for different values of n_0 (the boxed numbers) for $L_n = 0.5$ nm, 2 nm, and 20 nm, respectively. We see that there is an enhancement of the exciton formation rate when either L_n or a_0 is increased. A longer characteristic length results in a higher rate of exciton formation, even for smaller values of a_0. From Fig. 17.18(c), one can see a saturation effect occurring when both a_0 and n_0 are large. The maximum formation rate of 60 to 70% of R_{peak} occurs for n_0 between 10^{19} and 10^{21} m^{-3}.

Finally, Figs. 17.19(a)–17.19(c) show $\log(R/R_{\mathrm{peak}})$ as a function of x/d_0 for different values of a_0 (boxed numbers) for $n_0 = 10^{20}$ m^{-3} and $L_n = 0.5$ nm, 2 nm, and 20 nm. One can see the progressive localization of $\log(R/R_{\mathrm{peak}})$ near the NPB/$\mathrm{Alq_3}$ interface and the extent of the increase of this localization length for smaller values of L_n.

In the model one finds a couple of features often observed in organic electroluminescent devices, namely that the electroluminescent efficiency (assumed to be related to the exciton formation rate) increases with increasing carrier density and mobility (the latter is to be identified with the characteristic length). From the figures, one can see how R/R_{peak} changes with a_0 and L_n. In this study, a_0 is a parameter defining the optimal conditions for exciton formation and, as observed, it plays a crucial role in determining both the device efficiency

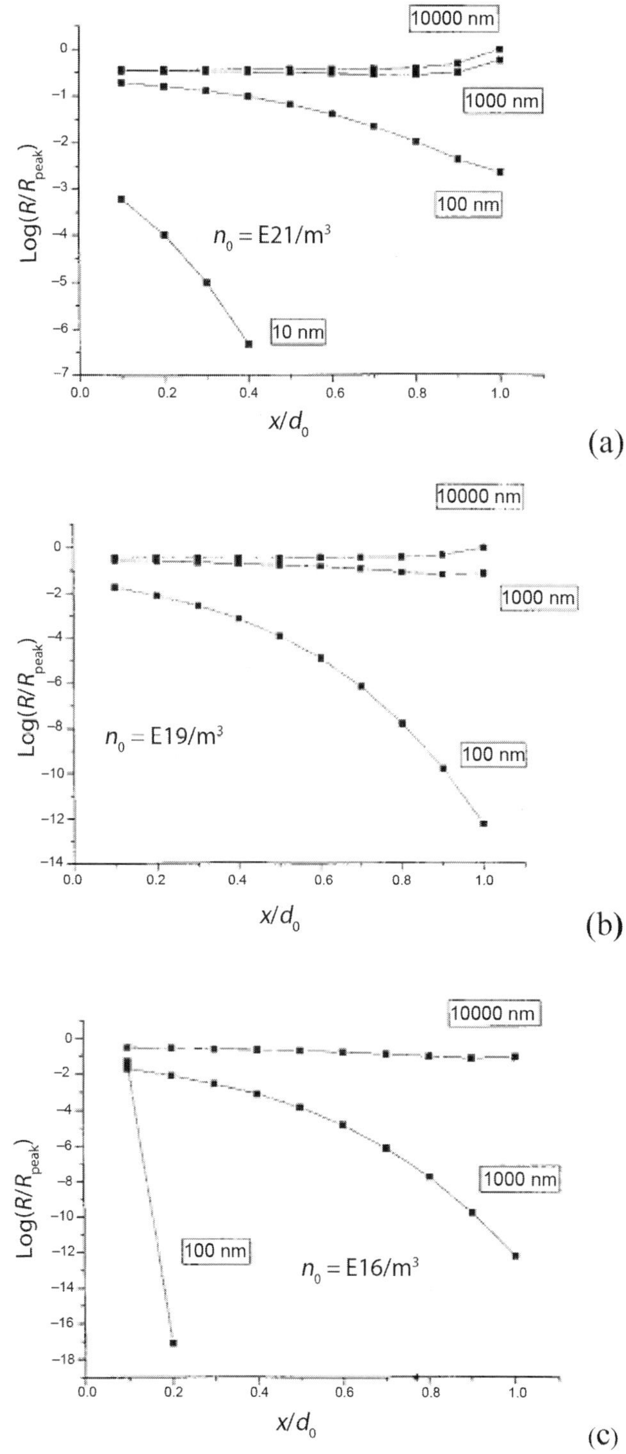

Figure 17.17 $(R/R_{peak})_{avg}$ versus log(a_0) for different values of n_0. (a) L_n = 0.5 nm; (b) L_n = 2 nm; and (c) L_n = 20 nm. (Reprinted from Ref. 21 with permission from Elsevier Science Ltd.)

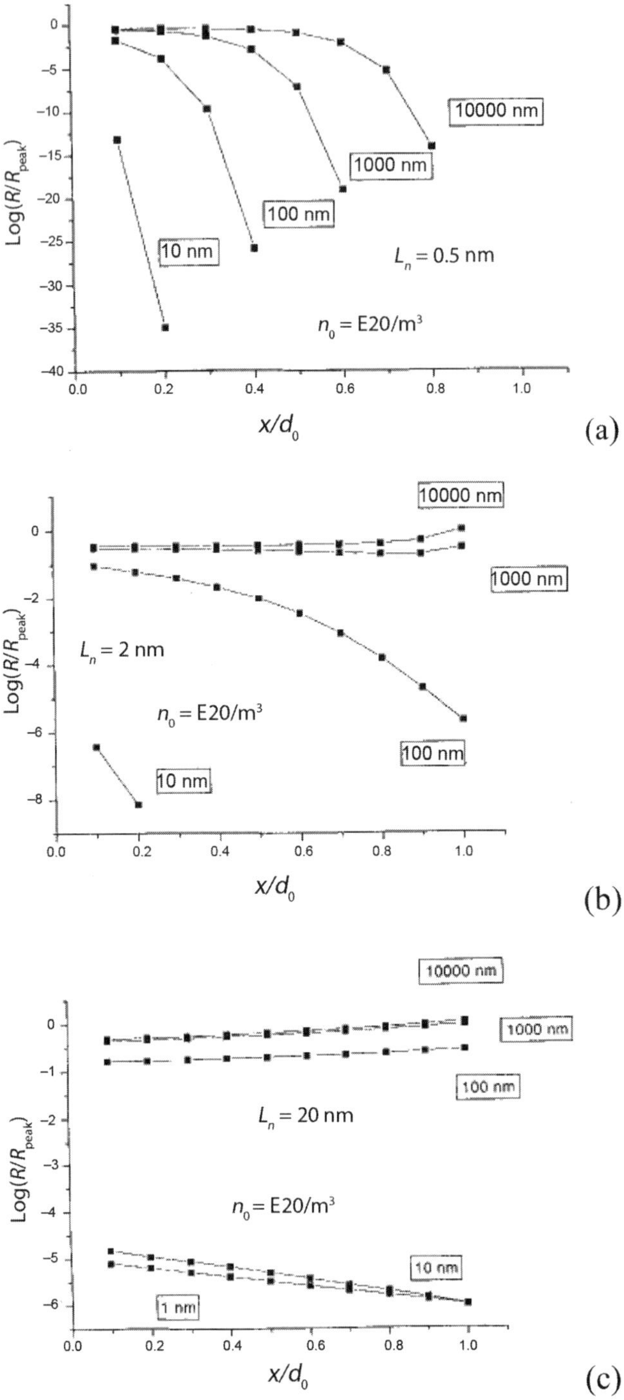

Figure 17.18 Log(R/R_{peak}) versus distance from the NPB/Alq_3 interface for different values of a_0. (a) $n = 10^{21}$ m^{-3}; (b) $n = 10^{19}$ m^{-3}; and (c) $n = 10^{16}$ m^{-3}. (Reprinted from Ref. 21 with permission from Elsevier Science Ltd.)

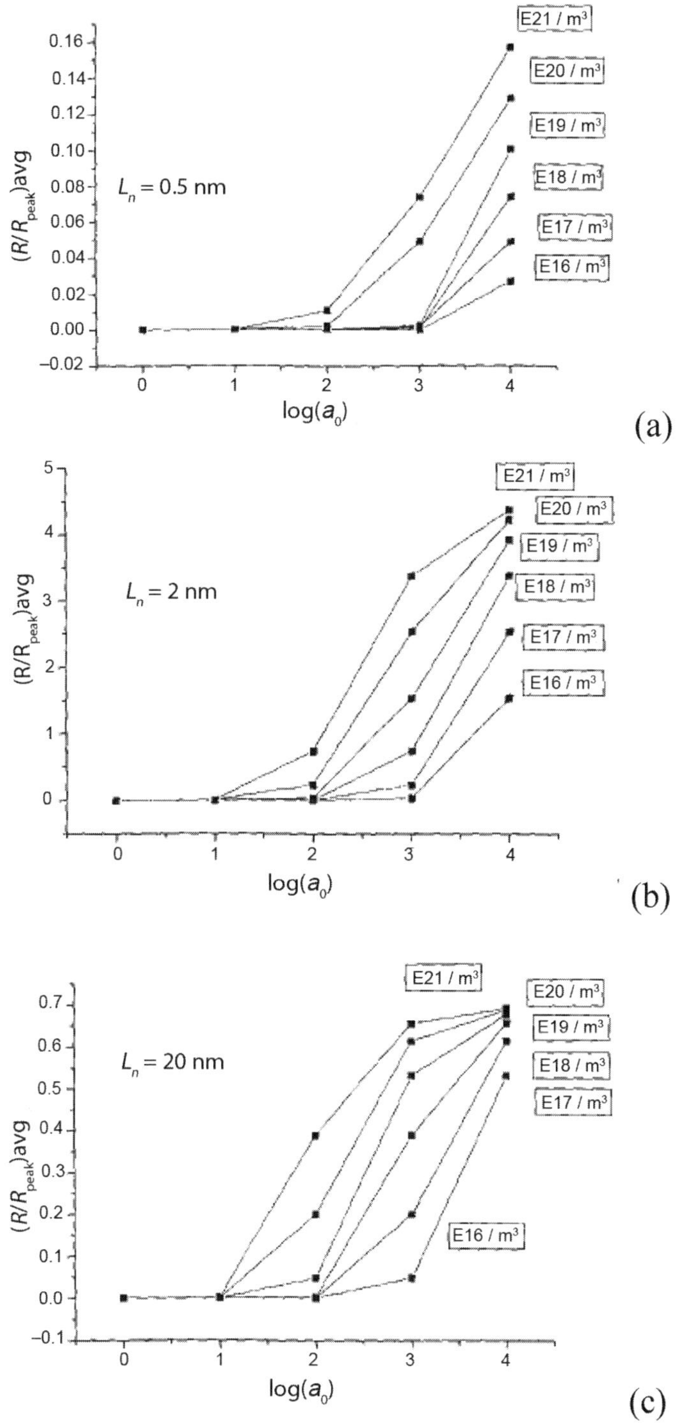

Figure 17.19 Log(R/R_{peak}) versus distance from the NPB/Alq_3 interface for $n_0 = 10^{20}$ m^{-3}. (a) L_n = 0.5 nm, (b) L_n = 2 nm, and (c) L_n = 20 nm. (Reprinted from Ref. 21 with permission from Elsevier Science Ltd.)

as well as the extent of localization of light emission. Using different combinations of a_0 and L_n, one may effectively shift the plateau of R/R_{peak} toward/away from the NPB/Alq_3 interface ($x = 0$). In a sense, this correlates very well with the observed sensitivity to degradation at/near the NPB/Alq interface in many electroluminescent devices. On the one hand, one may also move the plateau toward the Alq_3/Mg interface by decreasing the ratio n_0/p_0 or L_n/L_p. The sensitivity to L_n is particularly important because it is directly linked to the carrier mobility in the way one defines the model, and is in essential agreement with the perception that the carrier mobility plays an important role in determining electroluminescent efficiency. The parameter a_0 ought to be more or less dependent on the structure of the organic semiconductor and may be used to support the argument that, as far as light emission is concerned, a semiconductor/polymer with a high degree of regularity is preferred. Generally speaking, increasing n_0, p_0, L_n, and L_p will produce larger and more uniformly distributed values of R/R_{peak}. Based on the simulation results in Fig. 17.19(c), extremely high values of n_0 are prone to saturate R/R_{peak} when both L_n and a_0 are large. It is possible that there may not be any advantage in increasing n_0 above a certain limit while trying to increase the luminescence efficiency. On the contrary, it may produce a significant increase in the luminescence efficiency even for a small increase in n_0 when both a_0 and n_0 are small. [Compare for instance the values of R/R_{peak} in Figs. 17.19(b) through 17.19(c) when $\log(a_0) = 2$.] In some instances, it is likely that the luminescence efficiency will increase with little or no change in the carrier density. This can be attributed to a broadening of the R/R_{peak} profile due to an increase in the characteristic length caused by, for instance, an increase in the carrier mobility at high field. Such an effect would be more severe in situations where there exists a rapid spatial variation of R/R_{peak}.

17.7 Space-Charge Effect

Recently, there has been quite a bit of interest in studying current transport in organic light-emitting devices for the purpose of improving the luminescence efficiency.[24–26] Considerable effort has been spent identifying the transport mechanisms using *J-V* and *C-V* measurements and to provide a better understanding of the device physics. Unlike inorganic P-N junctions, organic light-emitting devices operate differently with respect to properties such as charge injection and recombination. This is partly due to the fact that the properties of the organic semiconductors are not the same as for inorganic semiconductors. In many ways organic semiconductors resemble insulators since they have low carrier mobilities and low dielectric constants. Unlike insulators, however, the carrier densities in organic semiconductors can be increased by doping. The observed field-dependent mobility in organic semiconductors, particularly in the thin films, raises the possibility of trap-related transport.[27] The presence of broken translational symmetry also contributes to difficulties in characterizing processes such as diffusion, generation, and recombination. Nonetheless, in some instances, concepts such as quasi-periodicity and super cells have been successfully used in models. Notwithstanding these issues, it is likely that fundamental properties such as drift and diffusion exist in organic light-emitting devices and "recombination" of carriers results in the formation of excitons, with relaxation (in the singlet states) responsible for light emission. In the previous section, we examined a model describing the process of exciton formation and showed that the process depended on material parameters such as the carrier densities, the characteristic lengths (analogous to diffusion lengths in inorganic semiconductors), as well as a parameter a_0 defining the optimal separation between the electrons and the holes for excitons to be formed. This model allowed us to quantify the recombination current and to examine spatially the distribution of excitons, which was used to explain regional degradation in devices due to selective doping. One can extend the model to include the space-charge effect and use it to explain the device *J-V* characteristics.[28] Earlier attempts made to explain the *J-V* characteristics of organic light-emitting diodes (OLEDs) included models based on space-charge-limited current, Fowler-Nordheim tunneling current, ohmic conduction, trap-related space-charge current, thermionic-emission current, etc. Transient effects were also analyzed. In the last section, we developed an equation for the rate of exciton formation based on bimolecular recombination. The model included parameters such as the characteristic lengths, which govern the distribution profile of the carriers and a parameter a_0 defining the most effective separation between the electrons and the holes for excitons to be formed. These characteristic lengths effectively dictated the spatial distribution of carriers in a manner similar to the diffusion lengths in inorganic semiconductors and led to an exponential decrease in the carrier densities away from their injection planes.

The existence of characteristic lengths in an organic semiconductor relies on the presence of a large quantity of traps (which slow down the carrier diffusion process) and the fact that the carrier capture rate is high when compared with the rate of exciton formation. In the model[21] the carrier densities essentially determine the rate of exciton formation $R(x)$; the peak rate R_{peak} ($= 1/8a_0^3\tau$) occurs when the carrier separation is equal to a_0. Equation (17.37), while valid for a bimolecular process, is not applicable in the presence of space charge. Normally space charge is not in equilibrium with the traps in the organic semiconductor. It is therefore necessary to modify this equation to include space charge. This gives rise to a space-charge-limited current. For the *p*-type organic semiconductor, one can replace the term

$$\exp\left[-\left|\frac{\exp\left(x/3L_p\right)}{\left(2a_0 p_0^{1/3}\right)-1}\right|\right] \text{ by } \exp\left\{-\left|\left[2a_0 p\left(x\right)^{1/3}\right]-1\right|\right\}.$$

In so doing, the integrity of the hole distribution function associated with space charge remains essentially unchanged provided that "recombination" is negligible. Equation (17.37) then becomes

$$R(x) = R_{peak} \exp\left\{-\left|\left[2a_0 p(x)^{1/3}\right] - 1\right|\right\} x \exp\left\{-\left|\frac{\exp\left[\frac{-(x-d_0)}{3L_n}\right]}{\left(2a_0 n_0^{1/3}\right)-1}\right|\right\}. \tag{17.38}$$

Under the space-charge-limited condition, the hole distribution is

$$p(x) = \frac{3e_s V}{4qd_0^{3/2} x^{1/2}}, \tag{17.39}$$

where ε_s is the semiconductor permittivity and q is the electron charge. To determine $R(x)$, one needs the values of a_0, τ, $p(x)$, n_0, d_0, and L_n. The recombination current density J_{rec} has the form

$$J_{\text{rec}} = q \int_0^{d_0} R(x)\, dx. \tag{17.40}$$

For most OLEDs, electrons and holes are injected separately from the contacts and they recombine throughout the active region (including any transport layers). Ideally, J_{rec} is directly proportional to the luminescence efficiency. According to Eq. (17.38), J_{rec} saturates at a large bias when both $n(x)$ and $p(x) >> 1/a_0^3$. The evaluation of the carrier profiles depends on the values of n_0 and p_0, which are linked to the properties of the contacts, including the interface materials. Previously we assumed the following expression for the electron density:

$$n_0 = n_{\text{eff}} \exp\left(\frac{qV}{\Phi_{01}}\right), \tag{17.41}$$

where n_{eff} is the effective electron density at zero bias and Φ_{01} is an "unknown" energy parameter. In practice, Φ_{01} includes any energy level mismatch and geometrical effects at the injection plane. Usually, carrier mobility in organic semiconductors has the following field dependence:

$$\mu = \mu_0 \exp\left(\alpha\sqrt{E}\right), \tag{17.42}$$

where μ_0 is the zero field mobility and α is a disorder parameter. Earlier, we mentioned that the characteristic lengths were analogous to the diffusion lengths with the presence of traps. Diffusion lengths are proportional to the square roots of diffusivities, which are assumed to have the same electric field dependence as the carrier mobility. This leads to

$$L_n = L_{n0} \exp\left(\frac{\alpha_1 \sqrt{E}}{2}\right), \tag{17.43}$$

where L_{n0} is the zero field characteristic length for the electrons. Combining Eqs. (17.38), (17.39), (17.40), and (17.43), J_{rec} can be found. At low bias, space-charge-limited current will dominate. For a *p*-type organic semiconductor, the space-charge distribution and the current density are given by

$$p(x) = \frac{3\varepsilon_s V}{4qd_0^{3/2} x^{1/2}} \tag{17.44}$$

$$J_{\mathrm{scl}} = \frac{9\theta\mu_h V^2}{8d_0^3}, \tag{17.45}$$

where μ_h is the hole mobility and θ is the ratio of free to trapped-plus-free carriers. It is possible to compare simulations with the reported experimental results. Figure 17.20 shows the *J-V* characteristics of an ITO/TPA-PPV (100 nm)/Al OLED[29] (TPA: triphenylamine-phenylene). These are compared to the simulated space-charge-limited current density given by Eq. (17.45) for the trap-free case ($\theta = 1$). Agreement with the experimental data at low bias indicates that space-charge-limited current is important when the bias voltage is small. As the bias voltage increases, trapping becomes more important and the extent of trapping is measured in terms of the ratio of the two current densities which is shown in Fig. 17.21(a) as a function of the bias voltage. Figure 17.21(a) can also be transformed into a plot of θ versus the bias voltage assuming that $\theta = 1 - J_{\mathrm{exp}}/J_{\mathrm{scl}}$. This is plotted in Fig. 17.21(b). At 3 V or higher, θ becomes meaningless as recombination current dominates the *J-V* characteristics.

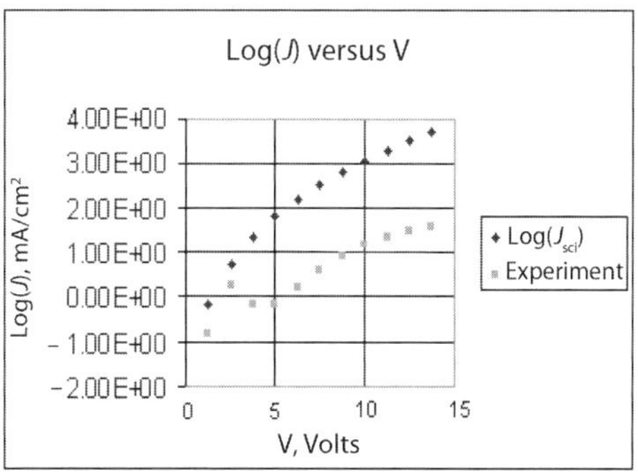

Figure 17.20 *J-V* characteristics and J_{scl} *-V* characteristics from Eq. (17.45). (Reprinted from Ref. 29 with permission from Elsevier Science Ltd.)

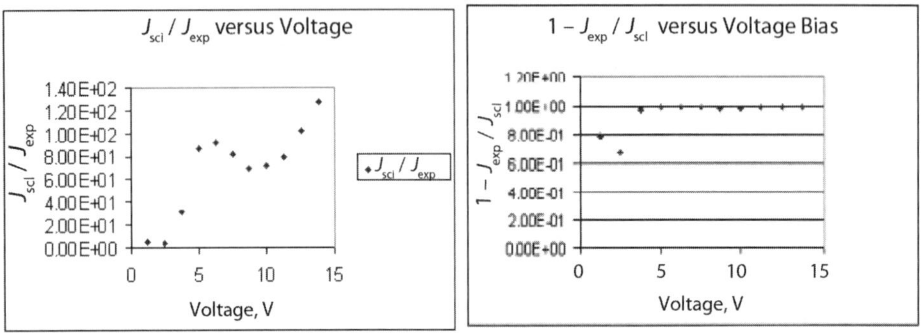

(a) (b)

Figure 17.21 (a) Ratio of free-to-trapped carriers (= J_{scl}/J_{exp}) versus the bias voltage. (b) Plot of $1 - J_{exp}/J_{scl}$ (= θ) versus the bias voltage. (Reprinted from Ref. 29 with permission from Elsevier Science Ltd.)

From Fig. 17.21(b) we see that the transition from trap free current conduction to trap-limited current conduction occurs at around 2 to 2.5 V. This voltage ought to mark the threshold at which trapping becomes important. If one were to associate trapping with recombination, then one may use the qualitative change in conduction mechanism to estimate the value of a_0. According to Eq. (17.39), $R(x)$ is reduced by about 40% from its nominal value if $a_0 p(x)^{1/3} \approx 1$. Using Eq. (17.45) to estimate the average value of $p(x)$ when the bias voltage is at 2.5 V, we found a_0 to be approximately 40 nm. This value of a_0 is used in our calculations described below unless otherwise specified. Computation of J_{rec} based on Eqs. (17.38) and (17.40) requires a knowledge of $p(x)$, a_0, d_0, τ, L_{n0}, α_1, μ_0(electrons), n_{eff}, and Φ_{01}. Values of $p(x)$ can be obtained from Eq. (17.44) [to be weighted by θ as shown in Fig. 17.21(b)]. One may also assume that d_0 = 100 nm; the other transport parameters can be found in the open literature for closely related organic semiconductors. The values of relevant parameters are listed in Table 17.3. Values of n_{eff} and Φ_{01} could not been taken from the literature as we expect them to be sensitive to processing. In our simulations, we have used them to fit the experimental data to our calculation at large bias for which the space-charge effect is negligible. The best-fit values were 2.5×10^{18} m^{-3} and 5.5 eV, respectively.

Using the above parameters, we computed $J_{scl} + J_{rec}$ and plotted the combined J-V characteristics in Fig. 17.22. There is a very close fit to the experimental data, suggesting that combined space-charge-limited current and recombination current adequately describes the observed J-V characteristics. To provide additional insight into the effect of a change in a_0, we plotted in Fig. 17.23 the J-V characteristics for a_0 = 100 nm. The upward shift in the calculated J-V curve is indicative of a broader range of recombination current. Electron injection is increased when Φ_{01} is lowered. The result of a reduction in Φ_{01} from 5.5 to 3 eV is illustrated in Fig. 17.24(a). The increase in the recombination current density is the result of more injected electrons. Further reduction of Φ_{01} to 1 eV

Table 17.3 Material parameters in the calculations of J_{rec}. (Data used from Ref. 29 with permission from Elsevier Science Ltd.)

τ	1 ns
α_1(holes)	5×10^{-4} m$^{0.5} \cdot$ V$^{-0.5}$
α_2(electrons)	8×10^{-4} m$^{0.5} \cdot$ V$^{-0.5}$
μ_0(holes)	3×10^{-11} m^2/V·s
μ_0(electrons)	3×10^{-13} m^2/V·s
L_{p0}	278 nm
L_{n0}	6.23 nm
n_{eff}	2×10^{18} m^{-3}
Φ_{01}	5.2 eV

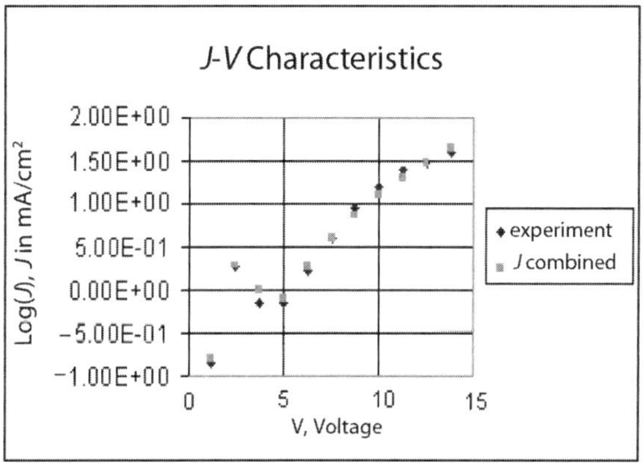

Figure 17.22 Experimental *J-V* characteristics and simulations after taking into account space-charge-limited current. (Reprinted from Ref. 29 with permission from Elsevier Science Ltd.)

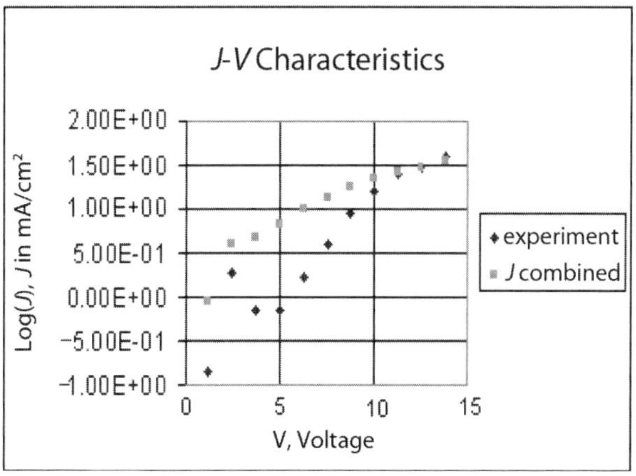

Figure 17.23 *J-V* characteristics for a_0 = 100 nm. (Reprinted from Ref. 29 with permission from Elsevier Science Ltd.)

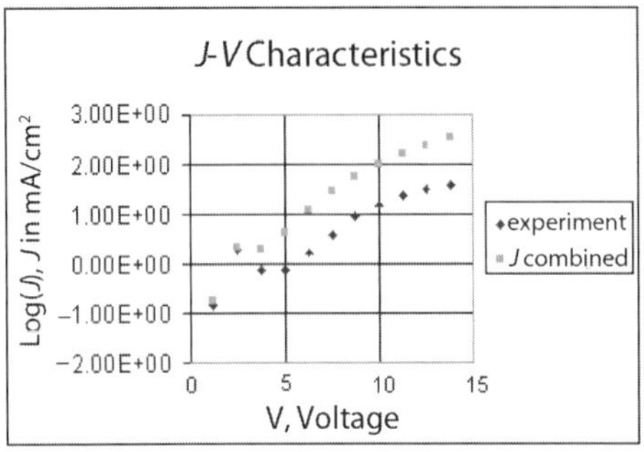

(a)

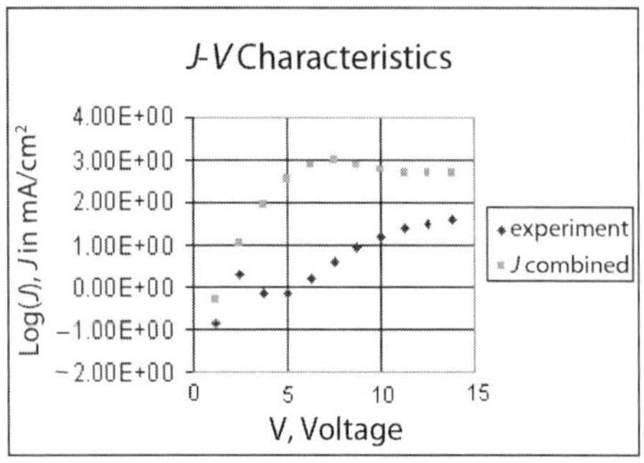

(b)

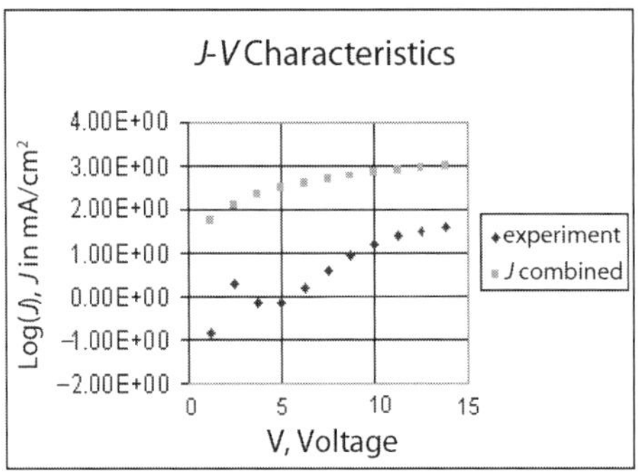

(c)

Figure 24 *J-V* characteristics when (a) Φ_{01} = 3 eV; (b) Φ_{01} = 1 eV; and (c) $n_{\text{eff}} = 2 \times 10^{20}$ m^{-3}. (Reprinted from Ref. 29 with permission from Elsevier Science Ltd.)

produces saturation in the *J-V* characteristics as shown in Fig. 17.24(b). The peak current density appears to be limited to ~ 1 A/cm^2, and a similar saturation is observed when n_{eff} is increased from 1×10^{18} m^{-3} to 2×10^{20} m^{-3}, as illustrated in Fig. 17.24(c).

Similarly to a_0, the thickness of the active layer d_0 also plays an important role in determining device performance. Figures 17.25(a) and (b) show the *J-V* characteristics as d_0 is changed from 40 nm to 200 nm. In both cases, the recombination currents are lowered, indicating that d_0 is no longer optimized with respect to a_0. The asymmetric drop in the device current density when $d_0 \neq 2a_0$ (= 80 nm) is in agreement with similar observations found in small molecule devices. Figure 17.26 shows the *J-V* characteristics when the electron mobility is increased by threefold from the value listed in Table 17.3. Only minor improvements in fit are observed.

In summary, carrier transport in OLEDs can be explained in terms of a recombination current at large bias and trap-free space-charge-limited current at low bias. The presence of a current peak around 3 V can be explained in terms of the transition between these two processes with trap centers serving as an

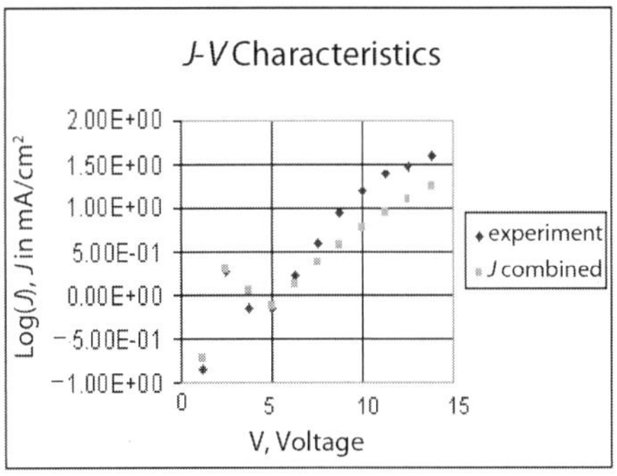

(a)

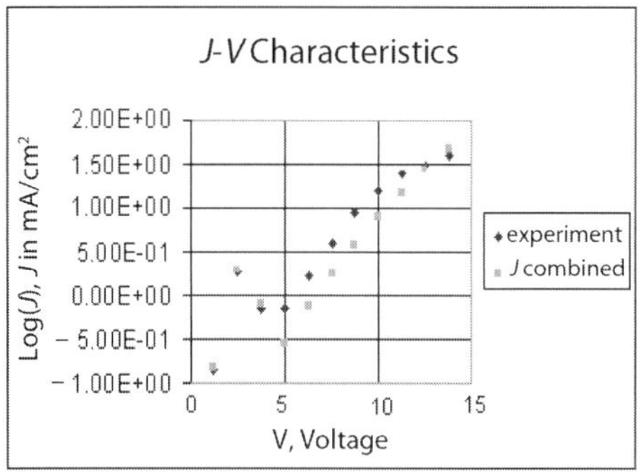

(b)

Figure 17.25 *J-V* characteristics when (a) d_0 = 40 nm and (b) d_0 = 200 nm. (Reprinted from Ref. 29 with permission from Elsevier Science Ltd.)

intermediary for exciton formation and subsequent light emission. The current peak, according to our simulations, exists when there is also a large disparity between the carrier densities [see Figs. 17.24(a)–(c)]. This happens when there is weaker injection for one of the electrodes. In the device we have examined, low electron injection appears to be linked to the interface properties between the PPV and the aluminium cathode. For exciton formation, the important device parameter is a_0, which is the optimal separation between the carriers. This parameter bears upon the physics of the recombination mechanism. Taking a simplistic view, a_0 may be considered as the capture parameter for carriers at a single trap level, or across a distribution of trap states. The Coulomb interaction has been implicitly assumed for the capture of the electrons and the holes. Based on the measured *J-V* characteristics of the device, a_0 has been found to be ~ 40 nm. The exact value is likely to vary depending on the properties of the organic semiconductor.

Other parameters affecting the *J-V* characteristics include d_0 and μ_0. Figures 17.25(a) and 17.25(b) show, for instance, that higher recombination currents are found near $d_0 = 2a_0$, an indication of the relationship between the two parameters as stipulated by the bimolecular recombination process. Similar observations have been reported for TPD/Alq_3 OLEDs (TPD: N,N'-diphenyl-bis-(3,3'-methylphenyl)-[4,4'-biphenyl]-1,1'diamine). We have also examined the effect of an increase in the electron mobility. A threefold increase in the electron mobility resulted in only a marginal increase in the recombination current at an intermediate bias, suggesting that the transit of carriers in the ultrathin active layers is not as important as the values of the carrier densities. Improvement is expected only when the electron injection level is moderate or low. Increasing the value of a_0 enhances the recombination current, as illustrated in Fig. 17.23. A moderate increase is observed. The overall result suggests that a major increase in "recombination" current occurs only if the injection properties (at the cathode) are improved.

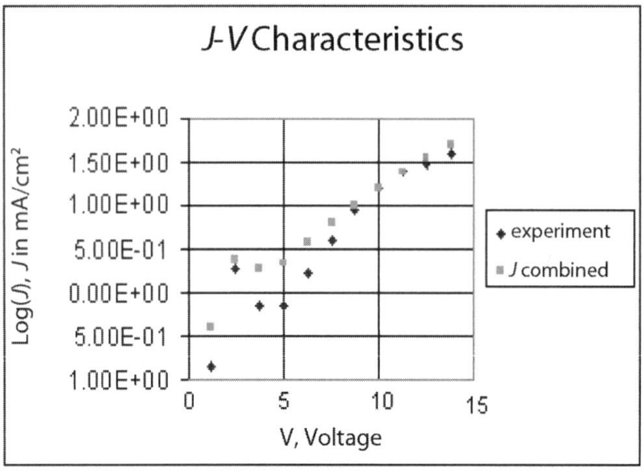

Figure 17.26 *J-V* characteristics when $\mu_0 = 3 \times 10^{-10}$ $m^2/V{\cdot}s$. (Reprinted from Ref. 29 with permission from Elsevier Science Ltd.)

17.8 Charge Transport in the Field-Effect Structure

Much effort has been expended to understand the conduction mechanisms in organic semiconductors because of the interest in using these materials in electronics. Previously, it has been established that organic semiconductors can be doped either *n*-type or *p*-type with high carrier densities. This opens up the possibility of using organic semiconductors in device applications. The mobility of the carriers in organic semiconductor thin films is much lower than that for inorganic semiconductors. For instance, the highest mobility reported for pentacene is only a few $cm^2/V{\cdot}s$, some two orders of magnitude below what is observed in silicon. This appears to support the opinion that devices made of pentacene could not operate at high speed.

Theoretical developments on the transport properties of organic semiconductors have been going on for a number of years. Early studies of the field-dependent carrier mobility relied on empirical formulae which did not provide much insight into the conduction mechanism. More recently, a few models have been more successful and are based on charge hopping in an environment of fluctuation potentials due to randomly oriented dipoles per atoms. These models have been used to explain measurements related to transit time dispersion and field-dependent carrier mobility. In addition, the voltage dependence of the field-effect mobility has been studied and was explained in terms of the gate-voltage-dependent activation energy for the carriers. In this work, we investigated the field-effect mobility in polycrystalline pentacene and compared the reported results with simulations based on the CDM. It has been observed that the different temperature dependence of the field-effect mobility for similarly prepared pentacene samples could be explained using the CDM with small values of σ, the rms width of the density of states (DOS). This implied that samples observed to have a higher mobility might have smaller disorder. These results led us to re-examine the validity of the CDM equation when applied to disordered solids and the physics that had given rise to variability in the field-effect mobility. It is also possible to explore the limit $\sigma \rightarrow 0$ and what might occur when the field-effect mobility has a negative exponential (electric) field dependence. One may consider the following equation for the current density J in an organic semiconductor:[30]

$$J = n_{\text{eff}} q v, \tag{17.46}$$

where n_{eff} is an effective carrier density, q is the electron charge, and v is the velocity of the carriers.

Since by definition mobility v equals μE, where E is the electric field, Eq. (17.46) can be combined with the field-dependent mobility equation commonly observed in disordered organic semiconductors $\left[\text{i.e., } \mu = \mu_0 \exp\left(\alpha\sqrt{E}\right)\right]$ to give

$$J = n_{\text{eff}} q \mu_0 \exp\left(\alpha\sqrt{E}\right) E, \tag{17.47}$$

where μ_0 is the zero-field mobility, α is the "disorder" parameter, and E is the electric field. To determine μ_0, we assume that the carriers move between hopping sites and that the amount of time they stay in these sites lowers the carrier mobility. Assuming that the delay time is directly linked to an effective barrier height Φ_B needed for the carriers to escape from the hopping sites (the hopping sites will be distinguished from trap sites as the former are correlated), the current density J has the form

$$J = n_{\text{eff}} q \mu' \exp\left(\frac{-\Phi_B}{E_\infty}\right) E, \tag{17.48}$$

where μ' and E_∞ are unknown constants. Equation (17.48) is, in fact, similar to other expressions linking the carrier density to barrier heights Φ_B associated with hopping. In some instances, E_∞ is given as kT_0, where T_0 is a parameter linked to the width of the energy distribution of the localized states. Physically, E_∞ (in our case) also includes a measure of the effectiveness of the "correlated" charge transfer. A large E_∞ therefore signifies a higher level of site correlation. In the presence of space charge (which includes injected carriers) near the hopping sites, it is possible to include a barrier height modulation term due to the Poole-Frenkel effect, which reduces Φ_B to $\Phi_B - \Delta\Phi$, where $\Delta\Phi = q\sqrt{qE/4\pi\varepsilon_s}$. Taking into account the barrier height reduction gives

$$J = n_{\text{eff}} q\mu' \exp\left(\frac{-\Phi_B}{E_\infty}\right) \exp\left[\frac{q\sqrt{\dfrac{qE}{4\pi\varepsilon_s}}}{E_\infty}\right]$$

$$= n_{\text{eff}} q\mu_0 \exp\left(\alpha\sqrt{E}\right)E, \tag{17.49}$$

where $\alpha = q\sqrt{q/4\pi\varepsilon}/E_\infty$, and $\mu_0 = \mu'\exp(-\Phi_B/E_\infty)$. For the calculations of the field-effect mobility, Φ_B is replaced by $\Phi_B + V_g'$, where V_g' is the value of the effective gate bias near the oxide-semiconductor interface (for a *p*-channel device, V_g' is negative). Equation (17.49) can also be compared to the known expression for the low-field carrier mobility ($T \rightarrow \infty$) during hopping

$$\mu_\infty = qa'^2 \nu_{\text{ph}} \exp \frac{\left(\dfrac{-2a'}{L}\right)}{\sigma}, \tag{17.50}$$

where a' is the site spacing, ν_{ph} is the attempt frequency for hopping to occur, L is the localization length, and σ is the rms width of the density of states (DOS). Note that $\mu_0 = \mu_\infty$ when $T \rightarrow \infty$. Within this limit one may compare Eqs. (17.49) and (17.50) to yield the following two equations:

$$\frac{\Phi_B}{E_\infty} = \frac{2a'}{L} \tag{17.51a}$$

$$\frac{qa'^2 \nu_{\text{ph}}}{\sigma} = \mu'. \tag{17.51b}$$

Physically, Eq. (17.51a) may be viewed as two different ways to express "obstacles" (either in the form of a physical barrier or a separation between neighboring localized states) presented to the carriers during hopping. The process is normally temperature dependent. Equation (17.51b), on the other hand, highlights the mechanism of the hopping process, which favors larger site spacing (longer mean free path) and less disorder (smaller σ). Making use of the relationship between α and E_∞ in Eq. (17.49), Eq. (17.51a) can be written as

$$\frac{\alpha\Phi_B}{q\sqrt{q/4\pi\varepsilon_s}} = \frac{2a'}{L}. \tag{17.52}$$

Note that Φ_B includes the gate bias dependence as well as the effect of site correlation. The outstanding parameter in Eq. (17.52) is α, which, according to the CDM, is

$$\alpha = -0.78\left[2-\left(\frac{\sigma}{kT}\right)^{3/2}\right]\left(\sqrt{\frac{qa'}{\sigma}}\right), \tag{17.53}$$

where k is the Boltzmann constant, and T is the absolute temperature. Combining Eqs. (17.52) and (17.53) gives

$$\frac{2a'}{L} = \frac{-0.78\ \Phi_B}{q\sqrt{\dfrac{4\pi\varepsilon_s a'}{\sigma}}}\left[2-\left(\frac{\sigma}{kT}\right)^{1.5}\right]. \tag{17.54}$$

The low-field carrier mobility now becomes

$$\mu_\infty = \frac{qa'^2 \nu_{\text{ph}}}{\sigma \exp\left\{\left(\dfrac{0.78\Phi_B}{q\sqrt{\dfrac{4\pi\varepsilon_s a'}{\sigma}}}\right)\left[2-\left(\dfrac{\sigma}{kT}\right)^{1.5}\right]\right\}}. \tag{17.55}$$

Equation (17.55) can be combined with Eq. (17.49) to give the full expression for the carrier mobility

$$\mu = \mu_\infty \exp\left[-\left(\frac{3\sigma}{5kT}\right)^2\right] \exp\left\{-0.78\left[2-\left(\frac{\sigma}{kT}\right)^{1.5}\right]\sqrt{\frac{qaE}{\sigma}}\right\}. \tag{17.56}$$

According to Eq. (17.56), ln(μ) can have either positive or negative field dependence depending on whether $[2 - (\sigma/kT)^{1.5}]$ is negative or positive. Potentially, this leads to negative values of α, which is not commonly observed in transport measurements. The key point here is that Eq. (17.56) arises from physical consideration of the microscopic hopping process based on the Gaussian DOS function, which has been shown to give the correct form of the mobility's temperature dependence [i.e., $\ln(\mu) \sim (T/T_0)^2$]. In addition, Eq. (17.56) does not require that $T \leq T_0$, as is the case in other theories. The fact that T can be less than T_0 is important if one attempts to analyze data where μ has been observed to be temperature independent. Several recent articles report measurements on the field-effect mobility in single-crystal and polycrystalline thin film organic transistors. In general, their results show temperature-dependent carrier mobility at low temperature and at low field, and temperature-independent carrier mobility otherwise. Such behavior had been attributed to the influence of traps that are active at low temperature and at low field, but are passive at high temperature. Polycrystalline field-effect transistors behave somewhat similarly to single-crystal field-effect transistors except for having lower values of the carrier mobility.

It is possible to develop an analysis of the mobility-temperature characteristics of polycrystalline field-effect transistors using the theory developed above. The reason for choosing these data is that they appear to cover the range of mobility-temperature characteristics observed in many organic field-effect transistors, and the variability in the data is of interest. One concern that arises in the application of the CDM to pentacene transistors is the fact that the model relies on the existence of permanent dipoles in the organic semiconductor, which apparently are absent in (nonpolar) pentacene. Our justification for using the CDM is based on the facts that: (a) only one to two pentacene layers are needed to support charge transport in a field-effect structure, and (b) these (active) layers are only a few nanometers away from the gate oxide which possesses permanent dipoles (in the case of silicon dioxide, the dipole moment is 1.6 Debye). In short, the polar background in the active layer arises from the permanent dipoles present in the nearby oxide layer. Figure 17.27 shows the temperature dependence of the field-effect mobility in pentacene reported in Ref. 30. The devices are labeled as samples A–C and are known to be grown under similar conditions. For comparison, we plotted in the same figure simulations of the carrier mobility using Eq. (17.56) with different values of σ chosen to provide a good fit. Table 17.4 lists the values of the material parameters (in addition to σ) used in the calculations. The electric field strength was 6.15 MV/m.

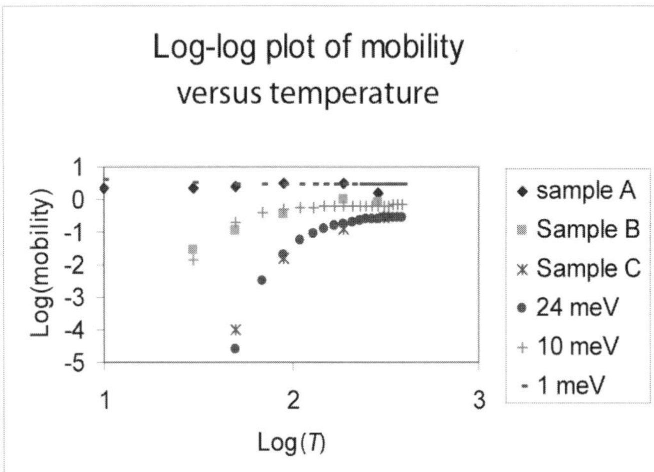

Figure 17.27 Log-log plots of the simulated and reported values on the carrier mobilities as a function of temperature for pentacene samples (A–C).The inset shows the values of σ used in the respective calculations. (Reprinted with permission from Ref. 30.)

Table 17.4 Values of the materials parameters for pentacene. (Reprinted with permission from Ref. 30.)

Parameters	a' (nm)	Φ_B (meV)	$\nu_{ph}(s^{-1})$	$\varepsilon_s/\varepsilon_0$
Values	1	40	1×10^{12}	2.5

As can be seen in Fig. 17.27, a close fit between data and simulations required $\sigma = 1$ meV, 10 meV, and 24 meV. These values are smaller than kT (~ 25.9 meV at room temperature), which required α to be negative over portions of the calculated mobility curves. To highlight this feature, we plotted in Fig. 17.28 log(μ) versus α using the values of σ and T used for Fig. 17.27. Here, one can observe the saturation of μ as α becomes negative. Increasing α (which corresponds to an increase in disorder) on the other hand reduces μ irrespective of the value of σ, the rms width of the DOS, and temperature. To explore the limiting value of μ when $\sigma \rightarrow 0$, we plotted in Fig. 17.29 the calculated values of log(μ) versus σ calculated using Eq. (17.56) and Table 17.4. As can be observed, the carrier mobility peaks for $\sigma \rightarrow 0$, but otherwise decreases rapidly with increasing σ. The peak value is ~ 2.8 $cm^2/V \cdot s$ (when $\sigma \rightarrow 0$), which is close to the best reported values of 1.5 to 3.2 $cm^2/V \cdot s$. It is useful to examine the physical implication as $\sigma \rightarrow 0$. With a reduced rms width of the density of states, one would expect the energy states of the organic semiconductor to approach those of the isolated atoms and a reduction in correlation between the hopping sites. This behavior resembles a low-dimensional semiconductor with a singular density of states. Observations have indeed been reported that in the poly (3-hexythiophene) lamella structure, the field-effect mobility should increase when the molecules are aligned head to tail.

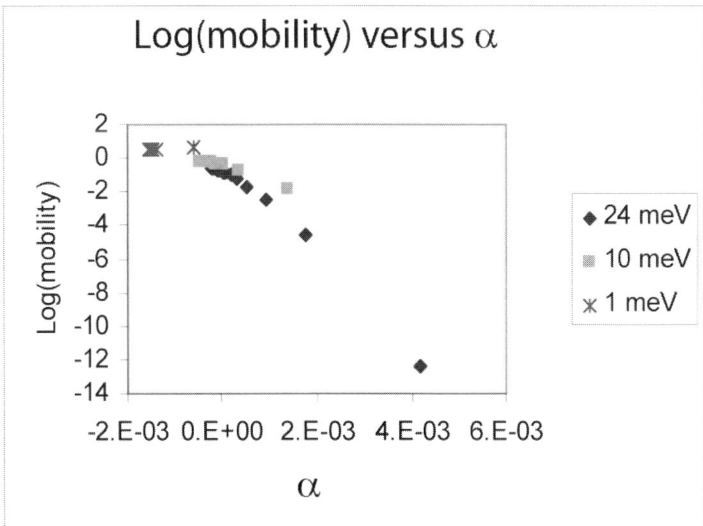

Figure 17.28 Semilog plot of the simulated carrier mobility versus α for different values of σ. (Reprinted with permission from Ref. 30.)

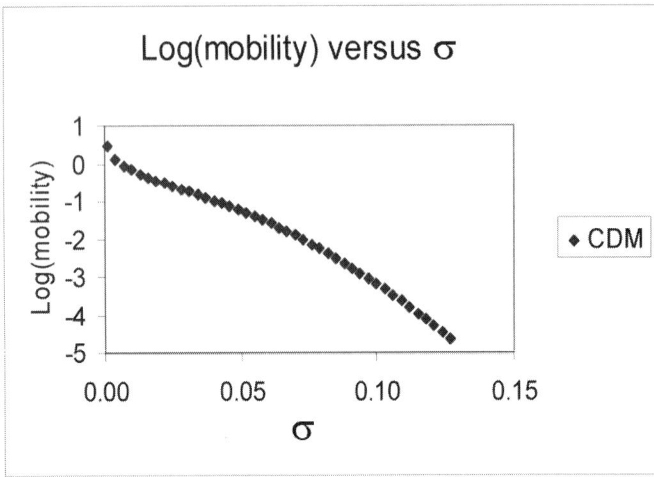

Figure 17.29 Plot of simulated log(μ) versus σ using the CDM. (Reprinted with permission from Ref. 30.)

Other parameters in Eq. (17.56) such as the site spacing a and the barrier height Φ_B also affect the value of the carrier mobility. Their calculated dependences are plotted in Figs. 17.30 and 17.31. As shown in Fig. 17.30, the mobility remains small (< 18 $cm^2/V{\cdot}s$) when $\sigma \rightarrow 0$, but can exceed 10^3 $cm^2/V{\cdot}s$ for larger values of σ and site spacing. Increasing the barrier height Φ_B reduces the carrier mobility for σ > 40 meV (as would be expected from theory) but such behavior is reversed for σ < 40 meV. These observations will be discussed next.

Figure 17.27 demonstrates the adequacy of Eq. (17.56) and the CDM to explain the temperature dependence of the carrier mobility in the samples reported. In the process, we have assumed that σ ranged between 1 to 24 meV. Such small values of σ are not commonly found in organic semiconductors and are presumably linked to the transistor structures and possibly the field effect imposed on the active layer.

Because of this observation, we plot in Figs. 17.28 and 17.29 the dependence of the calculated carrier mobility on α and σ. The change in the carrier mobility with σ is fairly straightforward, as it reflects the strong dependence of the carrier mobility on the broadening of the rms values of the density of states. The more important issue here is the mobility peak, which occurs for $\sigma \rightarrow 0$. This implies that, according to the CDM, carrier transport is enhanced only if the organic semiconductor is similar to an insulator with a narrow density of states. In this limit, the carrier mobility decreases with increasing electric field (which is often observed). Such behavior has been explained by a number of authors and was attributed to mechanisms such as: (a) a changeover from symmetrical hopping rate to asymmetric (Miller-Abrahams) hopping rate, (b) the influence of nonparabolic band structure for a small σ, and (c) velocity decline at high field due the emission of optical phonons. There is no direct evidence indicating that (b) or (c) played an important role in the data we analyzed.

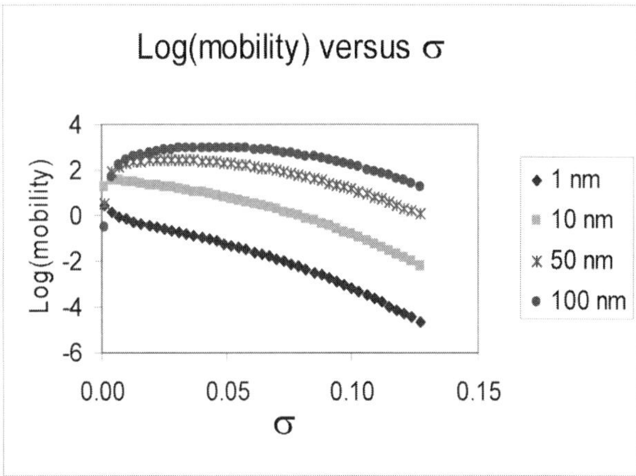

Figure 17.30 Log(μ) versus σ for different values of the site spacing. (Reprinted with permission from Ref. 30.)

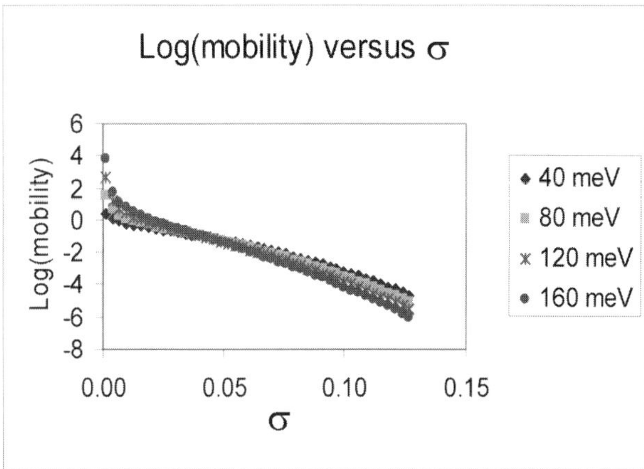

Figure 17.31 Log(μ) versus σ for different values of the barrier height. (Reprinted with permission from Ref. 30.)

The relationship between σ and α given by Eq. (17.53) is intriguing. First of all, α is a measurable quantity and is normally considered to be a marker of disorder in organic semiconductors, i.e., the larger α is, the greater the disorder. It is also related to σ, the rms width of the DOS, through Eq. (17.53). According to this equation, the term $[2 - (\sigma/kT)^{1.5}]$ determines whether α is positive or negative. What this means is that the CDM allows α to be negative (which implies a change in the transport mechanism) as the equation changes sign. We are of the opinion that this represents a transition from correlated hopping (large σ and low field) to uncorrelated hopping (small σ and high field). The former implies a strong coupling of the energy states at the hopping sites; the strength of the coupling depends on the barrier height and the site spacing. In uncorrelated hopping, carriers leaving the sites are subjected to band/lattice relaxations, which potentially lower the carrier velocity. Equations (17.55) and (17.56) show that other parameters, such as the site spacing and the barrier height Φ_B, also affect the carrier mobility. In Fig. 17.30 we plotted the simulated log(μ) versus σ for four different values of the site spacing (between 1 and 100 nm). Since the most frequently quoted value of the site spacing is ~1 nm, it is unlikely that higher values of the site spacing can be easily achieved (except perhaps in the case of ordered solids). According to the figure, for $\sigma \rightarrow 0$, the best mobility value is < 18 $cm^2/V{\cdot}s$. This implies that, in this limit, the carrier mobility remains low. In the case of a large σ, the carrier mobility increases rapidly and is found to exceed 10^2 $cm^2/V{\cdot}s$ when the site spacing approaches 50 nm. A closer examination of the relationship between σ and the site spacing reveals that this is unlikely to be the case, as σ is known to vary inversely with the square of the site spacing for a Gaussian energy distribution.

Finally, we plotted in Fig. 17.31 the simulated values of the carrier mobility for different barrier heights Φ_B (between 40 and 160 meV). The observed decrease of the carrier mobility with increasing Φ_B for $\sigma > 40$ meV (α positive) is in agreement with the expectation that the carriers are less likely to escape from the hopping sites when Φ_B is large. The opposite is observed when σ is small; this could have arisen from a changeover from symmetric to asymmetric hopping rates when the excess energy of the "released" carriers somehow affects the relaxation process. It ought to be pointed out here that the barrier height Φ_B is a parameter that is dependent on the correlation between the hopping sites and does not necessarily equate to the physical barrier height. We also wish to revisit the issue related to the dependence of the carrier mobility on the density of states function in disordered semiconductors. While other authors have successfully derived an expression for the carrier density dependence on the gate bias [Eq. (17.55)] and deduced the changes in the activation energy, the model we presented earlier has simply lumped the "activation" effect into the mobility parameter in the form of an effective barrier height $\Phi_B + V_g'$, which is bias dependent. Even though the two approaches may appear to be different, their effects on the current-voltage (I-V) characteristics of the field-effect transistor ought to be identical. We are therefore of the opinion that our model is very similar to the model presented; we have simply extended the applicability to cover theory and data recently reported in the literature.

Acknowledgments

H. Kwok wishes to express his appreciation to NSERC Canada for partial financial support. In addition, the authors would like to express their appreciation to Dr. Martin McCall for inviting a contribution to this book, to Prof. Mikhail Noginov and the editorial staff for reviewing and processing this submission, and to the permissions offices of Elsevier Science Ltd. and The Institution of Engineering & Technology for granting permissions to reuse the authors' published materials.

References

1. C. K. Chiang, C. R. Fincher. Jr., Y. W. Park, A. J. Heeger, H. Shirakawa, E. J. Louis, S. C. Gau, and A. G. MacDiarmid, "Electrical conductivity in doped polyacetylene," *Phys. Rev. Lett.* **39**, 1098–1101 (1977).
2. A. J. Heeger, "Nobel lecture: Semiconducting and metallic polymers: The fourth generation of polymeric materials," *Rev. Mod. Phys.* **73**, 681–700 (2001).
3. S. Forrest, P. Burrows, and M. Thompson, "The dawn of organic electronics," *IEEE Spectrum*, **37**, No. 8, 29–34 (2000).
4. K. Fesser, A. R. Bishop, and D. K. Campbell, "Optical absorption from polarons in a model of polyacetylene," *Phys. Rev. B* **27**, 4804–4825 (1983).
5. H. Bassler, "Charge transport in disordered organic photoconductors," *Physica Status Solidi* (*b*) **175**, 15–56 (1993).
6. S. V. Novikov, D. H. Dunlap, V. M. Kenkre, P. E. Parris and A. V. Vannikov, "Essential role of correlations in governing charge transport in disordered organic materials," *Phys. Rev. Lett.* **81**, 4472–4475 (1998).
7. N. F. Mott, and E. A. Davis, *Electronic Processes in Non-Crystalline Materials*, 2nd ed., Oxford University Press, Oxford (1979).
8. O. Tal, Y. Rosenwaks, Y. Preezant, N. Tessler, C. K. Chan, and A. Kahn, "Direct determination of the hold density of states in undoped and doped amorphous organic films with high lateral resolution," *Phys. Rev. Lett.* **95**, 256405 (2005). http://link.aps.org/doi/10.1103/PhysRevLett.95.256405
9. P. W. M. Blom, M. J. M. de Jong, and M. G. van Munster, "Electric-field and tempurature dependence of the hole mobility in poly(*p*-phenylene vinylene," *Phys. Rev. B* **55**, R656 (1997).
10. H. L. Kwok, *Electronic Materials*, PWS Publishing Co., Boston (1997).
11. H. L. Kwok, "Modeling negative capacitance effect in organic polymers," *Solid State Electron.* **47**, 1089–1093 (2003).
12. S. Berlab and W. Brutting, "Dispersive electron transport in tris(8-hydroxyquinoline) aluminum (Alq_3) probed by impedance spectroscopy," *Phys. Rev. Lett.* **89**, 286601 (2002).
13. H. L. Kwok, "Bias and temperature effect on the carrier transport parameters in impedance spectroscopy," *Proc. SPIE* **5508**, 105–112 (2004).
14. H. L. Kwok, "Understanding negative capacitance effect using an equivalent resistor-capacitor circuit," *Physica Status Solidi* (*c*) **5**, 638–640 (2008).
15. B. Sturman and E. Podivilov, and M. Gorkunov, "Origin of sketched exponential relaxation for hopping-transport models," *Phys. Rev. Lett.* **91**, 176602 (2003).

16. D. K. Ferry, *Semiconductors*, Macmillan Publishing Co., New York (1991).

17. Z. G. Yu, D. L. Smith, A. Saxena, R. L. Martin, and A. R. Bishop, "Molecular geometry fluctuations and field-dependent mobility in conjugated polymers," *Phys. Rev. B* **63**, 85202 (2001).

18. H. L. Kwok, Charge relaxation and dynamics in organic semiconductors," *Proc. SPIE* **6320**, 63200I (2006).

19. E. A. Hegmann, R. R. Tykwinski, K. P. H. Lui, J. E. Bullock, and J. E. Anthony, "Picosecond transient photoconductivity in functionalized pentacene molecular crystals probed by teraherz pulse spectroscopy," *Phys. Rev. Lett.* **89**, 227403 (2002).

20. A. B. Kaiser, "Electronic transport properties of conducting polymers and carbon nanotubes," *Rep. Prog. Phys.* **64**, 1–49 (2001).

21. H. L. Kwok and J. B. Xu, "A model for exciton formation in organic electroluminescent devices," *Solid State Electron.* **46**, 645–650 (2002).

22. M. Shur, *Physics of Semiconductor Devices*, Prentice Hall, Inc., Upper Saddle River, NJ (1990).

23. H. C. F. Martens, P. W. M. Blom, and H. F. M. Schoo, "Comparative study of hole transport in (*p*-phenylene vinylene) derivatives," *Phys. Rev. B* **61**, 7489–7493 (2000).

24. J. H. Burroughes, D. D. C. Bradley, A. R. Brown, R. N. Marks, K. Markey, R. H. Friends, P. L. Burn, and A. B. Holmes, "Light-emitting diodes based on conjugated polymers," *Nature* **437** 539–541 (1990).

25. H. C. F. Martens, H. B. Brown, and P. W. M. Blom, "Frequency-dependent electrical response of holes in poly(*p*-phenylene vinylene)," *Phys. Rev.* **60**, R8489–R8492 (1999).

26. P. W. M. Blom, M. J. M. de Jong, "Electrical characterization of light-emitting diodes," *IEEE J. Sel. Top Quantum Electron.* **4**, 105–112 (1998).

27. H. Meyer, D. Haarer, H. Naarmann, and H. H. Horhold, "Trap distribution for charge carriers in poly(paraphenylene vinylene (PPV) and its substituted derivative DPOP-PPV," *Phys. Rev. B* **52**, 2587–2598 (1995).

28. G. Yu, Y. Liu, S. Zhou, F. Bai, P. Zeng, M. Zeng, Z. Wu, and D. Zhu, "Anomalous current-voltage characteristics of polymer light-emitting diodes," *Phys. Rev. Lett.* **65**, 115211 (2002).

29. H. L. Kwok, "Modeling current transport in organic light-emitting diodes (OLEDs)," *Comput. Mater. Sci.* **33**, 200–205 (2005).

30. H. L. Kwok, Y. L. Wu and T. P. Sun, Investigation into the modeling of field-effect carrier mobility in disordered organic semiconductors," *IEE Proc. Circ. Dev. Syst.* **1153**, 124–128 (2006).

31. S. F. Nelson, Y. Y. Lin, D. J. Gundlach, and T. N. Jackson, Temperature-independent transport in high-mobility pentacene transistors," *Appl. Phys. Lett.* **72**, 1854–1856 (1988).

Biographies

Harry H. L. Kwok is a Professor in the Department of Electrical and Computer Engineering at the University of Victoria (1992–present) and is currently the acting director of CAMTEC (he also served as director and co-director of CAMTEC from 1997–2004). He graduated *magna cum laude* with a B.S. in engineering from University of California, Los Angeles, and obtained his Ph.D. from Stanford University. Kwok's academic research interests have been in the areas of electronic materials, devices, and circuits. In the past, he worked on projects related to ion implantation, gate array IC design, solar cells, organic thin film devices, gamma camera imaging systems (in collaboration with Vancouver Island Cancer Centre, BC), and high-speed GaAs CCD transient detection systems (in collaboration with TRIUMF, BC). He has been a visiting professor at many universities and has taught courses in areas such as solid state electronics, and circuits and devices. He wrote a textbook entitled *Electronic Materials*, published by PWS/Coles Publishing, Boston in 1997, which was translated into Korean in 2003. He has also supervised many M.S. and Ph.D. students and has published more than 100 journal and conference publications. His more recent areas of research are in thermoelectrics and DNA charge transport.

You-Lin Wu received his M.S. degree from National Tsing Hua University, Taiwan, in 1984 and his Ph.D. from National Taiwan University, Taiwan, in 1994, both in electrical engineering. From 1984 to 1991, he was a lecturer in the Department of Electronic Engineering at Hsinpu Junior College and served as head of the department from 1989 to 1990. In 1995, Dr. Wu joined the Powerchip Semiconductor Company where he was the manager of the Thin Film Department and was in charge of the development and fabrication of 16M dynamic RAM (DRAM). In 1998, he joined the faculty of the Electrical Engineering Department at National Chi-Nan University and served as chair of the department until 2001. He is currently a professor in the same department. His research interests involve ultrathin gate oxide reliability of metal-oxide-semiconductor (MOS) devices and the fabrication of biochemical sensors and organic thin-film transistors.

Tai-Ping Sun received his B.S. degree in electrical engineering from the Chung Cheng Institute of Technology, Taiwan, in 1974, his M.S. degree in material science engineering from the National Tsing Hua University, Taiwan, in 1977, and his Ph.D. degree in electrical engineering from the National Taiwan University in 1990. From 1977 to 1997, he worked at the Chung Shan Institute of Science and Technology on the development of infrared devices, circuits, and systems. He joined the Department of Management Information Systems at

Chung-Yu College of Business Administration in 1997 as an Associate Professor. Since 1999, he has been a Professor with the Department of Electrical Engineering, National Chi Nan University, Nantou, Taiwan. Since 2001, he has been Secretary General and currently serves as Dean of the College of Science and Technology at the same university. His research interests include infrared detectors and systems, analog/digital mixed-mode integrated circuit design, and special semiconductor sensors and their applications.

Chapter 18
Holography and Its Applications

H. John Caulfield and Chandra S. Vikram*
Fisk University, Nashville, TN, USA

18.1 Introduction

Several decades ago, one of the authors of this chapter was grading junior high school exams from a geographic science class taught by his wife, when he was rewarded with a wonderful answer to this seemingly straightforward question: "What is a rock?" The student's answer was 100% correct but not especially informative: "A rock is not a mineral." It is quite difficult to define something by the set of all things it is not. Yet it seems important to state at the beginning that this tutorial is not about holography as most people know it: a means to record and reproduce 3D images. In fact, we explicitly exclude image-forming holograms from the discussion.

---

*Professor Vikram died suddenly on August 17, 2007. His death was a great loss, and he will be remembered in many ways by the community.

This tutorial deals with holograms used as parts of more complex systems as ways of controlling or manipulating properties of that light—its direction, spectrum, polarization, speed of pulse propagation, etc. As there are other light manipulation means available, it will be important to understand when holograms offer advantages and what those advantages may be.

Like all of the materials discussed in this book, holograms are complex structures. Like most of the other complex media discussed, holograms are metamaterials. In electromagnetism (covering areas such as optics and photonics), a metamaterial is a structure that gains its (electromagnetic) material properties from the details of structure itself (and the properties of its components), rather than merely inheriting those properties directly from the materials of which it is composed. This term metamaterial is particularly used when the resulting material has properties not found in naturally formed substances. That is, metamaterials are "designer materials" created to have specific properties that the system or component designer needs.

Holographic metamaterials share one critical property with most other metamaterials: their behavior with respect to electromagnetic waves is related directly to the 3D periodic structure produced. Thus, we are forced to make an embarrassing confession (or boastful claim, depending on how you choose to look at it): most metamaterials are holograms.

Clearly we are in need of a definition for a hologram—something on which holographers themselves have not reached universal agreement. The broadest definition we tend to use is the following: "A hologram is a structure that uses periodic variations to convert one expected or easily created electromagnetic pattern (a wavefront) into another."

Most frequently, the easily obtained reconstruction wavefront derives from a point source, and the wavefront it produces upon interaction with the hologram is one that appears to come from a complex 3D scene. What makes a hologram a metamaterial (from our viewpoint) is that the wavefront produced by the hologram is not one that produces an image. It has far simpler functions, such as

- transmitting some light and reflecting the rest according to the wavelength of the light,
- deflecting light (forward or backward) in different directions according to its wavelength,
- canceling out the reflected light,
- diffusing the forward- or backward-propagating light,
- trapping light into a waveguide or detrapping it from a waveguide, and
- separating constant features of the input wavefront from changing features.

A rock is not a mineral. A holographic metamaterial is not an image-forming device; it is a hologram used to help construct a system. More generally, we will regard a holographic material as a material formed by recording one or more interference patterns. The recorded pattern may be the material itself or simply a step in making it. For example, a pattern in photoresist may, after etching, produce some other kind of material such as a photonic crystal.

18.2 Basic Information on Holograms

For several reasons, this is not an appropriate place for a detailed exposition on holography. First, many books and articles are available that do just that.[1–5] Second, most general discussions of holography are aimed more at display holography than at holographic metamaterials. Third, the detail can become so complex as to obscure the general concepts that are the focus of this tutorial. Nevertheless, several new developments on holography are available.[6]

Most, but not all holograms, are governed to a large degree by a single master equation that relates two complex wavefronts W_1 and W_2 at the hologram. That master equation is seldom perfectly implemented; however, it provides an easy way to discuss holographic fundamentals. The hologram is said to have a complex transmission

$$\begin{aligned} T &= \tau_0 + \beta|W_1 + W_2|^2 \\ &= \tau_0 + \beta(|W_1|^2 + |W_2|^2) + \beta W_1 W_2^* + \beta W_1^* W_2, \end{aligned} \tag{18.1}$$

where * indicates complex conjugation, τ_0 is the average transmittance, and β is the slope of the exposure-transmittance curve. For a common negative hologram, β is a negative quantity.

Suppose a wavefront W_1 is incident on such a hologram. It will produce five terms. Three of the terms, $W_1[\tau_0 + \beta(|W_1|^2 + |W_2|^2)]$, amount to merely the transmission of the incident wavefront. Usually, this is not particularly interesting. The other two terms do more interesting things. The term $\beta W_1 W_1 W_2^*$ produces the wavefront W_2^* modulated by $\beta W_1 W_1$. For obvious reasons, W_2^* represents the phase reversal of W_2. Usually this term, too, is uninteresting. Much effort goes into trying to force all of the incident light to go into the beam represented by the fourth term, $\beta|W_1|^2 W_2$. For a uniform intensity W_1, this is just a reproduction of W_2. The hologram has converted W_1 into W_2. By symmetry, it can also convert W_2 into W_1. But that is not all it can do. If W_1^* is incident on the hologram (that is, if W_1 is perfectly reversed and incident from the other side of the hologram), then the interesting term becomes $\beta|W_1|^2 W_2^*$. In other words, W_1 perfectly reverses W_2. Likewise, it can convert the perfectly reversed version of W_2 (optical scientists call this the phase conjugate of W_2) into the phase conjugate of W_1. Such considerations lead optical scientists to speak of holograms as wavefront transducers.

Removing the light corresponding to the uninteresting terms in the master equation is not easy. From this point on, we will assume so-called phase holograms that (ideally) conserve light and merely introduce periodic variations in the speed of light passing through them. The general approach is to use thick-phase (also called "volume" phase) holograms. A hologram is said to be thick if a ray in the reconstructing beam crosses multiple fringes in the periodic structure that is the hologram. Analyzing the effect of a thick hologram on the incident light can become extremely complex. The most popular way to do this is via "coupled wave theory,"[1–8] although alternate theories are available.[8] The idea is that an incident wave is scattered at every fringe it encounters, so that after the first fringe there are multiple waves in the thick hologram.

Consider a transmission hologram. The first fringe slightly reduces the incident light in the reconstructing beam by producing a small amount of light in the reconstructed beam. At the next fringe, more light is removed from the already-diminished reconstructing beam to add to the reconstructed beam. However, in addition, the reconstructed beam diffracts some light back into the reconstructing beam. A figurative battle for supremacy wages between the two. At some point if the hologram is thick enough and the phase modulation is strong enough, all of the light will go into the reconstructed beam. A hologram that can accomplish this is said to have 100% diffraction efficiency. If the hologram is thicker or the modulation is greater, then the struggle between reconstructing and reconstructed beams resumes. By symmetry, it is clear that 0% diffraction efficiency can also be reached. Indeed, the diffraction efficiency varies as sine squared of the thickness-modulation product. Thus, it becomes clear why this is called coupled-wave theory.

For a reflection hologram, some of the light leaves the hologram with each fringe encountered, so that as the reconstructing and reconstructed beams travel through more fringes, there is less light to "fight" over. A diffraction efficiency of 100% is approached asymptotically as the thickness-modulation product increases. Roughly speaking, the thickness multiplied by the index of refraction modulation needs to be of the order of half a wavelength in order to achieve essentially 100% diffraction efficiency in either a transmission or a reflection hologram.

As a general rule, reflection holograms tend to be particularly wavelength selective for a fixed angle of incidence. That is, they make good narrowband spectral filters that reflect the design wavelength and transmit the other wavelengths on either side. Maximum wavelength selectivity requires large thickness and therefore low modulation. On the other hand, reflection holograms are not particularly angularly selective. If we use extremely high modulation so that the effective thickness is small, then it is possible to make reflective holograms that reflect extremely broad spectral bands.

High-efficiency thick-transmission holograms tend to have high angular selectivity for a fixed wavelength. This refers to the angle in the plane defined by the reconstructed and reconstructing rays. A thick hologram is angularly tolerant in the orthogonal direction. Such holograms tend to transmit many wavelengths, but deflect each in a different direction.

Like the reflection hologram noted earlier, transmission holograms can have high efficiency and offer great tolerance of the restrictions that usually occur, if they are made extremely thin with high modulation; if this is the case, then a transmissive holographic grating can have high efficiency over a broad range of angles.

18.2.1 Hologram types

Most often, holographic metamaterials are made by exposing a high-resolution recording material to the interference pattern produced by two mutually coherent collimated beams of laser light, as shown in Fig. 18.1. The lines running perpendicular to the beam directions represent plane waves that are, by definition, spaced one wavelength λ apart. These plane waves can be viewed as moving at the speed of light. However, because the phase relationship between wavefronts in the two beams is constant, the plane waves form standing waves, or stable interference patterns. The spacing *S* between those interference patterns is

$$S = \frac{\lambda}{2\sin\theta_{1/2}}, \tag{18.2}$$

where $\theta_{1/2}$ is the half angle between the two directions of travel. Figure 18.2 shows the interference region in more detail.

Moreover, if the recording medium is thin, as shown in Fig. 18.3, we record a thin hologram—a diffraction grating. A hologram is said to be thick if a ray from either beam strikes multiple recorded fringes. Obviously, the ultimate thick hologram fills the entire interference region and looks like Fig. 18.2. Thick holograms are seldom used; rather, we use recording configurations, as are shown in Figs. 18.4 and 18.5.

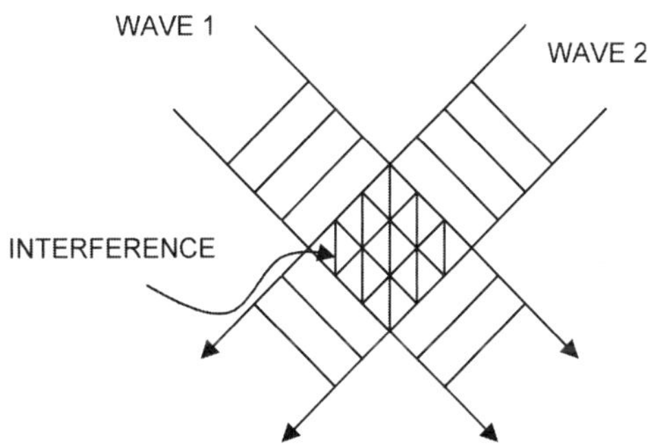

Figure 18.1 Two mutually coherent plane waves form interference fringes parallel to the bisector of the angle between the beams.

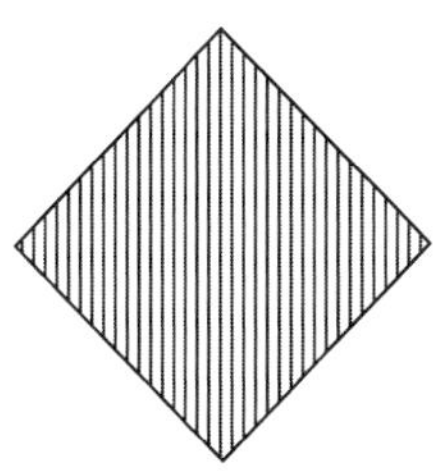

Figure 18.2 The overlap region from Fig. 18.1 has fringes that govern the holographic metamaterial structure.

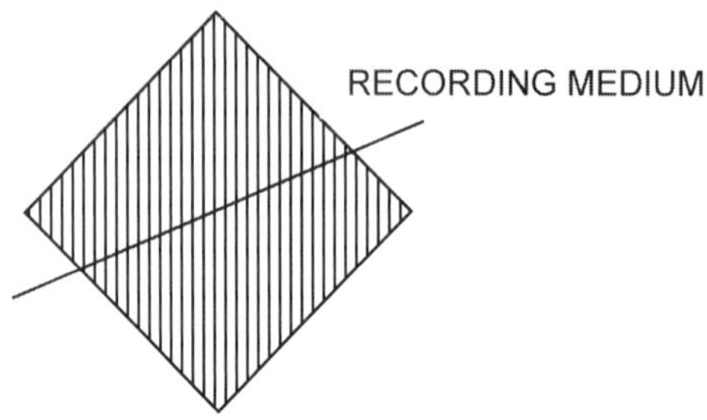

Figure 18.3 A thin recording medium will record a periodic pattern called a thin hologram or diffraction grating.

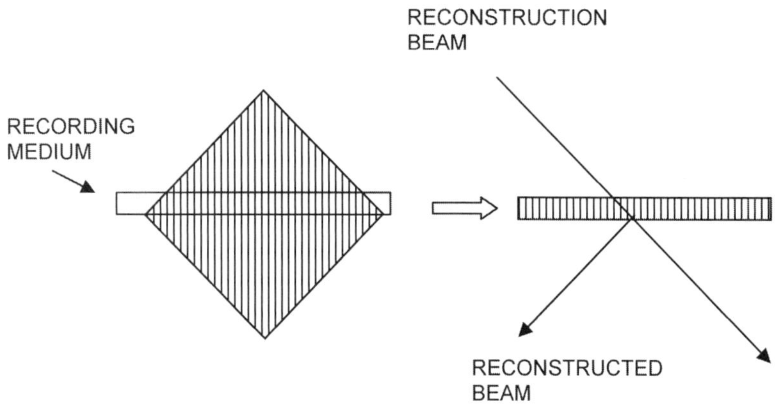

Figure 18.4 Configuration that leads to a transmissive thick hologram. In any kind of hologram, recreating one of the beams recreates the other, as suggested here. After the hologram is recorded, it can reconstruct one beam from the other beam that is transmitted through it.

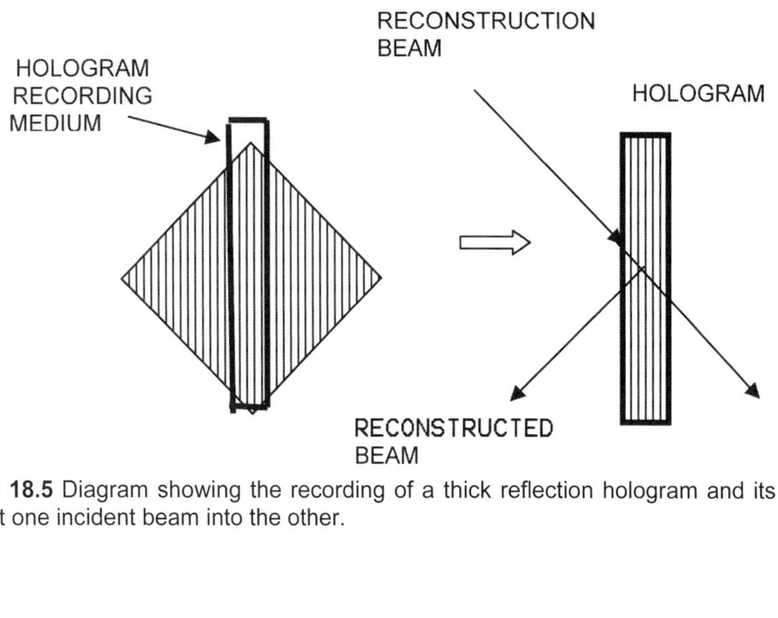

Figure 18.5 Diagram showing the recording of a thick reflection hologram and its use to convert one incident beam into the other.

Thus, the concepts of thick and thin holograms and transmissive and reflective holograms should be clear. However, both transmissive and reflective holograms also work in reverse, as shown in Figs. 18.6 and 18.7. Transmissive holograms remain transmissive, and reflective holograms remain reflective.

Looking down on the hologram formed as described above, one would see fringes such as those in the transmissive case, shown in Fig. 18.8. (The fringes are parallel to the surface in the reflective hologram, and not particularly interesting to look at). However, with two exposures, one can produce a pattern that looks similar to that shown in Fig. 18.9.

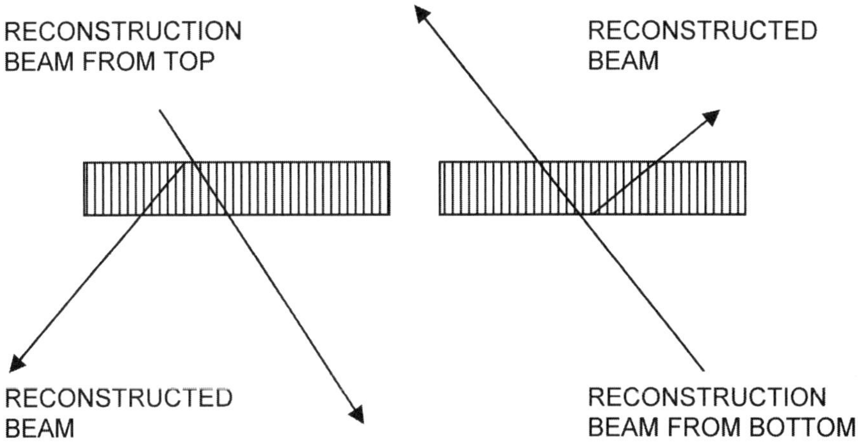

Figure 18.6 The same transmissive hologram can work with beams incident from either side.

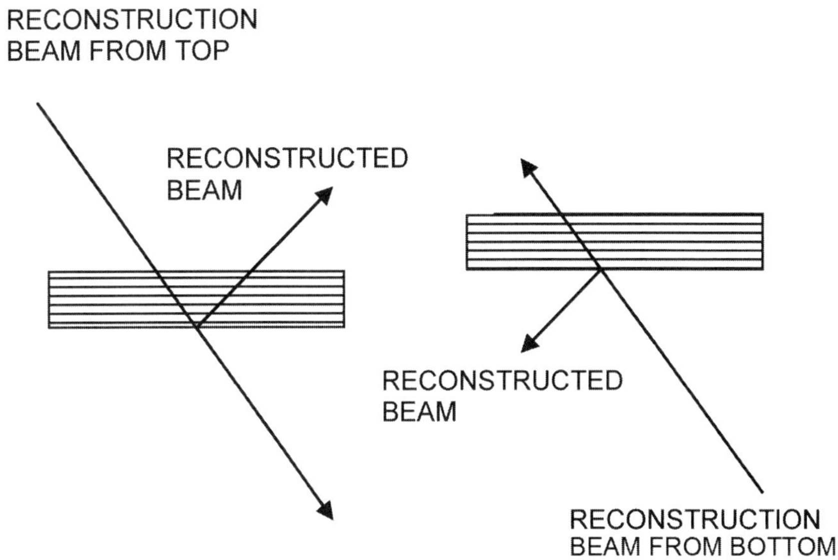

Figure 18.7 The same reflective hologram can work with beams incident from either side.

With three exposures, hexagonal patterns can be achieved, and so forth. Holograms are easy to form on curved surfaces as well. For instance, Fig. 18.10 shows how to record a reflective hologram on a curved surface.

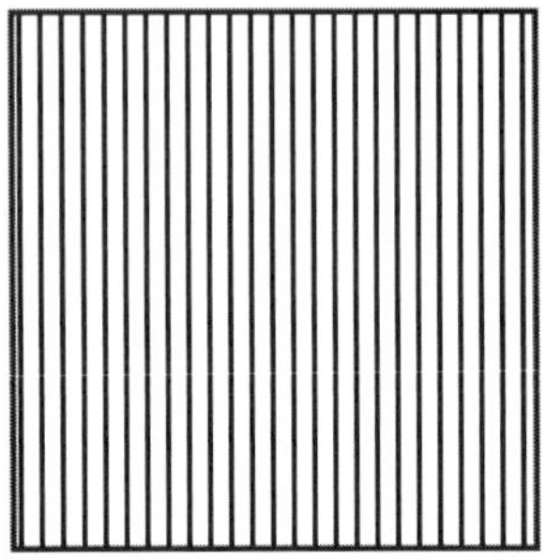

Figure 18.8 Viewed from its flat side, a transmissive hologram might look like this.

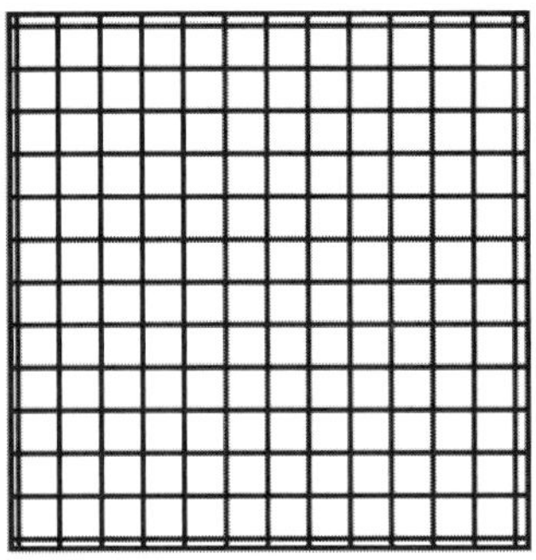

Figure 18.9 Multiple exposures can lead to more complex transmissive holograms, such as the squares shown here.

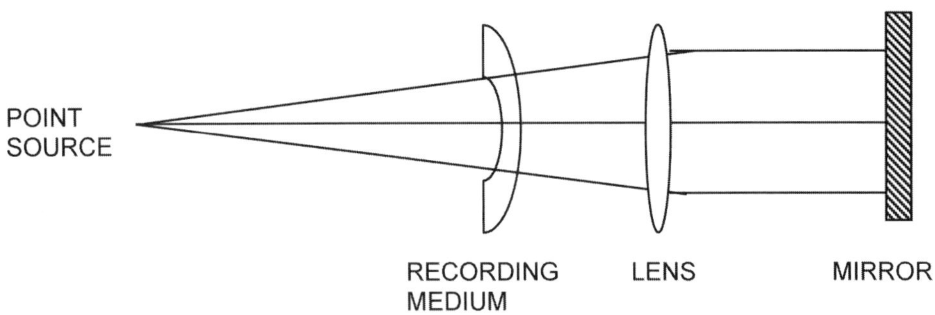

Figure 18.10 It is easy to reverse (phase conjugate) a spherical wavefront coming from a point source. This produces a backward-traveling wave that interferes with the forward-traveling incident wave to produce curved fringes ideally suited for making holograms on curved surfaces.

18.3 Recording Materials for Holographic Metamaterials

There are many requirements for recording materials. Here are a few of them:

1. They must be sensitive to light at the recording wavelength. This is a problem, especially for some popular materials that make good holograms but are sensitive only in the blue and green region.
2. They must modulate light effectively at the application or use wavelength (possibly distinct from the recording wavelength).
3. They must have enough resolution to record the fringes (most recording media do not).
4. They must be stable during the intended time of use in the expected conditions.

Here are some other useful and desirable properties:

1. Repeatability. Some materials make beautiful holograms every time, but the detailed hologram properties are extremely difficult and expensive to repeat.
2. Suitability for application to surfaces of various shapes. This can be a major issue. Some problems that can occur include: seams forming when the window, screen, or other structure is larger than the largest hologram we can make; ease of application by nonscientists; permanence (an issue that varies with circumstances), and ease of bubble removal.
3. Adjustability. If small adjustments in hologram properties can be made, such precise control over the recording and development processes is not needed.
4. Immunity to environmental degradation. For instance, some materials make wonderful holograms that slowly degrade with exposure to light, moisture, etc.

Here are some of the most widely used materials, along with some of their characteristics:

1. **Photographic materials.** These are attractive for many uses because of their extreme sensitivity and low cost; however, they are seldom used in holographic complex media. Sometimes photographic materials can be added into the emulsion to create useful photoactivated components.
2. **Photoresist.** After development, these can be used either directly (often with a reflective overcoat) or as a way to etch patterns into or deposit material onto another surface. Photoresists are available in positive and negative forms and are the critical component for photolithography. Unlike photographic materials, photoresists are grainless and therefore have essentially arbitrary resolution. Unfortunately, it is the grain structure that allows photographic materials to be so sensitive. For this reason, photoresists are much less sensitive than photographic materials.
3. **Photopolymers.** These are materials that exist as somewhat mobile monomers before exposure. Exposure to light polymerizes them, leaving

the material immobile. Indeed, monomers migrate from unexposed sites to the polymerization sites. The net result is a difference in index of refraction. Some photopolymers require wet chemical development; however, more and more dry developing materials are becoming available. These materials will undergo microscopic size changes when polymerization occurs unless they are somehow bound inside a rigid structure. Photopolymers in sol gel and other glassy materials effectively avoid such size changes.

4. **Photorefractives.** These are optically complex materials; they exhibit all sorts of behaviors, such as photoconductivity, charge migration, electro-optic effect, and so forth. These effects combine to create a phase pattern dictated by (and looking like) the interference pattern of the two beams of light striking them. However, this new phase pattern has an important difference: The spatial pattern is shifted so that the bright and dark regions are not in the same location as they would be if we simply photographed the interference pattern. This is called a spatial phase shift. Depending on the material and the power per unit area striking the material, the response time taken to produce the hologram can be extremely fast (nanosecond) or extremely slow (hours). These materials remain photosensitive unless a "fixing" operation is used.

Besides general descriptions[1–9] of the recording materials, specialized volumes are also available.[10–12]

18.4 Computer-Generated Holograms

Holograms need not be formed interferometrically. Many holograms (often not called holograms at all, but "kinoforms" or "diffractive optics") can be designed by computer, manufactured by conventional photolithography, and mass produced by thermoplastic pressing. Usually such holograms are on axis. They can achieve exceptionally high diffraction efficiency with a clever and totally unexpected method. The fringe spacings are below a wavelength, so all of the diffracted light is evanescent (nonpropagating in transmission or reflection). The only propagated light is zero order (undiffracted but phase and polarization modulated).

The general mode of generation is basic. A laser-illuminated object can be described by a large number of point scatters. Geometrical and physical parameters can then be used to compute the amplitude and phase reaching the hologram. Similarly, the reference beam parameters are more straightforward than those of interferometrically formed holograms. The hologram is thus plotted by a suitable display or storage device. Detailed descriptions on computer-generated holograms are generally available.[13–15] Some of the established applications are complex spatial filtering, laser-beam mode selection, and optical interconnections. Unique applications, such as diffusers for color mixing,[16] fabrication of high-frequency gratings,[17] waveguide gratings coupling with tailored spectral response,[18] and optical particle trapping[19] are frequently reported.

18.5 Simple Functionalities of Holographic Materials

Holographic materials can be used to produce desirable optical properties. In this respect, holograms are closely related to multilayer evaporated films, and both are governed by the well-known properties of interferometry.

One such property is high reflectivity in some bands and high transmission elsewhere. The reflected band can be very narrow for very thick holograms and very broad for very thin holograms. To accomplish high-diffraction efficiency, these holograms require that the product of the thickness of the material and the index of refraction modulation produced be roughly half a wavelength. That means that broadband reflection holograms require an extremely high index change, and narrowband reflection holograms require an extremely low index modulation.

The other simple property is extremely high transmission diffraction efficiency at a user-chosen angle of incidence. The angular tolerance can be great or small. Again, very thin holograms can have a large angular tolerance while very thick holograms can have very small angular tolerances. And again, high efficiency requires large index changes for thin holograms and small index changes for thick holograms.

The symmetries just noted go even farther. As a crude but useful "rule of thumb," we have[20]

$$\frac{\Lambda}{T} \approx \frac{\delta\theta}{\Theta} \approx \frac{\delta\lambda}{\Lambda}, \tag{18.3}$$

where Λ is the central wavelength, T is the physical thickness of the hologram, $\delta\theta$ is the angular spread over which high efficiency is achieved, Θ is the central angle of diffraction, and $\delta\lambda$ is the wavelength spread over which high efficiency can occur. Of course, $\delta\theta/\Theta$ applies to transmission holograms and $\delta\lambda/\Lambda$ applies to reflection holograms.

Another general rule concerns the solid angular field of view Ω of a hologram (or any other interferometer) and $\delta\lambda/\Lambda$. The most compact form of that law (that stems from diffraction limits) uses the resolving power R

$$R = \frac{\Lambda}{\delta\lambda} \approx \frac{T}{\Lambda}. \tag{18.4}$$

That rule is

$$R\Omega \geq 2\pi. \tag{18.5}$$

Thus,

$$W \geq \frac{2\pi}{R} \approx \frac{2\pi\Lambda}{T}, \tag{18.6}$$

which explains why we need very thin holograms to accomplish high diffraction efficiency with a large field of view.

There is a natural reflection loss at the surface of the hologram. These reflections are a result of rigorous computation of transmission and reflection of light at a boundary between two kinds of materials. The equations (called the Fresnel equations) are given in every book on optics. Often the small (about 4%) reflection is unimportant. When it is important to remove the reflection entirely, holography is very useful. It can create a beam identical in amplitude to the material that has the same amplitude but opposite phase. The two beams interfere in such a way as to remove the reflected beam altogether.

18.6 Phase Conjugation and Holographic Optical Elements

Made properly, holograms can overcome some of the imperfections in many optical systems. The phase conjugation property has been known since the early stages of holography.[21] For example, a reflection hologram can easily clean up the light reflected from an irregular object such as a distorting window. This is especially true if we use a particular type of dynamic hologram called double phase-conjugate mirrors (DPCM). Consider the situation presented in Fig. 18.11.

As shown in Fig. 18.12, a hologram of the highly complex light transmitted through or scattered from such an object can be recorded and perfectly reversed ("phase conjugated") so that when it is transmitted (Fig. 18.13) or scattered back, the object corrects the irregularities and restores the perfect phase conjugate (backward-traveling version) of the incident beam. The reimaging lens is used simply to capture more light for the hologram. It does not need to be a high-quality lens, as the use of a hologram compensates for imperfections. In that case, the system would aim to correct the effect of aberrations. In that sense the complex object and the lens are one combined unit. Permanent storage of the hologram is not necessary. Materials such as a photorefractive medium can be used for real-time processing. In fact, such phase-correction tools have been used to correct optical aberrations in particle field holography,[22–23] and high-resolution imaging, lithography, and communications.[24,25]

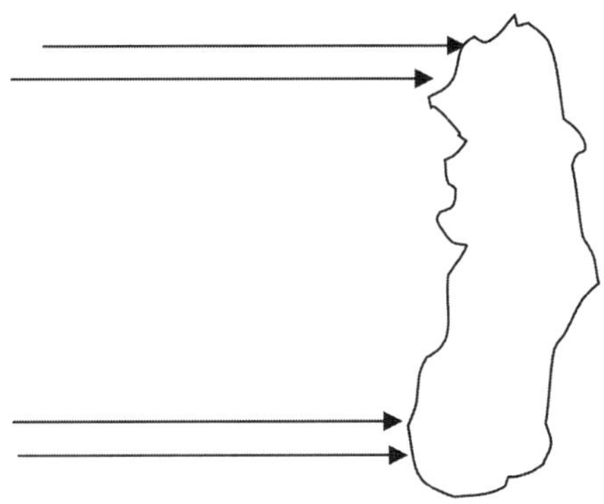

Figure 18.11 A collimated beam of light incident on a complex transmissive object is transmitted and reflected in many directions.

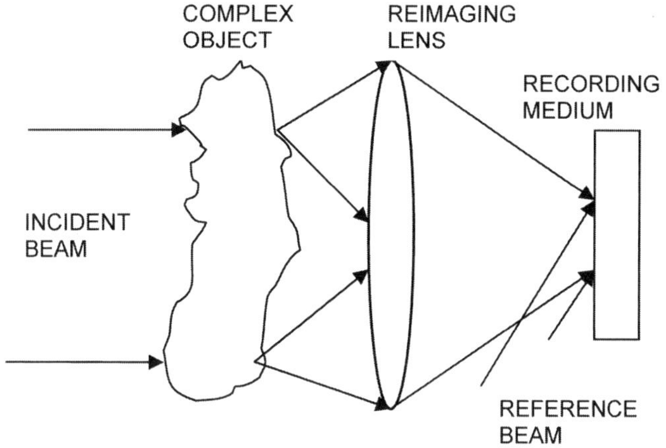

Figure 18.12 Diagram of how the hologram of a highly irregular wavefront can be recorded.

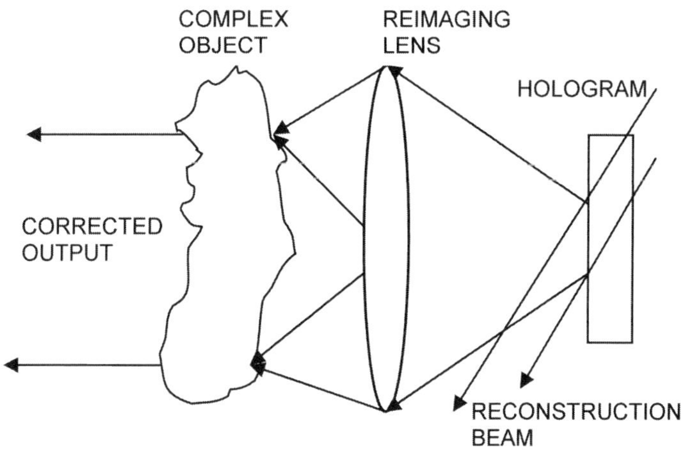

Figure 18.13 Reconstruction of the corrected wavefront using a hologram made by the arrangement of Fig. 18.12.

The above example of a hologram used to correct phase errors is, in fact, part of a larger group commonly known as a holographic optical element (HOE). A collection of reprints on the subject is available.[26] These elements have been applied to a large variety of applications such as optical interconnections, multifocusing, helmet-mounted display, fiber optic coupling, compact disc applications, automotive windshield displays, wideband phase-array antennas, confocal microscopy, spatial light modulation, and as optical elements in speckle metrology. These elements are flat, lightweight, and can be mass produced. In fact, as was explained in Sec. 18.2.1, a simple grating is a HOE. In the last few years, spectacular advances in optical materials have been made that allow the storage and user-selected replay of hundreds of independent bright holograms.[27]

Another interesting kind of HOE or holographic complex medium is the holographic diffuser.[28] Invented 35 years ago, these materials are made by exposure to laser speckle. Numerous companies around the world make affordable off-the-shelf holographic diffusers, and some manufactured products include them. The size of the speckles recorded is inversely proportional to the angle over which scattering takes place. Asymmetric speckles lead to asymmetric scattering. A HOE can be reflective or transmissive and has a number of other useful properties, such as

- controllable degree of diffusion (usually chosen to be 100%),
- memory—the diffusers remember the direction of the undiffused (specular) beam, and the scatter occurs around the unscattered beam,
- polarization preservation, and
- the ability to be formed in situ on flat or curved surfaces.

One such diffuser performance is represented in Fig. 18.14.[29]

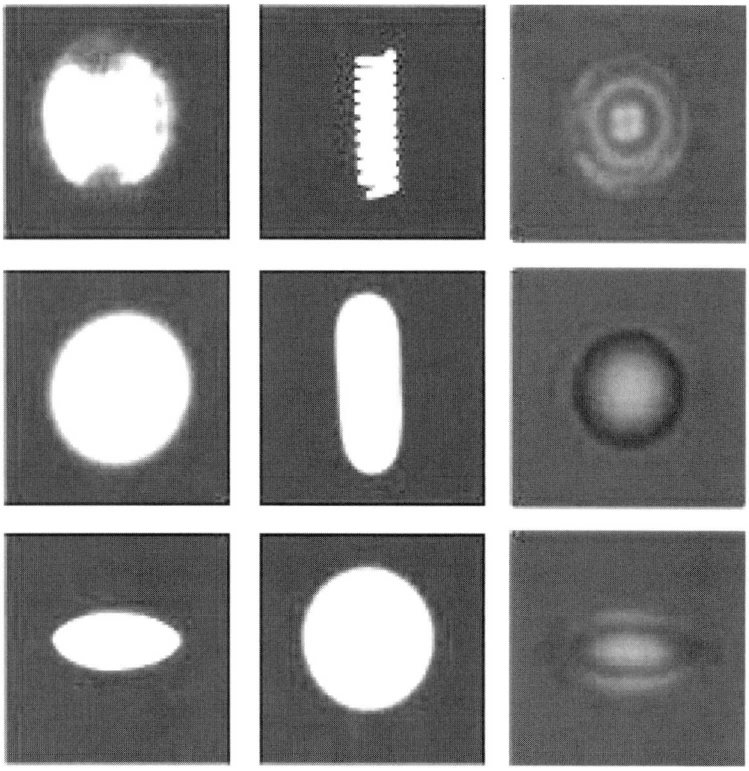

Figure 18.14 Different columns show "before" beams on the top row (flashlight, filament, and LED). Middle and bottom rows show effects of "circular" and "elliptically" scattering diffusers, respectively. (Reprinted from Ref. 29.)

18.7 Related Applications and Procedures

18.7.1 Holographic photolithography

Structured surfaces now play many roles in science and technology[30,31] Combined with photolithography developed for the manufacture of ultrasmall electronics, holography has played and will continue to play a role in recording patterns to be transferred to other materials to make structured surfaces.[32] Photonic crystals can thus be prepared,[33–36] including 3D versions.[33,37–40]

Matters of nomenclature immediately arise here. If a hologram is a recorded interference pattern, then "holographic lithography" means any photolithography performed using interference patterns. Thus, we could, if we so desired, call whatever is produced a result of a hologram. Arguing definitions is highly unprofitable, unless it is clear what the words mean in any context. Often it is useful to call the item produced something other than a hologram, as that same type of item might conceivably be produced in some other manner.

Producing large stable uniform deeply modulated interference patterns is by no means an easy task. Vibration is the major problem. This can be overcome by using sufficiently fast laser pulses so that vibration is eliminated, or with elaborate processes to minimize vibration, air currents, laser drift, and the like. Both approaches have been used.

The payoff of doing this well is a level of quality, repeatability, and uniformity of periodic patterns not obtainable in any other way. For its ability to produce something as simple as diffraction gratings and as complex as photonic crystals, holographic lithography is unexcelled. An example is seen in Fig. 18.15, which shows some results of holographic lithography. The simple experimental arrangement with a set of four beams is capable of 2D and 3D microstructures fabricated with submicrostructures in large areas. Potential applications are in microfluidic systems and templates for photonic crystal devices.[31]

More complex structures such as Bravais lattices can be conceived and fabricated, for example, by using multistep multibeam interference patterns. Figure 18.16 shows calculations of 3D patterns. In fact, Shoji and coworkers[39] experimentally demonstrated that such complex structures can be fabricated.

The above examples are just a few from a particularly active area of ongoing research. Active control for stable and adjustable (for tunable optical lattices) intensity patterns is one recent example.[41]

18.7.2 Copying of holograms

Mass copying[1,4,5,8] from a single master has economic and technical advantages. One obvious approach is to use a reconstructed image from the master to store hologram copies. Contact printing of the hologram itself is another option. However, one needs extremely close contacts (not more than Δ^2/λ where Δ is the fringe spacing and λ is the wavelength) with this mode.

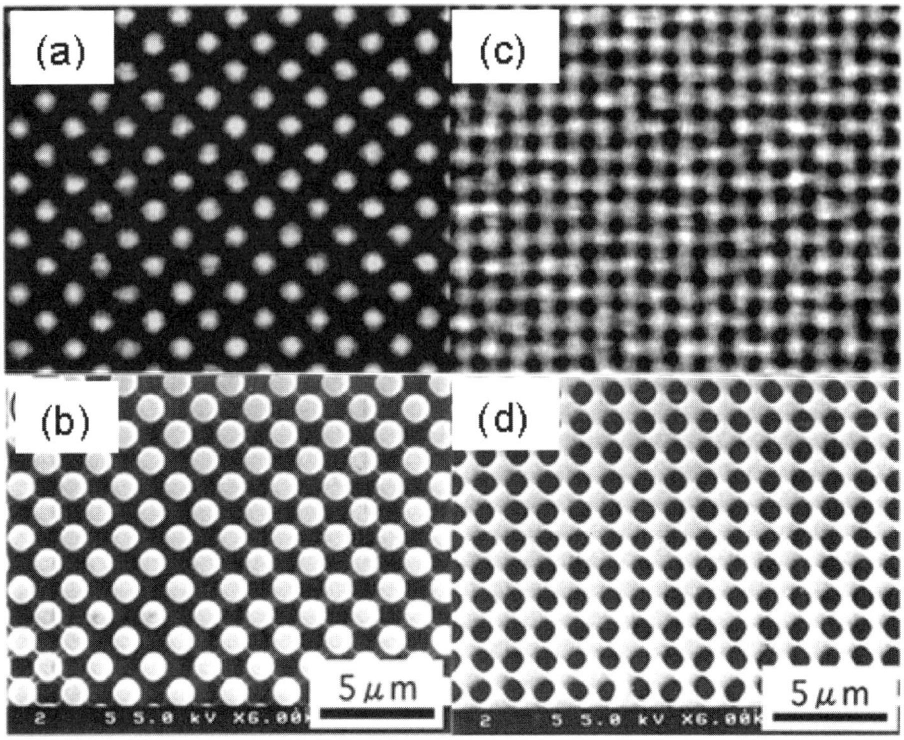

Figure 18.15 Four-beam holographic lithography in photoresist SU-8. Images (a) and (b) show process without phase difference. Images (c) and (d) are due to π/2 phase difference between diagonally paired beams. Images (a) and (c) are intensity patterns as seen by a CCD camera. Images (b) and (d) are SEM micrographs. (Reprinted from Ref. 32.)

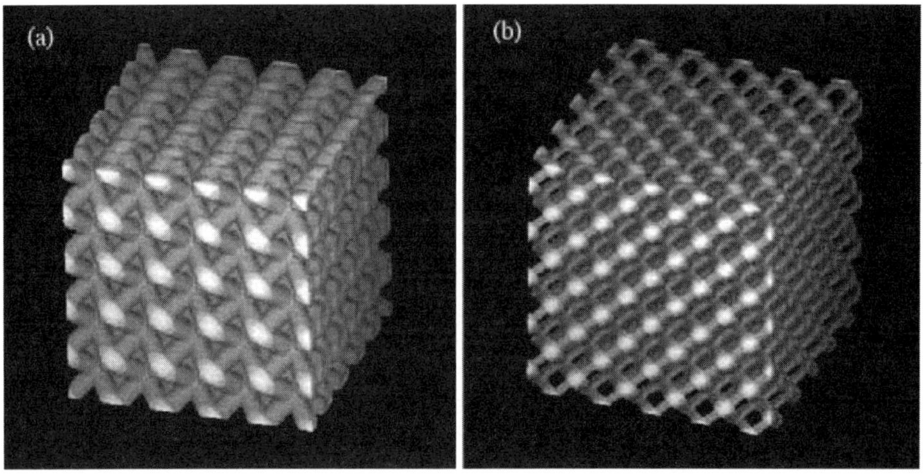

Figure 18.16. Computed 3D pattern created for three-step three-beam laser interferometry. (a) Body-centered cubic and (b) face-centered cubic lattices. (Reprinted from Ref. 39.)

Beyond the above historical approaches, the most common mass-producing technique is embossing, as is seen on credit cards, book covers, and product packets. A multistep process involves placing the master hologram on a photoresist media, metal vacuum depositing, and nickel electroforming; the hologram is then attached to an embossing die or a roller. Generally, holograms are hot rolled on plastic materials and coated with aluminum to make them reflective and protected. An interesting application was a hologram on chocolate or candy![42]

18.7.3 Holograms in nature and general products

Discussion on holograms is not complete without mentioning naturally existing structures. Parker[43] showed that the silver color of some fish scales is basically due to a reflection phenomenon caused by a multilayer dielectric structure. Parker did not call them thick-reflection holograms, but the scales of silvery fish are similar to such broadband holograms as described by Jannson et al.[44] Recently Vigneron et al.[45] found a similar case of blue-violet iridescence of the beetle *Hoplia coerulea* (Coleoptera). Earlier, Vigneron et al.[46] found that wings of the butterflies *Cyanophyrs remus* and *Lycaena virgaureae* consist of photonic-crystal films that can be termed thin-reflection holograms.

In gemology, *chatoyancy* is an optical reflectance effect in gemstones arising from structures such as inclusions or cavities. In woodworking, certain finishes result in a 3D appearance. Terms such as *scratch holograms*, *sandpaper holograms*, *wire-brush holograms*, and *phonograph holograms* have appeared in the past. Basically these natural and mechanical objects have complex structures representing a hologram of 'something' 3D. These examples show the vast potential of such structures in nature and common products.

References

1. H. J. Caulfield and S. Lu, *The Applications of Holography,* John Wiley & Sons, New York (1970).
2. R. J. Collier, C. B. Burckhardt, and L. H. Lin, *Optical Holography*, Academic Press, New York (1971).
3. H. M. Smith, *Principles of Holography*, 2nd ed., John Wiley & Sons, New York (1975).
4. H. J. Caulfield, *Handbook of Optical Holography*, Academic Press, New York, (1979).
5. P. Hariharan, *Optical Holography*, 2nd ed., Cambridge University Press, Cambridge (1996).
6. H. J. Caulfield and C. S. Vikram, *New Developments in Holography and Speckle*, American Scientific Publishers, Stevenson Ranch, California (2008).

7. W. T. Cathy, *Optical Information Processing and Holography*, John Wiley & Sons, New York (1974).

8. R. R. A. Syms, *Practical Volume Holography*, Clarendon Press, Oxford (1990).

9. H. M. Smith, *Holographic Recording Materials*, Springer-Verlag, Berlin (1977).

10. H. I. Bjelkhagen, *Silver Halide Materials for Holography & Their Processing,* 2nd ed., Springer-Verlag, Berlin (1995).

11. H. I. Bjelkhagen, *Selected Papers on Holographic Recording Materials*, SPIE Press, Bellingham, WA (1996).

12. R. A. Lessard and G. Manivannan, *Selected Papers on Photopolymers*, SPIE Press, Bellingham, WA (1995).

13. S. H. Lee, *Selected Papers on Computer-Generated Holograms and Diffractive Optics,* SPIE Press, Bellingham, WA (1992).

14. L. Yaroslavsky and M. Eden, *Fundamentals of Digital Optics*, Birkhäuser, Boston (1996).

15. V. Soifer, and M. Golub, *Laser Beam Mode Selection by Computer-Generated Holograms*, CRC Press, Boca Raton (1994).

16. W. Chao, S. Chi, C. Y. Wu, and C. J. Kuo, "Computer-generated holographic diffuser for color mixing," *Opt. Commun.* **151**, 21–24 (1998).

17. T. J. Suleski, B. Baggett, W. F. Dalaney, C. Koehler, and E. G. Johnson, "Fabrication of high-spatial-frequency gratings through computer-generated near-field holography," *Opt. Lett.* **24**, 602–604 (1999).

18. M. Li, B. S. Luo, C. P. Grover, Y. Feng, and H. C. Liu, "Waveguide grating coupler with a tailored spectral response based on a computer-generated waveguide hologram," *Opt. Lett.* **24**, 655–657 (1999).

19. M. Reicherter, E. Haist, E. U. Wagemann, and H. J. Tiziani, "Optical particle trapping with computer-generated holograms written on a liquid-crystal display," *Opt. Lett.* **24**, 608–610 (1999).

20. H. J. Caulfield, "Holographic spectroscopy," in *Advances in Holography:* Vol. 2, N. H. Farahat, Ed., 139–184, Marcel Dekker, New York (1976).

21. E. N. Leith and J. Upatnieks, "Reconstructed wavefronts and communication theory," *J. Opt. Soc. Am.* **52**, 1123–1130 (1962).

22. C. Ozkul and N. Anthore, "Residual aberration correction in far-field in-line holography using an auxillary off-axis hologram," *Appl. Opt.* **30**, 372–373 (1991).

23. D. H. Barnhart, R. J. Adrian, and G. C. Papen, "Phase-conjugate holographic system for high resolution particle-image velocimetry," *Appl. Opt.* **33,** 7159–7170 (1994).

24. G. Andersen, J. Munch, and P. Veitch, "Compact, holographic correction of aberrated telescopes," *Appl. Opt.* **36**, 1427–1432 (1997).

25. G. Andersen and R. J. Knize, "Holographically corrected microscope with a large working distance," *Appl. Opt.* **37**, 1849–1853 (1998).

26. T. W. Stone and B. J. Thompson, *Selected Papers on Holographic and Diffractive Lenses and Mirrors*, SPIE Press, Bellingham, WA (1991).

27. E. Fernández, C. García, I. Pascual, M. Ortuno, S. Gallego, and A. Beléndez, "Optimization of a thick polyvinyl alcohol-acrylamide photopolymer for data storage using a combination of angular and peristrophic holographic multiplexing," *Appl. Opt.* **45**, 7661–7666 (2006).

28. H. J. Caulfield, "Optically generated kinoforms," *Opt. Commun.* **4**, 201–202 (1971).

29. T. P. Janson, "Physical Optics at Physical Optics Corporation," in *Tribute to Emil Wolf*, T. P. Jannson, Ed., 115–139, SPIE Press, Bellingham, WA (2005).

30. E. Shamonina and L. Solymar, "Metamaterials: how the subject started," *Metamaterials* **1**, 12–18 (2007).

31. T. A. Leskova, A. A. Maradudin, E. E. García-Guerrero, and E. R. Méndez, "Structured surfaces as optical metamaterials," *Metamaterials* **1**, 19–39 (2007).

32. T. Kondo, S. Juodkazis, V. Mizeikis, H. Misawa, and S. Matsuo, "Holographic lithography of periodic two- and three-dimensional microstructures in photoresist SU-8," *Opt. Exp.* **14**, 7943–7953 (2006).

33. Z. V. Vardeny and M. Raikh, "Light localized on the lattice," *Nature* **446**, 37–38 (2007).

34. J. Feng, T. Okamoto, and S. Kawata, "Highly directional emission via coupled surface-plasmon tunneling from electroluminescence in organic light-emitting devices," *Appl. Phys. Lett.* **87,** 241109 (2005).

35. T. Okamoto, F. H'Dhili, and S. Kawata," Towards plasmonic band gap laser," *Appl. Phys. Lett.* **85**, 3968–3970 (2004).

36. J. Feng, T. Okamoto, and S. Kawata, "Enhancement of electroluminescence through a two-dimensional corrugated metal film by grating-induced surface-plasmon cross coupling," *Opt. Lett.* **30**, 2302–2304 (2005).

37. G. Zhou and F. S. Chau, "Three-dimensional photonic crystal by holographic contact lithography using a single diffraction mask," *Appl. Phys. Lett.* **90**, 181106 (2007).

38. Y. Lin, D. Rivera, and K. P. Chen, "Woodpile-type photonic crystals with orthorhombic or tetragonal symmetry formed through phase mask techniques," *Opt. Exp.* **14**, 887–892 (2006).

39. S. Shoji, R. P. Zaccaria, H.-B. Sun, and S. Kawata, "Multi-step multi-beam laser interference patterning of three-dimensional photonic lattices," *Opt. Exp.* **14**, 2309–2316 (2006).

40. Y. Zhong, L. Wu, H. Su, K. S. Wong, and H. Wang, "Fabrication of photonic crystals with tunable surface orientation by holographic lithography," *Opt. Exp.* **14**, 6837–6843 (2006).

41. X. S. Xie, M. Li, J. Guo, B. Liang, Z. X. Wang, A. Sinitskii, Y. Xiang, and J. Y. Zhou, "Phase manipulated multi-beam holographic lithography for tunable optical lattices," *Opt. Exp.* **15**, 7032–7037 (2007).

42. L. A. Peach, "Tasteful holograms adorn candy," *Laser Focus World* **35**, 50 (May, 1997).

43. A. Parker, *Seven Deadly Colours: The Genius of Nature's Palette and How it Eluded Darwin*, The Free Press, London (2005).

44. T. Jannson, I. Tengara, Y. Qiao, and G. Savant, "Lippmann-Bragg broadband holographic mirrors," *J. Opt. Soc. Am. A* **8**, 201–211 (1991).

45. J. P. Vigneron, V. Lousse, J.-F. Colomer, and N. Vigneron, "Natural layer-by-layer photonic structure in the scales of *Hoplia coerulea* (Coleoptera)," *Proc. SPIE* **6285**, 6285506 (2006).

46. J. P. Vigneron, V. Lousse, L. P. Biró, Z. Vértesy, and Z. Bálint, "Reflectance of topologically disordered photonic-crystal films," *Proc. SPIE* **5733**, 308–315 (2005).

Biographies

H. John Caulfield is the Distinguished Research Professor and director of the Conservative Optical Logic Devices (COLD) Project at Fisk University. He is also associated with several other universities and is Chief Technical Officer or Director of seven companies. He received his B.A. from Rice University in 1958 and his Ph.D. from Iowa State State in 1962, both in physics. He is author, coauthor, or editor of 14 books, 40 book chapters, 253 refereed journal papers, and numerous popular articles. He has received a number of major scientific awards. His main interests are cognitive science, soft computing, pattern recognition, and holography.

Chandra S. Vikram was a research professor at the COLD Project at FISK University. He held a B.Sc. from Agra University in physics, chemistry, and mathematics, a M.Sc. in physics, a M.Tech. in applied optics, and a Ph.D. in optics from the Indian Institute of Technology, Delhi. He authored or coauthored 3 books, 8 book chapters, and over 150 referred journal publications. His interests included several areas of modern optical science and technology, such as holography, speckle metrology, and optical logic. He received several scientific awards, and SPIE has created the annual Chandra S. Vikram Award in Optical Metrology to honor his work in the field.

In Memoriam:
Chandra S. Vikram
(1950–2007)

The last few years have been disastrous to our field. Great scientists such as Emmett Leith, Yuri Denisyuk, Steve Benton, Pal Greguss, Jacques Ludman, Roger Lessard, Steve Case, and my friend and colleague Chandra Vikram have died. I had the privilege of working closely with Chandra for perhaps a quarter of a century. He received many honors during his rather short life. Among them were the Dennis Gabor Award from SPIE and a distinguished alumnus award from the Indian Institute of Technology, Delhi. Chandra inspired those around him. At his funeral, many famous scientists from all over the United States took the time and ignored the expense to come to Huntsville to honor him one last time.

Chandra had a truly unique working style. When faced with a very difficult problem, he would study it, read articles on it, and then seemingly withdraw from it. A month or two later he would have the answer to the problem. In addition to possessing the "beautiful mind" of book and film, Chandra had a "beautiful subconscious," which was amazing to observe. We all miss him both as a scientist and as a friend.

H. John Caulfield

Chapter 19
Slow and Fast Light

Joseph E. Vornehm, Jr. and Robert W. Boyd
The Institute of Optics, University of Rochester, NY, USA

19.1 Introduction

In early 1999, a news article in the prestigious journal *Nature* led off with the announcement, "An experiment with atoms at nanokelvin temperatures has produced the remarkable observation of light pulses traveling at velocities of only 17 m/s." The review continued with the understatement, "Observation of light pulses propagating at a speed no faster than a swiftly moving bicycle... comes as a surprise."[1] These findings (and their review) marked the beginning of the current wave of interest in the field that has come to be called "slow light."[2]

When we refer to "the speed of light," we typically mean c, the phase velocity of light in a vacuum, or the speed of propagation of the phase fronts of monochromatic light. The phase fronts travel more slowly through a material, propagating at the speed c/n, where n is the index of refraction of the medium. However, this ordinary slowing of the *phase* velocity is *not* slow light. "Slow light" and "fast light" refer to changes in the *group* velocity of light in a medium.

A pulse of light can be decomposed mathematically into a group of monochromatic waves at slightly different frequencies, as in Fig. 19.1. In a dispersive material, these monochromatic waves travel at different speeds. When one views the propagation of the pulse as a whole, its apparent velocity depends on the extent of the spread of individual monochromatic-wave velocities. Each monochromatic wave travels at its own phase velocity, while the pulse travels at the group velocity.

Of course, the group velocity of a pulse of light is not a new concept. The field of slow and fast light has drawn on theory and developments from the work of Sommerfeld and Brillouin from 1907 to 1914,[3,4] experiments with early laser amplifiers in 1966,[5] and other work done through the end of the 20th century.[6,7] (For more of the history behind slow and fast light, see Refs. 8 and 9.)

19.1.1 Phase velocity

Consider the (complex) electric field of a monochromatic electromagnetic plane wave of amplitude E_0:

$$E(z,t) = E_0 e^{i(kz-\omega t)}. \tag{19.1}$$

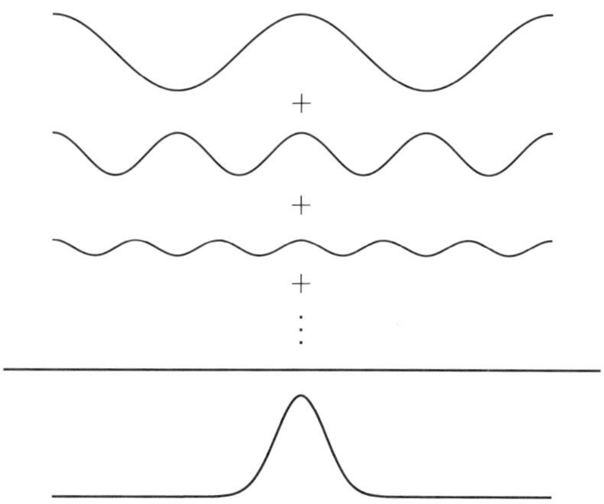

Figure 19.1 A pulse as the sum of its Fourier components. Each of the Fourier components (sinusoids, or monochromatic waves) that constitute the pulse must have the proper relative amplitude and phase in order to preserve the pulse shape. If the Fourier components become out of phase, the pulse will be distorted.

Here, ω is the angular frequency of the plane wave, k is the wavevector, z is the position in space, and t is the time. When we speak of the propagation of the plane wave, we mean the motion of the wave's phase fronts, or surfaces of constant phase. If we define

$$\phi = kz - \omega t \tag{19.2}$$

and rewrite the plane wave as

$$E(z,t) = E_0 e^{i\phi}, \tag{19.3}$$

we see that the phase fronts are defined by constant values of ϕ. For convenience, we observe the motion of the phase front located at the origin $z = 0$ at time $t = 0$ (such that $\phi = 0$). This motion is then defined by

$$kz - \omega t = 0. \tag{19.4}$$

To account for dispersion, the wavevector k and the refractive index of the medium n are written as functions of ω. We ignore absorption for the moment and write the wavevector as

$$k(\omega) = \frac{n(\omega)\omega}{c}. \tag{19.5}$$

By substituting Eq. (19.5) into Eq. (19.4), we find

$$z = \frac{\omega}{k(\omega)} t = \frac{c}{n(\omega)} t \equiv v_p t. \tag{19.6}$$

Thus, we have found the phase velocity v_p, or the speed of propagation of the phase front, to be

$$v_p(\omega) = \frac{\omega}{k(\omega)} = \frac{c}{n(\omega)}. \tag{19.7}$$

19.1.2 Group velocity

The propagation velocity of a pulse can be derived by requiring that the pulse shape remain constant under propagation, meaning its plane-wave Fourier components maintain their relative amplitudes and phases. Absorption needs to be included in the derivation, so we first define the complex wavevector $k_c(\omega)$ as

$$k_c(\omega) = k(\omega) + i\frac{\alpha(\omega)}{2} = \frac{n(\omega)\omega}{c} + i\frac{\alpha(\omega)}{2}. \tag{19.8}$$

Here, $n(\omega)$ is the (real) refractive index in the usual sense. The parameter $\alpha(\omega)$ is the linear intensity absorption per unit length experienced by a plane wave:

$$I(z) \propto |E|^2 = \left|E_0 e^{i\omega(nz/c-t)} e^{-(\alpha/2)z}\right|^2 = |E_0|^2 e^{-\alpha z}. \tag{19.9}$$

Assuming that the Fourier components are in a sufficiently narrow bandwidth $\Delta\omega$ around some center frequency (or carrier frequency) ω_0, we define $\omega = \omega_0 + \Delta\omega$ and expand the real and imaginary parts of $k_c(\omega)$ in a Taylor series about ω_0:

$$k(\omega) = k_0 + k_1\Delta\omega + \frac{1}{2}k_2\Delta\omega^2 + \frac{1}{6}k_3\Delta\omega^3 + \cdots = \sum_{m=0}^{\infty}\frac{k_m}{m!}(\Delta\omega)^m, \tag{19.10}$$

$$\alpha(\omega) = \alpha_0 + \alpha_1\Delta\omega + \frac{1}{2}\alpha_2\Delta\omega^2 + \frac{1}{6}\alpha_3\Delta\omega^3 + \cdots = \sum_{n=0}^{\infty}\frac{\alpha_n}{n!}(\Delta\omega)^n, \tag{19.11}$$

where we have defined

$$k_m \equiv \left.\frac{d^m k}{d\omega^m}\right|_{\omega=\omega_0}, \tag{19.12}$$

$$\alpha_n \equiv \left.\frac{d^n \alpha}{d\omega^n}\right|_{\omega=\omega_0}. \tag{19.13}$$

Let us now consider a pulse, broken down into a sum of plane-wave Fourier components as in Fig. 19.1. We write the Fourier components as

$$\begin{aligned} E(z,t) &= E_1 e^{i[k_c(\omega_1)z-\omega_1 t]} + E_2 e^{i[k_c(\omega_2)z-\omega_2 t]} + \cdots \\ &= E_1 e^{i[k(\omega_1)z-\omega_1 t]} e^{-\alpha(\omega_1)z/2} + E_2 e^{i[k(\omega_2)z-\omega_2 t]} e^{-\alpha(\omega_2)z/2} + \cdots \\ &= \sum_j E_j e^{-i\omega_j t}\exp[ik(\omega_j)z]\exp[-\alpha(\omega_j)z/2]. \end{aligned} \tag{19.14}$$

For each Fourier component, we define $\Delta\omega_j = \omega_j - \omega_0$. We then substitute Eqs. (19.10) and (19.11) into Eq. (19.14). Next, we take the ω_0, k_0, and α_0 terms outside the sums, and we take the k_1 term outside the sum over m, giving

$$E(z,t) = \sum_j E_j e^{-i(\omega_0+\Delta\omega_j)t} \exp\left[iz\sum_{m=0}^{\infty}\frac{k_m}{m!}(\Delta\omega_j)^m\right]$$
$$\times \exp\left[-\frac{z}{2}\sum_{n=0}^{\infty}\frac{\alpha_n}{n!}(\Delta\omega_j)^n\right]$$
$$= e^{-\alpha_0 z/2}e^{i(k_0 z-\omega_0 t)}\sum_j E_j e^{-i\Delta\omega_j(t-k_1 z)}$$
$$\times \exp\left[iz\sum_{m=2}^{\infty}\frac{k_m}{m!}(\Delta\omega_j)^m\right]\exp\left[-\frac{z}{2}\sum_{n=1}^{\infty}\frac{\alpha_n}{n!}(\Delta\omega_j)^n\right]. \quad (19.15)$$

In order for the pulse to propagate unchanged through a medium, the relative amplitudes and phases of the plane-wave Fourier components must remain constant. Propagation can at most be allowed to scale the entire pulse by a single complex constant. More specifically, if the pulse propagates with unchanged shape from the origin to the point z in a time t, we can write the propagation requirement formally as

$$E(z,t) = re^{i\phi}E(0,0). \quad (19.16)$$

Here, r and ϕ are free variables describing the change in amplitude and phase during propagation. By expanding the right-hand side of Eq. (19.16) according to the definition in Eq. (19.14), we find that

$$re^{i\phi}E(0,0) = re^{i\phi}\sum_j E_j e^{-i\omega_j 0}\exp[ik(\omega_j)\cdot 0]\exp[-\alpha(\omega)\cdot 0/2]$$
$$= re^{i\phi}\sum_j E_j. \quad (19.17)$$

In order for the pulse shape to remain unchanged upon propagation, Eq. (19.16) must be satisfied, meaning Eqs. (19.15) and (19.17) must be equal. In general, Eq. (19.16) can be satisfied only when a number of conditions hold true:

$$r = e^{-\alpha_0 z/2}, \quad (19.18)$$
$$e^{i\phi} = e^{i[k_0 z-\omega_0 t]}, \quad (19.19)$$
$$t = k_1 z, \quad (19.20)$$
$$k_2 = k_3 = \cdots = 0, \qquad \text{and} \quad (19.21)$$
$$\alpha_1 = \alpha_2 = \alpha_3 = \cdots = 0. \quad (19.22)$$

Equations (19.18) and (19.19) are automatically satisfied because g and ϕ are free variables that define the amplitude and phase change acquired by the pulse during propagation. It is shown below that Eq. (19.20) defines the group delay and group velocity, while Eqs. (19.21) and (19.22) have important consequences for pulse propagation and distortion.

The group delay τ_g, or the time required for the pulse to propagate a distance z, is given by Eq. (19.20) as

$$\tau_g \equiv t = k_1 z = z\frac{dk}{d\omega}. \tag{19.23}$$

The derivative is evaluated at $\omega = \omega_0$; we have dropped that notation for convenience. The group velocity v_g, or the speed at which the pulse envelope propagates, can then be defined as

$$v_g \equiv \frac{z}{\tau_g} = \left(\frac{dk}{d\omega}\right)^{-1} = \left[\frac{d}{d\omega}\left(\frac{n\omega}{c}\right)\right]^{-1} = c\left(n + \omega\frac{dn}{d\omega}\right)^{-1} = \frac{c}{n_g}, \tag{19.24}$$

where the quantity

$$n_g \equiv n(\omega) + \omega\frac{dn}{d\omega} \tag{19.25}$$

is called the group index, in analogy to Eq. (19.7). The first term in Eq. (19.25) is the refractive index, which may sometimes be termed the phase index (for contrast with the group index). The second term is a measure of the dispersion. In the literature, a positive value of $dn/d\omega$ is referred to as normal dispersion, and a negative value is called anomalous dispersion. Note that under normal circumstances, the second term in Eq. (19.25) is very small, so $n_g \approx n$ and $v_g \approx v_p$. The promise of slow and fast light lies in the fact that, because ω is so large in the optical regime, any method that increases the magnitude of the dispersion even slightly can produce astonishingly large or small group velocities.

Equations (19.21) and (19.22) have important consequences for slow light. In order for a general pulse shape to propagate wholly unchanged, second-order dispersion (group velocity dispersion, or GVD) and higher-order dispersion terms must be zero, and absorption must be spectrally flat. Any higher-order dispersion or absorption terms will introduce some amount of distortion into the pulse. As we shall see in Sec. 19.2.1, the Kramers–Kronig relations dictate that Eqs. (19.21) and (19.22) can never be precisely satisfied in real-world media. A great deal of experimental research in slow light is aimed at reducing the effects of pulse distortion.

The statement in Sec. 19.1 that the pulse travels at the group velocity does not give a complete picture. The pulse velocity is equal to the group velocity when distortion is minimal and the pulse more or less maintains its shape. However, severe distortion can cause the pulse to "break up" into a shape that has several peaks. In such situations, the group velocity is still defined according to Eq. (19.24), but there is no longer a single pulse for which to define a meaningful pulse velocity.

19.1.3 Slow light, fast light, backward light, stopped light

Slow light, then, is light that travels at an exceptionally slow group velocity, or in a medium with an exceptionally large group index.[2] Traditionally, slow light is defined as having $v_g \ll v_p$, which occurs when $n_g \gg n_p$. Slow-light techniques aim to increase n_g by increasing dispersion $(dn/d\omega)$. However, dispersion need not be positive. It can be zero, negative, or (nearly) infinite, leading to several other regimes of operation.

Fast light is light for which $v_g > c$. Fast light occurs in media for which $0 \leq n_g < 1$. Such propagation is also called "superluminal," owing to the fact that the group velocity is greater than the speed of light in vacuum. The limit of fast light is zero group index, or infinite effective group velocity, in the so-called critically anomalous dispersion regime;[10] in such a regime, the pulse appears to leave the medium at exactly the same time as it enters. It should be noted that fast-light propagation is entirely within the bounds of causality and special relativity.[11] No information can be communicated superluminally; for more discussion on this point, see Sec. 19.4.4. Fast light may also be defined as occurring when $v_g > v_p$ and $0 \leq n_g < n$.

Backward light occurs in media for which $n_g < 0$. A pulse travels in such a medium with $v_g < 0$, giving an apparent backward propagation of the pulse within the medium.[7,12,13] In this regime, shown in Fig. 19.2, the pulse appears to leave the medium before it enters, and the peak of the pulse within the medium propagates backward from the time the outbound pulse leaves to the time the inbound pulse

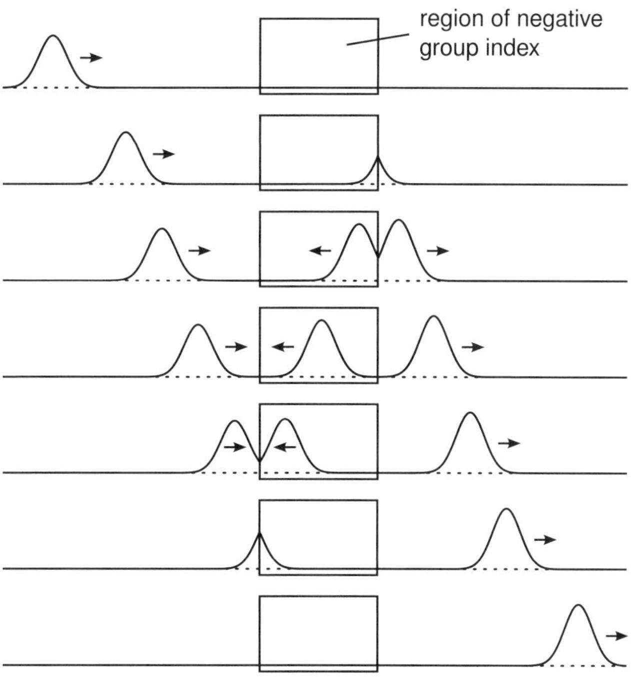

Figure 19.2 Backward pulse propagation in a medium with $n_g < 0$. Each plot shows a snapshot in time, with time advancing downward.

enters. Energy still flows in the forward direction.[14] Causality is not violated here, either; the effect can be seen as a form of pulse reshaping, selectively amplifying the long leading edge of a Gaussian pulse and attenuating the original peak.

Stopped light (or stored light) is related to slow light. Stopped-light techniques are used to "stop" or "trap" a pulse of light inside a medium for some amount of time. The pulse is slowed as it enters the medium, and then the medium is altered or stimulated so that the pulse is stored for a short time and then retrieved. The actual storage may happen by inducing the group velocity to become zero (and the group index to become infinite) or by some other means, such as by mapping the pulse into the spin coherence of a coherently prepared atomic medium.[15,16]

These different regimes of operation are all concerned with inducing strong dispersion and thus use very similar techniques. They are often all grouped together under the heading of "slow light" or "slow and fast light."

19.2 Slow Light Based on Material Resonances

19.2.1 Susceptibility and the Kramers–Kronig relations

Both the refractive index $n(\omega)$ and the absorption coefficient $\alpha(\omega)$ of a medium have their origin in the susceptibility of the medium $\chi(\omega)$. When an electric field is applied to a medium, the charged particles in the medium (the electrons and protons) shift their positions in response to the field. This shift in the positions of the charges creates an additional electric field. The electric dipole moment per unit volume of the material is known as the polarization density P, or simply the polarization. Some materials are more susceptible than others to being polarized by an incident electric field. The degree to which the material may be polarized by a given electric field is known as the electric susceptibility χ and is defined by

$$P = \epsilon_0 \chi E, \tag{19.26}$$

where E is the strength of the electric field and ϵ_0 is the permittivity of free space. The permittivity ϵ of the medium is defined as

$$\epsilon = \epsilon_0 \epsilon_r = \epsilon_0(1+\chi), \tag{19.27}$$

and the refractive index and absorption coefficient are defined respectively as

$$n(\omega) = \mathrm{Re}\left\{\sqrt{\epsilon_r}\right\} = \mathrm{Re}\left\{\sqrt{1+\chi(\omega)}\right\}, \tag{19.28a}$$

$$\alpha(\omega) = \frac{2\omega}{c}\,\mathrm{Im}\left\{\sqrt{\epsilon_r}\right\} = \frac{2\omega}{c}\,\mathrm{Im}\left\{\sqrt{1+\chi(\omega)}\right\}. \tag{19.28b}$$

When $\chi(\omega)$ is small, such as for dilute gases, Eqs. (19.28) simplify to

$$n(\omega) \approx 1 + \frac{1}{2}\,\mathrm{Re}\left\{\chi(\omega)\right\}, \tag{19.29a}$$

$$\alpha(\omega) \approx \frac{\omega}{c}\,\mathrm{Im}\left\{\chi(\omega)\right\}. \tag{19.29b}$$

One can thus think of the real part of the susceptibility as corresponding to the refractive index $n(\omega)$ and the imaginary part as corresponding to the absorption coefficient $\alpha(\omega)$. The correspondence is a useful conceptual device, although one must remember that Eqs. (19.29) are not strictly true for media with a strong electromagnetic response (a large susceptibility). In such cases, the more general forms in Eqs. (19.28) should be used.

The medium's electromagnetic response must be causal (it must obey causality). Any change in P at time t must be caused by changes in E that happen *before* time t. In other words, the cause must precede the effect. This may seem obvious, but the causality requirement has important consequences: It can be shown that the electromagnetic susceptibility of any causal medium also obeys the Kramers–Kronig relations

$$\mathrm{Im}\{\chi(\omega)\} = \frac{-2\omega}{\pi}\int_0^{\infty}\frac{\mathrm{Re}\{\chi(\omega')\}}{\omega'^2-\omega^2}\,d\omega', \tag{19.30a}$$

$$\mathrm{Re}\{\chi(\omega)\} = \frac{2}{\pi}\int_0^{\infty}\frac{\omega'\,\mathrm{Im}\{\chi(\omega')\}}{\omega'^2-\omega^2}\,d\omega'. \tag{19.30b}$$

These relations lead to several important results. First, any material that exhibits absorption must also possess dispersion. Conversely, any dispersive medium must also possess some spectral variation in absorption, meaning that $d\alpha/d\omega$ (and α_1) cannot be zero for all ω. Thus, distortion is ever-present in slow-light media. Additionally, the Kramers–Kronig relations dictate that $n(\omega)$ will be nearly linear in the neighborhood of a smooth peak or valley in the absorption spectrum. The value of $dn/d\omega$ (and k_1) can be either positive or negative, according to the concavity of the peak. As explained in the next section, one important slow-light technique is to induce sharp changes in the absorption spectrum. (For further discussion of the Kramers–Kronig relations, see Sec 1.7 in Ref. 17.)

19.2.2 Resonance features in materials

Many of the spectral features of a material's optical response come from material resonances. In many instances, the motion of bound charged particles in a material (such as electrons bound to atoms or molecules, or nuclei within a crystal lattice) is constrained to the form of a damped harmonic oscillator, similar to a mass on a spring. In this model, often called the Lorentz model, the charged particle tends to oscillate at a resonance frequency ω_r. The equation of motion of the charged particle can then be written as

$$\frac{d^2x}{dt^2} + 2\gamma_r\frac{dx}{dt} + \omega_r^2 x = \frac{eE}{m}, \tag{19.31}$$

where x is the particle's displacement from its equilibrium position, e is the charge carried by the particle ($e < 0$ for an electron), E is the magnitude of the applied electric field, m is the charged particle's mass, and γ_r is a damping coefficient. It can be shown that, under these conditions, the susceptibility χ of the medium due to the resonance has the form

$$\chi(\omega) \propto \frac{1}{\omega_r^2 - \omega^2 - 2i\omega\gamma_r}, \tag{19.32}$$

which gives the absorption spectrum $\alpha(\omega)$ a Lorentzian line shape centered at ω_r with linewidth γ_r. The Kramers–Kronig relations dictate that $n(\omega)$ and $\alpha(\omega)$ must have the relationship shown in Fig. 19.3(a). (For further discussion of material resonances, see Secs. 1.4 and 3.5 of Ref. 17 or Sec. 5.5 of Ref. 18.)

Of course, material systems have many different resonances, each with its own center frequency, linewidth, and relative strength. Beyond natural material resonances, resonances of a similar Lorentzian form can also be induced by certain optical processes. Lasing, for example, consists of creating an inverted population, such that a certain atomic or molecular transition (a resonance) experiences gain. In that case, ω_r is the frequency of the lasing transition, γ_r is its linewidth, and the value of $\alpha(\omega_r)$ is negative, indicating gain rather than absorption. The Kramers–Kronig relations then dictate essentially a reversed shape for $n(\omega)$.

It is easy to see from Fig. 19.3 how slow and fast light can be achieved in such systems. Near $\omega = \omega_r$, $n(\omega)$ varies nearly linearly with ω, making $d^2n/d\omega^2$ nearly zero. The absorption feature is only slightly dependent on ω near the resonance.

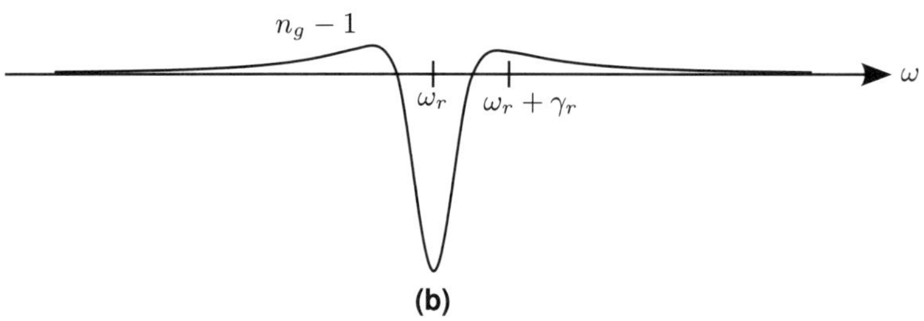

Figure 19.3 Dispersive features of an absorption resonance. (a) The absorption α and dispersion $n - 1$ of an absorption resonance at frequency ω_r with linewidth γ_r. Near the point $\omega = \omega_r$ (but not precisely at that point), α reaches a peak and $n - 1$ crosses the axis. (b) Group index of the same absorption resonance, using the same ω axis. Note that fast-light behavior ($|n_g| > c$) is exhibited near $\omega = \omega_r$, while slow-light behavior is exhibited around $\omega = \omega_r \pm \gamma_r$.

However, $|d\alpha/d\omega|$ becomes very large as the absorption drops off away from ω_r. One must be careful that the pulse spectrum does not extend too far into the region of large $|d\alpha/d\omega|$. Notice also that fast light occurs in regions of strong absorption; the fast-light experimenter may choose to deal with this absorption, to mitigate it somehow, or to create fast-light features using any of several other means that avoid absorption.

19.2.3 Spatial compression

It is clear that the reduced group velocity in slow light leads to spatial compression of the pulse by a factor equal to the group index. If a pulse of duration τ decelerates from speed c to speed c/n_g, its length L must likewise decrease from $L \propto c\tau$ to $L' = L/n_g$. Conservation of energy then dictates that if the pulse energy was distributed over length L but is compressed down to length L', the energy density u must increase by the same factor to $u' = n_g u$.

Interestingly, the intensity of a pulse $I = uv_g$ is unchanged upon entering a slow-light medium, because the increase in u is canceled exactly by the decrease in v_g. Likewise, the electric field strength $E \propto \sqrt{I}$ is unaffected by changes in the group velocity. Thus, although the pulse energy is spatially compressed, its peak electric field strength is unchanged.[8,19] In contrast to this result, slow light achieved in coupled resonators or photonic crystals is accompanied by an increase in electric field strength.[20]

19.2.4 Two-level and three-level models

Optical interactions with atoms (or molecules or other systems) can be described quantum-mechanically using a two- or three-level model. In the two-level model shown in Fig. 19.4(a), the incident light at some frequency ω is assumed to be resonant (or nearly resonant) with the atomic transition between energy levels $|1\rangle$ and $|2\rangle$. (By resonant, we mean that the photon energy $E = \hbar\omega$ is the same as the energy difference between states $|1\rangle$ and $|2\rangle$, and the optical transition is not forbidden by other issues such as parity.) Formally, the incident light has some probability of causing an atomic transition between any two energy levels; however, if the light is nearly resonant with the $|1\rangle$–$|2\rangle$ transition, the probabilities of other transitions become vanishingly small, and all other energy levels can be neglected. Thus, the atomic model is effectively reduced to two levels. Similarly, the three-level model shown in Fig. 19.4(b) is applicable when light fields are applied at two different frequencies ω_p and ω_c, which are respectively resonant (or nearly resonant) with the $|1\rangle$–$|3\rangle$ and $|2\rangle$–$|3\rangle$ transitions. (Note that the $|1\rangle$–$|2\rangle$ transition is typically forbidden by dipole selection rules.) The three-level model is often useful in experiments where the ω_c field is a strong coupling field or pump field that induces some optical effect and the ω_p field is a weak probe field that measures or "sees" the effect.

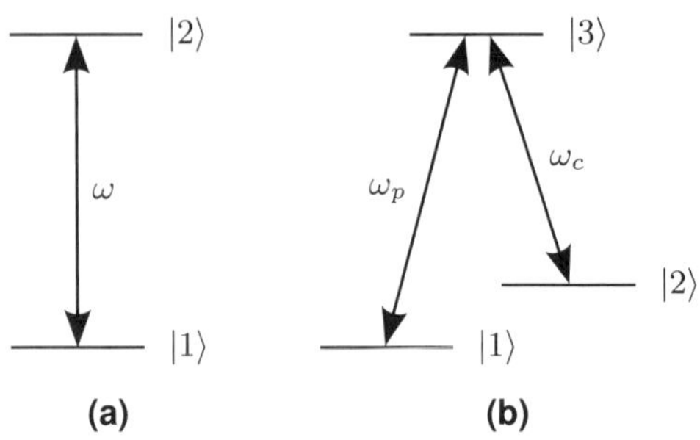

Figure 19.4 Energy-level models of an atom (or another quantum-mechanical system). The model can be reduced to include only those energy levels involved in resonant absorption or emission transitions (or nearly resonant transitions). All other levels can be neglected. (a) Two-level model. Incoming light at frequency ω is nearly resonant with the transition between energy levels $|1\rangle$ and $|2\rangle$. (b) Three-level model. Incoming light at frequency ω_p is nearly resonant with the transition between energy levels $|1\rangle$ and $|2\rangle$, and light at frequency ω_c is nearly resonant with the transition between energy levels $|3\rangle$ and $|2\rangle$. Typically, ω_c is a strong coupling beam or pump beam, while ω_p is a weaker probe beam.

19.2.5 Electromagnetically induced transparency (EIT)

Electromagnetically induced transparency (EIT) allows a very narrow window of increased transparency (a spectral hole) to be introduced in an absorption resonance.[21] The narrowness and depth of the spectral hole lead to a large group index, producing slow light. A three-level Λ-type model of an atom is shown in Fig. 19.4(b). The atom starts in state $|1\rangle$, the ground state. A strong pump field (also called a coupling field) at frequency ω_c is applied to the $|2\rangle$–$|3\rangle$ transition in such a way that a coherence is introduced between the $|1\rangle$–$|3\rangle$ and $|2\rangle$–$|3\rangle$ transitions. If a weak probe at frequency ω_p is later applied to the $|1\rangle$–$|3\rangle$ transition, it will undergo little or no absorption, whereas it would ordinarily be absorbed readily. The hole created in the probe absorption spectrum at ω_p is very narrow, so by the Kramers–Kronig relations, the medium has a large $dn/d\omega$ and a large group index at the same frequency.

As EIT is a quantum interference phenomenon, it is crucial that the EIT medium be maintained in an environment that preserves quantum coherence.[21] Typically, the medium is kept at cryogenic temperatures or, for vapors, at low pressure. A number of other requirements must also be met, and achieving EIT can be quite difficult experimentally. Despite these restrictions, EIT has been a popular experimental method for achieving slow light. The experiment that sparked recent interest in slow light was carried out using EIT in a Bose-Einstein condensate (BEC).[2] EIT-based slow-light experiments have been carried out in hot alkali vapor,[16] cold alkali vapor,[22] crystals,[15] semiconductor quantum wells,[23–25] and vapor confined within photonic band-gap fiber[26] and have been proposed in doped optical fiber[27] and semiconductor quantum dots.[28] Certain transparency effects similar to EIT have been predicted in resonator systems[29–32] and plasmas.[33–35]

EIT is also associated with an enhancement of the medium's optical nonlinearity.[19] When the fields applied to a resonant medium are tuned to a resonance, the optical nonlinearity of the medium reaches a local maximum. In the absence of EIT, linear absorption also reaches a local maximum, making the nonlinearity unusable. EIT allows access to these resonant nonlinearities that would otherwise be precluded by absorption. It would be incorrect to say that the nonlinearity enhancement in EIT is caused by the slow-light effect; however, the slow-light effect and the nonlinearity enhancement in EIT are inseparable.

19.2.6 Coherent population oscillation (CPO)

Coherent population oscillation (CPO) occurs in a two-level atom when a pump beam at frequency ω and a probe beam at frequency $\omega + \delta$ are applied to the same transition, such as the $|1\rangle$–$|2\rangle$ transition in Fig. 19.4(a). If ω and $\omega + \delta$ both lie within the natural linewidth $1/T_1$ of the transition, a portion of the atomic population oscillates between levels $|1\rangle$ and $|2\rangle$ at the beat frequency δ. The oscillating population produces a narrow hole in the absorption line centered at frequency ω. By the Kramers–Kronig relations, the narrow spectral hole results in a rapid index variation, producing slow light. (Of course, if the atomic population is initially in the excited state, the CPO effect produces a hole in the gain spectrum, giving a fast-light effect.)[36,37]

Equivalently, CPO can be viewed as a time-dependent saturable absorption or saturable gain effect. As part of the population moves from level $|1\rangle$ to level $|2\rangle$, that part of the population is unavailable to participate in the absorption process, leading to reduced absorption. A slow-light effect is achieved through pulse reshaping: The leading edge of the pulse is selectively attenuated, producing an effective delay of the peak. As part of the population moves from level $|2\rangle$ to level $|1\rangle$, the corresponding gain saturation occurs, giving a fast-light effect by selectively amplifying the leading edge of the peak. Note that the optimal pulse bandwidth is of the order of δ, so approximately one complete population cycle occurs.[38–41]

CPO as a slow- and fast-light method has several advantages over EIT. CPO is much easier to achieve at room temperature. Both methods can have similarly narrow linewidths, resulting in sharp dispersion and extreme group velocities. Additionally, CPO slow- and fast-light effects may be achieved with only a single beam by inducing a sinusoidal amplitude modulation at frequency δ on a pump beam at ω, or even a single pulse with approximate bandwidth δ. However, CPO typically suffers from a higher degree of residual absorption than EIT. Also, the bandwidth of CPO is limited to the natural linewidth $1/T_1$ of the atomic transition. CPO has been achieved in a variety of experimental setups, including in crystals,[37,42] erbium ions as dopants in an optical fiber,[14] semiconductor waveguides,[43] quantum dots in semiconductor waveguides,[44] and quantum dot semiconductor optical amplifiers (QD-SOAs).[45] CPO has also been used to generate slow light at cryogenic temperatures (10 K) in a semiconductor quantum well structure.[46]

19.2.7 Stimulated Brillouin and Raman scattering

Both stimulated Brillouin scattering (SBS) and stimulated Raman scattering (SRS) involve the scattering of light off a vibrational wave. If a pump field at frequency ω scatters off a vibrational wave at frequency Ω, scattered fields are generated at the Stokes frequency $\omega_S = \omega - \Omega$. The frequency Ω can be up to several gigahertz for SBS and up to several terahertz for SRS. If a probe field is applied to the medium at the Stokes frequency, it will experience gain as energy is scattered from the pump into the Stokes frequency. By the Kramers–Kronig relations, the SBS or SRS gain line induces a slow-light dispersion curve in the vicinity of the probe (Stokes) frequency.

Stimulated Brillouin scattering is based on the electrostrictive effect, whereby materials experience a slight change in density (and hence refractive index) in response to an applied optical field. In SBS, the pump and probe fields beat together and induce a traveling density modulation (a pressure wave or acoustic wave) in the medium at frequency Ω. Energy from the pump wave then scatters off the acoustic wave and into the probe field (since $\omega_S = \omega - \Omega$), which further enhances the acoustic wave, and so forth, creating a positive feedback loop. Absorption (or loss) increases at the anti-Stokes frequency $\omega_a = \omega + \Omega$. (If no Stokes field is applied initially, one can be generated by the scattering of the pump field off a thermal phonon at Ω. For further discussion of SBS, see Chaps. 8 and 9 of Ref. 17.) SBS can be readily induced in optical fibers.[47,48] Since SBS can be controlled via the pump, it can be tailored to minimize distortion[49] and to optimize gain bandwidth.[50]

Stimulated Raman scattering works by inducing molecular transitions between vibrational sublevels. As shown in Fig. 19.5, molecular vibrations at frequency Ω are excited by the beating of the pump and probe fields; these molecular vibrations scatter some energy from the pump into the Stokes sideband $\omega_S = \omega - \Omega$, and the probe field experiences gain. In a normal thermal distribution at room temperature,

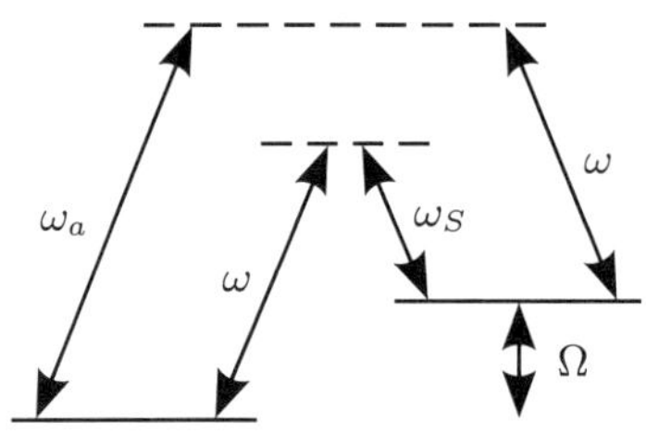

Figure 19.5 A typical level scheme for stimulated Raman scattering (SRS). The pump laser at frequency ω couples molecules (or atoms) from the vibrational ground state to the vibrationally excited state, emitting Stokes photons at frequency $\omega_S = \omega - \Omega$ and producing gain at the Stokes frequency (inner pair of slanted arrows). In the presence of the pump (and in the absence of four-wave mixing effects), anti-Stokes photons at frequency $\omega_a = \omega + \Omega$ can be absorbed by a stimulated Raman transition in ground-state molecules, resulting in loss at the anti-Stokes frequency (outer pair of slanted arrows). Note that the two upper levels are so-called virtual levels and need not correspond to real energy levels.

there are far more molecules in the ground state than in the excited state; these ground-state molecules can now absorb light in the anti-Stokes sideband $\omega_a = \omega + \Omega$, due to the presence of the pump to complete the stimulated Raman transition to the excited state. Thus, the anti-Stokes field, if present, experiences loss. (However, there is also a four-wave mixing component to SRS that can alter this balance. For further discussion of SRS, see Chapter 10 of Ref. 17.) SRS can be observed not only in molecules but also in atoms and crystals, i.e., in any system that can be vibrationally excited. Slow light based on SRS gain has been observed in both solids[51,52] and optical fibers.[53]

19.2.8 Other resonance-based phenomena

A number of other slow-light techniques based on resonance phenomena have been implemented successfully. Picosecond pulses were delayed by as many as 80 pulse widths by operating at the center frequency between two absorption lines (hyperfine ground states) of cesium.[54] Slow light was achieved using the gain of a vertical-cavity surface-emitting laser (VCSEL) configured as a semiconductor optical amplifier (SOA) rather than as a laser.[55] Slow light in semiconductors has also been achieved using a number of different mechanisms, including several excitonic mechanisms.[56–58] For more details on semiconductor effects, see Refs. 59 and 60 and their references.

19.3 Slow Light Based on Material Structure

19.3.1 Waveguide dispersion

In an optical fiber, a fraction of the energy in a guided electromagnetic mode propagates in the core, and the remainder propagates in the cladding. The effective refractive index of the mode depends on this fraction. For different wavelengths, the fraction changes, producing an index variation with frequency. This dispersion is known as waveguide dispersion or intramodal dispersion. Additionally, each mode of a multimode waveguide has its own group velocity; if a pulse coupled into a multimode waveguide propagates in several modes with different group velocities, intermodal dispersion results. For further discussion on dispersion in waveguides, see Chaps. 8 and 9 of Ref. 18 or Sec. 1.5 of Ref. 61.

Typical optical fibers do not have enough dispersion to be interesting for slow-light purposes. However, novel waveguides such as coupled-resonator structures can provide greatly enhanced waveguide dispersion (see Sec. 19.3.2). Semiconductor quantum well and quantum dot structures have also been used for slow-light experiments.[43,57]

19.3.2 Coupled-resonator structures

Slow-light effects have also been explored in coupled-resonator structures, alternately called coupled-resonator optical waveguides (CROWs) or coupled-cavity waveguides (CCWs). Examples of such structures are shown in Fig. 19.6. Low group velocities are observed in the propagation of light across the CROW, as a result of weak coupling and feedback between the resonators.[62] Here is a conceptual model of how the device works: Light couples evanescently into the first resonator. As the light resonates there, it couples evanescently into the second resonator, where it also resonates. It then couples evanescently into the third resonator, and so forth, until it has "leaked" across the entire waveguide.[63]

Photonic crystals (PCs or PhCs) are formed by introducing periodic refractive index changes in a dielectric medium. Because of the periodic index modulation, light within certain wavelength bands is unable to propagate in the photonic crystal. In analogy to semiconductor crystals, these bands are called forbidden bands or photonic band gaps. Often, photonic crystals are made by drilling rectangularly spaced air holes into a dielectric, with the index contrast between air and the dielectric providing the periodic index modulation. One-, two-, and three-dimensional photonic crystals can be formed in such a way. By convention, however, the term "photonic crystal" is typically reserved for two- and three-dimensional structures with high index contrast. Two-dimensional photonic crystals are perhaps more common since three-dimensional photonic crystals are quite difficult to fabricate.

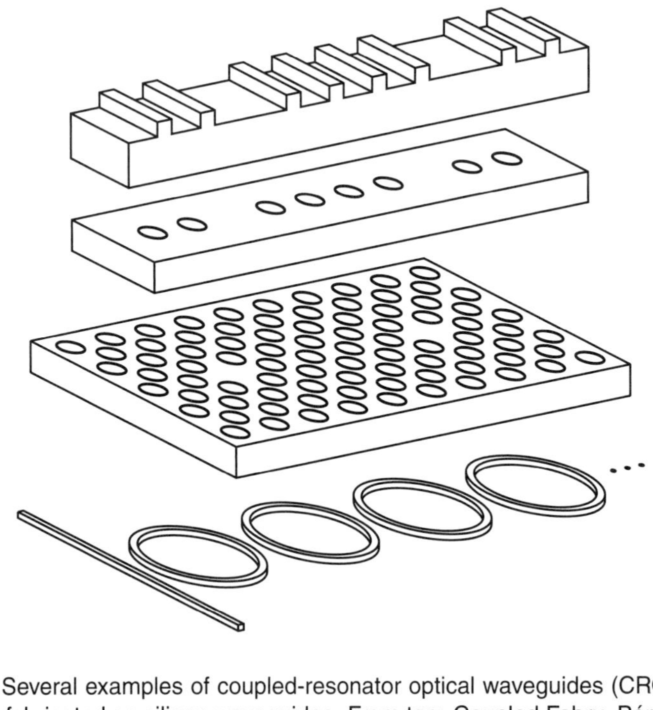

Figure 19.6 Several examples of coupled-resonator optical waveguides (CROWs). All devices can be fabricated as silicon waveguides. From top: Coupled Fabry–Pérot resonators, coupled one-dimensional photonic crystal defect resonators, coupled two-dimensional photonic crystal defect resonators, coupled ring resonators. (Adapted from Ref. 63; used by permission.)

A small defect introduced in the photonic crystal lattice will allow light to propagate in the vicinity of the defect, creating a resonator (see Fig. 19.6). If a periodic series of defects is introduced in the photonic crystal, the resonators can couple evanescently, forming a photonic crystal defect waveguide, another form of CROW.[63] Photonic crystals are particularly versatile, since they may be used to design many different optical devices and may be fabricated out of virtually any dielectric media. Photonic crystal devices may be created in highly nonlinear optical media as well. For example, the air holes in a photonic crystal lattice may be filled with a fluid having a high nonlinearity.[20]

In CROW devices, the electric field is enhanced within the resonators, leading to an enhancement of the optical nonlinearity.[20,63,64] Nonlinear effects typically depend on the strength of the incident electric field raised to some power. In a resonator, the resonating electric field builds up, leading to an enhancement of the nonlinear effect. Coupled-resonator optical waveguides typically see a nonlinearity enhancement that scales as the square of the slowing factor.[64]

Slow light has also been explored in certain optical filters, including fiber Bragg gratings[65] and Moiré fiber gratings.[66] These can be thought of conceptually (but not rigorously) as a series of coupled resonators. The dispersion and slow-light effects in optical filters are similar to those of coupled-resonator structures.

19.3.3 Band-edge dispersion

Photonic band-gap materials can be used to achieve slow light using another effect, one which they have in common with semiconductors. In the band gap, electromagnetic waves cannot propagate. Steep dispersion exists near the band edges, and precisely at the band edge, the group velocity goes to zero. At frequencies just outside the band gap (i.e., just inside the transmission band), light propagates at extremely slow group velocities.[67–71]

19.4 Additional Considerations

19.4.1 Distortion mitigation

The Kramers–Kronig relations dictate that any slow-light system will cause some degree of pulse distortion, as discussed in Sec. 19.2.1. In its simplest form, distortion takes the form of pulse broadening or compression, while more complicated forms of distortion can lead to pulse break-up. Workers in the field of slow light have developed a number of techniques for minimizing pulse distortion. For instance, Wang and coworkers[72] utilized the following scheme: Consider two resonances of the type shown in Fig. 19.3(a), spaced far apart. The refractive index will follow a smooth line between the resonances, producing a nearly linear refractive index profile (and hence a nearly constant group index) in the region directly between the two resonances. Such a configuration has since been used successfully by others.[54,73,74] One group suggested inducing two nearly overlapping resonances, which sum to produce a spectral region of nearly flat absorption and hence nearly flat group velocity.[49] Another group proposed doing the same with three resonances.[75] In coherent population oscillation, adding a continuous-wave background to the pulse can balance the competing effects of gain recovery and pulse spectrum broadening.[76] Many further techniques have been suggested.

19.4.2 Figures of merit

Several figures of merit are in common use among workers in slow light. The group index n_g may be quoted, either by itself or as the slowing factor $S = c/v_g$.[77] The most common figure of merit is simply the group delay τ_g, the time delay induced by propagation through the slow-light medium. This is nearly always the experimental quantity that is measured directly, so it is simple to report. However, it is generally easier to produce longer delays for longer pulses. Thus, a more meaningful measure is often the fractional delay, or the delay normalized by the pulse width.[78] Fractional delay coincides more closely with the particular application of slow-light delay lines and slow-light buffers for optical networking; fractional delay, then, becomes a measure of the number of bits that can be stored by such a delay line. Fast-light systems may be evaluated in terms of fractional advancement, or negative fractional delay. To offer a truer estimate of the technological value of experimental results, fractional delay is often quoted along with pulse width. Perhaps the most useful single figure of merit for optical delay lines is the delay–bandwidth product (DBP), which is also equal to the maximum possible fractional delay in a given slow-light system.[79] The delay–bandwidth product must also be quoted with the bit rate to be a definitive performance measure.

Many more figures of merit have been defined; indeed, the usefulness of a figure of merit depends on the merits it measures, which in turn depend on the intended application. The maximum possible delay can be represented in other ways, such as the length of a waveguide required to achieve a given time delay[71] or the ratio of a quantum memory's maximum storage time to the input pulse length.[80] Other figures of merit often include some measure of the absorption experienced by the pulse, such as the ratio of the group index or the delay–bandwidth product to the absorption coefficient,[81,82] or the time a signal can propagate in a slow-light buffer before needing regeneration or amplification.[58] Pulse distortion can be measured in several different ways, including the input–output pulse width ratio,[76] degree of dispersion near an absorption feature or a band edge,[71] or group velocity dispersion (GVD). The effects of pulse distortion on a telecommunications system are often the ultimate concern, so some experimenters use commercial telecommunications test equipment to test the bit-error rate (BER)[83] or the eye opening.[84]

19.4.3 Theoretical limits of slow and fast light

The most general theoretical limits of the performance of slow-light systems were already mentioned in Sec. 19.1.2: Group velocity dispersion k_2, frequency-dependent absorption α_1, and higher-order dispersion and absorption terms must be sufficiently small that the pulse is not distorted too much (though the degree of acceptable distortion is often application dependent). Generally, total linear absorption αL must also be sufficiently small that the signal can be detected. More specific limits than these depend on the particular slow-light technique in question. Some results are quoted here without further comment.

For many slow-light techniques, group velocity and bandwidth are proportional, requiring a trade-off between the two parameters.[82,85] Under many circumstances, the minimum spatial extent occupied by a single optical bit in a slow-light medium is roughly one vacuum wavelength.[82] Slow light using electromagnetically induced transparency (EIT) or coherent population oscillation (CPO) is most often limited by residual absorption on line center α_0 and by frequency-dependent absorption (α_1 and higher terms).[78,86] The excited state lifetime limits fractional delay in CPO in the short-pulse limit.[87] Group delay in EIT is partly limited by atomic collisions[88] and by nonlinear effects.[89] In slow-light systems using stimulated Brillouin scattering (SBS) in optical fibers, there is a trade-off between increased bandwidth and reduced pulse distortion.[84]

19.4.4 Causality and the many velocities of light

Fast light ($v_g > c$) and backward light ($v_g < 0$) seem at first to violate causality. However, careful analysis shows that this is not so. Causality is the requirement that any effect must be preceded by its cause. When combined with the special theory of relativity, causality requires that no information travel faster than the speed of light. (Otherwise, it would be possible to violate causality in certain frames of reference.) What does this mean for a group velocity greater than the speed of light? Extensive discussion of this question has occurred in the literature over the last century; see, for instance, Sec. 5.2 of Ref. 8, Ref. 11, Secs. 7.8–7.11 of Ref. 90, Ref. 91, and their references. An overview is presented here.

From 1907 to 1914, Sommerfeld and Brillouin examined the propagation of a discontinuous jump (like a step function) in the electric field. They examined the front velocity, or the speed of propagation of the first nonzero value of the electric field. They found that the front velocity can never exceed c and that no part of the waveform can overtake the front.[3,4] Their result was later extended to nonlinear media and to all functions with compact support, i.e., functions such as in Fig. 19.7 that are zero except over a finite range.[92]

Many fast-light experiments and theories use Gaussian-like pulses with long leading and trailing tails. The group velocity can then be used to describe the motion of the pulse envelope or the pulse peak. Using the presence or absence of a pulse as representing one bit (as in on-off keying), one may be tempted to think of the peak as carrying the information associated with the pulse, and hence conclude that information is propagating superluminally. However, the presence or absence of the long leading edge of the pulse carries the same information as the presence or absence of the peak. A true Gaussian pulse has infinite extent; in a sense, the pulse

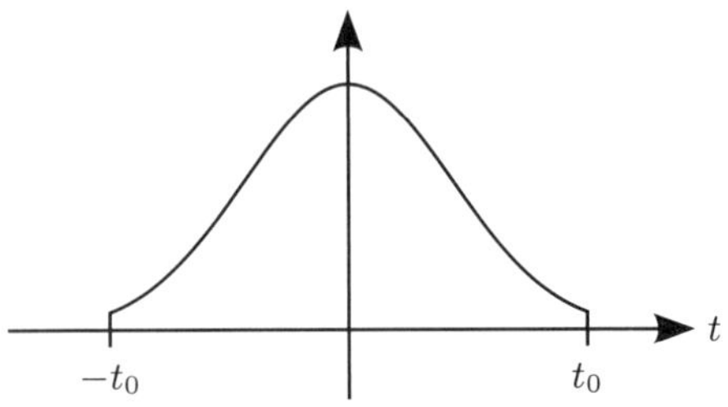

Figure 19.7 A truncated Gaussian pulse. The function has the value $\exp(-t^2/2)$ in the range $-t_0 \leq t \leq t_0$ and is zero outside this range, so it has compact support. A true Gaussian pulse would have infinite extent and therefore infinite support.

and its information have already "arrived" everywhere, irrespective of the motion of the peak. The superluminal peak velocity is therefore completely unrelated to the speed of information transfer.

For the more realistic case of a truncated Gaussian pulse (as in Fig. 19.7), the peak of the pulse may travel superluminally for a time, but the front of the pulse still propagates at or below the speed of light (since the pulse front is a discontinuity). None of the pulse energy can overtake the pulse front. For example, in an on-off keyed binary signal, fast light may shift the peak of a pulse within its bit slot but cannot advance the peak past the beginning of the bit slot. As the peak approaches the front, the pulse becomes highly distorted, often breaking up into a series of peaks or some other irregular shape. For further discussion, see Ref. 93 and Sec. 2.5 of Ref. 94.

A different, equally valid explanation of why superluminal propagation is consistent with causality follows from the formal description of fast-light effects. Fast light is predicted by the form of the susceptibility χ and by Maxwell's equations. The susceptibility is governed by the Kramers–Kronig relations, which are a consequence of requiring the medium in question to exhibit a causal response. In other words, the Kramers–Kronig relations are derived by requiring the electromagnetic response of the medium to occur strictly after the electromagnetic stimulus that causes it. Maxwell's equations, which govern the propagation of light within the medium, obey special relativity. It is easy to understand on these grounds that fast light could never be predicted to violate causality or special relativity.

It is sometimes convenient when working with relativistic considerations to consider the forward light cone, the region of space-time that is relativistically (and causally) accessible from a given point in space-time. In the example shown in Fig. 19.8, an event happens at the point $z = z_0$ at the time $t = t_0$. Only observers inside the forward light cone, or inside the boundaries $(z - z_0) = \pm c(t - t_0)$, could possibly observe the event. Any other observers would need a form of superluminal communication, or information transfer faster than the speed of light.

Imagine now that a long but truncated Gaussian pulse is transmitted through a superluminal medium, as shown in Fig. 19.9. A fast-light medium exists between

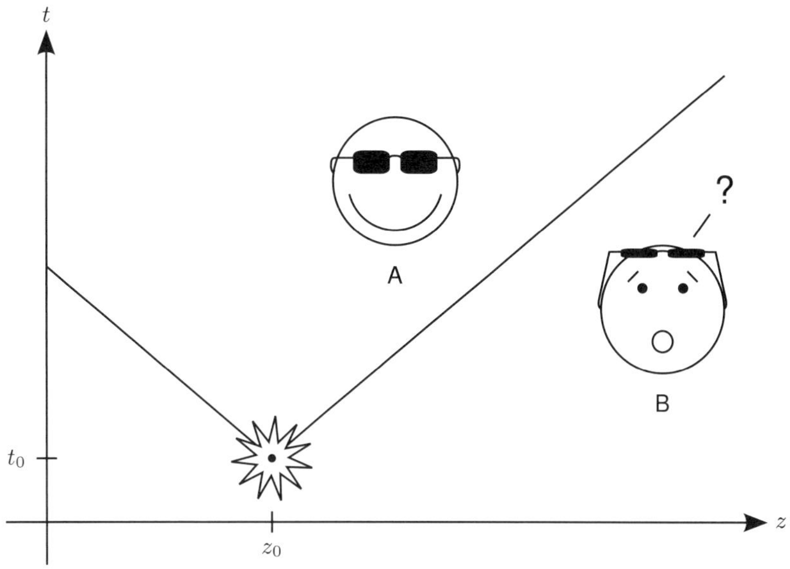

Figure 19.8 A representation of the forward light cone. If an event happens at the location $z = z_0$ at time $t = t_0$, then its forward light cone (the triangular region above the event) is bounded by the lines $(z - z_0) = \pm c(t - t_0)$. It is possible for observer A to observe the event without violating special relativity or causality, but not for observer B, who would need superluminal communication to gain information about the event.

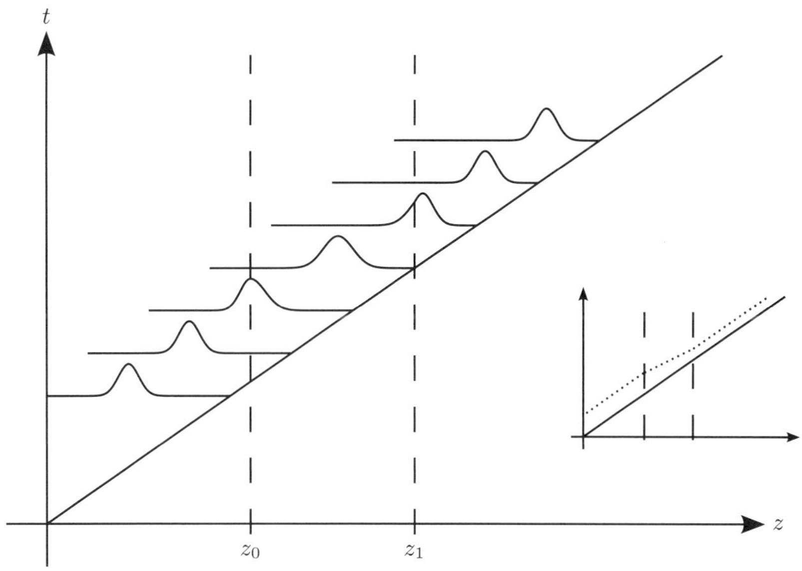

Figure 19.9 Superluminal (fast-light) propagation and the forward light cone. The front of the pulse propagates at c both in vacuum ($z < z_0$ and $z > z_1$) and in the medium ($z_0 \leq z \leq z_1$); the pulse peak travels at c in vacuum and at $v_g > c$ within the medium. Although the peak moves faster than c over a short space-time interval, it does not violate causality: The peak cannot escape the forward light cone of the beginning of the transmission event. Inset: The trajectory of the peak of the pulse (dotted line) and that of the pulse front (solid line). See text for details.

$z = z_0$ and $z = z_1$. A transmitter at location $z = 0$ begins transmitting a long but truncated Gaussian pulse at time $t = 0$. The forward light cone of the transmission event is denoted by the solid line, which also represents the pulse front traveling at its maximum velocity c. The peak of the pulse travels at c in vacuum and at $v_g > c$ in the medium. However, the peak can never overtake the front; instead, the pulse would become distorted. (Notice how the pulse expands inside the fast-light medium.) The inset to Fig. 19.9 shows the trajectory of the peak (dotted line) and the trajectory of the front (solid line). Inside the medium, the dotted line's slope is more horizontal than the solid line's slope, meaning that the speed of the peak is greater than the speed of the front and greater than c in that region. However, the dotted line may never cross the solid line. Fast-light propagation does not violate causality, because the peak of the pulse can never travel outside of the forward light cone of the event that began the transmission of the pulse. (This explanation is similar in spirit to the bit slot argument above.)

A number of other velocities have been defined in an attempt to understand the relationship between superluminal group velocities and causality. The original work of Sommerfeld and Brillouin defined five different velocities, including the front velocity and the group velocity, and showed that the front velocity is always less than or equal to c (always luminal), even when $v_g > c$. The group delay associated with an evanescent wave can appear superluminal, akin to the Hartman effect of quantum-mechanical tunneling through a barrier. However, the effect is a matter of energy storage and retrieval in the medium, rather than true propagation. The group delay of evanescent waves, then, should not be considered a propagation time.[95,96] For propagating waves, an "energy centroid" can be defined (similar to the center of mass for the total electromagnetic and stored material energy); its velocity is always luminal.[97–99] (Recall that, even in the case of backward light, the energy flow is still in the forward direction.[14]) One group attempted to quantify the information velocity by defining special pulses with definite points of distinguishability and then tracking the time at which the pulses could be distinguished. They found that the information velocity thus defined was always luminal, even for $v_g > c$.[73]

As a result of Brillouin's work, the special theory of relativity was reformulated slightly in the early 20th century: No information may be communicated faster than the speed of light.[73] There is no clear consensus on what constitutes the "information velocity" or the "signal velocity," perhaps due to the lack of a universal definition for "information" or for "signal." However, it has been well established that fast light and backward light cannot violate causality.

19.5 Potential Applications

Slow and fast light allow researchers to conduct many exciting fundamental studies of physics and light propagation, but they also have many potential practical applications. The applications proposed for slow and fast light cover many different areas, but they can be grouped into three main themes. Perhaps the most obvious use for a slow-light medium is as a tunable optical delay line, a compact device that can store or buffer light pulses for a time or perhaps even indefinitely. Tunable optical delay lines could find a number of applications within telecommunication networks, as well as in optical coherence tomography (OCT), ultrafast pulse metrology, and various kinds of optical signal processing.[100,101] Slow light can also be used to enhance the nonlinear effects of an optical material, leading to smaller devices and lower operating power in applications that require a high degree of nonlinearity. Finally, slow and fast light can enhance interferometry, producing more sensitive and more stable interferometers. (As discussed in Sec. 19.4.4, fast light may *not* be used to increase network data rates by increasing the propagation speed of light above c.)

19.5.1 Optical delay lines

19.5.1.1 Optical network buffer for all-optical routing

In a packet-based data network, such as the Internet, a router can be modeled as an $N \times N$ switch, as shown in Fig. 19.10. As packets arrive at the router's input ports, the router reads the destination information in the packets and sends them to the appropriate output ports. However, contention arises when two packets destined for the same output port arrive simultaneously at the router's inputs, as in Fig. 19.10(a). The router cannot simultaneously send both packets to the same output port. If it has no buffer, it can send one packet and must drop (discard) the other packet. Dropped packets must be retransmitted, causing increased network latency.

A much more desirable situation is shown in Fig. 19.10(b). If the router has an internal packet buffer, as in Fig. 19.10(c), it can send one packet and store the other in its buffer until the output port becomes available. Buffering enhances network robustness and throughput. (Of course, the actual router is much more complex, but this model suffices conceptually. For more details, see Chapter 5 of Ref. 102.)

At present, essentially all routing functions in an optical network are implemented in electronics. This requires converting the optical signal to an electrical signal for processing and then back to an optical signal for further transmission. This process, known as OEO conversion, adds delay to network transmissions and consumes additional power. An optical buffer would be the first step in all-optical networking, which many feel will soon become crucial to the increased performance of telecommunication networks.[103]

19.5.1.2 Network resynchronization and jitter correction

Tunable optical delay lines are ideal for all-optical jitter compensation.[104] In modern data networks, transmissions are synchronized to bit slots, regularly spaced time windows during which either a zero or one value is transmitted. For example, in traditional on-off keyed (OOK) transmission, zero or one is indicated respec-

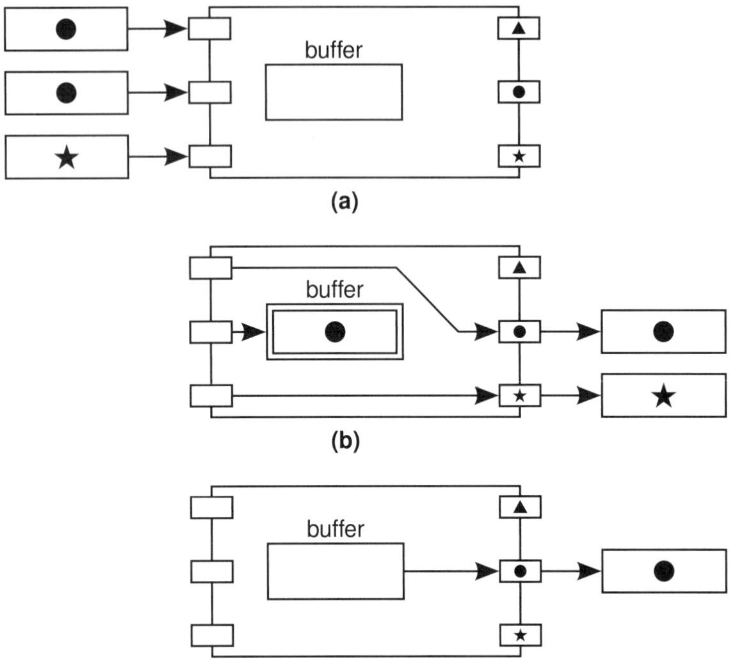

Figure 19.10 Contention resolution using a packet buffer in a 3×3 network switch. Symbols indicate packet destinations. (a) Three packets arrive simultaneously at the switch's input ports. The upper and middle packets are destined for the same output port. (b) The upper and lower packets are routed immediately to their destinations, while the middle packet is stored in a buffer. (c) When the appropriate output port becomes available, the buffered packet is released.

tively by the absence or presence of a light pulse in each bit slot. In a 10-Gbit/s transmission link, bit slots are 100 ps each. Every 100 ps, the transmitter transmits a pulse of light to indicate a one or transmits no pulse to indicate a zero.

Transmitters and receivers must agree on the size and start times of bit slots. Sometimes, parts of the data stream become slightly stretched, compressed, or shifted during the transmission process, resulting in a slight desynchronization between the transmitter and receiver. The receiver observes jitter in the data stream, meaning that each data bit arrives slightly before or after the expected time, shifted by a random amount. In other words, the data bits are not precisely aligned to the bit slots at the receiver.

Jitter is usually caused by random processes such as temperature changes, vibration, or pattern-dependent nonlinearities in the transmission medium or equipment. It is important that receiver equipment adjust its timing slightly to compensate for jitter, or else data corruption and increased bit-error rate (BER) can result. All-optical networks will require all-optical methods of jitter compensation, such as could be afforded by slow-light and fast-light delay lines.

19.5.1.3 Tapped delay lines and equalization filters

Optical delay lines can be used to implement tapped delay lines, as shown in Fig. 19.11(a). Tapped delay lines, in turn, can be used to implement certain optical signal processing elements, particularly filters.[105] Such filters are prevalent in electronics and allow reshaping of the signal spectrum in a well-defined manner. A typical filter is shown in Fig. 19.11(b). Optical filters may be particularly useful for equalization of an optical signal, whereby certain transmission effects can be mitigated and network robustness enhanced.

19.5.1.4 Optical memory and stopped light for coherent control

The group velocity of light can in fact be adjusted to zero, leading to so-called stopped light.[15,16,106] One stopped-light technique is to use electromagnetically induced transparency (EIT) to "map" a light pulse onto the spin coherence of a medium, effectively storing the pulse. The pulse can later be retrieved by performing the reverse operation. Other proposed techniques include using solitons in coupled-resonator structures.[63,107] A stopped-light system could be useful as an optical memory for storing pulses of light.

Many stopped-light techniques also preserve quantum coherence properties. EIT-based techniques, for example, can be used to preserve and store entanglement, or the coherence of two quantum-mechanical systems (such as two photons or a photon and an atom). A so-called quantum memory, onc that can store and re-

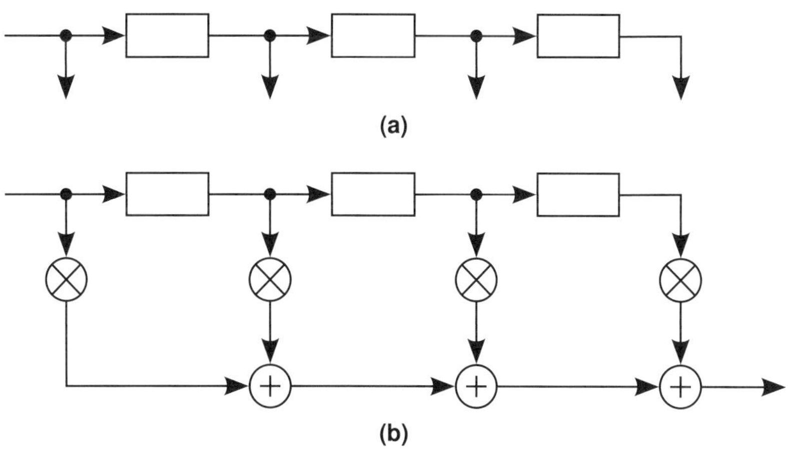

Figure 19.11 (a) An optical tapped delay line. Each block represents a delay element, such as a slow-light element. Each small dot represents a "tap," possibly implemented by a beamsplitter, that allows a delayed copy of the optical signal to be used. (b) An optical finite impulse response (FIR) filter. The cross symbols represent modulation of some kind (such as by variable attenuators), and the plus symbols represent combination of the signals (such as by a beamsplitter).

trieve entangled photons, could find applications in quantum computing, quantum cryptography, and any other technology that depends on entanglement.[108]

One drawback of stopped-light memories is the finite lifetime of the memory. EIT media gradually undergo decoherence (or dephasing). Other kinds of optical memories have different decay mechanisms, but all memories decay and lose their data over time. However, decay needs only to be slowed, not eliminated. A memory cell can be refreshed by reading out its value and rewriting the value into the cell, starting the decay cycle anew. In two 2001 stopped-light experiments using EIT, the coherence lifetime was 500 μs.[15,16] Thus, a stopped-light memory cell using similar techniques would need to be refreshed faster than every 500 μs. For comparison, modern electronic memory (DRAM) cells require a refresh every 7.8 μs.[109]

19.5.1.5 Optical image buffering

Slow-light media can buffer not only pulses of light, but in fact entire images.[110] Both amplitude and phase information in an image are preserved by slow-light media. Such an image buffer could have applications in optical image processing.

19.5.1.6 True time delay for radar and lidar

A phased-array radar antenna is a configuration of many individual antennas that radiate the same signal, only shifted in time relative to each other. Typically, the relative phases of the individual antenna elements are tuned to steer the radar beam in any direction. The rephasing acts like a time delay for narrowband radar. However, for wideband radar signals, using the signal phases to steer the beam will cause a phenomenon called beam squint, which results in directional inaccuracy. In these cases, a true time delay must be used to offset the signal of each antenna element.[111,112] There has been much interest in the past in the possibility of optical true time delay for radar, as optical signals can sustain the high bandwidth needed for modern radar. A radio frequency (RF) signal is impressed on an optical carrier, a tunable optical optical delay line creates the true time delay, and then heterodyne optical detection is used to reconstruct the delayed RF signal. Slow-light delay lines hold great promise in this area as a tunable source of delay for phased-array radar and lidar systems.[113]

19.5.2 Enhancement of optical nonlinearities

Slow light is sometimes associated with an enhancement of the optical nonlinearity. A reduced group velocity does not of itself increase the nonlinear response. However, in EIT, slow-light effects allow access to a kind of latent nonlinearity enhancement (see Sec. 19.2.5). In coupled-resonator structures, the field is enhanced by resonator effects, leading to a nonlinearity enhancement (see Sec. 19.3.2). Nonlinearity enhancements would let certain devices be miniaturized and consume less power.

19.5.2.1 Wavelength converter

In a wavelength-division multiplexed (WDM) optical network, each of a number of network channels is carried over the same optical fiber using a different wavelength range. One approach to routing the network data is to assign certain wavelengths to certain routes *a priori*, such as in Fig. 19.12. However, the network can be far more flexible and resilient if wavelength assignments can be modified dynamically.[114] A wavelength converter is a device that can convert one wavelength channel to another wavelength optically and programmably, without decoding and re-encoding the network data (as in an OEO conversion). The nonlinear optical process of four-wave mixing (FWM) allows just such a conversion to happen. The enhanced nonlinearity in a slow-light medium could enable more efficient FWM, allowing both smaller devices and reduced operating power requirements.[20]

19.5.2.2 Single-bit optical switching, optical logic, and other applications

Slow-light systems can be used for all-optical network switching at low light levels (i.e., using as little as or less than the equivalent energy of one data bit per switching operation).[115–117] The switching of light by light requires a nonlinear interaction, and the nonlinearity enhancement associated with slow light can make optical switching achievable at reasonable power levels. Optical logic gates may be implemented using a similar system.[20] Photonic crystal and coupled-resonator slow-light Mach–Zehnder interferometric modulators with favorable properties have already been fabricated.[63,118]

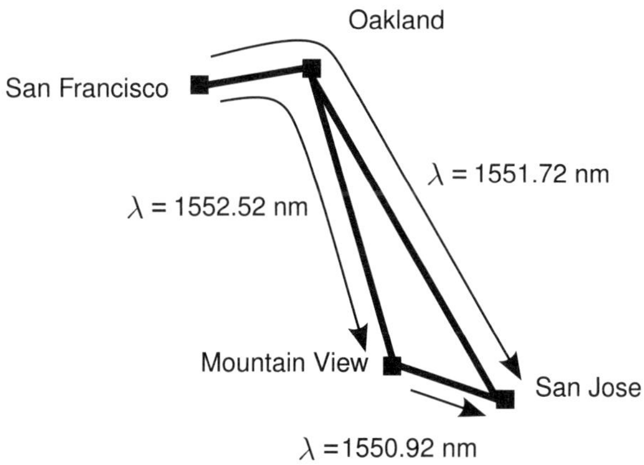

Figure 19.12 Wavelength routing in a hypothetical network in California, USA. Each filled square represents a router, and each thick line represents an optical fiber. Each arrow shows the allocation of a wavelength (a WDM channel) to a particular route. For example, only data sent from San Francisco to San Jose may use the 1551.72-nm band, regardless of actual network usage. In a complex network with many nodes and routes, wavelength converters allow dynamic wavelength allocation and make the network more robust.

Many other applications have been envisioned for slow-light devices with enhanced nonlinearities. Pulse generation and all-optical signal regeneration are possibilities. Traditional nonlinear optical processes, such as harmonic generation and the Kerr effect, can be implemented in slow-light devices.[19,119,120] All of these devices could be improved by using the nonlinearity enhancement associated with slow-light techniques.

19.5.3 Slow- and fast-light interferometry

19.5.3.1 Spectral sensitivity enhancement

The spectral properties of slow and fast light can be used to enhance the performance and robustness of spectroscopic interferometers. In such interferometers, a change in laser frequency causes a change in the optical path length difference (OPD) between the two interferometer paths. A slow-light medium placed in one path increases the sensitivity by a factor of the group index.[121] For example, if a slow-light medium of length L slows a pulse of light to a group velocity of $c/10^6$, it has the same effect on frequency sensitivity as would an OPD of $10^6 L$ (see Fig. 19.13). Of course, by using fast light, one could correspondingly decrease the frequency sensitivity of the interferometer. In theory, using stopped light with zero group velocity and zero group index, one could remove all sensitivity of the phase to frequency shifts, at least over the bandwidth of zero group velocity. Using a medium with a negative group index (in the backward-light regime), one could reverse the sign of the phase change with respect to frequency, although very few applications depend on the sign of the phase change. Slow and fast light have also been proposed for use in Fourier-transform interferometry[122] and interferometric rotation sensing.[123,124]

Similar sensitivity effects to those above are seen when a slow- or fast-light medium is placed inside a Fabry–Pérot cavity. The cavity linewidth is changed but the cavity storage time remains unaffected.[125,126] A slow-light medium narrows the cavity linewidth, while a fast-light medium broadens it. Thus, slow light can be

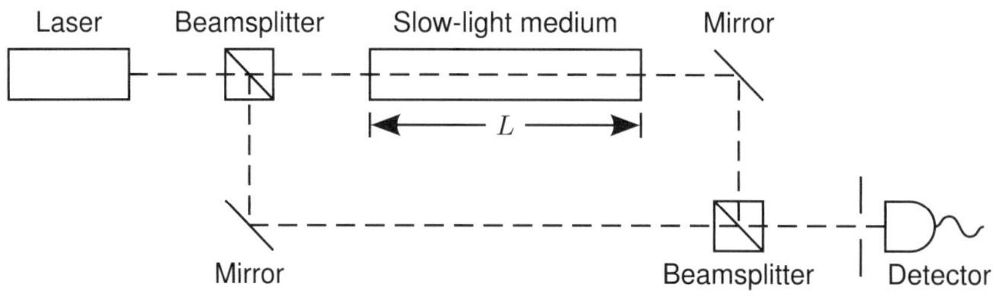

Figure 19.13 A slow-light Mach–Zehnder interferometer. (Adapted from Ref. 121; used by permission.)

used to enhance the spectral resolution of the cavity, while fast light can be used to decrease the sensitivity of the cavity to changes in length or laser frequency.[127]

19.5.3.2 White-light cavities

Fast light may also be used to construct a white-light cavity.[128] A white-light cavity is a Fabry–Pérot cavity that contains an anomalously dispersive (fast-light) element. In a normal Fabry–Pérot cavity, a slight detuning away from resonance will reduce the cavity transmission drastically: Each round trip acquires a slight phase shift, and the multiple round trips add destructively. However, in a white-light cavity, the fast-light element compensates for this slight phase shift, such that the cavity resonates across a range of wavelengths. In such a cavity, the electric field is greatly enhanced, making certain detection problems easier to solve.

Acknowledgments

The authors gratefully acknowledge financial support by the DARPA/DSO slow light program and through the NSF.

References

1. J. Marangos, "Slow light in cool atoms," *Nature* **397**, 559–560 (1999).
2. L. V. Hau, S. E. Harris, Z. Dutton, and C. H. Behroozi, "Light speed reduction to 17 metres per second in an ultracold atomic gas," *Nature* **397**, 594–598 (1999).
3. A. Sommerfeld, "Über die Fortpflanzung des Lichtes in dispergierenden Medien," *Ann. Phys.* **349**, 177–202 (1914).
4. L. Brillouin, "Über die Fortpflanzung des Lichtes in dispergierenden Medien," *Ann. Phys.* **349**, 203–240 (1914).
5. N. G. Basov, R. V. Ambartsumyan, V. S. Zuev, P. G. Kryukov, and V. S. Letokhov, "Nonlinear amplification of light pulses," *Sov. Phys. JETP* **23**, 16–22 (1966).
6. C. G. B. Garrett and D. E. McCumber, "Propagation of a Gaussian light pulse through an anomalous dispersion medium," *Phys. Rev. A* **1**, 305–313 (1970).
7. S. Chu and S. Wong, "Linear pulse propagation in an absorbing medium," *Phys. Rev. Lett.* **48**, 738–741 (1982).
8. R. W. Boyd and D. J. Gauthier, "'Slow' and 'fast' light," in *Progress in Optics*, E. Wolf, Ed., **43**, 497–530, Elsevier Science, Amsterdam (2002).

9. R. Y. Chiao and A. M. Steinberg, "Tunneling times and superluminality," in *Progress in Optics*, E. Wolf, Ed., **37**, 347–406, Elsevier Science, Amsterdam (1997).

10. P. G. Eliseev, H. Cao, C. Liu, G. A. Smolyakov, and M. Osiński, "Nonlinear mode interaction as a mechanism to obtain slow/fast light in diode lasers," in *Physics and Simulation of Optoelectronic Devices XIV*, M. Osiński, F. Henneberger, and Y. Arakawa, Eds., *Proc. SPIE* **6115**, 61150U (2006).

11. J. C. Garrison, M. W. Mitchell, R. Y. Chiao, and E. L. Bolda, "Superluminal signals: Causal loop paradoxes revisited," *Phys. Lett. A* **245**, 19–25 (1998).

12. A. M. Akulshin, S. Barreiro, and A. Lezama, "Steep anomalous dispersion in coherently prepared Rb vapor," *Phys. Rev. Lett.* **83**, 4277–4280 (1999).

13. M. González-Herráez, K.-Y. Song, and L. Thévenaz, "Optically controlled slow and fast light in optical fibers using stimulated Brillouin scattering," *Appl. Phys. Lett.* **87**, 081113 (2005).

14. G. M. Gehring, A. Schweinsberg, C. Barsi, N. Kostinski, and R. W. Boyd, "Observation of backwards pulse propagation through a medium with a negative group velocity," *Science* **312**, 895–897 (2006).

15. A. V. Turukhin, V. S. Sudarshanam, M. S. Shahriar, J. A. Musser, B. S. Ham, and P. R. Hemmer, "Observation of ultraslow and stored light pulses in a solid," *Phys. Rev. Lett.* **88**, 023602 (2001).

16. F. D. Phillips, A. Fleischhauer, A. Mair, R. L. Walsworth, and M. D. Lukin, "Storage of light in atomic vapor," *Phys. Rev. Lett.* **86**, 783–786 (2001).

17. R. W. Boyd, *Nonlinear Optics*, 3rd ed., Academic Press, Burlington, MA (2008).

18. B. E. A. Saleh and M. C. Teich, *Fundamentals of Photonics*, 2nd ed., Wiley–Interscience, New York (2007).

19. S. E. Harris and L. V. Hau, "Nonlinear optics at low light levels," *Phys. Rev. Lett.* **82**, 4611–4614 (1999).

20. M. Soljačić and J. D. Joannopoulos, "Enhancement of nonlinear effects using photonic crystals," *Nature Materials* **3**, 211–219 (2004).

21. S. E. Harris, J. E. Field, and A. Imamoğlu, "Nonlinear optical processes using electromagnetically induced transparency," *Phys. Rev. Lett.* **64**, 1107–1110 (1990).

22. H. Kang, G. Hernandez, and Y. Zhu, "Superluminal and slow light propagation in cold atoms," *Phys. Rev. A* **70**, 011801 (2004).

23. M. Phillips and H. Wang, "Spin coherence and electromagnetically induced transparency via exciton correlations," *Phys. Rev. Lett.* **89**, 186401 (2002).

24. S.-W. Chang, S. L. Chuang, C. J. Chang-Hasnain, and H. Wang, "Slow light using spin coherence and V-type electromagnetically induced transparency in [110] strained quantum wells," *J. Opt. Soc. Am. B* **24**, 849–859 (2007).

25. C. J. Chang-Hasnain and P.-C. Ku, "Variable semiconductor all-optical buffer using slow light based on electromagnetically induced transparency," U.S. Patent 7,038,827 (2006).

26. S. Ghosh, J. E. Sharping, D. G. Ouzounov, and A. L. Gaeta, "Resonant optical interactions with molecules confined in photonic band-gap fibers," *Phys. Rev. Lett.* **94**, 093902 (2005).

27. A. K. Patnaik, J. Q. Liang, and K. Hakuta, "Slow light propagation in a thin optical fiber via electromagnetically induced transparency," *Phys. Rev. A* **66**, 063808 (2002).

28. J. Kim, S. L. Chuang, P.-C. Ku, and C. J. Chang-Hasnain, "Slow light using semiconductor quantum dots," *J. Phys.: Cond. Matt.* **16**, S3727–S3735 (2004).

29. L. Maleki, A. B. Matsko, A. A. Savchenkov, and V. S. Ilchenko, "Tunable delay line with interacting whispering-gallery-mode resonators," *Opt. Lett.* **29**, 626–628 (2004).

30. D. D. Smith, H. Chang, K. A. Fuller, A. T. Rosenberger, and R. W. Boyd, "Coupled-resonator-induced transparency," *Phys. Rev. A* **69**, 063804 (2004).

31. D. D. Smith, N. N. Lepeshkin, A. Schweinsberg, G. Gehring, R. W. Boyd, Q.-H. Park, H. Chang, and D. Jackson, "Coupled-resonator-induced transparency in a fiber system," *Opt. Comm.* **264**, 163–168 (2006).

32. J. D. Franson and S. M. Hendrickson, "Optical transparency using interference between two modes of a cavity," *Phys. Rev. A* **74**, 053817 (2006).

33. S. E. Harris, "Electromagnetically induced transparency in an ideal plasma," *Phys. Rev. Lett.* **77**, 5357–5360 (1996).

34. G. Shvets and J. S. Wurtele, "Transparency of magnetized plasma at the cyclotron frequency," *Phys. Rev. Lett.* **89**, 115003 (2002).

35. M. Tushentsov, G. Shvets, A. Yu. Kryachko, and M. D. Tokman, "Undulator-induced transparency of magnetized plasma: New approach to electromagnetic energy compression," *IEEE Trans. Plasma Sci.* **33**, 23–31 (2005).

36. S. E. Schwarz and T. Y. Tan, "Wave interactions in saturable absorbers," *Appl. Phys. Lett.* **10**, 4–7 (1967).

37. M. S. Bigelow, N. N. Lepeshkin, and R. W. Boyd, "Observation of ultraslow light propagation in a ruby crystal at room temperature," *Phys. Rev. Lett.* **90**, 113903 (2003).

38. A. Schweinsberg, N. Lepeshkin, M. S. Bigelow, R. W. Boyd, and S. Jarabo, "Observation of superluminal and slow light propagation in erbium-doped optical fiber," *Europhys. Lett.* **73**, 218–224 (2006).

39. N. G. Basov, R. V. Ambartsumyan, V. S. Zuev, P. G. Kryukov, and V. S. Letokhov, "Propagation velocity of an intense light pulse in a medium with inverse population," *Sov. Phys. Dokl.* **10**, 1039 (1966).

40. A. C. Selden, "Pulse transmission through a saturable absorber," *Brit. J. Appl. Phys.* **18**, 743–748 (1967).

41. A. C. Selden, "Analysis of the saturable absorber transmission equation," *J. Phys. D* **3**, 1935–1943 (1970).

42. M. S. Bigelow, N. N. Lepeshkin, and R. W. Boyd, "Superluminal and slow light propagation in a room-temperature solid," *Science* **301**, 200–202 (2003).

43. J. Mřrk, R. Kjţr, M. van der Poel, and K. Yvind, "Slow light in a semiconductor waveguide at gigahertz frequencies," *Opt. Express* **13**, 8136–8145 (2005).

44. P. Palinginis, S. Crankshaw, F. Sedgwick, E.-T. Kim, M. Moewe, C. J. Chang-Hasnain, H. Wang, and S.-L. Chuang, "Ultraslow light (< 200 m/s) propagation in a semiconductor nanostructure," *Appl. Phys. Lett.* **87**, 171102 (2005).

45. H. Su and S. L. Chuang, "Room temperature slow and fast light in quantum-dot semiconductor optical amplifiers," *Appl. Phys. Lett.* **88**, 061102 (2006).

46. P.-C. Ku, F. Sedgwick, C. J. Chang-Hasnain, P. Palinginis, T. Li, H. Wang, S.-W. Chang, and S.-L. Chuang, "Slow light in semiconductor quantum wells," *Opt. Lett.* **29**, 2291–2293 (2004).

47. K. Y. Song, M. Herráez, and L. Thévenaz, "Observation of pulse delaying and advancement in optical fibers using stimulated Brillouin scattering," *Opt. Express* **13**, 82–88 (2005).

48. Y. Okawachi, M. S. Bigelow, J. E. Sharping, Z. Zhu, A. Schweinsberg, D. J. Gauthier, R. W. Boyd, and A. L. Gaeta, "Tunable all-optical delays via Brillouin slow light in an optical fiber," *Phys. Rev. Lett.* **94**, 153902 (2005).

49. M. D. Stenner, M. A. Neifeld, Z. Zhu, A. M. Dawes, and D. J. Gauthier, "Distortion management in slow-light pulse delay," *Opt. Express* **13**, 9995–10002 (2005).

50. Z. Zhu, A. M. C. Dawes, D. J. Gauthier, L. Zhang, and A. E. Willner, "12-GHz-bandwidth SBS slow light in optical fibers," in *Proceedings of the Optical Fiber Communication Conference*, paper PDP1, IEEE, Piscataway, NJ (2006).

51. K. Lee and N. M. Lawandy, "Optically induced pulse delay in a solid-state Raman amplifier," *Appl. Phys. Lett.* **78**, 703–705 (2001).

52. F. L. Kien, J. Q. Liang, and K. Hakuta, "Slow light produced by far-off-resonance Raman scattering," *J. Sel. Topics Quantum Electron.* **9**, 93–101 (2003).

53. J. Sharping, Y. Okawachi, and A. Gaeta, "Wide bandwidth slow light using a Raman fiber amplifier," *Opt. Express* **13**, 6092–6098 (2005).

54. R. M. Camacho, M. V. Pack, J. C. Howell, A. Schweinsberg, and R. W. Boyd, "Wide-bandwidth, tunable, multiple-pulse-width optical delays using slow light in cesium vapor," *Phys. Rev. Lett.* **98**, 153601 (2007).

55. X. Zhao, P. Palinginis, B. Pesala, C. J. Chang-Hasnain, and P. Hemmer, "Tunable ultraslow light in vertical-cavity surface-emitting laser amplifier," *Opt. Express* **13**, 7899–7904 (2005).

56. K. Ogawa, T. Katsuyama, and H. Nakamura, "Time-of-flight measurement of excitonic polaritons in a GaAs/AlGaAs quantum well," *Appl. Phys. Lett.* **53**, 1077–1079 (1988).

57. N. Shaw, W. Stewart, J. Heaton, and D. Wight, "Optical slow-wave resonant modulation in electro-optic GaAs/AlGaAs modulators," *Electron. Lett.* **35**, 1557–1558 (1999).

58. P.-C. Ku, C. J. Chang-Hasnain, and S. L. Chuang, "Variable semiconductor all-optical buffer," *Electron. Lett.* **38**, 1581–1583 (2002).

59. C. J. Chang-Hasnain, P.-C. Ku, J. Kim, and S.-L. Chuang, "Variable optical buffer using slow light in semiconductor nanostructures," *Proc. IEEE* **91**, 1884–1897 (2003).

60. P.-C. Ku, C. J. Chang-Hasnain, and S. L. Chuang, "Slow light in semiconductor heterostructures," *J. Phys. D* **40**, R93–R107 (2007).

61. G. P. Agrawal, *Lightwave Technology: Components and Devices*, Wiley-Interscience, New York (2004).

62. J. K. S. Poon, J. Scheuer, Y. Xu, and A. Yariv, "Designing coupled-resonator optical waveguide delay lines," *J. Opt. Soc. Am. B* **21**, 1665–1673 (2004).

63. J. Scheuer, G. T. Paloczi, J. K. S. Poon, and A. Yariv, "Coupled resonator optical waveguides: Toward the slowing & storage of light," *Opt. & Photon. News* **16**, 36–40 (2005).

64. Y. Chen and S. Blair, "Nonlinearity enhancement in finite coupled-resonator slow-light waveguides," *Opt. Express* **12**, 3353–3366 (2004).

65. G. Lenz, B. J. Eggleton, C. K. Madsen, and R. E. Slusher, "Optical delay lines based on optical filters," *J. Quantum Electron.* **37**, 525–532 (2001).

66. J. B. Khurgin, "Light slowing down in Moiré fiber gratings and its implications for nonlinear optics," *Phys. Rev. A* **62**, 013821 (2000).

67. E. Yablonovitch and T. J. Gmitter, "Photonic band structure: The face-centered-cubic case," *Phys. Rev. Lett.* **63**, 1950–1953 (1989).

68. S. John and J. Wang, "Quantum optics of localized light in a photonic band gap," *Phys. Rev. B* **43**, 12772–12789 (1991).

69. J. M. Bendickson, J. P. Dowling, and M. Scalora, "Analytic expressions for the electromagnetic mode density in finite, one-dimensional, photonic band-gap structures," *Phys. Rev. E* **53**, 4107–4121 (1996).

70. M. Scalora, R. J. Flynn, S. B. Reinhardt, R. L. Fork, M. J. Bloemer, M. D. Tocci, C. M. Bowden, H. S. Ledbetter, J. M. Bendickson, J. P. Dowling, and R. P. Leavitt, "Ultrashort pulse propagation at the photonic band edge: Large tunable group delay with minimal distortion and loss," *Phys. Rev. E* **54**, R1078–R1081 (1996).

71. M. L. Povinelli, S. Johnson, and J. D. Joannopoulos, "Slow-light, band-edge waveguides for tunable time delays," *Opt. Express* **13**, 7145–7159 (2005).

72. L. J. Wang, A. Kuzmich, and A. Dogariu, "Gain-assisted superluminal light propagation," *Nature* **406**, 277–279 (2000).

73. M. D. Stenner, D. J. Gauthier, and M. A. Neifeld, "The speed of information in a 'fast-light' optical medium," *Nature* **425**, 695–698 (2003).

74. R. M. Camacho, M. V. Pack, and J. C. Howell, "Low-distortion slow light using two absorption resonances," *Phys. Rev. A* **73**, 063812 (2006).

75. Z. Shi, R. Pant, Z. Zhu, M. D. Stenner, M. A. Neifeld, D. J. Gauthier, and R. W. Boyd, "Design of a tunable time-delay element using multiple gain lines for increased fractional delay with high data fidelity," *Opt. Lett.* **32**(14), 1986–1988 (2007).

76. H. Shin, A. Schweinsberg, G. Gehring, K. Schwertz, H. J. Chang, R. W. Boyd, Q.-H. Park, and D. J. Gauthier, "Reducing pulse distortion in fast-light pulse propagation through an erbium-doped fiber amplifier," *Opt. Lett.* **32**, 906–908 (2007).

77. J. K. S. Poon, L. Zhu, G. A. Rose, and A. Yariv, "Polymer microring coupled-resonator optical waveguides," *J. Lightwave Technol.* **24**, 1843–1849 (2006).

78. R. W. Boyd, D. J. Gauthier, A. L. Gaeta, and A. E. Willner, "Maximum time delay achievable on propagation through a slow-light medium," *Phys. Rev. A* **71**, 023801 (2005).

79. R. S. Tucker, P.-C. Ku, and C. J. Chang-Hasnain, "Delay-bandwidth product and storage density in slow-light optical buffers," *Electron. Lett.* **41**, 208–209 (2005).

80. M. Fleischhauer and M. D. Lukin, "Quantum memory for photons: Dark-state polaritons," *Phys. Rev. A* **65**, 022314 (2002).

81. P. Ginzburg and M. Orenstein, "Slow light and voltage control of group velocity in resonantly coupled quantum wells," *Opt. Express* **14**, 12467–12472 (2006).

82. R. S. Tucker, P.-C. Ku, and C. J. Chang-Hasnain, "Slow-light optical buffers: Capabilities and fundamental limitations," *J. Lightwave Technol.* **23**, 4046–4066 (2005).

83. B. Zhang, L. Yan, I. Fazal, L. Zhang, A. E. Willner, Z. Zhu, and D. J. Gauthier, "Slow light on Gbit/s differential-phase-shift-keying signals," *Opt. Express* **15**, 1878–1883 (2007).

84. E. Shumakher, N. Orbach, A. Nevet, D. Dahan, and G. Eisenstein, "On the balance between delay, bandwidth and signal distortion in slow light systems based on stimulated Brillouin scattering in optical fibers," *Opt. Express* **14**, 5877–5884 (2006).

85. M. D. Lukin and A. Imamoğlu, "Controlling photons using electromagnetically induced transparency," *Nature* **413**, 273–276 (2001).

86. S. E. Harris, J. E. Field, and A. Kasapi, "Dispersive properties of electromagnetically induced transparency," *Phys. Rev. A* **46**, R29–R32 (1992).

87. M. G. Payne, L. Deng, and K. J. Jiang, "Limitations of steady state solutions to a two-state model of population oscillations and hole burning," *Phys. Rev. A* **74**, 043810 (2006).

88. D. C. Roberts, T. Gasenzer, and K. Burnett, "Limitations of light delay and storage times in electromagnetically induced transparency experiments with condensates," *Phys. Rev. A* **66**, 023801 (2002).

89. A. Matsko, D. Strekalov, and L. Maleki, "On the dynamic range of optical delay lines based on coherent atomic media," *Opt. Express* **13**, 2210–2223 (2005).

90. J. D. Jackson, *Classical Electrodynamics*, 3rd ed., John Wiley & Sons, New York (1999).

91. G. Diener, "Superluminal group velocities and information transfer," *Phys. Lett. A* **223**, 327–331 (1996).

92. A. Icsevgi and W. E. Lamb, "Propagation of light pulses in a laser amplifier," *Phys. Rev.* **185**, 517–545 (1969).

93. D. J. Gauthier and R. W. Boyd, "Fast light, slow light and optical precursors: What does it all mean?" *Photon. Spectra* **41**, 82–90 (2007).

94. P. W. Milonni, *Fast Light, Slow Light, and Left-Handed Light*, Taylor & Francis, New York (2005).

95. H. G. Winful, "Energy storage in superluminal barrier tunneling: Origin of the 'Hartman effect'," *Opt. Express* **10**, 1491–1496 (2002).

96. H. G. Winful, "Nature of 'superluminal' barrier tunneling," *Phys. Rev. Lett.* **90**, 023901 (2003).

97. S. Glasgow, M. Ware, and J. Peatross, "Poynting's theorem and luminal total energy transport in passive dielectric media," *Phys. Rev. E* **64**, 046610 (2001).

98. M. Ware, S. Glasgow, and J. Peatross, "Role of group velocity in tracking field energy in linear dielectrics," *Opt. Express* **9**, 506–518 (2001).

99. J. Peatross, M. Ware, and S. Glasgow, "Role of the instantaneous spectrum on pulse propagation in causal linear dielectrics," *J. Opt. Soc. Am. A* **18**, 1719–1725 (2001).

100. J. E. Sharping, Y. Okawachi, J. van Howe, C. Xu, Y. Wang, A. E. Willner, and A. L. Gaeta, "All-optical, wavelength and bandwidth preserving, pulse delay based on parametric wavelength conversion and dispersion," *Opt. Express* **13**, 7872–7877 (2005).

101. Y. Okawachi, J. E. Sharping, C. Xu, and A. L. Gaeta, "Large tunable optical delays via self-phase modulation and dispersion," *Opt. Express* **14**, 12022–12027 (2006).

102. A. S. Tanenbaum, *Computer Networks*, 4th ed., Prentice Hall, Upper Saddle River, NJ (2003).

103. J. R. Lowell and E. Parra, "Applications of slow light: A DARPA perspective," in *Advanced Optical and Quantum Memories and Computing II*, H. J. Coufal, Z. U. Hasan, and A. E. Craig, Eds., *Proc. SPIE* **5735**, 80–86 (2005).

104. B. Zhang, L. Zhang, L.-S. Zhang, I. Fazal, J.-Y. Yang, and A. E. Willner, "Continuously-tunable, bit-rate variable OTDM using broadband SBS slow-light delay line," *Opt. Express* **15**, 8317–8322 (2007).

105. C. Yu, T. Luo, L. Zhang, and A. E. Willner, "Data pulse distortion induced by a slow-light tunable delay line in optical fiber," *Opt. Lett.* **32**, 20–22 (2007).

106. M. D. Lukin, S. F. Yelin, and M. Fleischhauer, "Entanglement of atomic ensembles by trapping correlated photon states," *Phys. Rev. Lett.* **84**, 4232–4235 (2000).

107. S. Mookherjea and A. Yariv, "Kerr-stabilized super-resonant modes in coupled-resonator optical waveguides," *Phys. Rev. E* **66**, 046610 (2002).

108. M. Fleischhauer and M. D. Lukin, "Quantum memory for photons: Dark-state polaritons," *Phys. Rev. A* **65**, 022314 (2002).

109. *JEDEC Standard No. 79-2E: DDR2 SDRAM specification*, 78. JEDEC Solid State Technology Association, Arlington, VA (2008).

110. R. M. Camacho, C. J. Broadbent, I. Ali-Khan, and J. C. Howell, "All-optical delay of images using slow light," *Phys. Rev. Lett.* **98**, 043902 (2007).

111. M. I. Skolnik, *Radar Handbook*, 3rd ed., 13.1–13.51. McGraw-Hill, New York (2008).

112. I. Frigyes and A. J. Seeds, "Optically generated true-time delay in phased-array antennas," *IEEE Trans. Microw. Theory Tech.* **43**, 2378–2386 (1995).

113. H. Su, P. Kondratko, and S. L. Chuang, "Variable optical delay using population oscillation and four-wave-mixing in semiconductor optical amplifiers," *Opt. Express* **14**, 4800–4807 (2006).

114. J. M. Yates, J. P. R. Lacey, D. Everitt, and M. Summerfield, "Limited-range wavelength translation in all-optical networks," in *INFOCOM '96: Fifteenth Annual Joint Conference of the IEEE Computer and Communications Societies*, **3**, 954–961 (1996).

115. H. Schmidt and R. J. Ram, "All-optical wavelength converter and switch based on electromagnetically induced transparency," *Appl. Phys. Lett.* **76**, 3173–3175 (2000).

116. D. A. Braje, V. Balić, G. Y. Yin, and S. E. Harris, "Low-light-level nonlinear optics with slow light," *Phys. Rev. A* **68**, 041801 (2003).

117. A. M. C. Dawes, L. Illing, S. M. Clark, and D. J. Gauthier, "All-optical switching in rubidium vapor," *Science* **308**, 672–674 (2005).

118. Yu. A. Vlasov, M. O'Boyle, H. F. Hamann, and S. J. McNab, "Active control of slow light on a chip with photonic crystal waveguides," *Nature* **438**, 65–69 (2005).

119. M. Soljačić, S. G. Johnson, S. Fan, M. Ibanescu, E. P. Ippen, and J. D. Joannopoulos, "Photonic-crystal slow-light enhancement of nonlinear phase sensitivity," *J. Opt. Soc. Am. B* **19**, 2052–2059 (2002).

120. Z. Yang, P. Chak, R. Iyer, A. D. Bristow, J. E. Sipe, J. S. Aitchison, H. M. van Driel, and A. L. Smirl, "Enhanced second harmonic generation using slow light in AlGaAs microring resonators," in *Slow and Fast Light*, paper MB6, OSA, Washington, DC (2006).

121. Z. Shi, R. W. Boyd, D. J. Gauthier, and C. C. Dudley, "Enhancing the spectral sensitivity of interferometers using slow-light media," *Opt. Lett.* **32**, 915–917 (2007).

122. Z. Shi, R. W. Boyd, R. M. Camacho, P. K. Vudyasetu, and J. C. Howell, "Slow-light Fourier transform interferometer," *Phys. Rev. Lett.* **99**, 240801 (2007).

123. M. S. Shahriar, G. S. Pati, R. Tripathi, V. Gopal, M. Messall, and K. Salit, "Ultrahigh enhancement in absolute and relative rotation sensing using fast and slow light," *Phys. Rev. A* **75**, 053807 (2007).

124. M. Salit, G. S. Pati, K. Salit, and M. S. Shahriar, "Fast-light for astrophysics: Super-sensitive gyroscopes and gravitational wave detectors," *J. Mod. Opt.* **54**, 2425–2440 (2007).

125. G. S. Pati, M. Salit, K. Salit, and M. S. Shahriar, "Demonstration of a tunable-bandwidth white-light interferometer using anomalous dispersion in atomic vapor," *Phys. Rev. Lett.* **99**, 133601 (2007).

126. G. S. Pati, M. Salit, K. Salit, and M. S. Shahriar, "Demonstration of tunable displacement-measurement-sensitivity using variable group index in a ring resonator," *Opt. Comm.* **281**, 4931–4935 (2008).

127. M. S. Shahriar, G. S. Pati, M. Messall, and K. R. Salit, "Slow-light based control of interferometer sensitivity," in *Frontiers in Optics/Laser Science XXII*, paper LTuJ4, OSA, Washington, DC (2006).

128. A. Wicht, K. Danzmann, M. Fleischhauer, M. Scully, G. Müller, and R.-H. Rinkleff, "White-light cavities, atomic phase coherence, and gravitational wave detectors," *Opt. Comm.* **134**, 431–439 (1997).

Biographies

Joseph E. Vornehm, Jr. received his B.S. degree in electrical engineering from Northeastern University (Boston, MA, USA) in 2001. After several years as an engineer in the Signal Processing Department of The MITRE Corporation (Bedford, MA, USA), he attended Northwestern University (Evanston, IL, USA) and received his M.S. degree in electrical engineering in 2005. His master's thesis, titled "Multi-spectral Raman gain in dual-isotope rubidium vapor," was supervised by Selim Shahriar. He is currently a Ph.D. student in the Nonlinear Optics group at the Institute of Optics of the University of Rochester, where he is a Sproull Fellow. His research interests include practical applications of slow and fast light. Joe is a student member of SPIE, the Optical Society of America, and the American Physical Society, and is a member of the IEEE.

Professor **Robert W. Boyd** received his B.S. degree in Physics from the Massachusetts Institute of Technology in 1969 and received his Ph.D. degree in Physics in 1977 from the University of California at Berkeley. His Ph.D. thesis was supervised by Professor Charles H. Townes and involved the use of nonlinear optical techniques for infrared detection for astronomy. Professor Boyd joined the faculty of the Institute of Optics of the University of Rochester in 1977 and is presently the M. Parker Givens Professor of Optics and Professor of Physics. His research interests include slow and fast light, optical physics, nonlinear optical interactions, nonlinear optical properties of materials, and applications of nonlinear optics including quantum and nonlinear optical imaging. Professor Boyd is a Fellow of the Optical Society of America and the American Physical Society and a member of SPIE and IEEE/LEOS. He is the author of *Radiometry and the Detection of Optical Radiation* (1983) and *Nonlinear Optics* (1992), and is the co-editor of *Optical Instabilities* (1986) and *Contemporary Nonlinear Optics* (1992). Professor Boyd has published more than 250 research papers, has been awarded seven U.S. patents, and has supervised the Ph.D. theses of 29 students.

About the Editors

Mikhail A. Noginov graduated from Moscow Institute for Physics and Technology (Moscow, Russia) in 1985 with a Master of Science degree in Electronics Engineering. In 1990 he received a Ph.D. degree in Physical-Mathematical Sciences from General Physics Institute of the USSR Academy of Sciences (Moscow, Russia). Dr. Noginov's affiliations include: General Physics Institute of the USSR Academy of Sciences (Moscow, Russia, 1985–1991); Massachusetts Institute of Technology (Cambridge, MA, 1991–1993); Alabama A&M University (Huntsville, AL, 1993–1997); and Norfolk State University (NSU) (Norfolk, VA, 1997–present).

Dr. Noginov has published one book, four book chapters, over 100 papers in peer-reviewed journals, and over 100 publications in proceedings of professional societies and conference technical digests (more than 20 of them being invited papers). He is a member of Sigma Xi, Optical Society of America, SPIE, and the American Physical Society and has served as a chair and a committee member on several conferences of SPIE and OSA. Since 2003, Dr. Noginov has been a faculty advisor of the OSA student chapter at NSU. His research interests include metamaterials, nanoplasmonics, random lasers, solid state laser materials, and nonlinear optics.

Graeme Dewar earned his Ph.D. in Physics from Simon Fraser University in 1980. After stints on the faculties of Princeton University and the University of Miami, he joined the University of North Dakota in 1989 where he is currently Professor and Chair of the Department of Physics and Astrophysics. Most of his research projects have involved the interaction of electromagnetic radiation with complex media. These have included experimental investigations of the radio frequency properties of ferromagnetic metals, with an emphasis on magnetoelastic effects, and photonic crystals. His current interests are primarily in metamaterials having a tailored permittivity and permeability.

Martin W. McCall graduated in Physics from Imperial College London in 1983. His doctoral thesis, completed while he worked for an electronics company, concerned development of vector coupled-wave theory describing anisotropic diffraction in photorefractives for use in real-time image processing. After a spell at the University of Bath, UK, where he worked on nonlinear dynamics in semiconductor lasers, McCall returned to Imperial College in 1988 to work on a range of optoelectronic themes, including nonlinear coupling and mixing in semiconductor amplifiers and laser arrays, optical interconnects, and Bragg grating physics. Sometimes referring to himself as a "reformed experimentalist," McCall's research is now purely theoretical. Broadly within the remit of describing the electromagnetics of complex media, he has specifically worked on chiral photonic films and negative index metamaterials. Recently he has specialized in the use of covariant methods in electromagnetism. He is currently Professor of Theoretical Optics at Imperial College London.

Nikolay I. Zheludev received his M.Sc., Ph.D., and D.Sc. from Moscow State University and joined the faculty of Southampton University in 1991. He is Deputy Director of the Optoelectronics Research Centre, University of Southampton and directs the EPSRC Centre on Nanostructured Photonic Metamaterials. His awards include a Senior Research Fellowship with the Leverhulme Trust (2001), a Senior Research Professorship of the UK Engineering and Physical Sciences Research Council (2002), and the Royal Society Wolfson Research Merit Award (2009). Prof. Zheludev is a Fellow of the Institute of Physics (London) and of the Optical Society of America. He is a member of the European Physical Society QEOD steering group and the Institute of Physics QEP steering group, and vice chaired the metamaterials group of the Physics and Engineering Research Council of the Optical Society of America. He is also Editor in Chief of the *Journal of Optics* (IOP Publishing).

Index

O

P

Q

R

S